The Chemistry Knowledge for Firefighters

Torsten Schmiermund

The Chemistry Knowledge for Firefighters

Torsten Schmiermund
Frankfurt a.M., Germany

ISBN 978-3-662-64422-5 ISBN 978-3-662-64423-2 (eBook)
https://doi.org/10.1007/978-3-662-64423-2

This book is a translation of the original German edition „Das Chemiewissen für die Feuerwehr" by Schmiermund, Torsten, published by Springer-Verlag GmbH, DE in 2019. The translation was done with the help of artificial intelligence (machine translation by the service DeepL.com). A subsequent human revision was done primarily in terms of content, so that the book will read stylistically differently from a conventional translation. Springer Nature works continuously to further the development of tools for the production of books and on the related technologies to support the authors.

This Springer imprint is published by the registered company Springer-Verlag GmbH, DE part of Springer Nature.
The registered company address is: Heidelberger Platz 3, 14197 Berlin, Germany

Foreword

Who does not remember with horror the chemistry lessons in school - all a bit suspect. Bang and blast was already interesting - but the backgrounds much too exhausting! And then you become a firefighter, hoping to hear no more about it all - and then it comes back! From burning and extinguishing to hazardous materials and radioactivity - again, chemistry is everywhere. But I want to be a firefighter - not a chemist! True, but understanding the craft of being a firefighter involves chemistry, and even more chemistry if you want to advance to the leadership ranks. Then you can't do it without a solid basic knowledge of chemistry! Now would be a good time to have a handy book to refer to! Everything that can happen to a firefighter with regard to chemistry should be in it, and if possible, then also in understandable language and of course related to the areas of application of the firefighter - and please of course technically correct!

I think, dear readers, you have such a thing in your hands! When I first leafed through it, I found almost every topic that has had any significance in my long firefighting life regarding chemistry! And I even understood what it said!

I hope you feel the same way! In your hand you have now a chemistry book for the firefighter that is good to use for learning and understanding, but also for reference later on. It would be nice if every instructor for the famous $^4/_5$ of additional knowledge to their $^1/_5$ material they have to teach could use this book as a guide, and if you could find it in the library of any fire department that takes their job seriously.

I hope you enjoy reading.

Berlin, Germany — Frieder Kircher

Foreword

What? Another chemistry book? There are enough chemistry books out there - aren't there?

Yes - and no.

There are books on general, analytical, inorganic, organic, physical and technical chemistry, textbooks, reference works, dictionaries, encyclopaedias, data and formula collections and table books, works on reaction mechanisms and theoretical chemistry, on principles, reactions, syntheses and much more. There are chemistry books for beginners, for trainees, for pupils, for "dummies", for students, for doctors, for biologists, for interested people and for engineers; in other words, for almost everyone.

But only almost: If a firefighter wants to/has to research "chemical topics/issues", he often has only two options: The use of pure firefighting literature or research in (chemistry) reference books or the Internet. The first option may not be able to explain everything as one would like - e.g. for the preparation of a lesson - and with the second option it takes a lot of time until one has found "the right thing".

That is why this book was created. It is intended to be a help and a reference book for all those who pursue the "most beautiful hobby in the world" throughout their entire firefighting career. From the basic fire brigade course to the platoon leader course and beyond that for all (specialist) courses on "NBC topics" or on "special dangers in civil protection".

The book is intended for the volunteer fire departments, the professional and plant fire departments and all interested parties of other aid organizations. It is suitable for the participants in a course and the instructors as well as an aid for the preparation of regular training and continuing education.

Frankfurt am Main, Germany
June 2018

Torsten Schmiermund

Foreword to the English edition

Originally, the book was written for the fire brigades/fire services in Germany and the German-speaking countries. It therefore contains some regulations based on German laws, regulations and standing orders.

Nevertheless, chemistry and physics apply throughout the whole universe.

Therefore, as the author, I am sure that this book has a benefit for all firefighters internationally. No matter whether they are at the beginning of their training or have been "in the business" for many years.

I wish all readers much enjoyment with this book and an always healthy return from their emergency operations.

Frankfurt am Main, Germany
September 2021

Torsten Schmiermund

Thank You

I would like to thank everyone who has helped me with this book project. First of all, to my test readers for their commitment and the valuable comments and suggestions I received. In particular, I would like to thank: Erika F., Karen H., Angelika K., Christian K., Kyra L., Bernd N., Martin R., Johanna St. and especially Lars H.

Special thanks to Beate Rocholz and Felicitas "Cean" Schmiermund for creating various illustrations.

From the publisher (Springer) I would like to thank Ms. Lerch for her patience, Ms. Fallert-Müller for her meticulous review of the manuscript, and Mr. Münz for his support.

The biggest thank you, however, goes to my wife Irene, who has supported me tirelessly and in many ways.

Torsten Schmiermund
June 2018
Frankfurt am Main

Contents

Part I Introduction

1 The Natural Sciences . . . 3
- 1.1 Differentiation of the Natural Sciences . . . 3
- 1.2 Differentiation of Physical and Chemical Processes . . . 4
- 1.3 What Are These "Substances"? . . . 5
- 1.4 Models . . . 6

2 Substances and Mixtures . . . 9
- 2.1 Substance Separation . . . 11
- 2.2 Element and Connection . . . 12
- 2.3 Substance Properties of Interest to the Fire Brigade . . . 15

Part II Forms of States of Matter

3 Aggregate States . . . 19
- 3.1 Heat Transport . . . 19
- 3.2 Change of the State of Aggregation . . . 29
- 3.3 Important Safety-Related Values . . . 35
- 3.4 Specific Heat Capacity and Latent Heats . . . 48

4 Gases . . . 59
- 4.1 Ideal Gas . . . 60
- 4.2 Pressure and Temperature . . . 60
- 4.3 Boyle-Mariotte Law . . . 61
- 4.4 Law of Amontons . . . 62
- 4.5 Law of Gay-Lussac . . . 63
- 4.6 General Gas Equation . . . 64
- 4.7 Avogadro Theorem . . . 65
- 4.8 Universal Gas Equation . . . 65
- 4.9 Standard Conditions . . . 66
- 4.10 Partial Pressures . . . 66
- 4.11 Diffusion . . . 67
- 4.12 Real Gases . . . 69

Part III Atomic Models and Periodic Table

5 Atoms and Atomic Shell 79
5.1 Development of the Atomic Theory 79
5.2 Structure of the Atomic Shell 82
5.3 Structure of the Atomic Nucleus 91
5.4 Particles in the Atom 94
5.5 Atomic Mass Units 96

6 The Periodic Table 101
6.1 Early Trials 101
6.2 Periodic Table According to Mendeleev & Meyer 102
6.3 Structure of the Periodic Table 103
6.4 Representation of the Electron Configuration 106
6.5 Periodic Properties 110
6.6 Main Groups of the Periodic Table 114
6.7 Subgroup Elements/d-Elements 119
6.8 Rare Earths/f-Elements 120
6.9 Oblique Relationship 120
6.10 Metals in the PTE 120

Part IV Molecules, Ions, Bonds

7 Introduction 125
7.1 Molecule Presentation 125
7.2 Notation 126
7.3 The Valence Stroke Formula 127
7.4 Other Formula Notations 133

8 Bonds 137
8.1 Strong Bonds 137
8.2 Weak Bondings 147
8.3 Other Types of Bonds in Solids 153

Part V Solutions and Chemical Reactions

9 Chemical Reactions: Fundamentals 159
9.1 Basic Laws 160
9.2 Reactions 162

10 Solutions 171
10.1 Basic Information on the Dissolving Behaviour 171
10.2 Composition of Mixed Phases 181
10.3 Reactions in Solution 188
10.4 Chemical Reactions During the Dissolving Process 191

11 Double Salts, Complexes and Dispersions . . . 193
11.1 Double Salts . . . 193
11.2 Complex Salts . . . 194
11.3 Disperse Systems . . . 199

Part VI Acids and Alkalis

12 Acid-Base Theories . . . 205
12.1 Definition According to Arrhenius . . . 205
12.2 Definition According to Brønsted and Lowry . . . 209
12.3 Definition According to Lewis . . . 212
12.4 HSAB Concept . . . 213

13 Acids and Alkalis . . . 217
13.1 Properties of Acids and Alkalis . . . 217
13.2 Important Acids . . . 221
13.3 Important Alkalis . . . 228

14 pH Value . . . 231
14.1 Explanation of the Value "pH value" . . . 231
14.2 The Neutral Point . . . 233
14.3 pH Value and pOH Value . . . 235
14.4 pH Value of Salt Solutions . . . 236
14.5 Acid and Alkali Strength . . . 238
14.6 Determination of the pH Value . . . 244

15 Neutralisation . . . 251
15.1 Basics of Acid-Base Neutralization . . . 251
15.2 Neutralisation Heat . . . 254
15.3 Buffer Solutions . . . 254
15.4 Neutralization and Emergency Operations . . . 257

Part VII Redox Reactions and Electrochemistry

16 Oxidation/Reduction Concept . . . 263
16.1 The Modern Redox Concept . . . 264
16.2 Oxidizing and Reducing Agents . . . 265
16.3 Redox Reactions . . . 266
16.4 Setting up Redox Equations . . . 269

17 Redox Pairs . . . 275
17.1 Half Cells . . . 275
17.2 Normal Potentials of Redox Couples . . . 277
17.3 Electrochemical Series . . . 278
17.4 Normal Potential and Reaction Course . . . 279

18 Calculation of the Electromotive Force 283
18.1 Calculation at Normal Conditions 283
18.2 Nernst's Equation 284

19 Galvanic Cells 287
19.1 Batteries 288
19.2 Accumulators 290
19.3 Fuel Cells 292
19.4 Electrochemical Corrosion 293

20 Electrolysis 295
20.1 Electrode Processes 295
20.2 Electrolysis of Aqueous Solutions 297
20.3 Applications of Electrolysis 300
20.4 Terms of Electrochemistry 303

Part VIII Radioactivity

21 Background Knowledge "Radiation" 307
21.1 Waves and Wave Radiation 307
21.2 Particle Radiation 315
21.3 Excursus: Ionizing Radiation 315

22 History of Radioactivity 319
22.1 Beginnings 319
22.2 Developments 322

23 Radioactivity: Terms and Notations 325
23.1 Notations of Nuclear Chemistry 325
23.2 Terms Relating to Nuclides and Isotopes 327

24 The Atomic Nucleus 331
24.1 Structure 331
24.2 Forces in the Core 333
24.3 Stability of the Atomic Nucleus 335

25 Radioactive Decay 339
25.1 Representation of Nuclear Reactions 339
25.2 α-decay 341
25.3 β-decay 343
25.4 γ-transition 348
25.5 Summary 352

26 Nuclide Cards 353
26.1 Current Nuclide Maps 353

27 Units of Measurement of the Radiation of Radioactive Substances 357
27.1 Radiation Protection Units 357
27.2 Other Units of Radiation Protection 367

28 Measuring Instruments for Radiation Emitted by Radioactive Substances 369
28.1 Measuring Principles 369
28.2 Measuring Instruments for Ionizing Radiation in Firefighting Operations 372

29 Radiation Exposure 379
29.1 Environmental Radioactivity 379
29.2 Civilisational Radiation Exposure 380
29.3 Total Load on Average 381

30 Biological Effects of Ionizing Radiation 383
30.1 Types of Radiation Damage 383
30.2 Radiobiological Reaction Chain 384
30.3 Factor dependence of the radiation effect 389

31 Use of Radioactive Substances 393
31.1 Application in the fire brigade 393
31.2 Applications in Technology 394
31.3 Applications in Science 396
31.4 Applications in medicine 397

32 Nuclear Reactions 399
32.1 Artificial Nuclear Transformations 399
32.2 Nuclear Fission 399
32.3 Nuclear Fusion 402
32.4 Nuclear Weapons 402

33 Labelling of Radioactive Substances and Areas 405
33.1 Transport Labeling 405

34 Protection Against Ionizing Radiation 411
34.1 Basic Behavior 411
34.2 Specific Protective Measures 412
34.3 Procedure According to Hazard Groups 420

Part IX Energy Conversion of Chemical Reactions

35 Energy 423
35.1 Law of Conservation of Energy 423
35.2 Definitions 423
35.3 Energies 425
35.4 Enthalpy H 426
35.5 Entropy S 429
35.6 The Driving Force of Chemical Reactions 432
35.7 Transitions 433

36 Catalysis 437
36.1 Introduction to Catalysis 437
36.2 Influences of a Catalyst 438
36.3 Aggregate States of Catalysts 441
36.4 Inhibition 444
36.5 Catalyst Poisons 445

Part X Burning and Extinguishing

37 Fire and Blazes 449
37.1 Fire – An Oxidation Process 449

38 The Process of Burning 453
38.1 Requirements for Burning 453
38.2 Combustible Substance 456
38.3 Oxygen 463
38.4 Ignition Energy 472
38.5 Mixing Ratio 475
38.6 Combustion Catalyst 479

39 Explosions 481
39.1 Differentiation of Explosion Processes 481
39.2 Explosion Indicators 486

40 The Chemistry of Combustion 491
40.1 Chain Reactions 491
40.2 Fire Course 497
40.3 Flue Gases 502
40.4 Energy Turnover During Fires 506

41 Extinguishing 515
41.1 Extinguishing Methods 515
41.2 Extinguishing Agent 520

42 Flame Retardants 533
42.1 Flame Retardant 533
42.2 Intumescent Coatings 537

Part XI Organic Chemistry

43 Indispensable Organic 541
43.1 Introduction 541

44 Hydrocarbons 545
44.1 Aliphatic Hydrocarbons 545
44.2 Excursus: Bonding Ratios at the C Atom 553
44.3 Annular Aliphatic Hydrocarbons 555
44.4 Aromatic Hydrocarbons 556

45 Organic Halogen Compounds . . . 561
45.1 The "Dirty Dozen" of Halogen Compounds . . . 561

46 Organic Oxygen Compounds . . . 563
46.1 Alcohols . . . 563
46.2 Aldehydes . . . 567
46.3 Ketones . . . 567
46.4 Ether . . . 567
46.5 Carboxylic Acids . . . 571
46.6 Ester . . . 573
46.7 Peroxides . . . 574

47 Organic Nitrogen Compounds . . . 577
47.1 Amines . . . 577
47.2 Nitro Compounds . . . 580
47.3 Nitrile . . . 581
47.4 Isocyanates . . . 581
47.5 Carboxylic Acid Amides . . . 582
47.6 Azo Compounds . . . 583

48 Organic Sulphur Compounds . . . 585
48.1 Thiols . . . 585
48.2 Thioethers and Disulphides . . . 586
48.3 Sulphonic Acids . . . 586
48.4 Sulphoxides and Sulphones . . . 587

49 Plastics . . . 589
49.1 Classification According to Thermal Properties . . . 589
49.2 Classification According to Education Mechanisms . . . 593
49.3 Fully Synthetic/Partially Synthetic Plastics . . . 595
49.4 Important Plastics . . . 595
49.5 Fire Behaviour of Plastics . . . 599

50 Surfactants . . . 605
50.1 Structure and Surfactant Groups . . . 605
50.2 Excursus: Fats and Oils . . . 606
50.3 Mode of Action of the Surfactants . . . 608
50.4 Surfactants in Fire Fighting . . . 610

Part XII CBRNE Hazards

51 Poisons . . . 615
51.1 General Information on Poisons . . . 615
51.2 Labelling and Classification of Toxic Substances . . . 617
51.3 Operational Measures . . . 619

52 Chemical Warfare Agents . . . 621
52.1 Introduction . . . 621

52.2 Physical, Chemical and Toxicological Properties of Warfare Agents . . . 625
52.3 Irritants . . . 630
52.4 Pulmonary Warfare Agents . . . 632
52.5 Blood Warfare Agents . . . 633
52.6 Skin Warfare Agents . . . 634
52.7 Nerve Agents . . . 637
52.8 Psychotoxic Warfare Agents . . . 641
52.9 Sabotage Poisons . . . 643
52.10 Strategic Ordnance . . . 644
52.11 Binary CWA . . . 645
52.12 Final Remark C-Weapons . . . 645

53 Biological Substances . . . 647
53.1 Introduction . . . 647
53.2 Classification of Biological Agents . . . 648
53.3 Classification of Biological Agents – Criteria . . . 650
53.4 Classification of Biological Agents – Risk Groups . . . 652
53.5 Labelling and Classification of Biological Agents . . . 654
53.6 Operational Measures . . . 656

54 Biological Agents . . . 659
54.1 Introduction . . . 659
54.2 Classification of Pathogens and Toxins . . . 662
54.3 Important Bacteria and Viruses . . . 663
54.4 Important Toxins . . . 668

55 Explosives . . . 671
55.1 Explosives Hazards . . . 671
55.2 Labelling of Explosive Substances and Goods . . . 671
55.3 Operational Measures . . . 674

Afterword . . . 677

Tables and Figures . . . 679

Bibliography . . . 701

Index . . . 711

Part I

Introduction

1 The Natural Sciences

Natural sciences are the fields of knowledge that deal with the (systematic) study of nature and the laws of nature.

The so-called “exact natural sciences” include astronomy, chemistry, geology, physics and the fields that connect them, such as astrophysics, meteorology or particle physics. “Exact” because they can be well described mathematically and physically.

In addition, there are the biological sciences as subfields of biology: Botany, genetics, ecology, zoology, etc.

Nowadays, not only regularities are investigated by means of corresponding experiments and (mathematical) models and described and explained in theories. The utilization of the natural sciences for technology, medicine, agriculture, materials science and in other areas of daily life occupies a large space today.

Therefore, we are constantly surrounded by the most diverse materials and substances with the most diverse properties to a greater extent than “in the past”.

New substances are constantly being produced, materials with improved properties developed, more modern medicines discovered. This leads to incidents, accidents and disasters to which the fire brigades are almost always called first. The members of the fire brigades must therefore have a sound basic knowledge of chemistry and physics - even outside the NBC platoons.

1.1 Differentiation of the Natural Sciences

As classical subjects in general education schools, a distinction is made in the natural sciences between:

Biology	Doctrine of life
Chemistry	The study of the transformation of substances
Physics	Science of the properties of substances

T. Schmiermund, *The Chemistry Knowledge for Firefighters*,
https://doi.org/10.1007/978-3-662-64423-2_1

In school, there is a clear, strict separation of these areas of knowledge; nevertheless, there are many subfields where these subjects overlap: Biochemistry (biology ↔ chemistry), physical chemistry (physics ↔ chemistry), physiology (physics ↔ biology).

Even though there is now a chemistry book here, we will also have to deal with physical conditions and properties. Many physical properties of substances are crucial to successful firefighting operations, so we will return to these again and again. These are also important for the understanding of "chemistry".

1.2 Differentiation of Physical and Chemical Processes

► **In** physical **processes, the** typical properties of **a substance are** retained.

Typical physical processes are, for example, changes in the state of motion (acceleration, free fall, ...), changes in the state of aggregation (melting, evaporation, ...) or deformation due to external forces.

The physical properties of the substances include:

- State of aggregation
- Focal point
- Vapour pressure
- Density
- Electrical conductivity
- Color
- Flash point
- Hardness
- Solubility
- Magnetizability
- Surface tension
- Radioactivity
- Melting point
- Heat of fusion
- Boiling point
- Temperature
- Heat of evaporation
- Malleability
- Viscosity
- Volume expansion
- Heat capacity
- Thermal conductivity

Each pure substance is characterized by a unique combination of these physical properties. This can be used to identify substances, as two different substances are not exactly the same in all properties.

▶ In **chemical** processes, the **typical** properties of the substance are **lost.** A new substance with other typical properties is created.

Please do not take the expression "*a* new substance" too literally. Often enough, two or more substances react with each other, which in turn form several new substances. But usually only one of them is considered to be the "target substance". In chemistry, we also speak of substance transformations.

For example, iron transforms into rust in moist air. An easily deformable metal that conducts electricity well becomes a reddish-brown, crumbly substance that is hardly electrically conductive.

The chemical properties of the substances include:

- Reactivity (to other substances, e.g. water, acid, alkali)
- Flammability (form of reactivity to oxygen)
- Corrosion behaviour (e.g. against moist air, acids, alkalis)
- Resistance (e.g. to acids, solvents)
- Acid or base constant
- Electronegativity (for elements)
- pH value

In the broadest sense, the so-called "physiological properties" also belong to the chemical substance properties. This refers to properties that are either perceptible to our senses or have effects on the environment. These are, for example, odour, taste, toxicity (poisonousness), ecotoxicity (harmfulness to the environment).
Some of the chemical and physical properties mentioned here will be encountered later.

1.3 What Are These "Substances"?

Everything that surrounds us (and of which we ourselves are made) is built up of the most diverse materials, substances, materials.

A sharp distinction between these three terms does not exist. Thus one speaks both of the (building) material and of the (building) material wood. A drinking glass and a window pane are both made of the same material: glass. Scissors, knives, drills and the nib of a fountain pen are very different in shape, size and purpose. Nevertheless, they are made of the same substance: steel.

In a nutshell:

▶ Everything that has a mass and occupies space is called a substance.

1.4 Models

In the following pages we will be confronted with a large number of different models: different atom and molecule models, models for the bonding of atoms with each other, different formulae (which are also based on corresponding models) and some others.

But what is a "model" in the chemical-physical sense? What purpose does it serve?

▶ A model always serves to simplify reality.

Many relationships - not only in the natural sciences - are very complex, sometimes even complicated. In order not to have to consider all aspects at the same time, one creates a 'simplified' or 'simplifying' model of certain sub-areas that appear to be interesting with regard to the question.

In these cases, models serve to provide a better overview and ensure that certain functions can be imagined without major problems.

These models are also called "functional models". They represent the basic process ("How does it work?") and thus enable us to influence the processes in a guiding way and to react to new conditions.

Sometimes you do not have the information about a subject in the depth of detail that you want. Here, too, models provide a good service. The interpretation and dissemination of the results of scientific research makes it necessary to take account of the need for clarity and imaginability. One likes to fall back on images (of meaning) and processes that are as close as possible to human perception and can be brought into harmony with everyday experience, so-called representational models.

"Good" models, however, are not a mere representation of previous results. They yield new information, raise new questions and also provide experimentally verifiable findings.

In both cases (functional and representational models) these models are "purpose-determined working tools". Their task is to make statements about the object - in such a way that one obtains statements about a certain area of objective reality. The model conception is based on general knowledge and/or theoretical concepts.

However, models must never be confused with reality or mistaken for 1: 1 images of reality. Models are quasi "representatives" of objective reality. They only represent certain aspects of reality that are essential for the problem in a symbolic form. Each model therefore only captures a certain section of reality.

▶ Models are aids that serve to illustrate unclear facts.

Some models have lost their model character in the course of time and have thus been elevated (at least in the collective memory) to reality itself.

However, this undermines the meaning and purpose of any model. The "section of reality" that a model represents is elevated to the status of the sole reality. With this, however, "real reality" can no longer be adequately described. Those who assume that they have grasped the full facts on the basis of a single model are mistaken.

▶ To explain a certain phenomenon, sometimes different models (explanations, explanatory approaches) are used side by side in order to be able to capture all aspects.

Each (!) model also has a range of validity. Only within these limits can the respective model represent and explain facts.

Substances and Mixtures 2

If we look at a glass of water, we cannot distinguish with our eyes whether it is pure (distilled) water, salt water or sugar water. It is different when we look at a piece of granite. Here we can recognize different components.

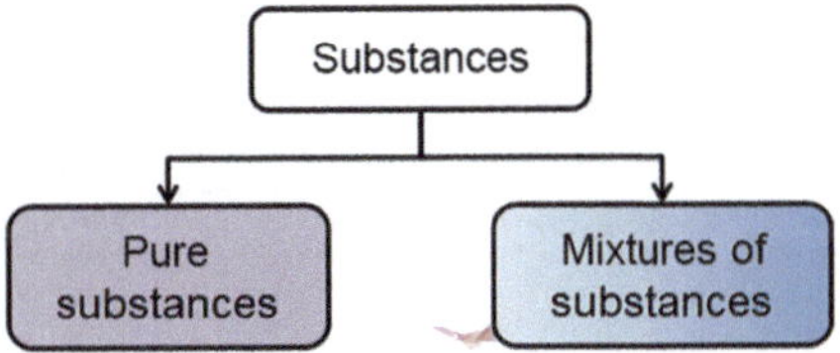

Salt water, sugar water and granite are mixtures of substances. Salt water and sugar water cannot be distinguished from each other, nor from pure water, by mere sight. They are uniform or homogeneous.

In the case of granite, on the other hand, the three main components can be distinguished even without aids:

Quartz (white, semi-transparent), feldspar (reddish, opaque) and mica (black, shiny).

▶ Mixtures that have a uniform appearance are called homogeneous mixtures; those with a non-uniform appearance are called heterogeneous mixtures.

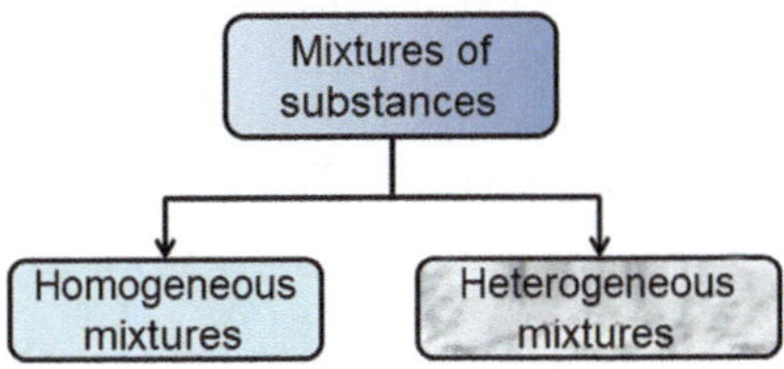

T. Schmiermund, *The Chemistry Knowledge for Firefighters*,
https://doi.org/10.1007/978-3-662-64423-2_2

Table 2.1 Heterogeneous mixtures (examples)

What in What	Designation of the mixture	Examples
Solid in solid	Mixture	Granite, potting soil, spice mix, cast iron, muesli (without milk)
Solid in liquid	Suspension	Yeast white beer, noodle soup, muesli (with milk)
Solid in gaseous	Smoke (aerosol)	Smoke, exhaust fumes
Liquid in solid	Gel or slurry	Wet clay, hair gel
Liquid in liquid	Emulsion	Milk, ointments, mayonnaise, skin cream
Liquid in gas	Mist (aerosol)	Clouds, fog
Gaseous in solid	Sponge	Pumice, aerated concrete
Gaseous in liquid	Foam	Whipped cream, shaving cream

Table 2.2 Homogeneous mixtures (examples)

Designation of the mixture	What in What	Examples
Solution	Solid in solid	Certain alloys
	Solid in liquid	Salt water, coffee, tea
	Gaseous in liquid	Mineral water, chlorinated water
	Liquid in liquid	Vinegar, apple juice, brandy
	Liquid in gaseous	Humidity
Gas mixtures	Gaseous in gaseous	Air, natural gas

In most cases, heterogeneous substances can already be recognized as such by eye (potting soil, smoke, fog), but sometimes it is necessary to use aids such as a magnifying glass or a microscope, e.g. to recognize the fat droplets in milk.

Depending on the state of aggregation, i.e. whether solid, liquid or gaseous (cf. Section 3), of the (intrinsically homogeneous) constituents involved, various heterogeneous systems can be distinguished from one another (Table 2.1).

In chemical engineering, substances that are finely dispersed in one another are often referred to as dispersions. The substance that is finely dispersed in another is then called a dispersed substance. The substance in which the substance is dispersed is called a dispersant.

▶ A heterogeneous mixture gaseous-in-gaseous cannot exist, since gaseous substances always (!) mix homogeneously with each other.

In the case of homogeneous mixtures, the individual components can no longer be identified (even with a microscope). These are gas mixtures or solutions (Table 2.2).

▶ Homogeneous mixtures still represent mixtures which can be separated into their constituents using suitable methods.

Alloys -Homogeneous or Heterogeneous?
Both in Table 2.1 (example "cast iron") and in Table 2.2 ("some alloys") metal-metal mixtures (= alloys) are listed.

Alloys have different structures depending on the proportions of the metals alloyed with each other:

- Alloys can consist of a structure of different "crystallites" (homogeneous, small crystals). This is the case, for example, with cast iron or brass. These are then heterogeneous mixtures.
- However, they can also be composed at the atomic level. These are then homogeneous alloys.
  - Many mercury alloys (amalgams) or, for example, an alloy of 50.8 % gold with 49.2 % copper (chemical formula: Cu_3Au) are "intermetallic phases" or "*solid* solutions". Here, a defined crystal structure of the defined compound exists (compare also Sect. 8.1).
  - Substitution alloys are formed when the foreign metal and the base metal are of comparable size. Here the foreign metal atoms are statistically distributed in the structure. This often occurs with elements that are adjacent to each other in the periodic table (e.g. Fe/Ni or Cu/Ag).
  - If the foreign atom is significantly smaller than the base metal, intercalation solid solutions are formed. Here, too, the foreign atoms are irregularly distributed in the structure. This often happens when the foreign atom is a non-metal (e.g. carbon, nitrogen) or a semi-metal (e.g. boron, silicon).

2.1 Substance Separation

A variety of methods can be used to separate homogeneous and heterogeneous mixtures. The selection of the suitable method depends on the separation task and the chemical-physical properties of the mixture or its components.

2.1.1 Separation of Heterogeneous Systems

Heterogeneous mixtures can often be separated into homogeneous mixtures, sometimes even into pure substances, by various methods. Frequently used methods are:

- Sieving
- Filter
- Decant
- Air classifying
- Magnetize
- Sublimate
- Sediment
- Centrifuge
- Segregation

2.1.2 Separation of Homogeneous Systems

A variety of methods are also used to separate homogeneous mixtures into their constituents, the so-called pure substances:

- Distillation methods (separation due to different boiling points)
- Crystallisation (exploitation of the varying solubility of different substances or solubility as a function of temperature)
- Sublimation ("evaporation" of a solid)
- Chromatographic methods
- Repricipation
- pH value changes
- Extraction

2.2 Element and Connection

Once a pure substance has been obtained by the above processes, it can be transformed in a variety of ways, e.g. by heat, light, electricity, other chemicals etc. In many cases, it is now possible to break down this pure substance into other, dissimilar components, even if "detours" have to be accepted in some cases.

Example 1

If mercury oxide is heated, it splits into a shiny silver liquid (mercury) and a gaseous substance (oxygen) (decomposition, analysis). Conversely, mercury oxide can also be produced again from mercury and oxygen (assembly, synthesis).

Mercury and oxygen cannot be further decomposed chemically. They are elements. ◀

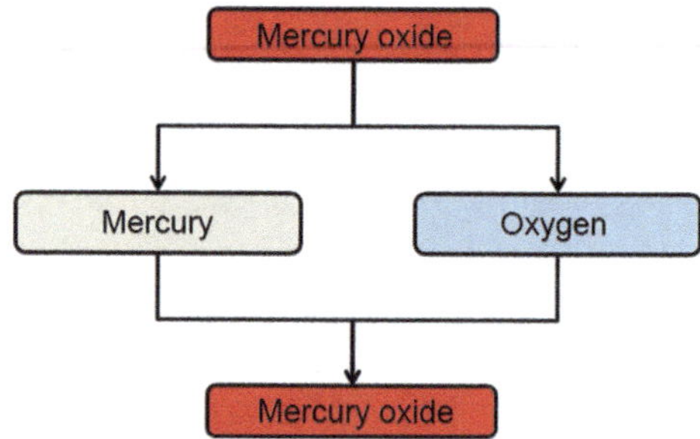

Example 2

When limestone is heated (to ~ 900 °C), two new substances are obtained: The gas carbon dioxide and burnt lime (calcium oxide CaO), a solid. These two

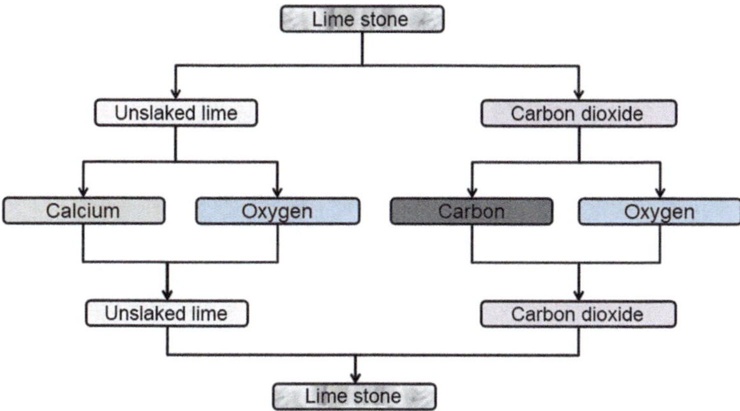

Fig. 2.1 Analysis and synthesis of limestone ($CaCO_3$)

substances can now be further decomposed. Burnt lime (also known as unslaked lime) into metallic calcium and oxygen, the carbon dioxide into carbon and oxygen.

Using suitable methods, limestone can be rebuilded (synthesized) from calcium, carbon and oxygen, see Fig. 2.1. ◀

Substances such as mercury, calcium, carbon and oxygen that cannot be broken down into further (simpler) substances are called basic substances or elements.

Substances that can be broken down by common physical and chemical methods are called compounds. The compounds in the two examples are: Mercuric oxide, limestone, quicklime, and carbon dioxide.

Historical Anecdote

When the Nobel Prize is awarded, the laureates receive not only a diploma but also a gold medal with the name of the laureate, an inscription and Alfred Nobel's picture. When Denmark was occupied by the Wehrmacht in April 1940, Niels Bohr feared for the gold medals entrusted to him by Max von Laue and James Franck. To prevent the confiscation of the medals by the Nazis, they were dissolved separately in aqua regia and placed with other chemicals in the laboratory. After World War II, the gold was recovered, the medals recast and returned to their owners.

Here, a compound ($AuCl_3$) was synthesized from an element (Au) and then the element was recovered.

(Source: Lit. [10–8], p. 328)

In summary, we can show the relationships between substances, mixtures, compounds and elements as in Fig. 2.2.

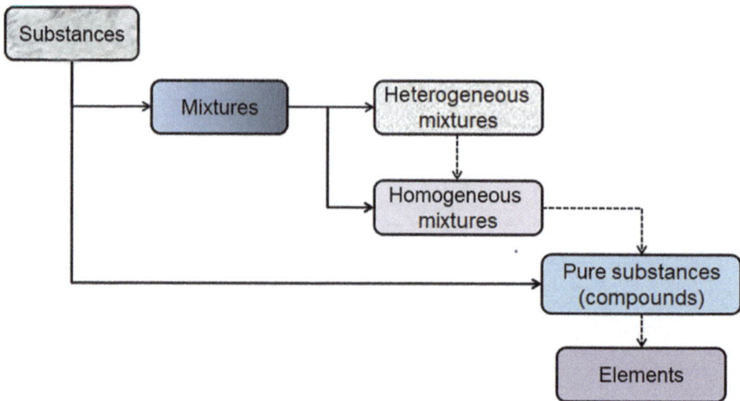

Fig. 2.2 Classification of substances

▶ **Mixtures of substances** can have a uniform (homogeneous) or non-uniform (heterogeneous) composition. They contain at least two pure substances in **any** mixing ratio.

▶ The properties of substance mixtures change with the mixing ratio of their individual components.

Melting and boiling points can change so much that it is better to speak of melting or boiling ranges. Other properties, such as the flash point, the pH value or the density, are also strongly influenced by the mixing ratio.

▶ **Pure substances** or **compounds** are uniform substances that can be described by **characteristic** properties.

Pure substances have defined and characteristic ("sharp") melting and boiling points. Other properties (density, electrical conductivity, colour, flash point, viscosity) are also specific to them. Compounds can be broken down into the elements of which they are composed by suitable methods.

▶ Pure substances that cannot be broken down further by chemical means are called **elements**.

The various elements are systematically listed and arranged in the periodic table of the elements (PTE, see Chap. 6). Of the 118 elements currently known, 91 occur naturally. The others were produced by man, i.e. they are "artificial elements".

2.3 Substance Properties of Interest to the Fire Brigade

A whole range of substance properties is of interest to the fire brigade. These can help to assess the behaviour of a substance or mixture of substances (e.g. fire smoke, hazardous goods) and then derive operational tactical measures from them. The most important properties are:

Flammability	Section 38.2
Vapour density ratio	Section 3.3.5
Vapour pressure	Sections 3.2.4 and 3.3.2
Density	Ratio of mass and volume
Diffusion coefficient	Section 4.11
Explosion pressure	Section 39.2.4
Explosion limits	Section 3.3.4
Flash point	Section 3.3.3
Calorific value	Section 40.4.2
Ionising radiation	Chapters 21 and 34
Linear expansion	Section 3.1.3
Miscibility with water	Section 10.1.8
pH value	Chapter 14
Melting point	Section 3.2.1
Boiling point	Section 3.2.2
Evaporation number	Section 3.3.1
Volume expansion	Section 3.1.4
Ignition temperature	Section 3.3.3

Part II

Forms of States of Matter

3 Aggregate States

Depending on its temperature, a substance exists in different forms, the aggregate states. In the case of gases, the pressure also plays a role. For example, everyone knows water as a liquid, as a solid (ice) and in a gaseous state (water vapour). These different forms are called states of aggregation. These states of aggregation differ significantly in their behaviour, as the overview in Table 3.1 shows.

Sometimes these states of aggregation are added as indices directly to the respective substance. The following abbreviations apply here:

$$(s) \rightarrow \text{solid}$$
$$(\ell) \rightarrow \text{liquid}$$
$$(g) \rightarrow \textit{gaseous}$$

Thus, $Fe_{(s)}$ denotes solid (metallic) iron, $H_2O_{(\ell)}$ liquid water and $CO_2(_g)$ gaseous carbon dioxide. $CO_2(_s)$ then stands for solid carbon dioxide ("dry ice") and $N_{2(\ell)}$ for liquid nitrogen. For the three states of water mentioned above, the following then applies analogously: Ice = $H_2O_{(s)}$, liquid = $H_2O_{(\ell)}$, water vapor = $H_2O_{(g)}$.

3.1 Heat Transport

The changes of the states of aggregation always have to do with changes of the temperature of the substance. They are therefore related to the temperature (and partly also to the pressure).

3.1.1 What Is "Heat"?

Phrases like "Warmth is when it's warm." or "Warmth is when it's not cold." show how difficult we find the concept of warmth in everyday life. Everyone knows what heat is, but we still find it difficult to formulate it in a simple sentence.

T. Schmiermund, *The Chemistry Knowledge for Firefighters*,
https://doi.org/10.1007/978-3-662-64423-2_3

Table 3.1 Overview of states of aggregation

	Solid	Liquid	Gaseous
Form fitting	Does **not** adapt to external, given shape	Adapts to external, predetermined shape	Adapts to external, predetermined shape
Surface	Depends on the degree of division	Surface available	No surface available
Compressibility	Incompressible	Incompressible	Compresses **easily**
Cohesion of particles	Stick together	Easily movable against each other	Freely movable
Particle arrangement	Dense, regular arrangement	Densely but irregularly arranged	Disorderly, evenly distributed in the room
Dispersal behaviour	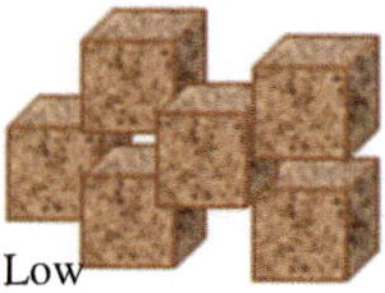 Low	 High, 2-dimensional	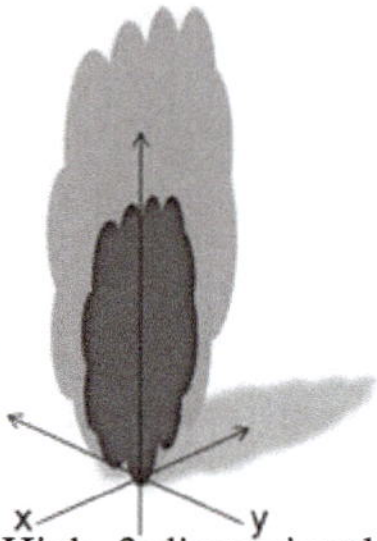 High, 3-dimensional
Particle image	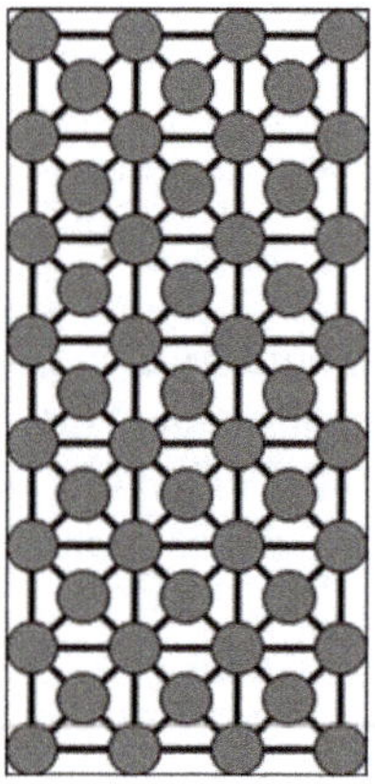	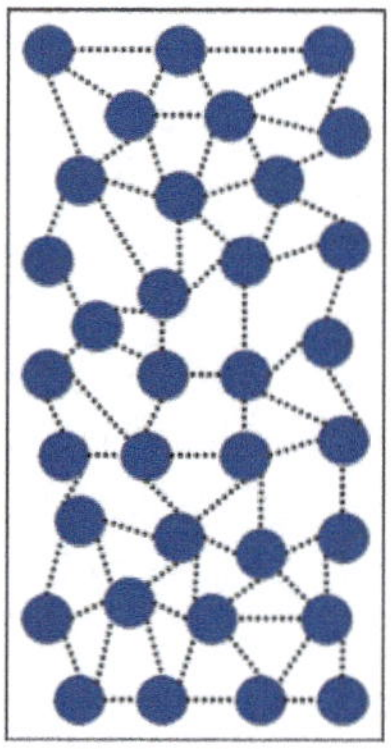	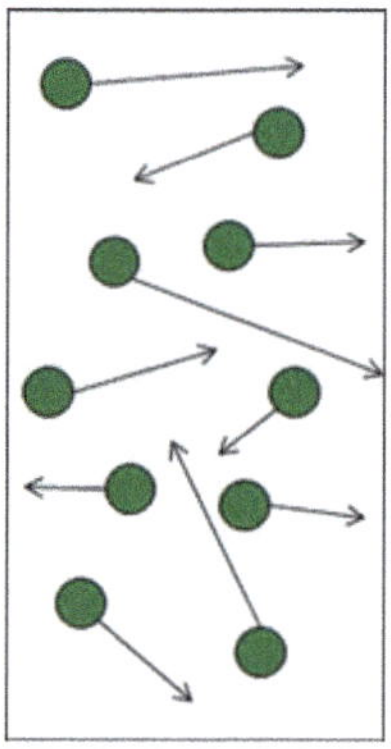

But even the natural scientists do not always have simple explanations in store. Here are a few examples:

- “When heat is added, the temperature of a substance generally increases.“
- “Heat is a measure of the energy content of a body.“
- “Thermal energy (= heat energy) is composed of the potential and kinetic energy of the particles of a system.”
- “Heat always flows from the warmer to the colder body.”

In Summary

Heat is a form of energy. This energy contained in the substance causes the particles that make up the substance to move minimally; you might say they "shake". The higher the amount of heat energy a particular substance carries, the more the particles that make it up move. Heat is therefore a measure of the kinetic energy of these particles.

If thermal energy is transferred, the colder substance heats up at the expense of the warmer substance, which cools down in the process. We will look at why this is the case.

The higher the thermal energy content, the faster the particles move and the higher the temperature we can measure.

3.1.2 What Is "Temperature"?

Temperature is a measure of the "thermal state" of a system. In the actual sense, the "mean kinetic energy" of the – incidentally disordered – particle motion is measured here.

All particles carry out an inherent motion caused by heat. However, this kinetic energy is not the same for all individual particles. Some move faster, others slower. We perceive the average value of this "movement" as temperature.

Let us compare the temperature with the wind force: An absolute calm corresponds to no particle movement or absolute zero. At some wind, leaves at trees move resp. particles vibrate around their resting position. At even more wind, already leaves are separated from tree. Just like this, bonds between particles within a solid will separate, it will melt. Finally, in a hurricane, trees break off or fall over. In terms of temperature, this means the breaking of chemical bonds.

3.1.2.1 Temperature Scales

Still most often used for temperature measurement thermometers filled with liquid (alcohol, mercury, etc.). Here a scale is attached, which allows the temperature to be read directly. The principle of operation is based on the volume expansion of liquids with an increase in temperature.

We (Center Europe) are most familiar with the **°C scale**. In 1742, Celsius[1] chose the freezing point of water (= 0 °C) and the boiling point of water (= 100 °C) as reference points and divided the range in between into 100 equal sections. He also proposed this division as a worldwide temperature scale; however, this was not implemented until 1948.

In the USA and Great Britain, the °F scale is very often used. Fahrenheit[2] chose three fixed points for his scale in 1714: A cold mixture (= 0 °F), the freezing point of water (= 32 °F) and the body temperature of humans (= 96 °F).

[1] Anders Celsius, 1701–1744, Swedish physicist.

[2] Daniel Gabriel Fahrenheit, 1686–1736, English physicist.

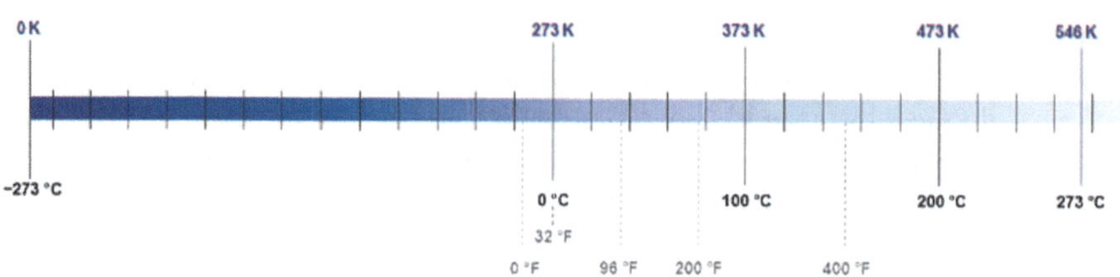

Fig. 3.1 Temperature scales

This results in the following comparative temperatures between the two scales:

$$0\ °C = 32\ °F \qquad 0\ °F = -17.8\ °C$$
$$100\ °C = 212\ °F \qquad 100\ °F = 38.8\ °C$$
$$35.5\ °C = 96\ °F \qquad 451\ °F = 232.8\ °C$$

- Conversion °C ↔ °F

°C → °F:	$t_F = \vartheta_C \cdot \frac{9}{5} + 32\ °F$	*(t_F = temperature in °F)*
°F → °C:	$\vartheta_C = t_F \cdot \frac{5}{9} - 17{,}8\ °C$	*(ϑ_C = temperature in °C)*

- Rule of thumb to get a feel for the order of magnitude °F to °C, the following rules of thumb can be used:

°C → °F:	$t_F = \vartheta_C \cdot 2 + 30\ °F$
°F → °C:	$\vartheta_C = t_F$: 2–15 °C

Another important temperature scale is the **Kelvin scale** (K; not [!] °K), which indicates the absolute temperature (see Sect. 4.6.1). The scale intervals °C and K are the same, therefore the conversion °C → K is simply done by adding 273. Compare Fig. 3.1.

3.1.3 Heat Conduction

What is the situation with heat transport? Somehow heat must (be able to) be transferred from a hot to a cold object.

In thermal conduction, the warmer particles (i.e. those that are moving more strongly) collide with the particles that are still cool, causing them to vibrate more strongly (Fig. 3.2). It is as if one billiard ball bumps into another and then both continue to move.

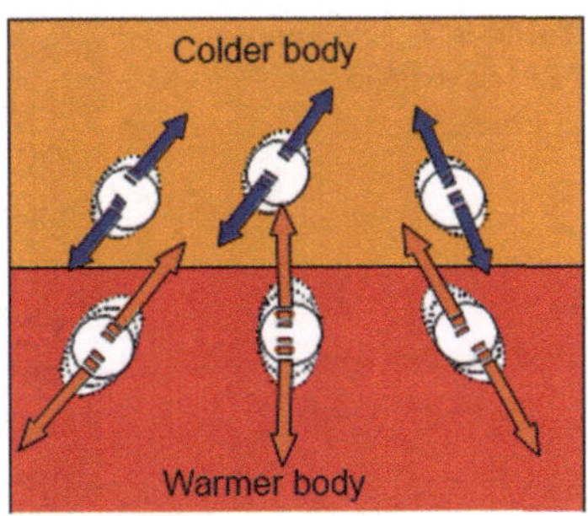

Fig. 3.2 Particle motion due to heat

The stronger vibrating particles need more space, the body expands. Of course, thermal expansion occurs in all three spatial directions, but many components (steel girders, railway tracks, power lines, ...) have a negligible width or height in relation to their length. Therefore, when considering heat conduction in solids, one usually also considers the resulting change in length, Fig. 3.3.

The so-called coefficient of linear expansion α (alpha) serves as a material-specific parameter for the temperature-related linear expansion. Examples are given in Table 3.2.

Calculation of Linear Expansion

To calculate the linear expansion of, for example, a steel beam, proceed as follows:

(a) Determine original length (ℓ_0) and temperature change (ΔT)
(b) Take the substance-specific coefficient of linear expansion from a table (e.g. Table 3.2).
(c) Calculation:

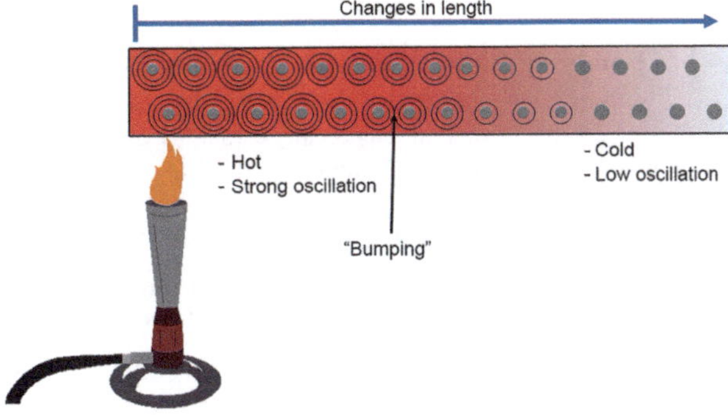

Fig. 3.3 Heat conduction

Table 3.2 Examples of linear expansion coefficients

Substance	Coefficient of linear expansion α
Aluminium	$23.8 \cdot 10^{-6}\,K^{-1}$
Lead	$29 \cdot 10^{-6}\,K^{-1}$
Bronze	$18 \cdot 10^{-6}\,K^{-1}$
Iron (pure)	$11.7 \cdot 10^{-6}\,K^{-1}$
Glass	$3.4...8.1 \cdot 10^{-6}\,K^{-1}$
Copper	$17 \cdot 10^{-6}\,K^{-1}$
Brass	$19 \cdot 10^{-6}\,K^{-1}$
Silver	$20 \cdot 10^{-6}\,K^{-1}$
Steel (stainless)	$10...16.5 \cdot 10^{-6}\,K^{-1}$
Zinc	$30 \cdot 10^{-6}\,K^{-1}$

$$\text{Formula}: \quad \Delta\ell = \ell_0 \cdot \alpha \cdot \Delta T$$

Original length of steel beam: 10 m;

Temperature change: 20 °C to 700 °C: $\Delta T = 700\ °C - 20\ °C = 680\ °C = 680\ K$

$$\text{Table value}: \quad \alpha \sim 14 \cdot 10^{-6}\,K^{-1}$$

$\Delta\ell = \ell_0 \cdot \alpha \cdot \Delta T = 10\ m \cdot 14 \cdot 10^{-6}\,K^{-1} \cdot 680\ K = 0{,}095\ m = 9{,}5\ cm$

Results:

The 10 m long steel beam expands by almost 10 cm (approx. 4 inch) when heated from room temperature to 700 °C. This can cause a load-bearing wall of a building to be literally pushed away. This can lead to a load-bearing wall of the building being literally pushed away, causing the building to collapse. ◀

Rule of Thumb

If the table values are given in the unit "$10^{-6}\,K^{-1}$", then the following applies approximately:

- For every 10 m of material length and every 10 °C increase in temperature, the substance expands by the table value in mm.

For steels, the change in length can be roughly estimated at 15 mm/100 m · °C, for copper at 18 mm and for aluminium at 24 mm.

3.1.3.1 Thermal Conductivity

The thermal conductivity λ (lambda) is a material property that is closely related to thermal conduction. It is a measure of how well a substance is able to transport thermal energy. Metals such as copper or silver are extremely good thermal conductors; they transport thermal energy quickly. Gases are very poor conductors of heat, so they act as thermal insulators. Naturally, a vacuum has the lowest thermal conductivity, since there are no particles to conduct the heat. Examples in Table 3.3.

Table 3.3 Thermal conductivities (examples)

Substance	Thermal conductivity λ
Aluminium	237 $Wm^{-1}K^{-1}$
Concrete	0.2...1.3 $Wm^{-1}K^{-1}$
Lead	353 $Wm^{-1}K^{-1}$
Iron (pure)	80.4 $Wm^{-1}K^{-1}$
Glass	0.7...0.9 $Wm^{-1}K^{-1}$
Copper	401 $Wm^{-1}K^{-1}$
Air	0.026 $Wm^{-1}K^{-1}$
Silver	429 $Wm^{-1}K^{-1}$
Rockwool	0.035 $Wm^{-1}K^{-1}$
Water	0.609 $Wm^{-1}K^{-1}$

The thermal conductivity of a wall is greater, the greater the quantity of heat Q that passes through the wall surface A in a given time t *for* the temperature difference $\Delta\vartheta$ (delta theta) and the thickness of the wall d. Here, the quotient $\Delta\vartheta/d$ is also referred to as the temperature gradient.

$$\lambda = \frac{Q}{A \cdot t \cdot \Delta\vartheta/d}$$

The amount of heat that is transferred is calculated according to:

$$Q = \lambda \cdot A \cdot t \cdot \frac{\Delta\vartheta}{d}$$

3.1.4 Heat Convection (Convection)

In liquids and gases, an increase in temperature causes the particles to move faster and more strongly. These now require more space – there is an increase in volume – and thus a decrease in the density of the substance.

The warmer, less dense (colloquially "lighter"), material rises to the top. The contained heat energy is transported along and can be released again at another place. Here, other substances with a lower energy content (lower temperature) can be heated.

As a result, the heat convection (short: convection) occurs: The substance is heated, rises, releases heat energy, sinks again, is heated again, ... (Fig. 3.4).

Central heating also works according to this convection principle (although here, too, it is usually supported by a pump).

However, the volume expansion of a liquid can also cause the container to "overflow" if it is already quite full before heating.

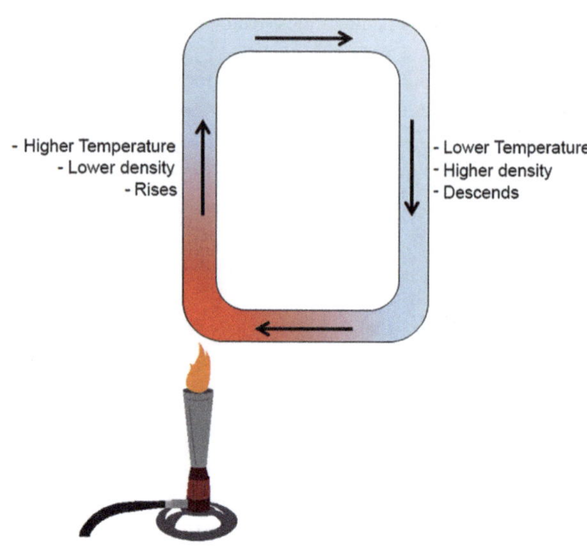

Fig. 3.4 Heat flow (convection)

3.1.4.1 Volume Expansion

The relevant parameter here is the **temperature-related volume expansion**, the so-called volume expansion coefficient γ (gamma). Examples of different volume expansion coefficients can be found in Table 3.4.

Calculation of the Volume Expansion

To calculate the volume expansion of a liquid, e.g. heating oil, proceed as follows:

Table 3.4 Coefficients of volume expansion (examples)

Substance	Coefficient of volume expansion γ
Gasoline	$10 \cdot 10^{-4}\ K^{-1}$
Diesel, heating oil	$9.6 \cdot 10^{-4}\ K^{-1}$
Diethyl ether	$16 \cdot 10^{-4}\ K^{-1}$
Ethanol	$11 \cdot 10^{-4}\ K^{-1}$
Machine oil	$9.3 \cdot 10^{-4}\ K^{-1}$
Methanol	$12 \cdot 10^{-4}\ K^{-1}$
Pentane	$16 \cdot 10^{-4}\ K^{-1}$
Mercury	$1.81 \cdot 10^{-4}\ K^{-1}$
Sulphuric acid	$5.7 \cdot 10^{-4}\ K^{-1}$
Water	$1.8 \cdot 10^{-4}\ K^{-1}$

1. determine original volume (V_0) and temperature change (ΔT)
2. Take the substance-specific volume expansion coefficient from a table (e. g. Table 3.4).
3. Calculation:

$$\text{Formula}: \quad \Delta V = V_0 \cdot \gamma \cdot \Delta T$$

Original volume of heating oil: 8 m^3 = 8000 L
Temperature change: 20 °C to 140 °C: ΔT = 140 °C – 20 °C = 120 = 120 K
Table value: $\gamma_{\text{heating oil}} = 9.6 \cdot 10^{-4}\ K^{-1}$

$$\Delta V = V_0 \cdot \gamma \cdot \Delta T = 8000\ L \cdot 9.6 \cdot 10^{-4}\ K^{-1} \cdot 120\ K = 921.6\ L$$
$$\approx 920\ L$$

Results:

The 8 m^3 of fuel oil take up almost 9 m^3 of volume after heating from 20 °C to 140 °C. A 10 m^3 container is filled to 80 % before heating and to 89 % afterwards. ◀

Rule of thumb
If the table values are given in the unit "$10^{-4}\ K^{-1}$", then the following applies approximately:

► For every m^3 of liquid volume and every 10 °C increase in temperature, the volume increases by the table value in litres.

For gasoline, diesel fuel and heating oil, the volume change can be roughly estimated at approx. 10 L/m^3 · 10 °C, for water at approx. 2 L.

3.1.4.2 Apparent Volume Expansion

Of course, when a liquid in a container is heated, the container also expands. This apparently reduces the volume expansion of the liquid.

To determine the apparent volume expansion, three times the coefficient of linear expansion (3α) of the container material must be subtracted from the coefficient of volume expansion of the liquid (γLiquid):

$$\gamma_{\text{appar}} = \gamma_{\text{liquid}} - 3\alpha_{\text{container}}$$

If we assume that the tank in the above example is made of steel with $\alpha = 12 \cdot 10^{-6}\ K^{-1}$, then we get:

$$\gamma_{appar} = \gamma_{liquid} - 3\,\alpha_{container}$$
$$\gamma_{appar} = 9.6 \cdot 10^{-4}\,K^{-1} - 3 \cdot 12 \cdot 10^{-6}\,K^{-1}$$
$$\gamma_{appar} = 9.24 \cdot 10^{-4}\,K^{-1}$$

Thus the fuel oil expands only by – apparent – 887 L. *(Please calculate yourself by replacing y by* γ_{appar} *calculated above).*

▶ However, the apparent volumetric expansion is of little importance during operation. It is better here ("safe side") to make estimates purely via the volume expansion of the liquid itself.

3.1.4.3 Volume Expansion in Completely Filled Containers

If a tank is completely filled with liquid (= no gas space present), then the expanding liquid very quickly exerts immense pressure on the tank when it is tightly closed. In the above example, this would be the case with a tank holding only 8500 L.

Since liquids are incompressible, i.e. cannot be compressed, high pressures are generated very quickly due to volume expansion.[3]

If, for example, you fill a 1-litre gas cylinder completely with water and close it, there is initially a pressure of 1 bar (= ambient pressure) in the container. When heated from 20 °C to 50 °C, the water expands by about 5 millilitres (mL). However, this expansion cannot take place in a container **completely** filled with liquid. As a result, the pressure increases massively: most containers will not survive the resulting 1000 bar (!) without damage.

3.1.5 Thermal Radiation

All substances give off energy in the form of so-called "heat radiation" (even if we are usually only aware of this when we are sitting too close to a campfire or can feel the warmth of a wall heated by the sun during the day in the evening). This constant loss of energy also causes any substance to always cool down to the ambient temperature. Here, the energy that is given off to or absorbed by the environment in the form of radiation (thermal radiation) is then in equilibrium.

Thermal radiation (also called infrared radiation or IR radiation for short) can be made visible with a thermal imaging camera, for example.

The higher the temperature of a body, the more heat is radiated. If the radiation hits a colder body (which emits less heat radiation), the particles are excited ("pushed") to vibrate more strongly. As a result, the body heats up, Fig. 3.5.

---

[3] An example calculation is not given here due to the complexity of the calculation. More details can be found in textbooks on physical chemistry.

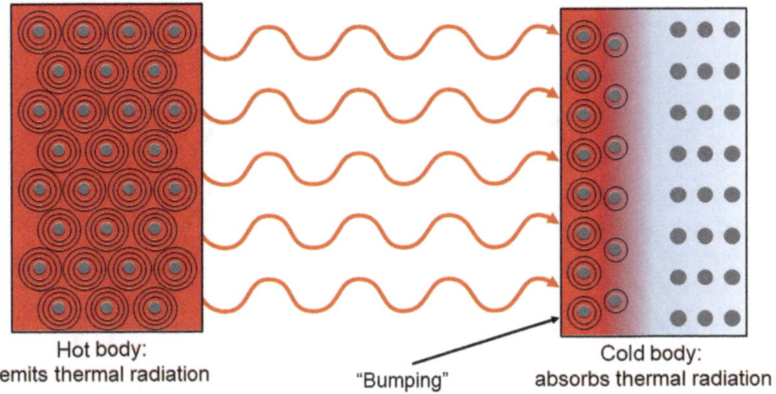

Fig. 3.5 Heat radiation

3.1.6 Heat Transport and Fire Occurrence

Heat transport is a major contributor to the spread of fire. An overheated chimney wall, for example (heat conduction), can ignite adjacent beams.

In the case of room fires, heat radiation and heat flow ensure that thermal energy is transferred to parts that are not yet burning and heats them up. This can then lead to a sudden ignition of the room furniture (*flash-over*). The rising of the hot fire gases, which collect under the ceiling of the room, creates a hot layer of smoke. This can also ignite (*roll-over*). Convection also causes stairwells to quickly fill with smoke, so that people on the upper floors can be literally trapped by the fire smoke (see Sect. 40.2.3).

Thermal radiation is often underestimated as a cause of fire spread. The large amount of energy transmitted here can even ignite relatively distant objects. Surrounding objects are therefore cooled to prevent them from heating up until the start of the fire.

3.2 Change of the State of Aggregation

In the following, we consider the changes in the states of aggregation (Fig. 3.6) and the associated differences in the "internal energy" of the respective states.

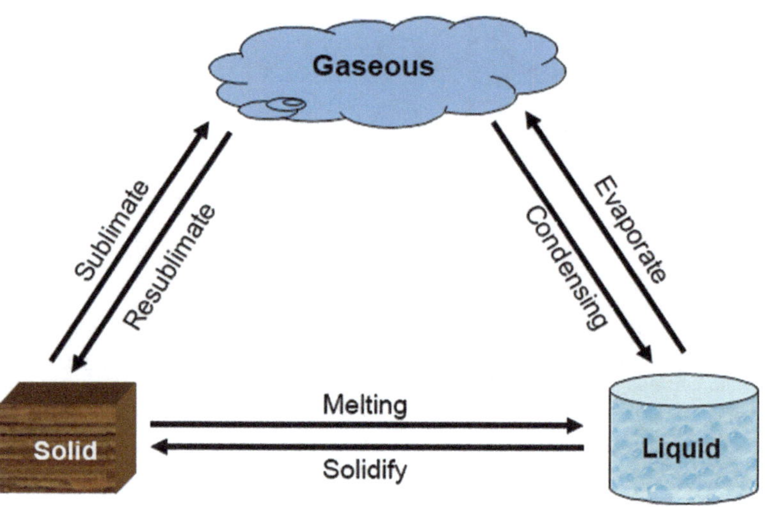

Fig. 3.6 Changes in the state of aggregation

3.2.1 Transitions Solid ↔ Liquid

3.2.1.1 Melting Point

▶ The transition from solid to liquid is called melting, and the associated temperature is called the melting point (mp).

A pure substance has a defined, substance-specific melting point. This can be used (as well as the boiling point) for identification and determination of purity. This works because contaminated substances always have a lower melting point than the pure substance.

In Germany sometimes the abbreviation "Fp" (*Festpunkt* = solid point) is used.

3.2.1.2 Solidification Point

▶ The transition from liquid to solid is called solidification, and the associated temperature is called the solidification point (sp) or freezing point.

For pure substances, the melting point and the solidification point are identical.

If a substance begins to melt, then the rigid bonds of the solid must first all be loosened. This process consumes energy. Therefore, the temperature remains constant until all bonds are loosened to such an extent that a liquid without solid content is present.

The reverse is true for solidification: the particles form rigid bonds with each other. Energy is released during this process. Despite further cooling, the temperature remains constant until the liquid has completely turned into a solid.

▶ During melting (solidification), the temperature remains constant for some time despite further energy supply (energy dissipation), since melting heat is consumed here (solidification heat is released).

In contrast to the boiling point (see Sect. 3.2.2), the melting/solidification point is independent of the ambient pressure.

3.2.2 Transitions Liquid ↔ Gaseous

3.2.2.1 Boiling Point

▶ The transition from liquid to gas is called boiling or boiling, the corresponding temperature is called boiling *point* (bp).

A pure substance has a defined, substance-specific boiling point. This (as well as the melting point) can be used for identification and determination of purity.

It should be noted here: If the impurity has a higher boiling point than the pure substance, the boiling point of the mixture increases. Conversely, if the impurity has a lower boiling point than the substance, the boiling point of the mixture is lowered.

For example, salt water boils at slightly above 100.0 °C, because the dissolved salt has a boiling point of over 1400 °C.

▶ The process when the vapor becomes a liquid is called condensation.

For pure substances, the boiling point and the condensation point are identical.

When a substance begins to boil, some of the liquid particles have absorbed enough energy to leave the "liquid bond". Other particles have not yet absorbed enough energy. The temperature therefore remains constant until all particles have passed into the gas phase. Only the resulting vapour can then heat up further and thus increase its temperature.

The reverse is true for condensation: the particles of the vapor are cooled and thus become a liquid. But only when all the particles of the vapor have become liquid again, the resulting liquid also cools down further.

▶ During boiling (condensation), the temperature remains constant for some time despite further energy supply (energy dissipation), since evaporation heat is consumed here (condensation heat is released).

In contrast to the melting point, the boiling point is strongly pressure-dependent (see Sect. 3.2.4).

3.2.2.2 Evaporation

The temperature is a measure for the average kinetic energy of the particles. I.e. slow, fast and extremely fast particles coexist at the same time. Some of the particles will therefore (almost) always be fast enough to leave the liquid. This process is called evaporation.

Of course, the particles that leave the liquid also take energy with them in the form of heat, so that the remaining liquid becomes colder. The cooling that occurs is called evaporative cooling.

The cooling of the human body during the evaporation of sweat in high heat is based on this principle. But also the cooling of containers and buildings by means of full jet or spray jet (or low expansion foam) makes use of this effect.

If you wear wet clothing for a longer period of time, hypothermia may occur. This is also a result of evaporative cooling.

3.2.3 Transitions Solid ↔ Gaseous

Just like liquids, solids can also "evaporate" or "vaporize" – although usually much more slowly. If this were not possible, some solids such as iodine, camphor or menthol would not have their own smell.

This process, known as sublimation, occurs when the most energetic particles of a solid overcome the binding forces due to their thermal motion (vibration). These particles then pass directly into the gas phase without passing through the liquid state. In the vapor phase, they exert a pressure called sublimation pressure. The sublimation pressure is, so to speak, the "vapor pressure of the solid".

The reverse process, the transition from the gas phase directly into the solid state, is called desublimation.

For most solids, however, the sublimation pressure is so small that it can be practically neglected. Important exceptions to this are, for example: Water (as ice), iodine, naphthalene (mothballs emit a characteristic odour), indigo or mercury (II) chloride ($HgCl_2$, "sublimate").

The sublimation ability of (frozen) water explains, for example, why laundry outdoors becomes dry even at temperatures below zero. Conversely, frost (= "solid water") forms directly from water vapour (= "gaseous water") at low temperatures.

In technology, this phenomenon is used for the purification of sublimable substances and to a large extent in freeze-drying.

In freeze-drying, the substance to be dried (e.g. an aqueous coffee extract) is frozen. The frozen substance is placed in a refrigerator and a vacuum is applied. The water sublimates and the remaining substance is anhydrous at the end – i.e. dry. In this way, instant coffee can be produced, sensitive medicinal substances can be dried and books and valuable documents that have been damaged by water can be saved.

3.2.4 Vapour Pressure Condition

In a sense, liquids represent the transition between solid and gas. A liquid consists of particles that are relatively free and always in motion. Unlike gases, however, the particles of a liquid are subject to mutual (intermolecular) forces of attraction. The particles move as close together as their own volume allows. These intermolecular forces of attraction are also called "cohesion" (Fig. 3.7).

The particles in a liquid have different velocities (at a given temperature), i.e. different kinetic energy. During heating, the average velocity of the particles in the liquid increases and more and more particles change into the gas phase. However, the more particles there are in the gas space above the surface of the liquid, the higher the pressure prevailing there (Fig. 3.8).

- The pressure created above the surface of a liquid by the liquid particles (atoms, molecules) that have become vaporous is called the vapor pressure. It is temperature-dependent and a substance-specific quantity.

- The higher the vapor pressure of a liquid, the more easily the substance volatilizes.

- If vapor pressure and ambient pressure are equal, then the liquid boils. In this process, the liquid suddenly begins to change to the gaseous state inside as well: Vapor bubbles are formed in the liquid.

In the case of an open liquid (e.g. water in a pot), the liquid boils when its vapour pressure is equal to the ambient pressure, i.e. reaches 1013 mbar at normal atmo-

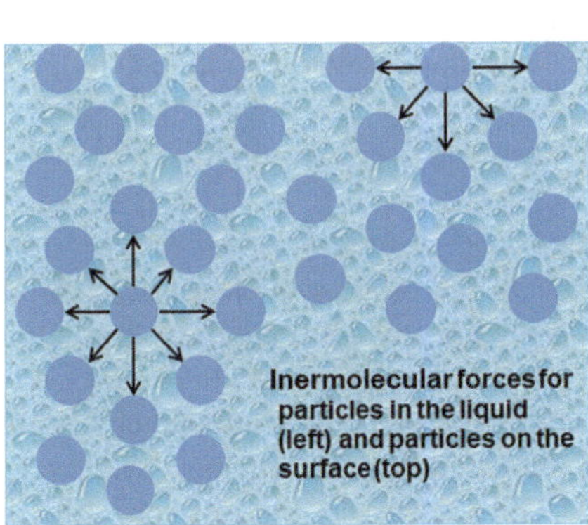

Fig. 3.7 Cohesive forces in a liquid

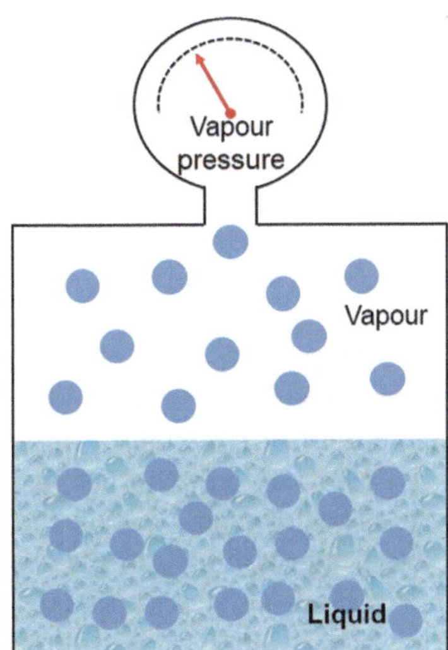

Fig. 3.8 Vapour pressure of a liquid

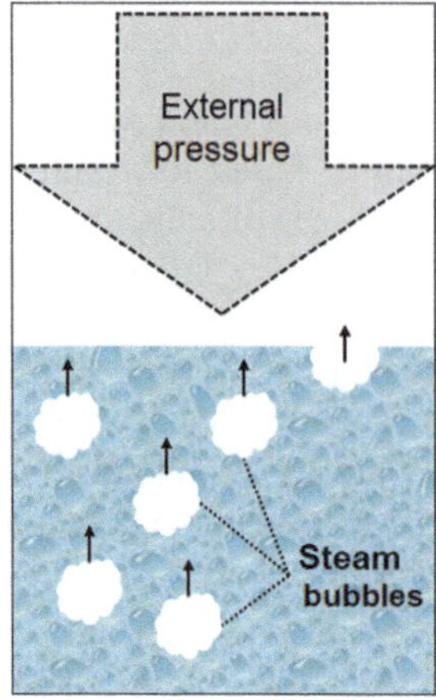

Fig. 3.9 Boiling process

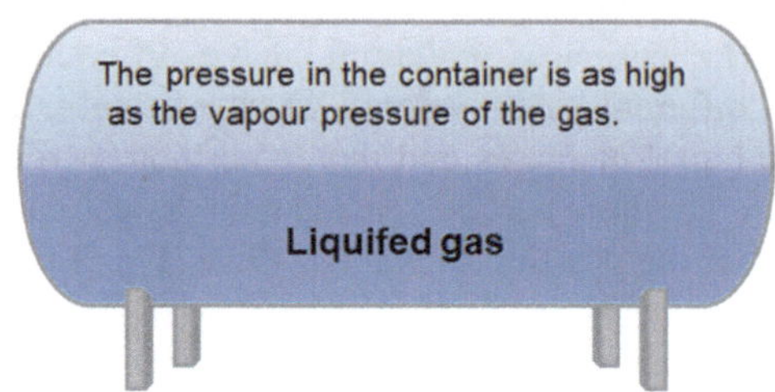

Fig. 3.10 Liquid gas tank

spheric pressure (Fig. 3.9). With falling air pressure (bad weather, low pressure), the boiling point then also falls. The effect is even more pronounced if you go to the high mountains. On the Zugspitze (Germany, 2962 m above sea level) the normal air pressure is only 693 mbar; water already boils at almost 90 °C. On Mount Everest (8000 m above sea level) the air pressure is only 363 mbar and water boils at 74 °C.

The situation is reversed in the pressure cooker. Here the pressure rises to approx. 1.8 bar and the water only boils when a temperature of 116 °C is reached.

In a liquefied gas tank (Fig. 3.10), the contained gas has been liquefied by pressure. The vapour pressure of the liquefied gas is higher than the ambient pressure outside the tank. At the same time, the vapor pressure of the gas is as high as can be expected based on the ambient temperature. However, the liquefied gas does not boil permanently. No additional energy is introduced, so that an equilibrium between condensation and evaporation is established.

The pressure dependence of the boiling point is also used in technology. Sensitive substances that would already decompose at their boiling point at normal pressure are subjected to vacuum distillation (i.e. distillation/evaporation at reduced pressure).

3.2.4.1 Water Steam Volatility

As we have already learned, a substance boils (at normal pressure) when its vapour pressure reaches 1013 mbar. With a trick, namely the process of steam distillation, which is often used in the chemical industry, a substance does not have to reach this vapour pressure itself.

Many high-boiling substances that are not very miscible with water can be distilled at temperatures around 100 °C if they are heated together with water (or if water steam is passed through them). The vapour pressure of the water is added to the small vapour pressure of the substance, so that a sum of vapour

Table 3.5 Examples of substances volatile in water vapour

Substance name	UN no.	Hazard no.	ERI Card	Hommel-MB	Features
Aniline	1547	60	6–09	31	Liquid, toxic, carcinogenic
p-Chloroaniline	2018	60	6–03	834	Solid, toxic, carcinogenic
Nitrobenzene	1662	60	6–03	210	Liquid, toxic, carcinogenic
p-toluidine	1708	60	6–09	329b	Solid, toxic, carcinogenic
Xylene	1307	30	3–05	208	Liquid, flammable
Naphthalene	1334	40	4–03	140	Solid

pressures of 1013 mbar or more is very easily reached. As a result, (much) water and (little) substance distils over (compare Sect. 4.10). The substance is thus "carried along" by the water vapour, so to speak.

If substances volatile in water steam (Table 3.5) are present at a fire event and come into contact with the water steam produced by the extinguishing measures, the same effect can occur here: The substance volatilises together with the water steam and then settles somewhere.

This can lead to unintentional contamination over a large area. This is particularly the case if toxic, water vapour volatile solids are affected.

These cases are certainly rare, but can lead to major problems if the "water vapour volatility" property is detected too late or not at all by the operations management.

3.3 Important Safety-Related Values

In the following, some important so-called "safety-related values" or "safety-related key figures" and their meaning are explained. Further key figures will be dealt with in later chapters.

In order to be able to evaluate operational situations, the boundary conditions of these key figures must always be kept in mind.

3.3.1 Evaporation Number

One parameter for liquids, especially flammable ones, is the evaporation number or evaporation rate, v_D.

Table 3.6 Evaporation number – significance as volatility

Meaning of the evaporation number	
Volatile	< 10
Medium volatile	10...35
Elusive	35...50
Highly volatile	> 50

Table 3.7 Evaporation figures (examples)

Evaporation figures	
Acetone	2
1-butanol	33
Iso-butanol	24
Butyl acetate	12
Diethyl ether	1
Ethanol	8.3
Ethyl acetate	2.9
Water	80
Xylene	17

► The evaporation number indicates how much slower a liquid evaporates than the reference liquid diethyl ether.

Examples of evaporation figures for some substances are given in Table 3.7. For the meaning of the numerical values/estimation of volatility on the basis of the evaporation number, see Table 3.6.

For evaporation intensity, i.e. the mass of substance that evaporates per minute in relation to the surface area, and the influence of the air flow velocity, compare Lit. [12-8].

► The greater the evaporation number, the slower the liquid evaporates.

3.3.2 Vapour Pressure

As already explained in Sect. 3.2.4, a high vapour pressure means a strong tendency of a liquid to change into the gaseous phase. A high vapour pressure is therefore associated with a low boiling point and usually also a low evaporation number.

The basic relationships are shown in Tables 3.8 and 3.9.

Naturally, no evaporation number can be given for gases. However, there is a clear correlation between the boiling point and vapour pressure of a gas.

For flammable gases and liquids it can be stated:

Table 3.8 Vapour pressure, boiling point, evaporation number (examples of liquids)

Substance	Vapour pressure (in mbar at 20 °C)	Boiling point (in °C)	Evaporation number
Iso-pentane	762	28	1
Diethyl ether	589,2	34.6	1,0
n-pentane	563	36.2	1
Petrol	~ 290	~ 35...200	~ 30...40
Acetone	246.6	56.1	2
Ethyl acetate	97	77.1	2.9
Ethanol	58.5	78.4	8
Iso-propanol	43.2	81	11
Toluene	8.8	110.6	6.1
Turpentine oil	5.86	~ 150...180	38
Glycerin	0.003	290	160

Table 3.9 Vapour pressure, boiling point (examples gases)

Substance	Vapour pressure (in bar at 20 °C)	Boiling point (in °CC)
Ethine (acetylene)	43.66	−83.6
n-propane	8.91	−42.3
n-butane	2.22	−0.5

▶ The higher the vapour pressure (the lower the boiling point), the easier it is for the particles to pass into the gaseous phase and the easier it is for a potentially explosive atmosphere to form on release.

3.3.3 Flash Point, Inflammation Point, Ignition Temperature

Flash point and inflammation point are key figures that only occur for flammable liquids. The ignition temperature, on the other hand, can be specified for practically all flammable substances (Fig. 3.11).

▶ The **flash point** is the lowest temperature at which a flammable liquid gives off enough vapors to cause it to ignite on contact with an ignition source.

The liquid does not continue to burn! There is only a short flare-up because the liquid cannot supply further vapours quickly enough.

However, other more flammable substances may be ignited by the flame. The liquid also gives off enough vapours to reach the LEL (lower explosion limit, see Sects. 3.3.4 and 38.5.2). This means that there is a risk of explosion.

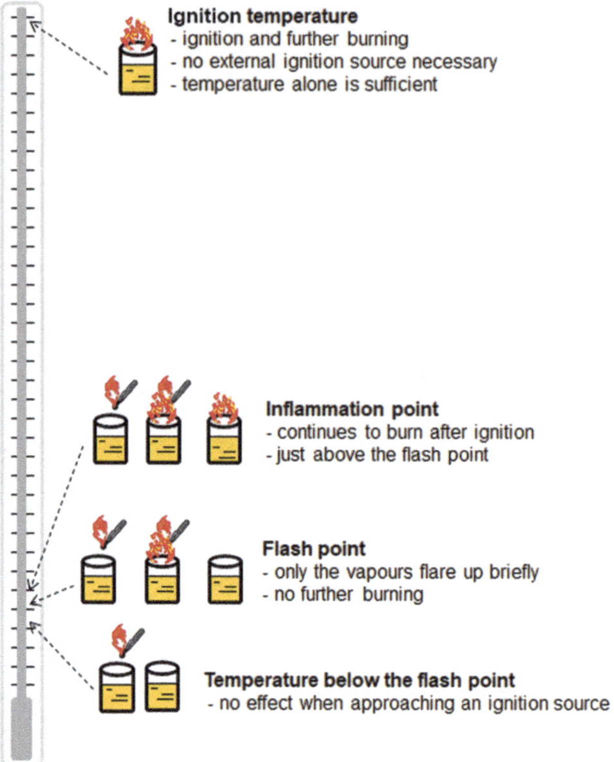

Fig. 3.11 Flash point, inflammation point, ignition temperature

▶ The lower the flash point, the more flammable the liquid.

The flash points generally increase with decreasing vapour pressure (i.e. with increasing molar mass, with increasing boiling point).

▶ The **inflammation point** is the temperature at which a flammable liquid gives off enough vapors to allow it to continue burning on its own after ignition of the vapors.

When the inflammation point is reached/passed, the liquid continues to burn after ignition.

However, vapours of flammable liquids and flammable gases do not only ignite by flames and sparks. They also ignite when the vapour/(gas) air mixture reaches a certain temperature: the ignition temperature. In the case of gases and vapours, the ignition temperature is the temperature that a heated wall must have in order to just bring the explosive mixture (with air) to ignition.

Table 3.10 Flash point, inflammation point, ignition temperature (examples liquids)

Substance	Flash point	Inflammation point	Difference	ignition temp.
Benzene	−11 °C	−9 °C	(+2 °C)	562 °C
Methanol	8 °C	13 °C	(+5 °C)	464 °C
n-octane	13 °C	19 °C	(+6 °C)	222 °C
Iso-propanol	14 °C	25 °C	(+11 °C)	400 °C
Pyridine	20 °C	25 °C	(+5 °C)	530 °C
n-propanol	23 °C	29 °C	(+6 °C)	371 °C

Table 3.11 Flash points and ignition temperatures (in comparison)

Substance	Flash point (in °C)	Ignition temp. (in °C)	Substance	Flash point (in °C)	Ignition temp. (in °C)
Toluene	**6**	535	*n*-octane	13	**222**
Methanol	**8**	464	*n*-propanol	23	**371**
n-octane	**13**	222	*Iso*-propanol	14	**400**
Iso-propanol	**14**	400	Methanol	8	**464**
Pyridine	**20**	530	Pyridine	20	**530**
n-propanol	**23**	371	Toluene	6	**535**
Heating oil	**>55**	~220	Benzene	−11	**562**

For combustible solids, it is the lowest temperature sufficient to ignite the solid.

▶ The **ignition temperature** is the lowest temperature that is sufficient to ignite a substance.

The ignition temperature is not a substance-specific property, but depends on the ambient conditions such as ambient pressure or oxygen content. Table 3.10 shows the values for normal air pressure and ambient air (21% by volume O_2).

It can be clearly seen that the burning point is only a few °C above the flash point. In addition, an explosive atmosphere can already arise when the flash point temperature is reached. The flash point, on the other hand, merely provides an indication of the fire hazard in the event of a spill of a flammable liquid. For these reasons, the flash point forms the basis for classifying flammable liquids according to their fire hazard.

Although benzene and toluene have low flash points, their ignition temperatures are unusually high (Table 3.11). Carbon disulfide and diethyl ether, on the other hand, not only have extremely low flash points. They also have very low ignition temperatures, so that they can easily be ignited by hot surfaces, such as those of a motor vehicle exhaust system.

For this reason, the classification into so-called temperature classes was made (Table 3.12). The temperature class indicates the maximum permissible surface temperature of an item of electrical equipment (e.g. a pump or a fan) and thus enables the selection of suitable equipment.

Table 3.12 Temperature classes

Temperature class	Max. Surface temperature	Ignition temperature of the substance	Examples (ignition temperature)
T1	450 °C	> 450 °C	Hydrogen (560 °C), acetone (535 °C) acetic acid (485 °C), methanol (464 °C)
T2	300 °C	> 300 °C	Ethanol (425 °C), *iso*-pentane (420 °C) Ethine (acetylene) (305 °C)
T3	200 °C	> 200 °C	Gasoline (~260 °C), diesel (~220 °C)
T4	135 °C	> 135 °C	Acetaldehyde (140 °C), diethyl ether (170 °C)
T5	100 °C	> 100 °C	–
T6	85 °C	> 85 °C	Carbon disulfide (95 °C)

If flammable liquids are involved in a fire, one possible fire-fighting method may be to cool the respective container to a temperature below the burning or flash point.

This method is a variant of the extinguishing method "smothering by reducing" (where reducing means reducing or completely stopping the fuel supply). In practice, however, this is only applicable to flammable liquids whose flash point is also significantly above the ambient temperature.

3.3.3.1 Water-Miscible Flammable Liquids

In the case of flammable liquids which are **infinitely** miscible with water, the flash point depends on the amount of water contained. The more water the liquid contains, the higher the flash point (compare Fig. 3.12).

This effect can be exploited in the course of fire-fighting operations under narrowly defined conditions:

- The liquid must be in a non-flammable container (e.g. steel tank).
- The volume of flammable liquid contained shall not exceed a few per cent of the total volume of the container.
- The liquid must not contain any components that are not miscible with water. These would otherwise "oil out" and possibly float.
- Dilution with water must be possible without danger (self-protection, accessibility).
- The instantaneous temperature of the flammable, water-miscible liquid must be below 100 °C (otherwise added water may evaporate and carry away the liquid).

There will only be a few cases in which these general conditions are completely fulfilled. In daily use, this special form of extinguishing effect "smother by reducing" will play a subordinate role, since the container will usually overflow before a noticeable effect is achieved. The potential danger of fire spreading due to overflowing of the container must be carefully weighed against the benefits.

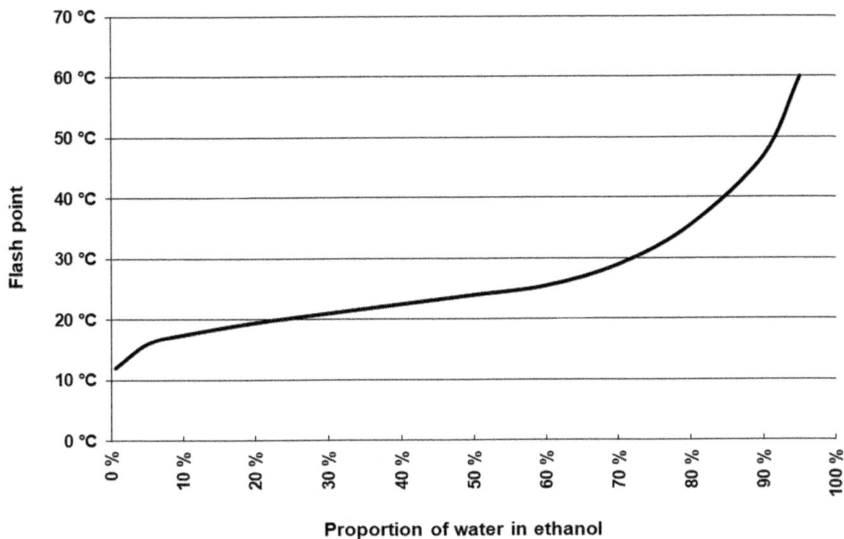

Fig. 3.12 Flash point of ethanol-water mixtures

3.3.4 Explosion Range

► An explosion is a particularly rapid combustion process which, once ignited, proceeds independently without any further supply of energy or air.

Explosions of flammable gases and vapours are only possible within a certain range (explosion range), which is limited by the lower explosion limit (LEL) and the upper explosion limit (UEL).

Below the LEL, the fuel concentration is too low; above the LEL, the oxygen concentration is too small to cause an explosion. However, vapour/air mixtures above the LEL can be ignited and burn extremely quickly (Fig. 3.13).

► The explosion range is the range between the LEL and the LEL. In this range, the mixture is explosible (i.e. capable of exploding).

► The lower explosion limit (LEL) and the upper explosion limit (UEL) are the concentration at which the gas/air or vapour/air mixture is just no longer explosive.

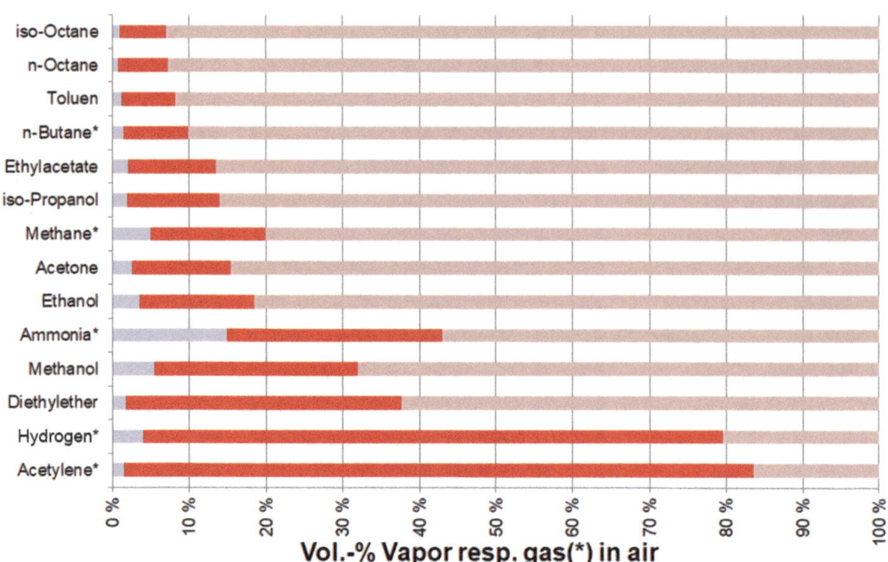

Fig. 3.13 Explosion limits

▶ The larger the explosion range, the greater the hazard posed by this substance.

In the case of flammable liquids, the quantity of fuel which must be completely vaporised can also be expressed in terms of liquid volume. To reach the lower or upper explosion limit, the quantities of liquid given in Table 3.13 must be vaporised in a 200 litre drum.

In short: Allowing a shot glass (= 20 mL) full of flammable liquid to evaporate in a 200-litre barrel filled only with air is usually sufficient to get within the explosion range. One part by volume of liquid thus forms 10 000 parts by volume of explosive vapour-air mixture.

If, however, a larger quantity of liquid than specified in Table 3.13 has evaporated in a drum, the UEL is exceeded. A spark inside the drum will therefore not cause an explosion. This is usually the case in drums that are empty of residue, as well as in the fuel tank of a vehicle.

The explosion limits are not purely substance-specific properties, they depend noticeably on the ambient conditions:

- An increase in the oxygen concentration (above the 21 % by volume of the ambient air) causes an extension of the explosion range (see Sect. 38.3.6).
- The addition of inert gases (nitrogen, CO_2, inert gases) narrows the explosion range.

Table 3.13 Volumes of evaporated liquid in the 200 L drum vs. explosion limits

Substance	Volume of liquid (for reaching the LEL)	Volume of liquid (for reaching the UEL)
Diethyl ether	15 mL	208 mL
Methanol	18 mL	88 mL
Ethanol	17 mL	72 mL
Acetone	15 mL	78 mL
Iso-propanol	13 mL	78 mL
Ethyl acetate	17 mL	92 mL
Toluene	10 mL	61 mL
n-octane	11 mL	87 mL

- A pressure reduction below the normal pressure (1013 hPa) reduces the explosion range.
- Rule of thumb for temperature dependence:
  - With a temperature increase of 100 °C the LEL decreases by approx. 10 %.
  - With a temperature increase of 100 °C the UEL increases by approx. 15 %.

▶ Within certain limits for the quantity ratios, all mixtures of easily combustible substances with air or oxygen are explosive.

For the relationship between explosion limits and explosion points, see Sect. 38.5.2.

3.3.4.1 About the Measurement Technology

In order to be able to determine at all whether a danger exists due to a potentially explosive vapour/air mixture, a corresponding measurement must first be carried out with an explosion limit warning device.

In the example in Fig. 3.14, the device was calibrated to nonane. Therefore, the device always displays the correct value for nonane only.

For ammonia, carbon monoxide, town gas, ethine (acetylene) and *n*-octane, reaching the threshold value "50 % LEL" is displayed prematurely. The real concentration (in % LEL) is lower than displayed.

For acetone, *iso*-octane, gasoline and carbon disulphide, on the other hand, the "50 % LEL" limit is already exceeded when the device displays the 50 %.

Instruments calibrated with nonane or toluene are more sensitive to most gases or vapors and are therefore better suited for measuring unknown mixtures than instruments calibrated to methane or carbon monoxide.

From an operational point of view, exceeding the OEG is often a problem. If flammable gases/vapours are driven out of a hazardous area by ventilation measures, the entire explosion range is passed through during these measures. The explosion hazard due to possible ignition therefore lasts for a relatively long time.

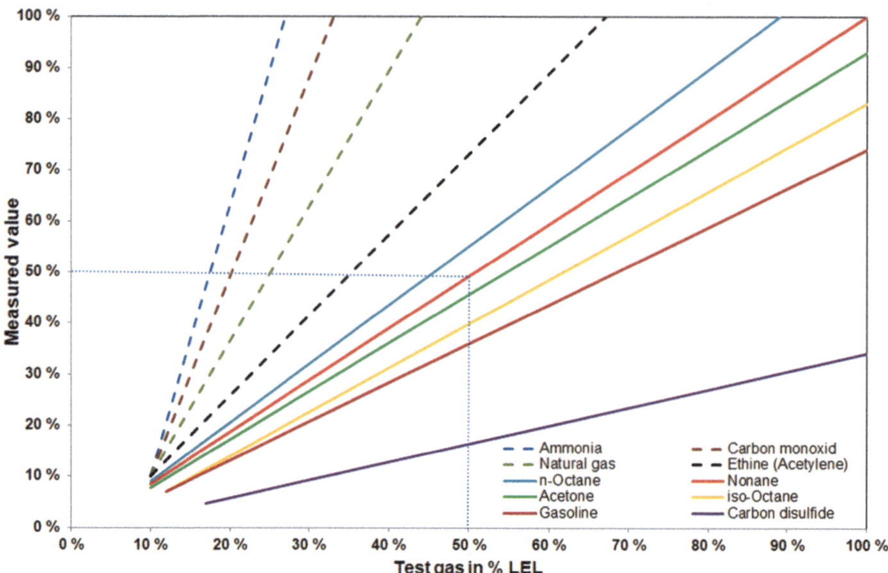

Fig. 3.14 Ex-meter display against calibration

3.3.4.2 Calculated Estimation LEL/UEL

If no data are available on the LEL or UEL for a gas or a vapour, the data can be estimated – at normal ambient pressure. For this purpose, the reaction equation of complete combustion is first established, the stoichiometric concentration of the fuel in air is determined from this and then multiplied by the given factors. The **factor 0.55** is used to calculate the **LEL** and the **factor 3.5 is** used to calculate the **UEL.**

For the combustion equation, see also Sect. 38.3.9.

Comparison of the display (measured value) of an explosion limit warning device with the real existing concentration (given in % LEL):

- Device is calibrated to Nonan → correct display over the entire range.
- The first 5 substances are displayed more sensitively. The device triggers earlier.
- For the last 4 substances, the device displays a lower measured value than the real concentration.

Example

LEL and UEL for 2-butene (C_4H_8) are to be determined.

4. Set up and solve the reaction equation for combustion

$$C_4H_8 + x\,O_2 \rightarrow a\,CO_2 + b\,H_2O \quad \Rightarrow \quad C_4H_8 + 6\,O_2 \rightarrow 4\,CO_2 + 4\,H_2O$$

5. Determination of the stoichiometric concentration in air The stoichiometric concentration is calculated according to

$$c_{st} = \frac{100}{1 + \frac{x}{0.21}}$$

where x is the number of oxygen molecules necessary for combustion. In our example, therefore:

$$c_{st} = \frac{100}{1 + \frac{6}{0.21}} = \frac{100}{1 + 28.57} = \frac{100}{29.57} = 3.38$$

The stoichiometric concentration of 2-butene in air is 3.4 % by volume.

6. Calculation of the LEL

The LEL is calculated according to: LEL = $c_{st} \cdot 0.55$.

$$\mathrm{LEL} = 3.4\ \mathrm{vol.}-\% \cdot 0.55 = 1.9\ \mathrm{vol.}-\%$$

7. Calculation of the UEL

The UEL is calculated according to: UEL = $c_{st} \cdot 3.5$.

$$\mathrm{UEL} = 3.4\ \mathrm{vol.}-\% \cdot 3.5 = 11.9\ \mathrm{vol.}-\%$$

Result

A swith literature values shows a relatively good agreement:

- LEL, calculated: 1.9 vol.-%; literature: 1.6 % –1.8 % by volume
- OEG, calculated: 11.9 vol.-%; literature: 9.4 % –11 % by volume ◀

3.3.5 Vapour Density Ratio

▶ The vapour density ratio indicates the ratio of the density of a gas/vapour to the density of air.

If the number is less than 1, the gas/vapour has a lower density than air (or colloquially: *"is lighter than air")*.

If the number is greater than 1, the gas/vapour has a greater density than air (is therefore *"heavier")*.

If the vapour density ratio is necessary to assess the overall situation (e.g. due to the question of whether vapours could possibly accumulate in lower-lying rooms), but the value is not available on site, it can be estimated from the ratio of the molar masses:

Air has an average molar mass of 28.8 g/mol (or 30 g/mol for rough calculations).

Methane (CH_4, the main component of natural gas) has a molar mass of 16 g/mol. That is, just under half. The vapour density ratio is therefore only a little more than 0.5. Methane is significantly lighter and rises.

The liquid gas butane (C_4H_{10}) has a molar mass of 58 g/mol and is therefore twice as dense ("heavier") than air. Butane therefore accumulates in lower-lying spaces.

Table 3.14 compares literature data with the rough calculation.

Some literature sources (e.g. Lit. [8-18]) do not give the ratio "vapour density to air", but the "density ratio of vapour-saturated air to air". A certain caution is required here, since it is not the vapors themselves that are considered, but a certain concentration of these vapors in the air. Therefore, completely different numerical values are obtained here.

Table 3.14 Relative density of vapours compared with molar masses (examples)

Substance	Formula	Molar mass (in g/mol)	Relative density (air = 1.0)	Rough calculation (air ~30 g/mol)
Methane	CH_4	16.04	0.55	~ 0.5
Ethine (acetylene)	C_2H_2	26.04	0.9	~ 0.85
n-propane	C_3H_8	44.1	1.56	~ 1.5
Ethanol	C_2H_5OH	46.07	1.59	~ 1.5
Acetone	$H_3C{-}CO{-}CH_3$	58.08	2.0	~ 2
n-butane	C_4H_{10}	58.12	2.05	~ 2
Iso-propanol	C_3H_7OH	60.1	2.07	~ 2
Iso-pentane	C_5H_{12}	72.15	2.49	~ 2.5
Diethyl ether	$(C_2H_5)_2O$	74.12	2.55	~ 2.5
Ethyl acetate	$CH_3COOC_2H_5$	88.11	3.04	~ 3
Toluene	$C_6H_5CH_3$	92.14	3.18	~ 3

Table 3.15 Relative densities of vapour-saturated air and of pure vapours (examples)

Substance	Relative density *vapour-saturated air* (air = 1)	Saturation of air at vol.-% substance	Relative density *pure vapour* (air = 1.0)
Toluene	1.067	3.1	3.18
iso-Propanol	1.049	4.5	2.07
Ethanol	1.036	6.1	1.59
Ethyl acetate	1.027	10.2	3.04
Acetone	1.260	25.7	2.0
Diethyl ether	1.958	61.4	2.55
iso-pentane	2.185	79.5	2.49

Some examples (sorted according to saturation in air) can be found in Table 3.15. It can be clearly seen that the values of the relative densities of vapour-saturated air (column 2 of the table) approach those of the relative density of pure vapour (column 4) with increasing concentration in the air (column 3).

3.3.6 Basic Tactical Rules

The key figures mentioned in Sects. 3.3.1, 3.3.2, 3.3.3, 3.3.4 and 3.3.5 provide some basic rules for the incident commander to assess the situation:

- Low boiling point = low evaporation number = high vapour pressure means easy evaporation of the liquid and thus leads to an extended danger area (= spatially extensive gas or vapour cloud).
- Flash points below ambient temperature increase the risk of explosion.
- The larger the explosion range, the greater the danger.
- Particularly dangerous are substances with
  - high vapour pressure / low boiling point / low evaporation number
  - low flash point
  - large explosion range
- Gases and vapours that are "heavier than air" are more problematic than "light" ones, as they travel along the ground and can accumulate in cellars, shafts, sinks, etc., where they can remain in dangerous concentrations for days (Fig. 3.15).

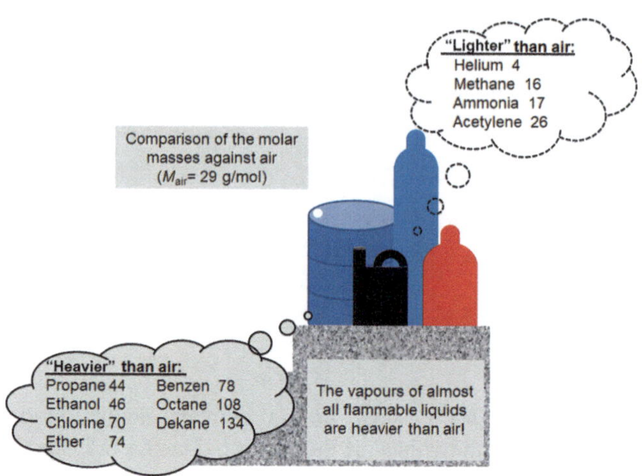

Fig. 3.15 Density of vapours/gases in relation to air

3.4 Specific Heat Capacity and Latent Heats

When a substance heats up, it absorbs energy from its surroundings. This can be a heater, a stove, a candle or any other warmer body. But also (thermal) radiation or flames and embers come into question as energy carriers.

3.4.1 Specific Heat Capacity

▶ The amount of energy required to heat a given mass of a substance (1 kg) by a given temperature (by 1 °C) is called the specific heat capacity.

The total heat energy that a substance absorbs when heated (or gives off when cooled) depends on the mass of the substance, its (substance-specific) heat capacity and the (measured) temperature change.

The heat quantity is calculated according to: $\Delta Q = m \cdot c \cdot \Delta\vartheta$.

Here are:

ΔQ	is the quantity of heat absorbed/discharged in the energy unit Joule (J).
m	is the mass of the substance in kg
c	is the specific heat capacity of the substance in kJ/ (kg · K)
$\Delta\vartheta$	the temperature change in °C

It should be noted that the same substance also has different heat capacities depending on its state of aggregation, since the specific heat capacity depends to a greater or lesser extent on the temperature.

Table 3.16 Specific heat capacities (examples). All values in kJ/(kg · K)

Solids		Liquids		Gases	
Aluminium	0.896	Gasoline	~ 2.0	Ammonia	2.060
Concrete	0.879	Benzene	1.74	Butane	1.658
Iron	0.452	Diethyl ether	2.37	Chlorine	0.502
Wood	~1.7	Acetic acid	2.05	Helium	5.193
Carbon dioxide$_{(s)}$	571.1	Ethanol	2.43	Carbon dioxide$_{(g)}$	0.833
Copper	0.382	Petroleum	2.14	Air	1.005
Magnesium	1.046	Mercury	0.139	Oxygen	0.920
Quartz glass	0.703	Sulphuric acid	1.39	Nitrogen	1.040
Water (ice)	2.060	Water	4.186	Steam	2.080

Examples of specific heat capacities of various substances are listed in Table 3.16.

The specific heat capacity (formula symbol "*c*") of a substance must not be confused with the heat capacity (formula symbol "*C*") of an object. The latter is the product of the specific heat capacity multiplied by the mass:

$$C = c \cdot m$$

In the past, the specific heat of water served as the unit of measurement for thermal energy. 1 calorie (cal) was defined as the amount of heat required to heat 1 g of water from 14.5 °C to 15.5 °C.

Conversion: 1 cal = 4.184 J.

3.4.2 Heat of Fusion

As already described in Sect. 3.2.1, when a sufficiently large portion of the substance melts, the temperature at the melting point does not change until the entire solid has melted. The energy that continues to be supplied is required to overcome attractive intermolecular forces (= forces between the particles of the substance) within the substance sample between the particles. This means that the temperature of the sample does not rise during the melting process.

▶ The specific heat of fusion (q) is a material constant and is given in the unit kJ/kg. The 'heat of fusion' is also called as 'heat of melting'.

Examples of specific heats of fusion of some substances are given in Table 3.17.

3.4.3 Heat of Evaporation

As already described in Sect. 3.2.4, the temperature at the boiling point of a sufficiently large portion of a substance does not change until the entire liquid has turned into vapour. In other words: The temperature of a boiling liquid does not rise even if further energy is added.

Table 3.17 Specific heats of fusion and evaporation

Specific heat of fusion q		Specific heat of evaporation r	
Aluminium	397 kJ/kg	Aluminium	10 500 kJ/kg
Lead	23 kJ/kg	Ammonia	1370 kJ/kg
Iron	277 kJ/kg	Benzene	394 kJ/kg
Copper	205 kJ/kg	Iron	6322 kJ/kg
Mercury	12 kJ/kg	Ethanol	840 kJ/kg
Silver	105 kJ/kg	Carbon dioxide	574 kJ/kg
Water	333 kJ/kg	Mercury	285 kJ/kg
Tin	60 kJ/kg	Water	2256 kJ/kg

The continuously added energy is needed to overcome the attractive intermolecular forces within the sample between the particles. This results in the temperature in the liquid not rising any further during the boiling process.

- The specific heat of vaporization (r) is a substance constant and is given in the unit kJ/kg.

- Heat of fusion and heat of evaporation are also referred to as latent (= hidden) heat quantities, since no increase in temperature can be observed despite a steady supply of energy.

Examples of specific heats of vaporization of some substances are given in Table 3.17.

3.4.4 Heat of Sublimation

The sum of the heat of fusion and the heat of vaporization is the heat of sublimation. However, it is only of significance for substances that change directly from the solid to the gaseous state, i.e. that sublimate.

For example, it is 569 kJ/kg for naphthalene, 226.5 kJ/kg for iodine and 568 kJ/kg for dry ice ($CO_{2(s)}$).

3.4.5 Heat Quantity Calculations

If one considers thermally insulated systems, then Richmann's rule of mixtures applies[4]:

- The warmer body releases as much energy as the colder body absorbs.

Written as a formula, the following applies

[4]Georg Wilhelm Richmann, 1711–1753, German-Russian physicist.

$$Q_{\text{off}} = Q_{\text{abs}}$$

Since for the quantity of heat Q holds:

$$Q = m \cdot c \cdot \Delta\vartheta$$

we can also write for the amount of heat given off:

$$Q_{\text{off}} = m_1 \cdot c_1 \cdot \Delta\vartheta_{\text{off}}$$

and for the absorbed heat quantity accordingly:

$$Q_{\text{abs}} = m_2 \cdot c_2 \cdot \Delta\vartheta_{\text{abs}}$$

Now, instead of the temperature difference $\Delta\vartheta$, we still insert the measured values (in °C). We denote the temperature of the warmer (energy emitting) substance as ϑ_1, that of the colder (energy absorbing) substance as ϑ_2 and the resulting mixing temperature as ϑ_M.

After substituting into above equations we get:

$$\begin{aligned} Q_{\text{off}} &= Q_{\text{abs}} \\ m_1 \cdot c_1 \cdot \Delta\vartheta_{\text{off}} &= m_2 \cdot c_2 \cdot \Delta\vartheta_{\text{abs}} \\ m_1 \cdot c_1 \cdot (\vartheta_1 - \vartheta_{\text{M}}) &= m_2 \cdot c_2 \cdot (\vartheta_{\text{M}} - \vartheta_2) \end{aligned}$$

3.4.5.1 Heat Mixtures without Changes of Aggregate State

Calculation Example 1

30 kg of water at 60 °C are mixed with 20 kg of water at 20 °C. What is the temperature? $c_{(\text{water})} = 4.19$ kJ/(kg K).

(For clarity, the units are omitted from the actual calculation here and in the following examples).

$$\begin{aligned} Q_{\text{off}} &= Q_{\text{abs}} \\ m_1 \cdot c_1 \cdot \Delta\vartheta_{\text{off}} &= m_2 \cdot c_2 \cdot \Delta\vartheta_{\text{abs}} \\ m_1 \cdot c_1 \cdot (\vartheta_1 - \vartheta_{\text{M}}) &= m_2 \cdot c_2 \cdot (\vartheta_{\text{M}} - \vartheta_2) \\ 30\ \text{kg} \cdot 4.19\ \text{kJ/(kg K)} \cdot (60\ ^\circ\text{C} - \vartheta_{\text{M}}) &= 20\ \text{kg} \cdot 4.19\ \text{kJ/(kg K)} \cdot (\vartheta_{\text{M}} - 20\ ^\circ\text{C}) \\ 30 \cdot 4.19 \cdot (60 - \vartheta_{\text{M}}) &= 20 \cdot 4.19 \cdot (\vartheta_{\text{M}} - 20) \\ 125.7 \cdot (60 - \vartheta_{\text{M}}) &= 83.8 \cdot (\vartheta_{\text{M}} - 20) \\ 7542 - 125.7\ \vartheta_{\text{M}} &= 83.8\ \vartheta_{\text{M}} - 1696 \\ 9218 - 125.7\ \vartheta_{\text{M}} &= 83.8\ \vartheta_{\text{M}} \\ 9218 &= 209.5\ \vartheta_{\text{M}} \\ \mathbf{44.0} &= \boldsymbol{\vartheta_{\text{M}}} \end{aligned}$$

Result: A temperature of exactly 44 °C results. ◀

Calculation Example 2

An iron cube of 750 grams, heated to 245 °C, is placed in a bucket of water given 4.5 liters of water of 18 °C.

What is the resulting temperature? Values: $c_{(water)} = 4{,}19$ kJ/ (kg K), $c_{(iron)} = 0{,}452$ kJ/(kg K)

$$\begin{aligned}
Q_{off} &= Q_{abs} \\
m_1 \cdot c_1 \cdot \Delta\vartheta_{off} &= m_2 \cdot c_2 \cdot \Delta\vartheta_{abs} \\
m_1 \cdot c_1 \cdot (\vartheta_1 - \vartheta_M) &= m_2 \cdot c_2 \cdot (\vartheta_M - \vartheta_2) \\
0.75\ \text{kg} \cdot 0.452\ \text{kJ/(kg K)} \cdot (245\ °\text{C} - \vartheta_M) &= 4.5\ \text{kg} \cdot 4.19\ \text{kJ/(kg K)} \cdot (\vartheta_M - 18\ °\text{C}) \\
0.75 \cdot 0.452 \cdot (245 - \vartheta_M) &= 4.5 \cdot 4.19 \cdot (\vartheta_M - 18) \\
0.339 \cdot (245 - \vartheta_M) &= 18.855 \cdot (\vartheta_M - 18) \\
83.055 - 0{,}339\ \vartheta_M &= 18.855\ \vartheta_M - 339.39 \\
422.445 - 0{,}339\ \vartheta_M &= 18.855\ \vartheta_M \\
422.445 &= 18.516\ \vartheta_M \\
\mathbf{22.8} &= \boldsymbol{\vartheta_M}
\end{aligned}$$

Result: The temperature rises only by almost 5 °C to almost 23 °C.

The relatively high heat capacity of water is clearly reflected in the relatively low heating. ◄

3.4.5.2 Heat Mixtures with Changes of Aggregate State

For calculations with a change of an aggregate state, the heat of fusion or evaporation must be entered into the equation on the "correct side". Thus, for cooling processes according to Q_{off}, for heating processes according to Q_{abs}. Of course, the different specific heat capacities solid – liquid – gaseous must also be taken into account.

Calculation Example

500 kg of steam at 108 °C, generated from waste heat, are to heat the water of a heating circuit to save energy. The volume in the circuit is 6400 L, the initial temperature 27 °C.

What is the resulting temperature? Values:

Values: $c_{(water)} = 4.19$ kJ/(kg K); $c_{(steam)} = 2.08$ kJ/(kg K); $r_{(water)} = 2256$ kJ/(kg K)

$$\begin{aligned}
Q_{off} &= Q_{abs} \\
m_1 \cdot c_1 \cdot \Delta\vartheta_{off} + m_1 \cdot r &= m_2 \cdot c_2 \cdot \Delta\vartheta_{abs} \\
m_1 \cdot c_1 \cdot (\vartheta_1 - \vartheta_M) + m_1 \cdot r &= m_2 \cdot c_2 \cdot (\vartheta_M - \vartheta_2) \\
500\ \text{kg} \cdot 2.08\ \text{kJ/(kg K)} \cdot (108\ °\text{C} - \vartheta_M) + 500\ \text{kg} \cdot 2256\ \text{kJ/kg} &= 6400\ \text{kg} \cdot 4.19\ \text{kJ/(kg K)} \cdot (\vartheta_M - 27\ °\text{C}) \\
500 \cdot 2.08 \cdot (108 - \vartheta_M) + 500 \cdot 2256 &= 6400 \cdot 4.19 \cdot (\vartheta_M - 27) \\
1040 \cdot (108 - \vartheta_M) + 1\ 128\ 000 &= 26\ 816 \cdot (\vartheta_M - 27) \\
112\ 320–1040\ \vartheta_M + 1\ 128\ 000 &= 26\ 186\ \vartheta_M - 724\ 032 \\
1\ 240\ 320–1040\ \vartheta_M &= 26\ 186\ \vartheta_M - 724\ 032 \\
1\ 240\ 320–1040\ \vartheta_M &= 26\ 186\ \vartheta_M \\
1\ 964\ 352 &= 28\ 266\ \vartheta_M \\
\mathbf{69.5} &= \boldsymbol{\vartheta_M}
\end{aligned}$$

Result: The temperature rises to almost 70 °C

3.4.6 Changes of Aggregate State and Extinguishing Agent Use

When using the extinguishing agents water, foam and carbon dioxide (CO_2) and even extinguishing powder, we also have to deal with changes of aggregate state (Table 3.18). These will be briefly considered here.

3.4.6.1 Water

Water is and remains the extinguishing agent par excellence. One of the great advantages is the large heat capacity of liquid water with around 4.19 kJ kg^{-1} K^{-1}. But the heat of evaporation and the heat capacity of steam are also high in relation.

For example, the energy required to heat a kilogram of liquid water from 0 °C to 100 °C is around 420 kJ. To convert this kilogram of water from 100 °C into water vapour at 100 °C, a further 2260 kJ of energy is required.

▶ Evaporating water removes five times more heat energy than heating it.

A large amount of energy can therefore be extracted from the fire with relatively little extinguishing agent, which in turn results in rapid extinguishing success. The best extinguishing effect is achieved by absorbing energy when the aggregate state changes from liquid to gaseous (Fig. 3.16).

Unfortunately, water is not suitable for fighting all damaging fires. Metal fires and fires of non-water-miscible flammable liquids cannot be extinguished with water.

Flammable liquids generally have a lower density (they are "lighter") than water. If water now penetrates into the burning liquid, it sinks. Sooner or later, the container will overflow and the fire will spread (Fig. 3.17).

Due to incomplete combustion, smear soot, gloss soot or flake soot can be deposited in chimneys. If this soot ignites, an ember fire occurs inside the chimney flue. In the case of **chimney fires**, the chimney heats up to several hundred °C. Temperatures of 1000 °C to 1500 °C can be reached. If one were to try to extinguish the fire with water, the result would be a steam explosion (see Sect. 39.1). One litre of water evaporates to produce 1700 litres of steam at 100 °C or 3000 litres of steam at 400 °C (see Sect. 4.8, Fig. 3.19). This abruptly released cloud of steam inevitably leads to the destruction of the chimney and can affect the entire building.

Table 3.18 Densities of solvents and extinguishing agents (comparison)

Densities of flammable solvents		Densities of extinguishing agents	
iso-pentane	0.62 kg/L	Water	1.0 kg/L
Diethyl ether	0.71 kg/L	Low expansion foam (LX)	~ 0.05 kg/L
Gasoline	~ 0.75 kg/L	Medium expansion foam (MX)	~ 0.015 kg/L
Heating oil	~ 0.86 kg/L	$CO_{2(s)}$	~ 1.56 kg/L
Toluene	0.87 kg/L	$CO_{2(g)}$	~ 0.002 kg/L

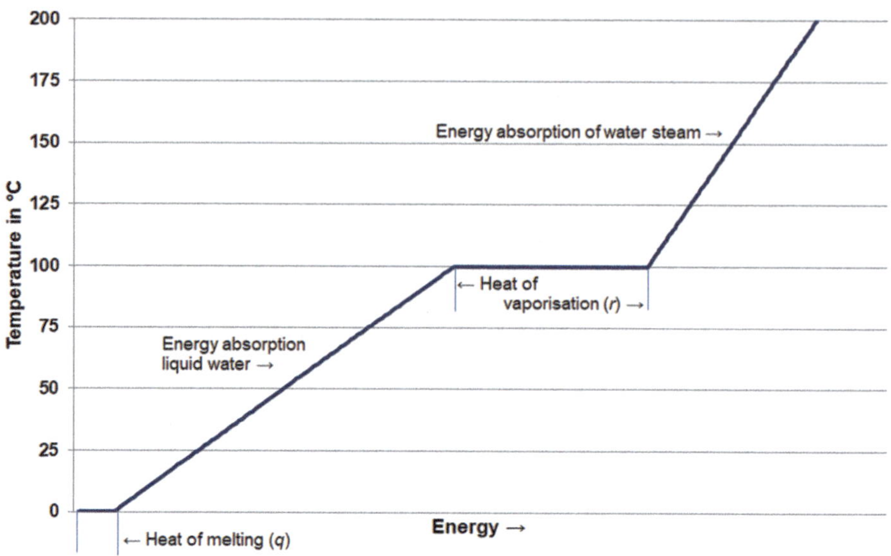

Fig. 3.16 Energy absorption of water

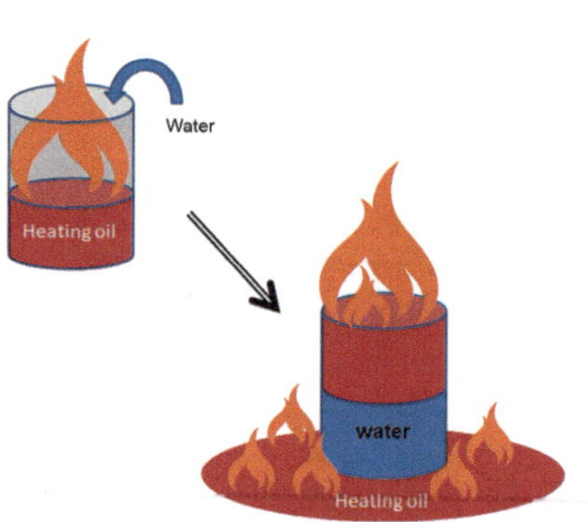

Fig. 3.17 Fire spread

Fig. 3.18 FCI-steam explosion between oil and water

> Chimney fires must never be extinguished with water.

The situation becomes even more precarious when the burning liquid has heated up to more than 100 °C. The penetrating water evaporates abruptly and takes the liquid with it. The penetrating water evaporates abruptly and carries the liquid with it. The mixture of water vapour and burning liquid swirls with the ambient air and ignites abruptly. This effect is known as a *boil-over*, as it often occurs in kitchens (Fig. 3.18).

Fig. 3.19 Water vapour evolution from 1 L water

A similar effect occurs when the burning liquid:

- heated to or above its boiling point (e.g. by heat conduction or flow), or
- an immiscible liquid of lower density and boiling point is present under the burning liquid.

Either the burning liquid boils itself or the liquid underneath comes to the boil. In both cases, the burning liquid is ejected *(boil over)* and the damage spreads. In the case of fires in tank farms, an area of 10 times the tank diameter can be reached by the ejection of burning liquid. The spreading speed of the burning oil can reach 9 m/s (32.5 km/h).

Water is also not suitable for fighting metal fires (Fig. 3.20). Due to the high combustion temperature, the water molecules are split – one also says "thermally dissociated". This produces a mixture of hydrogen and oxygen: oxyhydrogen gas.

Thus are dissociated:

at 700 °C: 0.000 03 % → approximately the temperature of a room fire
at 1000 °C: 0.003 %
at 1500 °C: 0.2 %
at 2000 °C: 2 % → ~ LEL of the mixture
at 2500 °C: 9 % → typical temperature for light metal fires
at 2700 °C: 15 %
at 3000 °C: 20 %
at 3200 °C: 40 %
at 3500 °C: 60 %

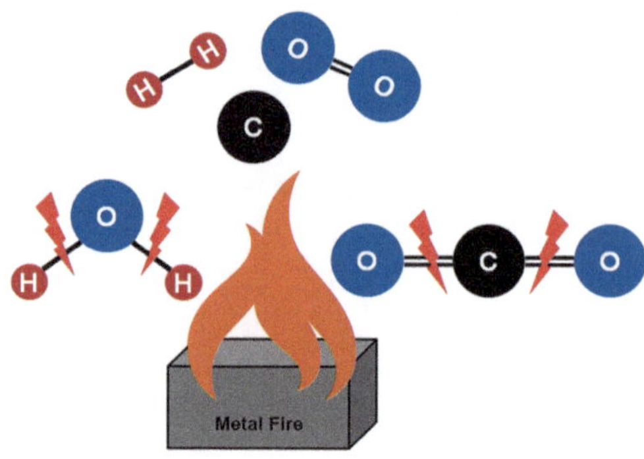

Fig. 3.20 Water and CO_2 vs. metal firing

State of aggregation / **Danger for:**	Solid	Liguid	Gaseous
Affected / injured persons	High	High	High
Emergency forces	low	high	high
Population	low	high to **medium**	high
Environment	low	**high** to medium	low

Fig. 3.21 "Hazard matrix aggregate states "chemical operations""

3.4.6.2 Foam

Unlike water, foam is also suitable for fighting fires involving burning liquids. The air trapped in the foam bubbles reduces the density immensely, so that the foam floats up and thus blocks the air supply.

Roughly speaking, you only have to divide the density of the water by the expansion ratio to get the approximate density of the extinguishing foam. The expansion ratio is the ratio of foam solution to foam.

With regard to the danger of a FCI-steam explosion between oil and water, however, the same applies as already mentioned for water, since the foam decomposes due to heat, among other things, and thus releases water again.

3.4.6.3 Carbon Dioxide

CO_2, which comes out of portable fire extinguishers as "dry ice snow", is well suited for fighting liquid fires, as it displaces the (air) oxygen. However, the energy absorption is very low, therefore almost no cooling effect on the fire material can be detected. As a gas, carbon dioxide then evaporates again relatively quickly. Here there is a risk of re-ignition, especially through hot surfaces.

CO_2 is also unsuitable for fighting metal fires. As with water, the CO_2 molecule is dissociated by the high temperatures (Fig. 3.20). Magnesium burns even when completely surrounded by dry ice according to:

$$2\,Mg + CO_2 \rightarrow 2\,MgO + C$$

3.4.6.4 Extinguishing Powder

When extinguishing ember fires, the extinguishing powder melts on the surface of the burning material and thus prevents further oxygen influx. Extinguishing powder for glowing fires, as used for substances of fire class A (fires of solid, ember-forming substances), already melts at temperatures below 100 ° C and also develops a low insulating effect, which protects the fire material from further thermal decomposition. The melting point of metal burning powder is much higher at approx. 800 °C. Particularly in the case of metal fires, care must be taken to ensure that the powder layer is completely tight. Exposure times of 10 minutes and more until a metal fire is extinguished are not uncommon. In order to control the extinguishing success, especially with metal fires, a thermal imaging camera can provide valuable services.

However, heat extraction by the extinguishing powder plays only a very minor role in the extinguishing effect with both types of powder.

3.4.7 Aggregate States in NBC Operations

The question of the physical state of the substances involved/spilled should always be one of the first questions asked when the emergency services arrive. Possible measures and hazards can already be derived at this stage.

Possible measures may include:

solids: cover, pick up and transfer to suitable containers.

Liquid substances: Dike, use binding agents, seal canal inlets

Gaseous substances: Warn population, check use of ventilators/water cannons.

In accordance with the (German) hazard matrix “Hazards at the operating scene”, the hazard matrix shown in Fig. 3.21 applies to the hazards arising purely from the physical state of the substances involved in a chemical incident.

▶ Please note that in the case of operations involving radioactive or biological hazards, it is not possible to make an estimate using the table (Fig. 3.21). Solids, especially powders, can represent a considerable hazard in A (atomic) and B (biological) operations.

Gases

4

The term "gas" probably goes back to Helmont,[1] who was concerned with the gas produced during fermentation, which he called "gas sylvestris" (lat. *Sylvestris* = wild) – carbon dioxide. Black[2] then called this gas "fixed air" in 1756, as he was able to extract it from inorganic substances (e.g. magnesium carbonate) in which it was bound.

Scheele[3] then discovered another component of air in 1774: the "fire air" (since it fsupported or made combustion possible in the first place) – known today as oxygen. Cavendish[4] determined the oxygen content of air (20.84%) and discovered hydrogen as a component of water.

Nitrogen as a component of air was discovered by D. Rutherford[5] in 1772.

However, extensive experiments with air had already been carried out around 100 years earlier:

Guerike,[6] who may be called the modern "inventor of the air pump", not only discovered the "pumpability" of air. He explained air pressure fluctuations as a weather phenomenon and showed that air can be weighed. The famous experiment with the "Magdeburg hemispheres" was probably carried out for the first time in 1657.

Torricelli[7] had already constructed a mercury barometer in 1644, and Pascal[8] continued these experiments. In the process, he realized that air pressure decreases with altitude. The "Pascal" (Pa) is today the unit of pressure. (Compare also Table A.4).

[1] Johan Baptista van Helmont, 1580–1644, Flemish naturalist.

[2] Joseph Black, 1728–1799, Scottish chemist and physicist.

[3] Carl Wilhelm Scheele, 1742–1786, Swedish pharmacist and chemist.

[4] Henry Cavendish, 1731–1810, English chemist and physicist.

[5] Daniel Rutherford, 1749–1819, Scottish botanist and chemist.

[6] Otto von Guericke, 1602–1686, German physicist and engineer.

[7] Evangelista Torricelli, 1608–1647, Italian mathematician and physicist.

[8] Blaise Pascal, 1623–1662, French mathematician and physicist.

T. Schmiermund, *The Chemistry Knowledge for Firefighters*,
https://doi.org/10.1007/978-3-662-64423-2_4

However, the first documented experiments with air and plans of pneumatically functioning machines were made by Philon[9] in the 2nd/third century B.C. He developed, among other things, pumps that were very similar to today's bucket pumps.

In all these early investigations, the pressure and temperature conditions were so moderate that deviations from the "ideal gas" were hardly observed.

Pressure Units

Today's legal unit of pressure is the "Pascal" (Pa). In weather reports, for example, the pressure of low and high pressure areas is given in "hectopascals" (hPa). In everyday use, the "bar" (bar) is more familiar as the so-called "SI-conform unit" for indicating pressures, e.g. for car tyres. The following applies: 1bar = 100,000 Pa; 1mbar = 100 Pa = 1 hPa.

The unit "Torricelli" (Torr) or also "mm mercury column" is today only of importance for the indication of the blood pressure (760 Torr = 1 bar = 10^5 Pa).

4.1 Ideal Gas

Ideal gases have certain properties that are favorable for models, but under certain conditions no longer exactly represent the real behavior of a gas.

Ideal gases have the following properties:

- Negligible mass of the individual gas particle
- Negligible intrinsic volume of the gas particles
- No interactions between the gas particles

These "conditions for an ideal gas" are most likely to be met at low pressures and high temperatures.

▶ All gas laws are based on the model of the ideal gas.

4.2 Pressure and Temperature

We have already learned about temperature as heat movement. The faster particles (atoms, molecules, ions) move, the higher the temperature of the system.

In the case of gases in particular, however, pressure is also an important variable. The pressure of an enclosed gas volume is caused by the gas particles moving freely and in a straight line and bouncing off the container wall. We perceive this "bouncing" as the "pressure" of a gas. It holds:

- The more particles per unit volume that hit the container wall, the higher the pressure.

[9] Philon of Byzantium, third-second century B.C., Greek inventor and author.

- The faster the gas particles are (i.e.: the more kinetic energy they possess = the higher the temperature of the gas), the more violently they collide with the wall, the higher the pressure.

4.3 Boyle-Mariotte Law

Gases, unlike liquids or solids, prove to be compressible. When a gas volume is compressed, the distances between the gas particles are reduced. As a result, the pressure increases because the particles in the smaller volume collide more frequently with the container wall.

If this takes place without a temperature change, it is called an isothermal process.

Independently of each other, Boyle[10] (1662) and Mariotte[11] (1676) recognized that the product of pressure *(p)* and volume (V) always remains the same:

$$p \cdot V = \text{const.}$$

This allows the pressure or volume resulting from the change to be calculated even in the case of changes in pressure or volume:

$$p_1 \cdot V_1 = p_2 \cdot V_2$$

Boyle-Mariott's law is sometimes just referred to as "Boyle's law".

$$p \cdot V = \text{const.}$$
$$p_1 \cdot V_1 = p_2 \cdot V_2$$

Volume and pressure change proportionally to each other (Fig. 4.1)

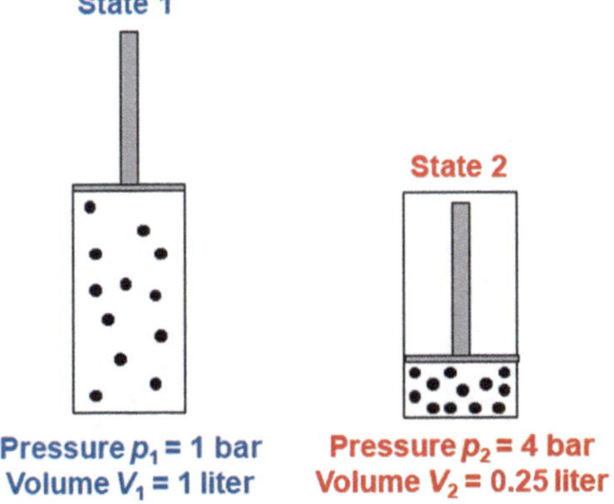

Fig. 4.1 Boyle-Mariotte law

[10] Robert Boyle, 1627–1691, English naturalist.

[11] Edme Mariotte, 1620–1684, French physicist.

4.4 Law of Amontons

If an enclosed gas volume is heated, the pressure increases by 1/273 of the initial pressure for each °C increase in temperature. Amontons[12] was already able to prove this by measurements at the end of the seventeenth century.

The gas particles move faster and therefore collide more frequently and with higher kinetic energy against the vessel wall. This is externally noticeable as an increase in pressure. Changes at constant volume are called *isochoric.*

The quotient of pressure and temperature always remains the same:

$$\frac{p}{T} = \text{const.}$$

This allows the pressure or temperature resulting from the change to be determined even in the event of changes in pressure or temperature.

$$p_1 : T_1 = p_2 : T_2$$

or

$$p_1 \cdot T_2 = p_2 \cdot T_1$$

The "Law of Amontons" is sometimes called the "second part of Gay-Lussac's law" (Fig. 4.2).

$$\frac{p}{T} = \text{const.}$$
$$\frac{p_1}{T_1} = \frac{p_2}{T_2}$$
$$p_1 \cdot T_2 = p_2 \cdot T_1$$

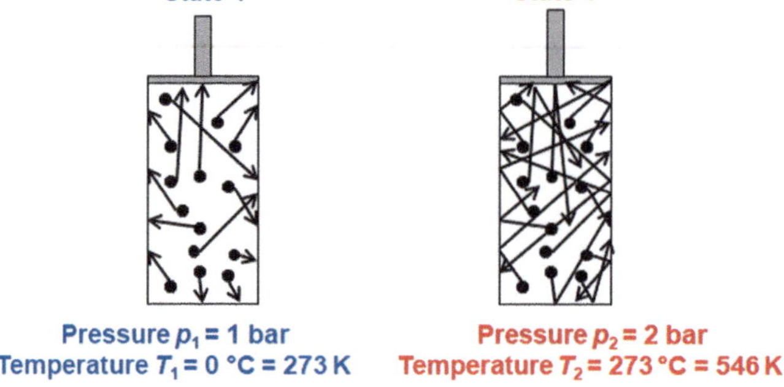

Fig. 4.2 Law of Amontonss

[12]Guilliaume Amontons, 1663–1705, French physicist.

4.5 Law of Gay-Lussac

Charles[13] investigated the volume changes at constant pressure (isobaric change of state) and different temperatures as early as 1787, but did not initially publish the results. Gay-Lussac[14] carried out similar investigations in 1802.

If an enclosed quantity of gas is heated at constant pressure, the volume increases by 1/273 of the initial volume for each °C increase in temperature.

If you increase the temperature of a gas, the kinetic energy of the particles increases. If the volume remains constant, the pressure increases. But if you leave the pressure unchanged, the volume increases.

The quotient of pressure and temperature remains the same:

$$\frac{V}{T} = \text{const.}$$

This allows the temperature or volume resulting from changes in volume or temperature to be determined.

$$V_1 : T_1 = V_2 : T_2$$

$$\frac{V}{T} = \text{const.}$$
$$\frac{V_1}{T_1} = \frac{V_2}{T_2}$$

or

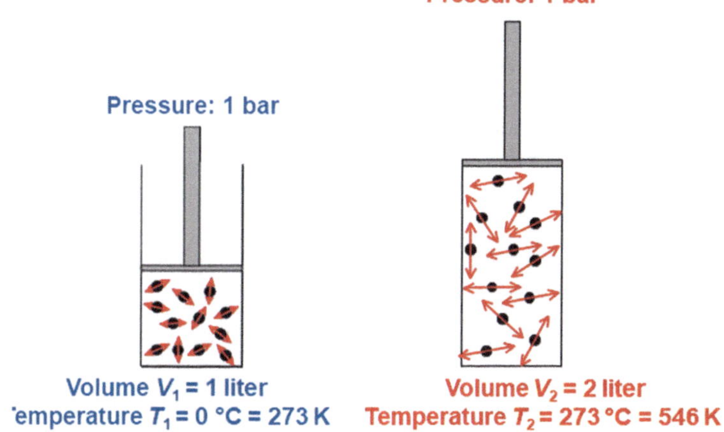

Fig. 4.3 Gay-Lussac law

[13] Jaques Charles, 1746–1823, French physicist.

[14] Joseph Louis Gay-Lussac, 1778–1850, French physicist and chemist.

$$V_1 \cdot T_2 = V_2 \cdot T_1$$

The "Law of Gay-Lussac" is sometimes referred to as the "Law of Charles" (Fig. 4.3).

4.6 General Gas Equation

By combining the three aforementioned gas laws, one arrives at the general gas equation:

$$\frac{p \cdot V}{T} = \text{const.}$$

or

$$\frac{p_1 \cdot V_1}{T_1} = \frac{p_2 \cdot V_2}{T_2}.$$

This equation can already be used to perform a great many calculations for gases.

4.6.1 Absolute Zero

From Amontons' law (Sect. 4.4) it follows that at a temperature of −273 °C the pressure must be zero. At even lower temperatures, this would result in a **negative pressure**. This is **not possible!**

Similarly, it follows from Gay-Lussac's law (Sect. 4.5) that at a temperature of −273 °C the volume of a gas is zero. This is not practically possible (intrinsic volume of the gas particles), but it follows from the rule. At even lower temperatures, a **negative volume** would result according to this calculation. A negative volume is also **not possible!**

Consequently, the temperature cannot fall below −273 °C (more precisely: −273.15 °C). This temperature is called "absolute zero". In honour of the physicist W. Thomson[15], the later Lord Kelvin, "Kelvin" (K) was chosen as the unit of absolute temperature.

The temperature scales of the °C-division and the K-scale differ only in the choice of the zero point. Therefore:

$$0\ ^\circ\text{C} = 273.15\ \text{K} \quad \text{and} \quad 0\ \text{K} = -273.15\ ^\circ\text{C}$$

The conversion is therefore done by adding or subtracting the difference |273|.

Equal volumes always contain the same number of gas particles under the same conditions (pressure, temperature).

[15]William Thomson, 1824–1907, English physicist, ennobled in 1892, then Lord Kelvin of Largs.

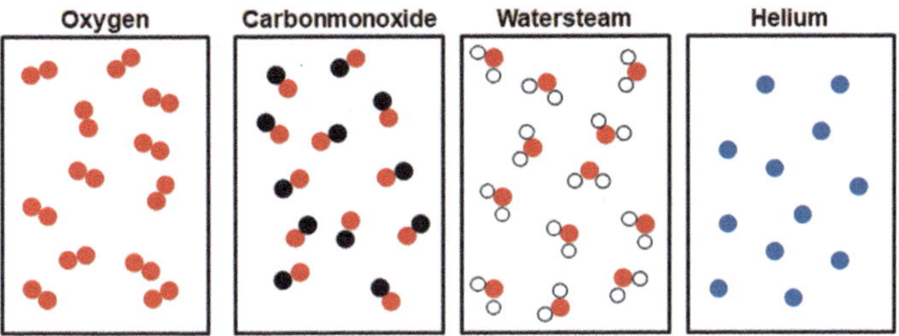

Fig. 4.4 Avogadro's theorem

4.7 Avogadro Theorem

After it had been established that all gases behave in the same way with respect to temperature changes and that in chemical reactions with gases their volumes exhibit simple integer ratios, Avogadro[16] concluded in 1811 that equal volumes also contain an equal number of particles (Fig. 4.4).

So the sentence reads:

▶ Equal volumes of gases contain the same number of particles (atoms, molecules) at the same pressure and temperature.

Avogadro's considerations made it possible to introduce a new quantity: the amount of substance n (cf. Sect. 5.5.3). If we now link the amount of substance with the general gas equation, we obtain the universal gas equation (Sect. 4.8).

4.8 Universal Gas Equation

The universal gas equation links pressure, temperature and volume of a gas with the amount of substance. For this, a certain (additional) factor is necessary: The universal gas constant R. In contrast to the "general gas equation", one is now independent of the portion of substance. The universal gas equation is:

$$p \cdot V = n \cdot R \cdot T$$

The universal gas constant depends on the chosen system of units. Its value is:

- $8.314\ \mathrm{Pa \cdot m^3 \cdot K^{-1} \cdot mol^{-1}}$ (pressure in Pascal, volume in m^3)
- $0.083\ 14\ \mathrm{bar \cdot L \cdot K^{-1} \cdot mol^{-1}}$ (pressure in bar, volume in L)
- $83.14\ \mathrm{hPa\ L \cdot K^{-1} \cdot mol^{-1}}$ (pressure in mbar or hPa, volume in L)

[16] Amedeo Avogadro, 1776–1856, Italian physicist.

Calculation Example "Universal Gas Equation"

We are looking for the volume of water vapour at 100 °C and at 400 °C that is produced from one litre of water.

The amount of substance *n of* water is calculated according to:

Amount of substance = mass/molar mass $n = m/M$. With the molar mass $M(H_2O) = 18$ g/mol and 1 L water = 1 kg = 1000 g, the amount of substance is $n(H_2O) = 55.56$ mol.

The volume of water vapour at 100 °C (= 373 K) is then calculated according to:

$$V = \frac{n \cdot R \cdot T}{p} = \frac{55.56 \text{ mol} \cdot 0.083\,14 \text{ bar} \cdot \text{L} \cdot 373 \text{ K}}{1.013 \text{ bar} \cdot \text{mol} \cdot \text{K}} = 1700 \text{ L}$$

Similarly, a volume of about 3 m^3 results for steam at 400 °C· ◀

4.9 Standard Conditions

In order to be able to compare the volumes of gases, it is necessary to define "comparison conditions" in terms of temperature and pressure.

As a standard, the so-called standard conditions (also called "normal condition" – German abbreviation: i. N., in English *"STP"* from *standard temperature and pressure)* and set these arbitrarily:

- Pressure: p_n = 1013.25 hPa = 1013.25 mbar = 1.013 25 bar
- Temperature: T_n = 273.15 K = 0 °C

In the further one refers to the amount of substance of 1 mol and the molar volume of a gas:

- Amount of substance: n_n = 1 mol
- Volume: $V_{m,0}$ = 22.414 L/mol = 0.022 414 m^3/mol (compare Sect. 5.5.4)

4.10 Partial Pressures

Dalton's partial pressure law[17] (also called Dalton's law) states that the total pressure of a mixture of ideal gases is equal to the partial pressures that each of the constituents would exert if it filled the gas volume alone (Fig. 4.5).

[17] John Dalton, 1766–1844, English chemist and physicist.

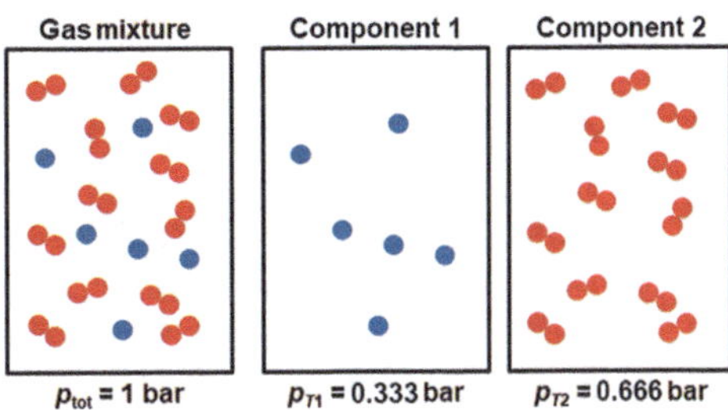

Fig. 4.5 Partial pressure law

In other words, if a cubic meter of gas consisting of 20 % oxygen and 80 % nitrogen exerts a pressure of 1.0 bar:

- The amount of oxygen (20%) contained in the mixture exerts the pressure of $1.0 \text{ bar} \cdot 20\,\% = 0.2 \text{ bar}$ in 1 m^3. The partial pressure of the oxygen is 0.2 bar.
- The quantity of nitrogen (80%) contained in the mixture exerts the pressure of $1.0 \text{ bar} \cdot 80\,\% = 0.8 \text{ bar}$ in 1 m^3. The partial pressure of the nitrogen is 0.8 bar.

Dalton's law also applies to the partial vapour pressure of a vapour.
The carbon dioxide and oxygen partial pressure is important in medicine and in the rescue service for the respiration of patients. The calculation of partial pressures is part of the basic training for scuba divers and nitrox divers.

A great advantage of this law is that the volume ratio – just like the mass ratio – of the individual components can be used.

4.11 Diffusion

From experience, we know that a "scent" or a "stench" spreads throughout a room over time. This process is called diffusion.

> Diffusion is an independent process of concentration equalization in which particles (atoms, molecules, ions) move from the region of higher concentration to the region of lower concentration as a result of their thermal movement.

Diffusion is an irreversible (non-reversible) process. It continues until the concentrations of all substances involved have balanced out; i.e. until a dynamic equilibrium prevails.

Diffusion occurs in gases, in liquids and even in solids. The speed at which diffusion occurs depends on the size of the particles involved.

Diffusion proceeds much more slowly than mixing by flow processes. Therefore, it has no significance outdoors or in ventilated/ventilatable rooms. Inside pits or shafts in which no air flow occurs, however, diffusion is the determining variable for the dispersion of gases and vapours. Here, it also determines the passage of the explosion limits.

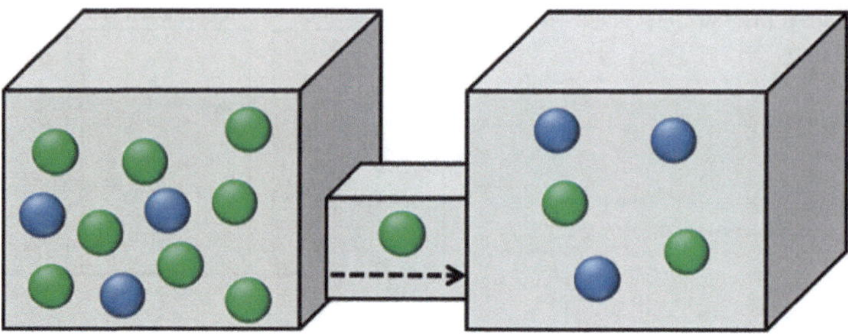

Fig. 4.6 Diffusion coefficient (illustration)

4.11.1 Diffusion Coefficient (Diffusion Constant)

The diffusion coefficient is a quantity that serves to compare the diffusion capacity or the diffusion speed of gases and vapours. It indicates the quantity of a gas that diffuses into another gas over an area of one cm^2 per second (compare Fig. 4.6). Liquid gases and many hydrocarbon vapors have values around ~ 0.09 cm^2/s. Hydrogen has a very high value of 0.96 cm^2/s, so diffuses quite rapidly. Ammonia and ethyne (acetylene) are in a medium range with 0.22 cm^2/s.

4.11.2 Brownian Molecular Motion

The previously mentioned "thermal motion" of the particles is nothing other than Brownian molecular motion.

Around 1827, the botanist Brown[18] observed a zigzag movement of very small particles suspended in liquid or gas – in his case: pollen. This movement is caused by the liquid molecules bumping particles from all sides. Depending on which side/ point the number of collisions predominates, the particle moves away.

The movement can be observed, for example, under a microscope. Figure 4.7 shows the path of a single particle when looking through a microscope.

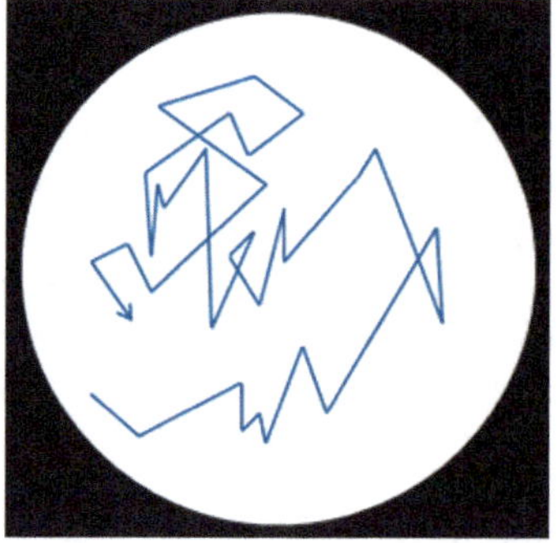

Fig. 4.7 Brownian motion of a particle under the microscope

[18] Robert Brown, 1773–1858, British botanist.

4.12 Real Gases

So far, ideal gases have been considered. In reality, however, the gas particles have an inherent volume. Also, at sufficiently high pressure, they influence each other by attracting and repelling intermolecular forces.

Hydrocarbons (i.e. primarily the liquid gases propane and butane, in addition to methane and ethane) already show noticeable deviations at pressures of 50 bar. For most other gases, the application of the universal or general gas equation results in negligible deviations up to a pressure of approximately 200 bar.

There are several mathematical models of varying complexity (e.g. the "Van der Waals equation")[19] to calculate the behaviour of real gases. In daily firefighting practice, however, the real behaviour of a gas need only interest us in a few cases. But it is precisely these cases that will be briefly considered here.

4.12.1 Breathing Air – A Real Gas

The extent to which the real behaviour of a gas differs from that of an ideal gas is explained in Table 4.1 using the example of breathing air.

As can be seen from Table 4.1 (column "real gas"), today's breathing apparatus with their 6 L gas cylinders and 300 bar filling pressure contain the same quantity of breathing air as the 2-cylinder apparatus used in the past with 200 bar filling pressure. The deviation of the "ideal gas" (calculation according to the general gas equation) from the "real gas" (calculation according to Van der Waals) amounts to approximately 4 % at 200 bar and is thus negligible in practice. At 300 bar, on the other hand, there is a difference of about 15 %. This is already noticeable.

Table 4.1 Ideal and real behaviour of breathing air

Breathing air bottle	Advertisement duck	Breathing air volume at normal pressure, calculated according to:		
		General gas equation	Universal gas equation	Real gas (van der Waals)
2 · 4 L	**200 bar**	**1600 L**	**1500 L**	**1540 L**
	100 bar	800 L	750 L	750 L
1 · 6 L	**300 bar**	**1800 L**	**1680 L**	**1540 L**
	200 bar	1200 L	1100 L	1150 L
	100 bar	600 L	600 L	600 L
	50 bar	300 L	300 L	300 L

[19]Example calculations are not given here. More details can be found in textbooks of "Physical Chemistry".

At a pressure of 400 bar(!), the 1800 L of breathing air of the general gas equation can also be accommodated in a 6-liter cylinder. The gain of only 260 L of breathing air is offset by a pressure increase of 100 bar (or a plus of 33 %).

4.12.2 Critical Pressure and Critical Temperature

In order to liquefy a real gas, a certain pressure is required at a given temperature. Just as when a gas cools down sharply, the velocities of the particles decrease and the gas particles constantly come closer to each other. At some point, they are so close that intermolecular forces of attraction take effect. The mutual proximity of the particles causes the mutual attractive forces to eventually become strong enough to cause the gas to become liquid.

By increasing the pressure/decreasing the temperature, a real gas deviates more and more from the ideal behaviour until it finally liquefies. The higher the temperature of the gas, the higher the pressure to be applied.

However, a gas above its **critical temperature** (T_{crit} or ϑ_{crit}) cannot be liquefied by pressures, no matter how high. The pressure that must be applied to liquefy at the critical temperature is called the **critical pressure** (p_{crit}).

The point where the two critical quantities meet is called the **critical point.**

Nitrogen or oxygen, for example, can therefore only be stored in liquid form at correspondingly low temperatures.

Below the critical temperature, a pressure lower than the critical pressure is already sufficient to convert the gas into its liquid state.

Above the critical temperature, only one phase exists, a so-called "**supercritical fluid**" (compare Fig. 4.8). A distinction between liquid and gas is no longer possible here. The fluid has properties of both phases: Density and solvent properties of a liquid, low viscosity of a gas.

Supercritical carbon dioxide, for example, is used to decaffeinate coffee.

- When heated strongly, supercritical fluids behave like liquids.

Some gases (compare Table 4.2) can only be liquefied if they are cooled to temperatures well below room temperature.

To achieve these low temperatures, one makes use of the effect that compressed gases cool down when they expand. This behavior, known as the **Joule-Thomson effect,** comes from the fact that the intermolecular forces of attraction must be overcome. Energy is consumed in this process (as it is when a solid melts) and this energy is taken from the gas itself. As a result, the gas cools down strongly.

In technology, this effect is exploited in the Linde air liquefaction process.

In everyday firefighting, we encounter it when breathing air from a cylinder. The cylinder also cools down during this process, so that drops of condensation often appear on the outside in the valve area. If the breathing air is not absolutely dry (= free of water vapour residues), the valve can even freeze up inside the cylinder, as an ice plug forms.

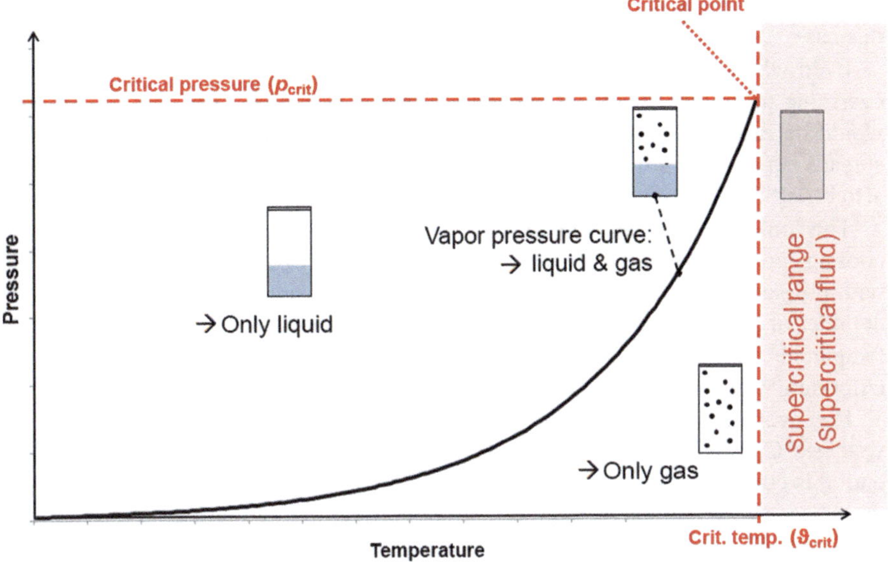

Fig. 4.8 Critical data of a gas

Explanation

- Above the vapour pressure curve *(black line)* only liquefied gas exists.
- Below the vapor pressure curve, only the gas form exists (up to the critical temperature).
- On the vapor pressure curve, gas and liquid coexist.
- Above the critical temperature only a supercritical fluid occurs.

When extinguishing with a portable CO_2 fire extinguisher, this effect is so noticeable that part of the extinguishing agent carbon dioxide escapes from the

Table 4.2 Critical data of gases (examples)

Substance	Formula	Boiling point (at p_n in °C)	Critical temperature (ϑ_{crit} in °C)	Critical pressure (p_{crit} in bar)
Helium	He	−269	−268	2.3
Hydrogen	H_2	−253	−240	13
Ammonia	NH_3	−33.5	132.1	112.6
Air	–	–	−140.7	37.7
Oxygen	O_2	−183	−118.5	50.6
Carbon dioxide	CO_2	−78	31.1	73.9
Propane	C_3H_8	−42	96.8	40.6
Chlorine	Cl_2	−34	144	77.1
Butane	C_4H_{10}	−0.5	152	38

extinguisher as a frozen gas (dry ice snow). The gas cools down to below −78 °C in this case.

Even when gases flow out unintentionally/uncontrolled, they cool down. However, due to its low intrinsic temperature, this cold gas has a higher density than would be expected according to the ambient temperature. The cold gas therefore spreads out near the ground. This increased gas density due to the previous cooling is also called the **heavy gas effect.**

The Joule-Thomson effect can cause the escaping gas to liquefy by itself due to cooling and thus stop the gas escaping. This has already been observed, for example, with pressurised chlorine gas cylinders in swimming pools. With a full 50 kg gas cylinder of chlorine, only about $^1/_3$ the total quantity escapes. In the meantime, the temperature drops to −34 °C (= boiling point of Cl_2) and practically no more chlorine gas escapes.

However, the pressurized gas cylinder must not be subjected to a water jet, as the water (8 °C to 18 °C) adds heat and the full 50 kg of chlorine is present as a gas at the end; this corresponds to a gas cloud of a good 16 m^3 – without taking into account the dilution by the ambient air!

Gases whose critical temperature is significantly below normal ambient temperatures are also referred to as **permanent gases** or **compressed gases.** These are present in compressed gas containers exclusively in gaseous form (N_2, O_2, CH_4, H_2, ...).

Gases that have been at least partially liquefied by pressure and are transported in this way are also called **liquid** gases or **liquefied gases.**

Another special case is ethine (acetylene), which is a **gas dissolved under pressure**, a so-called **dissous gas.** (The pressurized gas cylinder is filled with diatomaceous earth soaked with acetone. Acetylene is dissolved in acetone because, as a pure gas, it tends to self-decompose. See also Sect. 44.1.3).

Rule of Thumb for Liquid Gases

- 1 kg liquid gas = 2 L liquid
- 1 kg liquid gas = 500 L gas
- 1 kg liquid gas = 30,000 L explosible gas-air mixture

4.12.2.1 BLEVE

If a gas cylinder containing LPG (or a LPG tank) is heated to a high temperature (e.g. due to a fire), the pressure in the container will rise as well as the temperature. This leads to the associated safety valve responding at some point and gas escaping. At the same time, the liquid phase expands and reduces the available gas space, which acts as a pressure cushion. If the escaping gas ignites, this results in a more or less long or high flame (depending on the internal pressure in the container).

However, the effect of temperature also influences the material from which the container is constructed: the strength decreases. In particular, the area above the liquid phase is weakened, since the gaseous phase has a lower heat capacity than the liquid phase. The liquid gas therefore dissipates heat from the vessel wall faster than the gaseous phase can. If the pressure reduction via the safety valve is not sufficient, the internal pressure can increase to such an extent that a pressure vessel rupture occurs.

If the vessel is completely filled with liquid due to the volume expansion of the liquid phase, a pressure increase of approximately 8 bar per °C temperature increase can be expected. If the internal temperature rises above the critical temperature of the liquid gas, the pressure build-up is extreme. It is almost certain that a pressure vessel rupture will occur very quickly.

Explanation of the Extreme Pressure Build-Up

Liquids are practically not compressible (they are "incompressible"). If you increase the pressure of a volume of 50 L water from 1 bar to 1000 bar (!), the reduction in volume is only ($^1/_4$ L or 0.5 %). But after exceeding the critical temperature, the now supercritical fluid behaves like a liquid. And this fluid tries to expand when heated further. However, since it no longer finds any "free space" in the container, the increase in volume becomes noticeable through an immense increase in pressure.

The consequence of container rupture in the case of liquefied gases is the same in all cases:

The container tears open (danger from flying parts). Only approach the container from the side. End faces are particularly at risk.), the gas ignites and at the same time almost the entire quantity of gas escapes. A fireball occurs.

This effect is also referred to as "BLEVE". This is the abbreviation of "***b**oiling **l**iquid **e**xpanding **v**apour **e**xplosion*".

According to the ILO *(International Labour Office, "Major Hazard Control", Geneva, 1988),* the fireball generated by a BLEVE can be estimated as follows:

$$r_{\mathrm{FB}} = 29 \cdot \sqrt[3]{m_{\mathrm{LG}}} \qquad h_{\mathrm{FB}} = 2 \cdot r_{\mathrm{FB}}$$

Meaning:

r_{FB}: Radius of the fireball in meters

h_{FB}: Height of the fireball in meters

m_{LG}: Mass of the liquefied gas in tonnes

This means that even a full 11 kg propane gas cylinder will produce a fireball around 6.5 m in diameter and 13 m in high.

4.12.3 Solubility of Gases

Many gases can be dissolved in water. Some are dissolved "physically", i.e. without reacting with the water (N_2, O_2), others react at least partially (cf. Chap. 13). Examples of the solubility of gases in water can be found in Table 4.3.

- ▶ The solubility of gases in liquids increases as the temperature of the liquid decreases.

- ▶ The solubility of gases in liquids increases with increasing pressure.

When diving, the solubility of the air in the blood increases. With increasing depth, the total pressure acting on the diver increases and so does the solubility of the

Table 4.3 Temperature-dependent solubility of gases (examples, in g per litre of water)

Gas	0 °C	10 °C	20 °C	50 °C
Hydrogen bromide (HBr)	2212 g	~ 2060 g	~ 1765 g	1695 g
Ammonia (NH_3)	907 g	765 g	542 g	281 g
Hydrogen chloride (HCl)	842 g	~ 780 g	720 g	589 g
Chlorine (Cl_2)	14.6 g	10 g	7.3 g	3.88 g
Hydrogen sulfide (H_2S)	7.19 g	5.23 g	4.0 g	2.1 g
Carbon dioxide (CO_2)	3.38 g	2.36 g	1.73 g	1.31 g
Ethine (acetylene) (C_2H_2)	2.03 g	1.53 g	1.21 g	–
Oxygen (O_2)	0.070 g	~ 0.057 g	0.044 g	~ 0.031 g
Nitrogen (N_2)	0.030 g	~ 0.025 g	0.019 g	0.0137 g

breathing gas. If the pressure is reduced too quickly by ascending too fast, the gas, which is now no longer soluble, "bubbles" out of the blood. These air bubbles in the blood, the so-called "diver's disease", can even lead to death.

When a bottle of mineral water is opened, gas, in this case carbon dioxide, also bubbles out. The cause is the reduction of the pressure in the supernatant gas phase. Due to the now reduced pressure, the solubility decreases and "excess" CO_2 for this pressure bubbles out.

Solubility of ammonia and carbon dioxide (per litre of water)

Pressure:	1 bar	2 bar	5 bar
Ammonia (NH_3)	542 g	717 g	1593 g
Pressure:	1 bar	24.5 bar	49 bar
Carbon dioxide (CO_2)	1.73 g	28.5 g	50.7 g

4.12.3.1 Solubility of Gases During Firefighting Operations

The high solubility of HCl, HBr, HI and NH_3 in water is due to the fact that these gases react with the water. In the former cases, an acid is formed, and when ammonia is dissolved, an alkali is formed (cf. Chap. 13). Also, the solubilities given in Table 4.3 can only be achieved under laboratory conditions.

In a real outdoor application, an efficiency of only about 1 ‰ of the values in Table 4.3 can be achieved. The following rule of thumb applies in practice:

► 1 L (1 m^3) of water dissolves about 1 g (1 kg) of leaked gas that is readily soluble in water.

This means that 50 000 litres of water are needed to "bind" just 50 kg of leaked ammonia. Using 2 B-branch pipes of 400 L/min each, the time required to wash out the gas from the ambient air is around one hour.

Although the gases, which are extremely soluble in water, can be contained in their spread by a wall of water or by a spray jet/water mist, the fact that this produces a large quantity of highly diluted acid or alkali must be taken into account by the

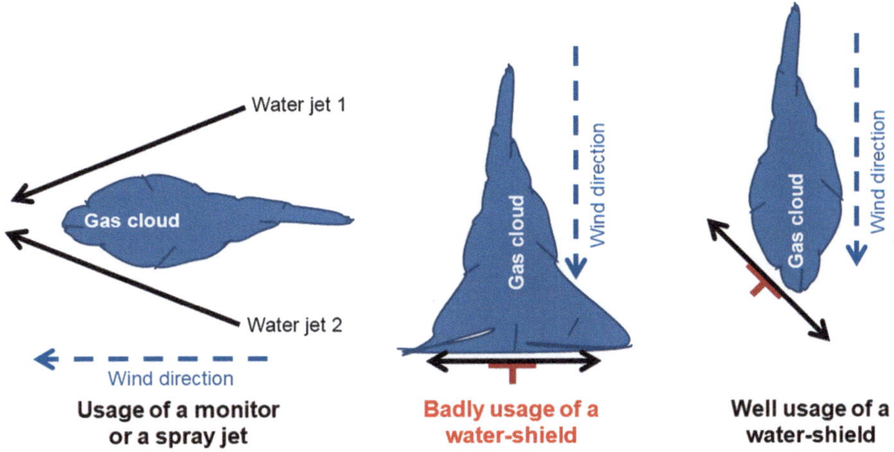

Fig. 4.9 Use of water cannon or hydro shield

incident commander. In particular, measures to protect surface waters must not be disregarded.

The use of spray jets, water cannons or so-called "hydro shields" to knock down or contain gas clouds requires the correct deployment tactics (Fig. 4.9). The jets from water cannons or similar should always be directed with the gas cloud to increase the contact time of the water with the gas and thus achieve greater success. Hydro shields are to be set up at an angle of approximately 45° to the cloud in order to prevent an uncontrolled distribution of the gas cloud "in the width" and to be able to "steer" the gas cloud.

In the case of gases that are poorly soluble in water, such as chlorine, only a mechanical effect can be achieved with spray jets or hydroshields. One can only try to prevent the gas cloud from spreading or to push it back.

In the case of a **punctual** gas leak, e.g. in a pipeline with a small cross-section or the bunghole of a barrel, consideration can be given to using a medium expansion foam branch pipe to absorb the escaping gas, which is readily soluble in water.

A medium expansion foam branch pipe draws in a relatively large amount of air and mixes it intensively with the water flowing through it. If the air inlet openings of the foam branch pipe are brought as close as possible to the outlet point, the gas is literally sucked in (injector principle, water suction pump effect) and dissolved relatively well by the strong mixing with the water. The contaminated water must, of course, still be disposed of in an orderly manner.

Part III

Atomic Models and Periodic Table

Atoms and Atomic Shell

5

One of the early questions that natural scientists asked themselves was about the "structure of everything". While the Greek philosophers were still convinced that everything was made up of indivisible particles, they later became convinced that everything was made up of the four "elements" of earth, fire, water and air. It was not until the seventeenth century that the "atom theory" was taken up again and then scientifically confirmed.

5.1 Development of the Atomic Theory

Already Democritus[1] explained that all substances are made up of "atoms" (Greek: *atomos* = indivisible). He extended the view of his teacher Leukip[2] by assuming that the atoms of the different substances differ in form, size and mass. Furthermore, these atoms could assume a certain position in relation to each other, collide with each other, unite and separate again. In this way Democritus was able to explain the diversity of the substances that surround us.

Decades after Democritus, Epicurus[3] developed the atomic theory further and wrote ten theorems about the basic building blocks of the world:

1. Nothing arises from what is not.
2. Nothing dissolves into that which is not.
3. The whole thing is infinite.
4. The whole thing has always been the way it is now, and always will be.
5. The whole is made up of the bodies and the void.
6. There are two types of bodies: atoms and aggregates (atomic compositions).

[1]Democritus of Abdera, 460–370 B.C., Greek philosopher.

[2]Leucipp of Miletus, c. 450 BC, Greek philosopher.

[3]Epicurus, 341 to c. 270 B.C., Greek philosopher.

T. Schmiermund, *The Chemistry Knowledge for Firefighters*,
https://doi.org/10.1007/978-3-662-64423-2_5

7. The atoms move without stopping.
8. Atoms have only three things in common with sensual things: shape, volume and weight.
9. The atoms are infinite in number, the void is infinite in extent.
10. The atoms of identical form are infinite in number, but their forms are indeterminate in number, but not infinite.

Lucretius[4] then explained some natural phenomena by means of Epicurus' theorems in the first century BC.

After that, the atomic theory fell into oblivion until it was renewed and improved by Gassendi.[5] Boyle[6] took this up again in his seminal book "The Sceptical Chymist" (published in 1661) and developed a further improved theory of atoms.

Here Boyle rejects the "doctrine of the four elements" and the "doctrine of the three principles" which were based on the writings of Paracelcus'[7] and dominated the ideas of chemistry/alchemy for a long time.

Boyle introduced a new concept of elements that excluded fire, earth, water, and air, among others, as elements. He attributed the chemical changes to changes in the structure of differently shaped particles ("corpuscles") and assumed that the different substances are formed by the coming together of differently shaped corpuscles into different forms.

5.1.1 Dalton's Atomic Model – Sphere Model

It was only with Dalton[8] that we can speak of a modern, scientific view of the concept of the atom. On the basis of his investigations into gases (Sect. 4.10) and their solubility in liquids, he developed the first "scientific" model of the atom (cf. Table 5.1).

The central implications of Dalton's atomic model, published in 1805, are:

▶
- Substances consist of smallest, not further decomposable particles, the atoms.
- The atoms of different elements have different masses and have different properties.
- The atoms of an element are equal to each other in mass and chemical properties.

[4]Titus Lucretius Carus (called Lucretius), c. 98 to c. 54 B.C., Roman poet.

[5]Pierre Gassendi, 1592–1655, French philosopher, physicist, and astronomer.

[6]Robert Boyle, 1627–1692, English chemist.

[7]Actually Theoprastus Bombastus von Hohenheim, 1494–1541, physician, philosopher and naturalist.

[8]John Dalton, 1766–1844, English chemist.

Table 5.1 Overview of early atomic models

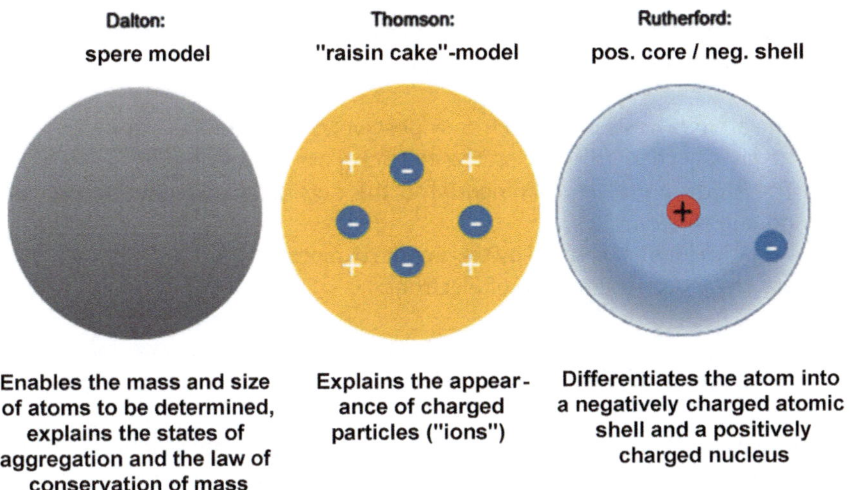

Dalton: spere model	Thomson: "raisin cake"-model	Rutherford: pos. core / neg. shell
Enables the mass and size of atoms to be determined, explains the states of aggregation and the law of conservation of mass	Explains the appear-ance of charged particles ("ions")	Differentiates the atom into a negatively charged atomic shell and a positively charged nucleus

- Atoms of different elements can combine with each other, in the ratio of simple integers, to form compounds.
- When a compound decomposes, the atoms that remain unchanged can form the same or other compounds again.
- The atoms can be conceived as massive, matter-filled spheres.

Dalton laid the foundation for the laws of "constant and multiple proportions" with his investigations on the relative masses of atoms (with the reference value hydrogen = 1, since he identified its atoms as those with the lowest mass) and a table of the relative atomic masses of 18 elements and compounds published in 1805. With his composite "atomic compounds" he already anticipated the concept of molecules, which was not introduced until 1860.

5.1.2 Thomson's Atomic Model – Raisin Cake Model

Around 1900, Lenard[9] carried out investigations on cathode rays and already pointed out that it must be possible *"to obtain information on the nature of molecules and atoms with the help of these rays"*. His research brought him to the conclusion that the atom could not be a compact sphere, but had to consist of very small centres of force and large interstices.

[9] Philipp Lenard, 1862–1947, German physicist, Nobel Prize in Physics in 1905.

In 1910, Thomson[10] was able to show that the cathode rays have a mass more than a thousand times smaller than the hydrogen atoms and calculated the relationship between the mass and charge of these particles. His investigations resulted in a modified atomic model (compare Table 5.1):

- The atom consists of an electrically positively charged sphere in which negatively charged particles, the electrons, are embedded.
- The atoms are electrically neutral to the outside, but can accept or donate electrons.
- Negative ions are created by this uptake of electrons, and positive ions are created by the release of electrons.

5.2 Structure of the Atomic Shell

Rutherford's scattering experiment (cf. Sect. 5.3.1) differentiated the atom, which until then had been a solid, impenetrable structure, into a nucleus and a shell. As a consequence, there was an almost rapid further development of atomic models, especially of the structure of the atomic shell.

5.2.1 Rutherford Atomic Model

Only a few years after Thomson, E. Rutherford[11] carried out his famous scattering experiment. Rutherford had already begun research on radioactivity in 1896 and then established in scattering experiments that the atom must essentially consist of "empty space", a small nucleus and a large "shell" (compare Table 5.1).

His atomic model (set up in 1911) states:

- The atom consists of an atomic nucleus and an atomic shell.
- The atomic nucleus is positively charged, carries almost the entire mass of the atom and is located in its centre.
- The atomic nucleus has only about 1/10 000 of the diameter of the whole atom.
- The atomic shell determines the size of the atom.
- The atomic shell is negatively charged. In it are the electrons of the atom.

[10] Joseph John Thomson, 1856–1940, English physicist, Nobel Prize in Physics in 1906.

[11] Sir Ernest Rutherford, 1871–1937, English physicist, Nobel Prize in Chemistry in 1908.

5.2.2 Bohr's Atomic Model

Niels Bohr[12] took up Rutherford's results and reinterpreted them.

According to the theories valid at that time, the electrons would have had to crash into the nucleus by giving off energy (in shape of radiation). So no stable atoms could exist at all ...

Bohr was thus forced to rethink Rutherford's interpretation. In doing so, he also included Planck's[13] findings from 1900:

Planck's Constant

Planck had found that energy in the atomic range is not absorbed or emitted in any size, but only in the form of energy packets (energy quanta) or multiples thereof. The energy results from the frequency of the radiation and the so-called Planck's quantum of action.

Bohr's Postulates

With these findings, Bohr postulated basic conditions for the structure of atoms in 1913:

- Electrons can only stay on certain orbits (also called energy levels, energy states or shells) in the atomic shell. Each orbit has a defined distance from the atomic nucleus and consequently a defined energy.
- These orbits are arranged concentrically around the atomic nucleus. Each orbit is designated from the inside to the outside with a letter (K, L, M, N, ...) or a number (1, 2, 3, 4, ...) (compare Fig. 5.1).
- An electron that is on such a circular orbit is on a "stationary" orbit and thus in an "allowed" energy state.
- Energy is absorbed or released by an electron only when it changes from one allowed energy state to another allowed energy state (i.e., from one orbit to another orbit).
- This energy is absorbed or emitted in the form of electromagnetic radiation (e.g. light).
- The energy difference between the allowed orbits corresponds to integer multiples of Planck's quantum of action.

The "quantization" of energy, i.e. the existence of "energy packets" (which have a fixed size) was a completely new idea, but it fitted in well with all the observations. Together with the "stationary trajectories", some measured values could now be better interpreted or even understood.

[12] Niels Bohr, 1885–1962, Danish physicist, Nobel Prize in Physics in 1922.

[13] Max Planck, 1858–1947, German physicist, Nobel Prize in Physics in 1918.

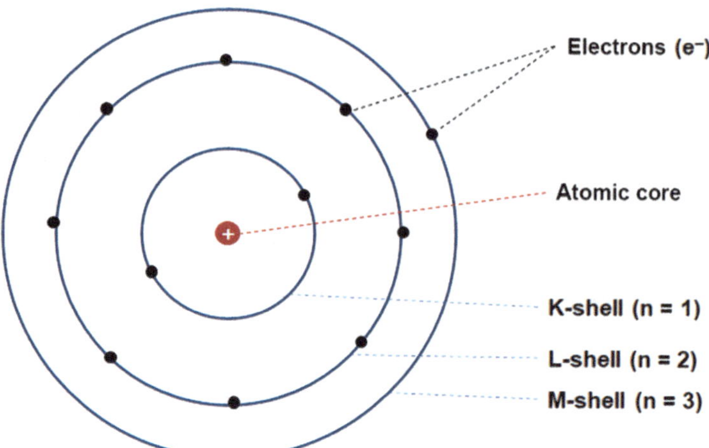

Fig. 5.1 Bohr's atomic model (using sodium as an example)

- **For Bohr's atomic model holds (in addition to Rutherford):**
  - The atom consists of an atomic nucleus and an atomic shell.
  - The atomic nucleus is positively charged, carries almost the entire mass of the atom and is located in its centre.
  - The atomic nucleus has only about 1/10 000 of the diameter of the whole atom.
  - The atomic shell determines the size of the atom.
  - The atomic shell is negatively charged. In it are the electrons of the atom.
  - **The electrons move only on certain ("allowed") paths around the atomic nucleus.**
  - **The further out the orbits are, the more energetic the electron is.**
  - **When changing from one orbit to another, energy is only absorbed or released in the form of "energy packets" (energy quanta).**

5.2.3 Bohr-Sommerfeld Atomic Model

The Bohr atomic model marks the turning point to the quantum mechanical atomic model (= orbital model, Sect. 5.2.4). Sommerfeld[14] generalized the Bohr model and extended it to include elliptical orbits. Similarly, various so-called "quantum numbers" were introduced here:

---

[14] Arnold Sommerfeld, 1868–1951, German physicist.

Principal quantum number "n" is equal to Bohr's orbit and includes *nearly coincident* energy states of the electron.

The secondary quantum number "nn the secondary quantum number designation" characterizes the energy states of the electron *within* each shell.

Like Bohr's orbits ("main shells"), the secondary quantum numbers ("subshells") are also designated by letters:

$$\ell = 0 \rightarrow \mathrm{s} \qquad \ell = 1 \rightarrow \mathrm{p} \qquad \ell = 2 \rightarrow \mathrm{d} \qquad \ell = 3 \rightarrow \mathrm{f}$$

The magnetic quantum number "m" refines the splitting of the energy states by taking into account influences of (strong) magnetic fields (possible values: $-\ell$...0... $+\ell$).

The spin quantum number "s" indicates the "intrinsic angular momentum" of the electrons. The spin can assume the values +1/2 and −1/2.

Explanation of the Spin

In the model, the electrons rotate around their own axis while moving around the nucleus (just as the earth rotates around itself while moving around the sun).

If the direction of rotation of the electron (the earth) is the same as the direction of movement around the atomic nucleus (the sun), i.e. in both cases e.g. counterclockwise, then one speaks of a positive spin (+1/2).

If the directions of rotation are different (once clockwise, once counterclockwise), then you get a negative spin (−1/2).

The electron spin as "rotation of the electron around its own axis" must not be taken literally. It is only a simplified model conception.

Pauli Principle

Pauli[15] postulated the "spin" of electrons in 1924 and likewise formulated his "exclusion principle".

The rule states that no two electrons within an atom may match in the four quantum numbers. At least one of the quantum numbers (n, l, m or s) must be different. It also follows that there can never be more than two electrons in an atomic orbital (see Sect. 5.2.4).

From this it follows that, according to Bohr, the main shells can accept a maximum number of electrons: $\mathbf{2n^2}$ (see Table 5.2).

For the notation of the electron distribution:

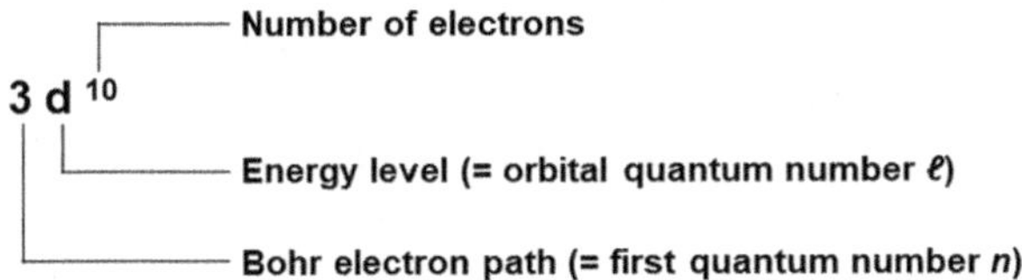

[15] Wolfgang Pauli, 1900–1958, Austrian theoretical physicist, Nobel Prize in Physics in 1945.

Table 5.2 Bohr orbits and number of electrons

Bohr's shell		Number of energy levels	Designation of the energy levels	Calculation max. Number of electrons	Electron distribution
K-Shell	1. Shell	1	1 s	$2 \cdot 1^2 = 2 \cdot 1 = \mathbf{2}$	1 s^2
L-Shell	2. Shell	2	2 s, 2p	$2 \cdot 2^2 = 2 \cdot 4 = \mathbf{8}$	$2s^2$ $2p^6$
M-Shell	3. Shell	3	3 s, 3p, 3d	$2 \cdot 3^2 = 2 \cdot 9 = \mathbf{18}$	$3s^2$ $3p^6$ $3d^{10}$
N shell	4. Shell	4	4 s, 4p, 4d, 4f	$2 \cdot 4^2 = 2 \cdot 16 = \mathbf{32}$	$4s^2$ $4p^6$ $4d^{10}$ $4f^{14}$

Why Do 10 Electrons Fit on the d Energy Level?

It is true that if the spin quantum number designation is "d", then $\ell = 2$. The magnetic quantum number m then has the possible values -2, -1,0, +1, +2; that is, 5 different values. And to each of the values there can exist an electron with positive and with negative spin quantum number. These are then electrons. $5 \cdot 2 = 10$ same applies to the f-energy level with its maximum of 14 electrons.

With the help of the Bohr-Sommerfeld atomic model, not only the structure of the electron shell of the atoms can be understood, but also the structure of the periodic table of the elements.

This model can also be used to explain the X-ray spectra characteristic of each element and how a laser works.

The described electron orbits, the main and subshells, however, are to be understood merely as descriptive images and terms for the energy states of the electrons in the atomic shell.

5.2.4 Orbital Model

Today, the so-called "orbital model" is the model that describes the conditions in the atomic shell in the most detail. For this purpose, however, the Bohr-Sommerfeld model had to be extended.

Matter Waves

In 1924, in his doctoral thesis, de Broglie[16] introduced the concept of "matter waves". According to him, all objects have wave properties. However, this only comes into play at atomic dimensions, since for larger things the wavelengths become immeasurably small.

According to de Broglie, the atomic and subatomic building blocks of matter thus possess particle and wave properties in equal measure. Sometimes electrons behave as if they were particles and in other cases as if they were waves. *The question "Is the*

[16]Louis Victor de Broglie, 1892–1987, French physicist, Nobel Prize in Physics in 1929.

electron a particle or a wave?" is therefore as meaningless as the question "Are you a CDU- or SPD-voter?

Without matter waves, for example, there could be no electron microscopes at all. One of the proofs for the correctness of the theory.

Schrödinger Equation

Schrödinger[17] then formulated his wave equation in 1926, connecting the energy (E) and spatial coordinates (H) to the wave function (Ψ, Psi) of the electron:

$$E \cdot \Psi = H \cdot \Psi$$

The solutions of the Schrödinger equation provide the so-called wave functions, in which all information about the system is contained.

The Schrödinger equation is a differential equation of second order, which theoretically has infinitely many solution functions. Therefore, a deeper consideration is omitted here.

Uncertainty Relation

Heisenberg[18] realized in 1927 that it is not possible to determine the location **and** momentum (motion) of a particle with absolute **accuracy.**

If you want to determine the location and momentum of a particle as small as an electron, for example, you can use light beams to measure it. If one uses light with a long wavelength (= small energy), then one cannot measure the location exactly, because the wavelength of the light is larger than the spatial expansion of the electron. With small wavelengths (= high energy) one "taps" the electron by the measurement, like a billiard ball hits another one. Consequence: One can no longer measure the momentum accurately.

Orbital

Therefore, it is only possible to state the probability with which an electron is located at a certain distance from the atomic nucleus.

At the same time, a "picture" of the spatial distribution of the electrons – a "residence probability range" – is obtained. These areas are called orbitals.

- The principal quantum number n tells us something about the relative size of the orbital and indicates the Bohr orbit.
- The secondary quantum number' indicates the number of nodal levels and thus determines the (basic) shape of the respective orbital (s, p, d, f).
- The number of possible magnetic quantum numbers m gives information about the number of different arrangements of the orbital in space.

[17] Erwin Schrödinger, 1887–1961, Austrian physicist, Nobel Prize in Physics in 1933.

[18] Werner Heisenberg, 1901–1976, German physicist, Nobel Prize in Physics in 1932.

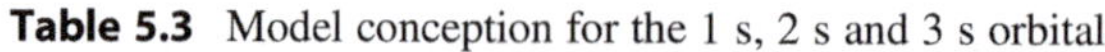

Table 5.3 Model conception for the 1 s, 2 s and 3 s orbital

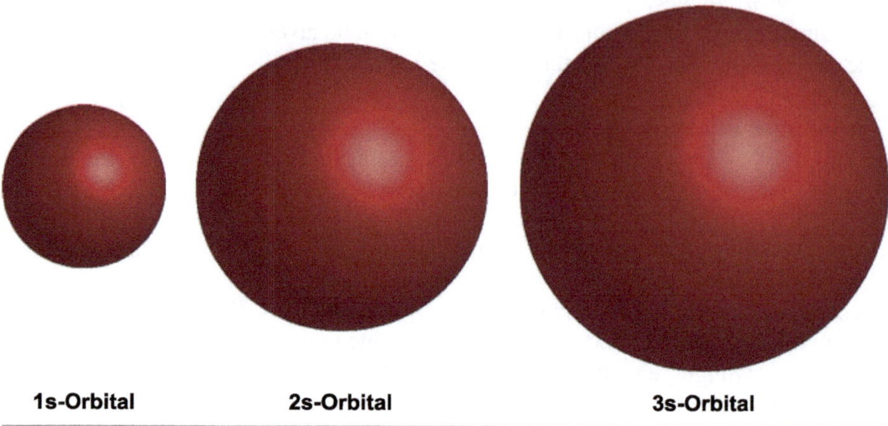

- The spin quantum number s gives information about the spin of the electron.
- Each orbital can be occupied by a maximum of two electrons.

The orbitals are therefore "spatial structures" in which an electron resides with a "certain probability". The geometric form of these "residence probability spaces" can be determined by calculating the probability of an electron being in a certain position relative to the atomic nucleus.
Tables 5.3 and 5.5, Figs. 5.3 and 5.2 show the space in which an electron is located with a 90% probability. In part, two-colour representations were chosen to provide a better overview.

Model Presentation "Balloon

For a better idea, consider a round balloon that we inflate. We get a sphere which has no nodal planes ($\ell = 0$). We have an s orbital. If we blow up the balloon more, we turn the 1 s orbital into the 2 s orbital and so on. (Table 5.3). The atomic nucleus is always in the centre of the ball.

Now we put a string around the balloon and pull it together (Table 5.4). We get two hemispheres and a nodal plane (the string), which corresponds to the minor quantum number $\ell = 1$. This gives the shape of the p-orbital.

This p-orbital can be arranged vertically, horizontally and in depth. This results in three differently arranged p-orbitals (p_{x-}, p_{y-} and p_{z-}orbital, Table 5.5) which are energetically equivalent (same secondary quantum number). The difference in spatial arrangement is expressed in the three possible magnetic quantum numbers ($m = -1, 0, +1$). The atomic nucleus is always located in the centre of the nodal plane.

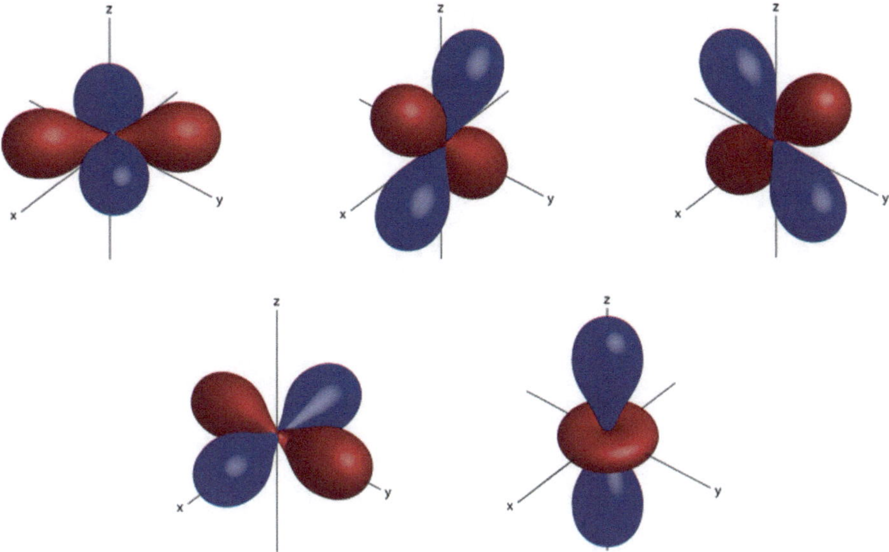

Fig. 5.2 The five d-orbitals © Beate Rocholz

Table 5.4 Formation of a p-orbital (balloon model)

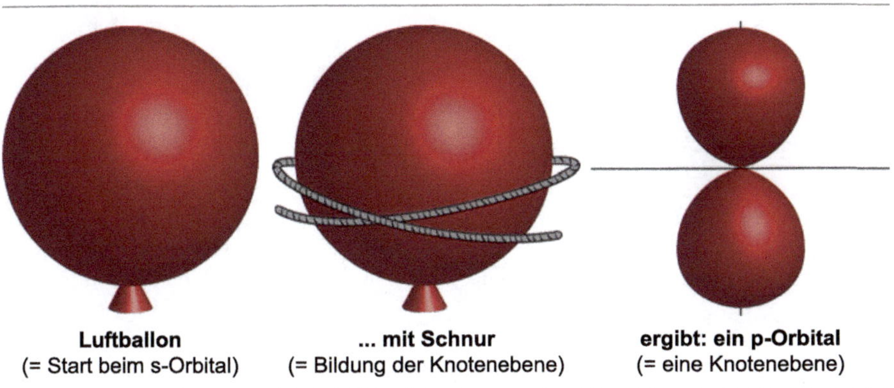

Luftballon (= Start beim s-Orbital)	**... mit Schnur** (= Bildung der Knotenebene)	**ergibt: ein p-Orbital** (= eine Knotenebene)

To form the d-orbitals (Fig. 5.2), we mentally place a second string around the balloon (= formation of a second nodal plane, since $\ell = 2$). In the simpler case we get a kind of cloverleaf (the two strings cross at right angles). In the other case, two oval structures connected by a bead (the two strings are parallel to each other).

Here, too, different spatial orientations are possible, so that we obtain five arrangements (corresponding to the magnetic quantum numbers −2, −1, 0, +

Table 5.5 Model conception for px, py and pz orbitals

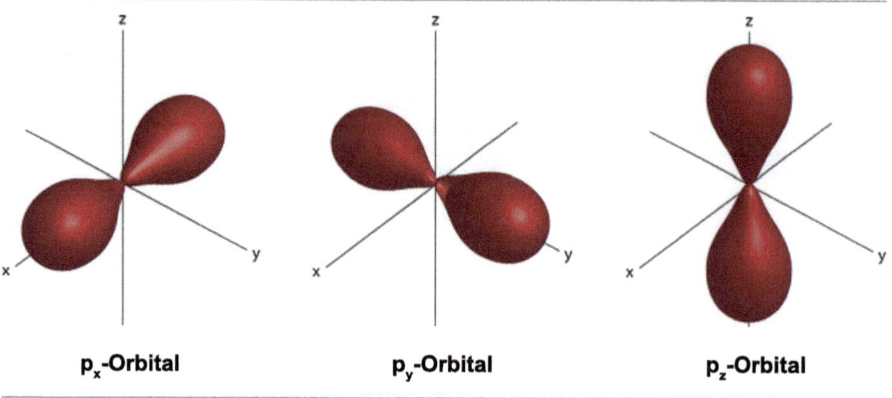

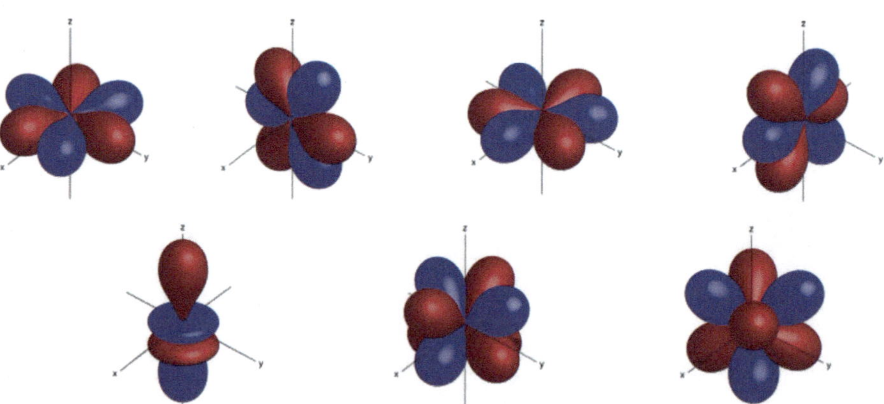

Fig. 5.3 The seven f-orbitals © Beate Rocholz

1, +2). In sum, a maximum of 10 electrons can be accommodated at the d-energy level.

Here, too, we find the atomic nucleus in the center of the structure.

To – last but not least – now also form the f-orbitals, a third string $(\ell = 3)$ is required.

The seven possible shapes (corresponding to the magnetic quantum numbers $m = -3, \dots 0, \dots +3$) are shown in Fig. 5.3.

In the atom itself, the is-orbital is naturally enclosed by the 2 s-orbital and the 2p-orbitals penetrate these two s-orbitals. The same applies to the d and f orbitals. Since graphics quickly become very confusing here, a pictorial representation has been omitted.

► *The orbitals are neither observable as such, nor in their form. They are mathematically determined areas whose geometric shape is calculated on the basis of mathematical rules.*

5.2.5 The Electron

The atomic shell consists of electrons. Here are a few words about the electron itself:

G. J. Stoney[19] had already postulated the existence of a "carrier of electric charge" in 1874 and suggested the name "electron" (Greek: "amber"). In 1897, J. J. Thomson succeeded in determining the mass of the electron and finally, in 1909, R. Milikan[20] was also able to measure the elementary charge of this particle.

Final Sentence on the Atomic Shell

The electrons move within certain regions called orbitals. The shape of these orbitals and their distance from the atomic nucleus can be described by so-called quantum numbers.

In connection with chemical bonds and chemical reactions (Chaps. 8 and 9), Bohr's atomic model will usually serve us as a "picture", but certain facts (e.g. several valence states in one element, "equality" of the bonds in the carbon atom, similarities of the groups of the periodic table, etc.) can only be understood well if one is familiar with the basic features of the orbital model.

5.3 Structure of the Atomic Nucleus

The first insights into a structure of atomic nuclei of some kind were provided by studies of natural radioactivity (Bequerel, M. Curie, etc.; cf. also e.g. Chap. 22).

After Lenard had already assumed "centres of force" in atoms in 1900 and Thomson was able to prove positive and negative charges in the atom in 1910, many physicists were concerned with "the atom". In 1911, Rutherford published a model of the atom in which he assumed a positively charged nucleus and a negatively charged shell.

5.3.1 Rutherford's Scattering Test

In 1894, Lenard had bombarded thin aluminium foils with electrons ("cathode rays") and found that these penetrated the foils (0.002 mm, approx. 7500 atomic layers thick). However, according to Dalton's atomic model ("atoms are balls of mass"),

[19] George Johnstone Stoney, 1826–1911, Irish physicist.

[20] Robert Millikan, 1868–1953, US physicist, Nobel Prize in Physics in 1923.

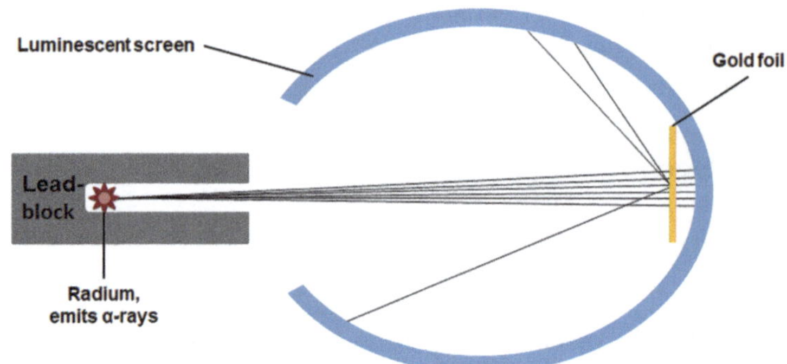

Fig. 5.4 Rutherford's scattering test (setup)

this should not have happened. He therefore assumed that there must be a "centre of force" in the atom.

Rutherford's scattering experiment (1909) provided further insights.

His colleague E. Marsden[21] had placed a radioactive source (radium) in a block of lead. Through a hole, so-called α-radiation (whose positive charge was already known) could now hit a very thin gold foil (0.0005 mm = approximately 2000 atomic layers thick). In order to be able to observe the paths of the α-rays, a luminescent screen was placed around this structure (Fig. 5.4). If a particle hits the screen (at that time a glass plate coated with ZnS), it produces a tiny flash of light (of course, the experiments were carried out in a darkened room).

It was now observed that almost all particles penetrated the gold foil and only a few were deflected. And this despite the fact that the α-particle is 7300 times heavier than the electron and has a double positive charge.

As Rutherford described it, *"It was like firing a cannonball at a piece of tissue paper and the bullet bounced back."*

Further investigations revealed that the largest part of the atoms consists of empty space. The (electron) shell has a diameter 10,000 times larger than the atomic nucleus, but in which practically the entire mass of the atom (> 99.9%) is concentrated (Fig. 5.5). Even in uranium, the 92 electrons contribute less than 0.5 ‰ to the total mass.

The atoms (and thus the atomic shells) have diameters in the range of about 100 to 300 pm (pm = picometer = 10^{-12} m). The diameter of the atomic nuclei is about 10 fm (fm = femtometer = 10^{-15} m).

So if you imagine the atomic nucleus to be the size of a ping-pong ball Ø = 3 cm), then the atom has a diameter of 300 m! Even if one were to imagine the atom as a small pinhead (ø = 1 mm) the atom still has a diameter of 10 m!

[21] Ernest Marsden, 1889–1970, English physicist.

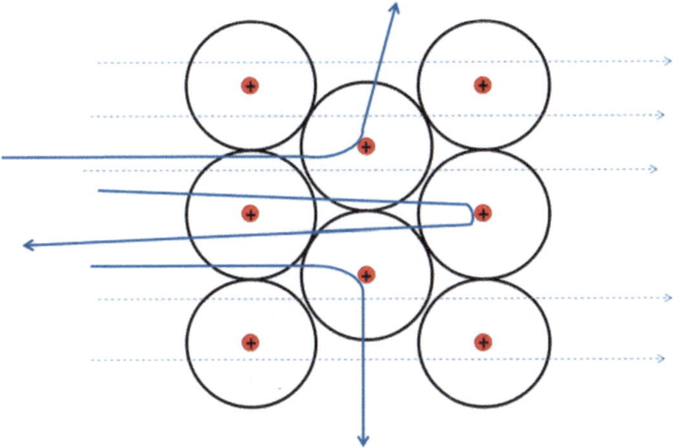

Fig. 5.5 Interpretation of the scatter plot

The alpha particles mostly penetrate the empty space between the atoms. Since the same charges repel each other, those that come too close to the nucleus are deflected or even thrown back by the nucleus.

5.3.2 The Proton

Eugen Goldstein[22] discovered the so-called "channel rays" as early as 1886. In contrast to the already mentioned "cathode rays", however, no electrons are released here, but positive particles. Using hydrogen as the gas filling of the "Geissler[23] tube" (after the developer of the "channel ray tube"), Wilhelm Wien[24] was able to obtain the lightest electrically positively charged particles ever in 1898. The charge of these particles is the same as that of electrons, but with the opposite sign.

In 1906, J. J. Thomson succeeded in determining the mass of this particle, and in 1919, E. Rutherford proposed to call this particle the "proton" (from Greek "the first") – after having detected it in the nucleus of the nitrogen atom.

5.3.3 The Neutron

As a rule, atoms are electrically neutral towards the outside. So they must have as many electrons within atom-shell as protons within nucleus. Consequently, the

[22]Eugen Goldstein, 1850–1930, German physicist.

[23]Heinrich Geißler, 1814–1879, German glassblower and instrument maker.

[24]Wilhelm Wien, 1864–1928, German physicist, Nobel Prize in Physics in 1911.

Table 5.6 Particles in the atom

Name	Symbol	Loading	Mass in [kg]	in [u]	in [e^- -mass]
Electron	e^-	−1	$9.109\,383 \cdot 10^{-31}$	0.000 548 58	1
Proton	p	+1	$1.672\,621 \cdot 10^{-27}$	1.007 276	1836
Neutron	n	0	$1.674\,927 \cdot 10^{-27}$	1.008 665	1839

masses of the atoms should also be corresponding multiples of the proton (or hydrogen).

However, the actual masses of the atoms (hydrogen excepted) are significantly greater. To explain this, Rutherford already in 1920 assumed another particle in the atomic nucleus. It should have the mass of the proton, but be electrically neutral. The name of this neutral particle was given by W. D. Harkins[25] in 1921.

James Chadwick[26], who had already developed a method for determining the number of positive charges of the various atoms, was then able to detect this particle in 1932.

5.4 Particles in the Atom

Table 5.6 gives an overview of the particles that make up the atoms.

Explanations

The charge of the electron is also called the "elementary charge" (e_0). Its value is coulombe$_0$ = 1, 602 177 · 10^{-19}(C). Here, 1 C = 1 As (ampere-second).

The charges of the proton and the electron are exactly the same. They differ only in their sign.

Since protons and neutrons build up the atomic nucleus, they are also called "nukleons" (from Greek *nukleus* = nucleus).

The neutron has a slightly larger mass than the proton.

1 u = 1/12 of the mass of the ^{12}C atom (see: Sect. 5.5).

Cohesion of the Atomic Nucleus

We know from experience with magnets that unlike poles attract and like poles repel.

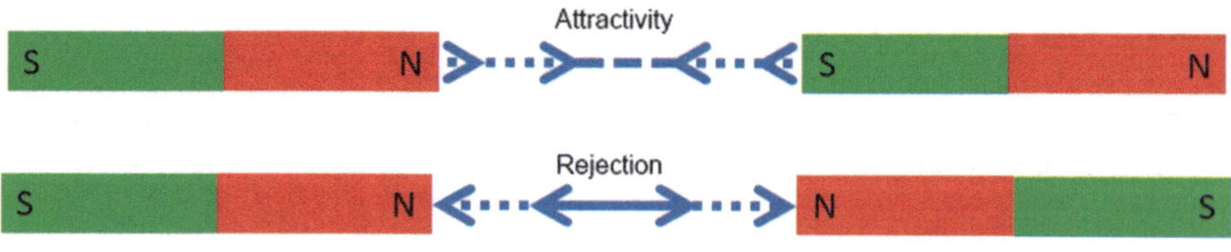

[25] William Draper Harkins, 1873–1951, American chemist and nuclear physicist.

[26] James Chadwick, 1891–1974, English physicist, Nobel Prize in Physics in 1935.

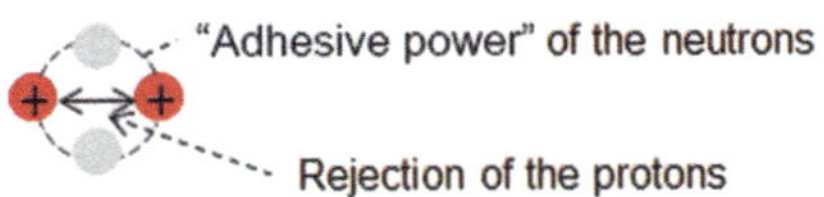

Fig. 5.6 Neutrons and protons in the atomic nucleus. The neutrons act like "glue" so that the atomic nucleus holds together despite exclusively positive charges

Magnetic attraction and repulsion: Unequal poles attract each other, equal poles repel each other

The same applies to electric charges: Unequal charges (sign + : −) attract each other, equal charges (sign + : + or − : −) repel each other.

So no atomic nucleus, with the exception of hydrogen, should be stable, right?

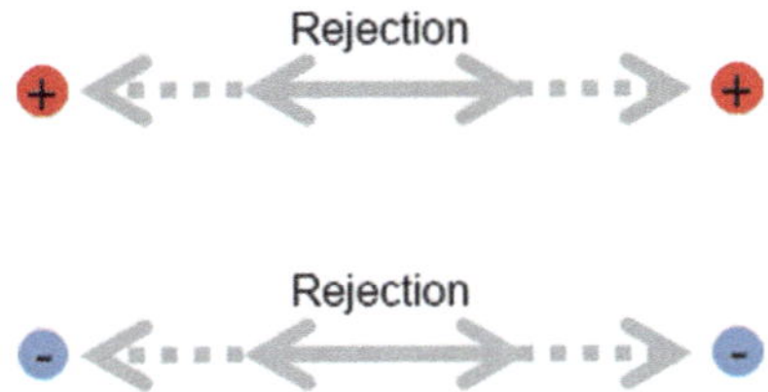

Equal electric charges repel each other

For now, it may suffice to say that the additional neutrons contained in the nucleus act as a kind of "glue" and thus hold the positive charges of the protons together (Fig. 5.6).

The neutrons contribute to the mass of the atom practically as much as the protons themselves. Elements with different numbers of neutrons are called isotopes. For example, there are three isotopes of hydrogen: "normal" hydrogen (H; 1 proton), deuterium ("heavy hydrogen", D; 1 proton, 1 neutron) and "superheavy hydrogen" (tritium, T; 1 proton, 2 neutrons).

A more detailed consideration is given in Sect. 24.2.

Atomic Density

Since almost all the mass is concentrated in the nucleus and the nucleus is very, very small, the nucleus itself has a very high density. This density is about the same for all atomic nuclei. It is about $1.4 \cdot 10^{17}$ kg/m^3. Thus, the nucleus of an atom is $6.5 \cdot 10^{12}$ times (6 500 000 000 000 times) denser (or colloquially, *"heavier")* than the element platinum (which is one of the three densest elements). Put another way: An atomic nucleus that has a volume of just over 40 mL has about the same mass as our Earth: about $6 \cdot 10^{21}$ tons. Such extremely high densities are otherwise only found in nature in so-called neutron stars.

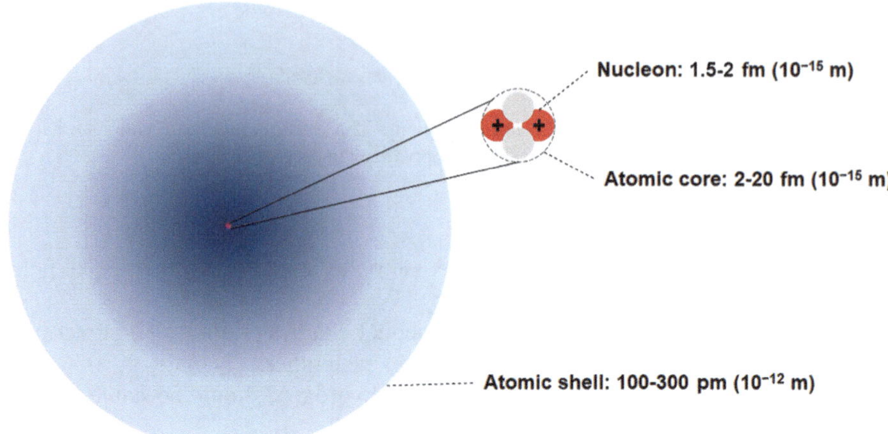

Fig. 5.7 Size ratios in the atom. The figure is **not** to scale. The atomic nucleus in the left part is the red dot. With a nucleus diameter of only 1 mm, the shell would have to have a diameter of 10 m!

Other Elementary Particles

The protons and neutrons are made up of other, even smaller particles (so-called "quarks"). Also other subatomic particles are known. However, this is irrelevant for the understanding of chemistry *at this point.*

Ratio Atom: Atomic Nucleus

One more word about the diameter of the atoms in relation to the atomic nucleus:

In chemistry- and physics-books, but also in internet-sources, different values of size-ratios of atomic nucleus to atom can be found. The values given range from 1 : 10000 to 1 : 100 000 (sometimes up to 150 000).

This often leads to confusion. Therefore, the origin of these different indications shall be briefly explained here:

The diameter of the atoms is 100 pm to 300 pm, the diameter of a neutron about 1.7 fm, that of the proton about 1.4 fm. The diameters of atomic nuclei are ~2 fm to ~20 fm, thus can differ by a factor of 10. Now, in some books the ratio of hydrogen nucleus to atomic shell (~ 2 fm : ~ 200 pm = 1 : 100 000), in other books the "average" ratio (~ 15 fm : ~ 150 pm = 1 : 10 000) is given (compare Fig. 5.7).

5.5 Atomic Mass Units

The different atomic models also constantly stimulated to get information about the mass of the single atom. The results obtained here led to various comparative figures, which will now be considered in more detail.

Atoms and Molecules Are Tiny

To illustrate the tininess of atoms and molecules, here are two examples:

- If you were to pour 20 cm^3 of alcohol (= 1 shot glass) into the sea and distribute it evenly over all the seas (= 1370 million km^3 of water), you would still find 153 alcohol molecules in every litre of sea water.
- The amount of iron atoms contained in a pinhead (1 mm^3) is about 100 trillion (10^{20}), the diameter of an iron atom is about 20 nm (nanometer = trillionth of a meter). Placed in a chain, the distance between the earth and the moon could be measured more than fifty times. In other words, a single millimetre would consist of five million iron atoms.

5.5.1 Absolute Atomic Mass (m_A) and Absolute Molecular Mass (m_M)

The absolute atomic mass is given in kg. But since atoms and molecules are extremely small, these masses are also extremely small. They are (depending on the element) 10^{-27} to 10^{-25} kg. A single hydrogen atom weighs $1.674 \cdot 10^{-27}$ kg. Or written differently:

$$m(\mathrm{H}) = 0,000\,000\,000\,000\,000\,000\,000\,000\,001\;674\,\mathrm{kg}.$$

The absolute mass of individual molecules is almost as tiny. A single sugar molecule ($C_6H_{12}O_6$) weighs only 0.000 000 000 000 000 000 000 000 299 156 kg and is thus a good 180 times heavier than a hydrogen atom.

It is probably only too understandable that it is difficult to work with such “unwieldy” numbers on a permanent basis.

5.5.2 Relative Atomic Mass (A_r), Relative Molecular Mass (M_r), and “u”

In order not to have to work with the absolute atomic masses, the so-called “relative atomic mass” was introduced, which tells how much heavier an element or a compound is compared to a reference point.

At first, this reference point was hydrogen as the lightest of all elements. Its relative mass was arbitrarily set to “1”. Later, one took the oxygen as the most common earthly element with the mass 16 as reference. Today, a certain carbon isotope (the ^{12}C, which consists of 6 protons, 6 neutrons and 6 electrons) is the reference.

The relative atomic and molecular masses are without unit, as they are purely comparative figures.

However, one can also specify the “atomic mass units, u” (from *unified atomic mass unit;* from obsolete amu: *atomic mass unit).* The numerical values do not change in this case.

- One unit (i.e.: 1 u) corresponds to exactly 1/12 of the mass of the carbon isotope C-12.

- The relative molecular masses (M_r) or the mass (given in u) are calculated by adding the relative atomic masses (A_r) of the atoms involved.

Example

Methane

$$M_r(CH_4) = A_r(C) + 4 \cdot A_r(H) = 12{,}011 + 4 \cdot 1{,}008 = 16.043$$

Methane has a relative molecular mass of 16.043; $M_r(CH_4) = 16.043$.

Water

$$u(H_2O) = 2 \cdot u(H) + 1 \cdot u(O) = 2 \cdot 1.008\ u + 1 \cdot 15.999\ u = 18.015\ u$$

The mass of a water molecule is 18.015 u.

Sugar

$$M_r(C_6H_{12}O_6) = 6 \cdot 12.011\ u + 12 \cdot 1.008\ u + 6 \cdot 15.999\ u = 180.16\ u$$

The relative molecular mass is 180.16 u. ◀

5.5.3 Amount of Substance *n*

The relative atomic masses or u correspond to certain gram quantities (masses) or certain volumes. An example may clarify this: The relative atomic mass of hydrogen is 1.0 u, that of carbon is 12.0 u. If we now take any number of hydrogen atoms, the total mass of the same number of carbon atoms will always be 12 times greater than that of the hydrogen atoms.

As long as this mass ratio of 1 : 12 is fulfilled, there are also the same number of atoms of the two elements.

Or in general:

- The quantity in grams of an element corresponding to the numerical value of the relative atomic mass always contains the same number of atoms.

This number of atoms is called the “Avogadro number”. The corresponding portion of substance is called a mole.

▶ A mole of a substance is the amount of substance of a system (an amount of matter) that consists of as many smallest particles (atoms, molecules, ...) as are contained in exactly 12.0 g of the carbon isotope ^{12}C.

The symbol for **mole** is "**mol**", the number of moles – better: the amount of substance – is abbreviated with the formula symbol "$\boldsymbol{n}$".

A wide variety of methods can be used to determine how many atoms or molecules are contained in one mole of pure substance: from the kinetic theory of gases, from Brownian molecular motion, from the surface tension of dilute solutions, from the electrical charge of oil droplets (Millikan experiment), and many more.

With these very different methods the same numerical value was found everywhere: 602 214 076 000 000 000 000 000 particles (molecules or atoms) are contained in the quantity of substance of one mole.

This number of particles is called "Avogadro's constant" or "Avogadro's number" after the physicist Avogadro and is abbreviated N_A:

$$N_A = 6.022 \cdot 10^{23}\ \text{mol}^{-1}.$$

Avogadro's law states that the same number of particles (atoms or molecules) are present in the same volumes of all gases at the same pressure and temperature (see Sect. 4.7).

▶ The substance portion 1 mole contains $6.022 \cdot 10^{23}$ particles (atoms, molecules) of a substance. This number is called the Avogadro constant (N_A).

However, this material portion "mole" is not at all as abstract as this huge, hardly imaginable number may first appear.

It is the same kind of "abbreviation" as the dozen (= 12 pieces), the gross (= 12 dozen), the hundredweight (= 50 kg) or the light-year (9.46 trillion km).

A dozen is always 12 pieces – no matter if eggs, nails, cars or houses. One mole is always $6.022 \cdot 10^{23}$ particles – no matter of which substance.

Of course it is nonsensical to talk about "x moles of rice" – no one buys his Sunday rolls in hundredweight ...

For the "handling" of atoms and molecules, however, the "mole" has proven to be just as useful as the "dozen" when buying eggs.

A major advantage is that the relative mass (without unit), the mass in atomic mass units (i.e. in u) and the molar mass (to be given in g/mol) for a given substance always have the same numerical value (see Table 5.7).

5.5.4 The Molar Volume (V_m)

Studies of gases have shown that one mole of a gas not only consists of the $6.022 \cdot 10^{23}$ gas particles already mentioned, but that this number of particles also always occupies the same space independently of the gas – assuming the same conditions of pressure and temperature.

Table 5.7 Comparison: absolute, relative and molar masses

Fabric X	Symbol/ Formula	Absolute mass $m_A(X)$, $m_M(X)$	Relative mass $A_r(X)$, $M_r(X)$	Mass in u	Molar mass
Neon	Ne	$3.351 \cdot 10^{-26}$ kg	20.18	20.18 u	20.18 g/mol
Chlorine	Cl	$5.887 \cdot 10^{-26}$ kg	35.45	35.45 u	35.45 g/mol
Water	H_2O	$2.992 \cdot 10^{-26}$ kg	18.015	18.015 u	18.015 g/mol
Ammonia	NH_3	$2.828 \cdot 10^{-26}$ kg	17.031	17.03 u	17.03 g/mol
Table salt	NaCl	$9.705 \cdot 10^{-26}$ kg	58.443	58.44 u	58.44 g/mol
Sugar	$C_6H_{12}O_6$	$2.992 \cdot 10^{-25}$ kg	180.156	180.16 u	180.16 g/mol

A pressure of 1013.25 mbar and a temperature of 0 °C (= 273.15 K) were set as standard conditions. Under these conditions, one mole of a gas occupies the volume of exactly 22.413 837 L.

▶ The molar standard volume of a gas is: $V_{m,0} = 22.4$ L/mol (at 0 °C, 1.013 bar)

5.5.5 Loschmidt Number

In the older (mainly German) literature, the Loschmidt number (N_L) is erroneously found as a synonym for the Avogadro constant ($N_L = N_A = 6.022 \cdot 10^{23}$ 1/mol).

However, according to a *more recent definition,* the Loschmidt number (also: Loschmidt constant, n_0) is the number of particles contained in 1 m³ of a gas (under standard conditions). It holds:

$$N_L = n_0 = N_A / V_{m,0} = 2.687 \cdot 10^{25} \frac{1}{m^3}$$

The Periodic Table

6

The periodic table of the elements, also known as the periodic table or PTE for short, is something most people will remember from their school days. How this system has developed historically and what structure it is based on will be described in Sects 6.1, 6.2, 6.3 and 6.4.

6.1 Early Trials

Attempts were made early on to bring order into the confusion of different substances and their properties. But neither the alchemists nor the early natural scientists succeeded in this.

Between 1817 and 1829, Döbereiner[1] published different combinations of three elements each ("triads"), in which the properties of these elements were compared.

- Lithium – Sodium – Potassium
- Calcium – Strontium – Barium
- Sulphur – Selenium – Tellurium
- Chlorine – Bromine – Iodine

The elements of each triad are similar in their chemical properties and the atomic mass of the middle element is approximately equal to the average of the other two. Between 1863 and 1866, Newlands[2] then proposed the 'octave law':

If one orders the elements according to increasing atomic mass, then the chemical properties repeat themselves in every eighth position. In this way he even succeeded in predicting a missing element between silicon and tin, which should have an

[1] Johann W. Döbereiner, 1780–1849, German chemist.

[2] John A. R. Newlands, 1837–1898, English chemist.

T. Schmiermund, *The Chemistry Knowledge for Firefighters*,
https://doi.org/10.1007/978-3-662-64423-2_6

atomic mass of 73 (in 1886 germanium, atomic mass 72.6 u, was discovered). However, this rule of eight proved to be too rigid and was therefore discarded.

Many other chemists also endeavored to establish a system: Falckner (1824), Gmelin (1843), von Pettenkofer (1850), Odling (1857), Lenssen (1857), Dumas (1858), Chancourtois (1862), Hinrichs (1866) and others.

The breakthrough was achieved independently by Mendeleev[3]and Meyer[4] in 1869. As a rule, Mendeleev is regarded as the "father of the periodic table", since he published his ideas first.

6.2 Periodic Table According to Mendeleev & Meyer

Mendeleev and Meyer arranged the elements according to increasing atomic mass and similarities in chemical behavior. Mendeleev initially arranged similar elements in rows, but later restructured this so that similar elements were placed in vertical columns below one another.

Here gaps arose which for Mendeleev simply represented missing, i.e. not yet discovered, elements. He was so convinced of the correctness of his system that he even dared to make predictions about these missing elements.

Gallium (still called "Eka-aluminum" by Mendeleev, 1875), scandium ("Eka-boron", 1879), germanium ("Eka-silicon", 1886) and polonium ("Eka-tellurium", 1898) were discovered.

It was the discovery of the predicted elements that helped Mendeleev and Meyer's periodic table achieve its breakthrough.

The noble gases, which were only discovered between 1892 and 1898, had not yet been envisaged by Mendeleev, but they fitted into the system without any problems.

From Mendeleev's point of view, the only blemishes were three pairs of elements (argon/potassium, cobalt/nickel and tellurium/iodine) which had to be exchanged in relation to their atomic mass.

Note that in the diagram in Fig. 6.1 the basic formula of the respective oxide is also given for each group.

Moseley's Law

In 1913/14 Moseley[5] solved the problem of "switched" atomic masses:

He studied the X-ray spectra of 38 elements and found a relationship between the frequency of a particular line in the spectrum and the atomic number.

It also turned out that Moseley's atomic number agreed well with Rutherford's determinations of nuclear charge numbers:

[3]Dimitri Ivanovich Mendeleev, 1834–1907, Russian chemist.

[4]Julius Lothar Meyer, 1830–1895, German chemist.

[5]Henry G. J. Moseley, 1887–1915, British physicist.

Reihen	Gruppe I. — R^2O	Gruppe II. — RO	Gruppe III. — R^2O^3	Gruppe IV. RH^4 RO^2	Gruppe V. RH^3 R^2O^5	Gruppe VI. RH^2 RO^3	Gruppe VII. RH R^2O^7	Gruppe VIII. — RO^4
1	H=1							
2	Li=7	Be=9,4	B=11	C=12	N=14	O=16	F=19	
3	Na=23	Mg=24	Al=27,3	Si=28	P=31	S=32	Cl=35,5	
4	K=39	Ca=40	—=44	Ti=48	V=51	Cr=52	Mn=55	Fe=56, Co=59, Ni=59, Cu=63
5	(Cu=63)	Zn=65	—=68	—=72	As=75	Se=78	Br=80	
6	Rb=85	Sr=87	?Yt=88	Zr=90	Nb=94	Mo=96	—=100	Ru=104, Rh=104, Pd=106, Ag=108
7	(Ag=108)	Cd=112	In=113	Sn=118	Sb=122	Te=125	J=127	
8	Cs=133	Ba=137	?Di=138	?Ce=140	—	—	—	— — — —
9	(—)	—	—	—	—	—	—	
10	—	—	?Er=178	?La=180	Ta=182	W=184	—	Os=195, Ir=197, Pt=198, Au=199
11	(Au=199)	Hg=200	Tl=204	Pb=207	Bi=208	—	—	
12	—	—	—	Th=231	—	U=240	—	— — — —

Fig. 6.1 Periodic table of Mendeleev (1869)

"*In the atom there is a fundamental quantity which increases in regular steps from one element to another. This quantity can only be the positive charge of the atomic nucleus,*" Moseley stated.

Thus the atomic number (which until then had only been a kind of serial number for each element) became the number of positive nuclear charges (the protons) and thus a fundamental ordering principle.

6.3 Structure of the Periodic Table

The periodic table of the elements follows a clear structure principle. If the elements were first arranged on the basis of their "chemical relationship", an order then followed according to the structure of the electron shell. Only after it was recognized that the "numerator atomic number" was directly related to the number of protons in the atomic nucleus itself, did the atomic number also become a true order characteristic.

6.3.1 Display Mode

In the modern representation, the elements are ordered by increasing atomic number in the rows (called periods) from left to right. Similarly, the total number of electrons also increases by one from element to element.

A first classification is based on the so-called main groups and subgroups. Let us first take a look (Fig. 6.2) at the so-called "short period system" and thus at the so-called "main group elements".

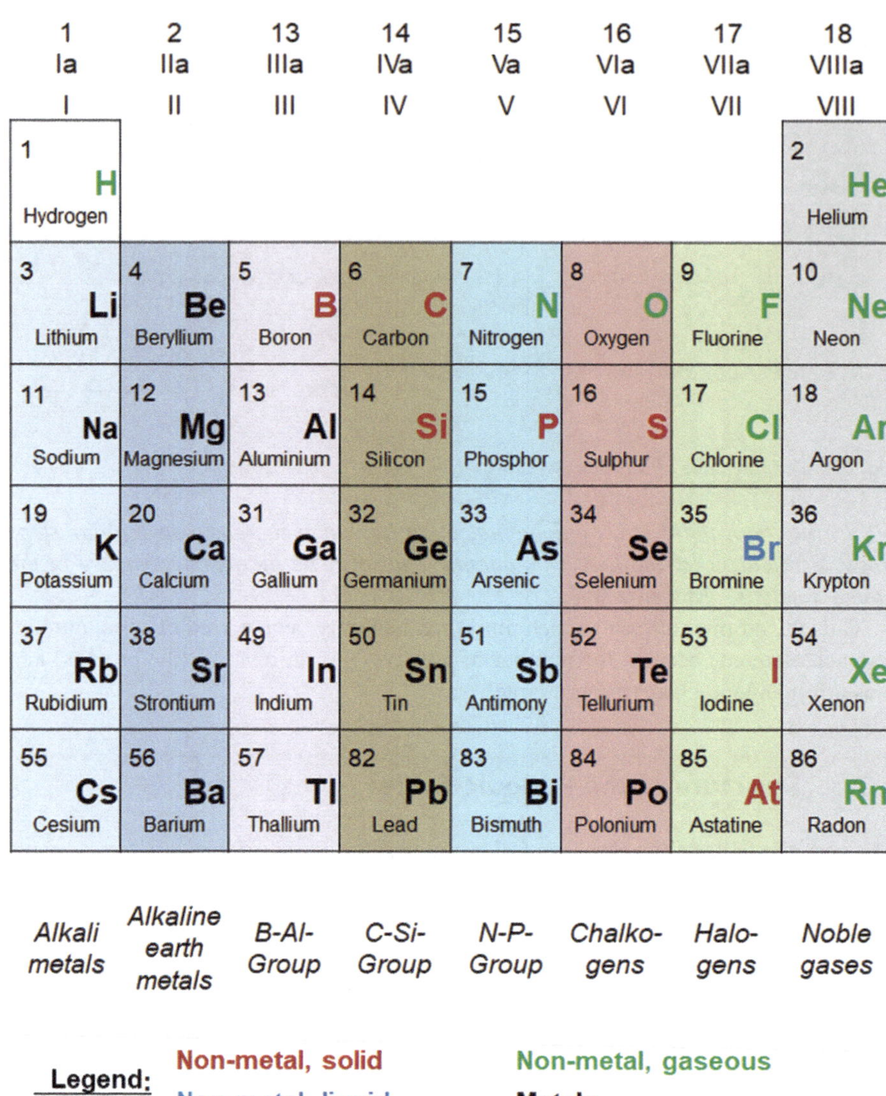

Fig. 6.2 Short period system

For the Numbering of the Groups

The old group number (third numbering line) is indicated with Roman numerals and corresponds to the number of outer electrons.

In a later variant, the main groups were given the suffix "a", the subgroups (not shown here) the suffix "b" (second line). The roman numeral of the subgroup elements (Ib – VIIIb) indicates the number of "available" outer electrons.

Currently the groups are simply numbered from 1 to 18 (first line).

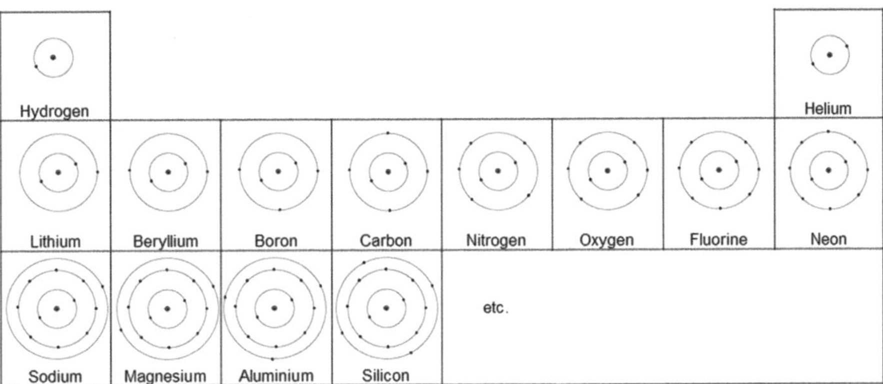

Fig. 6.3 Bohr's atomic model in the periodic table

In all three variants, the "inner transition metals" are not numbered.

If we now compare Fig. 6.2 with the structure of the elements according to Bohr's atomic model, we obtain the representation shown in Fig. 6.3.

It can be seen that the elements of the 1st main group have one electron in the outermost shell, those of the second main group have two electrons, and so on.

Only with the noble gases (8th main group = 8 outer electrons) helium makes an exception. It has only two outer electrons, because the 1s orbital cannot accept more electrons.

Due to the periodic arrangement in groups, elements with the same (outer) electron configuration come to stand among each other. This same composition of the outer electrons then also explains the almost identical chemical behaviour within a group.

6.3.2 Casting Order

As supplementary material to the book, there are various periodic tables of the same structure but with different information. These are freely available at https://www.springer.com/en/book/9783662566060. In one of these periodic systems, the occupation of the electron shells is highlighted in color so that it can be easily understood. By the way, elements with the same electron configuration of the outer shell (read from the right, Sect. 6.4.3) are always placed one below the other (are thus "divided" into groups).

The order in which the orbitals are occupied by electrons can be seen in Fig. 6.4 (read line by line).

But: How does it explain that before the 3d-orbital first the 4s-orbital is occupied? And, after the 3d-orbital is completely filled, only then the 4p-orbital is occupied?

As is so often the case, this is related to energy. The 4s orbital has a somewhat lower energy than the 3d orbital, which in turn is somewhat lower in energy than the 4p orbital.

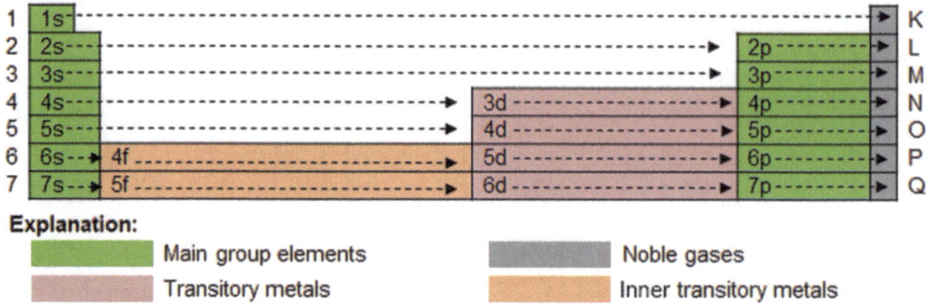

Fig. 6.4 Overview of the cast sequence

Fig. 6.5 Energy levels of the orbitals

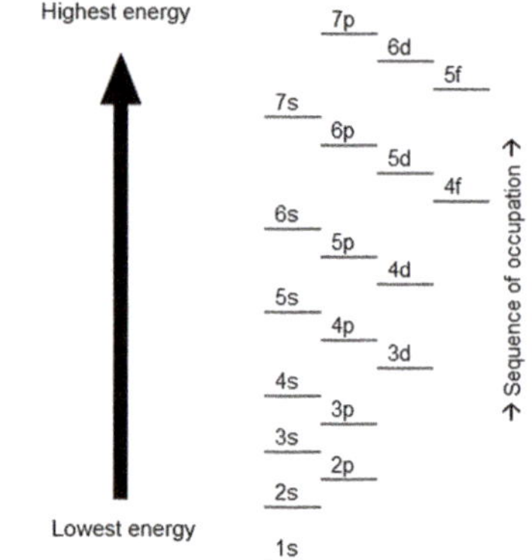

For this reason, the M-shell (the third Bohr orbital) is filled up to the end only in the fourth period. Analogously it behaves also with the f-orbitals.

Therefore, the representation can also be chosen according to the "amount of energy" (Fig. 6.5).

However, in order to remember the order of occupation, another method is more helpful: the "chessboard method" (Table 6.1).

6.4 Representation of the Electron Configuration

Sometimes it is necessary to represent the electron configuration. Two different notations are commonly used for this purpose, which can also be used side by side depending on the question.

Observing the Hund rule, the configuration of all its electrons can be written or represented for each element.

Table 6.1 Orbital occupation according to the “chessboard method

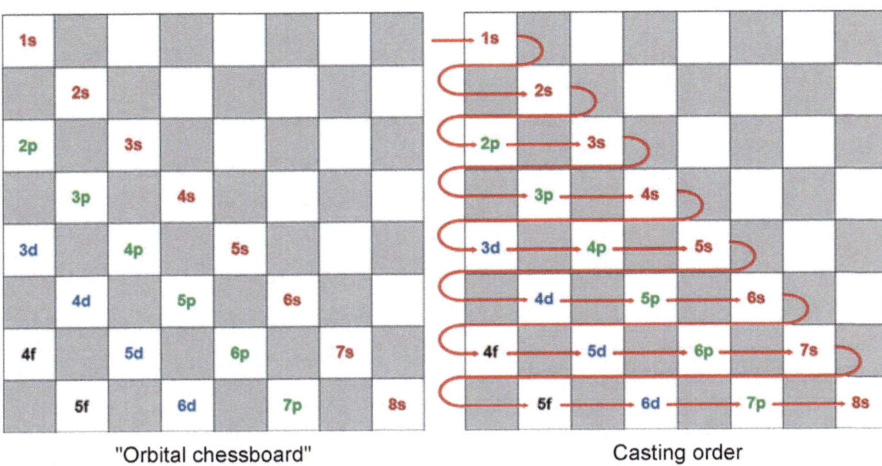

6.4.1 Hund’s Rule

According to the rule established by F. Hund[6] in 1927, when the atomic shell is filled up with electrons, first all energy levels (corresponding to the minor quantum number ℓ) within a shell are occupied with an electron. Only when all orbitals belonging together are simply occupied, they are filled up with electrons with opposite spin.

The negative charge of the electrons causes their repulsion among each other. In the case of energetically equivalent orbitals, they therefore first distribute themselves to the different orbitals. Only then does pairing occur in the same orbital.

The mutual influence of the electrons not only causes the initial single occupation of the orbitals; it also ensures that a maximum number of electrons with parallel spin ($+^1/_2$ *or* $-^1/_2$) is preferred. This state is lower in energy than spin alternating from electron to electron.

Equal-energy orbitals, such as the three 2p orbitals, are also called “degenerate”.

6.4.2 Orbital Diagram

One way of indicating the electron configuration of an element is the orbital diagram, also known as the “Pauling notation” or casually as the “box notation” after its developer (Fig. 6.6).

[6] Friedrich Hund, 1896–1997, German physicist.

Fig. 6.6 Orbital diagram (1)

	1s	2s	2p		
Hydrogen: $1s^1$	↑				
Helium: $1s^2$	↑↓				
Lithium: $1s^22s^1$	↑↓	↑			
Beryllium: $1s^22s^2$	↑↓	↑↓			
Boron: $1s^22s^22p^1$	↑↓	↑↓	↑		
Carbon: $1s^22s^22p^2$	↑↓	↑↓	↑	↑	
Nitrogen: $1s^22s^22p^3$	↑↓	↑↓	↑	↑	↑
Oxygen: $1s^22s^22p^4$	↑↓	↑↓	↑↓	↑	↑
Fluorine: $1s^22s^22p^5$	↑↓	↑↓	↑↓	↑↓	↑
Neon: $1s^22s^22p^6$	↑↓	↑↓	↑↓	↑↓	↑↓

Here, one draws a box for each orbital (more precisely: for each magnetic quantum number per orbital). The occupation with electrons is done by arrows pointing upwards or downwards. The direction of the arrow symbolizes the electron spin ($+^1/_2$ or $-^1/_2$). Here the "Hund's rule" is to be considered.

In another representation, the "orbital boxes" are slightly offset in height to show the different energies of the orbitals in a "stepped" manner.

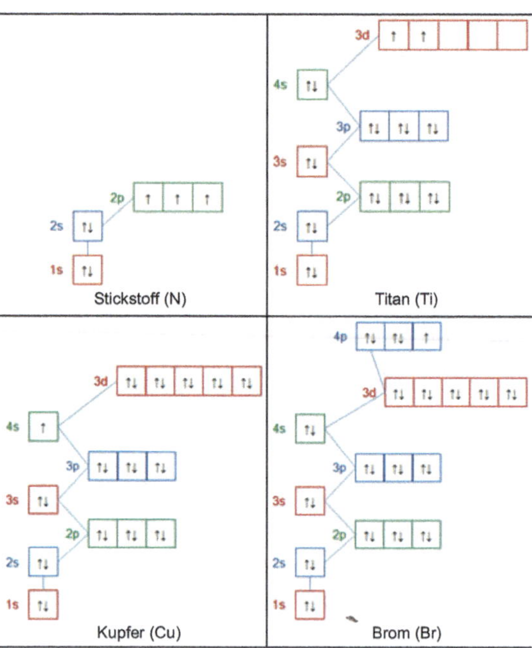

This method of representation sometimes has its advantages, e.g. if one wants to compare energy levels (orbitals) with each other.

6.4.3 Term Notation

The term notation of the electron configuration has already been briefly described in the Bohr-Sommer-feld atomic model (Sect. 5.2.3).

One writes in the order: shell – energy level – number of electrons → next shell – energy level(s) – number of electrons → next shell – energy level(s) – number of electrons → etc.

It should be noted here that the order of the orbitals follows the shell numbers. Thus, the 3d orbital is written before the 4s orbital, although the 3d orbital is occupied after the 4s orbital. The p orbitals follow the s orbitals, since the s orbitals are occupied first.

Table 6.2 shows the term notation of the first 20 elements.

Table 6.2 Term notation of the electron configuration of the first 20 elements

Bohr's shell	Highest occupied orbital	Item	Total number of electrons	Number of outer electrons	Electron configuration
1	1s	Hydrogen	1	1	$1s^1$
		Helium	2	2	$1s^2$
2	2s	Lithium	3	1	$1s^2\ 2s^1$
		Beryllium	4	2	$1s^2\ 2s^2$
	2p	Boron	5	3	$1s^2\ 2s^2\ 2p^1$
		Carbon	6	4	$1s^2\ 2s^2\ 2p^2$
		Nitrogen	7	5	$1s^2\ 2s^2\ 2p^3$
		Oxygen	8	6	$1s^2\ 2s^2\ 2p^4$
		Fluorine	9	7	$1s^2\ 2s^2\ 2p^5$
		Neon	10	8	$1s^2\ 2s^2\ 2p^6$
3	3s	Sodium	11	1	$1s^2\ 2s^2\ 2p^6\ 3s^1$
		Magnesium	12	2	$1s^2\ 2s^2\ 2p^6\ 3s^2$
	3p	Aluminium	13	3	$1s^2\ 2s^2\ 2p^6\ 3s^2\ 3p^1$
		Silicon	14	4	$1s^2\ 2s^2\ 2p^6\ 3s^2\ 3p^2$
		Phosphorus	15	5	$1s^2\ 2s^2\ 2p^6\ 3s^2\ 3p^3$
		Sulphur	16	6	$1s^2\ 2s^2\ 2p^6\ 3s^2\ 3p^4$
		Chlorine	17	7	$1s^2\ 2s^2\ 2p^6\ 3s^2\ 3p^5$
		Argon	18	8	$1s^2\ 2s^2\ 2p^6\ 3s^2\ 3p^6$
4	4s	Potassium	19	1	$1s^2\ 2s^2\ 2p^6\ 3s^2\ 3p^6\ 4s^1$
		Calcium	20	2	$1s^2\ 2s^2\ 2p^6\ 3s^2\ 3p^6\ 4s^2$

Table 6.3 Shorthand notation of the electron configuration (examples)

Preceding noble gas	Item	Short form
Helium	Beryllium	[He] $2s^2$
	Nitrogen	[He] $2s^2$ $2p^3$
Neon	Aluminium	[Ne] $3s^2$ $3p^1$
	Sulphur	[Ne] $3s^2$ $3p^4$
	Chlorine	[Ne] $3s^2$ $3p^5$
Argon	Calcium	[Ar] $4s^2$
	Scandium	[Ar] $3d^1$ $4s^2$
	Titanium	[Ar] $3d^2$ $4s^2$
	Chrome	[Ar] $3d^5$ $4s^1$
	Iron	[Ar] $3d^6$ $4s^2$
	Arsenic	[Ar] $3d^{10}$ $4s^2$ $4p^3$
Xenon	Cerium	[Xe] $4f^2$ $6s^2$
	Platinum	[Xe] $4f^{14}$ $5d^9$ $6s^1$
	Lead	[Xe] $4f^{14}$ $5d^{10}$ $6s^2$ $6p^2$

Since this “long notation” becomes very quickly confusing, one has developed a shortened form. Here the *preceding* noble gas is written in square brackets (= [NG]), only then the actual term notation follows. For further simplification, the Bohr orbit (principal quantum number) is no longer written in front of each orbital. Examples see Table. 6.3.

We will return to this notation as needed.

The configuration of all elements can be found in the periodic table (additional material, available at https://www.springer.com/de/book/9783662566060).

6.5 Periodic Properties

The properties of the elements, which are related (although not always easy to identify) to the electron configuration, change periodically with increasing atomic number. These are, for example:

Atomic radius	Ionization energy
Density	Oxidation numbers
Electrical conductivity	Melting point
Electron affinity	Boiling point
Electronegativity	Heat conductivity
Ion radius	*et al.*

As early as 1870, L. Meyer worked out an “atomic volume curve” (Fig. 6.7).

However, this atomic volume only gives an approximate indication of the size of an atom, since the “packing density” of the atoms is not taken into account here.

Some of the periodic properties will be considered in more detail in Sects. 6.5.1, 6.5.2, 6.5.3 and 6.5.4.

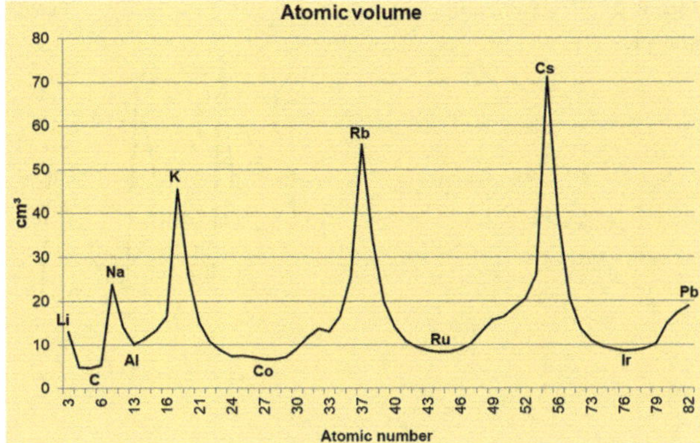

Fig. 6.7 Atomic volume curve (after L.Meyer)

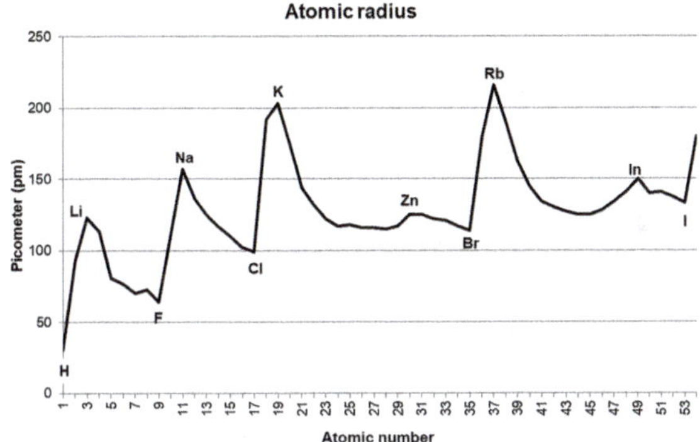

Fig. 6.8 Atomic radii

6.5.1 Atomic Radius

With the beginning of each new period, the atomic radius increases as expected. In the course of the respective period, the atom itself becomes steadily smaller, since the "more" protons in the nucleus pull the entire shell more strongly towards the nucleus. The period minimum is then reached with the halogens. Within a group, the atomic radius increases because new shells are constantly being added (Fig. 6.8).

6.5.2 Ionization Energy

► The ionization energy is the energy that must be expended to remove an electron from the electron shell of an atom or ion.

The "first ionization energy" is therefore the energy required to turn an atom into a single positively charged ion.

This first ionization energy is shown in Fig. 6.9.

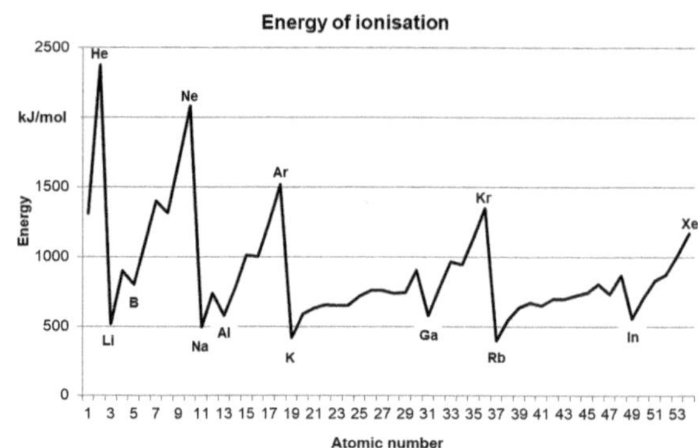

Fig. 6.9 First ionization energies

The following statements can be made based on the graph:

- The elements of the 1st main group (Li, Na, K, Rb) have very low ionization energies. The single outer electron (electron configuration: s^1) is easy to separate.
- For the elements of the second main group, the energy is somewhat higher. Here, an electron must be removed from the fully occupied s-orbital (electron configuration s2).
- In the third main group (B, Al, Ga, In) the ionization energy is lower again. The single electron of the p-orbital (electron configuration s2 p1) is again easier to remove.
- Finally, for the noble gases, the ionization energy is unusually high. The electron configuration s2 p6 represents a very stable state.

6.5.3 Electron Affinity

> Electron affinity is the energy converted by a neutral atom when it accepts an electron.

Energy that is released by the absorption of an electron is represented in Fig. 6.10 by positive values. Energy that must be expended is shown accordingly as a negative value.

- The halogens (F, Cl, Br, I) take up electrons very easily. They then form the more stable noble gas configuration s^2 p^6 from their original electron configuration s^2 p^5.
- The slightly higher affinity of chlorine to fluorine is due to the fact that the increasing nuclear charge from fluorine to chlorine (from 9 p^+ to 17 p^+) exceeds the effect of the increasing atomic radius (from 64 pm to 99 pm).
- The alkaline earth metals (Be, Mg, Ca, Sr) can only with great difficulty add another electron to their atomic shell, since a p-orbital must be formed in addition to the fully occupied s-orbital.

Fig. 6.10 Electron affinities

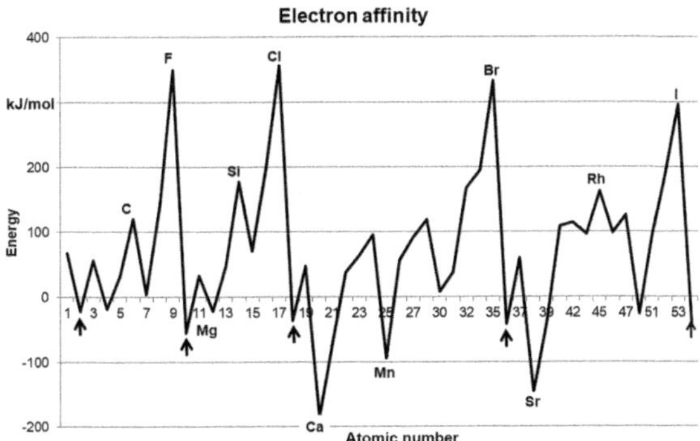

- The energies of the noble gases *(arrows)* are all about the same, since the formation of a "new shell" (the next s-orbital is occupied by an electron) requires a comparably high energy input.
- Manganese (Mn) has the configuration [Kr] $4s^2$ $4p^6$ $4d^5$ $5s^1$ before ionization. Both the 4d and 5s orbitals are half occupied. Overcoming this rather stable state in the electron configuration is noticeable in the low electron affinity.
- The situation is different for rhodium (Rh). The electron configuration before ionization is [Kr] $4s^2$ $4p^6$ $4d^8$ $5s^1$. By adding an electron, the 5s orbital can be completed. This is relatively easy and brings an energetic gain for the atom.

6.5.4 Electronegativity

L. Pauling[7] developed a system in 1932 to estimate the polarity of a bond. For this purpose, he introduced a measure, the electronegativity.

► Electronegativity (EN) is a measure of the tendency of an atom to attract electrons to itself in a bond.

By fixing the EN of fluorine to the value 4.0, it was possible to specify a closed electronegativity scale for all elements. The EN plays an important role in the bonds between the atoms and thus in the formation of ions and molecules (cf. Chap. 2).

- The halogens (F, Cl, Br, I) have the highest electronegativities. Within the group the EN decreases.
- The alkali metals (Li, Na, K, Rb, Cs) consistently have the lowest EN values. Here, too, the value decreases further with increasing period.
- Manganese (Mn) and technetium (Tc) with their electron configuration [NG] d^5 s^2, as well as zinc (Zn) and cadmium (Cd) with [NG] d^{10} s^2 have particularly low-energy states, so that the EN drops significantly within the period.

[7]Linus Carl Pauling, 1901–1994, US chemist, Nobel Prize in Chemistry 1954, Nobel Prize in Peace 1963.

H						
2.1						
Li	Be	B	C	N	O	F
1.0	1.5	2.0	2.5	3.0	3.5	4.0
Na	Mg	Al	Si	P	S	Cl
0.9	1.2	1.5	1.8	2.1	2.5	3.0
K	Ca	Ga	Ge	As	Se	Br
0.8	1.0	1.6	1.8	2.0	2.4	2.8
Rb	Sr	In	Sn	Sb	Te	I
0.8	1.0	1.7	1.8	1.9	2.1	2.5

Since noble gases do not form compounds in the "classical sense", no electronegativities can be determined for them. In Fig. 6.11 the values were arbitrarily set to "0".

Summary: Periodic Properties of the Elements

The most important periodic properties of the elements based on their position in the PSE are listed in Table 6.4.

6.6 Main Groups of the Periodic Table

As shown in the short period system (Fig. 6.2), the PTE consists of eight so-called main groups. These main groups will be briefly characterised in Sects. 6.7.1, 6.7.2, 6.7.3, 6.7.4, 6.7.5, 6.7.6, 6.7.7 and 6.7.8.

6.6.1 First Main Group – Alkali Metals

The elements of this group (Li, Na, K, Rb, Cs, Fr; with the exception of hydrogen) represent light metals that can be cut with a knife and are combustible because they react readily with atmospheric oxygen. They also react readily with water. This

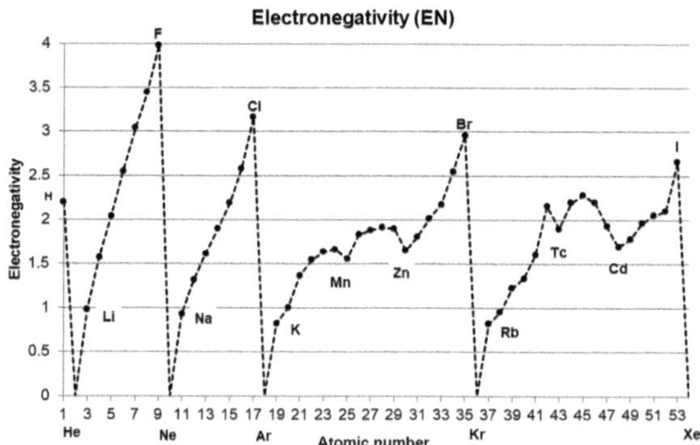

Fig. 6.11 Electronegativity

Table 6.4 Summary of periodicity in the PSE

Atomic radius, atomic volume	Increases from top to bottom. Decreases from left to r8Tabight.	Atomic volume, Atomic radius
Ionization energy	Decreases from top to bottom. Increases from left to right.	Lonisation energy
Electron affinity	Very high for the halogens. Very low for the alkaline earth metals.	Electron affinity
Electronegativity	Decreases from top to bottom. Increases from left to right. Most electronegative element: Fluorine *(most non-metallic non metal")* Most electropositive element: Francium *(most metallic metal)*	Electronegativity
Metal character	Increases from top to bottom. Decreases from left to right.	Metal character

produces (flammable) hydrogen and the corresponding alkali (corrosive!). This "alkaline" behavior was also the inspiration for the group name. Alkali metals and alloys with a high proportion of these must therefore never be brought into contact with water.

As a combustible gas, hydrogen (H) occupies a special position in this group. It is by far the most common element in the universe. Metallic hydrogen is formed at pressures of 2.7 megabar ($2.7 \cdot 10^6$ bar).

Since all elements of this group have only one outer electron, they are "monovalent", i.e. they form only one "bond".

- **Sodium (Na)**: Element essential for life, contained as common salt in seawater, used in glass, soaps, yellow street lamps (sodium vapour lamps), as sodium hydroxide (NaOH) an important basic chemical substance.
- **Potassium (K)**: Essential, most common use: Fertilizer; but also found in soft soap, glass, and black powder; Used as potassium hyperoxide in regeneration devices (respirators).
- **Rubidium (Rb)**: Use in photocells
- **Caesium (Cs)**: Is the most reactive metal, forms as caesium hydroxide (CsOH) the strongest base of all elements; used in photocells, photomultipliers and light barriers – and in atomic clocks
- **Francium (Fr)**: Radioactive, extremely short-lived

6.6.2 Second Main Group -Alkaline Earth Metals

The alkaline earth metals (Be, Mg, Ca, Sr, Ba, Ra) are found in many minerals and ores ("earths") and form hydroxides that react slightly alkaline (hence **alkaline** earth metals). The oxides and hydroxides also have an "earthy" feel (naming: alkaline **earth** metals).

They are all divalent metals with low density and a silvery-white surface. They are not quite as reactive as the alkali metals and can therefore be stored in air.

The alkaline earth metals also do not occur elementally in nature. They are mainly found in minerals – the magnesium also in the leaf dye chlorophyll.

- **Beryllium (Be)**: Mainly used as an alloying element, makes alloys hard, strong, corrosion resistant and low sparking.
- **Magnesium (Mg)**: For the production of light alloys, essential for all living beings, also used in fireworks and in so-called "magnesium torchs", which continue to burn even under water.
- **Calcium (Ca)**: Essential for all living organisms (skeleton), contained in gypsum, marble, mortar and cement
- **Strontium (Sr)**: Used for red flares and in permanent magnets, and as radioactive Sr-90 in nuclear batteries.
- **Barium (Ba)**: As sulphate in white wall paints and as X-ray contrast agent, burns with green paint (fireworks), in magnets and high-temperature superconductors
- **Radium (Ra)**: Much more radioactive than uranium, formerly used in luminous paints.

6.6.3 Third Main Group – Boron Group

This group consists of the elements B, Al, Ga, In, Tl.

Boron is a hard nonmetal, aluminum and the other elements in the group are soft metals.

Aluminium is involved in a large number of compounds (minerals) but, like the other elements in the group, does not occur elementally in nature.

In contrast to the other elements of the third main group, boron is rather inert.

- **Boron (B)**: As perborate in detergents, for heat-resistant borosilicate glasses (Duran®, Pyrex®, Jenaer Glas®), boron nitride has a similar hardness to diamond (Widia®, e.g. for drills)
- **Aluminium (Al)**: Most common metal in the earth's crust, versatile use (cans, foils, tubes, car and aircraft construction, components such as windows, profiles, ...) due to its low density.
- **Gallium (Ga)**: Melts already at 29.8 °C, used in semiconductor technology (microchips, diodes, microwaves and lasers), therefore contained in every PC and mobile phone.
- **Indium (In)**: Used for solar cells, transistors, displays, touch screens, etc.

- **Thallium (Tl)**: Formerly used for rat poison, today used for infrared detectors (thermal imaging) and as a colouring component in green signal flares

6.6.4 Fourth Main Group - Carbon Group

The elements C, Si, Ge, Sn and Pb belong to the fourth main group. Carbon and silicon are nonmetals, germanium is a semimetal, tin and lead have a distinct metal character.

Carbon is the "element of life", it is an omnipresent component of the substances of living nature. It is also a component of many minerals such as limestone or marble.

Silicon is extremely abundant in the earth's crust, as many minerals contain silicon.

Lead and lead compounds have a high density. Also, lead is the end product of radioactive decay series.

- **Carbon (C)**: Found as graphite and diamond; used in iron and steel production, printing inks, tires and paints. Petroleum and coal as carbon sources form one of the foundations of modern civilization.
- **Silicon (Si)**: Basis for microchips and photovoltaics, for silicones in medicine, technology and household (baking moulds), second most common element in the earth's crust (sand!)
- **Germanium (Ge)**: Used in transistors and semiconductors.
- **Tin (Sn)**: Alloy with copper = bronze; as soldering material ("solder"), for coins, jewellery, bells, organ pipes, statues, etc.
- **Lead (Pb)**: Lead in letterpress printing, in lead accumulators ("car battery", see also Sect. 19.2.1), as an additive in lead crystal glass and in radiation protection (shielding of X-rays and gamma rays).

6.6.5 Fifth Main Group - Nitrogen Group

The nitrogen group includes N, P, As, Sb and Bi. Here the increase in metal character (from top to bottom) becomes particularly clear: nitrogen is a typical (gaseous) non-metal. Phosphorus occurs in two non-metallic and one metallic modification. Arsenic also has a non-metallic form and antimony and bismuth are typical metals.

- **Nitrogen (N)**: Main component (78 % by volume) of air, in compounds (ammonia, nitric acid) as fertiliser and important raw material for industry; also contained in amino acids, nitrous oxide and urea; used as inert gas and liquid coolant
- **Phosphorus (P)**: Used as fertilizer, detergent and for flame retardant products; vital element (bones, teeth, DNA, cell metabolism)

- **Arsenic (As)**: Arsenic oxide is a potent poison; industrially, As is used to dope semiconductor materials and for LEDs.
- **Antimony (Sb)**: For hardening alloys and for flame retardants
- **Bismuth (Bi)**: For fuses, e.g. in sprinkler systems

6.6.6 Sixth Main Group - Chalcogens

The elements of the sixth main group are found particularly often as components of ores (formerly also called "earths"). Chalcogenes actually means "earth formers".

The group includes O, S, Se, Te, Po. Oxygen and sulfur are typical nonmetals, selenium, tellurium and polonium are semimetals.

- **Oxygen (O)**: Contained in the air at 21 % by volume; essential for life (respiration!); is the most common element in the earth's crust; the most important oxygen compound is also essential for life: Water
- **Sulphur (S)**: Yellow solid, used in the production of sulphuric acid, in fertilisers and for the vulcanisation of rubber (e.g. for the production of tyre rubber); also contained in some proteins
- **Selenium (Se)**: Used for photocells, copiers and as semiconductor material
- **Tellurium (Te)**: Used in optoelectronics, infrared cameras, solar cells and for alloys.
- **Polonium (Po)**: Is highly radioactive and is therefore used as a long-life energy source for satellites.

6.6.7 Seventh Main Group - Halogens

The elements of this group form salts particularly easily (Greek *halos* = salt and *genos* = to form). They are reactive nonmetals. Fluorine and chlorine are gases, bromine is liquid and iodine is solid. Fluorine is the most reactive element and has the highest electronegativity.

- **Fluorine (F)**: Toxic, corrosive gas, extremely reactive; used for chemically and thermally highly stable plastics (e.g. Teflon©), in UF_6 for isotope separation, essential for bones and teeth; hydrogen fluoride attacks glass (glass etching)
- **Chlorine (Cl)**: Toxic, corrosive, yellow-green gas, very reactive; essential for life as table salt; important basic chemical substance for all kinds of products; also used in bleaching agents and disinfectants.
- **Bromine (Br)**: Brown, smelly, easily evaporable liquid; used in roll films and for flame retardants
- **Iodine (I)**: Vital element with a metallic sheen, occurs in thyroid hormone; used as an antiseptic and in halogen lamps
- **Astat (At)**: Has no practical meaning; rarest element on earth

6.6.8 Eighth Main Group -Noble Gases

He, Ne, Ar, Kr, Xe and Rn are inert gases (inert → noble; hence the name). They occur in traces in the air and are enclosed in tiny bubbles in rocks. They are used as fillings for incandescent lamps (xenon lamps in motor vehicles) and for fluorescent tubes (neon tubes) and as inert gases in welding (argon or inert gas welding).

Helium is the second most abundant element in the cosmos after hydrogen and was first detected in the sun (*helios* = sun) before it was discovered on earth.

- **Helium (He)**: Is the second most abundant element in the universe; used as a coolant (e.g. in magnetic resonance tomographs) and as a balloon filler.
- **Neon (Ne)**: Glows red and was the first gas ever used for fluorescent tubes; hence the synonym for fluorescent tubes: Neon tubes
- **Argon (Ar)**: Used as a filling for incandescent lamps, fluorescent tubes and double-glazed windows, and as a shielding gas in welding.
- **Xenon (Xe)**: **Used in** floodlights, in the UV tubes of sunbeds, in high-power lasers and in car headlights; used as an anaesthetic gas when mixed with oxygen.

6.7 Subgroup Elements/d-Elements

The (classical) subgroup elements are also called "*transition* metals" or "d-elements" (occupation of the d-orbitals). These are the elements of groups 3 to 12 or Ib to VIIIb.

This also includes the most important commodity metals: iron, cobalt, nickel, manganese, chromium, etc.; the coinage metals gold and silver; and the only liquid metal: mercury.

The difference to the main group elements is caused by the existing d-orbitals, which are now -in the second outermost shell- filled up and also partly participate in bonds.

This results in various effects:

- High melting and boiling points (exception: mercury)
- High hardness (exception: copper and gold)
- High densities (only titanium and yttrium are light metals)
- The formation of complex compounds (see Sect. 11.2)
- A catalytic efficiency
- The occurrence of several oxidation states
- The appearance of colored compounds

For the "electron-deficient" subgroup elements of groups 3–7 (IIIb-VIIb), the maximum oxidation state results from the sum of the electrons of the s- and d-orbitals. The number of stable oxidation states increases with increasing group number.
In the "electron-rich" subgroup elements (groups 8–12 or VIIIb, Ib, IIb), the d-electrons are relatively tightly bound and therefore participate less and less in reactions until, in the end, only the s-electrons further out form bonds.

The d-elements are used in a wide variety of materials, e.g. for steels (Fe, Cr, Ni, Co, Mo, Mn, W, V), for catalysts (Pt, Ir, Ni, Ru, Rh, Pd), in electrical engineering (Ni, Cd, Cu, Ag, Au, Pt, Ir), for alloys (Cu, Zn, Ni), in medicine for instruments and implants (Ti, Nb, Ta), as jewellery metals (Au, Ag, Ti, Rh, Pt) and many more.

6.8 Rare Earths/f-Elements

The "rare earths", also called "inner transition metals" or "f-elements", consist of the 14 elements following lanthanum ("lanthanoids", old: "lanthanides") and actinium ("actinoids", old: "actinides") respectively.

Here, the respective f-orbital – and thus the third outermost shell – is filled with electrons. This causes that these elements behave chemically extremely similar and are also difficult to separate from each other.

The f-elements are also catalytically active and in some cases form coloured compounds. The tendency to form complexes is significantly lower than with the d-elements.

Most people will have encountered cerium (Ce) – probably the only one of the f-elements: Lighter flint is made of an alloy (cerium iron) that consists of 70 % Ce.

6.9 Oblique Relationship

Elements that are "oblique" to each other in the periodic table sometimes show strong similarities in their behaviour. This is also called "oblique relationship".

Due to the atomic radius increasing from top to bottom, the ionic part of the bonds becomes larger. However, it decreases from left to right due to the increasing charge. This leads to the fact that the following element pairs are very similar in their chemical behaviour: Li – Mg / Be – Al / B – Si / Ag – Hg.

6.10 Metals in the PTE

Among the more than 112 elements are less than 20 classical nonmetals. All other elements are metals or semi-metals. In this context, a wide variety of terms are used again and again, which will be explained briefly:

Non-Ferrous Metals

Non-ferrous metals are all heavy metals that are coloured by themselves or in their alloys (e.g. brass), with the exception of iron (Fe), silver (Ag), gold (Au) and the platinum metals (Ru, Os, Rh, Ir, Pd, Pt). These are: Cadmium (Cd), Cobalt (Co), Copper (Cu), Nickel (Ni), Lead (Pb), Tin (Sn) and Zinc (Zn).

Precious Metals

Metals which are more "noble" than hydrogen (see Sect. 17.3, Table 17.1). They are not dissolved by dilute hydrochloric acid and do not corrode in air. These include the six platinum metals, the three coinage metals and mercury (Hg).

Semi-Metals

Semi-metals are the elements that occupy an intermediate position between metals and non-metals. These are: Boron (B), Silicon (Si), Germanium (Ge), Arsenic (As), Antimony (Sb), Bismuth (Bi), Selenium (Se), Tellurium (Te) and Polonium (Po).

As elements, they usually occur in a metallic and a non-metallic modification. Due to the partial occupation of the outermost electron shell, they can both donate electrons ("metallic" behaviour) and accept them ("non-metallic" behaviour).

Light Metals

These are the 15 metals whose density is $< 5\ g/cm^3$. All other metals are heavy metals. The density of the metallic elements varies from lithium ($0.5\ g/cm^3$) to osmium ($22.5\ g/cm^3$) in a ratio of 1 : 45.

The 15 light metals are: Lithium (Li), Sodium (Na), Potassium (K), Rubidium (Rb), Caesium (Cs), Francium (Fr), Beryllium (Be), Magnesium (Mg), Calcium (Ca), Strontium (Sr), Barium (Ba), Aluminium (Al), Scandium (Sc), Yttrium (Y) and Titanium (Ti).

Coin Metals

This refers to the three metals copper (Cu), silver (Ag) and gold (Au), which also occur in nature in "native" (elemental) form and have been valued since time immemorial. As precious metals they have a high corrosion resistance and are (especially Cu) easy to extract from ores. The mass ratio in the earth's crust is about 20 000 Cu : 20 Ag : 1 Au. They are the best conductors of electric current among the elements.

Non-Ferrous Metals (NF metals)

Non-ferrous metals are defined as all non-alloyed metals (with the exception of iron) as well as alloys in which a metal other than iron makes up the largest mass proportion.

Platinum Metals

The six platinum metals are ruthenium (Ru), osmium (Os), rhodium (Rh), iridium (Ir), palladium (Pd) and platinum (Pt). They are precious metals that are often used in catalytic converters.

Part IV

Molecules, Ions, Bonds

7 Introduction

In chemical processes, no "element transformation" takes place. The atoms of the starting molecules rearrange themselves to form the molecules of the end products. In this process, bonds between the atoms are first broken and then reestablished.

In order to better understand these processes, we must first agree on a uniform, suitable notation and learn the basic rules for compounds before we take a closer look at the different types of chemical bonding.

7.1 Molecule Presentation

The term "molecule" was coined as early as *1618* by Sennert[1] *(molecula* = small mass), although here he made no distinction between atom and molecule. In *1805*, Dalton introduced the term "atomic compounds" as the "smallest particles of compounds", thus anticipating today's concept of molecule. It was not until around *1860 that* the term became established in its present form. In older literature, the term "molekel" can also be found instead of the plural "molecules" used today.

Only the noble gases occur as individual atoms. All other nonmetals combine with each other to form larger structures in which the atoms are linked together, called molecules. In order to bond together, two or more atoms must form bonds with each other. The resulting molecule behaves as a single unit in chemical-physical processes. Many elements also form molecules, for example: O_2, N_2, Cl_2, Br_2, P_4, S_8, ...

[1] Daniel Sennert, 1572-1637, German physician.

T. Schmiermund, *The Chemistry Knowledge for Firefighters*,
https://doi.org/10.1007/978-3-662-64423-2_7

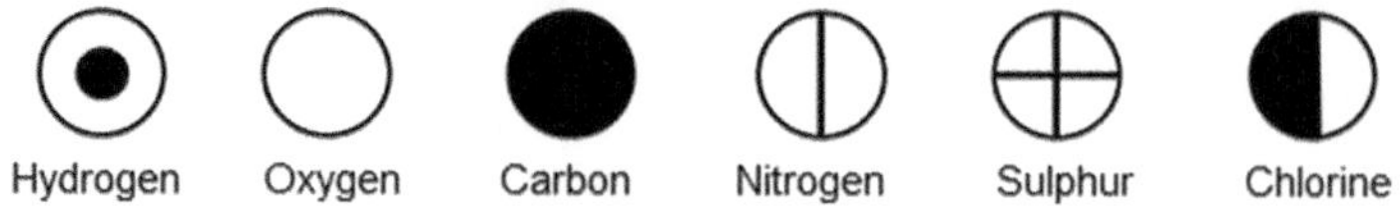

Fig. 7.1 Element symbols according to Dalton (examples)

7.2 Notation

The alchemists of the Middle Ages still each had their own system, in which the various substances were represented with different symbols. When the "secret science of alchemy" began to grow more and more into a "natural science", the need for a "common" notation for the exchange of information became more and more urgent.

For example, Dalton introduced uniform symbols for the various elements (Fig. 7.1).

Dalton's notation represents not only the proposal of unification. Because of his atomic model, his symbols also simultaneously represented *1* atom, or a standard mass containing a standard number of atoms. Since then, chemical formulas and reaction equations also contain a quantitative, a quantitative, statement.

7.2.1 Element Symbols

The present representation of the elements by "letter symbols" (element symbols) was introduced by Berzelius[2] in *1814.* He replaced the pictorial symbols with the initial letters of the Latin names of the elements. He also simplified Dalton's notation by introducing indices (subscripted digits) for the number of atoms within a compound.

The element symbols of some elements appear to us directly obvious:

B – Boron, **Al** – Aluminum, **Cl** – Chlorine, **He** – Helium, **Mn** – Manganese, **Mg** – Magnesium, **Ds** – Darmstadium, etc.

In others, the atomic symbol is only revealed by the Latin/Greek designations, as introduced by Berzelius:

H – Hydrogen *(hydrogenium* = water former); **C** – Carbon *(carbonium* = charcoal); **N** – Nitrogen *(nitrogenium);* **O** – Oxygen *(oxygenium);* **Fe** – Iron *(ferrum),* **Cu** – Copper *(cuprum);* **Ag** – Silver *(argentum);* **Sn** – Tin *(stannum);* **Sb** – Antimony *(stibium);* **Au** – Gold *(aurum);* **Hg** – Mercury *(hydrargyros* – "liquid silver") and **Pb** – Lead *(plumbum).*

These element symbols must – willy-nilly – be learned by heart.

However, in most cases it is sufficient to limit yourself to the most common/important elements. These are:

[2] Jöns Jakob Berzelius, 1779–1848; Swedish chemist and physician.

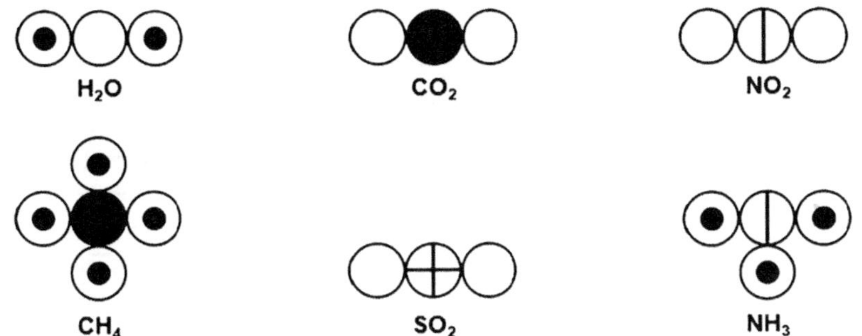

Fig. 7.2 Comparison of the molecular notation according to Dalton and according to Berzelius

- from the first main group: H, Li, Na, K
- from the second main group: Mg, Ca, Ba
- from the third main group: B, Al
- from the fourth main group: C, Si, Sn, Pb
- from the fifth main group: N, P
- from the sixth main group: O, S
- from the seventh main group: F, Cl, Br, I
- from the eighth main group: He, Ne, Ar, Kr, Xe, Rn
- from the subgroups: Ti, Cr, Mn, Fe, Co, Ni, Cu, Zn, Mo, Ag, Pt, Au, Hg
- from the lanthanoids and actinoids: Ce, U, Pu

7.2.2 Comparison of Molecular Notations

We want to draw a comparison between Dalton and Berzelius concerning the notation of molecules. In Fig. 7.2 the (corrected) Dalton's molecules are shown pictorially, below them the (modern) notation of Berzelius. Even in these few uncomplicated examples, the captivating simplicity of Berzelius' proposal is apparent.

7.3 The Valence Stroke Formula

The valence bar formula is probably the most common formula representation of compounds because of its clarity. At the same time, it is also one of the ways of representing structures, i.e. a variant of the structural formulae. It is also called the "Lewis notation" after its inventor.

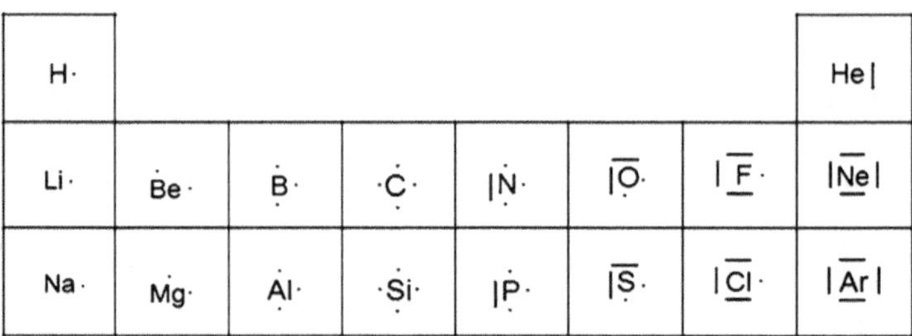

Fig. 7.3 Lewis notation of the first 18 elements

- **Rules for the formation of the valence stroke formula**
  - Only the outer electrons (valence electrons) are considered. Electrons on more inner shells are not considered.
  - Single (unpaired) electrons are represented as a dot (•) on the element symbol.
  - Paired electrons are represented as a colon (:) or a dash (|) (Fig. 7.3).
  - Electron pair bonds (as they consist of two electrons each) are also drawn as a line (–) between the atoms.
  - Double bonds (=) or triple bonds (≡) are symbolized by the corresponding number of parallel bars.
  - In compounds, valence electrons are distributed throughout the molecule in such a way that electron pairs are formed.

7.3.1 Noble Gas Rule

As an aid to understanding the formation of bonds, Lewis introduced the so-called "noble gas rule" or "octet rule" as early as 1916. Since the noble gas atoms are the only ones that do not form bonds with each other, the noble gas state (more precisely: the arrangement of the outer electrons in the case of noble gases, ns^2 np^6 or $1s^2$ in the case of helium) must represent a particularly stable state.

Lewis found that the atoms of the remaining elements, by changing their electron shells, tend to develop this "noble gas configuration" (fully occupied outer shell). This is the lowest-energy (and thus the most stable) state that the electrons in the compound can assume.

This configuration can be achieved if two neighbouring atoms share the electrons that form the bond between them, or if a (metal) atom gives up its outer electrons or a (non-metal) atom takes up a corresponding number of electrons.

Example

- Hydrogen has one outer electron. In the hydrogen molecule (H_2), two atoms share two electrons each; both hydrogen atoms "possess" 2 electrons each (= helium configuration).

$$H\bullet + \bullet H \rightarrow H{-}H$$

$$H + H \rightarrow H_2$$

- Nitrogen has 5 outer electrons. In the nitrogen molecule, the two nitrogen atoms share three electrons each: Both nitrogen atoms thus have 8 outer electrons each (= neon configuration).

$$|N\vdots + \vdots N| \rightarrow |N{\equiv}N|$$

$$N + N \rightarrow N_2.$$ ◀

About 95 % of all chemical compounds form a noble gas configuration during chemical bonding. This means that most compounds can also be represented well using the Lewis notation.

7.3.1.1 Exceptions to the Noble Gas Rule

Some compounds, such as boron trifluoride (BF_3) do not fulfil the noble gas rule. BF_3 has only three bond electron pairs, i.e. 6 electrons – a so-called electron deficient compound. This explains that BF_3 reacts very easily with substances that have a free electron pair (e.g. NH_3). Here, the nitrogen of the ammonia provides its entire (free) electron pair for the bond.

$$BF_3 + NH_3 \rightarrow BF_3NH_3$$

Another exception is hydrogen boride. The BH_3 molecule is only hypothetical. Investigations showed that it is in reality a B_2H_6 molecule. In this context, one also speaks of a "three-centre bond".

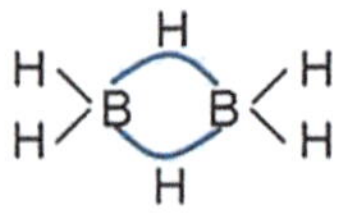

7.3.1.2 Bonds of Heavy Elements (From Third Period)

In addition to the valence electrons in the s- and p-orbitals, the elements from the third period onwards, i.e. from atomic number 11 (sodium), also have unoccupied d-orbitals available. Including these d-orbitals, with simultaneous "loss" of the s- and p-orbitals, all outer electrons can degenerate and thus form bonds up to the total number of outer electrons.

For example, both sulfur in sulfuric acid and phosphorus in phosphoric acid have expended their maximum possible number of outer electrons to bond.

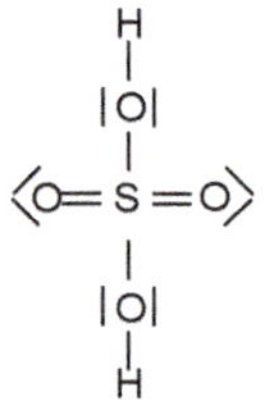

Sulphuric acid (H_2SO_4)

Phosphoric acid (H_3PO_4)

S
6. main group
= 6 valence electrons
= max. 6 bonds

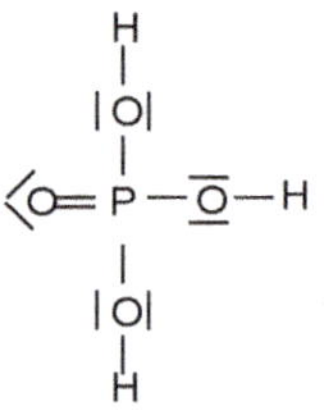

P
5. main group
= 5 valence electrons
= max. 5 bonds

7.3.1.3 Formal Charge

Bonds can also be formed when atoms with free electron pairs (N, P, O) provide the bonding partner with an electron pair rather than an electron: We have already learned about this in the compound BF_3NH_3. The bonding electrons of the boron-nitrogen bond both originate from nitrogen.

This special form of the electron pair bond, sometimes called a dative bond or coordinative bond, is sometimes written as $H_3N \rightarrow BF_3$ instead of BF_3NH_3. The arrow indicates which atom provides an electron pair for the bond.

However, if we divide the electrons of the bonds between the individual atoms, we find that the B atom has one electron more and the nitrogen atom one electron less than in the elementary state. Therefore, the B is assigned a formal charge of -1, the N a formal charge of $+1$.

Similarly, the carbon and oxygen atoms in the carbon monoxide molecule can also be assigned formal charges.

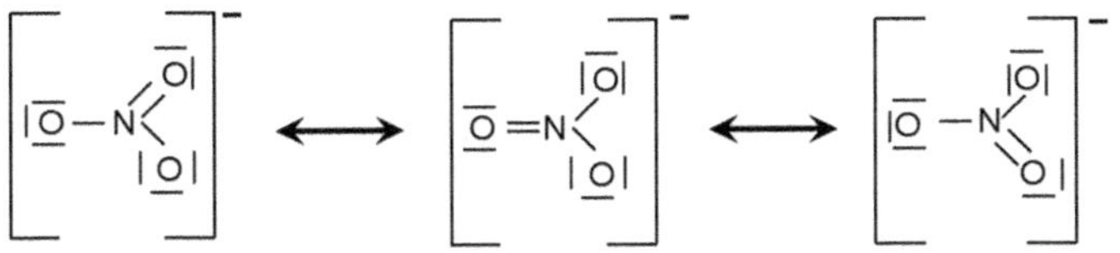

Resonance formula 1 ⟷ Resonance formula 2 ⟷ Resonance formula 3

Fig. 7.4 Mesomerism of the nitrate ion (NO_3^-)

$$\overset{\oplus}{H_3N} - \overset{\ominus}{BF_3} \qquad |\overset{\ominus}{C} \equiv \overset{\oplus}{O}|$$

Formal charges are always written "circled". Under no circumstances should they be confused with the actual charge of an atom.

The formal charge can be calculated as follows:

valence electrons of the atom
  – electrons not involved in bonds
  – binding electron pairs
  formal charge of the atom

7.3.1.4 Mesomerism

For some compounds, the Lewis notation gives several ways to record the formula. These variants are called "limit formulas". So as formula pictures, which represent an extreme. The "true" bond state always lies between these boundary formulas, because these extremes are always more energetic than the molecules in reality. This is also the case, for example, with the nitrate ion (Fig. 7.4).

The "true" bonding state cannot be represented in a Lewis formula in most cases. Therefore one uses these boundary formulas and marks this so-called mesomerism with a mesomerism arrow. *In the case of the boundary formulas, think of the individual atoms numbered consecutively. You can then see that the bonds flip back and forth. The molecule itself is neither rotated nor mirrored.*

In a few cases, "special notations" of the Lewis formulae exist. For example, in the case of benzene or the acid residue ions of organic acids (Fig. 7.5).

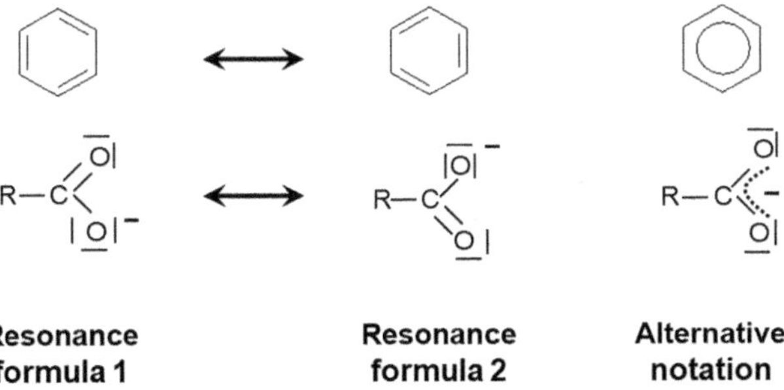

Fig. 7.5 Alternatives to mesomeric notation (examples)

Example

Examples of mesomerism and boundary formulas are given in Figs. 7.4 and 7.5, respectively. ◀

7.3.2 Electronegativity (EN)

As we will see in Sects. 8.1 and 8.2, the electronegativity (EN), or more precisely the difference in electronegativities (Δ EN) of the atoms involved, has a significant influence on the bond type. It also helps to determine which elements involved receive a negative or positive oxidation number. The element with the higher electronegativity always receives the negative sign. (For EN in PSE, see Sect. 6.5.4).

7.3.3 Oxidation Number

The oxidation number, also known as "valency", "bonding" or "valence", indicates how many formal, mathematical bonds the individual element has in the compound. Figure 7.6 shows an example of three sulfur compounds for which the oxidation number of the S atom has been given. Compare here with the number of bonds taking into account the electronegativities of O or H with respect to S.

Other (computational) examples are:

- NaCl → the Na could be replaced by an H (HCl). Na has the oxidation number + I.
- NaCl → the Cl could take an H instead of the Na: Oxidation number Cl: –I.
- MgO → the Mg could be replaced by two H (H_2O): Oxidation number: +II.
- MgO → the oxygen could be replaced by two Cl: Oxidation number: –II.
- NH_3 → the nitrogen carries 3 H → oxidation number: –III.
- NH_3 → 3 hydrogens are present. Oxidation number: +I.
- P_2O_5 → the two P could be replaced by 10 H, 10: 2 = 5, P: +V.

One can also formulate:

- The oxidation number of a nonmetal indicates by how many hydrogen atoms the bonding partner could be replaced.
- The oxidation number of a metal indicates by how many chlorine atoms the bonding partner could be replaced.

The older term "valence" comes from this *comparison*. Hydrogen and chlorine are each monovalent; they each form a bond in the hydrogen chloride molecule.

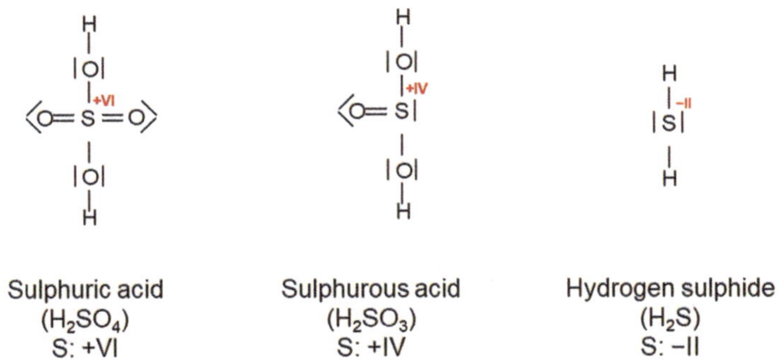

Fig. 7.6 Example: Oxidation numbers for S-compounds

In order to avoid confusion between oxidation numbers and charges, we write the oxidation numbers with Roman numerals and preceded by a sign.

An overview of the possible oxidation numbers of the second and third period elements can be found in Figs. A.1 and A.2.

7.4 Other Formula Notations

We have already become acquainted with the valence line formula, as an example of a structural formula representation. But chemistry would not be what it is, had not other, for the respective purpose useful, representation variants developed in the course of the years.

7.4.1 Ratio Formula

The ratio formula merely reflects the ratio of the atoms involved. It must **not** be confused with the molecular formula. For example, the ratio formulae of benzene (molecular formula C_6H_6) and ethyne (acetylene, molecular formula: C_2H_2) are identical: C_1H_1.

Many substances, such as salts, do not exist as "independent" molecules. Here, either the ratio formula (e. g.: NaCl) or the constitutional formula (e. g.: $Al_2(SO_4)_3$) is used instead of the molecular formula.

7.4.2 Sum Formula

The molecular formula only provides information about the number of elements involved in the compound. It is useful for calculating the molar mass, for example.

Aluminum sulfate	Water	Indigo
$Al_2S_3O_{12}$	H_2O	$C_{16}H_{10}N_2O_2$
Dimethyl ether	**Ethanol**	**iso-octane**
C_2H_6O	C_2H_6O	C_8H_{18}
n-Propanol	**iso-Propanol**	**Toluene**
C_3H_8O	C_3H_8O	C_7H_8

Unfortunately, the molecular formula does not give us any clues as to how the atoms of this substance are arranged. It is not possible to see from the molecular formula why dimethyl ether (a gas) and ethanol (a liquid) differ in their properties.

7.4.3 Constitutional Formula

The constitutional formula, sometimes called the partial-sum formula, is a variety of the valence-stroke formula and the preferred form for simpler molecules. Here, important structural elements are written separately. In the field of inorganic chemistry, these are often acid residues; in organic chemistry, they are the so-called "functional groups". In a way, this notation represents a kind of "half-structure formula".

Aluminum sulfate	Water	Indigo
$Al_2(SO_4)_3$	H_2O	–
Dimethyl ether	**Ethanol**	**iso-octane**
$(CH_3)_2O$	C_2H_5-OH	$C_7H_{15}-CH_3$
or	*or*	
CH_3-O-CH_3	CH_3-CH_2-OH	
n-Propanol	**iso-Propanol**	**Toluene**
C_3H_7-OH	$C_2H_5(OH)CH_3$	$C_6H_5-CH_3$
or	*or*	
$CH_3-CH_2-CH_2-OH$	$CH_3-CH_2(OH)-CH_3$	

Now more can be seen:

- Aluminium sulphate consists of two Al atoms and three SO_4 groups.
- In dimethyl ether, the oxygen that connects the two CH_3 groups is singled out.
- In ethanol, the OH group is important and is highlighted separately.
- The two isomeric propanols show a difference with respect to the position of the OH group.

7.4.4 Structural Formulas

By means of the partial sum formula, individual structural elements can already be identified and delimited. But this is not yet a "real" structure.

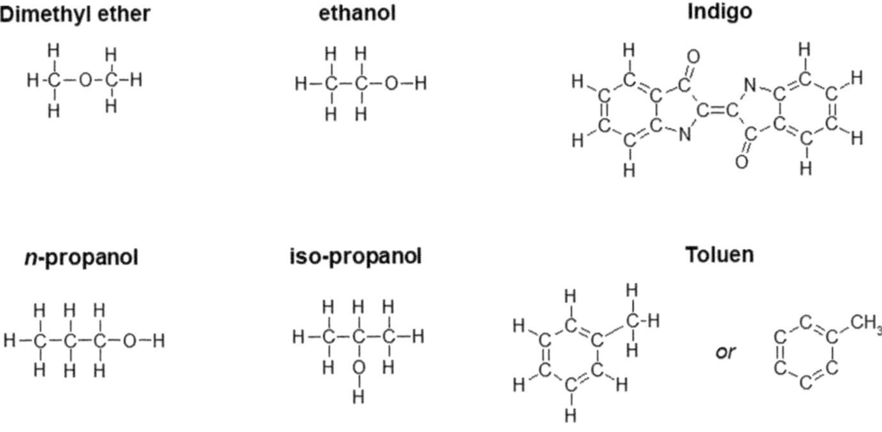

Fig. 7.7 Example: Structural formulae

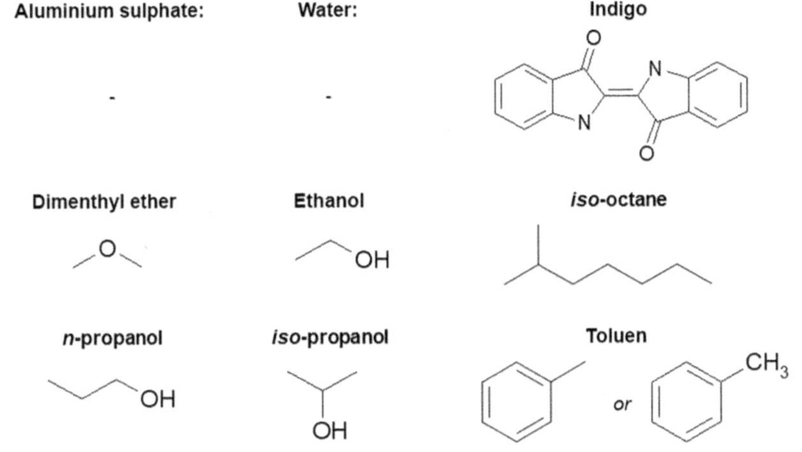

Fig. 7.8 Example: Skeleton formulas

In the structural formula (Fig. 7.7), the individual atoms and their direct links to each other are drawn. It is a simplified form of the valence line formula, in which the free electron pairs are usually omitted. Please take into account that the molecules are "flattened" on the paper plane. Only limited statements can be made here about the three-dimensional structure.

7.4.5 Skeleton Formulas

In the so-called skeleton formula (Fig. 7.8), the carbon atoms are drawn as corners or ends. Heteroatoms (= non-carbon atoms) are shown as element symbols. The hydrogen is generally not drawn. The number and position of the hydrogens results from the number of free bonds.

Especially in the field of organic chemistry this representation is often used.

8 Bonds

At this point we want to get an overview of the binding forces. Chemical bonds can be divided into "strong bonds" and "weak bonds".

Strong bonds are also called valence bonds or main valence bonds. The bonding forces always conform to the valences in these bonds. They are: the metal bond, the electron pair bond and the ionic bond.

Weak bonds are also called interactions or secondary valence bonds. In these bonds, any number of particles can approach each other; at least as long as they have the space to do so. They are: The hydrogen bond, the Van der Waals forces and the dipole-dipole interaction.

Wawra et al. have found a very nice model for chemical bonds in their book "Chemie verstehen" [Ref. 3-13, page 28, translated]:

"Think of a chemical bond as an interaction between two hedgehogs. Suppose two hedgehogs are sitting somewhere and both are cold. To get warm, these hedgehogs will move as close as possible. But if they get too close to each other, they will sting each other. So they will move away from each other again (and quickly!!). But now they are freezing again! After some moving back and forth, the two hedgehogs will finally have found a distance where they are warm enough, but not yet stinging each other. And they will keep this distance of maximum comfort (which corresponds to the minimum energy of two atoms) as far as possible. That's exactly how atoms do it in a chemical bond."

8.1 Strong Bonds

8.1.1 Metal Binding

More than ¾ of all elements are metals. Since metals have few outer electrons and are electropositive, they cannot form electron pair bonds or ionic bonds with each other to obtain a noble gas configuration.

However, if they "release" all their outer electrons, positively charged atomic hulls are formed which are held together by the delocalized electrons. These electrons can move freely through the overall structure, so-called "electron gas"

T. Schmiermund, *The Chemistry Knowledge for Firefighters*,
https://doi.org/10.1007/978-3-662-64423-2_8

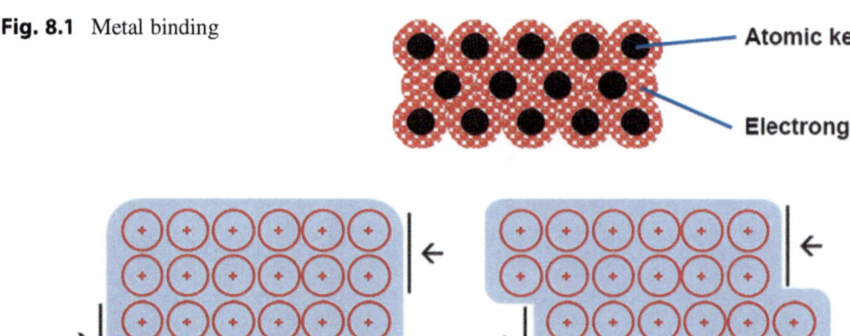

Fig. 8.1 Metal binding

Fig. 8.2 Deformation of the metal grid

(Fig. 8.1). The electrostatic forces that occur are undirected, so that the metal cations assemble in a regular metal lattice.

In this process, the metal atoms (according to the Lewis model) give up their outer electrons, so to speak, so that these become "freely available".

$$\text{Na} + \text{Na} \rightarrow 2\,\text{Na}^{+} + 2\,\text{e}^{-}$$

► Metal atoms form metal lattices. The metal atoms release their valence electrons into a common electron gas that moves freely between the atomic trunks. The bond is formed by the electrostatic forces between the atomic trunks and the electron gas.

This type of bonding explains (among other things) the good electrical conductivity, the metallic luster, the good thermal conductivity and the relatively high boiling points (compared to the molar mass) of the metals.

Metals are electrical "1st order conductors" or "electron conductors". The freely movable valence electrons emitted into the electron gas are also called conduction electrons, since they enable the current to flow. One calculates with 10^{20} conduction electrons per cm^3 metal. Therefore, a considerable flow of electrons occurs even at low voltages.

The "heating up" of a current-carrying metal wire can be imagined as "frictional heat" of the electrons at the atomic trunks.

Metals are ductile, i.e. easily deformable. Only the atomic trunks are displaced, the electron gas provides the permanent cohesion (Fig. 8.2).

8.1.2 Ion Binding

If a sodium atom is reacted with a chlorine atom, both atoms can reach a "noble gas shell" in that the sodium atom donates an electron and the chlorine atom accepts this electron (Fig. 8.3).

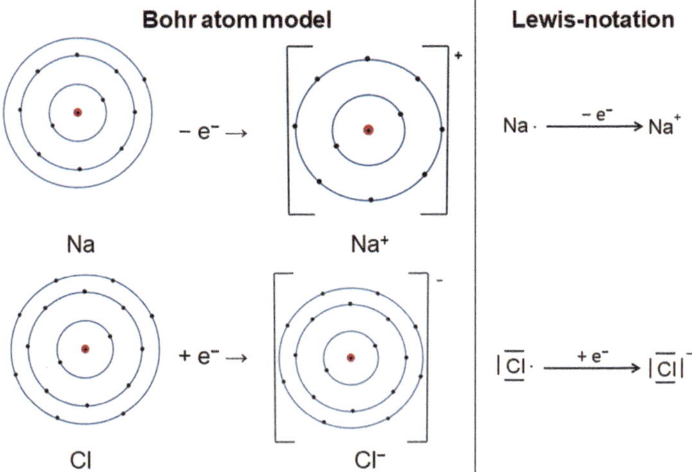

Fig. 8.3 Formation of Na^+ and Cl^-

The sodium atom becomes the positive sodium ion (Na^+) and acquires the outer shell of neon. The chlorine atom becomes the negative chloride ion (Cl^-) and acquires the outer shell of argon.

The number of protons in the nucleus does not change, of course. The Na has one more proton than total electrons and is therefore positively charged. The Cl has one electron more than it has protons in the nucleus, so it is negatively charged.

The cohesion in the resulting sodium chloride molecule comes from the electrostatic attraction (unequal charges attract each other due to Coulomb forces) of the resulting ions.

This type of bond is called an ionic bond (or heteropolar bond, electro-valent bond, electrostatic bond or ionogenic bond) and is characteristic of salts. Since the attraction between the ions acts uniformly in all directions, salts form lattice structures. In NaCl, for example, each Na^+ ion is surrounded (in the crystal lattice) by six Cl^- ions, and each Cl^- ion by six Na^+ ions. The lattice structure explains the high melting and boiling points of salts.

► In ionic bonding, elementary atoms donate electrons or accept electrons to achieve the outer electron configuration of a noble gas. Thus, positively charged cations or negatively charged anions are formed. The cohesion is achieved by electrostatic attraction.

According to the ion model, the complete separation of one or more electrons from the metallic component (then: the cation) and their absorption by the non-metal atom (then: the anion) is assumed. The number of electrons released/absorbed depends on the oxidation number of the element.

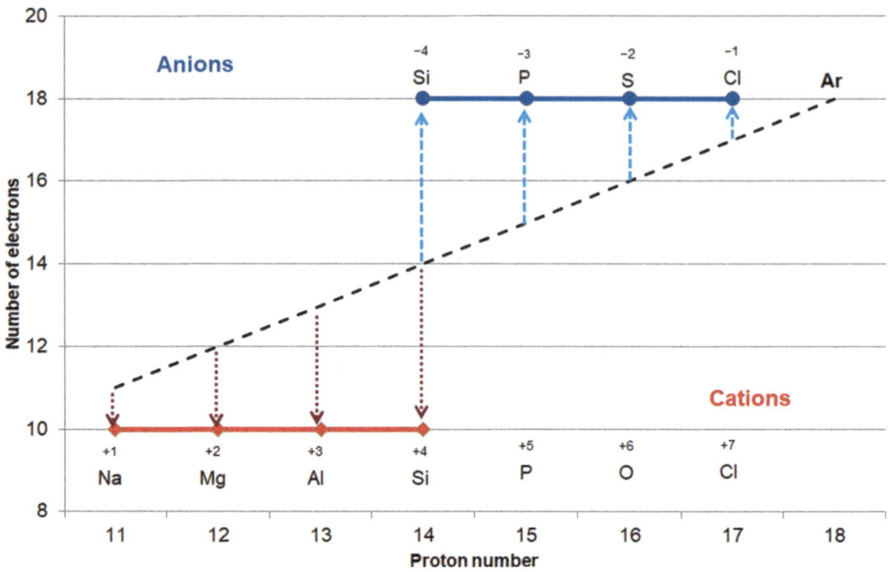

Fig. 8.4 Ion formation in the ion model

▶ Compounds that are made up of cations and anions are also called salts.

In Fig. 8.4 (using the third period as an example), pale red arrows pointing downward symbolize electron donation and pale blue arrows pointing upward symbolize electron acceptance.

The higher the ionic charge and the smaller the ionic radii, the greater the electrostatic forces. In addition, they act spherically symmetrically, i.e. uniformly in all spatial directions.

This leads to several conclusions:

- Each ion tries to arrange as many counter ions around itself as possible. The result is an unlimited, three-dimensional lattice with a sphere packing as dense as possible.
- The repulsion of equally charged ions forces them to be as far away from each other as possible. The ions are distributed as symmetrically as possible around the central ion. Ion lattices are therefore highly symmetrical.
- The total system "salt" satisfies the law of electrical neutrality. The total number of positive charges of the cations must be equal to the number of negative charges of the anions.

For example, the general stoichiometric compositions according to Table 8.1 or the charges of individual ions according to Table 8.2 are obtained.

In common salt, the smaller sodium ions are packed into the spaces between the larger chloride ions (Fig. 8.5a). Each Na^+ ion (grey) is surrounded by six Cl^- ions (green) - and vice versa (Fig. 8.5b).

Table 8.1 Composition of ionic compounds (examples)

Formula	Examples
$Z^{+}A^{-}$	NaCl, KCl, NaOH, CsF, NH_4Br
$Z^{2+}A^{2-}$	CaO, $MgSO_4$,
$Z^{2+}A_2^{-}$	CaF_2, $Pb(NO_3)_2$, $MgBr_2$
$Z_2^{+}A^{2-}$	Na_2O, Na_2SO_4, K_2CrO_4
$Z_2^{3+}A_3^{2-}$	Al_2O_3, $Al_2(SO_4)_3$
$Z_3^{2+}A_2^{3-}$	$Ca_3(PO_4)_2$

Table. 8.2 Common ions

Charge of the ion	Main group elements	Subgroup elements	Loaded connections
1+	H^+, Na^+, K^+, Rb^+, Cs^+	Ag^+, Au^+, Cu^+	H_3O^+
2+	Mg^{2+}, Ca^{2+}, Sr^{2+}, Ba^{2+}, Sn^{2+}, Pb^{2+}	Mn^{2+} Fe^{2+},Co^{3+}, Ni^{2+}, Pt^{2+}, Cu^{2+}, Zn^{2+}, Cd^{2+}, Hg^{2+}	–
3+	Al^{3+},Bi^{3+}	Cr^{3+}, Au^{3+},Fe^{3+},Ce^{3+}	–
4+	$Pb4^+$	Mn^{+4}	–
1-	H^-,F^-,Cl^-,Br^-,I^-	–	NO_3^-, ClO_4^-, ClO_3^-, OH^-
2-	O^{2-}, S^{2-}, Se^{2-}	–	SO_4^{2-}, SO_3^{2-}
3-	N^{3-}, P^{3-}	–	PO_4^{3-}-

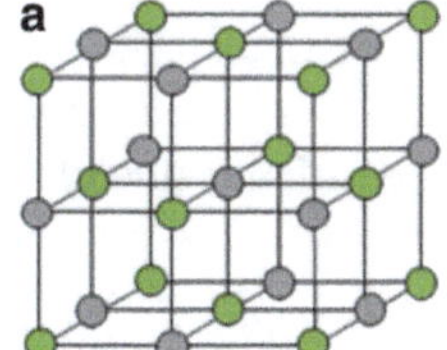

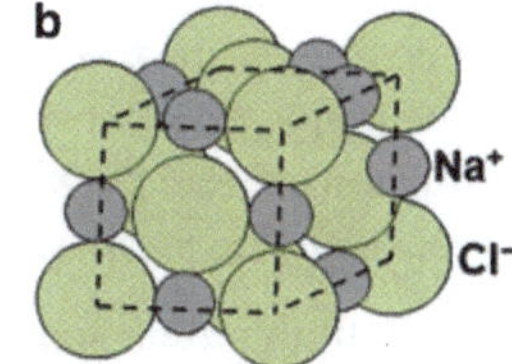

Fig. 8.5 (**a**) NaCl grid (structure), (**b**) NaCl grid (space filling)

In ionic bonding, a distinction is made between the positively charged cations (usually metal ions) and the negatively charged anions (usually "acid residue ions", cf. Chap. 12).

▶ Cations are positively charged: "t" as in "⊕" → Ka ⊕ ion. **Anions** are negatively charged.

Ionic compounds are second order conductors (ionic conductors). They conduct electric current only as a melt or in (mostly aqueous) solution. The conductivity increases with temperature, as the ionic mobility also increases with temperature. The addition of salts massively increases the electrical conductivity of water. (On the subject of foam additives and conductivity, see Sect. 50.4.2).

If anions donate electrons or cations accept electrons, they discharge. This means that the original compound is decomposed (see Chap. 20).

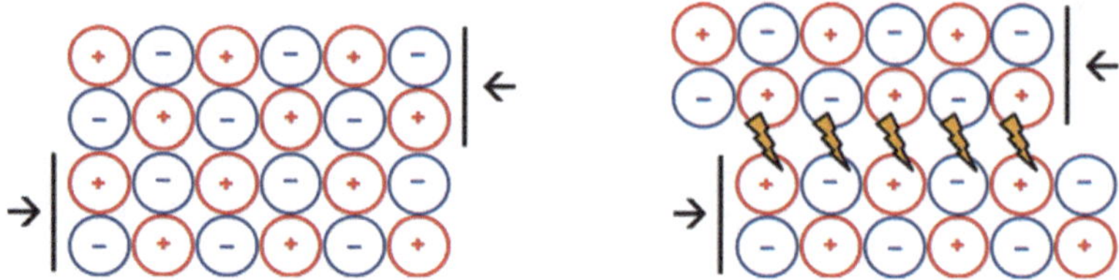

Fig. 8.6 Deformation of ion crystals

When attempting to deform ion crystals, ions with the same charge come to rest on top of each other. As a result of the electrostatic repulsion of equal charges, the crystal breaks apart (Fig. 8.6). Ionic compounds are therefore brittle.

8.1.3 Electron Pair Bond

The electron pair bond (EPB) is the most common bond, not least because of the huge number of organic compounds that are predominantly constructed from this type of bond.

It is also called homeopolar bond, covalent bond, nonpolar bond, unitary bond or atomic bond.

EPB occurs when nonmetal atoms with unpaired electrons approach each other. When they come close enough, the nucleus of one atom attracts the electrons of the other atom. The two atoms then continue to approach each other until finally the repulsion of the positive charges of the nuclei determines the minimum distance (Fig. 8.7). Since all valence electrons are now in the sphere of influence of both nuclei, both atoms share electrons. *Think of the freezing hedgehogs!*

The sharing of electrons also gives rise to the multiple bonds (Fig. 8.8).

For molecules with electron pair bonding, the melting and boiling points depend only on the molecular mass, the molecular size and the Van der Waals forces. They are therefore naturally very low. Also, the molecules can be easily displaced with respect to each other, so they are often gases, liquids, and soft solids. The fact that the electrons are "trapped" in the electron pair bond explains the poor electrical conductivity.

8.1.4 Polarized Electron Pair Bond

The polarized or polar EPB, also called "electron pair bond with ionic character" or "electron pair bond with ionic part", represents the step between the pure electron pair bond and the ionic bond.

In ionic bonding, the difference in electro-negativities (EN) is large enough to snatch an electron from an atom, thus forming a cation. On the other hand, this

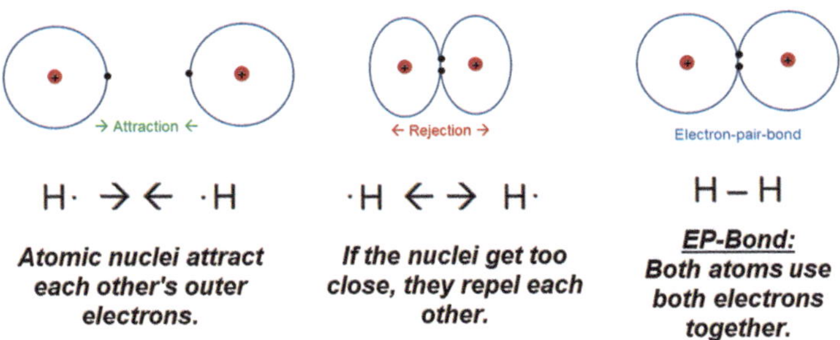

Fig. 8.7 Electron pair bond

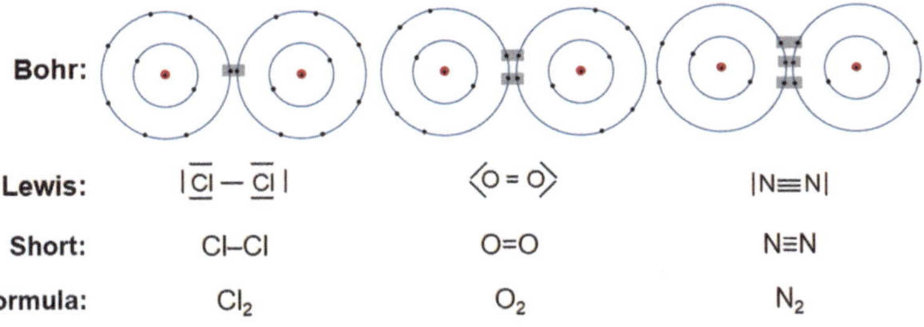

Fig. 8.8 Single, double, triple binding

difference ensures that an anion is formed, since the more electro-negative atom in question accepts this electron into its valence shell.

In polarized EPB, the atoms share the two bonding electrons, analogous to the pure electron pair bond. However, due to the relatively high EN difference, the probability of one or both electrons being closer to the more electronegative atom is very high. As a result, this more electronegative atom is also "a bit" negative. It carries a partial charge.

However, this partial charge is not to be understood as a "kind of ionic bond". Rather, the bonding electrons, which are soon with one atom, soon with the other, soon between the two, simply stay longer on average with the more electronegative partner. This is responsible for the formation of these partial charges.

As Fig. 8.9 shows, there are actually two transitions between the electron pair bond and the ionic bond: the polarized electron pair bond and the "distorted ionic bond". However, these distorted ions are still ions - and therefore belong to the ionic bond.

To represent the polarized EPB, sometimes small deltas (with ± symbol), sometimes "wedges" are written (where the broad side of the wedge represents the more electron-rich side); sometimes both representation variants are combined (Fig. 8.10).

A measure of the ionic bond fraction is the ionic bond fraction or ionic character of the respective bond due to the EN difference. This can be calculated and compared with the differences of the electro-negativities (Table 8.3).

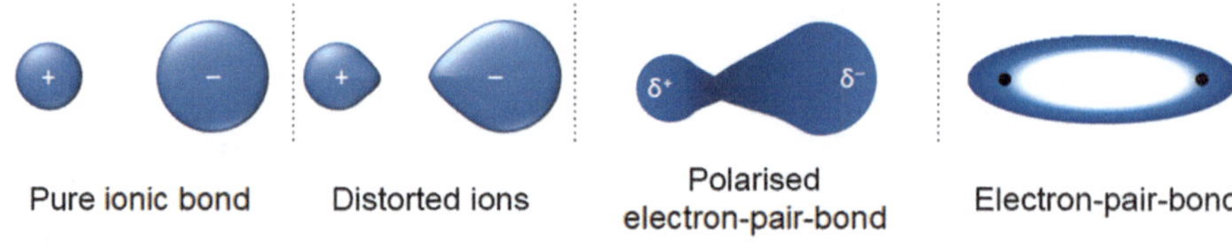

Fig. 8.9 Transitions ionic bond - electron pair bond

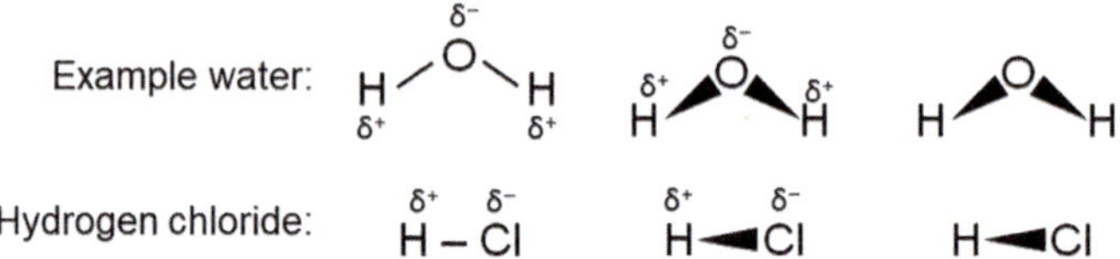

Fig. 8.10 Representation of the polarized electron pair bond

Table 8.3 EN difference and ion character (extract, extended list in Table A.7)

Δ EN	Ionic character	Δ EN	Ionic character
0.0	0 %	**2.0**	63 %
0.4	4 %	**2.2**	70 %
0.8	15 %	**2.4**	76 %
1.2	30 %	**2.6**	82 %
1.6	47 %	**2.8**	86 %
1.7	51 %	**3.0**	89 %
1.8	55 %	**3.2**	92 %

Table 8.4 Ionic character and bond type of selected compounds as examples

Compound	**Formula**	**Δ EN**	**Ion character**	**Binding type**
Nitrogen	N_2	*0.0*	*0 %*	*Covalent*
Methane	CH_4	*0.4*	*4 %*	*Covalent*
Hydrogen chloride	**HCl**	**0.9**	**19 %**	**Polar**
Water	H_2O	**1.4**	**39 %**	**Polar**
Table salt	NaCl	**2.1**	67 %	Ionic
Caesium fluoride	CsF	3.3	96 %	Ionic

The comparison between ionic character and bond type of some substances can be seen in Table 8.4.

8.1.5 Transitions Between the Bond Types

To get a better idea of the smooth transition between the electron pair bond and the ionic bond, please refer to Fig. 8.11.

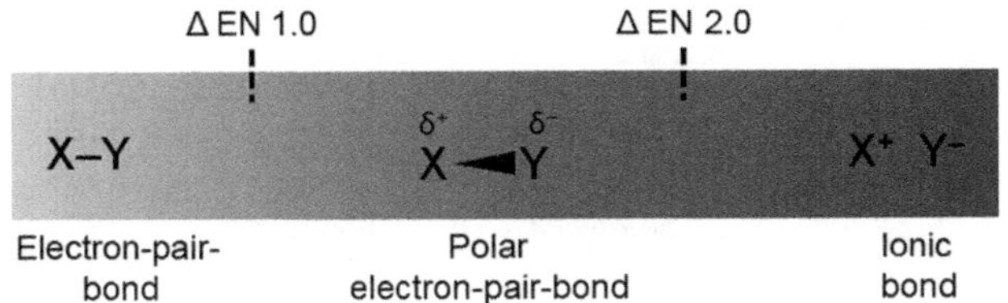

Fig. 8.11 Transition electron pair bond - ionic bond

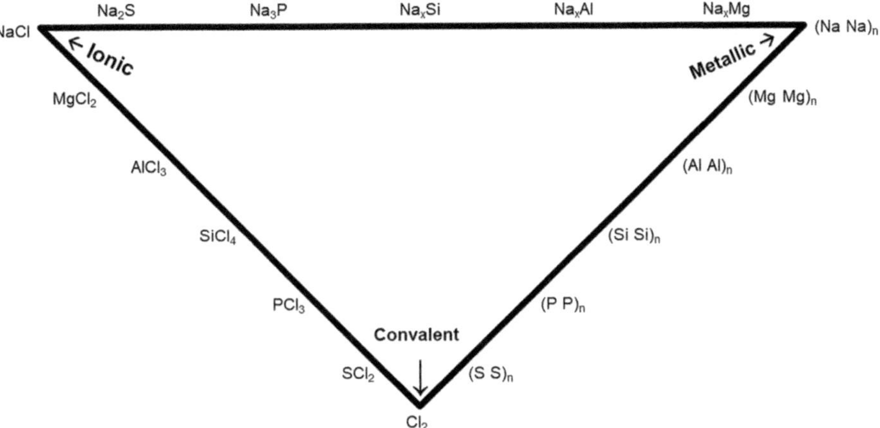

Fig. 8.12 Transitions of the bond types

There are not only transitions between the electron pair bond and the ionic bond, but also between the electron pair bond and the metal bond. This becomes clear if we consider the elements and their interconnections from one period in the PSE. This is illustrated in Fig. 8.12 using the example of the elements of the third period.

Typical transition compounds may include:

- ionic ↔ covalent:
  oxides and sulfides of heavy metals
  (e.g.: ZnS, ZnO)
- metallic ↔ covalent:
  semimetals (e.g.: Sb, Bi, In)
- ionic ↔ metallic:
  intermetallic compounds
  (e.g.: LaBi, Mg_2Ge, Mg_3Sb_2)

An overview of the chemical bonds is shown in Table 8.5.

Table 8.5 Comparison of binding types

Designation	Metal binding	Ionic Bond	Electron pair bond	Polarized electron pair bond
Binding partner	Metal - Metal	Metal - non-metal	Similar non-metals	Different Non-metals
EN-Difference	0	> 1.6	< 0.5	0.5...1.6
Examples	Iron, platinum	NaCl, CaF_2	nh_3, ch_4	H_2O, HCl
Image				δ+ δ− H–Cl
	Metal atoms emit their valence electrons into an electron gas.	Metal atoms give up their valence electrons to non-metal atoms. Positive and negative ions are formed.	Atoms share electron pairs to reach the noble gas state.	Atoms share electron pairs to reach the noble gas state. Due to the EN difference, the electrons are shifted to the more EN negative partner.
Starch	~ 150 to ~ 680 kJ/mol	~ 200 to > 1000 kJ/mol	~ 120 to ~ 650 kJ/mol	
Melting and boiling points	High, due to the strong attraction between metal ions and electron gas	High, due to the strong attraction of the anions and cations	Very low, due to weak intermolecular attraction	Low, due to dipole forces; slightly higher than pure EPB
State of aggregation	Solids	Mostly solids	Mostly gases or liquids	Mostly liquids
Malleability	Good deformability (lattice sites shift, electron gas holds everything together).	Brittle, break apart when deformed	Liquids or gases	Liquids or gases
electrical conductivity	Very good due to the electron gas	Only in the liquid (molten) or dissolved state	None Power line	None Power line
Heat conduction	Very good due to the electron gas	Very poor, only possible by heat movement of the ions around their lattice sites	Poor, only possible through the thermal movement of the molecules	Poor, only possible through the thermal movement of the molecules
Special features	–	–	Cohesion of the molecules practically only through van der Waals forces	(Very) frequent dipoles. Dipole-dipole interactions and hydrogen bonds occur

8.2 Weak Bondings

8.2.1 Dipole-Dipole Interaction

All polar substances are first of all dipoles, i.e. they consist of two (= *di)* different (electric) poles.

With regard to dipole-dipole interactions, however, the spatial structure of the respective molecule must first be considered. This results in either a mutual amplification or a mutual cancellation of the partially charged atoms of the molecule. Free electron pairs must also be taken into account here (Table 8.6).

Ammonia (NH_3), hydrogen fluoride (HF), chloromethane (CH_3Cl) and water (H_2O) are permanent dipoles. They are capable of dipole-dipole interactions. In the case of carbon dioxide (CO_2) and in the case of carbon tetrachloromethane (CCl_4), the dipole moments cancel each other out, so that the respective molecule does not represent a permanent dipole towards the outside.

Table 8.6 Dipoles: charge distribution and molecular geometry

Formula	Charge distribution	Dipole direction	Repeal?	Permanent Dipole?
H_2O	$^{\delta+}H-O^{\delta-}-H^{\delta+}$	$^{\delta+}H-O-H^{\delta+}$ (lone pairs δ^-)	No	Yes
CO_2	$O^{\delta-}=C^{\delta+}=O^{\delta-}$	$O^{\delta-}=C^{\delta+}=O^{\delta-}$	Yes	No
CCl_4	$C^{\delta+}(Cl^{\delta-})_4$	$C^{\delta+}(Cl^{\delta-})_4$	Yes	No
CH_3Cl	$H_3C^{\delta+}-Cl^{\delta-}$	$H_3C^{\delta+}-Cl^{\delta-}$	No	Yes
HF	$H^{\delta+}-F^{\delta-}$	$H^{\delta+}-F^{\delta-}$	No	Yes
NH_3	$H^{\delta+}-N^{\delta-}(-H^{\delta+})-H^{\delta+}$	$H^{\delta+}-N(-H^{\delta+})-H^{\delta+}$ (lone pair δ^-)	No	Yes

Fig. 8.13 Dipole-dipole interaction

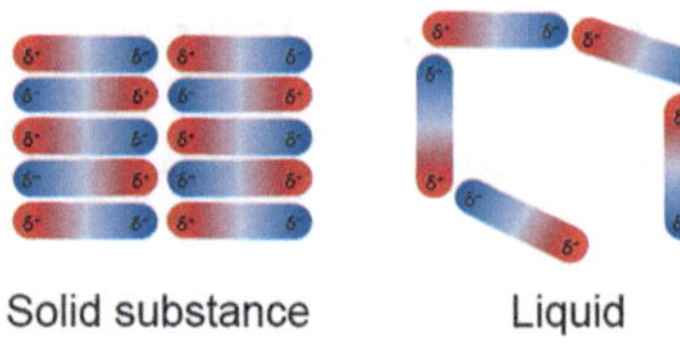

Table 8.7 Boiling point change due to dipole-dipole interaction

Stronger polar connection	bp	Δ bp	bp	Less polar connection
ortho-Dinitrobenzene	329 °C	**17 K**	312 °C	*para*-dinitrobenzene
ortho-Xylene	143.6 °C	**5.2 K**	138.4 °C	*para*-Xylene
para-nitrotoluene	239 °C	**17 K**	222 °C	*ortho*-nitrotoluene

► The dipole-dipole interaction occurs when dipoles are permanent. The cohesion occurs through the electrostatic attraction of the partially differently charged "molecule ends".

The mutual attraction of permanent dipoles results in a mutual alignment of these dipoles. This results in an overall lower-energy state. This alignment effect is temperature dependent, since the thermal motion of the particles disturbs this alignment.

Due to the relatively small effect of the interaction, it is often also assigned to the Van der Waals forces (Sect. 8.2.3), so that the chemical literature is inconsistent here.

In the solid state, the dipoles arrange themselves alternately. In the liquid state, the arrangement is loose but still directed (Fig. 8.13).

8.2.1.1 Effects of the Dipole-Dipole Interaction

Pure dipole-dipole interactions are difficult to grasp on their own. On the one hand, they are superimposed by the Van der Waals forces, on the other hand, they lead to hydrogen bonds in extreme cases.

As examples of the boiling point increase to be expected, for example, three substance pairs with their boiling points (bp) are listed in Table 8.7, where the differences (Δ bp, middle of table) are practically only due to the arrangement of atomic groups within the respective molecule.

8.2.2 Hydrogen Bond

The hydrogen bond is a directed weak bond and can be seen as a special case of the dipole-dipole interaction. The intermolecular forces of attraction are much stronger in some molecules than could be explained by the dipole-dipole interaction alone.

In all cases, the hydrogen atoms of the compound are bonded to small, strongly electronegative atoms (N, O, F, Cl). These exert a strong attraction on the bonding electrons and pull them towards them (polar atomic bond; compare Sect. 8.1.4). This

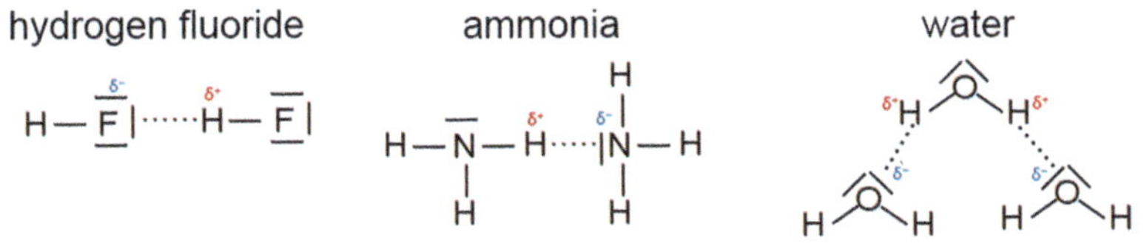

hydrogen fluoride ammonia water

Fig. 8.14 Hydrogen bonds (examples)

Fig. 8.15 Intramolecular and intermolecular hydrogen bonds

Intramolecular hydrogen bond (salicylic acid)

Intermolecular hydrogen bond (mixture of acetone (l.) and chloroform (r.)

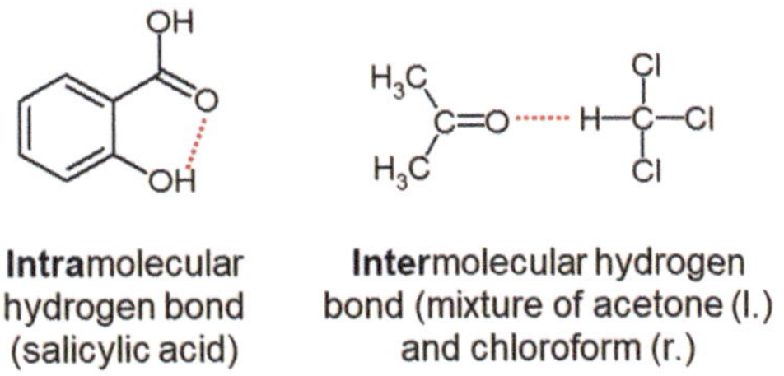

results in a high partial positive charge on the hydrogen atom, which is then hardly shielded from other atoms/molecules by an electron cloud.

► The mutual attraction of the positive hydrogen atom of one molecule and the free electron pair of another molecule creates a water-matter bond.

In formulas, this H-bridge binding is usually represented by a dotted or dashed line (such as in Fig. 8.14).

Larger atoms than those mentioned above, such as sulphur or bromine, do not tend to form hydrogen bonds despite their electronegativity. This is well illustrated by the boiling points of substances of similar composition (Table 8.8).

Apart from the compounds mentioned above, hydrogen bonds occur mainly with amines (e.g. methylamine $H_3C{-}NH_2$), inorganic and organic acids (e.g. nitric acid, sulphuric acid, perchloric acid, formic acid, acetic acid etc.) and with alcohols (e.g. methanol, ethanol, *n*-propanol, *iso*-propanol etc.).

The effect of hydrogen bonds can be seen, for example, in the comparison of ethanol and ethyl acetate. Although ethyl acetate, at 88.1 g/mol, has almost twice the molar mass of ethanol (46 g/mol), both have almost the same boiling point (77. . .78 °C). Another time the hydrogen bonds stand out is when comparing the evaporation figures of the two substances. Ethanol evaporates more than 2.5 times slower than ethyl acetate (for boiling point and evaporation figures, see Table 3.8).

Hydrogen bonds can also occur intramolecularly (= within a single molecule) or only when two substances are mixed (intermolecularly = between several molecules).

Effects of Hydrogen Bonding

- In ice, water molecules are held together by hydrogen bonds. Here, each O atom is tetrahedrally surrounded by 4 H atoms (Fig. 8.16). Cavities are formed which cause the low density of the ice. During melting, some of the hydrogen bonds

Table 8.8 Boiling points vs. H-bridge binding

Substance	Δ EN	Molar mass	Dipole	H-Bridges	Boiling point
CH_4	0.4	16.0 g/mol	no	no	−162 °C
SiH_4	0.3	32.1 g/mol	no	no	−112 °C
PH_3	0.0	34.0 g/mol	weakly	no	−88 °C
HCl	0.9	36.5 g/mol	weakly	yes	−85 °C
H_2S	0.4	34.1 g/mol	weakly	no	−62 °C
NH_3	0.9	17.0 g/mol	strongly	yes	−33 °C
HF	1.9	20.0 g/mol	strongly	yes	19 °C
H_2O	1.4	18.0 g/mol	strongly	yes	100 °C

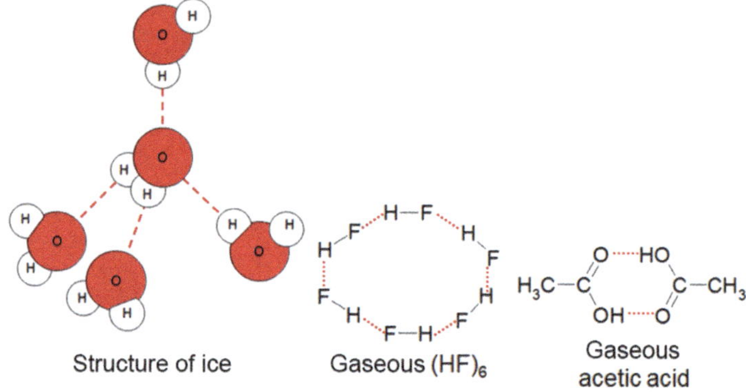

Fig. 8.16 Structures of dipole molecules

dissolve, the cavities collapse and the density of the liquid water initially increases.

This phenomenon is called the "anomaly of water". Because ice has a lower density and floats on top of the water, lakes do not freeze completely (which they would if the ice sank to the bottom of the water). This allows fish to survive in the lake.
Water has its greatest density at 4 °C. At higher temperatures it expands again due to thermal molecular motion and at 100 °C (and normal pressure) overcomes the hydrogen bonds: It boils. According to literature [3-3] (there page 273), even at 90 °C only a few percent of the hydrogen atoms are not involved in hydrogen bonds.

- Even in the gaseous state, certain substances exist as multiple molecules which are connected to each other by hydrogen bonds (Fig. 8.16).
- Hydrogen bonds are essential for life itself. DNA and proteins can only carry out their biological function if they are present in the correct form, the so-called tertiary structure. This is ensured, among other things, by hydrogen bonds.

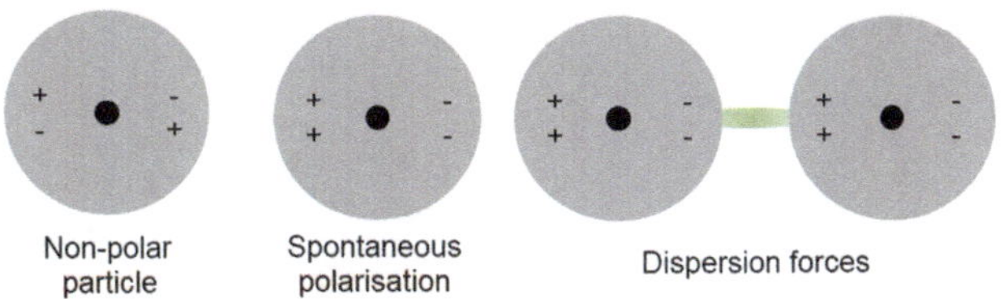

Fig. 8.17 London dispersion forces

8.2.3 Van der Waals Forces

▶ Van der Waals forces are nonspecific, undirected interactions that are always effective regardless of any other type of bonding.

They consist of the induction effect and London's dispersion forces (after F. London[1]). Often the dipole-dipole interactions (Sect. 8.2.1) are added to the Van der Waals forces.

The weak dispersion forces arise from a spontaneous polarization of a particle, which in turn induces a dipole in neighboring particles (Fig. 8.17). On time average, however, these fluctuating dipoles in each particle are equal to zero.

The induction effect is based on a similar principle:

Electrons can only move within certain boundaries (orbitals) in an atom. This leads to a constantly changing charge distribution within the individual atom. This creates temporary dipoles. If two molecules come close enough to each other, then one molecule induces a dipole in the other molecule; they attract each other. Of course, a permanent dipole in a nonpolar molecule can also produce such a temporary dipole when it comes close.

Long-chain compounds, in particular, can be deposited against each other here almost like a zipper (Fig. 8.18).

8.2.3.1 Effects of the Van der Waals forces

- The very low Van der Waals forces can add up very strongly under certain circumstances:
  - Thus, some substances decompose when an attempt is made to distil them at normal pressure. Here, the sum of the Van der Waals forces exceeds the binding energy of the (individual) electron pair bonds.
  - The gecko adheres to smooth surfaces only due to Van der Waals forces by means of very fine hairs and can therefore walk on the ceiling or on glass surfaces. The Van der Waals forces of these animals add up to approximately 40 Newton [Lit. 13-1].

[1] Fritz London, 1900-1954, German-American physicist.

Fig. 8.18 Induced dipoles

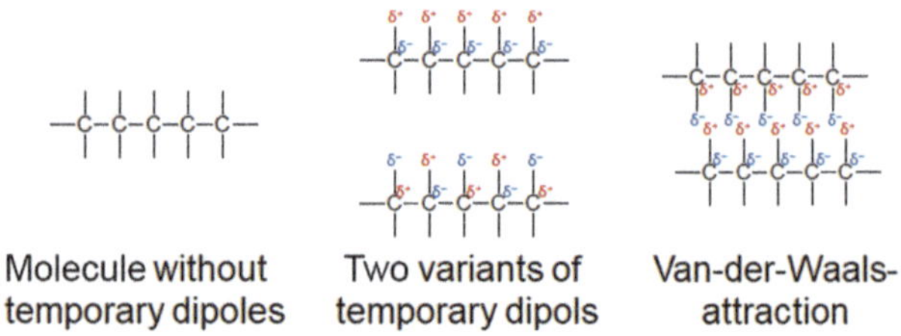

Molecule without temporary dipoles

Two variants of temporary dipols

Van-der-Waals-attraction

Table 8.9 Structure. Van der Waals forces and boiling point

Name	*n*-pentane	*iso*-pentane	*neo*-pentane
Systematic name	Pentane	2-methylbutane	2.2-dimethylpropane
Structural formula	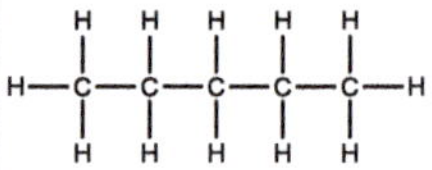	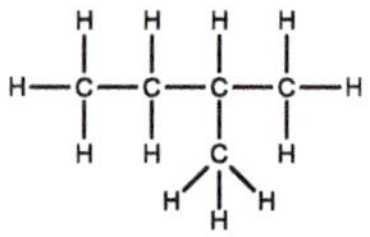	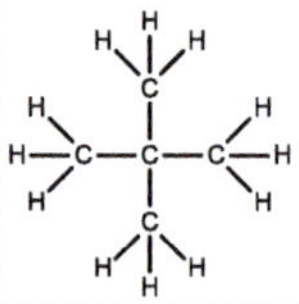
Model		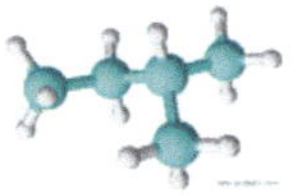	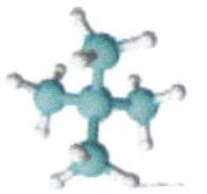
Van-der-Waals forces	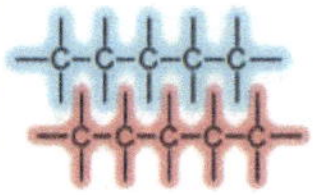	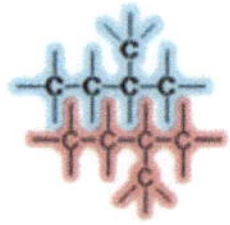	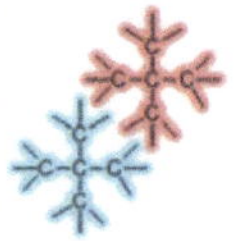
Boiling point	36 °C	28 °C	9,5 °C

- The boiling points of the three isomeric pentanes (all C_5H_{12}) differ markedly due to the different strength of the van der Waals forces (Table 8.9). They decrease markedly from the linear *n*-pentane via the less branched *iso*-pentane to the almost spherical *neo*-pentane.
- The Van der Waals forces are also clearly noticeable at the transition from gas to liquid. While the average distance of the particles in gases is determined by pressure and temperature, the particles in liquids have defined molecular distances due to mutual attraction. In liquids, therefore, a certain order occurs in a very small range, the so-called "near order". This is constantly broken down and built up again by the thermal movement of the particles.
- The close order caused by the Van der Waals forces not only influences melting and boiling points, but also, for example, viscosity. The non-polar hydrocarbon mixtures gasoline (shorter chains) and motor oil (longer chains) differ significantly in this aspect.
- In supercritical fluids (see Sect. 4.12.2), the high temperature prevents the Van der Waals forces from forming. The high pressure severely restricts the space

Table 8.10 Comparison of interactions

Designation	Hydrogen bond	Dipole-dipole interaction	Van der Waals forces
Binding is...	Directed	Directed	Undirected
Image	H—F|······H—F|		
Examples	Water (H_2O), hydrogen fluoride (HF)	Carbon monoxide (CO), Sulphur trioxide (SO_3)	Pentane (C_5H_{12}), Benzene (C_6H_6)
	Partial charges (H: δ^+/ X: δ^-)	Permanent dipole	Polarizability
Occurrence	Alcohols, acids, amines	Polar substances and their mixtures	All substances, all solutions and mixtures
Binding energy	~ 10 to ~ 50 kJ/mol	~ 1 to ~ 20 kJ/mol	~ 0.5 to ~ 5 kJ/mol
Temperature dependence	Strongly	Strongly	Strongly
Influence on melting and boiling points	High, due to the relatively strong	Medium, attraction is weaker than for hydrogen bonds	Very low, due to the weak intermolecular attraction.
State of aggregation	Mostly liquids	Solids and liquids	Solids and liquids
explained:	– anomaly of water – high boiling points of certain substances	– Molecular lattice of polar molecules	– supercritical fluids – Molecular lattice of nonpolar molecules

available to the molecules. Consequence: Supercritical fluids behave like liquids with the viscosity of gases.

- In particular, the dispersion forces can explain the liquid phases or the crystals (existing at extremely low temperatures) of the noble gases.

However, the range of these forces is very small, in fact it is limited to the nearest neighbour, as it decreases much faster over distance than the ion-ion interaction of the ionic bond.
A comparison of the interactions can be found in Table 8.10.

8.3 Other Types of Bonds in Solids

We have already learned about the metal bond and the ionic bond. In this section, we will explain the molecular lattices, which are formed by the van der Waals forces and the dipole-dipole interactions already described.

In addition, the atomic lattice will be discussed and a special property of some elements, allotropy, will be explained.

Table 8.11 Molecular lattices (examples)

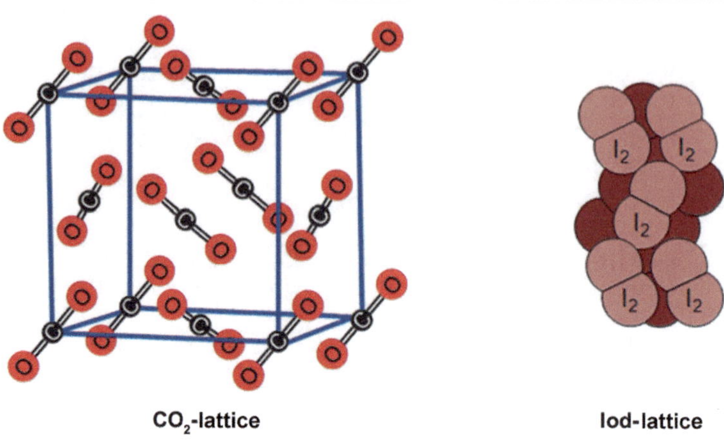

8.3.1 Molecular Lattice

Solids differ from liquids primarily in that they not only have a near order, but also a defined far order. In the case of metals, this results in different types of alloys, and in the case of ionic lattices, it results in a wide variety of crystal forms. Solid substances that do not have a long-range order are called amorphous (Greek: *amorphous* = without shape), because they do not form defined structures, but only an irregular pattern.

In molecules, the solid state usually consists of a crystal lattice whose lattice building blocks are molecules (Table 8.11). These are held together only by dipole-dipole interactions and/or van der Waals forces. The molecular crystal is then formed by a large number of such lattice components.

Compared to salts or metals, the melting points are naturally relatively low. However, the melting and boiling points of polar molecules are higher than those of non-polar molecules, even if the size of the molecule is comparable. Some substances that are made up of molecular lattices do not even pass through the liquid state when heated: they sublime, such as iodine or carbon dioxide.

8.3.2 Atomic Lattice

Atoms also arrange themselves in lattice structures. In an atomic lattice, the atoms of the element or compound are linked to each other by electron pair bonds. This sometimes results in high-melting and very hard compounds.

Thus boron nitride (BN), aluminium nitride (AlN) and silicon carbide (SiC) arrange themselves in atomic lattices, but also the elements sulphur (S), phosphorus (P), carbon (C), silicon (Si) etc. A substance can crystallise in different structures that can be distinguished from each other. This phenomenon is called polymorphism

Table 8.12 Comparison of graphite and diamond grids

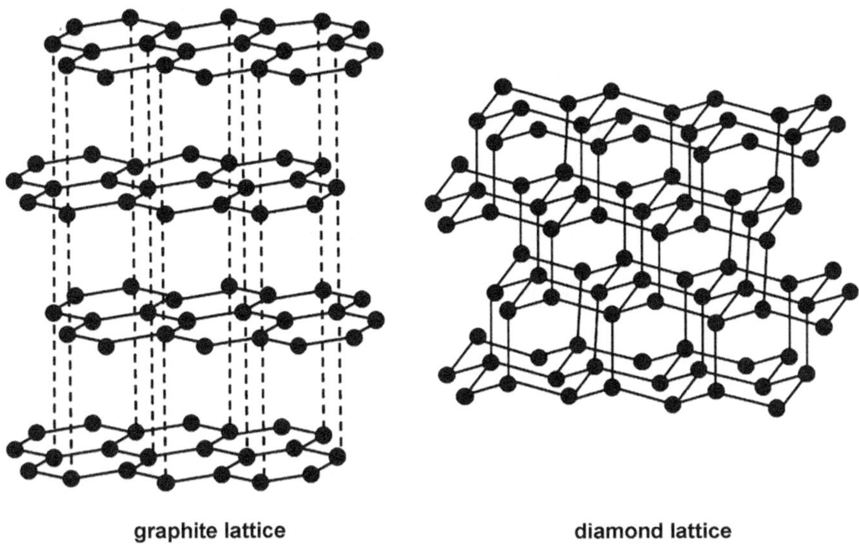

(Greek *poly* = many; *morphos* = shape; together "multiform") or, in the case of the elements, allotropy (from Greek *allotropos* = in a different way). A typical example is carbon (Table 8.12).

In the diamond lattice, all carbon atoms are connected to each other with electron pair bonds in a three-dimensional network. This results in the great hardness and transparency of diamond. In graphite, on the other hand, the carbon atoms are only connected in a two-dimensional layer lattice with electron pair bonds. The individual layers are held together by Van der Waals forces. As a result, the individual layers can be easily displaced against each other. Graphite is therefore used as a lubricant or in pencils (Table 8.12).

8.3.2.1 Modification

If substances differ in their physical properties, but not in the composition of their simplest molecular unit, then these different forms of state are also called modifications (from Latin *modus* = kind, way; *facere* = to make, to do; meaning: change, transformation).

Graphite and diamond are therefore different modifications of the element carbon. Phosphorus occurs in four modifications: white, red, black and violet. Red phosphorus is amorphous, white consists of P_4 molecules and the other two have complicated lattice structures. Tin has a metallic β-modification that slowly changes to the powdery α-modification below 13.2 °C. (This worsened the situation for Napoleonic troops in the winter of 1812 during the Russian campaign: Their tin uniform buttons literally crumbled to dust).

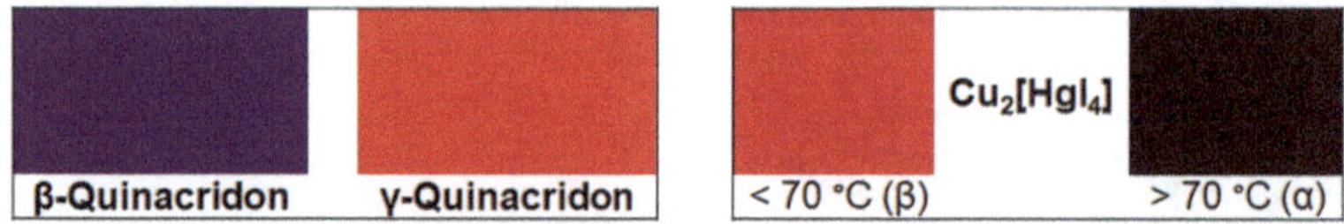

Fig. 8.19 Color and modification (examples)

Table 8.13 Overview of solid grids

Designation:	Metal grid	Ion grid	Molecular lattice	Atomic lattice
Formed from:	Metal ions and electrons	Cations and anions	Atoms or molecules	Atoms
Image				
Binding type	Metal binding	Ionic Bond	Van der Waals forces and dipole-dipole interactions	Electron pair bonsd
Melting points	Medium to high	High	Low (< 300 °C)	Particularly high
Hardness	Low to high	Large	Low	Very large
Solubility	Only in liquid metals	In polar liquids	In non-polar or polar liquids	Insoluble in almost all substances
Electr. conductivity	Very large (electrons)	Large (due to ions)	Very little	Very little
Examples	Cu, Fe, Ni, W	NaCl, KBr, KNO_3	I_2, S_8 CCl_4, SO_2	Diamond, AlN

However, the modifications resulting from the change in the crystal structure also show significant color differences in some cases:

- The color of the β-modification of quinacridone (a pigment) is reddish purple, while that of the γ-modification is red (Fig. 8.19).
- Copper-(I)-tetraiodomercurate-(II), formula: $Cu_2[HgI_4]$, is red at room temperature. On heating to ~ 70 °C, the structure changes and it becomes dark brown (Fig. 8.19). On cooling it reverts to the red colour. This effect is called thermochromism.

In both examples, the pairs are chemically identical and do not undergo any chemical change. The differences in hue (Fig. 8.19) are a purely physical property, which is explained by the different spatial structure of the molecules in the crystal lattice. Thermochromic substances are used, for example, in thermal paints, temperature-measuring paints, coffee cups whose motif changes when a hot drink is poured ("magic cups"), or film-like wine and clinical thermometers. An overview of solid grids is given in Table 8.13.

Part V

Solutions and Chemical Reactions

This section is about solutions, the dissolving process itself, concentration data and reactions in solution. Most chemical conversions take place between substances that have been dissolved - in a solvent usually not involved in the actual reaction.

Before that we deal with the basics of chemical reactions or reaction equations and the naming of simple inorganic compounds.

Preliminary remark: mass and weight

The weight (also weight force), of a substance or object depends on the location, because the force exerted by the mass is calculated according to: Weight Force = Mass multiplicated with Gravitational Acceleration ($G = m \cdot g$). It therefore makes sense to choose a quantity (i.e. mass) that is independent of location.

For clarification: One liter of water has a weight (!) of 1 kg · 9.81 m/s^2 = 9,81 kg m/s^2 = 9.81 N (N = Newton = unit of weight force) on earth. On the moon, however, the weight force is only 1 kg · 1.64 m/s^2 = 1.64 N. The difference comes from the fact that the moon is much smaller than the earth and therefore "attracts" objects and substances weaker than the earth (only ! $^1/_6$).

The mass (unit: kg) is a property of the respective body itself. The weight (unit: N) is the result of external influences on the body. The units "gram", "kilogram", "hundredweight", "ton" etc. always denote a mass. The daily used *"has a weight of 12 kilos."* is colloquial language and factually incorrect. Weights are given in the unit Newton. The terms "mass" and "weight" are used in this book exclusively in the scientific sense.

9 Chemical Reactions: Fundamentals

- A process in which new substances – with new properties – are created from given starting materials is called a chemical process or a chemical reaction.

Physical processes, such as the dissolution of sugar in water, are not chemical reactions. Even if a "new" substance (here: sugar water) appears to be created here, the substance-specific properties of the individual components do not change. It is merely a mixture of the individual components.

- The reaction of several substances to form one (or more) new substances is called **synthesis.** The decomposition of a substance into several new substances is called **analysis** (Fig. 9.1).

Neither gold, nor other precious metals or even diamonds can be produced with a chemical reaction. One can only convert existing starting materials into new end materials.

- In chemical reactions, no transformation of matter takes place. Only a combination or regrouping of the atoms of the elements involved takes place.
- The substances that are present before the reaction starts are called starting substances, **educts** or reactants. The substances formed during the reaction are called final substances or **products.**

Chemical reactions always take place in a certain numerical ratio between the educts and the products. Therefore, a reaction equation can only be correct if the number of atoms of each element involved is also the same on both sides of the reaction equation.
Here however real relations must not be disregarded completely. Oxygen, for example, occurs as a diatomic gas molecule. Therefore, "O_2" must always be

T. Schmiermund, *The Chemistry Knowledge for Firefighters*,
https://doi.org/10.1007/978-3-662-64423-2_9

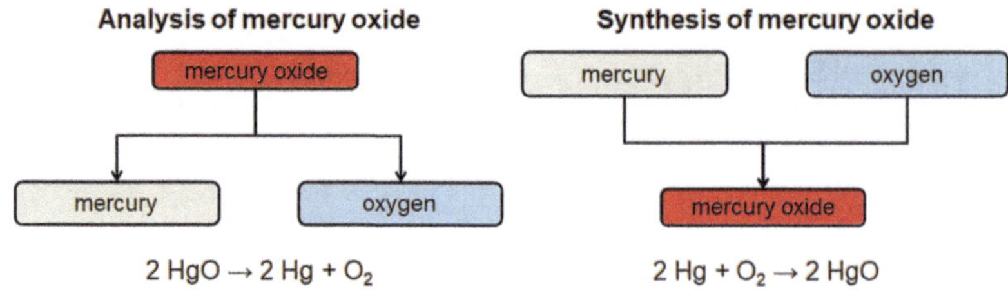

Fig. 9.1 Analysis and synthesis of mercury oxide

written. The notation "½ O_2" has to be omitted – and is equally nonsensical – as e.g. "½ NH_3". If necessary, simply double the reaction equation.

9.1 Basic Laws

The following laws can be derived well from Dalton's atomic hypothesis and represent "basic pillars" for chemical reactions.

9.1.1 Law of Conservation of Mass

If a candle is burned, it becomes lighter. If an iron nail rusts, it becomes heavier. There seems to be a contradiction here in chemical reactions: Sometimes the mass decreases, sometimes it increases.

However, if the reactions are carried out in a closed vessel, it can be seen that there is no change in the total mass of all the substances involved.

After Lomonosov[1] had already postulated in 1748 that the total mass would remain constant, Lavoiser[2] verified this assumption in 1774 with the help of precise scales. Both findings led to the "law of conservation of mass":

- In all chemical processes, the total mass of the substances involved remains unchanged.

This statement seems to contradict Einstein's principle of energy-mass equivalence. However, the energies that are converted in chemical processes are so low that they have no influence on the mass here (compare Sect. 22.1.4).

[1] Mikhail Vasilyevich Lomonosov, 1711–1765, Russian naturalist and poet.

[2] Antoine Laurent de Lavoisier, 1743–1794, French chemist and jurist.

9.1.2 Law of Equivalent Proportions

For example, 1 g of hydrogen reacts with 3 g of carbon (1 : 3) to form 4 g of methane (CH_4), but with 8 g of oxygen (1 : 8) to form 9 g of water.

This law was established by J. B. Richter[3] in 1791 and includes the two laws that follow.

- Elements always combine in the ratio of certain "compound masses" or their integral multiples to form chemical compounds.

Please bear in mind that at Richter's time the amount of substance (cf. Sect. 5.5.3) had not yet been introduced as a "measure". Richter worked exclusively with mass ratios. Nowadays it is much easier to understand this – on the basis of the mass ratios.

9.1.3 Law of Constant Proportions

Furthermore, it was found that the mass ratio of two elements that combine to form a compound is always constant with respect to each other.

For example, 1 g of hydrogen always reacts with 8 g of oxygen to form 9 g of water. If more hydrogen or more oxygen is available than corresponds to this ratio, a residue of the gas in excess remains.

Mercury oxide (HgO) consists of 92.6 % mercury and 7.4 % oxygen. Therefore, there are always 100 g of mercury for every 8 g of oxygen.

- The mass ratios of the elements in a compound are constant.

This rule was recognized by Proust[4] in 1797 and is sometimes referred to as "Proust's Law."

9.1.4 Law of Multiple Proportions

Lead can react with oxygen to form lead oxide (PbO) or lead dioxide (PbO_2). In this process, 10 g of lead combine with 0.77 g or with 1.54 g of oxygen. Similarly, carbon can react to form carbon monoxide (so-called "incomplete combustion") or carbon dioxide ("complete combustion"). 1 g of carbon reacts with 1.33 g of oxygen to form 2.33 g of carbon monoxide (CO) or with 2.66 g of oxygen to form 3.66 g of carbon dioxide (CO_2).

The oxygen ratio is 1 to 2 for both the PbO/PbO_2 pair and the CO/CO_2 pair.

---

[3] Jeremias Benjamin Richter, 1762–1802, Polish chemist and mining expert.

[4] Joseph-Louis Proust, 1754–1826, French chemist.

- The mass ratios of two elements that can combine to form different compounds are in simple, integer ratios to each other in these compounds.

This law was recognized by Dalton in 1808 and supported his atomic hypothesis. It builds on the law of constant proportions.

9.1.5 Humboldt's Gas Law

During the combustion of hydrogen, two volumes of hydrogen react with one volume of oxygen to form two volumes of water vapour. If water is decomposed (e.g. by means of electric current in so-called electrolysis), two parts by volume of hydrogen are always produced for one part by volume of oxygen.

Written as reaction equations:

$$2\,H_{2(g)} + O_{2(g)} \xrightarrow{\text{Combustion}} 2\,H_2O_{(g)}$$

$$2\,H_2O_{(\ell)} \xrightarrow{\text{Electrolysis}} 2\,H_{2(g)} + O_{2(g)}$$

- In gas reactions, simple and integer volume ratios always occur.

This relationship was established around 1810 by von Humboldt[5], who worked closely with Gay-Lussac.

9.2 Reactions

Chemical conversions, i.e. "reactions", are expressed in the form of reaction equations. In many cases, this appears to be a kind of "secret code". However, it is only a fixed system of abbreviations. In this way, one saves paperwork and has certain facts easier, faster and clearer "in view".

But how does one arrive at a certain reaction equation? How do we know "how" and "whether" certain substances react with others? Here, there is a set of rules on how substances behave towards other substances. In addition, there is the experience of the individual natural scientist. He simply "knows", partly by analogy, how certain substances behave towards others.

[5] Alexander von Humboldt, 1769–1859, German naturalist.

9.2.1 Basic Reactions

There are a few basic, largely inorganic, reactions that actually always take place in this form. These will be presented briefly.

A. Metal and acid react to form salt and hydrogen.

$$Fe + 2\,HCl \rightarrow FeCl_2 + H_{2(g)}$$

$$Mg + H_2SO_4 \rightarrow MgSO_4 + H_2 \uparrow$$

B. Metal oxide and acid react to form salt and water.

$$FeO + 2\,HCl \rightarrow FeCl_2 + H_2O_{(\ell)}$$

$$MgO + H_2SO_4 \rightarrow MgSO_4 + H_2O_{(\ell)}$$

C. Lye and acid react to form salt and water.

$$Mg(OH)_2 + H_2SO_4 \rightarrow MgSO_4 + 2\,H_2O$$

$$NaOH + HCl \rightarrow NaCl + H_2O$$

D. Metal and oxygen react to form metal oxide.

$$4\,Na + O_2 \rightarrow 2\,Na_2O$$
$$2\,Ca + O_2 \rightarrow 2\,CaO$$

E. Metal oxide and water react to form metal hydroxide.

$$CaO + H_2O \rightarrow Ca(OH)_2$$
$$Na_2O + H_2O \rightarrow 2\,NaOH$$

F. Non-metal and oxygen react to form non-metal oxide.

$$S + O_2 \rightarrow SO_2$$
$$N_2 + 2\,O_2 \rightarrow 2\,NO_2$$

G. Non-metal oxide and water react to form acid.

$$SO_2 + H_2O \rightarrow H_2SO_3$$
$$SO_3 + H_2O \rightarrow H_2SO_4$$

H. Strong acids displace weak acids from their salts.

$$2\,NaCl + H_2SO_4 \rightarrow Na_2SO_4 + 2\,HCl$$
$$2\,KNO_3 + H_2SO_4 \rightarrow K_2SO_4 + 2\,HNO_3$$

I. Carbonates react with acids to form salt, water and carbon dioxide.

$$CaCO_3 + 2\,HCl \rightarrow CaCl_2 + H_2O_{(\ell)} + CO_{2(g)}$$

$$Na_2CO_3 + H_2SO_4 \rightarrow Na_2SO_4 + H_2O + CO_2\uparrow$$

J. **Complete combustion of** organic substances produces water and carbon dioxide.

$$C_3H_8 + 5\,O_2 \rightarrow 3\,CO_2 + 4\,H_2O_{(g)}$$

$$2\,C_2H_2 + 5\,O_2 \rightarrow 4\,CO_2 + 2\,H_2O$$

(For the reaction equation of complete combustion, see also Sect. 38.3.8*).*

K. **Incomplete combustion of** organic substances produces water and carbon monoxide.

$$C_8H_8 + 6\,O_2 \rightarrow 8\,CO + 4\,H_2O$$
$$C_2H_4 + 2\,O_2 \rightarrow 2\,CO + 2\,H_2O$$

9.2.2 Reaction Equations

Knowing the formulas of the starting materials and those of the reaction products, an equation can be written to form the products: The reaction equation. If we follow a few simple rules, we can make a whole series of statements based on a reaction equation.

The following conventions have been agreed upon:

- The starting materials (educts) are on the left.
- The final substances (products) are on the right.
- A reaction arrow (→) is used instead of an equal sign (=).
- The number (sum) of the different atoms must be the same on both sides.
- The number of atoms/molecules involved is written in front of the formula.

Examples

Hydrogen	+	Chlorine	=	= Hydrogen chloride
	+		=	
$\mathbf{H_2}$	+	$\mathbf{Cl_2}$	→	$\mathbf{2\ HCl}$
1 mole hydrogen	+	1 mole chlorine	→	2 moles hydrogen chloride
2 g hydrogen	+	70.9 g chlorine	→	72.9 g hydrogen chloride
22.4 L hydrogen	+	22.4 L chlorine	→	44.8 L hydrogen chloride

Hydrogen	+	Nitrogen	=	Ammonia
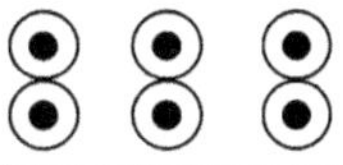	+		=	
$\mathbf{3\ H_2}$	+	N_2	→	2 NH_3
3 mol hydrogen	+	1 mol nitrogen	→	2 mol ammonia
6 g hydrogen	+	28 g nitrogen	→	34 g ammonia
67.2 L hydrogen	+	22.4 L nitrogen	→	44.8 L ammonia

Thus, by means of a reaction equation and the molar masses, statements can be made about the mass ratios of a reaction – in the case of gas reactions also about the volume ratios. ◄

9.2.3 Rules for Setting Up Reaction Equations

- Formulas of the compounds must not be changed.
- Write down the substances involved according to the scheme "educts → products"
- Balancing the number of atoms of each element on both sides of the equation by coefficients.
  - Coefficients ("priors") have the meaning of multipliers for the substance below.
  - Example: 5 H_2O = five molecules of water

$5 \cdot 2 = 10$ atoms of hydrogen
$5 \cdot 1 = 5$ atoms oxygen

  - The coefficients should appear in the final reaction equation as the smallest possible integers.
- Checking the number of atoms of each element involved on both sides of the equation.
  - The number must be identical on the right and left sides of the equation.
  - It is calculated by multiplying the coefficient by the index.

Example 1

Description: **Nitrogen and hydrogen react to form ammonia**

$$\text{Formula equation}: \quad \mathbf{N_2} + \mathbf{H_2} \rightarrow \mathbf{NH_3}$$

On the left there are 2 N-atoms, on the right only 1. By balancing you get:

$$\text{Equalization 1} \quad \mathbf{N_2} + \mathbf{H_2} \rightarrow \mathbf{\underline{2}\,NH_3}$$

Now we balance the hydrogens (left 2, right 6):

$$\text{Equalization 2} \quad \mathbf{N_2} + \mathbf{\underline{3}\,H_2} \rightarrow \mathbf{2\,NH_3}$$

$$\text{Reaction equation}: \quad \mathbf{N_2} + \mathbf{3\,H_2} \rightarrow \mathbf{2\,NH_3}.$$ ◀

Example 2

Description: Pentane & oxygen react to form carbon dioxide & water

$$\text{Formula equation}: \quad \mathbf{C_5H_{12}} + \mathbf{O_2} \rightarrow \mathbf{CO_2} + \mathbf{H_2O}$$

On the left there are 5 C-atoms, on the right only 1. By equalization you get:

$$\text{Equalization 1}: \quad \mathbf{C_5H_{12}} + \mathbf{O_2} \rightarrow \mathbf{\underline{5}\,CO_2} + \mathbf{H_2O}$$

On the left there are 12 H-atoms, on the right only 2. After equalizing:

$$\text{Equalization 2}: \quad \mathbf{C_5H_{12}} + \mathbf{O_2} \rightarrow \mathbf{5\,CO_2} + \mathbf{\underline{6}\,H_2O}$$

Now we balance the oxygen atoms (left 2, right 16):

$$\text{Equalization 3}: \quad \mathbf{C_5H_{12}} + \mathbf{\underline{8}\,O_2} \rightarrow \mathbf{5\,CO_2} + \mathbf{6\,H_2O}$$

$$\text{Reaction equation} \quad \mathbf{C_5H_{12}} + \mathbf{8\,O_2} \rightarrow \mathbf{5\,CO_2} + \mathbf{6\,H_2O}$$

◀

The following special features should also be noted:

- For elements, all metals, all noble gases, and almost all solid nonmetals are considered "monatomic". Since the index "1" is omitted, one thus writes, for example: Fe, Cu, Ag, Au, C, Zn, Sn, S, P, Ne, Ar, ...
- Seven nonmetals exist as diatomic molecules and are invariably represented as such. They are: H_2, N_2, O_2, F_2, Cl_2, Br_2, I_2.

9.2.4 Stoichiometry

The emergence of stoichiometry was one of the important prerequisites for the development from alchemy to the modern natural science of chemistry. J. B. Richter is regarded as the founder. Stoichiometry (Greek *stoicheia* = basic substance and *metron* = measure) deals with the quantitative description of chemical reactions. The laws underlying stoichiometry have already been discussed in Sects. 9.1.1, 9.1.2, 9.1.3, 9.1.4 and 9.1.5.

The calculations on this basis are extremely helpful:

- In the case of a synthesis, the theoretical yield can thus be calculated and compared with the real yield.
- It is possible to determine the quantity of reactants required to obtain a certain quantity of target product.
- In analyses, the quantitative relationships can be calculated and thus, for example, the proportion of iron in an ore can be determined.
- In electrochemistry, the mass fractions of the deposited elements can be determined from the amount of current consumed.

The formula symbol ν (Greek letter *ny)* for the stoichiometric factors indicates the relationship to the amount of substance (formula symbol n). In the case of reactions, it has been agreed that a negative sign should always be written for the reactants, i.e. for the "consumed substances". However, this is only the case when the reaction stoichiometry number is given "purely"; in a reaction equation there are always positive numbers on the right and left sides.
The **coefficients** that ***precede*** a compound or element are also called the stoichiometric factor or *stoichiometric number. For* example, the stoichiometric factors of the reaction are $N_2 + 3\ H_2 \rightarrow 2\ NH_3$: $\nu(N_2) = 1$, $\nu(H_2) = 3$ and $\nu(NH_3) = 2$.

The **indices *in*** compounds are also called *particle stoichiometric number.* Thus, the stoichiometric factors for NH_3 $\nu_N = 1$ and $\nu_H = 3$ or for $CaCl_2$ $\nu_{Ca} = 1$ and $\nu_{Cl} = 2$.

Stoichiometry and Oxidation Number
A great help to be able to balance reaction equations more quickly is to know the oxidation numbers. These are given, for example, in various periodic tables (see also supplementary material to this book, available at https://www.springer.com/en/

book/9783662566060). A method for calculating oxidation numbers within compounds and in the course of chemical reactions is shown in Chap. 16.

9.2.5 Naming Connections

Of course, the starting and final substances of chemical reactions do not only want to be represented as a formula, they also need unique names or substance designations. Some substances have historically determined designations, so-called "trivial names", which have to be learned more or less by heart. The "systematic names", on the other hand, follow rules.

The composition of a compound is often indicated by prefixing the respective element with Greek numerical words and combining them to form a substance name. The *mono* for 1 is usually omitted.

P_2O_5	Diphosphorus pentoxide
SiO_2	Silicon dioxide (sand, quartz)
CO	Carbon monoxide
CO_2	Carbon dioxide
PCl_5	Phosphorus pentachloride
$NaCl$	Sodium chloride (table salt)
$SiCl_4$	Silicon tetrachloride
SF_6	Sulphur hexafluoride
SO_2	Sulphur dioxide
SO_3	Sulphur trioxide
MnO_2	Manganese dioxide

If there is only one possible stoichiometric composition of the substance, the number words are often omitted:

$AlCl_3$	Aluminium chloride
$Al_2(SO_4)_3$	Aluminium sulphate
H_2S	Hydrogen sulphide
H_2O	Hydrogen oxide (water)
$POCl_3$	Phosphorus oxychloride
$CaBr_2$	Calcium bromide

If an element occurs in different compounds which are composed of the same elements, the oxidation numbers differ. For this reason, the oxidation number is given as a Roman numeral in the substance name (trivial names in brackets):

Cu_2Cl_2	$\mathbf{Cu}_2^{+\mathbf{I}}Cl_2^{-I}$	Copper(**I**)-chloride
$CuCl_2$	$\mathbf{Cu}^{+\mathbf{II}}Cl_2^{-I}$	Copper(**II**)-chloride
$FeSO_4$	$Fe^{+II}(SO_4)^{-II}$	Iron(**II**)-sulphate
$Fe_2(SO_4)_3$	$\mathbf{Fe}_2^{+\mathbf{III}}(SO_4)_3^{-II}$	Iron(**III**)-sulphate
MnO	$Mn^{+II}O^{-II}$	Manganese(**II**)-oxide
MnO_2	$\mathbf{Mn}^{+\mathbf{IV}}O_2^{-II}$	Manganese(**IV**)-oxide
H_2SO_4	$H_2^{+I}\mathbf{S}^{+\mathbf{IV}}O_4^{-II}$	Sulphuric(**VI**)-acid (sulphuric acid)
H_2SO_3	$H_2^{+I}\mathbf{S}^{+\mathbf{IV}}O_3^{-II}$	Sulphuric(**IV**)-acid (sulphurous acid)
$HClO_4$	$H^{+I}\mathbf{Cl}^{+\mathbf{VII}}O_4^{-II}$	Chloric(**VII**)-acid (perchloric acid)
$HClO_3$	$H^{+I}\mathbf{Cl}^{+\mathbf{V}}O_3^{-II}$	Chloric(**V**)-acid (chloric acid)
$HClO_2$	$H^{+I}\mathbf{Cl}^{+\mathbf{III}}O_2^{-II}$	Chloric(**III**)-acid (chlorous acid)
$HClO$	$H^{+I}Cl^{+I}O^{-II}$	Chloric(I)-acid (hypochlorous acid)

There are extended rules for the naming of complex salts and organic compounds, which will be discussed later (Sect. 11.2.7).

Solutions 10

As we learned in Chap. 2, practically all homogeneous mixtures are "solutions". They therefore have the same properties everywhere. Even gas mixtures such as air are sometimes regarded as solutions of several gases (O_2, CO_2, ...) in a "dissolving gas" (nitrogen).

A solution is always composed of a solvent and the dissolved substance(s). As a rule, the liquid parts are referred to as the solvent. If this is not possible, e.g. with alloys, then the substance that is present in the highest quantity is the solvent.

In the following we consider the dissolution of a substance in a solvent. Reactions under dissolution, such as the dissolution of a metal in acid, are not the subject of this chapter, as this is a chemical reaction in the narrower sense.

10.1 Basic Information on the Dissolving Behaviour

Sugar and salt dissolve in water, but not in oil. Conversely, grease stains cannot be treated with pure water. This is where petroleum ether, which was often used in grandma's day, comes in handy. From this we can conclude that there are certain substances that dissolve in water and water-like compounds, others do not. Conversely, some substances dissolve in oil and oil-like compounds, but not in water.

More detailed consideration of these soluble or insoluble substances and the solvents used culminates in a short sentence:

- "Same dissolves in same" or "like dissolves in like".

This statement refers primarily to the polarity of solvent or solute and is intended to express the following: Polar substances, such as salts, generally dissolve well in polar solvents, such as water. Non-polar substances (grease) dissolve in non-polar or less polar solvents (petrol, chloroform).

T. Schmiermund, *The Chemistry Knowledge for Firefighters*,
https://doi.org/10.1007/978-3-662-64423-2_10

Table 10.1 Solubility of potassium chloride (KCl) and sulphur (S_8) in various solvents

*L** of KCl at 0 °C in			*L** of S_8 at 20 °C in		
Waters	H_2O	28.2	**Carbon disulfide**	CS_2	42.4
Heavy water	D_2O	21.2	**Carbon tetrachloride**	CCl_4	0.72
Hydrogen peroxide	H_2O_2	63.3	**Benzene**	C_6H_6	1.73
Sulphur dioxide	$SO_{2(\ell)}$	0.041	**Toluene**	$C_6H_5CH_3$	1.82

In order to be able to make comparisons regarding the maximum amount of a substance that can be dissolved in a certain solvent, the term "solubility" (*L**) was introduced:

- The solubility (L*) of a substance is defined as grams of substance per 100 grams of solvent.

In addition to the temperature (see Sect. 10.1.5), the specification of the solvent is also of decisive importance (Table 10.1).

10.1.1 Dissolving Process: Polar Substances in Polar Solvents

Within a polar solid, such as a common salt crystal, the charges of the ions are balanced. At the surface, however, the positive/negative charges of the ions act as corresponding forces in space. The outermost layer of particles is therefore not quite as firmly anchored in the solid lattice as the layers further inwards. In a salt crystal surrounded by water, the water molecules accumulate around the ions. Here the water molecule aligns itself as a dipole (Sect. 8.2.1) correspondingly opposite to the charge of the ion. The ions are surrounded by a "water envelope" ("hydrate envelope") and pass into the liquid (Fig. 10.1). This process is promoted by the molecular movement of the particles (diffusion, Sect. 4.11) or by simple stirring.

At the end of the dissolution process, the solution consists of water molecules and hydrated chloride and sodium ions.

The number of water molecules in the hydrate shell of an ion depends on the particular ion. For example, a Li^+ ion is surrounded by 12, a Na^+ ion by 8 and a K^+ ion by 4 H_2O molecules.

Polar solvents such as water or ethanol predominantly dissolve:

- Ionic compounds (salts)
- polar inorganic compounds (e.g. HCl, NH_3, H_2SO_4)
- Organic compounds with hydrophilic (water-friendly) groups, such as: –OH (alcohols, sugars), –COOH (carboxylic acids), –CHO (aldehydes), $-NH_2$ (amines), $-SO_3H$ (sulphonic acids), which are also polar.

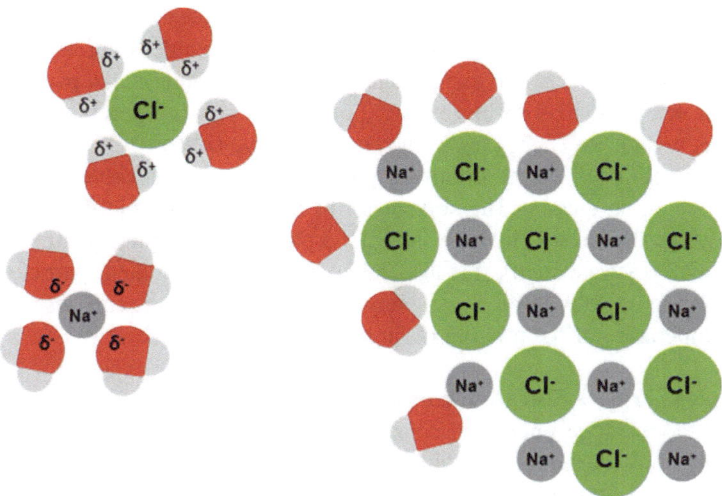

Fig. 10.1 Dissolution of NaCl in water (schematic)

Dissolving NaCl in water:

$$NaCl + H_2O \rightarrow Na^+_{(aq)} + Cl^-_{(aq)}$$

The index (aq) denotes the (hydrated) ion surrounded by the hydrate shell.

10.1.2 Dissolving Process: Non-polar Substances in Non-polar Solvents

Non-polar or less polar solids are held together in their molecular lattices only by the weak Van der Waals forces (Sect. 8.2.3). As in an ion crystal, the particles in the outermost layer are less tightly bound than the particles in the interior. Analogous to the dissolution process in polar liquids, a solvate shell (solvent shell) forms around the individual molecule as it leaves the molecular lattice.

10.1.3 Energy Consumption During the Dissolving Process

In order for a substance to be considered "dissolved" at all, its individual components, i.e. the molecules or ions of which the substance consists, must be individually surrounded by solvent molecules. This process is called "**solvation**" or, in the case of water as a solvent, "**hydration**".

Here, two different energies compete: The energy that holds the solid together, i.e. the lattice energy, which must be overcome, and the solvation or hydration energy that is released during dissolution.

The lattice energy is first of all the energy that is released during the formation of the crystal lattice. However, it must also be expended in order to break up the lattice and transfer the particles into the gas phase. During dissolution, the particles have to be "separated" just like during evaporation. Therefore, the process of dissolution consumes this lattice energy and removes it from the solvent.

The "enveloping" of the dissolved particles with the molecules of the solvent releases solvation energy. The solvent forms a solvate shell (solvent shell), in the case of water also called "hydrate shell". If a particle is completely surrounded by a solvate shell (hydrate shell), it is also called a solvated (hydrated) particle.

Depending on which of the two energies predominates, one can observe changes in the temperature of the mixture during the dissolution process:

- **Lattice energy > Solvation energy**
- Since more lattice energy is consumed than solvation energy is released, the mixture cools during the dissolution process.
- Typical examples are: NH_4NO_3, $Ba(NO_3)_2$, KNO_3, $CaCl_2 \cdot 6\ H_2O$, urea
- **Solvation energy > Lattice energy**
- If the solvation energy released is greater than the required lattice energy, the mixture heats up during the dissolution process.
- Typical examples are: $CaCl_2$, $Al_2(SO_4)_3$, $MgCl_2$, NaOH, KOH
- **Lattice energy ≈ Solvation energy**
- Lattice energy and solvation energy are approximately equal. There is little or no detectable temperature change in the mixture during the dissolution process. Typical examples are: NaCl, NaBr, Na_2SO_4, $Al_2(SO_4)_3 \cdot 27\ H_2O$.

In the case of some salts, a temperature effect can even be determined as a function of the crystal water content of the salt. Anhydrous calcium chloride ($CaCl_2$), for example, dissolves when heated. Calcium chloride containing water of crystallization ($CaCl_2 \cdot 6\ H_2O$), on the other hand, dissolves when cooled.
In the latter, the Ca^{2+} ion is already surrounded by water molecules in the crystal. Therefore, only the hydration energy of the two Cl^- ions is released. However, this is less than the lattice energy to be expended: the solution cools down.

10.1.4 Unsaturated, Saturated and Supersaturated Solutions

An **unsaturated** solution is a solution that can dissolve other substances if you add some. However, this dissolving capacity is exhausted at some point. A sediment of undissolved substance remains. You have obtained a **saturated** solution.

This can be decanted or filtered off the sediment. However, if the solution remains above the sediment, there is a dynamic equilibrium between the sediment and the saturated solution: particles constantly leave the undissolved phase to enter the solution. At the same time, dissolved particles that get close enough to the sediment due to molecular motion are attracted by the sediment and reintegrated into the crystal (Fig. 10.2).

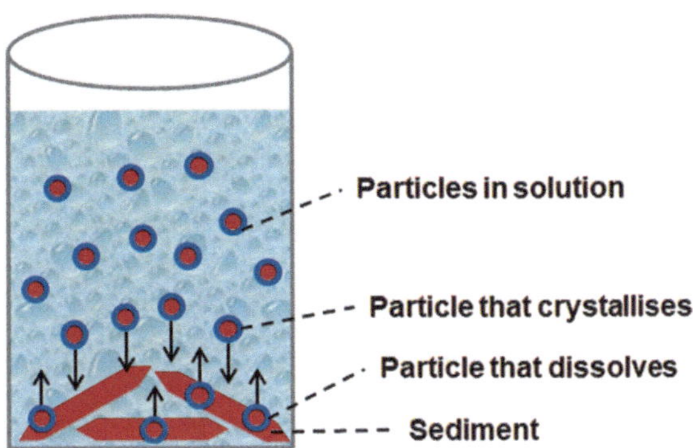

Fig. 10.2 Dynamic solution equilibrium

A solution that is saturated with respect to *one* substance can, however, also dissolve *other* substances. For example, it is possible to dissolve additional sugar in a saturated sodium chloride solution - without any part of the sodium chloride crystallizing out of the solution.

Under certain circumstances, it is even possible to produce solutions in which more material is dissolved than would be expected in terms of solubility - at this temperature. This is called a **supersaturated solution**. But this solution is not stable. Dust grains that get into it, intentionally introduced small crystals of the dissolved substance (as a so-called "seed crystal") or vibrations then ensure that the "too much" of dissolved substance precipitates/crystallizes out - and thus a saturated solution with a bottom body is obtained again. The lattice energy released during crystallization heats the solution in many cases.

Practical Application
The sodium salt of acetic acid (sodium acetate) forms supersaturated solutions particularly easily. The salt has a solubility of 61 g per 100 g water at room temperature. If 500 g of the salt containing water of crystallization is mixed with 50 mL of water and heated to above 60 °C, a supersaturated solution is obtained. This can be filled into so-called "heat pads" and serves as a "latent heat storage". By "cracking" a metal plate contained in the bag, seed crystals are formed and the salt crystallizes out. The released crystallization energy provides temperatures around 58 °C. By reheating in a water bath until the salt is completely dissolved at temperatures of 65 °C – 70 °C, the heat pad can then be regenerated and is available again.

10.1.5 Temperature Dependence of the Solubility

Until a solution is saturated, this substance will also dissolve if further substance is added. At some point, however, a proportion of substance remains which no longer dissolves. The solution is now saturated.

If the already saturated solution is now heated, it is found that in the vast majority of cases the soil body dissolves. The solubility therefore usually increases with the

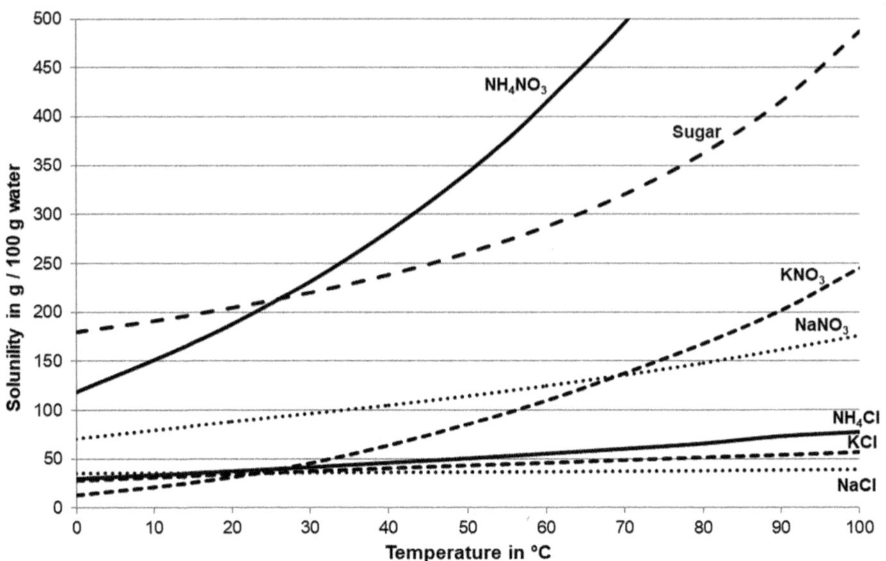

Fig. 10.3 Temperature dependence of solubility

temperature. As is so often the case in chemistry, there are exceptions: Calcium sulfate (gypsum, $CaSO_4$) is probably still familiar to most. Another exception is the rather exotic holmium sulfate ($Ho_2(SO_4)_3$).

Some examples of solubility and its temperature dependence are shown in Fig. 10.3.

It is interesting to convert the solubility of sugar into "sugar cubes per cup of coffee":

At 20 °C 82 pieces of sugar cubes, at 50 ° C 104 pieces, at 80 ° C 145 and at 100 ° C even 195 (!) pieces of sugar cubes dissolve completely in a small cup of coffee á 125 mL.

10.1.6 Crystal Water

If the reformation of the solid salt from the solution is initiated by evaporation or evaporation at not too high temperatures, then the ions can partly take over their hydrate shell into the crystal lattice. Water bound in the crystal in this way is called crystal water.

Also in formulas the water of crystallization is usually given, e. g.: $CaCl_2 \cdot 6\ H_2O$. The dot (·) is to indicate that there is no direct chemical bond here (Table 10.2).

Salts containing crystal water often form very beautifully formed and large crystals. The largest crystals found so far consist of $CaSO_4 \cdot 2\ H_2O$ and were

Table 10.2 Examples of salts with water of crystallization

$Al_2(SO_4)_3 \cdot 18\ H_2O$	$BaCl_2 \cdot 2\ H_2O$	$Na_3PO_4 \cdot 12\ H_2O$
$KAl(SO_4)_2 \cdot 12\ H_2O$	$CaCl_2 \cdot 6\ H_2O$	$Na_2HPO_4 \cdot 12\ H_2O$
$Na2SO_4 \cdot 10\ H_2O$	$FeCl_2 \cdot 4\ H_2O$	$Ni(NO_3)_2 \cdot 6\ H_2O$
$CaSO_4 \cdot 2\ H_2O$	$ZnCl_2 \cdot 2\ H_2O$	$Ba(OH)_2 \cdot 2\ H_2O$
$FeSO_4 \cdot 7\ H_2O$	$NiCl_2 \cdot 2\ H_2O$	$Pb(CH_3COO)_2 \cdot 3\ H_2O$
$CuSO_4 \cdot 5\ H_2O$	$CoCl_2 \cdot 6\ H_2O$	$CaBr_2 \cdot 6\ H_2O$

found in a mine in northern Mexico. They are over 4 m long and more than ½ meter in diameter.

However, water of crystallization also plays a role in the coloration of the salts of the subgroup elements. For example:

- $CuSO_4 \cdot 5\ H_2O$ forms beautiful blue crystals. Anhydrous $CuSO_4$ is almost colourless.
- $CoCl_2 \cdot 6\ H_2O$ is pink, anhydrous $CoCl_2$ is dark blue.

10.1.7 Application-Related Dissolving Behaviour of Solids

It is relatively unimportant in the field of application whether any spilled solid has a solubility in water of 23.5 or 32.7 g per 100 g of water (cf. also Sect. 10.2.1). A semi-quantitative statement of solubility is quite sufficient. This must then be related to the chemical-toxicological properties of the substance (toxicity, pH of the solution, effects on the environment, etc.) so that the necessary measures can be coordinated.

Example

- Lime ($CaCO_3$) is a non-toxic, non-corrosive substance that is not very soluble in water. Here, after picking up and filling into suitable containers, a "post-cleaning" with water can be considered.
- Lead cyanide ($Pb(CN)_2$, UN No. 1620) is a slightly soluble but toxic solid. If the substance enters the sewage system, for example, it will only dissolve over a longer period of time. Undissolved parts may be washed away. Therefore, a long-lasting release into water bodies and a widespread spreading (dispersion or spreading of the contamination) is to be expected. Long-term effects ("creeping poisoning") in rivers and lakes cannot be excluded. The same applies to contamination of the soil.
- Sodium arsenate (Na_3AsO_4, UN No. 1685) is a highly soluble ($L^* = 10.5$) toxic substance. An acute toxic effect is to be expected here. If the substance enters the sewage system or surface waters, acute poisoning may occur, e.g. of fish stocks. The operators of waste water treatment plants must be informed at an early stage. If the substance reaches the soil as a solution, a high contamination in acute toxic concentrations is to be expected. ◀

Table 10.3 Solubility of solids

Corresponds to: Statement:	kg water/kg substance (for complete dissolution)	L^* [g per 100 g water]	~ w % (solution)
Good / very good soluble	< 1...10	> 10	> 10 %
Soluble	10...50	2...10	2...10 %
Hardly soluble	50...100	1...2	1...2 %
Difficult/very difficult to dissolve	100...10 000	0.01...1	0.01...1 %
Practically insoluble	> 10 000	<0.01	< 0.01 %

Table 10.4 Miscibility of liquids with water

Corresponds to: Statement:	kg water/kg substance (for complete dissolution)	L^* [g per 100 g water]	~ w % (solution)
Fully miscible	0...∞	∞	0...100 %
Partially miscible	1...10	> 10	> 10 %
Moderately miscible	10...100	1...10	1...10 %
Slightly miscible	100...1000	0.1...1	0.1...1 %
Very slightly miscible	> 1000	< 0.1	< 0.1 %

Solubility of Solids

Table 10.3 can be used to assess the solubility of solids. The first column indicates how much water is necessary to dissolve the substance completely. The second column gives possible literature values of solubility. Finally, in the third column, the approximate content (as a mass fraction) of the resulting solution is estimated.

10.1.8 Water Miscibility of Liquids

In the case of liquids, one often speaks colloquially of "miscibility with water" instead of solubility (Table 10.4). This miscibility is often given in reference works and tables (e.g. in "Hommel") and describes the ability of a liquid to mix with water. Strictly speaking, however, this is also the solubility of a - liquid - substance in water.

Solubility of Liquids in Use

In the case of hazardous substance operations involving the escape of liquid substances, the density of the liquid must be taken into account in addition to its miscibility with water. The combination results in four main groups:

I. Slightly / very slightly water-miscible substances, density < 1 g/mL
   - Substance floats on top of the water and spreads out as a thinner and thinner layer on the water surface as the amount of water increases.

- Sometimes carried very far with the water. In the case of flammable liquids, an explosive vapour-air mixture can develop far from the point of discharge. Can usually be separated from the water by oil separators or oil barriers.
- Examples: Gasoline, diesel fuel, heating oil, toluene.

II. Slightly / very slightly water-miscible substances, density > 1 g/mL
- Substance sinks below the water surface, remains on the bottom in watercourses, canals and sewage treatment plants and seeps into the bottom of the watercourse. Oil barriers and oil separators have no effect. Substances are often environmentally hazardous and/or toxic.
- Examples: Tetrachloroethene, chlorobenzene, chloroform, 1,2-dichlorobenzene, nitrobenzene.

III. Substances completely miscible with water
- Form a homogeneous mixture substance: Water.
- Substances are often readily biodegradable, but in large quantities can quickly overload wastewater treatment plants. Oil separators and oil barriers are ineffective.
- The degree of dilution determines the properties of the mixture. This becomes apparent e.g. in the change of the flash point (compare Sect. 3.3.3) or the "corrosivity" (compare e.g. Chap. 13).
- Typical liquids that can be mixed with water indefinitely are
  - Short-chain organic acids (formic acid, acetic acid, acrylic acid)
  - Lower alcohols (methanol, ethanol, propanol, glycol, glycerol)
  - Inorganic acids (e.g. sulphuric acid, phosphoric acid, perchloric acid) and alkalis (e.g. sodium hydroxide solution, potassium hydroxide solution)

IV. Partially / moderately water miscible substances
- Here, an individual consideration of the substance and its chemical-physical-toxicological properties is required in order to be able to make an assessment.

10.1.8.1 Mixture Gap for Liquids

Some liquids absorb water only to a limited extent, while at the same time part of the substance dissolves in the water. If there is little water and much of the substance in question, a water-saturated substance phase is obtained. If there is a large amount of water and little of the substance, a substance-saturated aqueous phase is obtained.

Example

- Phenol
  - water-saturated phenol: 72.5 % phenol / 27.5 % water
  - phenol-saturated water: 8 % phenol / 92 % water
- *iso*-Butanol
  - water-saturated *iso*-butanol: 82 % *iso*-butanol / 18 % water
  - *iso*-butanol-saturated water: 8 % *iso*-butanol / 92 % water

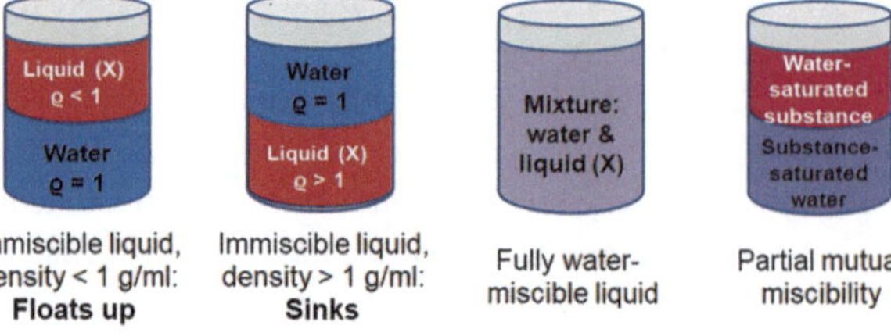

Fig. 10.4 Different miscibilities of liquids with water

For all other mixing ratios, two liquid phases form which have at most the specified composition. By increasing the temperature, this effect, known as the miscibility gap, can be eliminated. For example, above 68 °C phenol mixes with water in any ratio. Depending on the density of the respective substance, the water-saturated phase floats up **or** sinks down (Fig. 10.4).

10.1.8.2 Simple Detection Options

Oil test paper can be used to decide whether an oily-looking liquid is really oil or whether oil is in water or in the soil. The pale blue paper turns deep blue on contact with hydrocarbons, especially petrol, fuel oil, lubricating oil etc. (Fig. 10.5b).

The following can be detected without any problems (according to manufacturer's specifications): Petrol (25 mg/L), heating oil (10 mg/L), lubricating oil (1 mg/L).

But be careful: All liquids that can dissolve fats or oils give a positive result, even completely water-miscible substances such as acetone. However, these usually wet the paper extremely strongly. Purely aqueous solutions, on the other hand, do not wet the paper.

Under certain circumstances, even slightly water-miscible substances with a density above 1 g/mL can be detected with the oil test paper. In this case, dark spots appear on the water droplet on the paper.

In order to be able to determine the height of a water layer *under a* floating oily liquid in a container, so-called **water detection paste can be** used. This is a grey paste which is applied to a suitable object, e.g. a wooden slat. You then dip the slat into the container, wait a few seconds and then pull it out again. A pinkish discoloration indicates the level of the water layer under the oil (Fig. 10.5a).

With acids and acidic liquids a foaming of the paste occurs. At the same time, no colour change takes place at all. Substances which are only partially miscible with water, i.e. which have a miscibility gap, can only be examined to a limited extent for water underneath. If the water content is too high, false-positive detection of both phases occurs.

10.1.9 Rules for the Solubility of Salts in Water

Unfortunately, there is no simple rule for solubility. The mutual influences and interactions of the anions and cations with each other and with the solvent are too diverse. Nevertheless, a few basic rules can be derived from the observations.

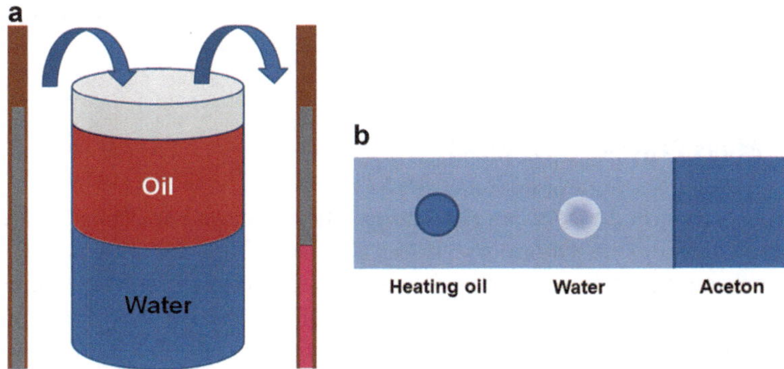

Fig. 10.5 (**a**) Water detection paste (application, schematic), (**b**) Oil test paper (display, schematic)

- Predominantly form soluble salts:
  - All nitrates $\left(NO_3^-\right)$, chlorates $\left(ClO_3^-\right)$, perchlorates $\left(ClO_4^-\right)$and acetates (CH_3COO^-)
  - Chlorides (Cl^-), bromides (Br^-), and iodides (I^-); exceptions: Ag^+, $Hg_2^{2+}$and Pb^{2+}
  - Alkali metals (Li^+, Na^+, K^+) and ammonium$\left(NH_4^+\right)$; Exception: $KClO_4$
  - All sulfates $\left(SO_4^{2-}\right)$; exceptions: Ca^{2+}, Sr^{2+}, Ba^{2+}, Hg^{2+}, Pb^{2+}
    Hydrogencarbonate $\left(HCO_3^-\right)$
- Predominantly form poorly soluble salts:
  - Carbonates $\left(CO_3^{2-}\right)$and phosphates $\left(PO_4^{3-}\right)$; exceptions: Na^+, K^+, NH_4^+
  - Oxalates [$(COO^-)_2$] and cyanides (CN^-); exceptions: Li^+, Na^+, K^+, NH_4^+
  - Hydroxides (OH^-); exceptions: Li^+, Na^+, K^+, NH_4^+, Ba^{2+}
  - Oxides (O^{2-}) and sulfides (S^{2-}); exceptions: Alkali and alkaline earth metal compounds

10.2 Composition of Mixed Phases

There are several variants for specifying how much substance is contained in a certain solution, depending on the respective task in the laboratory or chemical production. Since applications in laboratories or in a chemical plant cannot be ruled out, an overview of the most important concentration indications that are used will be given here.

Preliminary Note

- Fractions always refer to the sum of the individual components
- Concentrations always refer to the "finished" solution/mixture
- ratios always refer to the ratio of substance A to substance B

- Molality is the exception, as it is effectively a concentration based on mass rather than volume.

10.2.1 Mass Fraction (*w*, *w* %)

The mass fraction is the concentration information that we usually encounter most often, even outside the laboratory. It is calculated from the ratio "mass of the component sought" divided by "total mass of the mixture" and can therefore be conveniently applied to all kinds of mixtures. The total mass is the sum of the individual components.

$$w(\mathrm{X}) = \frac{m(\mathrm{X})}{m_{\mathrm{total}}} \qquad w\%(\mathrm{X}) = w(\mathrm{X}) \cdot 100\% \qquad m_{\mathrm{total}} = m_1 + m_2 + m_3 \ldots$$

In the past, the mass fraction was also referred to as "weight percentage" (wt.-%). This designation should not be used, since mass and weight ("weight force") are fundamentally different.

10.2.1.1 Conversion Solubility (L^*) ↔ Mass Fraction (*w*)

To convert mass fraction to solubility or solubility to mass fraction, use the following formulas:

Conversion from mass fraction (w) to solubility ($L*$) :

$$L^*(\mathrm{X}) = \frac{100 \cdot w(\mathrm{X})}{1 - w(\mathrm{X})}$$

Conversion from solubility ($L*$) to mass fraction (w) :

$$w(\mathrm{X}) = \frac{L^*(\mathrm{X})}{L^*(\mathrm{X}) - 100.}$$

The solubility must be entered here in grams of substance per 100 g of solvent. The mass fraction **must** not be given in % but as a decimal number.

10.2.2 Volume Fraction (φ,φ %, vol.-%)

In analogy to the mass fraction, the volume fraction φ (phi) is calculated according to "volume of the sought substance" divided by "sum of all individual volumes"; i.e. in the case of the volume fraction, the individual volumes are already added **before** mixing.

$$\varphi(\mathrm{X}) = \frac{V(\mathrm{X})}{V_{\mathrm{total}}} \qquad \varphi\%(\mathrm{X}) = \varphi(\mathrm{X}) \cdot 100\,\% \qquad V_{\mathrm{total}} = V_1 + V_2 + V_3 \ldots$$

Volume fraction and volume concentration are identical if no change in volume occurs during the mixing process of the substances involved. However, this is not always the case. A mixture of 50 mL water and 50 mL pure ethanol only results in a total volume of 96 mL (instead of 100 mL). This behavior is called volume contraction and is due to the fact that the strongly polar water molecules can occupy spaces between the polar ethanol molecules.

For the indication of the alcoholic strength by volume of alcoholic beverages, the alcoholic strength by volume is expressed as "vol.-%" (percent by volume).

10.2.3 Mass Concentration (β)

The mass concentration β *(beta) is* calculated by dividing the mass of the respective substance by the total volume of the resulting mixture.

$$\beta(\mathrm{X}) = \frac{m(\mathrm{X})}{V_{\text{total}}} \qquad V_{\text{total}} = \sum (V_1 + V_2 + V_3 \ldots)$$

The unit of mass concentration is grams per litre (g/L).

10.2.4 Volume Concentration (σ, σ %)

Similarly, the volume concentration σcalculated from the volume of the sought substance to the total volume of the mixture.

$$\sigma(\mathrm{X}) = \frac{V(\mathrm{X})}{V_{\text{total}}} \qquad \sigma\%(\mathrm{X}) = \sigma(\mathrm{X}) \cdot 100\% \qquad V_{\text{total}} = \sum (V_1 + V_2 + V_3 \ldots)$$

Volume concentration and volume fraction are identical if no volume change occurs during the mixing process of the substances involved (compare Sect. 10.2.2).

10.2.5 Mass Concentration (c)

The concentration of a substance, i.e. the amount of substance (in mol) contained in a certain volume (usually: 1 L), is one of the most frequently used concentration values in laboratories. Solutions with a defined substance concentration are used, among other things, to determine the content of acids or alkalis. The unit is mol/L.

$$c(\mathrm{X}) = \frac{n(\mathrm{X})}{V_{\text{total}}}$$

10.2.6 Mass Ratio (r)

The substance quantity ratio is simply calculated according to the ratio of the substance quantity of substance X to the substance quantity of substance Y:

$$r_{\mathrm{XY}} = \frac{n_{\mathrm{X}}}{n_{\mathrm{Y}}}$$

Analogously, volume ratios (psi, ψ) or mass ratios (zeta, ζ) can also be specified. A conversion into % values is absolutely unusual here.

10.2.7 Molality (b)

In this series, the molality is more or less the exception to the rule. This specification is often important in technical chemistry. This is due to the fact that liquids are known to expand when heated. A content quantity specification as "concentration", i.e. as a unit related to the volume, is not always practical in chemical engineering due to heating or cooling processes.

The mass of the solvent, on the other hand, does not change. Therefore, one uses the quotient of the amount of substance of the sought substance (n_X) and the mass of the solvent (m_{LM}) - the molality.

$$b(\mathrm{X}) = \frac{n_{\mathrm{X}}}{m_{\mathrm{LM}}}$$

The unit of molality is mol/kg. Please note: The mass of solvent is taken as reference, **not the** mass of the final solution.

Concentration Data

120 mL = 100 g = 2 mol of substance A are mixed with 6 L = 6 kg of substance B (the solvent).

It amounts to:

Mass fraction:	$w(\mathrm{A}) = 0.0164 \rightarrow w\%(\mathrm{A}) = 1.64\ \%$
Mass concentration:	$\beta(\mathrm{A}) = 16.34$ g/L
Substance concentration:	$c(A) = 0.327$ mol/L
Volume concentration:	$\sigma(\mathrm{A}) = 0.0196 \rightarrow \sigma\%(\mathrm{A}) = 1.96\ \%$
Molality:	$b(A) = 0.333$ mol/kg

10.2.8 Small Concentrations (‰, ppm, ppb, ppt)

In analysis, especially in the field of environmental analysis and occupational health and safety, concentrations far below the percentage or per mille limits are routinely determined. The symbols or abbreviations and their meaning are listed in Table 10.5.

Table 10.5 %, ‰, ppm, ppb, ppt

Abbreviation	Designation	Fracture	Potency	Unit examples
%	Percent Parts per hundred	$^1/_{100}$ (hundredths)	10^{-2}	–
‰	Per mil Parts of a thousand	$^1/_{1000}$ (thousandths)	10^{-3}	g/kg; L/m^3, mL/L
ppm	Parts per million Parts of a million	$^1/_{1\ 000\ 000}$ (millionths)	10^{-6}	mg/kg; mL/m^3
ppb	Parts per billion Parts of a billion	$^1/_{1\ 000\ 000\ 000}$ (billionths)	10^{-9}	mg/t; μL/m^3
ppt	Parts per trillion Parts of a trillion	$^1/_{1\ 000\ 000\ 000\ 000}$ (trillionth)	10^{-12}	–

– Please note: 1 vol.% corresponds to 10,000 ppm or 1 ppm corresponds to 0.0001 vol.-%.

Figure 10.6 shows the ratios to scale. The blue square is the reference, the black = 1 % and the yellow 1 ‰ of the blue. If you increase the size of the yellow square by a factor of 1000, then the purple square ends up being 1 ppm of the original blue square. A ppb is then one thousandth of the area of the violet square and a ppt is then again one thousandth of it.

There Is Another Way to Put This

One sugar cube weighs about 3.1 grams. If dissolved in one kilogram of water, the concentration is 0.31 % or 3.1 ‰. Dissolved in a ton (1 m^3) of water, the sugar concentration is 3.1 ppm. If we take an Olympic swimming pool with 2500 m^3 of water as the "dissolving tank", the concentration is still 1.24 ppb. So to get to the order of magnitude of about 1 ppt, the sugar cube must be dissolved in 1000 Olympic-size swimming pools.

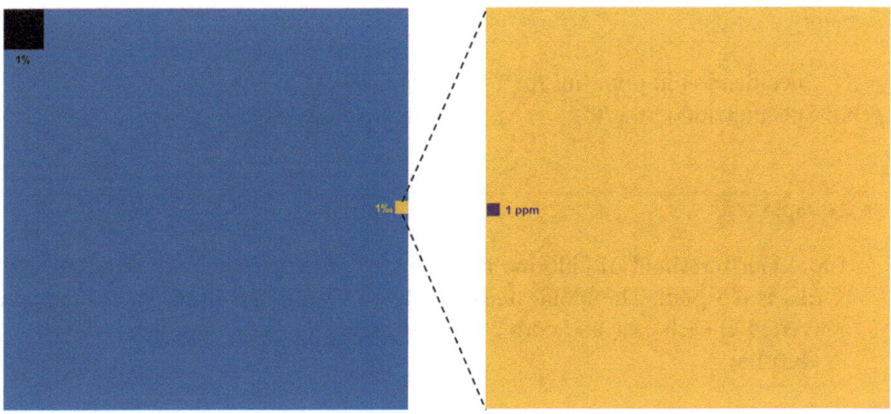

Fig. 10.6 Ratio %, ‰, ppm

10.2.8.1 Conversion Mass Concentration ↔ Volume Concentration

Particularly in the field of NBC operations, it is always necessary for the command and/or control centre to convert literature/measured values from the mass concentration (mg/m^3) to the volume concentration (mL/m^3).

This can be done with the following formulas:

Conversion β [mg/m3] to ppm [mL/m3] $$\sigma(X) = \frac{\beta(X) \cdot 22.4}{M(X)}$$

Conversion ppm [mL/m3] to β [mg/m3] $$\beta(X) = \frac{M(X)}{\sigma(X) \cdot 22.4}$$

The above conversion refers to the standard molar volume of a gas of 22.4 L mol^{-1} at 0 °C and 1.013 bar.

In order to take into account the ambient temperature, the molar volume of gas corrected to a different temperature according to the following table shall be used instead of the value "22.4":

Ambient temperature	−10 °C	−5 °C	**0 °C**	10 °C	15 °C	20 °C	25 °C	30 °C
Formula correction	21.6	22.0	**22.4**	23.2	23.6	24.1	24.5	24.9

The temperature correction factor $f_{T(cor)}$ can also be calculated temperatures that are not specified:

$$f_{T(cor)} = \frac{273 + \vartheta}{273}$$

The values corrected for the respective temperature are then obtained according to:

$$\sigma_\vartheta(X) = \sigma_0(X) \cdot f_{T(cor)} \qquad \beta_\vartheta(X) = \beta_0(X) / f_{T(cor)}$$

It means:

$\sigma(X)$: Specification in ppm (mL/m^3) M: molar mass in g/mol
$\beta(X)$: Specification in mg/m^3 ϑ: Temperature in °C

Example

The odour threshold of chlorine is 0.1 mg/m^3, the occupational exposure limit (OEL) is 0.5 ppm. The molar mass $M(Cl_2) = 70.9$ g/mol. Both values must be converted in each case and corrected to 20 °C.

Results:

- Odour threshold 0.1 mg/m_3 corresponds to 0.034 ppm = 34 ppb
- OEL = 0.5 ppm corresponds to 5.9 mg/m^3 ◀

10.2.9 ppm Values in Firefighting Operations

Especially in the context of dispersion calculations, e.g. according to MET ("Model for Effects with Toxic Gases and Vapours"), but also for the determination of threshold limits, as well as of danger and exclusion zones, a large number of different specifications are in use. These often originate from occupational health and safety, but also in part from incident considerations.

The most frequently used values (abbreviations) and their meaning can be found in Table A.14. A comparative overview is given in Fig. A.3. For the conversion of the most frequently used concentration values see Table 10.6.

Only the AEGL-2 value for an exposure time of 4 hours (AEGL-2 (4), Table 10.7) is considered in more detail, since this corresponds to the operation tolerance value (ETW = German: *Einsatztoleranzwert*) of the vfdb (Association for the Promotion of German Fire Protection = German: *Vereinigung zur Förderung des deutschen Brandschutzes e. V.*) according to vfdb Guideline 10/01.

If this value is not exceeded, no permanent damage will occur after exposure over a period of 4 hours.

All substances listed in the table as examples - with the exception of hydrogen chloride ("hydrochloric acid gas") - can be detected with the equipment of the (German) NBC exploration vehicle. Note the sometimes considerable differences between the odour threshold and the AEGL 2(4) value.

Table 10.6 Conversion of frequent concentration data

	vol.-%	ppm	ppb
Vol.-% = $\frac{10\ \text{L per m}^3}{1\ \text{cL per L}}$	1	10 000 (10^4)	10 000 000 (10^7)
ppm = $\frac{\text{mL per m}^3}{\mu\text{L per L}}$	0.000 1 (10^{-4})	1	1000 (10^3)
ppb $\frac{\mu\text{L per m}^3}{\text{nL per L}}$	0.000 000 1 (10^{-7})	0.001 (10^{-3})	1

Table 10.7 ETW values and odour threshold (examples)

Substance	UN No.	AEGL-2(4)/ETW	Odour threshold
Ammonia	1005	110 ppm	0.27 ppm
Chlorine	1017	1 ppm	0.1 ppm
Hydrogen chloride	1050	11 ppm	9 ppm
Hydrogen cyanide	1051	3.5 ppm	0.75 ppm
Acetic acid	2789	20 ppm	0.24 ppm
Ethanol	1170	3000 ppm	0.17 ppm
Methyl mercaptan	1064	14 ppm	0.002 ppm
Sulphur dioxide	1079	0.75 ppm	3.1 ppm
Carbon disulfide	1013	100 ppm	0.21 ppm
Toluene	1294	310 ppm	0.15 ppm

10.3 Reactions in Solution

Most chemical reactions take place "in solution". No matter whether inside a cell, in the laboratory or in the production plant. In the dissolved state, the particles are similarly isolated as in the gaseous state, but packed many times more densely. They are not/no longer present as aggregates or agglomerates - and thus as solid particles. This facilitates "favorable" collisions, i.e. collisions of the various reactants that also lead to a chemical reaction. In addition, the particles in the solution are in constant motion anyway. Especially in aqueous solution, chemical reactions usually proceed quickly.

10.3.1 Exchange Reactions, General

Many reactions in aqueous solution involve one of these three processes:

- Precipitation formation
- Formation of a gas
- Formation of a weak electrolyte

These reactions run according to the scheme

$$AX + BY \longrightarrow AY + BX$$

from. Therefore, these reactions are also called exchange or *metathesis* reactions (Greek *metathesis* = conversion). Since in most cases salt solutions are reacted with each other here, one can also say that the cations and anions exchange their partners.

Based on the rules for the solubility of salts (Sect. 10.1.9), the conditions for the formation of gases (Sect. 10.3.3) and the formation of weak electrolytes (Sects. 10.3.4 and 14.5), the course of these reactions can be predicted.

10.3.2 Precipitation Formation

First, let us look at an example reaction:

$$AX + BY \longrightarrow AY + BX$$

$$NaCl + AgNO_3 \longrightarrow NaNO_3 + AgCl$$

Common salt and silver nitrate react to form sodium nitrate and silver chloride. The silver ion virtually exchanges its anion for the anion brought by the sodium - and vice versa.

Since the reaction takes place in aqueous solution, we better write:

$$NaCl_{(aq)} + AgNO_{3(aq)} \rightarrow NaNO_{3(aq)} + AgCl_{(s)}$$

However, we also know that all dissolved ions are each surrounded by a hydrate shell, so it is only correct to formulate the reaction as a complete ionic equation:

$$Na^+_{(aq)} + Cl^-_{(aq)} + Ag^+_{(aq)} + NO^-_{3(aq)} \rightarrow Na^+_{(aq)} + NO^-_{3(aq)} + AgCl_{(s)}$$

At this point, at the latest, it should be obvious that the silver chloride is not present in dissolved form, but - with the loss of the hydrate shell of the individual ions - as a solid.

So let us look at the reaction equation again, but focus on the formation of the silver chloride:

$$Na^+_{(aq)} + \mathbf{Cl}^-_{(aq)} + \mathbf{Ag}^+_{(aq)} + NO^-_{3(aq)} \rightarrow Na^+_{(aq)} + NO^-_{3(aq)} + \mathbf{AgCl}_{(s)}$$

From this we can formulate what is called the **net ionic equation:**

$$Ag^+_{(aq)} + Cl^-_{(aq)} \rightarrow AgCl_{(s)}$$

From this it follows inevitably: This reaction always takes place when dissolved silver ions and dissolved chloride ions meet. Therefore, this reaction can be used as a detection for the presence of chloride ions (by adding $AgNO_3$ solution) as well as for the detection of silver ions (by adding NaCl solution).

In the laboratory, this reaction is often written in the form:

$$NaCl + AgNO_3 \rightarrow NaNO_3 + AgCl \downarrow$$

Here, the downward arrow indicates that the protruding substance is failing.

Silver chloride is one of the substances that are very difficult to dissolve. Therefore, the silver content of a silver salt solution can be determined by adding chloride ions (e.g. common salt solution or diluted hydrochloric acid) to precipitate the silver ions completely, filtering them off, drying them and weighing them out. This form of analysis is known as "gravimetric analysis" or "gravimetry".

Whether a precipitate is formed at all, i.e. the reaction in the sense of the scheme

$$AX + BY \rightarrow AY + BX$$

can be estimated using the solubility rules (Sect. 10.1.9).

If no reaction takes place, as, for example, when common salt solution and potassium nitrate solution are added together, one often writes:

$$NaCl_{(aq)} + KNO_{3(aq)} \not\rightarrow$$

The crossed-out reaction arrow indicates that no reaction is taking place.

However, metathesis reactions are not limited to the presence of aqueous solutions. Solid reactions can even take place. If barium hydroxide containing water of crystallization is mixed with ammonium nitrate, the two solids react with each other:

$$Ba(OH)_{2(s)} + 2\,NH_4NO_{3(s)} \rightarrow Ba(NO_3)_{2(s)} + 2\,NH_4OH_{(\ell)}$$

10.3.3 Formation of Gases

Many reactions in which gases are formed are also exchange reactions. In the simpler case, a less stable acid or alkali is formed first, which then immediately decomposes again or has only a rather low solubility per se.

A typical example is the reaction of sodium carbonate solution with hydrochloric acid:

$$AX + BY \longrightarrow AY + BX$$

$$Na_2CO_3 + 2\,HCl \longrightarrow 2\,NaCl + H_2CO_3$$

The carbonic acid formed immediately decomposes into water and carbon dioxide:

$$H_2CO_{3(aq)} \longrightarrow H_2O_{(\ell)} + CO_{2(g)}$$

As a net ionic equation, it can be written:

$$CO^{2-}_{3(aq)} + 2\,H^+_{(aq)} \rightarrow H_2CO_3 \rightarrow H_2O + CO_2\uparrow$$

The following gases can be produced in this way:

- CO_2 from carbonates (CO_3^{2-})or hydrogen carbonates (HCO_3^-) with acids
- H_2S from sulfides (S^{2-}) or hydrogen sulfides (HS^-) with acids: $H_2S\uparrow$
- SO_2 from sulfites (SO_3^{2-}) or hydrogen sulfites (HSO_3^-) with acids. The sulphurous acid formed decomposes according to: $H_2SO_3 \rightarrow H_2O + SO_2\uparrow$
- NH_3 from ammonium salts and lye ($NH_4OH \rightarrow H_2O + NH_3\uparrow$)

Here, too, the reaction is not bound to aqueous solutions. For example, gaseous, anhydrous HCl gas can be produced by reacting can be produced. HBr, for example, can also be obtained in a similar way.

$$NaCl_{(s)} + H_2SO_{4(\ell)} \rightarrow NaHSO_{4(s)} + HCl_{(g)}$$

10.3.4 Formation of Weak Electrolytes

An electrolyte is first of all a substance that increases the conductivity of electric current when dissolved in water. This can be salts, acids or alkalis. Sugar dissolved in water does not increase the conductivity and is therefore not an electrolyte.

So-called "strong electrolytes" and "weak electrolytes" have been defined. Strong electrolytes are practically completely present in the form of hydrated ions. Weak electrolytes, on the other hand, are largely dissolved as molecules and only a small proportion of them have decomposed into hydrated ions. (For more details see Chap. 14).

Now, if there is a possibility of a weak electrolyte forming in a reaction, this will happen. As an example, let us consider the reaction of sodium acetate solution with dilute hydrochloric acid:

$$AX + BY \longrightarrow AY + BX$$

$$Na\,CH_3COO + HCl \longrightarrow Na\,Cl + CH_3COO\,H$$

The ion equation is:

$$Na^+_{(aq)} + CH_3COO^-_{(aq)} + H^+_{(aq)} + Cl^-_{(aq)} \rightarrow Na^+_{(aq)} + Cl^-_{(aq)} + CH_3COOH_{(\ell)}$$

And from this the net ion equation

$$CH_3COO^-_{(aq)} + H^+_{(aq)} \rightarrow CH_3COOH_{(\ell)}$$

Typical weak electrolytes are: Acetate CH_3COO^- (from acetic acid, CH_3COOH), nitrite NO_2^- (from nitrous acid, HNO_2), fluoride F^- (from hydrofluoric acid, HF), sulfide S^{2-} (from hydrogen sulfide, H_2S), and ammonium NH_4^+ (from ammonia, NH_3).

- If a sparingly soluble salt can be formed, this reaction proceeds preferentially to the formation of a weak electrolyte.

10.4 Chemical Reactions During the Dissolving Process

Sometimes substances react with the solvent in the sense of a chemical reaction. Here, too, the term "dissolving" is often used incorrectly.

10.4.1 Reactions with Acids

When magnesium reacts with dilute hydrochloric acid, the metal seems to disappear. It "dissolves" with the formation of gas. Similarly, calcium hydroxide, which is sparingly soluble in water, "dissolves" readily in dilute hydrochloric acid. The same applies to copper oxide, for example.

Even if one colloquially speaks of a "dissolving process", no dissolving in the actual sense takes place here. In all cases, it is a chemical reaction in which the corresponding salts are formed.

10.4.2 "Solutions" of Gases

Some gases that dissolve extremely well in water are not solutions of these gases in the sense discussed so far. It is rather the case that these gases react with the water during the dissolving process, form ions and these ions are then dissolved (hydrated).

Typical examples are the aqueous solutions of HCl, HBr and HI. H^+ ions and X^- ions are formed. The gas itself is practically not present in the solution, but its reaction products with water are. The same applies to the basic solution of ammonia in water.

10.4.3 Hydrolysis

If electron pair bonds of substances are dissolved and newly linked during the reaction with water, this is called hydrolysis. Here, too, no dissolving process (formation of a hydrate shell) takes place. It is a chemical reaction.

Example

$$P_2O_5 + 3\,H_2O \rightarrow 2\,H_3PO_4$$
$$PCl_5 + 4\,H_2O \rightarrow H_3PO_4 + 5\,HCl$$ ◀

Double Salts, Complexes and Dispersions 11

Even if no chemical reaction takes place initially, new substances can still be formed when salt solutions are mixed: Double salts. In other cases, stable compounds are formed with the formation of electron pair bonds: Complexes. If "large" particles are present in the solution, these are dispersions. These groups should not be confused with each other and will be briefly addressed.

11.1 Double Salts

If an aqueous solution of aluminium sulphate is mixed with a solution of potassium sulphate, no reaction takes place.

$$Al_2(SO_4)_{3(aq)} + K_2SO_{4(aq)} \nrightarrow$$

When the solution is evaporated, one would expect the less soluble of the two salts to crystallize first. But this is not the case. Instead, a new salt is obtained: potassium aluminium sulphate, $KAl(SO4)_2$ ("alum"), more precisely: $KAl(SO_4)_2 \cdot 12\ H_2O$.

As a solid, this salt has different properties than the two solid starting materials. In solution, however, it behaves in exactly the same way as a mixture of the solutions of the starting materials. In these cases, one speaks of double salts. The ionic equation of the dissolution process makes this clear:

$$KAl(SO_4)_2 + n\,H_2O \rightarrow K^+{}_{(aq)} + Al^{3+}{}_{(aq)} + 2\ SO_4{}^{2-}{}_{(aq)}.$$

The solution contains potassium ions, aluminium ions and sulphate ions – just like the mixture of a potassium sulphate solution with an aluminium sulphate solution.

Typical double salts are the "alums": potassium aluminium sulphate ("alum", "potassium alum", $KAl(SO_4)_2$), ammonium aluminium sulphate ("ammonium alum", $(NH_4)Al(SO_4)_2$), potassium chromium (III) sulphate ("chromium alum", $KCr(SO_4)_2$). They are used, for example, in tanning and fabric dyeing. Potassium

T. Schmiermund, *The Chemistry Knowledge for Firefighters*,
https://doi.org/10.1007/978-3-662-64423-2_11

aluminium sulphate is used as a so-called "blood-stopping stick" (for wet shaving) and as a "deodorant crystal". All alums contain 12 mol of water of crystallization.

11.2 Complex Salts

Complex salts do **not** behave in solution like mixtures of the solutions of the individual components. Normally, for example, silver ions react with chloride ions to form poorly soluble silver chloride: $Ag^{+}{}_{(aq)} + Cl^{-}{}_{(aq)} \rightarrow AgCl\downarrow$. After addition of ammonia solution this reaction no longer takes place. The silver ions have formed a complex with the dissolved ammonia which prevents this precipitation reaction:

$$Ag^{+}_{(aq)} + 2\,NH_3 \rightarrow [Ag(NH_3)_2]^{+}$$
$$[Ag(NH_3)_2]^{+} + Cl^{-} \nrightarrow$$

Iron (II) ions react with sulfide ions to form insoluble $Fe^{2+} + S^{2-} \rightarrow FeS\downarrow$. If cyanide ions are present in the solution, a complex compound is formed which no longer interacts with the sulfide ions: $[Fe(CN)_6]^{4-}$.

These observations can be explained by the fact that the complexes formed are so stable that they do not exist as single ions: The complex does not dissociate.

11.2.1 Important Complexes/Complexing Agents

Without us really realising it, we use complexing agents in detergents, dishwashing products and rinse aids. Complexing agents are used here to make the water "softer". Easily soluble Ca^{2+} or Mg^{2+} complexes are formed, which keep these ions as easily soluble compounds in the washing suds throughout the entire cleaning process. In this way, greying of the laundry, but also limescale stains on glasses, can be effectively prevented.

However, the most important complex for us might be the "heme" in hemoglobin. Hemoglobin consists of 4 heme units that are bound to a globin molecule (globin is a protein). In the center of the heme molecule is an Fe^{2+} ion. Four of the six coordination sites of the iron are linked to the heme, the fifth to another protein. Only the sixth coordination site can take up O_2 and thus provide transport to the cells.

Since the O_2 molecule is complexed, other molecules or ions capable of forming complexes can also occupy this ligand site. CO and CN^{-}, for example, pose a particular danger, as relatively stable complexes are formed here. As a result, oxygen transport can be restricted to such an extent that an extremely severe undersupply occurs, which leads to death.

11.2.2 Complexions

Complex compounds, also known as coordination compounds, consist of a central building block (coordination centre, an atom or ion) around which so-called ligands (molecules or ions) are arranged. The central atom is surrounded by a ligand shell, so to speak.

The actual complex ion is written in square brackets, the charge outside the brackets. Here, the charge of the complex is calculated from the sum of the charges of all the particles of which it consists. The number of ligands that a coordination center can bind is called the coordination number (C.N.) or ligancy.

Central ion	Ligand	Complex	Coordination number
Al^{3+}	F^{-}	$[AlF_6]^{3-}$	6
Cr^{3+}	NH_3	$[Cr(NH_3)_6]^{3+}$	6
Fe^{3+}	H_2O	$[Fe(H_2O)_6]^{3+}$	6
Cu^{2+}	NH_3	$[Cu(NH_3)_4]^{2+}$	4
Ag^{+}	CN^{-}	$[Ag(CN)_2]^{-}$	2

The complex compounds are especially important for compounds of the metals of the d- and the f-series. An entire branch of chemistry, "complex chemistry", is devoted to complex compounds. Some complexes are important as technical catalysts.

11.2.3 Structure of Complexes

The spatial structure of a complex depends strongly on the coordination number. Since the odd coordination numbers (3, 5, 7, 9) and the C.N. 8 are rare, they are not addressed here. It is also not possible here to deal with the different spatial arrangements (isomerism).

- Coordination number 2 forms linearly arranged complexes, e.g. $[Ag(CN)_2]^{-}$ or $[AuCl_2]^{-}$
- Coordination number 4:
  (a) tetrahedral complexes, e.g. $[BeF_4]^{-}$, $[FeCl_4]^{-}$, $[Co(SCN]_4]^{2-}$
  (b) planar-square complexes, e.g. $[Pt(NH_3)_2Cl_2]$ or $[Cu(NH_3)_4]^{2+}$; very common with Rh^{+}-, Ir^{+}-, Pt^{2+}-, Pd^{2+}-, Au^{2+} – and Cu^{2+}-compounds
- Coordination number 6 is found very often in transition metals and forms an octahedral structure, examples are Cr^{3+}-, Co^{3+}-, Fe^{3+}- or Sn^{4+}- compounds.

Figure 11.1 shows these structures. "M" is the coordination center, "L" are the ligands. The electron pair bonds are shown with arrows, the spatial shape is highlighted with dashed lines.

Fig. 11.1 Spatial structure of complexes

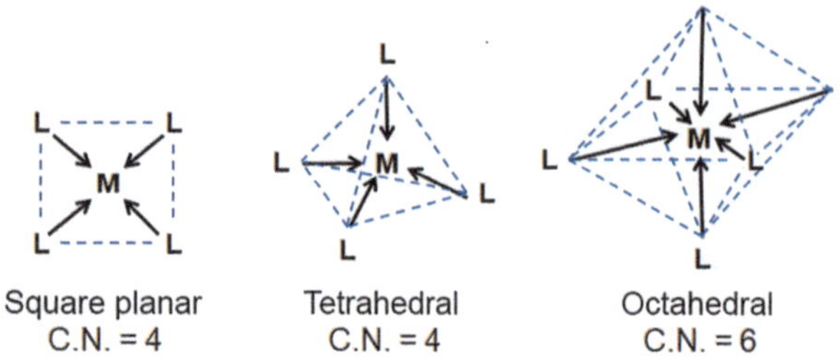

11.2.4 Denticity of the Ligands

Ligands such as NH_3, H_2O, Cl^-, F^-, CN^- can only form one bond with the central atom. To emphasize this, one also speaks of "monodentate" ligands. Other ligands can occupy more than one coordination site. They can coordinate 2, 3, 4, 5 or even 6 positions of the central atom, i.e. be "bidentate", "tridentate", etc.

Examples of bidentate ligands are the oxalate ion (abbreviated "ox", $^-OOC{-}COO^-$) or ethylenediamine (abbreviated "en", $H_2N{-}CH_2{-}CH_2{-}NH_2$). Hexadentate ligands, such as ethylenediamine-tetraacetate (EDTA) are even able to bind the alkaline earth metal ions – which generally have an extremely low tendency to form complexes – in extremely stable complexes.

Bidentate and multidentate ligands are also called chelating agents (Greek *cheles* = crab claws). They occupy the coordination positions as if a crab claw had grasped them.

11.2.5 Stability of Complexes

The ligands must have at least one free electron pair to be effective as such; they are Lewis bases (compare Sect. 12.3.2). The central atom is the corresponding Lewis acid. The free electron pair of the ligand now forms an electron pair bond to the central atom. In this case, however, the binding electron pair originates only from the ligand. To distinguish it from the classical electron pair bond (= one electron from each of the two bonding partners), this special form of the EPB is also called a "dative bond".

Different ligands form different strength bonds with the central atom. Therefore, some complexes are more stable than others. By adding the "stronger" ligand, the "weaker" ligand is displaced from the complex; the more stable complex is formed. In general, complexes with bidentate or polydentate ligands are considerably more stable than complexes with monodentate ligands.

More detailed considerations of the stability and the binding theory of complexes would go beyond the scope of this book. Here we refer to the literature appendix.

Table 11.1 Designations of ligands (examples)

Anionic ligands		Nonionic ligands	
F^-	Fluorido	H_2O	Aqua
Cl^-	Chlorido	NH_3	Ammine
OH^-	Hydroxido	CO	Carbonyl
CN^-	Cyanido		

11.2.6 Colourfulness of Complexes

Most complex compounds are coloured. This is used, for example, in analysis. For example, copper can be detected as a deep blue $[Cu(NH_3)_4]^{2+}$ complex or iron as a red $[Fe(SCN)_6]^{3+}$ complex even in relatively low concentrations. The coloration of some salts containing water of crystallization is just as much due to complex bonds as the coloration of many aqueous solutions of d-metal salts.

11.2.7 Nomenclature of Complex Compounds

11.2.7.1 Ligands

When naming complexes, the number and name of the ligands are given first. Here, the number – as usual in chemistry – is designated by Greek numeral words. Then follows the name of the central atom. The oxidation number of the central atom is indicated at the end by Roman numerals in brackets.

In the case of negatively charged ligands, the fact that it is a ligand is indicated by an appended "o" to the root name of the ion (compare Table 11.1).

11.2.7.2 Cationic Complexes

Analogous to "normal" salts, the cation is always designated first, followed by the anion in the case of complex salts.

Cationic complex				–	Anion
Number of ligands	Ligand	Central ion	Oxidation number	–	Anion
$[Ag(NH_3)_2]Cl$					
Di	ammin	silver	(I)	–	chloride
$[Co(H_2O)_6]Cl_2$					
Hexa	aqua	cobalt	(II)	–	chloride
$[Cr(H_2O)_6]Cl_3$					
Hexa	aqua	chrom	(III)	–	chloride

11.2.7.3 Anionic Complexes

For anionic complexes, the name of the central particle ends in "-ate". In some cases, the name of the central particle is derived from the Latin name (e.g.: Ag = argentate; Au = aurate, Fe = ferrate). The rule "cation before anion" remains.

Cation	–	Anionic complex			
Cation	–	Number of Ligands	Ligand	Central ion	Oxidation number
$Na[Ag(CN)_2]$					
Sodium	–	di	cyanido	argentate	(I)
$K_2[CoCl_4]$					
Potassium	–	tetra	chlorido	cobaltate	(II)
$Na_2[PtCl_4]$					
Sodium	–	tetra	chlorido	platinate	(II)

11.2.8 Water Hardness and Extinguishing Water Supply

The so-called "hardness formers" are Ca^{2+} and Mg^{2+} ions. The more of these ions are dissolved in the water, the "harder" the water is. The majority of water hardness is caused by dissolving lime by adding CO2 as easily soluble hydrogen carbonate:

$$CaCO_3 + CO_2 + H_2O \rightarrow Ca(HCO_3)_2$$

To indicate degrees of hardness, the Ca^{2+} and Mg^{2+} content is converted to CaO. 1 °dH (=1 degree of German hardness) corresponds to a content of 10 mg CaO per litre of water. Recently – based on the Washing and Cleaning Agents Act (Wasch- und Reinigungsmittelgesetz, WRMG) – the content is stated as "calcium carbonate in mmol L^{-1}". Conversion: 1°dH = 5.6 mmol L^{-1}. This is used to classify the following hardness ranges:

°dH	Hardness range	mmol $CaCO_3$ per L
<8.4	Soft	<1.5
8.4...14	Medium	1.5...2.5
>14	Hard	>2.5

A previously specified hardness range "very hard" (>21.3 °dH equivalent to >3.8 mmol L^{-1}) is no longer common following adaptation to European standards.

Total, Temporary and Permanent Hardness

The total hardness of the water is made up of the temporary hardness and the permanent hardness. The temporary hardness – also known as carbonate hardness –

"disappears" when heated. The easily soluble hydrogen carbonates decompose into the poorly soluble carbonates with CO_2 splitting off, which precipitate out of the water and form the so-called "scale".

$$Mg(HCO_3)_2 \overset{\text{Heat}}{\rightarrow} MgCO_3 \downarrow + CO_2 \uparrow + H_2O$$

Permanent hardness is the "residue" of dissolved alkaline earth metal ions, which is predominantly present as chloride and sulphate.

Extinguishing Water Supply

In areas with very hard water, lime deposits, similar to scale, can already occur in the pipelines for the water supply. This process, known as incrustation, can cause the pipe cross-sections to become severely restricted over time. The process itself is gradual and often only becomes noticeable over many years. What may still be annoying in the case of a domestic water pipe may jeopardise the success of extinguishing in the case of a hydrant pipe required for fire-fighting purposes, as the necessary quantities of extinguishing water can no longer be drawn off. In consultation with the local water supply company, the flow rates of the hydrant lines should therefore be checked regularly, if necessary. This applies in particular to pipes leading to farms, etc., or other branch pipes that branch off from the ring main system.

11.3 Disperse Systems

At the end of the chapter "Solutions" we come back to the inhomogeneous mixtures of substances (compare Chap. 2). Between the real solutions (molecules or ions surrounded by a solvate shell, so-called "molecular disperse systems") and the sometimes very easily recognizable suspensions or emulsions (e.g. noodle soup or milk), there are still some "intermediate forms" which we would like to briefly illuminate.

First of all, dispersions are mixed phases of two or more individual substances. In one phase, which is usually liquid or gaseous, one or more other substances are finely dispersed ("dispersed"). In this case, the dispersed substance is distributed in the dispersing agent. Dispersions are thus also an umbrella term for emulsions, suspensions and aerosols.

Classification of Dispersions

Dispersions can be conveniently classified according to the particle size of the dispersed ("distributed") substance:

• Coarse dispersed systems	Particle size 10 μm to 1 mm
• Finely dispersed systems	Particle size 100 nm to 10 μm
• Colloid dispersed systems	Particle size 1...100 nm
• Molecular dispersed systems	Real solutions

In the following, we will only briefly discuss the finely dispersed and the colloid dispersed systems. The coarse dispersed systems can already be easily recognized as suspensions or emulsions and do not need to be discussed further here.

11.3.1 Finely Dispersed Systems

- Probably the most common application of finely dispersed systems is in so-called dispersion paints. These are paints that contain dispersed plastic particles (100 nm to 5 μm in size) as a binder. After drying, the plastic component holds the fillers and pigments to the object, providing a hard-wearing coating.
- Another group are the dispersion adhesives. These are also plastic dispersions in the broadest sense, which form an adhesive film through evaporation of the dispersion agent (usually water). With this, a wide variety of materials (wood, cardboard, leather, paper, wallpaper, felt, cork, ...) can be glued. The best known example is probably the classic wood glue.

11.3.2 Colloid Disperse Systems

These are collections of atoms in the order of $10^3...10^9$ atoms. Due to their size of 1...100 nm, they can neither be defined as clearly heterogeneous nor as clearly homogeneous. Colloids are not filterable, so that a separation into dispersed substance and dispersant is difficult.

Important colloids are e.g. the red gold ruby glass (also 'cranberry glass'), purple of Cassius (for dyeing glass and porcelain) and silver colloids (antimicrobial effect, water disinfection).

11.3.3 Tyndall Effect

The Tyndall[1] effect describes the scattering of light by dispersed particles whose size lies in the range of light wavelengths. In particular, when the particles are dispersed in gas (air) or liquid (usually water) (Fig. 11.2a).

Light hits the finely dispersed or colloidal particles and is scattered to the side. Thus, for example, the course of a laser beam can be made visible by "blowing" on it

[1] John Tyndall, 1820–1893, British naturalist.

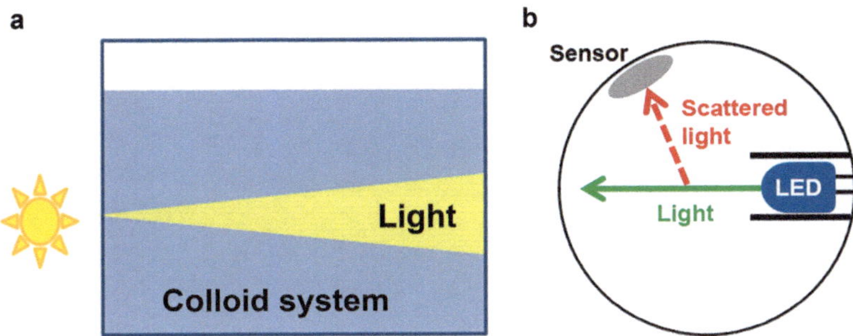

Fig. 11.2 (**a**) Tyndall effect (schematic), (**b**) optical smoke detector

with smoke or fog. The clusters of rays from the sun in haze, dust or in faint fog are also based on this effect.

Optical smoke detectors exploit the Tyndall effect: Smoke particles scatter light from a small light source onto a light-sensitive sensor. Smoke-free air does not cause scattering, so the sensor does not receive a signal. Smoke, on the other hand, scatters part of the light to the sensor, which then triggers an alarm at a defined threshold value (Fig. 11.2b).

Part VI

Acids and Alkalis

"An acid is when it tastes sour." This statement is - in principle - correct at first. But seriously, who wants to taste every spilled substance in the field? What if the substance is toxic? Does a dilute acid taste less sour than a concentrated one? And besides, what do alkalis taste like then? Sweet? Bitter?

In this chapter we will take a closer look at acids and bases and their properties. We will learn about their strength, deal with strong and weak electrolytes, neutralization and the pH value.

12 Acid-Base Theories

Some acids were already known in antiquity and the Middle Ages: "vitriol oil" (today: sulphuric acid), "aqua fortis" (nitric acid), "aqua regia" (a mixture of hydrochloric acid and nitric acid), formic acid and acetic acid, to name but a few.

So the alchemists and the early chemists were looking for common ground. Lavoisier assumed that all acids must contain oxygen in the molecule, thus giving oxygen its name. He supported his assumption by the fact that eventually all solutions of non-metal oxides react acidically. Today, however, we know that there are also acids that do not carry oxygen in their molecule. Davy[1] already regarded hydrogen as the essential characteristic of acids and was better off with his theory (as we know today). von Liebig[2] defined acids as hydrogen compounds that can be converted into salts by metals. All three, however, gave no definition of bases.

The Terms "Base" and "Alkali

A base is virtually the "counterpart" to an acid. The solution of a base is called lye. Alkali metals react with water to form a base (or lye). The solution then reacts "basic" or "alkaline".

In contrast, an acid is always called an acid and always reacts acidically.

12.1 Definition According to Arrhenius

Arrhenius[3] defined acids in 1887 as "substances that release H^+ ions when dissolved in water". Bases (lyes) are substances that "give off OH^- ions when dissolved in water." "When acids and bases are combined, water is formed."

[1] Humphry Davy, 1778–1829, English chemist.

[2] Justus von Liebig, 1803–1873, German chemist.

[3] Svante August Arrhenius, 1859–1927, Swedish physicist and chemist, Nobel Prize in Chemistry 1903.

T. Schmiermund, *The Chemistry Knowledge for Firefighters*,
https://doi.org/10.1007/978-3-662-64423-2_12

Arrhenius considers the dissolution process and the ions that are formed. Since one molecule of acid or base becomes two particles, this is also referred to as dissociation. The original particles dissociate ("decay") to new (dissolved) particles.

12.1.1 Acids

- Acids are substances that release H^+ ions when dissolved in water. They dissociate into H^+ and acid residue ions.

$$\text{Acid molecule} \rightleftharpoons \text{Hydrogen ion} + \text{Acid residue ion}$$

$$HA \rightleftharpoons H^+ + A^-$$

Characteristic is the formation of a positively charged hydrogen ion and a negatively charged anion, the so-called acid residue ion. The double arrow ($\rightleftharpoons$) indicates that the reaction proceeds in both directions and an equilibrium is reached.

Examples

Hydrochloric acid	$HCl \rightleftharpoons H^+_{(aq)} + Cl^-_{(aq)}$
Acetic acid	$H_3C{-}COOH \rightleftharpoons H^+_{(aq)} + H_3C{-}COO^-_{(aq)}$
Hydrogen cyanide	$HCN \rightleftharpoons H^+_{(aq)} + CN^-_{(aq)}$

◀

12.1.2 Bases

- Bases are substances that release OH^- ions when dissolved in water. They dissociate into OH^- ions and (metal) cations.

The most important inorganic bases are metal hydroxides.

$$\text{Metal hydroxide} \rightleftharpoons \text{Cation} + \text{Hydroxyde ion}$$

$$M(OH) \rightleftharpoons M^+ + OH^-$$

Characteristic is the formation of a hydroxide ion (OH-) and a cation, which is then called base residue, rarely base residue ion.

Example

Sodium hydroxide	$NaOH \rightleftharpoons Na^{+}{}_{(aq)} + OH^{-}{}_{(aq)}$
Barium hydroxide	$Ba(OH)_2 \rightleftharpoons Ba^{2+}{}_{(aq)} + 2\ OH^{-}{}_{(aq)}$
Aluminium hydroxide	$Al(OH)_3 \rightleftharpoons Al^{3+}{}_{(aq)} + 3\ OH^{-}{}_{(aq)}$

◀

12.1.3 Neutralization

- When acids and bases react, salt and water are formed.

$$\text{Acid} + \text{Base} \rightarrow \text{Salt} + \text{Water}$$

$$HA + M(OH) \rightarrow MA + H{-}OH$$

Example

$$HCl + NaOH \rightarrow NaCl(aq) + H_2O.$$

$$2\ H_3C{-}COOH + Ba(OH)_2 \rightarrow Ba(CH_3COO)_{2(aq)} + 2\ H_2O.$$

Let us now consider the first example as a complete ionic equation

$$H^{+}_{(aq)} + Cl^{-}_{(aq)} + Na^{+}_{(aq)} + HO^{-}_{(aq)} \rightarrow Na^{+}_{(aq)} + Cl^{-}_{(aq)} + H_2O$$

and transform into the net ionic equation

$$H^{+}_{(aq)} + OH^{-}_{(aq)} \rightarrow H_2O$$

it can be seen that the actual reaction is the formation of water molecules. ◀

- Neutralization is the formation of water molecules from H^{+} and OH^{-} ions.

12.1.4 Hydrolysis

If salt and water are formed from an acid and a base, then it can be assumed that when a salt is dissolved (at least in part) the reverse reaction also occurs: Salt and water become acid and base.

$$\text{Salt} + \text{Water} \rightarrow \text{Acid} + \text{Base}$$

$$MA + H{-}OH \rightarrow HA + M(OH)$$

$$Na_2CO_3 + H{-}OH \rightarrow NaHCO_{3(aq)} + NaOH_{(aq)}$$

$$KCl + H{-}OH \rightarrow HCl_{(aq)} + KOH_{(aq)}$$

It follows from this purely formal consideration:

- Hydrolysis is the reverse of neutralization.

The effects of hydrolysis on the pH value of salt solutions are explained in Sect. 14.4.

12.1.5 Salts

According to Arrhenius, salts can be distinguished into:

- Neutral salts that contain neither H^+ nor OH^- ions: $NaCl$, $AlCl_3$, $CuSO_4$
- Acidic salts that can still release $\mathbf{H^+}$ ions: $K\mathbf{H}SO_4$, $Na\mathbf{H}CO_3$, $Na\mathbf{H_2}PO_4$
- Basic salts that can still release $\mathbf{OH^-}$ ions: $Ca(\mathbf{OH})Cl$, $Mg(\mathbf{OH})Br$, $Na[Al(\mathbf{OH})_4]$

12.1.6 Limitations of the Model and Outlook

The acid-base model according to Arrhenius is limited to water as solvent. Hydrated hydrogen ions ($H^+_{(aq)}$) must be present. However, more detailed observations show that "in reality" H_3O^+ ions (old: hydronium ions, new: oxonium ions) are present and that water is thus an important reaction partner of the acids.

The basic behaviour of substances that do not contain OH groups cannot be explained. It remains unclear in this model, for example, why an aqueous ammonia solution reacts alkaline.

The process of neutralization and the formation of acidic or basic salt solutions (which this model also explains) will be discussed in more detail in Sect. 14.4.

12.2 Definition According to Brønsted and Lowry

In 1923, Brønsted[4] and Lowry[5] independently published an extended acid-base theory. Both assume that an acid reacts with a partner (the so-called base). The acid/base definitions therefore differ somewhat from Arrhenius'.

12.2.1 Acids and Bases

Acids are particles that can release protons (H^+). They are therefore also called proton donors.

All Arrhenius acids are therefore also Brønsted acids. Furthermore, all compounds that are capable of giving off an H^+ ion are also Brønsted acids. Thus, for example, the acidic salts of Arrhenius:

$$H_2PO_4^- \rightarrow H^+ + HPO_4^{2-}$$

Bases are particles that can accept protons (H^+). They are therefore also called proton acceptors.

All acid residue ions (Arrhenius) are therefore Brønsted bases:

$$Br^- + H^+ \rightarrow HBr$$

In the Brønsted-Lowry definition, free protons (H^+) do not exist because they are not stable in solution. The reaction of a substance as an acid necessarily requires the existence of a reaction partner, the base. The base (2) takes up the proton of the acid (1) and thus becomes a positively charged particle: The acid (2); the acid anion thus becomes the base (1). So acid and base always belong together as a pair. This is also called ***corresponding acid-base pairs***.

Since all acid-base reactions according to this model represent proton transfer reactions, they are also called protolysis.

Example

$$\text{Acid(1)} + \text{Base(2)} \rightleftharpoons \text{Acid(2)} + \text{Base(1)}$$

$$HX + Y \rightleftharpoons HY^+ + X^-$$

$$HCl + H_2O \rightleftharpoons H_3O^+ + Cl^-$$

[4]Johannes Nicolaus Brønsted, 1879–1947, Danish chemist.

[5]Thomas Martin Lowry, 1874–1936, English chemist.

- Acid (1) is the HCl molecule. It can donate one proton: $HCl \rightarrow H^+ + Cl^-$
- Base (2) is the water because it can hold a proton: $H_2O + H^+ \rightarrow H_3O^+$
- Acid (2) is the H_3O^+ ion. It can donate one proton: $H_3O^+ \rightarrow H_2O + H^+$
- Base (1) is the chloride ion because it can accept a proton: $Cl^- + H^+ \rightarrow HCl$ ◀

We therefore have ***two*** corresponding acid-base pairs:

(a) Acid: $HCl \leftrightarrow$ corresponding base: Cl^-
(b) Base: $H_2O \leftrightarrow$ corresponding acid: H_3O^+

The formation of the oxonium ion H_3O^+ (older name: hydronium ion) is characteristic for acid-base reactions in aqueous solution.

In addition, acid-base pairs can be formulated for other solvents. For example, the acid-base reaction in the dissolution of hydrogen chloride in methanol:

$$\text{Acid}(1) + \text{Base}(2) \rightleftharpoons \text{Acid}(2) + \text{Base}(1)$$

$$HCl_{(g)} + H_3C{-}OH_{(\ell)} \rightleftharpoons H_3C{-}OH_2{}^+{}_{(lq)} + Cl^-{}_{(lq)}$$

The index *(lq)* denotes the liquefied particles.

Since the corresponding base of an acid is formed by releasing ***a*** proton from it, a distinction is also made according to:

- Neutral acids, e.g. HCl, H_2SO_4, HNO_3, $H_3C{-}COOH$
- Cationic acids, e.g. NH_4^+
- Anionic acids, e.g. HSO_4^-, $H_2PO_4^-$, HPO_4^{2-}

Thus, during protolysis, the neutral acid H_2SO_4 becomes the corresponding base HSO_4^-, which then protolyses as anionic acid to its corresponding base SO_4^{2-}.

$$H_2SO_4 + H_2O \rightleftharpoons HSO_4{}^- + H_3O^+$$

$$HSO_4{}^- + H_2O \rightleftharpoons SO_4{}^{2-} + H_3O^+$$

Some important acid-base pairs are listed in Table 12.1.

12.2.2 Salts

In this model, all substances that form ion lattices in the solid state are referred to as salts. Besides NH_4Cl, NaCl, Na_2SO_4, KBr, for example, also NaOH and $KHSO_4$. But: During protolysis in water, the base OH^- is released from the salt NaOH, and the acid HSO_4^- is released from the salt $KHSO_4$.

Table 12.1 Acid-base pairs according to Brønsted-Lowry

Acid	Base	Acid	Base
HCl	Cl^-	**H_2S**	**HS^-**
H_2SO_4	HSO_4^-	**H_2PO_4**	**HPO_4^{2-}**
H_3O^+	**H_2O**	**NH_4^+**	**NH_3**
HNO_3	**NO_3^-**	**HCN**	**CN^-**
HSO_4	SO_4^{2-}	**HPO_4^{2-}**	PO_4^{3-}
H_3PO_4	**$H_2PO_4^-$**	**HS^-**	**S^{2-}**
HF	F^-	**H_2O**	**OH^-**
H_2CO_3	**HCO_3^-**	**NH_3**	**NH_2^-**

12.2.3 Amphoteric Substances

If a substance can react both as an acid (= release of protons) and as a base (= absorption of protons), then this substance is amphoteric; it is a so-called ampholyte. The best known substance with amphoteric properties is probably water:

$$H_2O + H_2O \rightleftharpoons H_3O^+ + OH^-$$

Two corresponding acid-base pairs are also present here:

(a) Acid: $H_2O \leftrightarrow$ corresponding base: OH^-
(b) Base: $H_2O \leftrightarrow$ corresponding acid: H_3O^+

Other amphoteric compounds include: $HSO_4^-, H_2PO_4^-, HPO_4^{2-}, Al(OH)_3$.

12.2.4 Improvements Against Arrhenius

In addition to being transferable to a variety of solvents, the Brønsted-Lowry theory can also explain the reactions of ammonia:

$$\begin{aligned}
&\text{Acid}(1) + \text{Base}(2) \rightleftharpoons \text{Acid}(2) + \text{Base}(1)\\
&H_2O + NH_3 \rightleftharpoons NH_4^+ + OH^-\\
&HCl_{(g)} + NH_{3(g)} \rightleftharpoons NH_4^+Cl^-{}_{(s)}
\end{aligned}$$

- The base NH_3 forms ammonium and hydroxide ions in the first equation with the acid H_2O. The assumption of the intermediate stage "$NH_4OH_{(aq)}$" can be dispensed with.
- The second reaction equation shows the reaction of the two gases HCl and NH_3 to the solid salt NH_4Cl (ammonium chloride). This could not be explained by Arrhenius' theory, because no water is present in the reaction of the two gases.

12.2.5 Limitations of the Model and Outlook

The Brønsted-Lowry model is limited to compounds that contain hydrogen atoms in the molecule (acids) or can accept protons (bases).

The process of autoprotolysis of water, the connection with the pH value and the concept of acid/base strength will be considered in more detail in Sect. 14.5.

12.3 Definition According to Lewis

In 1923, Lewis[6] introduced a new acid-base concept based not on protons but on electron pairs. For this purpose, he used the Lewis notation that we are already familiar with.

12.3.1 Lewis Acids

- Lewis acids are substances that can attach an electron pair. They are electron pair acceptors.

Lewis acids include:

- Substances that have an octet gap; e.g.: BF_3, $AlCl_3$, SO_2, $B(OH)_3$
- Substances that have "unsaturated coordination"; e.g.: PF_5, $FeBr_2$, $TiCl_4$, Ag^+.
- Cations that can act as a central atom in complex compounds; e.g.: $\underline{Fe}^0$ in $[Fe(CO)_5]$, Cu^{2+} in $[Cu(NH_3)_4]^{2+}$

12.3.2 Lewis Bases

- Lewis bases are substances that can provide electron pairs. They are electron pair donors.

Lewis bases include:

- Substances with free electron pairs with large expansion; e.g.: $H{-}\overline{\underline{O}}{-}H$, $|NH_3$
- Molecules with double/triple bonds
- Substances with a free pair of electrons that can occur as ligands in complexes; e.g.: CN^-, H_2O, NH_3, F^-, CO All bases according to the Brønsted-Lowry definition.

[6] Gilbert N. Lewis, 1875–1946, American chemist.

12.3.3 Acid-Base Reaction According to Lewis

$$\text{Lewis-Acid} + \text{Lewis-Base} \rightleftharpoons \text{Lewis-Acid-Base-Adduct}$$

$$AlCl_3 + Cl^- \rightleftharpoons AlCl_4^-$$

$$Cu^{2+} + 4\,NH_3 \rightleftharpoons [Cu(NH_3)_4]^{2+}$$

The Lewis acid-base adducts are also called "acceptor-donor complexes". Characteristic is the formation of an electron pair bond between Lewis acid and Lewis base. As a rule, the oxidation number of the reaction partners involved does not change.

12.3.4 Limits of the Concept According to Lewis

In order to be able to make good predictions regarding the position of complexation or precipitation equilibria, Lewis' acid-base concept must be combined with the HSAB concept.

12.4 HSAB Concept

In 1963 Pearson[7] published his concept of *"**H**ard and **S**oft **A**cids and **B**ases"*. This concept is used in many areas of chemistry to estimate the stability of compounds or their reactivity.

12.4.1 Overview HSAB Concept

Pearson distinguishes between hard and soft acids or hard and soft bases and their mutual affinity:

- Hard acids combine preferentially with hard bases, soft acids combine preferentially with soft bases.

"Hard" refers to particles that have small ionic radii, possess high charges, and have low polarizability (though they themselves are strongly polarizing). "Soft" refers to particles that have large ionic radii, have low charges, and are easily polarizable. So this concept is more or less a kind of "like and like go together" principle.
With additional consideration of the electronegativity (acids: low, bases: high) Table 12.2 results.

[7] Ralp G. Pearson, born 1919, American chemist.

Table 12.2 Hard and soft acids and bases

Type	Features	Examples
Hard acid	• Low electronegativity • Low polarizability • Small ionic radius	H^+, Na^+, K^+, Mg^{2+}, Ca^{2+}, Al^{3+}, Fe^{3+}, Co^{3+}, BF_3
Hard base	• High electronegativity • Low polarizability • Small ionic radius	OH^-, F^-, Cl^-, O^{2-}, SO_4^{2-},PO_4^{3-}, CO_3^{2-}, NH_3, H_2O
Soft acidity	• Low electronegativity • High polarizability • Large ionic radius	Cu^+, Ag^+, Au^+, Pt^{2+}, Cd^{2+}, Pb^{2+}, Hg^{2+}, Fe^{2+}, BH_3
Soft base	• High electronegativity • Low polarizability • Large ionic radius	I^-, SCN^-, CN^-, S^{2-}, R_2S

Example

Let us compare a few connections based on this concept:

- FeS (Fe^{2+}: soft acid; S^{2-}: soft base) and Fe_2O_3 (Fe^{3+}: hard acid; O^{2-}: hard base) are found in natural ore deposits. The compounds are quite stable. Fe_2S_3 and FeO are found extremely rarely in nature because they are less stable.
- Ag^+ is a soft acid. The compound with the hard base F^- (AgF) is readily soluble, whereas the compound with the soft base I^- (AgI) is insoluble.
- HF (hard/hard) should be a weak acid. In fact, HI (hard/soft) is about 10^{13} times (= 10 000 000 000 000 times) "more acidic" than HF.
- Lead sulphide (PbS) proves to be difficult to dissolve (Pb^{2+}: soft acid, S^{2-}: soft base). ◄

12.4.2 Limitations of the HSAB Concept

There are hardly any "absolutely hard" or "absolutely soft" particles. The distinction between hard and soft acids and bases should therefore not be regarded as absolute. Tendencies, on the other hand, can usually be estimated well. If other effects occur, such as a high hydration energy, then these effects often predominate:

According to HSAB, NaOH (hard/hard), for example, should be a stable compound with low solubility. However, NaOH dissolves very well in water. The high hydration energy (NaOH dissolves in water when heated) outweighs the HSAB prediction here.

An overview of various substances and their classification in the different acid-base theories is given in the table in Fig. 12.1.

Overview acid-base definitions.

Definition by	Arrhenius	Brensted	Lewis
Acid ...	Releases H^+ in water	Releases H^+ ions	Has electron gaps
Gen. Reaction acid	$HA \rightleftharpoons H^+ + A^-$	$HX + Y \rightleftharpoons X^- + HY^+$	–
Example of an acid	$HCl \rightleftharpoons H^+ + Cl^-$	$HCl + H_2O \rightleftharpoons Cl^- + H_3O^+$	–
Base ...	Releases OH in water	Absorbs H^+ ions	Has a free electron pair
Gen. Reaction base	$M(OH) \rightleftharpoons M^+ + OH^-$	$HX + Y \rightleftharpoons X^- + HY^+$	–
Example of a base	$NaOH \rightleftharpoons Na^+ + OH^-$	$H_2O + NH_3 \rightleftharpoons OH^- + NH_4^+$	–
Special features	– "free" H^+ *ion*	- Always corresponding acid-base pairs - H_3O^+ ion	- Formation of covalent bonds
Keywords	- Neutral salts - Acid salts - Basic salts - Neutralization - Hydrolysis	- Neutral acid - Cationic acid - Anionic acid - Amphoteric substances - Protolysis	- Complex
Advantages	- Simple model - Explains reactions in water - Explains neutralization - Explains salt formation - Explains acid/basic salts	- Solvent independent - Defines the strength of an acid/base - Explains acidic/basic reaction of neutral salts in solution	- Dispenses entirely with protons
Disadvantages	- Restricted to water - cannot explain all acids/bases - cannot explain all acid-base reactions	- More complex than Arrhenius	- HSAB concept additionally necessary for assessment

Fig. 12.1 Different substances and acid-base theories

Subst.	Acid-base-definition acc. to:		
	Arrhenius	Brønsted	Lewis
H^+	n. def.	Acid	Acid
HNO_3	Acid	Acid	n.def.
HSO_4^-	Acid	Acid/Base	n. def.
NaCl	Salt	Salt	Salt
NaOH	Base	Salt	Salt
OH^-	n. def.	Base	Base
NH_3	n. def.	Base	Base
$AlCl_3$	Salt	Salt	Acid
Cl^-	*Ac. residue*	Base	Base
NH_4^+	n. def.	Acid	n. def.
K^+	n. def.	n. def.	Acid

13 Acids and Alkalis

Before we turn to the pH value and the strength of acids and bases, first a few basic properties of these compounds and an overview of important acids and alkalis.

13.1 Properties of Acids and Alkalis

We want to get a brief overview of the most important properties of acids and alkalis, identify similarities and point out differences.

13.1.1 Shift of the pH Value

Probably the most obvious property of acids and alkalis is that they change the pH of substances with which they are mixed. Acids change the pH value towards the acidic, alkalis shift into the alkaline (basic) range.

This can be determined, for example, using litmus paper as an acid-base indicator (Fig. 13.1).

It is precisely this change in pH value that can be a major problem, as it can also be noticeable in large dilutions. If acid or lye penetrates the soil or gets into surface waters via canals or rain, a massive pH change may occur. For many microorganisms and small animals, but also for fish, this creates a life-threatening, sometimes even deadly environment from their traditional habitat.

13.1.2 Corrosivity

Acids and alkalis have a corrosive and irritant effect on the skin, mucous membranes and eyes. The degree of damage depends on the substance itself, its concentration and the duration of exposure.

T. Schmiermund, *The Chemistry Knowledge for Firefighters*,
https://doi.org/10.1007/978-3-662-64423-2_13

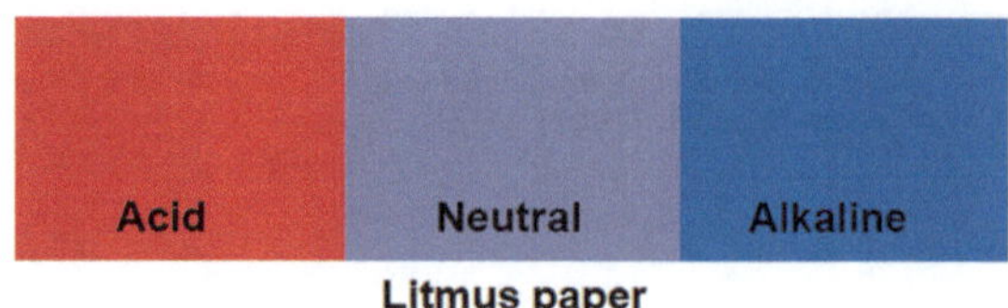

Fig. 13.1 Litmus paper (schematic)

An irritant effect is defined as redness or inflammation occurring within minutes to several hours after contact, which disappears completely after 24–72 hours. Irritation of the respiratory tract is often expressed by a "scratchy throat", irritation of the eyes by burning and tearing.

The corrosive effect of acids and alkalis is fundamentally different. In the case of acid burns, the proteins of the cells are denatured – they solidify, "flocculate". (This can be easily demonstrated, for example, by adding a dash of vinegar to a glass of milk). This clumping of the proteins prevents the acid from penetrating as deeply into the tissue. In medicine, this is called ***coagulation necrosis*** because the affected cells die due to clotting (coagulation).

In the case of a caustic burn, on the other hand, the damaged tissue (skin, mucous membrane, eye) is liquefied. The lye can thus penetrate much deeper and causes much more extensive damage. Even small amounts of lye cause much more severe burns than acids. The medical term here is ***colliquative necrosis***.

Measures in Case of Contact with Corrosive Liquids

- General measures
  - Remove wetted clothing immediately.
  - In case of contact with concentrated acids/alkalis, dab off first if possible (kitchen roll, toilet paper, etc.). First aiders: Do not forget gloves!
  - Rinse with plenty of water. Do not allow diluted acid/alkaline to run over uninjured parts of the body.
- Eye contact:
  - Rinse with water when the eyelid is open. To do this, the eyelid must be held open with the fingers (eyelid closure reflex!).
  - Physiological saline solution ("Ringer's solution") directly from the infusion bag can also be used for rinsing.
  - Always flush the eye from the root of the nose towards the ear so that the previously uninjured eye is not affected.
  - If available: Use eye wash bottle.
- Incorporation (swallowing, inhalation):
  - Consultation with a poison control center is absolutely necessary.
  - Do not induce vomiting! Do not administer fluids without medical instruction.

13.1.3 Corrosion Effect

Corrosion is the reaction of a material with substances in its environment. In the context of acids and alkalis, this is understood to mean whether and to what extent a material is attacked by them. The corrosion effect of acids and alkalis is different.

Non-oxidizing acids (e.g. hydrochloric acid, phosphoric acid, citric acid, acetic acid) "dissolve" base metals such as iron, zinc, magnesium with hydrogen evolution. Oxidizing acids (e.g. nitric acid, perchloric acid) also attack more noble metals such as copper or silver. With the exception of alkali metals and alkaline earth metals, most metals are resistant to alkalis.

Aluminium (fittings, couplings) is an important exception among the materials. Aluminium is hardly attacked by oxidising acids, as a very dense, protective oxide layer forms. However, aluminium is destroyed by sodium hydroxide or potassium hydroxide with the formation of hydrogen:

$$2\,NaOH + 2\,Al + 6\,H_2O \rightarrow 2\,Na[Al(OH)_4] + 3\,H_2 \uparrow$$

In fires involving the plastic PVC (polyvinyl chloride), noticeable quantities of hydrogen chloride ($HCl_{(g)}$) are released. In moist air, large quantities of diluted, finely distributed hydrochloric acid are thus formed. This may not only have an irritating effect on humans and animals. It can also attack the building stock. High remediation costs can then be the result.

The classifications according to the Chemicals Act in "corrosive" (C) and "irritant" (Xi) were replaced in 2012 by the "Globally Harmonised System of Classification and ***Labelling*** of Chemicals" (GHS) based on the CLP (**C**lassification, ***Labelling***, **P**ackaging) Regulation. Within the framework of labelling according to GHS, not only substances that have a corrosive effect on the skin, eyes and mucous membranes are marked with the corresponding symbol. All substances that have a corrosive effect on metals must also be labelled with the "corrosive" symbol. Some solids, on the other hand, are marked with the "exclamation mark" symbol. An overview of the labeling can be found in Fig. 13.2.

A hydrochloric acid with the concentration of $c(HCl) = 0.1\ mol\ L^{-1} \approx 0.4\ \%$ *used to* be given the symbol "irritant" according to the Chemicals Act. According to GHS – due to the fact that even this strongly diluted hydrochloric acid has a corrosive effect on metals – a "Corrosive" symbol must now be affixed.

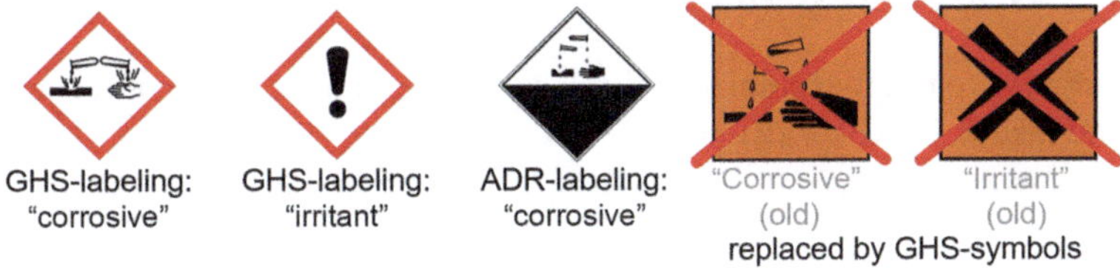

Fig. 13.2 Labeling of acids and alkalis

13.1.4 Change in Electrical Conductivity

The formation of H^+ or OH^- ions massively increases the electrical conductivity of water. Even relatively small quantities of strong acids/alkalis can cause high conductivity in contaminated water – and thus significantly increase the risk of electric shock for the emergency services.

13.1.5 Suitable Binders

Small quantities of leaked or spilled acids and alkalis are best absorbed with a suitable binder. But: Which binding agent is "suitable"?

Trade and industry offer a variety of different chemical binders. They are available in the form of fleeces, granules and so-called "snakes". They are made from the most diverse, but always absorbent materials.

In principle, type III binders based on inorganic substances (e.g. magnesium aluminosilicates or diatomaceous earth), which have *not* been waterproofed (= not hydrophobicized), are the most suitable for absorbing acids or alkalis. For this, compare Lit. [13-2].

Organic base materials such as peat, bark, cork, corn spindle granulate, cotton fibres etc. are generally not recommended as binders for acids/alkalis. Acids such as nitric acid may react dangerously with them (fire!). On contact with alkalis, the mixture may become "slimy" and can then no longer be absorbed properly.

In most cases, dry sand or expanded clay (from gardening supplies) can be used as a "makeshift binder". Cement is only suitable for alkaline solutions, as the alkaline components in cement cause strong heating when acids are added. Temperatures well above 100 °C can be reached.

If it is not possible to place a suitable receptacle underneath in the case of drip leaks, the liquid can often be covered with newspaper or kitchen roll, for example, to prevent the coarsest spread. But be careful: These are also organic materials.

For all measures, it must be noted that the wetted binder must be treated in the same way as the absorbed hazardous substance itself. Due to the increase in the surface area of the substance on the binder, greater evaporation is to be expected (wicking effect). Suitable personal protective equipment (PPE) and orderly disposal should be a matter of course.

13.1.6 Possible Reaction Hazards

We would like to briefly summarize the possible reaction hazards that can arise from acids and alkalis in NBC operations.

Table 13.1 Reaction possibilities of acids

Acid mixes in	Possible dangers
Acid	• Heating/Spraying • Gas evolution (e.g.: $H{-}COOH + H_2SO_4 \rightarrow CO \uparrow$)
Lye	• Neutralisation heat: heating/spraying
Water	• Heat of solution/hydration: • Heating/Spraying (e.g. H_2SO_4 & H_2O)
Metal	• Corrosion • Gas evolution (always: $H_2 \uparrow$; for HNO_3: $NO_x \uparrow$)
Organic substances	• Destruction/corrosivity • Charring (contact with H_2SO_4) • Fire (contact with HNO_3)
Salt	• Gas evolution possible; depending on the salt, the following can be generated: HCl, NO_x, SO_2, CO_2, HCN and others.

Basically

- Emergency personnel shall wear substance-related, adequate PPE.
- All devices and aids must be tested for their resistance to the respective hazardous substance before use.
- The corrosive effects of gases, vapours and aerosols should not be underestimated. Equipment (breathing apparatus!), fittings, even vehicles downwind can be affected under certain circumstances.
- Gas developments ("formation of bubbles") must be reported immediately to the incident command by the squadron on site. Particularly in closed rooms, this can result in a considerable additional hazard potential for the forces deployed:
  - H_2: Explosion hazard
  - CO_2: oxygen displacement
  - SO_2, HCl, HCN, NH_3 and other gases: Toxic, irritant and corrosive effects
- Gases, vapours and aerosols can cause serious damage up to pulmonary oedema in persons without adequate respiratory protection.
- Reactions which can potentially occur with certain groups of substances are listed in Table 13.1 (acids) and Table 13.2 (alkalis).

13.2 Important Acids

Here, some of the technically most important acids will be briefly considered with regard to their properties, production and use.

13.2.1 Hydrochloric Acid ($HCl_{(aq)}$)

Features

- Hydrochloric acid is the aqueous solution of the gas hydrogen chloride (HCl) and belongs to the strong acids. Its salts are called chlorides (Cl^-).

Table 13.2 Reaction possibilities of alkalis

Lye mixes in	Possible dangers
Acid	• Neutralisation heat: heating/spraying
Lye	• Rarely chemical reactions
Water	• Heat of solution/hydration: Heating/Spraying (e.g. $NaOH_{(s)}$ & H_2O)
Metal	• Corrosion possible • Gas evolution in Al materials ($H_2 \uparrow$)
organic substances	• Destruction/corrosivity • Dissolution (cotton, skin, ...)
Salt	• Gas evolution ($NH_3 \uparrow$) possible with ammonium salts
Oxidizing agent	• In case of contact of e.g. hydrazine with H_2O_2 or HNO_3 there is a risk of fire and explosion

- It dissolves many metals, with the exception of the precious metals and some "special cases" (e.g.: Ta, Ti, Ge) with hydrogen evolution: $Fe_{(s)}$ + 2 H
- Metal oxides dissolve to form salt and water, e.g.: $CuO + 2\,HCl \rightarrow CuCl_2 + H_2O$.
- It is commercially available as concentrated hydrochloric acid with approx. 33. . .38 % (also called "fuming" hydrochloric acid, as it tends to form mist in air) or as dilute hydrochloric acid with 10. . .25 % HCl. Solutions above 25 % are classified as highly corrosive.
- Warning sign

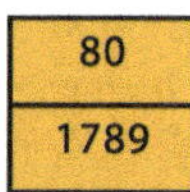

Production
- On a large scale, HCl can be obtained by gas phase reaction of chlorine with hydrogen according to $H_2 + Cl_2 \rightarrow 2\,HCl$.
- During the chlorination of hydrocarbons, HCl is formed as a by-product: $R{-}H + Cl_2 \rightarrow R{-}Cl + HCl$.

Use
HCl is an important basic chemical in industry:
- Production of metal chlorides (e.g.: $FeCl_3$, $CaCl_2$, $BaCl_2$)
- For neutralizations, precipitations, acid hydrolyses
- For the processing of ores and phosphates
- In pharmacy for the preparation of readily water-soluble hydrochlorides of active substances
- In analytics for the determination of the content of alkalis
- In metal processing for pickling, etching and soldering
- 1 % hydrochloric acid is authorised as a food additive (E 507).

Special Features

- Hydrochloric acid is contained in the gastric juice of mammals (i.e.: also in humans) in a concentration of approx. 0.3...0.4 %.
- A hydrochloric acid of just under 20 % can be distilled without changing the concentration (w(HCl) = 0.204; 1.013 bar; bp: 108.6 °C). Gaseous HCl first escapes from more highly concentrated solutions, while water is distilled off first from lower concentrated solutions.

13.2.2 Sulphuric Acid (H_2SO_4)

Features

- Sulphuric acid is a diphotonic, strong, oxidising acid. It forms two series of salts: the hydrogen sulphates (HSO_4^-) and the sulphates (SO_4^{2-}).
- Pure (96...100%) sulfuric acid is an oily, odorless liquid with a density of 1.84 g/mL that mixes indefinitely with water. When mixing sulfuric acid and water, a large amount of hydration energy is released, which can lead to evaporation of the water and thus splashing of the acid. Therefore, when diluting, always pour the acid into the water in small portions.

▶ "First the water, then the acid – otherwise the monstrous will happen."

- Sulphuric acid attacks all metals that are less noble than hydrogen.
- It is commercially available as concentrated sulphuric acid (96 %), as "monohydrate" (100 %) and with additionally dissolved SO_3 (fuming sulphuric acid, oleum).
- Warning signs

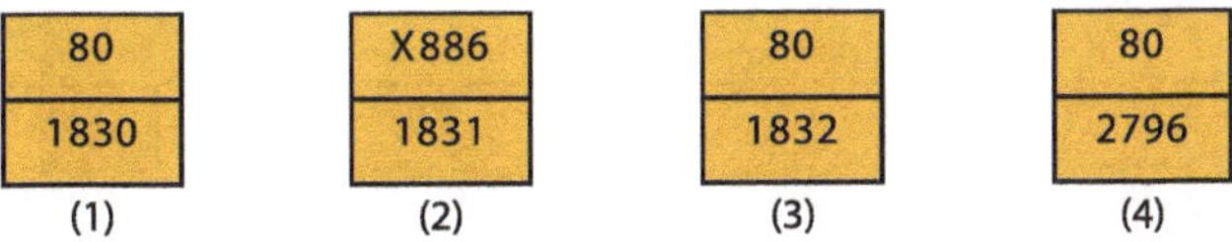

It means:

1. Sulphuric acid, containing more than 51 % acid
2. Sulphuric acid, fuming
3. Sulphuric acid, used
4. Sulphuric acid, containing not more than 51 % acid

Production

- SO_2 is produced from ores containing sulphides. This is oxidized to SO_3 (catalyst: V_2O_5) and then dissolved in water: $SO_3 + H_2O \rightarrow H_2SO_4$.

Use

$H_2\ SO_4$ is the most important inorganic acid in the chemical industry:

- Production of fertilisers (superphosphate, ammonium sulphate, etc.)
- Production of other acids (HF, H_3PO_4, ...)
- Extraction of titanium dioxide (e.g. for white wall paints)
- As accumulator acid ("battery acid")
- For the production of sulfonic acids (dyes, pigments, drugs)

Special Features

- Concentrated sulphuric acid can be transported in iron or steel containers without destroying them. A protective layer forms in the container, which protects the metal from damage.
- Concentrated sulfuric acid has a strong hygroscopic effect (water-attracting). It can be used as a desiccant, e.g. for gases. However, it also destroys organic substances (wood, paper, sugar, skin) by dehydration and slow carbonization.

13.2.3 Nitric Acid (HNO_3)

Features

- Nitric acid is a strong, oxidizing acid. Its salts are called nitrates (NO_3^-).
- Pure (100%) nitric acid is a liquid which fumes strongly in air (density 1.52 g/mL), is coloured brown to deep red by dissolved NO_2 and gives off brown vapours (NO_2). It mixes indefinitely with water.
- HNO_3 is able to dissolve silver, but not gold.
- It is commercially available as "red fuming nitric acid" with 100 % or as "concentrated nitric acid" with 65 % HNO_3.
- "Warning signs"

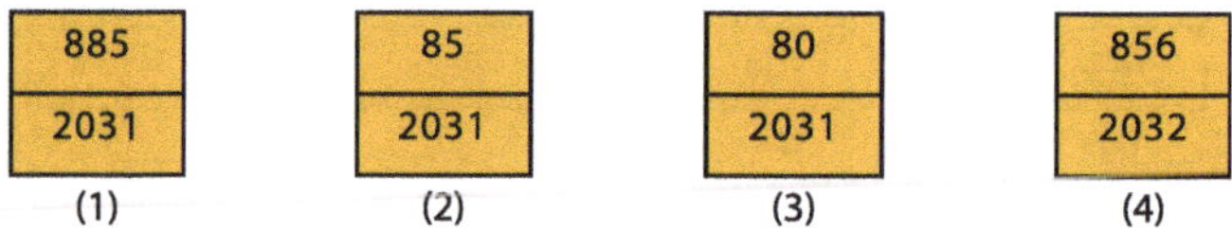

It means:

1. Nitric acid with more than 70 % acid
2. Nitric acid with 65. . .70 % acid
3. Nitric acid with less than 65 % acid
4. Nitric acid, red fuming (~100 % strength)

Production

- By combustion of ammonia on a Pt/Rh catalyst. The overall reaction is: $NH_3 + 2\ O_2 \rightarrow HNO_3 + H_2O$.

Use

- Nitric acid is one of the 10 most widely used industrial products.
- Nitrates and nitric acid are used in the fertilizer industry, in the production of explosives (nitroglycerine) and in pyrotechnics.
- It is used to introduce the nitro group (-NO_2) in the production of amines, drugs, dyes and pigments.

Special Features

- 65...100 % nitric acid reacts so strongly with organic substances such as wood, clothing or e.g. turpentine that they may ignite; therefore: Never pick up even small quantities with sawdust or similar!
- A mixture of hydrochloric and nitric acid is called “aqua regia” because it is even able to dissolve gold (the “king of metals”).

13.2.4 Phosphoric Acid (H_3PO_4)

Features

- H_3PO_4 is a medium strength acid that forms colorless, transparent crystals. It is commercially available mostly as an 85% solution. It forms three series of salts: Dihydrogen phosphates ($H_2PO_4^-$), hydrogen phosphates (HPO_4^{2-}) and phosphates(PO_4^{3-}).
- Warning signs

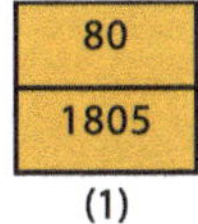

(1)

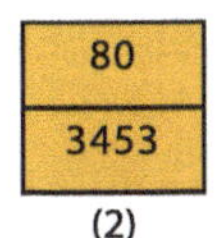

(2)

It means:

1. Phosphoric acid, solution
2. Phosphoric acid, solid

Production

- By digestion of the mineral apatite according to:

$$Ca_3(PO_4)_2 + 3\,H_2SO_4 \rightarrow 3\,CaSO_4 \downarrow + 2\,H_3PO_4$$

- By burning phosphorus and reacting the diphosphorus pentoxide with water:

$$P_4 + 5\,O_2 \rightarrow P_4O_{10}$$
$$P_4O_{10} + 6\,H_2O \rightarrow 4\,H_3PO_4$$

Use

- Phosphoric acid is used on a large scale for the production of fertilizers.
- Phosphates are contained in many detergents and cleaning agents.
- Phosphoric acid is a component of industrial cleaners and agents for metal treatment (rust removers).
- It is used in the food industry to acidify soft drinks (food additive E 338).

13.2.5 Other Inorganic Acids

13.2.5.1 Hydrofluoric Acid ($HF_{(aq)}$)

The aqueous solution of hydrogen fluoride is a weak acid. It is mainly used for the production of fluorides. It must not be stored in glass bottles, as it dissolves glass (simplified: $SiO_2 + 4\ HF \rightarrow SiF_4 \uparrow + 2\ H_2O$) and is therefore also used for glass etching.

Hydrofluoric acid causes severe skin burns that heal very poorly. Rapid treatment with calcium gluconate is required to "bind" the acid. This produces sparingly soluble calcium fluoride ("fluorspar"): $Ca^{2+} + 2\ F^{-} \rightarrow CaF_2 \downarrow$.

- Warning signs

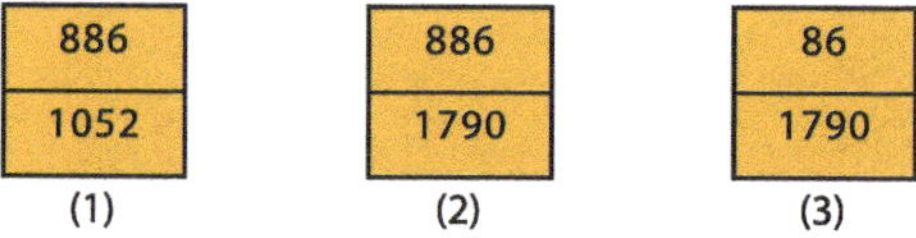

(1) (2) (3)

It means:

1. Hydrogen fluoride, anhydrous
2. Hydrofluoric acid with more than 60 % HF
3. Hydrofluoric acid with not more than 60 % HF

Further information on HF, hydrofluoric acid and fluorides can be found in Lit. [13-18].

13.2.5.2 Hydroiodic Acid ($HI_{(aq)}$)

The aqueous solution of hydrogen iodide is the strongest acid whose molecule does not contain oxygen. It is used, among other things, for the production of iodine-containing pharmaceuticals (e.g. thyroid preparations).

It is prepared by synthesis from the elements: $H_2 + I_2 \rightarrow 2\ HI$

- Warning signs

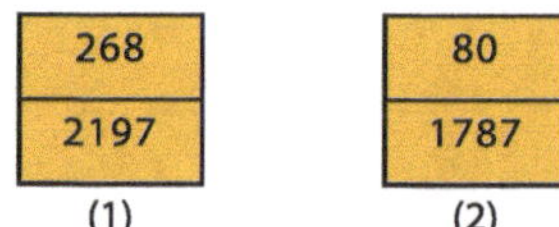

(1) (2)

It means:

1. Hydrogen iodide, anhydrous
2. Hydrogen iodide, aqueous solution

13.2.6 Organic Acids

13.2.6.1 Formic Acid (H−COOH)

Colourless liquid with a pungent odour. Mixes in any ratio with water and many organic solvents, such as ethanol, chloroform or toluene. It is produced from carbon monoxide at approx. 300 bar:

$$CO + H_2O \rightarrow H - COOH$$

It is the simplest carboxylic acid and much more acidic than acetic acid. Base metals are dissolved with hydrogen evolution, the salts are called formates. With conc. Sulphuric acid CO is released.

It is used in the textile, leather and animal feed industries. Since it has a bactericidal effect, it is also used for preservation and disinfection in medicine and food production. Calcium formate ($Ca(H{-}COO)_2$) is highly soluble in water, so it can also be used as an agent for decalcification.

- Warning signs

(1)

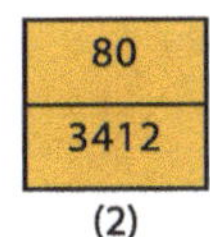

(2)

It means:

1. Formic acid with more than 85% acid
2. Formic acid, with 5...85% acid

13.2.6.2 Acetic Acid ($H_3C{-}COOH$)

Acetic acid is a weak acid with a characteristic odour which mixes indefinitely with water and various solvents (e.g. ethanol, diethyl ether). The pure acid solidifies at 16 °C to ice-like crystals (“glacial acetic acid”) and is flammable.

Production used to be by “acetic acid fermentation”, now by the catalytic oxidation of acetaldehyde: $2\ H_3C{-}CHO + O_2 \rightarrow 2\ H_3C{-}COOH$.

It is sufficiently known for flavour adjustment and as a preservative (vinegar ~5 %, vinegar essence ~25 % acid). A large number of other basic chemicals (in particular acetic acid esters as solvents and for plastics production) are obtained from it. In the form of acetic essence, it is often used as a descaler.

- Warning signs

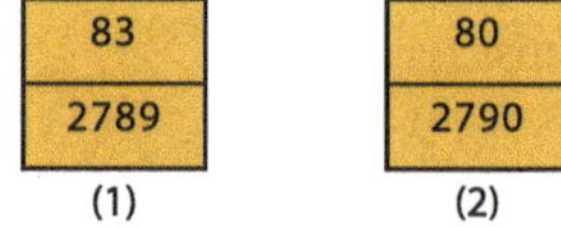

It means:

1. Acetic acid with more than 80 % acid
2. Acetic acid, with 10...80 % acid

13.2.6.3 Peroxoacetic Acid ($H_3C-COOOH$)

It is also known as peroxyacetic acid or peracetic acid (abbreviation: PAA) and is a colourless, pungent-smelling liquid. It is produced from hydrogen peroxide and acetic acid:

$$H_3C - COOH + H_2O_2 \rightarrow H_3C - COOOH + H_2O$$

It is used as a bleaching agent (TCF bleach; totally chlorine-free) for paper and textiles and for disinfection (approx. 1 % solution). Due to the equilibrium (see above), storage of diluted residues makes little sense.

- Warning sign

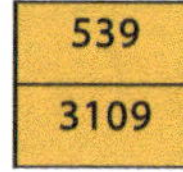

Organic peroxide, type F, liquid

PAA tends to self-decomposition when contaminated (→ Never pour residues back into the storage vessel!) and when heated (→ storage conditions!). The decomposition is exothermic and self-accelerating. This can lead to the liquid boiling up or even to a "deflagration".

13.3 Important Alkalis

13.3.1 Sodium Hydroxide Solution ($NaOH_{(aq)}$)

Features

- Solid NaOH ("caustic soda") comes in the form of cookies, flakes or beads. In air, it absorbs water and liquefies in the process. NaOH dissolves in water under strong heat development.
- Caustic soda is usually sold as a 33...50 % solution. It is an oily liquid that mixes indefinitely with water. In the process, "heat of dilution" is released.

- Even in diluted form, the solution is still strongly alkaline and has a corrosive effect. Aluminium and its alloys are attacked by caustic soda.
- Warning signs

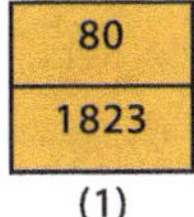

(1)

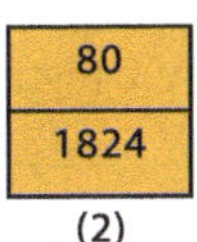

(2)

It means:
1. Sodium hydroxide, solid
2. Sodium hydroxide, solution

Production
- Caustic soda is produced during chlorine production ("chlorine-alkali electrolysis"). The simplified reaction equation is:

$$2\ NaCl + 2\ H_2O \rightarrow 2\ NaOH_{(aq)} + Cl_{2(g)} + H_{2(g)}.$$

Use
Caustic soda is the most important industrial lye:
- Alkalisation and neutralisation of any kind
- In detergents and soap manufacture
- Metalworking
- Preparation of sodium salts
- Aluminium extraction
- Food additive E 524 (for lye pastry)
- and much more.

13.3.2 Caustic Potash Solution ($KOH_{(aq)}$)

Features
- The properties are similar to those of NaOH/soda lye (see Sect. 13.3.1).
- Warning signs

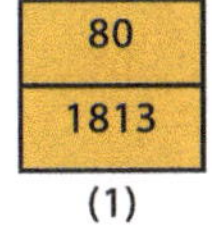

(1)

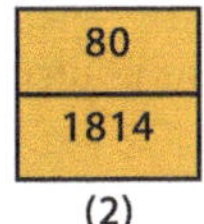

(2)

It means:
1. Potassium hydroxide, solid
2. Potassium hydroxide, solution

Production

- Electrolysis process, analogous to sodium hydroxide.

Use

- The use is similar to that of caustic soda (see Sect. 13.3.1).

13.3.3 Ammonia/Ammonia Solution (NH_3/$NH_{3(aq)}$)

Features

- Ammonia is a colorless, flammable, pungent-smelling gas that dissolves readily in water. The solution is also called "ammonia solution" or "ammonia water". It is a weak base.
- Salts contain the ammonium cation (NH_4^+), which is very similar in chemical behavior to the potassium ion (K^+).
- The commercial solution has a content of 25 % NH_3 and is corrosive.
- Warning signs

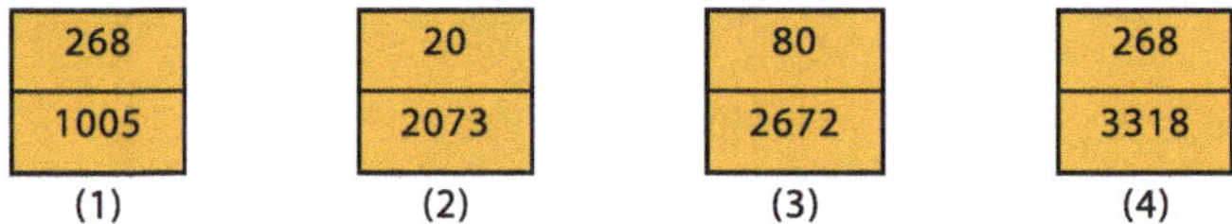

It means:

1. Ammonia, anhydrous
2. Ammonia solution, in water, 35...50 % NH_3
3. Ammonia solution, in water, 10...35 % NH_3
4. Ammonia solution, in water, more than 50 % NH_3

Production

- Ammonia synthesis ("Haber-Bosch process") from the elements at approx. 500 °C and approx. 200 bar: $N_2 + 3\ H_3 \rightarrow 2\ NH_3$.

Use

- Production of nitric acid
- Fertilizer production (ammonium salts and urea)
- Detergent
- Anhydrous as refrigerant in appropriate systems.

Special Features

- Liquid (i.e. anhydrous!) ammonia shows similar properties to water. It dissolves many salts, organic compounds and even the alkali metals.

14 pH Value

The theorem "The pH value is the negative decadic logarithm of the hydrogen ion concentration. I.e.: {pH = $-\log c(H^+)$}" should be familiar to some readers.

But: What exactly does that actually mean? And: What can you do with this value?

Let us look at a few everyday examples of pH in Fig. 14.1. We see here that we can certainly come into contact with substances that cover practically the entire range of the pH value scale (0–14).

14.1 Explanation of the Value "pH value"

Let us take the theorem apart:

The pH value is the negative decadic logarithm of the concentration of H_3O^+-ions.

- negative: $\cdot (-1)$
- decadic: Base = 10
- logarithm: Exponent
- concentration of H_3O^+-ions: Specification

So that means: first, pH is once other way of indicating the "amount" of hydrogen ions contained in a liter of the acid solution.

Thus, the specification in the "standard unit" mol/L does not always seem appropriate.

Example

The amount of H^+ ions (more precisely: of H_3O^+ ions) in a dilute acid is $^1/_{100\,000}$ mol/L = 0.000 01 mol/L. Both specifications are rather unwieldy. Therefore, one likes to write here $1 \cdot 10^{-5}$ mol/L.

T. Schmiermund, *The Chemistry Knowledge for Firefighters*,
https://doi.org/10.1007/978-3-662-64423-2_14

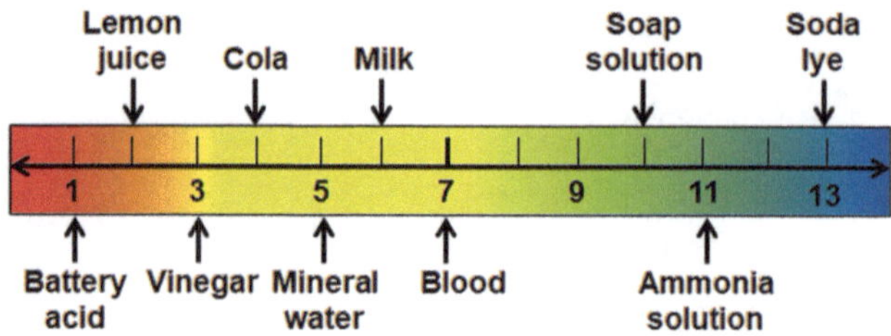

Fig. 14.1 pH values of everyday substances

If we now simply leave out the base, i.e. the "10", then we get "−5" as the remainder. Now we follow the instruction "· (−1)" (i.e. we reverse the sign) and get "5".

The pH of a dilute acid solution that has exactly $1 \cdot 10^{-5}$ mol/L H_3O^+-ions is pH = 5.0.

So we can deduce several things right now:

- The pH value is another way of indicating the concentration of H_3O^+ ions ("acid ions").
- An acid with pH 1 (equivalent to $^1/_{10}$ mol H_3O^+ per litre) is 10 times more "acidic" than a solution with pH 2 ($^1/_{100}$ mol H_3O^+ ions per litre), because it is 10 times more concentrated in H_3O^+ ions.
- An alkaline solution with pH 11 has only a vanishingly small amount of H_3O^+ ions per liter ($1 \cdot 10^{-11}$ mol/L).

Note on Calculations

From the mathematician C. F. Gauß[1] comes the quotation: "The lack of mathematical education reveals itself by nothing so conspicuously as by immoderate acuteness in numerical calculation." [Lit. 8-9]

Gauß wants to express here that one must neither calculate with 24 digits after the decimal sign, nor give a result with so many digits. If a balance only shows two decimal places, then it is nonsensical to state the calculation of the yield to 5 decimal places. A good pH meter displays two decimal places. It is therefore not very useful to state the result of a pH value calculation with more decimal places. I would therefore like to appeal to the reader to round calculation results with common sense and not to copy all the digits of the calculator display without thinking.

[1] Johann Carl Friedrich Gauss, 1777–1855, German physicist, mathematician, and astronomer.

14.1.1 pH Value Calculation with the Calculator

Not all concentrations of acid or alkali can be calculated as easily "in the head" as in the example above. For example, hydrochloric acid with a concentration of c(HCl) = 0.025 mol/L is not so easy to calculate.

Let us first consider the protolysis of hydrochloric acid:

$$HCl + H_2O \rightarrow H_3O^+ + Cl^-$$

Since the number of HCl molecules and the number of H_3O^+ ions are equal according to the reaction equation, the following applies in this case: $c(HCl) = c(H_3O^+) = 0.025 \text{ mol L}^{-1}$.

To calculate the pH value, we enter into the calculator:

[0] [,] [0] [2] [5] **0,025** [log] **−1,60205999...** [±] **1,60205999...**

The pH of a hydrochloric acid solution with $c(H_3O^+) = 0.025$ mol/L is pH = 1.6.

The reverse case, calculating the H_3O^+ ion concentration from the pH value, is also easy to do with a calculator.

The pH of a dilute hydrochloric acid is pH = 2.5. To calculate, we type:

[2] [,] [5] **2,5** [±] **−2,5** [10^x] **0,003162277...**

The concentration of H_3O^+ ions – and thus also of hydrochloric acid – is 0.003 16 mol L^{-1} = $3.16 \cdot 10^{-3}$ mol L^{-1} = 3.16 mmol L^{-1}.

Mnemonic for the Calculator If the concentration is given, press the logarithm key ([log]).

If the **pH** value is given, use the "High x" key ([10^x]).

14.2 The Neutral Point

In Sect. 12.2.3 we have already seen that water is an ampholyte, it reacts amphoterically. It can act as an acid or as a base – depending on the reactant. However, water is also subject to what is called "autoprotolysis", i.e. it reacts with itself. Let's look at the reaction equation again:

$$H_2O + H_2O \rightleftharpoons H_3O^+ + OH^-$$

This is initially an equilibrium reaction. That is, the reaction from left to right proceeds just as quickly/frequently as the reaction from right to left. And: Both partial reactions run continuously.

Using various methods, it is now possible to determine how many water molecules have dissociated into their ions (H_3O^+ and OH^-). The value obtained is referred to as the "ion product of water" (K_W). It is at 22 °C:

$$K_W = c(H_3O^+) \cdot c(OH^-) = 1.0 \cdot 10^{-14} mol^2/L^2$$

Since in pure water the concentration of H_3O^+ ions is equal to the concentration of OH^- ions, also holds:

$$c(H_3O^+) = c(OH^-)$$
$$c^2(H_3O^+) = c^2(OH^-) = 1.0 \cdot 10^{-14} mol^2/L^2$$
$$\sqrt[2]{K_W} = c(H_3O^+) = 1.0 \cdot 10^{-7} mol/L$$

Even without a calculator, the pH value of this solution can be determined: pH = 7. This is therefore the so-called neutral point of pure water.

- The neutral point of ***pure*** water is pH 7.
- Acids or acidic solutions have pH values < 7.
- Alkaline or basic solutions have pH values > 7.

14.2.1 Temperature Dependence of K_W

The autoprotolysis of water is a dynamic equilibrium that lies largely on the left side of the equation (small numerical value of K_W). However, by increasing the temperature, this equilibrium is shifted slightly to the right (the numerical value of K_W becomes larger). The increase in temperature simply causes more particles to dissociate. Conversely, dissociation is made more difficult at low temperatures.

The pK_W value given in Table 14.1 is nothing other than the "negative decadic logarithm of the K_W value". It is always twice the pH value.

However, it is **not** the case that pure water becomes alkaline when cooled or acidic when heated. Rather, the size of the scale changes. At low temperatures the scale is wide, at higher temperatures it becomes narrower.

Table 14.1 Temperature dependence of K_W

Temperature	K_W	pK_W	pH
0 °C	$0.13 \cdot 10^{-14}$ mol²/L²	14.88	7.44
20 °C	$0.85 \cdot 10^{-14}$ mol²/L²	14.06	7.03
22 °C	$1.01 \cdot 10^{-14}$ mol²/L²	14.00	7.00
25 °C	$1.26 \cdot 10^{-14}$ mol²/L²	13.90	6.95
50 °C	$5.95 \cdot 10^{-14}$ mol²/L²	13.23	6.61
100 °C	$7.4 \cdot 10^{-14}$ mol²/L²	12.13	6.07

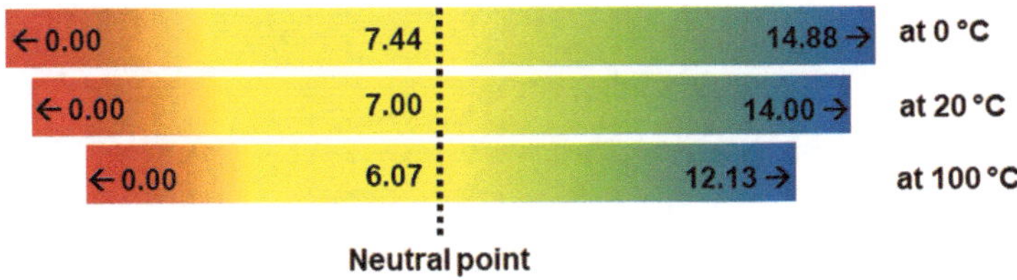

Fig. 14.2 pH value scale at different temperatures

The neutral point of the scale is always (!) the centre of the scale (= the pH value of the pure water given in Table 14.1). Compare Fig. 14.2.

For **accurate** pH measurements, the temperature must therefore be taken into account. Modern pH measuring instruments (pH meters) have the option of temperature compensation and indicate the pH value corrected to 20 °C in the display.

14.3 pH Value and pOH Value

For the following considerations we always assume that the acidic or alkaline solutions are present at room temperature, i.e. no temperature-dependent corrections have to be made.

14.3.1 Relationship Between pH and pOH Value

If the concentration of the H_3O^+ ions and that of the OH^- ions is the same, then the solution reacts neutrally. At room temperature $c(H_3O^+) = 1 \cdot 10^{-7}$ mol/L and pH = 7.0. But then also $c(OH^-) = 1 \cdot 10^{-7}$ mol/L. The negative decadic logarithm of the OH^- ion concentration is the pOH value. For $c(OH^-) = 1 \cdot 10^{-7}$ mol/L, then pOH = 7.0. Accordingly:

$$\mathrm{pH} + \mathrm{pOH} = 14$$

This also results from the ion product of the water:

$$K_\mathrm{W} = c(\mathrm{H_3O^+}) \cdot c(\mathrm{OH^-}) = 1.0 \cdot 10^{-14} \mathrm{mol^2/L^2}$$

$$\mathrm{p}K_\mathrm{W} = \mathrm{pH} + \mathrm{pOH} = 14$$

We can thus set up the overview as shown in Fig. 14.3.

$c(H_3O^+)$ (in mol/L)		pH	$c(OH^-)$ (in mol/L)		pOH
1	1	0	$^1/_{100\,000\,000\,000\,000}$	$1 \cdot 10^{-14}$	14
$^1/_{10}$	$1 \cdot 10^{-1}$	1	$^1/_{10\,000\,000\,000\,000}$	$1 \cdot 10^{-13}$	13
$^1/_{100}$	$1 \cdot 10^{-2}$	2	$^1/_{1\,000\,000\,000\,000}$	$1 \cdot 10^{-12}$	12
$^1/_{1\,000}$	$1 \cdot 10^{-3}$	3	$^1/_{100\,000\,000\,000}$	$1 \cdot 10^{-11}$	11
$^1/_{10\,000}$	$1 \cdot 10^{-4}$	4	$^1/_{10\,000\,000\,000}$	$1 \cdot 10^{-10}$	10
$^1/_{100\,000}$	$1 \cdot 10^{-5}$	5	$^1/_{1\,000\,000\,000}$	$1 \cdot 10^{-9}$	9
$^1/_{1\,000\,000}$	$1 \cdot 10^{-6}$	6	$^1/_{100\,000\,000}$	$1 \cdot 10^{-8}$	8
$^1/_{10\,000\,000}$	$1 \cdot 10^{-7}$	7	$^1/_{10\,000\,000}$	$1 \cdot 10^{-7}$	7
$^1/_{100\,000\,000}$	$1 \cdot 10^{-8}$	8	$^1/_{1\,000\,000}$	$1 \cdot 10^{-6}$	6
$^1/_{1\,000\,000\,000}$	$1 \cdot 10^{-9}$	9	$^1/_{100\,000}$	$1 \cdot 10^{-5}$	5
$^1/_{10\,000\,000\,000}$	$1 \cdot 10^{-10}$	10	$^1/_{10\,000}$	$1 \cdot 10^{-4}$	4
$^1/_{100\,000\,000\,000}$	$1 \cdot 10^{-11}$	11	$^1/_{1\,000}$	$1 \cdot 10^{-3}$	3
$^1/_{1\,000\,000\,000\,000}$	$1 \cdot 10^{-12}$	12	$^1/_{100}$	$1 \cdot 10^{-2}$	2
$^1/_{10\,000\,000\,000\,000}$	$1 \cdot 10^{-13}$	13	$^1/_{10}$	$1 \cdot 10^{-1}$	1
$^1/_{100\,000\,000\,000\,000}$	$1 \cdot 10^{-14}$	14	1	1	0

Fig. 14.3 Relationship between pH and pOH value

14.3.2 Calculation of the pOH and pH Values of an Alkaline Solution

The pH of a sodium hydroxide solution with a molar concentration of *c*-(NaOH) = 0.001 mol/L is to be calculated. We first determine the pOH value:

$$c(\text{NaOH}) = c(\text{OH}^-) = 1 \cdot 10^{-3}\text{mol/L} \longrightarrow \text{pOH} = 3$$

and from that the pH value:

$$\text{pH} = 14 - \text{pOH} \longrightarrow \text{pH} = 11$$

In this way the pH or pOH values of all strong acids and alkalis can be determined with sufficient accuracy. (For the calculation of weak acids and bases, see Sect. 14.5.6).

14.4 pH Value of Salt Solutions

From Arrhenius' acid definition, acid salts result. Namely, those salts that contain ions from which an H^+ ion can still be released. Thus also $KHSO_4$ forms according to

$$\text{KHSO}_{4(\text{aq})} \rightarrow \text{K}^+_{(\text{aq})} + \text{H}^+_{(\text{aq})} + \text{SO}^{2-}_{4(\text{aq})}$$

a rather acidic solution. The pH value is about 1.2. Analogously, one would now also expect a pH value <7 from a $NaHCO_3$ solution. However, the pH is ~ 8.3. When dissolving the neutral salt NaCl, a solution with pH 7 is obtained. The neutral salt NH_4Cl gives pH 4.8 in solution.

According to Arrhenius, these correlations cannot be explained.

14.4.1 Salt of a Strong Acid and a Strong Base

According to the theory of Brønsted and Lowry, strong acids (bases) are completely protolyzed. Weak acids (bases), on the other hand, are not. (More on the strength of acids and bases in Sect. 14.5).

Let us consider the example of NaCl:

$$Na^+_{(aq)} + 2H_2O \leftarrow NaOH_{(aq)} + H_3O^+_{(aq)}$$

$$Cl^-_{(aq)} + H_2O \leftarrow HCl_{(aq)} + OH^-_{(aq)}$$

The solution reacts neutrally. This is true for the aqueous solutions of all salts formed from a strong acid and a strong base. For our example, water is a stronger base than Cl^- or a stronger acid than Na^+.

14.4.2 Salt of a Weak Acid and a Strong Base

A weak acid is only incompletely protolyzed. When the salt is dissolved, this protolysis equilibrium is re-established. Therefore, for example, the acetate ion of a sodium acetate solution reacts according to

$$H_3C{-}COO^- + H_2O \rightarrow H_3C{-}COOH + OH^-$$

Since the strong base behaves neutrally to water, the pH value is >7. The solution reacts alkaline.

14.4.3 Salt of a Strong Acid and a Weak Base

Here, the strong acid behaves neutrally towards the water. The weak base, e.g. NH_3 (from NH_4Cl) forms in the aqueous solution due to the protolysis equilibrium according to:

$$NH_4^+ + H_2O \rightarrow NH_3 + H_3O^+$$

As a result, the solution reacts acidly.

14.4.4 Salt of a Weak Acid and a Weak Base

No general rule can be given for this case. The acid strength and the base strength must always be compared directly with each other. Ammonium acetate solution reacts neutrally, since acetic acid and ammonia are about equally strong. Ammonium cyanide solution reacts basic, since NH_3 is more basic than HCN is acidic.

In order to be able to compare the acid strengths (base strengths) directly with each other, the pK_A (pK_B) value was introduced (compare Sect. 14.5).

Summary

- Only ions that can form a weak electrolyte by reacting with water affect the pH of the solution.
- Anions derived from weak acids behave basic in aqueous solution (e.g.: AcO^-, NO_2^-, SO_3^-)
- Cations derived from weak bases behave acidly in aqueous solution (e.g.: NH_4^+, Al^{3+})
- Salts of strong acids and strong bases do not affect the pH of an aqueous solution (e.g.: NaCl, KNO_3, $Ba(ClO_3)_2$, Na_2SO_4). It is ~ 7.
- Salts of weak acids with strong bases give alkaline solutions (pH > 7). E.g.: KNO_2, NaCN, Na_2CO_3
- Salts of strong acids with weak bases give acid solutions (pH < 7). E.g.: NH_4Cl, NH_4NO_3, $AlCl_3$
- Salts of weak bases with weak acids can form both alkaline and acidic solutions. The pH value depends on whether the acidic character of the cation or the basic character of the anion is more pronounced.

14.5 Acid and Alkali Strength

In the previous sections, we have already talked about strong and weak acids or bases. At this point we want to take a closer look at this.

Dissociations (Arrhenius, compare Sect. 12.1) or protolyses (Brønsted, compare Sect. 12.2) do not always proceed completely. Rather, they are equilibrium reactions. The reaction of the acid with water to H_3O^+ and the acid residue occurs **simultaneously** with the reverse reaction. Here, a dynamic equilibrium is formed in which the concentrations no longer change.

Whether the equilibrium in question is more on the side of the reactants (left) or the products (right) depends on how easily the acid releases protons and how easily the base binds protons.

- Acid strength is a measure of how easily an acid gives off protons.
- Base strength is a measure of the ease with which a base accepts protons.

14.5.1 Degree of Dissociation α

Dissociation, or the "decay" of an acid into its ions, can be expressed as the degree of dissociation α (alpha):

$$\alpha = \frac{\text{Number of dissociated particles}}{\text{Number of dissolved particles}} = \frac{c_{\text{diss}}}{c_0}$$

For e.g., acetic acid (= AcOH) then applies:

$$\alpha(\text{AcOH}) = \frac{c(\text{H}^+)}{c_0(\text{AcOH})} = \frac{c(\text{AcO}^-)}{c_0(\text{AcOH})}$$

For an acetic acid solution of concentration $c(\text{AcOH}) = 2$ mol/L, a dissociation of $2 \cdot 10^{-3}$ mol/L was determined experimentally. The degree of dissociation a is 0.001 or 0.1%.
The degree of dissociation can take the values from 0 (not dissociated at all) to 1 (completely dissociated). The following applies to the strength of acids/bases:

- strong acids/bases: $\alpha = 0.6\ldots1$
- medium strength acids/bases: $\alpha = 0.01\ldots0.6$
- weak acids/bases: $\alpha < 0.01$

- Strong acids and bases are almost completely dissociated into their ions. Weak acids and bases dissociate less than 1%.

14.5.2 Degree of Protolysis β

Already in the discussion of the acid-base theories it was stated that free protons (H^+) do not actually occur. So, in order to compare the ease of proton release of an acid or the ease of proton uptake of a base with other acids/bases, one must choose the same reaction partner. We are of course looking at the reaction in water.

In analogy to the degree of dissociation, we first define the degree of protolysis β (beta) as the quotient of the concentration of protolysed particles to the concentration of the acid as a whole:

$$\beta = \frac{c(\text{H}_3\text{O}^+)}{c(\text{Acid})}$$

About the general reaction equation of protolysis

$$\text{HA} + \text{H}_2\text{O} \rightleftharpoons \text{H}_3\text{O}^+ + \text{A}^-$$

results in a degree of protolysis β_A for the acid and a degree of protolysis of β_B for the corresponding base:

$$\beta_{\mathrm{A}} = \frac{c(\mathrm{H_3O^+})}{c(\mathrm{HA})} \qquad \beta_{\mathrm{B}} = \frac{c(\mathrm{A^-})}{c(\mathrm{H_2O})}$$

14.5.3 Acid Constant: K_A and pK_A

In dilute solutions, the concentration of the water is to be considered constant (see below). We can therefore set this equal to 1. In addition, $c(H_3O^+) = c(A^-)$. For the overall reaction, we combine the two degree of protolysis equations into a new constant: The acid constant K_A.

$$K_{\mathrm{A}} = \frac{c(\mathrm{H_3O^+}) \cdot c(\mathrm{A^-})}{c(\mathrm{HA})}$$

If we also apply the "trick" of the negative decadic logarithm here, analogous to the pH value, we obtain the pK_S value:

$$pK_S = -logK_A$$

A large value of the acid constant means that the equilibrium of the reaction is shifted to the right side:

$$\mathrm{AH} + \mathrm{H_2O} \rightleftharpoons \mathrm{H_3O^+} + \mathrm{A^-}$$

Constant Concentration of Water

One liter of water contains 1000 g: 18 g/mol = 55.55 mol H_2O. A hydrochloric acid solution with $c(HCl) = 0.1$ mol/L "consumes" 0.1 mol water to form 0.1 mol H_3O^+ ions. This leaves 55.45 mol of water (or about 99.8%). In even more dilute (acid) solutions, the "consumption" is correspondingly less. Therefore, the concentration of the water may be considered constant.

14.5.4 Base Constant: K_B and pK_B

The same applies to the base constant K_B and its negative decadic logarithm pK_B:

$$\mathrm{B} + \mathrm{H_2O} \rightleftharpoons \mathrm{BH^+} + \mathrm{OH^-}$$

$$K_{\mathrm{B}} = \frac{c(\mathrm{BH^+}) \cdot c(\mathrm{OH^-})}{c(\mathrm{B})}$$

$$\mathrm{p}K_{\mathrm{B}} = -\log K_{\mathrm{B}}$$

Again, a large value of the base constant means that the equilibrium is shifted to the right side.

Since in acidic solution the H_3O^+ ion and in basic solution the OH^- ion are the actual carriers of the acidic and basic properties respectively, a corresponding acid-base pair applies analogously to the pH value:

$$K_S \cdot K_B = 1 \cdot 10^{-14} = K_W$$
$$pK_S + pK_B = 14 = pK_W$$

14.5.5 Strong Acids and Strong Bases

Strong acids have pK_S values <1 (i.e.: $pK_B > 13$) and strong bases have pK_B values <0 (i.e.: $pK_S > 14$). Strong acids (bases) are practically completely dissociated in water, so that the pH value (pOH value) can easily be calculated from the total concentration of the acid (base).

Examples

Acid

$$c(\text{HCl}) = 0.01\ \text{mol/L} \qquad c(\text{HCl}) = c(\text{H}_3\text{O}^+) = 1 \cdot 10^{-2}\text{mol/L} \rightarrow \text{pH} = 2$$

Lye

$$c(\text{NaOH}) = 0.01\ \text{mol/L} \quad c(\text{NaOH}) = c(\text{OH}^-) = 1 \cdot 10^{-2}\text{mol/L} \rightarrow \text{pOH} = 2$$

$$\text{pH} = 14 - \text{pOH} \rightarrow \text{pH} = 12$$

Restriction for Higher Concentrations

Strong electrolytes are largely to almost completely dissociated in aqueous solution ($\alpha > 0.6$). These include almost all salts and all strong acids and bases. These are very ion-rich solutions. The total concentration of all ions thus reaches very large values. An undisturbed, orderless movement of the ions is thus no longer possible: The particles are so close to each other that they influence each other.

This mutual influence now causes an apparently lower mass concentration of ions and thus an apparently larger proportion of undissociated molecules. The activity coefficient f_a serves as a correction factor.

Multiplying the concentration c by the activity factor f_a gives the activity a *of* the solution. The activity coefficient is strongly concentration-dependent and approaches the value 1 with increasing dilution.

This explains the deviations between the calculated and measured pH value in more concentrated solutions. However, the error is usually not greater than ~0.2 pH units.

Hydrochloric Acid

c(HCl)	Activity coefficient f_a	a(HCl)
$1 \cdot 10^{-1}$ mol/L	0.796	$0.796 \cdot 10^{-1}$ mol/L $\approx 0.8 \cdot 10^{-1}$ mol/L
$1 \cdot 10^{-2}$ mol/L	0.904	$0.904 \cdot 10^{-2}$ mol/L $\approx 0.9 \cdot 10^{-2}$ mol/L
$1 \cdot 10^{-3}$ mol/L	0.966	$0.966 \cdot 10^{-3}$ mol/L $\approx 1.0 \cdot 10^{-3}$ mol/L

14.5.6 Weak Acids and Weak Bases

The calculation of the pH values of weak acids and bases is not quite as simple as that of strong acids and bases. In addition to the substance concentration of the acid (base), the acid strength (base strength) must also be included in the calculation. The derivation of the calculation formulas shall be omitted here.

Weak Acids
The following applies to the pH value of weak acids:

$$\mathrm{pH} = \frac{1}{2} \cdot (\mathrm{p}K_\mathrm{S} - \log c_\mathrm{Acid})$$

Weak Bases
The same applies to the pOH value of weak bases:

$$\mathrm{pOH} = \frac{1}{2} \cdot (\mathrm{p}K_\mathrm{B} - \log c_\mathrm{Base})$$
$$\mathrm{pH} = 14 - \mathrm{pOH}$$

Calculation Examples
Weak Acid
Acetic acid with a concentration of 0.1 mol/L; pK_S(AcOH) = 4.76.

$$c(\mathrm{AcOH}) = 0.1 \mathrm{mol/L} = 10^{-1} \mathrm{mol/L} \rightarrow \log c_\mathrm{Acid} = -1$$
$$\mathrm{pH} = \frac{1}{2} \cdot (4.76 - (-1)) = \frac{1}{2} \cdot (4.76 + 1) = \frac{1}{2} \cdot 5.76 = 2.88$$

The pH value is approximately 3.
Weak Base
Ammonia solution with a concentration of 0.05 mol/L; pK_B (NH_3) = 4.75.

$$c(\mathrm{NH_3}) = 0.05 \mathrm{mol/L} \rightarrow \log c_\mathrm{Base} = -1.3$$
$$\mathrm{pOH} = \frac{1}{2} \cdot (4.75 - (-1.3)) = \frac{1}{2} \cdot (4.75 + 1.3) = \frac{1}{2} \cdot 6.05 = 3.03$$
$$\mathrm{pH} = 14 - \mathrm{pOH} = 14 - 3.03 = 10.97$$

The pH value is approximately 11.

The two examples clearly show that an estimation of the pH value for weak acids and bases purely on the basis of the mass concentration can lead to large errors.

14.5.7 Acid and Base Strength

Table 14.2 gives an overview of the strength of various acids and bases. The sorting is done according to the acid strength, i.e. the strongest acids are at the top of the table, the strongest bases at the bottom.

Summary

- The strength of an acid (base) is expressed in its K_A value (K_B) or pK_A value (pK_B).
- Strong acids (bases) have pK_A values < 1 (pK_B values < 0). They are almost completely dissociated.
- Weak acids (bases) have pK_A values > 3.5 (pK_B values > 3.5). They are only dissociated to a small extent.
- pH values can be determined by calculation. Different formulas must be used for strong and for weak acids (bases).

Table 14.2 Acid/base strengths

Acid	Corresponding base	pK_A value	pK_B value
$HClO_4$	ClO_4^-	–9	23
HI	I^-	–8	22
HBr	Br^-	–6	20
HCl	Cl-	–6	20
H_2SO_4	HSO_4^-	–3	17
HNO_3	NO_3^-	–1.3	15.3
$HClO_3$	ClO_3^-	0	14
H_3O^+	H_2O	0	14
HSO_4^-	SO_4^{2-}-	1.92	12.08
H_3PO_4	$H_2PO_4^-$	1.96	12.04
H_2SO_3	HSO_3^-	1.96	12.04
HF	F^-	3.1	11.9
HNO_2	NO_2^-	3.35	10.65
H_3C-COOH	$H_3C - COO^-$	4.75	9.25
H_2CO_3	HCO_3^-	6.5	8.5
HSO_3^--	SO_3^{2-}	7.2	6.8
$H_2PO_4^-$	HPO_4^{2-}	7.2	6.8
NH_4^+	NH_3	9.2	4.8
HCN	CN^-	9.4	4.6
HPO_4^{2-}-	PO_4^{3-}	12.3	1.7
H_2O	OH^-	14	0
OH^-	O_2^-	24	–10

- The pH value (pOH value) makes **no** statement about the strength of an acid (base). It only says something about the concentration of H_3O^+ ions (OH^- ions) in the solution.
- The pH value does **not** indicate the hazardousness or corrosiveness of an acid or base.

14.6 Determination of the pH Value

In order to determine the pH value of an aqueous liquid, various possibilities are available. We will look at the most important ones here. Before we do so, we will explain what acid-base indicators (pH indicators) are and how they work.

14.6.1 pH Indicators

An indicator is an "marker". A pH indicator shows whether a liquid is acidic or alkaline. We have already become familiar with the litmus indicator (compare Fig. 13.1). However, there are many other indicators. The exciting thing is that different indicators change their colour at different pH values (compare Fig. 14.4).

pH indicators are usually organic dyes. Their color in solution is in turn dependent on the pH value. These dyes are themselves extremely weak acids. Since they are very color-intensive, only a few drops of an indicator solution are needed to achieve a clear coloration of the sample solution. There is no influence on the pH value of the sample solution. Table 14.3 lists some of the indicators and their turnover range.

14.6.2 Functioning of a pH Indicator

Indicators, more precisely: pH indicators, are themselves weak acids (HIn). Their conjugate base is the indicator anion (In^-). The reaction proceeds according to

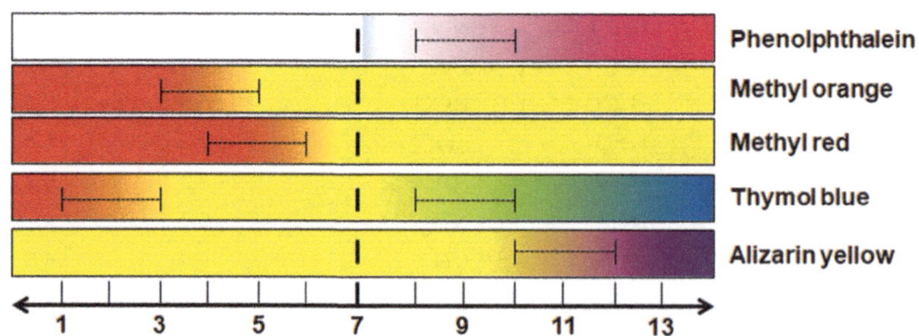

Fig. 14.4 Colour change of pH indicators

Table 14.3 pH indicators and turnover ranges

Indicator	Color at low pH	Handling area	Colour at high pH
Thymol blue	red	1.2...2.8	yellow
Methyl orange	red	3.1...4.5	yellow
Bromocresol green	yellow	3.8...5.5	blue
Methyl red	red	4.2...6.3	yellow
Litmus	red	5.0...8.0	blue
Bromothymol blue	yellow	6.0...7.6	blue
Thymol blue	yellow	8.0...9.6	blue
Phenolphthalein	colorless	8.3...10.0	raspberry red
Alizarin yellow	yellow	10.0...12.1	blue-violet

$$HIn + H_2O \rightleftharpoons In^- + H_3O^+$$

The indicator acid and its conjugate base have different colors, so that depending on the position of the equilibrium the respective color HIn or In^- becomes visible:

- By adding acid, the equilibrium is shifted to the left (towards the indicator acid). This increases the concentration of HIn and the colour of the indicator acid becomes visible.
- By adding base, the equilibrium is shifted to the right (towards the indicator base). The H_3O^+ ions react with the base to form water and are thus removed from the equilibrium. This makes the colour of the In^- ions visible.

For one of the two colours to be visible individually, the concentration HIn (In^-) must be approximately 10 times higher than the concentration In^- (HIn). This explains why pH indicators do not have a very sharp transition point but a transition range.

Indicators are preferably used for acid-base titration. Due to the different turnover ranges, care must be taken that the turnover range of the indicator used corresponds to the pH value at the end point of the neutralization. For example, the indicator phenolphthalein {pronounced fenolf-ta-le-in} is well suited for the titration of acetic acid with sodium hydroxide solution. For the titration of hydrochloric acid with caustic soda, all three indicators listed in Fig. 14.5 can be used.

14.6.3 Detect: Track, Measure, Analyze

At this point, the differences between "tracking", "measuring" and "analysing" should be briefly pointed out (cf. Fig. 14.6).

The generic term is "detection". Detection can be qualitative (*What* is it?) or quantitative (*How much* is it?) and depends on the (measurement) method used. The result of hazardous substance detection is typically referred to as a "measured value" or "measurement result".

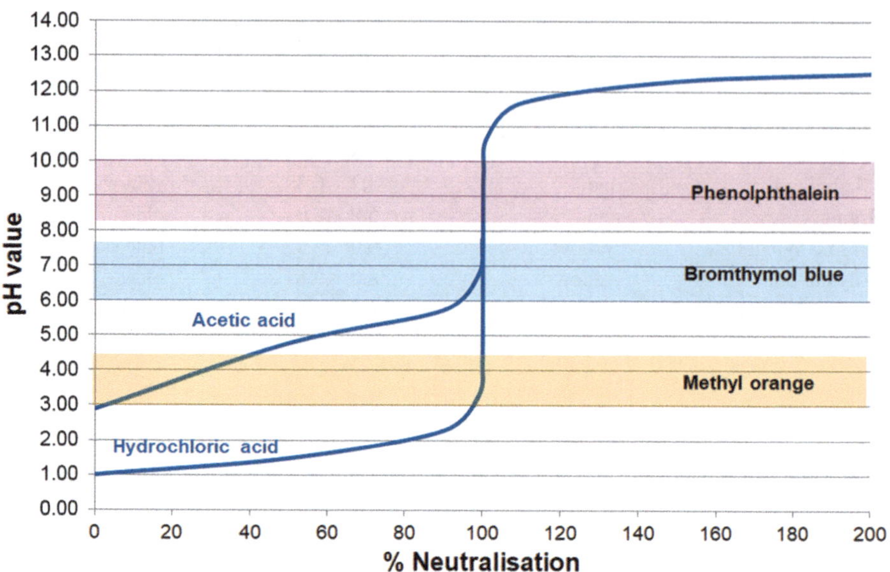

Fig. 14.5 Titration with different indicators

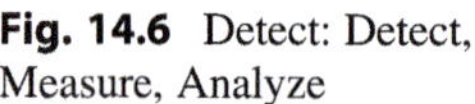

Fig. 14.6 Detect: Detect, Measure, Analyze

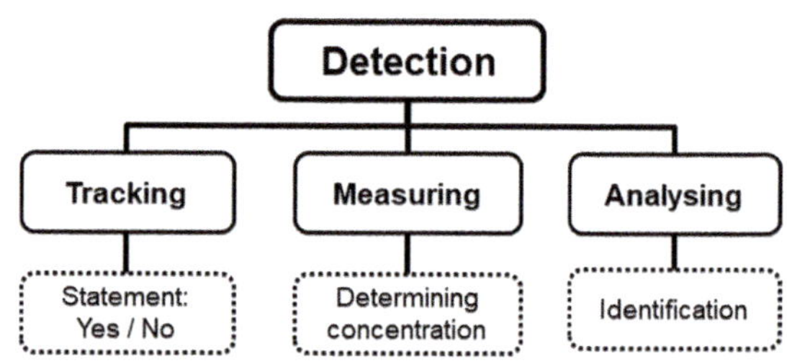

Tracking

Tracking is the search for released hazardous substances and the hazards they pose. For this purpose, simple detection methods are used which allow a yes/no statement. Substance-specific statements are not possible, but the equipment required is low.

"Tracking"

Oil test paper, water detection paste, CWA detection paper, CWA detection powder, leak detection spray, litmus paper (indicator paper).

Measurement

Measurement is the determination of the concentration of a hazardous substance or the danger posed by a hazardous substance. Single measurements or continuous measurements can be carried out.

Special care is required when selecting the measuring instruments, as the respective instruments are only suitable for certain substances or substance groups.

"Measuring"
Explosion limit warning devices, test tubes, electrochemical measuring devices, photoionization detectors (PID), substance-specific test kits, pH paper.

Analyze
Analysis is the identification of an unknown hazardous substance. An analysis can only be carried out with complex equipment on site or after sample preparation in the laboratory. The analysis is complicated by the fact that hazardous substances can occur in all three aggregate states and as pure substances, preparations or mixtures.

The necessary equipment is often cost-intensive and places high demands on the operating personnel and their training.

"Analysing"
Mobile mass spectrometers (MS), gas chromatographs (GC), ion mobility spectrometers (IMS), hazmat detector array.

Sensory Perceptions
Our sensory perceptions are **not** part of the detection methods. It would also be grossly negligent to send an firefighter into the danger zone with the task of "sniffing out" a hazardous substance.

Nevertheless, we perceive – even if unintentionally, quasi "alongside" – our environment by seeing, smelling, hearing and tasting. Caution is required here, because our senses can easily be deceived. Acid gases such as SO_3, HCl and HBr can easily be confused with each other, as all three have a "pungent smell". Ammonia and methylamine smell virtually the same to the "untrained" nose. Brown plumes, such as may be encountered in certain smoldering fires, are easily mistaken for nitrous gases. At high concentrations, hydrogen sulfide anesthetizes the olfactory cells – the gas is no longer perceived.

Added to this is the health hazard. Chlorine, hydrogen chloride and phosgene, for example, can lead to pulmonary oedema, which can take hours to appear. Carbon monoxide is odourless, but a strong blood poison.

Of course, a "odd" smell can indicate a hazardous substance. The same applies to the "strange" discoloration of a body of water. In such cases, the presence or absence of hazardous substances must be determined beyond doubt using suitable detection and measurement methods, or an analysis if necessary. In doing so, the possible danger to oneself must not be disregarded.

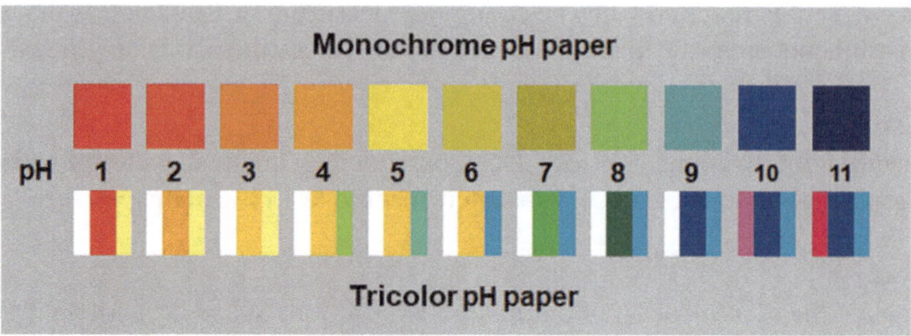

Fig. 14.7 pH papers (examples, schematic)

14.6.4 pH Value Measurement with pH Paper

Litmus paper, which we have already encountered, is actually an acid-alkali tracing paper. With this, only the statement "acid" (pH < 5) or "basic" (pH > 8) can be made.

However, the trade offers a large number of other so-called pH papers. Rolls or strips with an indication of pH 1 to 11 or 12 and a gradation of one pH unit are perfectly adequate for use (see Fig. 14.7). In terms of accuracy, these pH papers are thus somewhere between litmus paper and a pH meter.

pH papers are offered in the form of rolls, tear-off pads and paper strips. In addition, there are so-called test strips. These are plastic strips with usually several test fields (which must all be immersed in the liquid to be tested). The associated scale is always shown on the package, so that a direct comparison of wetted paper to the scale can be used to make a statement about the pH value of the liquid.

Whether you use a single color or a multi-color pH paper is up to you.

Sources of Error

- Concentrated (i.e.: anhydrous) acids do not give any indication (missing H_3O^+ ions!). If necessary, carefully (!) dilute the liquid to be tested with water.
- Strong oxidizing agents destroy the indicators: The paper bleaches out. Possible problem solution: Short immersion, quick reading.
- Strongly coloured solutions make reading difficult. Sometimes it helps to let the liquid rise into the paper via capillary forces.

Tips for Use

- pH paper (roll or strip) can be attached to a rod with e.g. adhesive tape or drawing pins. This facilitates handling for the preceding squad, especially if liquid is wetting the floor or leaks are to be located. The disadvantage is that several rods must be prepared for several locations to be examined.
- pH paper can be fastened to the clothing, e.g. to the chemical protective suit, with transparent adhesive tape. In this case, only a part of the pH paper is covered with adhesive tape. If the squad notices a change from "free end" to "protected paper", corresponding hazardous substances are present in the ambient air.

- Shifts into the acidic range can be caused by e.g.: HCl, SO_3, NO_x,. . .
- Shifts into the basic range can occur, for example, with: Ammonia (NH_3) or amines
- Fading/colour loss can be an indication of e.g. ozone (O_3) or chlorine (Cl_2).

This process works better the higher the humidity and the higher the water solubility of the acidic or basic reacting gas.

14.6.5 pH Value Measurement with pH Meter

In laboratories, pH values are very often determined with corresponding measuring instruments called pH meters. These are mostly portable devices that indicate the pH value to one or two decimal places. They consist of a pH electrode and the device itself.

The pH electrode is a so-called combination electrode. This means that the actual measuring electrode and a reference electrode are installed together in one housing. In the actual sense, the voltage difference between the reference electrode and the measuring electrode is then measured and displayed as the pH value.

pH electrodes require regular care and maintenance. They must be calibrated and must not dry out. Calibration is carried out with at least two different buffer solutions, which in turn must be "OK". Calibration, or at least a check of the instrument, must be carried out fairly regularly. If this is not done, measured value deviations are practically pre-programmed.

If a new electrode is connected to a pH meter, calibration is mandatory.

The electrodes must be protected against drying out as they have a glass membrane. This ensures the exchange of H_3O^+ or OH^- ions between the measuring electrode and the liquid to be tested. If this membrane dries out, it is usually damaged to such an extent that it can no longer be regenerated. The pH electrode is then unusable.

A potassium chloride solution with $c(KCl) = 3\ mol\ L^{-1}$ is usually used as storage medium (follow manufacturer's instructions). Storage in distilled water or drinking water will render the electrode unusable in the long term.

In the case of portable battery-powered devices, it must also be ensured that the batteries in the device cannot leak and that replacement batteries are available.

15 Neutralisation

In the following, we will consider the reactions of acids with bases, which are generally referred to as neutralisation or neutralisation reactions. Here, an acid present is "neutralized" by a base or a given alkali is "neutralized" by an acid or, to put it crudely, rendered harmless.

15.1 Basics of Acid-Base Neutralization

This "rendering harmless" is due to the fact that the actually "corrosive" part of an acid (H_3O^+) or an alkali (OH^-) is converted to water.

- Neutralization is the formation of water (H_2O) from oxonium ions (H_3O^+) and hydroxide ions (OH^-).

$$H_3O^+ + OH^- \rightarrow 2H_2O$$

The remaining base residue and acid residue ions then form the corresponding salt. Therefore, it can be summarized in a simplified way:

$$\text{Acid} + \text{Base} \rightarrow \text{Salt} + \text{Water}$$

A neutralization can be considered complete when all H_3O^+ ions have been converted to water by OH^- ions.

- A solution with pH = 7 is neutral compared to ***pure*** water.

As we have seen, not all solutions of salts react neutrally (compare Sect. 14.4). Therefore, it must be stated:

T. Schmiermund, *The Chemistry Knowledge for Firefighters*,
https://doi.org/10.1007/978-3-662-64423-2_15

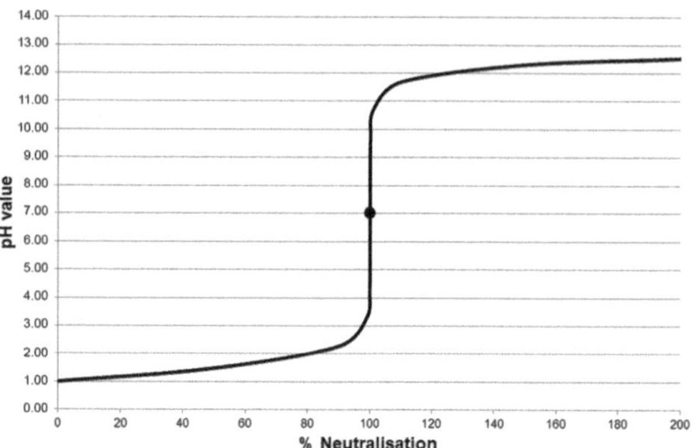

Fig. 15.1 Neutralization of hydrochloric acid with sodium hydroxide solution

- The solution of a completely neutralised acid (base) may have a pH value different from 7.

The differences that arise here with strong/weak acids and bases will now be examined in more detail. The graphs show the addition of lye with a concentration of 0.1 mol/L to one litre of acid with a concentration of 0.1 mol/L. The horizontal axis shows the volume of lye added. The horizontal axis shows the volume of lye added, the vertical axis the corresponding pH value.

15.1.1 Neutralization Strong Acid ↔ Strong Base

Let us first consider the neutralization of the strong acid HCl with the strong base NaOH.

The reaction equation is: $HCl_{(aq)} + NaOH_{(aq)} \rightarrow Na^+_{(aq)} + Cl^-_{(aq)} + H_2O$. We can easily see that to neutralize one liter of a hydrochloric acid with the molar concentration of $c(HCl) = 0.1$ mol/L, exactly (!) one liter of a sodium hydroxide solution with the concentration of $c(NaOH) = 0.1$ mol/L is also required. The course of the neutralisation is shown in Fig. 15.1.

On the other hand, for the neutralization of the strong sulfuric acid with sodium hydroxide solution, the following results:

$$H_2SO_{4(aq)} + NaOH_{(aq)} \rightarrow Na^+_{(aq)} + HSO^-_{4(aq)} + H_2O$$
$$HSO^-_{4(aq)} + NaOH_{(aq)} \rightarrow Na^+_{(aq)} + SO^{2-}_{4(aq)} + H_2O$$
$$H_2SO_{4(aq)} + 2NaOH_{(aq)} \rightarrow 2Na^+_{(aq)} + SO^{2-}_{4(aq)} + 2H_2O$$

The neutralisation of the sulphuric acid thus proceeds in two stages. It is shown in Fig. 15.2. The point at which all H_2SO_4 molecules have been converted to HSO_4^--ions is at pH 1.7 – i.e. in the strongly acidic range, and is not recognisable as a separate “stage”.In both cases, the neutralisation point is apparently not so easy to

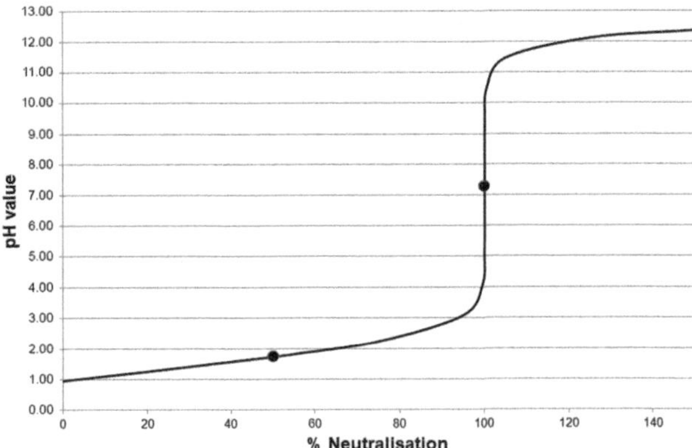

Fig. 15.2 Neutralization of sulfuric acid with sodium hydroxide solution

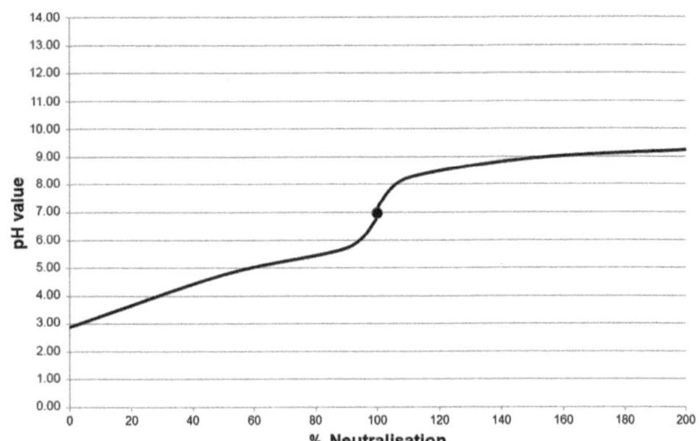

Fig. 15.3 Neutralization of acetic acid with ammonia solution

"hit". Even the smallest amounts of additional lye cause a massive increase in the pH value into the alkaline range.

15.1.2 Neutralisation Weak Acid ↔ Weak Base

Let us now consider the neutralization of the weak acetic acid with the weak base ammonia solution: $AcOH_{(aq)} + NH_4OH_{(aq)} \rightarrow NH^+_{4(aq)} + AcO^-_{(aq)} + H_2O$.

The curve (see Fig. 15.3) is flatter overall than in the examples given in Sect. 15.1.1. Here, too, the smallest amounts of further alkali are sufficient at the neutralisation point to cause the pH value to rise to over 8.

15.1.3 Neutralisation Strong Acid (Base) ↔ Weak Base (Acid)

Finally, we consider the curve of the neutralization of hydrochloric acid with ammonia solution: $HCl_{(aq)} + NH_4OH_{(aq)} \rightarrow NH^+_{4(aq)} + Cl^-_{(aq)} + H_2O$.

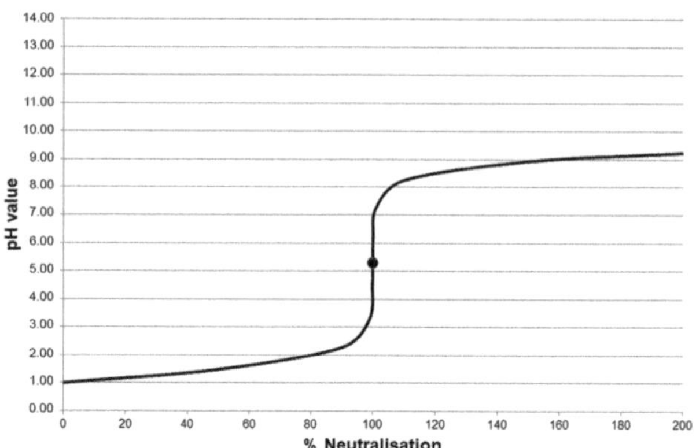

Fig. 15.4 Neutralization of hydrochloric acid with ammonia solution

In the acidic range, the curve is similar to the neutralization of hydrochloric acid with NaOH, but does not increase as much in the alkaline range. The neutralization point is pH 5.3 (Fig. 15.4).

15.2 Neutralisation Heat

During the actual neutralization, i.e. the reaction of H_3O^+ ions with OH^- ions to form water, the same amount of energy of 57.3 kJ is always released per mole.

$$H_3O^+ + OH^- \rightarrow 2H_2O \qquad \Delta H = -57.3\text{kJ}$$

When 1 L hydrochloric acid with $c(\text{HCl}) = 1$ mol/L and 1 L sodium hydroxide solution with $c(\text{NaOH}) = 1$ mol/L are added together, the temperature rises by approximately 6–7 °C.The neutralisation heat has a much greater effect when concentrated acids and alkalis react with each other, as can happen, for example, as a result of an accident. Thus, during the "rapid neutralisation" of 1 L of 33% sodium hydroxide solution with concentrated hydrochloric acid, temperatures around 100 °C (212 °F) and above are reached.

If acids ***and*** alkalis are involved in an incident, they must be prevented from running into each other if possible.

15.3 Buffer Solutions

If small amounts of acids or bases are dissolved in water, the pH value changes drastically. The pH value of water is very sensitive to acidic/basic additions and is therefore easily changed. In contrast, so-called buffer solutions have a relatively stable pH value that changes only slightly even when noticeable amounts of acid or alkali are added.

15.3.1 Functionality of Buffer Solutions

In order to be able to capture ("buffer") the H_3O^+ ions that are released when acid is added, the buffer solution must contain a base. On the other hand, in order to intercept the OH^- ions when an alkali is added, the solution must also contain an acid. Buffer solutions are often prepared from a weak acid and its alkali salt (e.g. acetic acid and sodium acetate). Depending on the acid-base combination chosen, the solution buffers in a certain pH range. The amount of acid (base) that a buffer solution can absorb without significantly changing its pH depends on the concentration of the buffer solution. This absorption capacity for acid or alkali, called buffer capacity, increases with the concentration of the buffer solution.

Buffer System Acetic Acid/Sodium Acetate
The solution contains sodium acetate as base. When acid is added, the weak, hardly dissociated acetic acid is formed:

$$H_3C{-}COO^- + H_3O^+ \rightarrow H_3C{-}COOH + H_2O$$

The acid in this buffer is acetic acid. This reacts with bases according to:

$$H_3C{-}COOH + OH^- \rightarrow H_3C{-}COO^- + H_2O$$

Due to the formation of water, the influence on the pH value is only slight in both cases. Both H_3O^+ and OH^- ions are intercepted; the solution buffers.

15.3.2 Calculation of the pH Value of a Buffer

The Henderson-Hasselbalch[1] equation (also known as the buffer equation) is used to calculate the pH value of a buffer solution; its derivation is omitted here. It reads:

$$\mathrm{pH} = \mathrm{pK_S} + \log \frac{c_{(\mathrm{Salt})}}{c_{(\mathrm{Acid})}}$$

For the case $c(\text{acid}) = c(\text{salt})$ the result is $\mathrm{pH} = \mathrm{pK_S}$. This point also represents the inflection point in the buffer curve.

Example Calculation

A buffer solution is prepared by dissolving 1 mol of acetic acid and 1 mol of sodium acetate in 1 L of water. The *pK_S value* of acetic acid is 4.75. The pH value is calculated according to:

[1]Lawrence Joseph Henderson, 1878–1942, American chemist and biologist, Karl Albert Hasselbalch, 1874–1962, Danish physicist and chemist.

$$pH = pK_S - \log\frac{c_{(Salt)}}{c_{(Acid)}} = 4.75 - \log\frac{1\ mol/L}{1\ mol/L} = 4.75$$

Now add 0.1 mol/L of HCl. The acid concentration changes to: 1 mol/L + 0.1 mol/L = 1.1 mol/L. At the same time, the acetate concentration decreases: 1 mol/L − 0.1 mol/L = 0.9 mol/L. This results in:

$$pH = 4.75 - \log\frac{1.1\ mol/L}{0.9\ mol/L} = 4.75 - 0.0872 = 4.66$$

The pH value only changes by about one tenth of a unit. Pure water would have reached pH 1 and thus changed by a full 6 pH units.If 0.01 mol of lye is added to the above buffer solution, the acid concentration is: 1 mol/L − 0.01 mol/L = 0.99 mol/L. The acetate concentration is then: 1.0 mol/L + 0.01 mol/L = 1.01 mol/L. Substituted into the formula:

$$pH = 4.75 - \log\frac{0.99\ mol/L}{1.01\ mol/L} = 4.75 - (-0.087) = 4.75 + 0.087 = 4.76$$

◀

The pH value remains practically constant. With pure water, on the other hand, the pH value would have increased from 7 to 12, i.e. by 5 units.

15.3.3 Important Buffer Systems

In addition to a wide variety of buffer solutions across the entire pH spectrum, which have been designed for a wide range of applications, buffers are used as calibration solutions for pH measuring instruments. Buffer solutions also play an important role in many technical processes, such as electroplating, dye production or leather tanning or the manufacture of eye drops.

Important Buffer Systems Are

- Acetate buffer (AcOH/NaAcO), buffers at pH 4.5...5.0; technical use
- Phosphate buffer (NaH_2PO_4/Na_2HPO_4), buffers at ~pH 7; technical use
- Ammonium buffer (NH_4Cl/NH_3), buffers at pH 9...9.5; technical use
- Carbonate buffer (H_2CO_3/HCO_3^-); regulates CO_2 concentration in atmosphere/biosphere
- Blood, pH ≈ 7.4
- Gastric juice, pH ≈ 2

15.4 Neutralization and Emergency Operations

Time and again, the question is raised as to whether spills of acids or alkalis should not be neutralised "on site". We have already seen that "neutralization" (i.e. achieving an environmentally compatible pH value of between 5.5 and 7.5), "carried out on the road" in the context of an incident, is likely to be extremely difficult.

Nevertheless, various possibilities are cited again and again. These are now to be considered in more detail – from a chemical point of view.

15.4.1 Dilution up to pH 7

Sometimes it is suggested to simply dilute leaked acids or alkalis generously with water in order to finally achieve an environmentally compatible pH value via the dilution.

As an example, consider the dilution of one liter of a hydrochloric acid of the molar concentration of 0.1 mol L^{-1} (~ 0.36 % strength; corresponding approximately to the concentration of our stomach acid) in Fig. 15.5.

As can be seen from Fig. 15.5, ***one*** litre ultimately becomes ***one million*** L (or 1000 m^3, approx. 6290 bbl {US}). The argument that it is not necessary to dilute to pH 7, since pH 5 is already environmentally compatible, reduces the total volume of water, but 10 m^3 (~ 63 bbl) of water are needed to reach pH 5 – for one litre of approximately 0.4 % hydrochloric acid! This is to be illustrated again by the scaled Fig. 15.6.

To dilute even one litre of ***concentrated*** hydrochloric acid to pH 5 (weakly acidic), more than 1200 m^3 of water are required. Conversely, to dilute one litre of 33 % sodium hydroxide solution to pH 7.5 (weakly alkaline) requires around 57 200 m^3 of water.

These numerical examples may illustrate the immense dilutions that would have to be spoken of here in the event of an operation, if the operational command decided in favour of this measure.

	1 Liter hydrochloric acid $c(HCl)$ = 0.1 mol/L	pH 1
+	9 Liter water	↓
=	10 Liter hydrochloric acid $c(HCl)$ = 0.01 mol/L	pH 2
+	90 Liter water	↓
=	100 Liter hydrochloric acid $c(HCl)$ = 0.001 mol/L	pH 3
+	900 Liter water	↓
=	1 000 Liter hydrochloric acid $c(HCl)$ = 0.000 1 mol/L	pH 4
+	9 000 Liter water	↓
=	10 000 Liter hydrochloric acid $c(HCl)$ = 0.000 01 mol/L	pH 5
+	90 000 Liter water	↓
=	100 000 Liter hydrochloric acid $c(HCl)$ = 0.000 001 mol/L	pH 6
+	900 000 Liter water	↓
=	1 000 000 Liter hydrochloric acid $c(HCl)$ = 0.000 000 1 mol/L	pH 7

Fig. 15.5 Dilution of hydrochloric acid

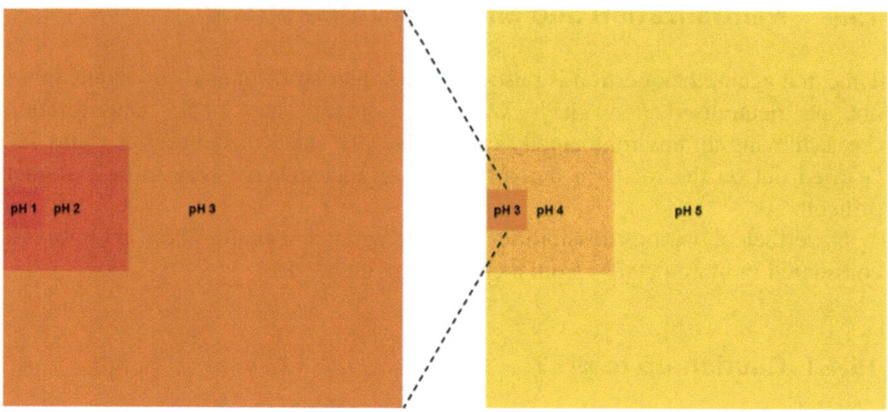

Fig. 15.6 Dilution to pH 5 (to scale)

15.4.2 Neutralisation with Diluted Acids/Alkalis

The use of concentrated acids or alkalis for the neutralisation of a spilled hazardous material is generally ruled out. The heat development (heat of neutralization!) is in no way controllable. The liquid mixture immediately comes to the boil and the resulting water vapour carries droplets of acid or alkali with it. Contamination of the emergency personnel involved in the operation and of the surrounding area can hardly be avoided.

Diluted neutralising agents would therefore have to be used. A lower concentration also means a less violent reaction. The water contained in the dilution absorbs a good part of the neutralisation heat.

Procedure

The leaked acid (lye) must be collected in a sufficiently large container, neutralised there with constant mixing with the diluted lye (acid) and can then be discharged into a channel, if necessary with further dilution.

Estimation of Quantities

- To bring 25 L of a leaked 85 % phosphoric acid (corrosive!) with a 10 % sodium hydroxide solution (corrosive!) into the pH range given above, 146 L must be used. So the final volume has increased about sevenfold!
- Similarly, 50 L of sodium hydroxide solution, 33 % (corrosive!), is to be neutralized with a 5 % hydrochloric acid (corrosive!): Just under 410 L of the dilute hydrochloric acid is needed. Eight times the amount of lye that has leaked out!

Table 15.1 Solids quantities for neutralisation of 1 L acid

Acid	Amount	Density (g/mL)	Ca $(OH)_2$	Na_2 CO_3	$CaCO_3$	Litres of CO_2 (generated)
Hydrochloric acid (HCl)	37 %	1.19	0.45 kg	0.64 kg	0.60 kg	145
Sulphuric acid (H_2SO_4)	96 %	1.84	1.33 kg	1.91 kg	1.80 kg	433
Nitric acid (HNO_3)	65 %	1.40	0.53 kg	0.77 kg	0.72 kg	174
Phosphoric acid (H_3PO_4)	85 %	1.68	1.08 kg	1.54 kg	1.46 kg	350
Formic acid (H–COOH)	100 %	1.22	0.98 kg	1.41 kg	1.33 kg	319
Acetic acid (H_3C–COOH)	100 %	1.05	0.65 kg	0.93 kg	0.88 kg	210

Table 15.2 Solids quantities for neutralisation of 1 L lye

Lye	Amount	Density g/mL	Citric acid monohydrate	Amido-sulphonic acid
Sodium hydroxide solution ($NaOH_{(aq)}$)	33 %	1.36	0.79 kg	1.1 kg
Potassium hydroxide solution ($KOH_{(aq)}$)	30 %	1.29	0.48 kg	0.67 kg
Ammonia solution ($NH_{3(aq)}$)	25 %	0.907	0.93 kg	1.3 kg

15.4.3 Neutralisation with Solid Acids/Alkalis

Again and again the use of solid substances for neutralisation is suggested. Let us first have a look at Table 15.1. The respective amount of solid material required for the neutralisation of the acid is given. For the use of carbonates for acid neutralisation, the volumes of CO_2 formed are also given. In Table 15.2 the quantities required for the neutralisation of alkalis are given analogously.

Procedure

The leaked acid (lye) must be collected in a sufficiently large container. The respective solid is then added here with constant mixing. With further dilution, the neutralised solution can then be fed into a channel if necessary.

Estimation of the Moment of Danger

- The quantities to be used (Tables 15.1 and 15.2) seem moderate at first sight. However, one must be aware that here (with the exception of $CaCO_3$) one is again dealing with corrosive or at least highly irritating hazardous substances, which require appropriate protective equipment.

- The temperature development must also be taken into account when using solids, because in the end you are neutralising with a highly concentrated base or acid. Under laboratory conditions (i.e. good agitation and good heat dissipation) were measured during the neutralization of 0.1 L (!):
  - H_3PO_4 with Na_2CO_3: 95 °C
  - Formic acid with Na_2CO_3: 90 °C.
  - NaOH solution with citric acid: 90 °C.
- In all three cases, considerable water vapor evolution takes place. The water vapour can entrain micro-droplets of acid (lye) and thus distribute them in the surrounding atmosphere. If the mixing is not ideal or if the neutralising agent is added too quickly, the acid (lye) can boil at certain points. Hot acid (lye) then splashes around and endangers the emergency personnel entrusted with the task.
- If acids are to be neutralized with carbonates, **CO_2 evolution** is added. Carbon dioxide formed entrains acid and disperses it in the form of very fine droplets in the surrounding atmosphere. At the same time, the effect of **CO_2** in higher concentrations must be taken into account:
  Neutralization of 1 L acetic acid with carbonate already produces 210 L **CO_2**. In order to dilute to 1 vol.-% **CO_2**, 20.8 **m^3** fresh air must be supplied (compare Table A.10.).

15.4.4 Result of the Observations

- Neutralisation tests “on site” are logistically hardly feasible and/or involve additional dangers for the emergency services.

Almost independent of the amount of acid (lye) that has leaked out, there is a sequence of measures to be taken:

I. Collection, decanting, transfer
II. Absorb residues with binder
III. Rough cleaning after consultation with authority/advisor/operator
IV. Handing over the operation site to the competent authority or the operator

Part VII

Redox Reactions and Electrochemistry

16 Oxidation/Reduction Concept

The terms oxidation and reduction have changed over the years – similar to what we have already seen with the acid-base term. Lavoiser was able to demonstrate that oxygen (French: *oxygène*) is consumed in every combustion, and thus coined the term “oxidation”.

An oxidation is the reaction of a substance with oxygen. The substance takes up oxygen, it is oxidized.

A typical example is the combustion of carbon:

$$C + O_2 \rightarrow CO_2 \uparrow$$

Carbon absorbs oxygen and is oxidized to carbon dioxide.Reduction was initially understood as the “return” of a metal from its ore (lat. *Reducere* = to return). Later this became the splitting off of oxygen from a compound.

Reduction is the removal of oxygen from a compound. The compound is reduced. Here are two examples:

1. Copper oxide is reduced with carbon to copper. Carbon dioxide is formed in the process.

$$2\ CuO + C \rightarrow 2\ Cu + CO_2 \uparrow$$

2. When heated, potassium nitrate is reduced to potassium nitrite with the release of oxygen.

$$2\ KNO_3 + \text{Energy} \rightarrow 2\ KNO_2 + O_2 \uparrow$$

T. Schmiermund, *The Chemistry Knowledge for Firefighters*,
https://doi.org/10.1007/978-3-662-64423-2_16

16.1 The Modern Redox Concept

When copper oxide reacts with carbon, not only is the copper oxide reduced to copper, but the carbon is also oxidized to carbon dioxide. Oxidation and reduction take place simultaneously. Other reactions also take place under the appearance of fire, e.g. the reaction of metallic sodium in a chlorine gas atmosphere to NaCl.

Closer examination revealed that:

a. the oxidation numbers of the elements involved change during both oxidation and reduction,
b. oxidation does not take place without another substance being reduced,
c. it is the real or formal transfer of electrons.

This leads to a new, more modern definition of the two terms:

Oxidation is the release of electrons. Reduction is the acceptance of electrons.

16.1.1 Clarification

Similar to the acidic effect of an acid, which can only develop when it reacts with a base (remember: $HCl + H_2O \rightarrow H_3O^+ + Cl^-$; H_2O is the base with which the acid HCl reacts), an oxidation can only take place together with a reduction. If one substance gives up electrons, another substance must also be able to take up these electrons.

Oxygen is one of the most electronegative elements, along with fluorine and chlorine. When a substance reacts with oxygen, the oxygen attracts the electrons of this substance. The substance that is oxidized gives up these electrons to the oxygen (which is reduced in the process).

$$\overset{0}{C} + \overset{0}{O_2} \rightarrow \overset{+IV-II}{C\,O_2}$$

Carbon (C^0) is oxidized to C^{+IV}, oxygen ($O_2{}^0$) is reduced to O^{-II}.

- Redox reactions are characterized by the displacement of electrons. In this process, the oxidation numbers of the elements involved change.

Let's look again at the reduction of copper oxide:

$$2\,CuO + C \rightarrow 2\,Cu + CO_2$$

and taking into account the oxidation numbers.

$$2\,\overset{+II-II}{Cu\,O} + \overset{0}{C} \rightarrow 2\,\overset{0}{Cu} + \overset{+IV-II}{C\,O_2}$$

Table 16.1 Redox terms

Oxidation		Reduction
Oxygen uptake Release of hydrogen Takes place through the oxidizing agent Release of electrons ($- e^-$) Oxidation = Ox number high	–	Oxygen release Absorption of hydrogen Is carried out by the reducing agent Absorption of electrons ($+ e^-$) Reduction = Ox number down
Oxidizing agent		Reducing agent
Accepts electrons ($+ e^-$) Is reduced during the reaction Oxidation number becomes smaller	–	Emits electrons ($- e^-$) Is oxidized during the reaction Oxidation number becomes larger

The divalent copper (Cu^{+II}) accepts two electrons (e^-) and is thus reduced to elemental copper (Cu^0). At the same time, the carbon gives up 4 e^- and is oxidized to C^{+IV}. The oxidation number of the oxygen (O^{-II}) does not change in this example.

As can be seen from these examples, the electrons do not have to be transferred "completely". The change of the oxidation number is the decisive factor (compare Sect. 7.3.3).

16.2 Oxidizing and Reducing Agents

- Substances that can easily accept electrons are called oxidizing agents. The oxidizing agent is reduced during the redox reaction.
- Substances that easily give up electrons are called reducing agents. The reducing agent is oxidized during the redox reaction.
- Neither oxidation nor reduction can occur by itself.

Let us use Table 16.1 to get a better overview of the terms.

- **Attention:** The pairs of terms oxidation/reduction, electron donation/electron acceptance, oxidant/reductant must not be confused or mixed.

16.2.1 Comparison Redox ↔ Acid-base

A certain similarity between corresponding acid-base pairs and corresponding redox pairs has already been pointed out. Let us compare these pairs in Table 16.2.

Like acid-base reactions, redox reactions are also reversible. However, the equilibrium in the redox pairs is usually on the side of one partner. The reduced (Red) and the oxidized (Ox) form are related as a corresponding redox pair (also called redox system) analogous to a corresponding acid-base pair according to Brønsted.

There are even more analogies:

- Compounding (see Sect. 16.3.1) can be understood in the same way as neutralisation.

Table 16.2 Redox ↔ acid-base comparison

Redox reactions	Acid-base reactions
Red(1) + Ox(2) ⇌ Red(2) + Ox(1)	Acid(1) + Base(2) ⇌ Acid(2) + Base(1)
Sn^{2+} + 2 Fe^{3+} ⇌ 2 Fe^{2+} Sn^{4+}	HCl + H_2O ⇌ H_3O^+ + Cl^-
Corresponding redox pairs	Corresponding acid-base pairs
Red ⇌ Ox + e^- (Oxidation, Electron donation →; ← Electron acceptance, reduction)	Acid ⇌ Base + H^+ (Proton donation →; ← Proton acceptance)
Na ⇌ Na^+ + e^-	NH_4^+ ⇌ NH_3 + H^+
Fe^{2+} ⇌ Fe^{3+} + e^-	$HClO_4$ ⇌ ClO_4^- + H^+
Sn^{2+} ⇌ Sn^{4+} + 2 e^-	H_2SO_4 ⇌ SO_4^{2-} + 2 H^+
2 Cl^- ⇌ Cl_2 + 2 e^-	H_2O ⇌ OH^- + H^+

- Elements that are present in medium oxidation numbers act as oxidizing or reducing agents, depending on the reaction partner. This analogue to the amphoteric behaviour of acids and bases is called redoxamphoteric.

16.3 Redox Reactions

Most inorganic reactions that are neither direct acid-base reactions nor complex formation are redox reactions. These include the dissolution of a metal in acid, the combustion of a metal/non-metal or the reaction of an alkali metal with water.

Let us first consider the reaction of zinc with dilute hydrochloric acid:

$$Zn + 2\ HCl \rightarrow ZnCl_2 + H_2 \uparrow$$

In the ion equation, the reaction is represented like this:

$$Zn + 2\ H^+ \rightarrow Zn^{2+} + H_2 \uparrow$$

The zinc apparently supplies two electrons to the protons contained in the solution and discharges them: Hydrogen is released as a gas.If we look at the reaction equation by writing the oxidation numbers, we see that the oxidation number of the zinc increases, so it is oxidized. The oxidation number of the hydrogen, on the other hand, decreases; it is reduced.

$$\overset{0}{Zn} + 2\ \overset{+I\,-I}{HCl} \rightarrow \overset{+II\,-I}{ZnCl_2} + \overset{0}{H_2}\uparrow$$

We can now formulate in two partial equations that account for electron transfer:

- Oxidation $Zn - 2\ e^- \rightarrow Zn^{2+}$ *or* $Zn \rightarrow Zn^{2+} + 2\ e^-$
- Reduction $2\ H^+ + 2\ e^- \rightarrow H_2$ *or* $2\ H^+ \rightarrow H_2 - 2\ e^-$

In the following, the first variant is retained, since the reference to the definitions of terms is better recognizable here.

For ions consisting of only one atom, the oxidation number is equal to the charge. In the case of polyatomic ions, the oxidation number must be determined and written down separately for each element involved. Only in this way can the changes in the oxidation numbers – and thus the number of electrons transferred – be accurately determined.

16.3.1 Types of Redox Reactions

It is observed that in redox reactions the oxidation numbers of the elements involved in the actual reaction always change. On closer examination, however, it is found that not all redox reactions also change the charges of the respective elements, i.e. there is not always a change in the charge state. Some reactions only take place in acidic or alkaline solutions. Therefore, redox reactions are divided into three different types.

16.3.1.1 Redox Reactions Under Electron Exchange

A real exchange of charges takes place here. Uncharged substances receive a charge and thus become ions. Ions are discharged and thus (mostly) become an element:

$$Cu^{2+}_{(aq)} + Zn_{(s)} \rightarrow Zn^{2+}_{(aq)} + Cu_{(s)}$$

In other cases, the charge of the ions involved is increased or decreased by the exchange of electrons:

$$2\ Fe^{3+} + Sn^{2+} \rightarrow 2\ Fe^{2+} + Sn^{4+}$$

16.3.1.2 Redox Reactions Under Partial Electron Transfer

Only the formation of polar electron pair bonds occurs. Therefore, only the oxidation number of the elements involved changes. Charges do not occur.

$$\overset{0}{Si} + 2\ \overset{0}{Cl_2} \rightarrow \overset{+IV}{Si}\ \overset{-I}{Cl_4}$$

16.3.1.3 Protolysis-coupled Redox Reactions

Many redox reactions only take place in an acidic or basic environment. In part, the acidic/alkaline reactions also differ in the number of electrons that are transferred. What they have in common is that water is involved in the form of H^+ or OH- ions. (For the sake of clarity, we will again write H^+ instead of H_3O^+).

+VII – II		+II		+I		+II		+III		+I – II
MnO_4^-	+	$5\ Fe^{2+}$	+	$8\ H^+$	→	Mn^{2+}	+	$5\ Fe^{3+}$	+	$4\ H_2O$

Table 16.3 Redox reactions (examples)

$2\ Mg + O_2 \rightarrow 2\ MgO$	Oxidation: $Mg - 2\ e^- \rightarrow Mg^{2+}$ Reduction: $O_2 + 4\ e^- \rightarrow 2\ O^{2-}$	Reducing agent: Mg Oxidizing agent: O_2
$2\ CO + O_2 \rightarrow 2\ CO_2$	Oxidation: $\overset{+II}{C} - 2e^- \rightarrow \overset{+IV}{C}$ Reduction: $\overset{0}{O_2} + 4e^- \rightarrow 2\ \overset{-II}{O_2}$	Reducing agent: CO Oxidizing agent: O_2
$2\ KI_{(aq)} + Cl_2 \rightarrow 2\ KCl_{(aq)} + I_2$	Oxidation: $2I^- - 2\ e^- \rightarrow I_2$ Reduction: $Cl_2 + 2\ e^- \rightarrow 2\ Cl^-$	Reducing agent: I^- Oxidizing agent: Cl_2
$2\ Na + H_2O \rightarrow 2\ NaOH + H_2$	Oxidation: $2\ Na - 2\ e^- \rightarrow 2\ Na^+$ Reduction: $2\ \overset{+I}{H} + 2e^- \rightarrow H_2$	Reducing agent: Na Oxidizing agent: H^+
$S + O_2 \rightarrow SO_2$	Oxidation: $\overset{0}{S} - 4e^- \rightarrow \overset{+IV}{S}$ Reduction: $\overset{0}{O_2} + 4e^- \rightarrow 2\ \overset{-II}{O_2}$	Reducing agent: S Oxidizing agent: O_2
$4\ P + 5O_2 \rightarrow P_4O_{10}$	Oxidation: $\overset{0}{P} - 5e^- \rightarrow \overset{+V}{P}$ Reduction: $\overset{0}{O_2} + 4e^- \rightarrow 2\ \overset{-II}{O}{}^{2-}$	Reducing agent: P Oxidizing agent: O_2
$2\ P + 3\ Cl_2 \rightarrow 2PCl_3$	Oxidation: $\overset{0}{P} - 3e^- \rightarrow \overset{+III}{P}$ Reduction: $\overset{0}{Cl_2} + 2e^- \rightarrow 2\ \overset{-I}{Cl^-}$	Reducing agent: P Oxidizing agent: Cl_2
$2\ F_2 + O_2 \rightarrow 2\ F_2O$	Oxidation: $\overset{0}{O_2} - 4e^- \rightarrow 2\ \overset{+II}{O}{}^{2+}$ Reduction: $\overset{0}{F_2} + 2e^- \rightarrow 2\ \overset{-I}{F}{}^-$	Reducing agent: O_2 Oxidizing agent: F_2

16.3.2 Examples of Redox Reactions

Table 16.3 shows some examples of different redox reactions. Only a small selection of the many possibilities is to be shown. At the same time, by studying the examples, the terminology can be deepened.

16.3.3 Disproportionation

Reactions in which molecules or ionic compounds with an average oxidation number change into those with higher and lower oxidation numbers are called disproportionations.

		0		+I		−I
2 NaOH	+	Cl_2	→	NaClO	+	NaCl

+V			+VII	−I
$4\ KClO_3$	→	$3\ KClO_4$	+	KCl

16.3.4 Comproportioning

Comproportionation, sometimes also called synproportionation, is the not too often used name for the opposite process to disproportionation: compounds of higher and lower oxidation numbers react with each other to form compounds of intermediate oxidation numbers.

+VII		+II				+IV		
2 MnO_4^-	+	3 Mn^{2+}	+	4 OH^-	$\rightarrow$	5 MnO_2	+	2 H_2O

This type of reaction can be understood analogously to the neutralization reaction:

+VII	+II	+IV	
2 MnO_4^- + 3 $Mn^{2+} \rightleftharpoons$ 5 MnO_2			

$$H_3O^+ + OH^- \rightleftharpoons 2\,H_2O$$

16.4 Setting up Redox Equations

Setting up redox reaction equations is divided into three parts:

1. Determination of the oxidation numbers before and after the reaction
2. Determination of the transferred electrons and compensation of these
3. Balancing of charges, water balance if necessary

16.4.1 Determination of the Oxidation Numbers

Oxidation numbers are imaginary charges of atoms. They are obtained by decomposing the electron pair bonds in molecules or complex ions in such a way that the electronegative element is *mathematically* assigned the shared electrons.

From the electronegativity relations and the rule for the conservation of charges some rules are derived.

Rules for Determining the Oxidation Numbers

1. Elements always have the oxidation number 0.
2. Metals always have positive oxidation numbers.
3. Non-metals generally have negative oxidation numbers.
4. Fluorine always has the oxidation number -I.
5. Oxygen always has the oxidation number -II; exceptions:
   - in peroxides –I (e.g. in H_2O_2)
   - in F_2O, here +II

6. The halogens Cl, Br, I always have oxidation number – I, except in compounds that carry oxygen in the molecule.
7. Hydrogen always has the oxidation number CI; except in hydrides, there –I.
8. For monatomic ions, the oxidation number is equal to the charge.
9. In molecules, the sum of the oxidation numbers must equal zero. (Molecules are electrically neutral).
10. In polyatomic ions, the sum of the oxidation numbers is equal to the charge of the ion (charge of the ion must be conserved). Important oxidation numbers of some elements are listed in Table 16.4.

For redox reactions, it is sufficient to determine which elements have changed their oxidation numbers and then to set up one partial equation each for oxidation and reduction.

16.4.2 Determination and Compensation of Transferred Electrons

After determining which elements change their oxidation number during the reaction, partial equations can be created for oxidation and reduction respectively. This then gives the number of electrons transferred.

Table 16.4 Common oxidation numbers of important elements (examples)

Oxidation number	Important elements with this oxidation number	Example connections
+VIII	Os	OsO_4, OsF_8
+VII	Cl, I	$HClO_4$
+VI	S, U	H_2SO_4, UF_6
+V	N, P, As, Cl	HNO_3,$HClO_3$
+IV	C, Si, Sn, Pb, S	CO_2, SO_2, PbO_2
+III	B, Al, La, Fe, Au, N, P, Cl	PCl_3, $FeCl_3$, $HClO_2$
+II	Mg, Ca, Ba, Sr, Fe, Ni, Cu, Zn, Cd, Hg, Pb	$MgCl_2$, $BaSO_4$, $FeCl_2$, CuO, PbS
+I	H, Li, Na, K, Ag, Cu, Au, Cl	HClO, NaCl, KOH
0	Elements	He, Ar, C, N_2, O_2, P_4, S_8
–I	F, Cl, Br, I, H	NaCl, HF, KBr, NaH
–II	O, S	H_2S, H_2O
–III	N, P	NH_3, PH_3
–IV	C	CH_4

Example

$$\_\_\ K\overset{+VII}{Mn}O_4 + \_\_\ H\overset{-I}{Cl} \rightarrow \_\_\ KCl + \_\_\ \overset{+II}{Mn}Cl_2 + \_\_\ \overset{0}{Cl_2} + \_\_\ H_2O$$

- Oxidation (electron release):

$$\overset{-I}{Cl} - e^- \rightarrow \overset{0}{Cl}$$

Consideration of "gas generation":

$$2\,\overset{-I}{Cl} - 2e^- \rightarrow \overset{0}{Cl_2}$$

- Reduction (electron absorption)

$$\overset{+VII}{Mn} + 5e^- \rightarrow \overset{+II}{Mn}$$

Changes in the oxidation number occur only in manganese and chlorine. Two electrons are transferred during oxidation, five during reduction. To compensate for this, use the "least common multiple" (LCM) and complete the partial equations to the LCM as the electron count. (In this example, the LCM of 2 and 5 = 10).

- Oxidation (electron release):

$$10\,\overset{-I}{Cl} - 10e^- \rightarrow 5\,\overset{0}{Cl_2}$$

- Reduction (electron absorption)

$$2\,\overset{+VII}{Mn} + 10e^- \rightarrow 2\,\overset{+II}{Mn}$$

By determining the total electrons transferred, the "redox part" of the reaction equation is already solved. ◀

16.4.3 Charge Balance, Water Balance

We insert the determined coefficients into the reaction equation

1. $\underline{2}\ \mathrm{K}\overset{+\mathrm{VII}}{\mathrm{Mn}}\mathrm{O_4} + \underline{10}\ \mathrm{H}\overset{-\mathrm{I}}{\mathrm{Cl}} \rightarrow \_\ \mathrm{KCl} + \underline{2}\ \overset{+\mathrm{II}}{\mathrm{Mn}}\mathrm{Cl_2} + \underline{5}\ \overset{0}{\mathrm{Cl_2}} + \_\ \mathrm{H_2O}$

Now balance the remaining cations (here: K, 2 times) and the remaining anions (here: Cl, 2 times for KCl, 4 times for 2 $MnCl_2$ = 6).

2. $\underline{2}\ \mathrm{K}\overset{+\mathrm{VII}}{\mathrm{Mn}}\mathrm{O_4} + \underline{16}\ \mathrm{H}\overset{-\mathrm{I}}{\mathrm{Cl}} \rightarrow \underline{2}\ \mathrm{KCl} + \underline{2}\ \overset{+\mathrm{II}}{\mathrm{Mn}}\mathrm{Cl_2} + \underline{5}\ \overset{0}{\mathrm{Cl_2}} + \_\ \mathrm{H_2O}$

At the end you check hydrogen and oxygen atoms (= "water balance"). In this example there are 8 oxygen atoms (2 times $KMnO_4$) and 16 hydrogen atoms (16 times HCl) on the left. This results in exactly 8 water.

3. $\underline{2}\ \mathrm{K}\overset{+\mathrm{VII}}{\mathrm{Mn}}\mathrm{O_4} + \underline{16}\ \mathrm{H}\overset{-\mathrm{I}}{\mathrm{Cl}} \rightarrow \underline{2}\ \mathrm{KCl} + \underline{2}\ \overset{+\mathrm{II}}{\mathrm{Mn}}\mathrm{Cl_2} + \underline{5}\ \overset{0}{\mathrm{Cl_2}} + \underline{8}\ \mathrm{H_2O}$

16.4.4 "Strikeout Trick"

Redox equations are sometimes a bit confusing. In order to get a better overview of the steps that have already been carried out, it is sometimes useful to cross out the atoms that have already been "completed" by calculation.

The easiest way to do this by hand is to use a slash (/), which is supplemented by a second slash (\) to form an "X". Quite analogous to the marking of searched rooms in SCBU operations: Simply searched = "/" or thoroughly searched = "X".

This notation then appears like this:

1. $\underline{2}\ \mathrm{KMnO_4} + \underline{10}\ \mathrm{HCl} \rightarrow \_\mathrm{KCl} + \underline{2}\ \mathrm{MnCl_2} + \underline{5}\ \mathrm{Cl_2} + \_\mathrm{H_2O}$
2. $\underline{2}\ \mathrm{KMnO_4} + \underline{16}\ \mathrm{HCl} \rightarrow \underline{2}\mathrm{KCl} + \underline{2}\ \mathrm{MnCl_2} + \underline{5}\ \mathrm{Cl_2} + \_\mathrm{H_2O}$
3. $\underline{2}\ \mathrm{KMnO_4} + \underline{16}\ \mathrm{HCl} \rightarrow \underline{2}\mathrm{KCl} + \underline{2}\ \mathrm{MnCl_2} + \underline{5}\ \mathrm{Cl_2} + \underline{8}\mathrm{H_2O}$

The final reaction equation is:

$$2\ KMnO_4 + 16\ HCl \rightarrow 2\ KCl + 2\ MnCl_2 + 5\ Cl_2 + 8\ H_2O$$

Let us now turn to the electrochemical processes. These influence us in many areas of life – often without us being aware of it: Batteries, rechargeable batteries, chrome-plated fittings in bathrooms and kitchens, but also food cans and anodic corrosion protection. Electrochemistry plays a role everywhere.

Redox Pairs

17

As already stated in Chap. 16, oxidation and reduction can only take place in parallel. If one reactant gives up electrons (oxidation), these electrons must be taken up by the other reactant (reduction).

17.1 Half Cells

If a zinc rod is placed in a copper sulphate solution, elemental copper is deposited on the zinc (Fig. 17.1). During this process, known as cementation, a redox reaction takes place:

Oxidation:	$Zn - 2\,e^- \rightarrow Zn^{2+}$
Reduction:	$Cu^{2+} + 2\,e^- \rightarrow Cu$
Redox-operation:	$Cu^{2+} + Zn \rightarrow Cu + Zn^{2+}$

We are dealing with two redox pairs here: $Zn_{(s)}/Zn^{2+}$ and $Cu^{2+}/Cu_{(s)}$. The combination of **one** redox couple with a conductor (an "electrode"), e.g. a submerged wire for electron conduction, is called a half-cell. For metal/metal ion redox couples, the metal in question serves as the electrode.

The combination of two half cells is called cell, chain, galvanic element, galvanic cell or voltaic element. These cells are used in a wide variety of ways as location-independent power sources, e.g. in batteries or accumulators.

If the two redox pairs are separated and the two half-cells thus produced are conductively connected to each other, the voltage generated can be measured. The flow of electrons takes place via the metal electrodes Zn or Cu, which are conductively connected to each other with wires. A voltmeter is interposed to

T. Schmiermund, *The Chemistry Knowledge for Firefighters*,
https://doi.org/10.1007/978-3-662-64423-2_17

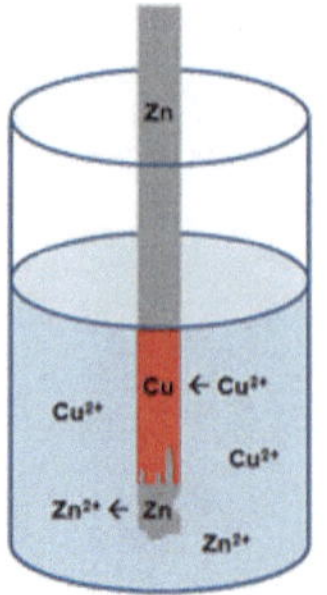

Fig. 17.1 Cementation

measure the voltage. The electrolytic conduction between the separate half-cells takes place either through a so-called diaphragm or a current key.

With reference to the above example, this arrangement, called the Daniell element, is also written $Zn_{(s)}/Zn^{2+}//Cu^{2+}/Cu_{(s)}$. The double slash separates the two half cells, the single slash symbolizes the phase boundary.

17.1.1 Diaphragm

Diaphragm is the name for a porous, partially permeable partition (Greek *dia* = through and *phragma* = separation). The diaphragm serves to separate the anode compartment from the cathode compartment. It prevents the mixing of the electrolytes, but allows the passage of current. Sometimes diaphragms are also called semi-permeable membranes (Greek *semi* = half, *permeo* = to pass through; Latin *membrana* = skin), because only certain particles can pass through them, but not all particles (Fig. 17.2).

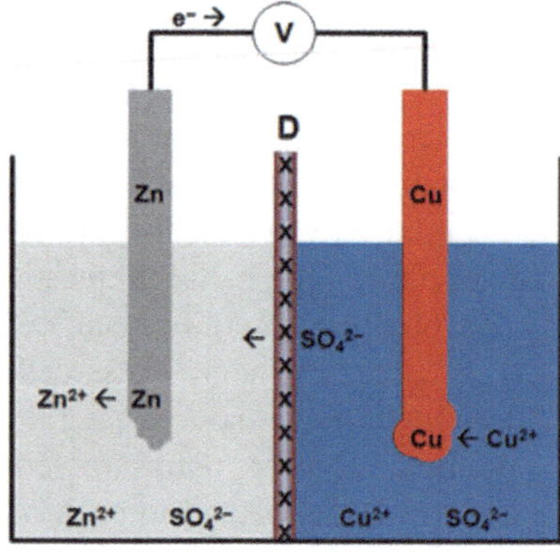

Fig. 17.2 Half cells with diaphragm (D)

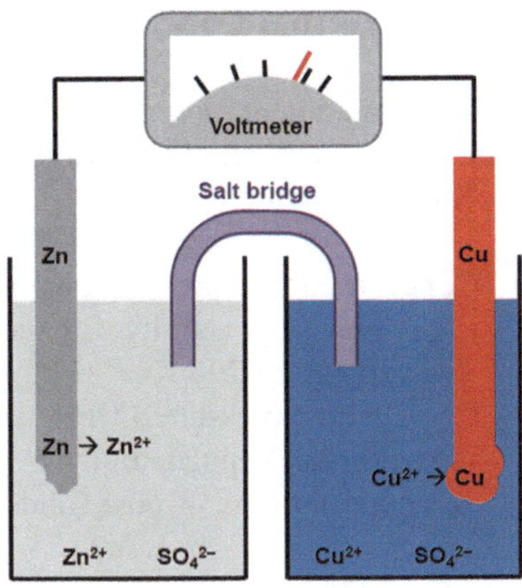

Fig. 17.3 Half cells with salt bridge

17.1.2 Salt Bridge

Salt bridges (in Germany sometimes called "current keys"), electrolyte bridges or ion bridges, are U-shaped tubes filled with a concentrated salt solution (usually KCl or KNO_3 or NH_4NO_3). They serve as an ion-conducting connection between two half cells, see Fig. 17.3.

17.2 Normal Potentials of Redox Couples

A redox pair, such as Zn/Zn^{2+}, Fe^{2+}/Fe^{3+}, H_2O/H_3O^+ or $Cl_2/2\ Cl^-$ has a very specific electrochemical potential, the redox

potential, under precisely defined conditions. Since it is not possible to measure the absolute potentials between, for example $Zn_{(s)}$ and $Zn^{2+}{}_{(aq)}$ individually, the potential differences are determined.

If any half-cell is now combined with a standardised half-cell, the individual voltage of the half-cell can be determined in relation to this standard half-cell. The standard selected is the normal hydrogen electrode and the potential zero is assigned to it.

17.2.1 Standard Hydrogen Electrode

The standard hydrogen electrode (SHE) is a half cell. It consists of a platinized platinum electrode which is flushed with hydrogen gas at a constant pressure of 1.013 bar at 25 °C. The electrode is immersed in an aqueous acid solution whose H_3O^+-concentration is 1 mol/L. (More precisely: the activity $a(H_3O^+) = 1$ mol/L; compare Sect. 14.5.5).

17.2.2 Normal Conditions

If the measurements are carried out against the normal hydrogen electrode at normal conditions, the **normal potentials** $\boldsymbol{E^0}$ of the redox pairs concerned are obtained.

Normal conditions are:

- Concentration of the reactants of 1 mol L^{-1} (more precisely: activity = 1 mol L^{-1})
- Temperature of 298 K = 25 °C
- Gases with a pressure of: 1013.25 hPa = 1013.25 mbar = 1.01325 bar
  It was further stipulated:
- The concentration of pure solids and pure liquids is always 1.

▶ **Please note:** The normal conditions refer to 25 °C (298 K). The standard conditions (gases) refer to 0 °C (273 K). Do not confuse with each other!

17.2.3 Signing

Redox pairs that donate electrons when combined with the normal hydrogen electrode are given a negative sign. They act as reducing agents with respect to the redox couple $H_2/2\ H^+$.

Redox couples that accept electrons opposite the normal hydrogen electrode receive a positive sign. They act as oxidizing agents.

17.3 Electrochemical Series

With the standard potentials determined against the normal hydrogen electrode, a table can be drawn up according to increasing normal potential. This is called the electrochemical series or electromotive (force) series, for short: emf series (Table 17.1).

If we consider only metals, we can differentiate between noble and base metals:

Li K Ca Na Mg Al	Mn Zn Cr Fe Cd Co Ni Sn Pb	$\mathbf{H_2}$	Cu Ag Hg	Au Pt
Light metals (base metal)	Heavy metals (base metal)		Semi-precious metals	Precious metals

Table 17.1 Electrochemical series (detail)

Redox-pair reduced form ⇌ oxidized form	Number of transferred e^-	Normal potential E^0
$Li \rightleftharpoons Li^+$	1	−3.04 V
$K \rightleftharpoons K^+$	1	−2.93 V
$Ca \rightleftharpoons Ca^{2+}$	2	−2.87 V
$Na \rightleftharpoons Na^+$	1	−2.71 V
$Mg \rightleftharpoons Mg^{2+}$	2	−2.37 V
$Al \rightleftharpoons Al^{3+}$	3	−1.66 V
$Cr \rightleftharpoons Cr^{2+}$	2	−0.91 V
$H_2 + 2\ OH^- \rightleftharpoons 2\ H_2O$	2	−0.83 V
$Zn \rightleftharpoons Zn^{2+}$	2	−0.76 V
$Cr \rightleftharpoons Cr^{3+}$	3	−0.74 V
$Fe \rightleftharpoons Fe^{2+}$	2	−0.45 V
$Cr^{2+} \rightleftharpoons Cr^{3+}$	1	−0.43 V
$H_2 \rightleftharpoons 2\ H^+$	2	0.00 V
$Sn^{2+} \rightleftharpoons Sn^{4+}$	2	+0.15 V
$Cu \rightleftharpoons Cu^{2+}$	2	+0.34 V
$Ag \rightleftharpoons Ag^+$	1	+0.80 V
$Pt \rightleftharpoons Pt^{2+}$	2	+1.18 V
$2\ H_2O \rightleftharpoons O_2 + 4\ H^+$	4	+1.23 V
$2\ Cl^- \rightleftharpoons Cl_2$	2	+1.36 V
$Au \rightleftharpoons Au^{3+}$	3	+1.50 V
$Co^{2+} \rightleftharpoons Co^{3+}$	1	+1.92 V
$2\ F^- \rightleftharpoons F_2$	2	+2.87 V

▶ Metals with a negative normal potential (e.g. Mg, Zn, Pb) are called base metals. Metals with positive normal potential (e.g. Cu, Ag, Au) are called noble metals.

17.4 Normal Potential and Reaction Course

The normal potential E^0 of a redox couple characterizes its oxidation or reduction capacity in aqueous solution. The lower (more negative) the potential is, the stronger it acts as a reducing agent. The higher (more positive) the potential of the redox couple, the stronger its effect as an oxidizing agent.

The oxidizable particle (reducing agent) can only be oxidized by the oxidizing agent if the potential of the redox couple of the oxidizing agent is greater (more positive) than that of the reducing agent. Similarly, a reducing agent can only have a reducing effect if its potential is smaller (more negative) than that of the reactant.

▶ From the knowledge of the normal potentials it can be predicted whether a certain redox reaction is possible.

Of particular interest are the redox pairs of metals and the solution of their salts ($Me_{(s)}/Me^{n+}$):

a. All metals with an $E^0 < 0$ (i.e. above hydrogen) dissolve as base metals in non-oxidizing acids with hydrogen evolution:

$$Fe + 2\,H^+ \rightarrow Fe^{2+} + H_2 \uparrow$$

b. Metals with negative E^0 can reduce metals with positive E^0. That is, they can separate the metal ions as an element from their solutions:

$$Zn + Cu^{2+} \rightarrow Zn^{2+} + Cu \downarrow$$

▶ In order to determine whether a redox reaction can take place at all, the following scheme is used:

▶ One writes down the involved redox pairs as in the emf series (= more negative pair upwards) one below the other. The reaction then always takes place from top left to bottom right.

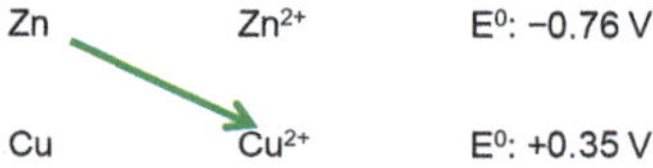

▶ In words: Zn atoms react with Cu^{2+} ions to form metallic Cu and Zn^{2+} ions.

Certain **inhibition phenomena** can occur which nevertheless prevent an electrochemically possible reaction from taking place or noticeably delay it. These include passivation (formation of a dense protective layer of the oxide or hydroxide) and overvoltage (e.g. caused by gas bubbles on the surface).

This is the only way to explain why magnesium, for example, does not react with pure water. According to the voltage series, the same reaction would be expected as for sodium, for example ($2\,Na + 2\,H_2O \rightarrow 2\,NaOH + H_2 \uparrow$); both metals are above hydrogen. The immediate formation of an *insoluble* hydroxide layer ($Mg + 2\,H_2O \rightarrow Mg(OH)_2 + H_2 \uparrow$) causes the reaction to stop immediately. The protective layer of $Mg(OH)_2$ formed prevents further reaction. However, when an acid is added, the hydroxide dissolves ($Mg(OH)_2 + 2\,HCl \rightarrow MgCl_2 + 2\,H_2O$); the reaction inhibition is no longer present.

Aluminium has a very dense oxide layer which protects the metal to a large extent. In strongly basic solutions, the Al-oxide dissolves with complex formation. Aluminium therefore dissolves in alkaline solutions with hydrogen evolution.

The noble metals do not dissolve in acids such as HCl or H – COOH, but partially in oxidizing acids such as concentrated nitric acid or sulfuric acid.

$$Cu + 2\,HCl \nrightarrow Cu + 4\,HNO_3 \rightarrow Cu(NO_3)_2 + 2\,H_2O + 2\,NO_2 \uparrow$$

18 Calculation of the Electromotive Force

The difference between the redox potentials of two half cells is the *maximum* voltage that a galvanic cell can deliver. It is called the electromotive force (EMF).

Potential Difference
For an electric current to flow, there must be a difference in the "level" of the two electric poles. A water "current" flows from the higher level to the lower level due to the difference in height. Electric current flows from the more electron-rich pole (= high level) to the less electron-rich pole (= low level). In the case of water, the force with which the water hits the lower level increases as the height of the waterfall increases. The same applies to electric current: the greater the difference in level, the greater the potential difference.

This potential difference is also called electric voltage. Its unit is the Volt (V).

18.1 Calculation at Normal Conditions

If two different half-cells are used under normal conditions, the EMF can be determined relatively easily. The potential difference (= the maximum voltage) of the cell is determined by the difference of the potentials of the half cells involved:

$$\begin{aligned}
\Delta E^0_{\text{Cell}} &= E^0_{\text{Cathode}} - E^0_{\text{Anode}} && \text{C} - \text{A} \\
&= E^0_{\text{Reduction}} - E^0_{\text{Oxidation}} && \text{Red} - \text{Ox} \\
&= E^0_{\text{Ox-Medium}} - E^0_{\text{Red-Medium}} && (+) - (-) \\
&= E^0_{\text{nobel}} - E^0_{\text{base}} && \text{n} - \text{b}
\end{aligned}$$

These are four identical formulas, with only different indices and therefore with different mnemonic devices.

T. Schmiermund, *The Chemistry Knowledge for Firefighters*,
https://doi.org/10.1007/978-3-662-64423-2_18

▶ A positive ΔE means that the reaction is spontaneous.

Example

An element is composed of the two half-cells tin in tin sulphate solution and silver in silver nitrate solution. The concentration of the solutions is 1 mol L^{-1} in each case.

In which direction does the reaction take place and what voltage is the cell capable of delivering?

The redox pairs are: Sn/Sn^{2+} ($E^0 = -0.136$ V) and Ag/Ag^+ ($E^0 = +0.8$ V).

Sn → Sn²⁺ E⁰: −0.136 V

Ag ← Ag⁺ E⁰: +0.8 V

Tin is oxidized to Sn^{2+} (anode process), Ag^+ is reduced to silver (cathode process).

The electromotive force is calculated according to:

$$\Delta E^0_{\text{Cell}} = E^0_{\text{Cathode}} - E^0_{\text{Anode}}$$
$$\Delta E^0_{\text{Cell}} = 0.8\ \text{V} - (-0.136\ \text{V}) = 0.8\ \text{V} + 0.136\ \text{V} = 0.936\ \text{V}.$$

Similarly, the EMF of the Daniell element (see Sect. 17.1.1) can be calculated as 1.1 V. ◀

18.2 Nernst's Equation

If the reaction partners of a cell or half-cell are not present under normal conditions, this results in a different value for the electromotive force. This EMF can be calculated with an equation developed by W. Nernst[1] in 1889. *(The derivation of Nernst's equation is omitted here.)*

According to Nernst, the EMF of a redox couple is calculated as follows:

$$E = E^0 + \frac{0.059}{n} \cdot \log \frac{c_{\text{Ox}}}{c_{\text{Red}}}$$

To explain:

n Number of electrons transferred
c_{Ox} Product of the concentration of all oxidizing agents
c_{Red} Product of the concentration of all reducing agents

[1] Walther Nernst, 1864–1941, German physicist and chemist.

Stoichiometric factors occur as exponents of the concentrations.

► **Please note:** Nernst's equation does **not** refer to the oxidation/reduction **processes,** but **always** to the oxidizing/reducing agents. The nobler element is always at the top, since it is itself reduced and thus represents the oxidizing agent.

18.2.1 EMF and Concentration Variations

When operating a $Zn_{(s)}/Zn^{2+}//Cu^{2+}/Cu_{(s)}$ cell (Daniell element), the concentrations of both half-cells change with time. The concentration of Cu^{2+} ions decreases and that of Zn^{2+} ions increases. However, this means that the normal conditions ($c = 1$ mol/L) are no longer given. As a result, the EMF must change.

Putting this into Nernst's equation gives the following for the Daniell element:

$$E = E^0_{\text{Cu}} - E^0_{\text{Zn}} + \frac{0.059\text{V}}{n} \cdot \log \frac{c(\text{Cu}^{2+})}{c(\text{Zn}^{2+})}$$

After some time, the copper ion concentration has halved from 1 mol/L (normal conditions) to, for example, 0.5 mol/L. For the concentration of the Zn^{2+} ions this means an increase of 0.5 mol/L and thus a "current" concentration of 1.5 mol/L.If we insert these values into the equation, we calculate

$$E = 0.34\text{ V} - (-0.76\text{ V}) + \frac{0.059\text{ V}}{2} \cdot \log \frac{0.5\frac{\text{mol}}{\text{L}}}{1.5\frac{\text{mol}}{\text{L}}}$$
$$E = 0.34\text{ V} - (-0.76\text{ V}) + 0.0295 \cdot (-0.477\text{ V})$$
$$E = 1.10\text{ V} + (-0.014\text{ V})$$
$$E = 1.086\text{ V}$$

As expected, the voltage decreases because Cu^{2+} ions are "consumed" and are therefore no longer available. If after a certain time all Cu^{2+} ions are reduced to copper, the battery is empty – the redox reaction can no longer take place, electron transport no longer takes place.

18.2.2 Concentration Chains

Since the potential of a half cell depends on the concentration of the ions, a galvanic element can also be constructed from two half cells which differ only in the concentration of the salt solution. Such cell combinations are called concentration chains.

Let us consider two silver/silver nitrate half cells. Let the concentration of half-cell (1) be $c(AgNO_3) = c(Ag^+) = 0.5\ \text{mol L}^{-1}$, and that of half-cell (2) be $c(AgNO_3) = c(Ag^+) = 0.01\ \text{mol L}^{-1}$. Since solutions of different concentration have the tendency to equalize their concentration, the following reactions take place:

Half cell (1)	Half cell (2)
$Ag^+ + e^- \rightarrow Ag$	$Ag - e^- \rightarrow Ag^+$
Cell of higher concentration: Ag^+ concentration becomes lower $\rightarrow$ Ag^+ is oxidizing agent	Cell of lower concentration: concentration Ag^+ increases $\rightarrow$ Ag^+ is reducing agent

Since for the potential of a half cell holds:

$$E_{\text{Ag}} = E^0 + 0.059\ \text{V} \cdot \log c(\text{Ag}^+)$$

results in the combination:

$$\Delta E = 0.059\ \text{V} \cdot \log \frac{c(Ag^+)(1)}{c(Ag^+)(2)}$$

$$\Delta E = 0.059\ \text{V} \cdot \log \frac{0.5\,\frac{\text{mol}}{\text{L}}}{0.01\,\frac{\text{mol}}{\text{L}}} = 0.059\ \text{V} \cdot 1.7 = 0.1\ \text{V}$$

The concentration chain assumed here therefore delivers a maximum voltage of 0.1 Volts.

19 Galvanic Cells

The combination of two half-cells to form a galvanic element (also called a galvanic cell) leads to redox reactions, which supply current due to the flow of electrons.

► In all galvanic elements, chemical energy is converted into electrical energy.

Even today, they are indispensable components of our everyday lives. In flashlights, calculators, watches, mobile phones, motor vehicles, radios, etc. They are used in many ways, as batteries, accumulators or fuel cells.

Galvanic cells are divided into three groups:

I. Primary elements
II. Secondary elements
III. Fuel cells

Historical

Galvani[1] conducted anatomical experiments with frogs' legs in 1780. Their muscles contracted when they came into contact with copper and iron at the same time, whereby the two metals also had to be connected. With the publication of this observation, he laid the foundation for the development of electrochemical cells.

In 1792, Volta[2] learned of Galvani's experiments. The dispute as to whether the frog's leg was a kind of detector (Volta) or rather a kind of capacitor ("current collector", Galvani) prompted Volta to spend many years investigating.

[1] Luigi Galvani, 1737–1798, Italian physician, anatomist, and biophysicist.

[2] Allessandro G. A. A. Volta, 1745–1827, Italian physicist.

T. Schmiermund, *The Chemistry Knowledge for Firefighters*,
https://doi.org/10.1007/978-3-662-64423-2_19

19.1 Batteries

Batteries are intended for single use and are called primary elements. The chemical processes that take place when current is drawn cannot be reversed. On the contrary:

> **Warning** Attempting to charge batteries is extremely dangerous. Often the batteries burst apart.

19.1.1 Historical Batteries

19.1.1.1 Voltaic Column

Around 1799/1800, Volta developed the first battery – the Voltaic column. It consists of many copper and zinc plates alternately layered on top of each other. Pieces of cardboard soaked in saline solution or diluted acid are placed between them at regular intervals.

The arrangement Volta chose (series connection) produced a clearly measurable voltage from the small potential difference of the single Volta element.

As the first power source capable of supplying continuous current, it had a great influence on many technical fields. It enabled the discovery of electrolysis, which in turn led to the production of many base elements (e.g. Na, K, Ca, Mg). It also paved the way for electroplating (the deposition of metal coatings on objects).

19.1.1.2 Daniell Element

In 1836, Daniell[3] developed his zinc-copper element, which has already been discussed in detail (Sect. 17.1). Here, for the first time, the two half-cells were housed in separate reaction chambers. The Daniell element provides a fairly constant voltage over relatively long periods of time.

Unloading Process

Cathode: $Cu^{2+} + 2\,e^- \rightarrow Cu$
Anode: $Zn - 2\,e^- \rightarrow Zn^{2+}$

Cell voltage: ~ 1.1 V

19.1.1.3 Leclanche Element

Leclanché[4] patented the first battery in the modern sense in 1866. At the World's Fair in Paris in the same year, the invention caused a sensation and won a bronze medal.

[3] John Frederic Daniell, 1790–1845, British chemist.

[4] Georges Lechlanche, 1839–1882, French chemist.

In its original form, it was a "wet battery", since the electrolyte (NH_4Cl solution) was housed as a liquid in the case. As a first improvement, the electrolyte solution was thickened by sawdust or starch.

Unloading Process

Cathode: $2\ MnO_2 + 2NH_4^+ + 2\ e^- \rightarrow Mn_2O_3 + H_2O + 2\ NH_3$
Anode: $Zn - 2\ e^- \rightarrow Zn^{2+}$

Ammonia, by the way, is not released as a gas. The NH_3 reacts with the $ZnCl_2$ to form the complex salt $[Zn(NH_3)_2]Cl_2$.

19.1.2 Zinc-carbon Battery

The zinc-carbon battery is a further development of the Leclanche element. The electrolyte (NH_4Cl solution) is thickened so that a "real" dry battery is present.

The actually expected H_2 evolution is prevented by the presence of oxygen-saturated activated carbon (next to the MnO_2). H_2 is oxidized to H_2O. If this is no longer possible, then the battery may inflate and "run out".

Unloading Process

Cathode: $2\ MnO_2 + 2\ NH_4^+ + 2\ e^- \rightarrow Mn_2O_3 + H_2O + 2\ NH_3$
Anode: $Zn - 2\ e^- \rightarrow Zn^{2+}$

Cell voltage: ~ 1.5 V

19.1.3 Alkaline Manganese Cell

It is also a further development of the Leclanche element. The electrolyte is a KOH solution, zinc flakes form the anode, manganese dioxide the cathode. Due to the reversed electrode arrangement compared to the zinc-carbon battery and the two-stage conversion of MnO_2 to $Mn(OH)_2$, up to 50 % better battery performance is achieved.

Unloading Process

Cathode: $MnO_2 + H_2O + e^- \rightarrow MnO(OH) + OH^-$
$MnO(OH) + H_2O + e^- \rightarrow Mn(OH)_2 + OH^-$
Anode: $Zn - 2\ e^- \rightarrow Zn^{2+}$

Cell voltage: ~ 1.5 V

19.1.4 Zinc Mercury Battery

The most common type of this battery is the so-called "button cell". It is often used in small devices, e.g. in watches, pocket calculators, hearing aids, etc.

Unloading Process

Cathode: $HgO + H_2O + 2\ e^- \rightarrow Hg + 2\ OH^-$
Anode: $Zn - 2\ e^- + 2\ OH^- \rightarrow Zn(OH)_2$

Cell voltage: ~ 1.35 V.

Increasingly, the toxicologically questionable HgO of button cells is being replaced by silver oxide, Ag_2O.

19.2 Accumulators

Accumulators are secondary elements. With them, the reactions that take place when the current is drawn can be reversed. The current drain is called the discharge process, the reverse reaction is called the charge process.

However, due to changes in the electrodes, the charging of a battery is not 100 %, so that a accumulator cannot be regenerated (charged) as often as desired.

The term accumulator comes from the Latin *accumulare* = to accumulate, to collect. In English, accumulators are also called rechargeable battery, storage battery or storage cell.

19.2.1 Lead Accumulator

The lead accumulator finds its most common application in motor vehicles and is usually mistakenly referred to as a "battery" (car battery).

Unloading.

Cathode: $PbO_2 + SO_4^{2-} + 4\ H_3O^+ + 2\ e^- \rightarrow PbSO_4 + 6\ H_2O$
Anode: $Pb + SO_4^{2-} - 2e^- \rightarrow PbSO_4$

Cell voltage: ~ 2 V.

By connecting the cells in series within a housing, lead-acid batteries with 6 V, 12 V and 24 V are commercially available.

As the sulphuric acid concentration drops during discharging, the density of the so-called "battery acid" changes. The charge state of the battery can be checked by measuring the density. If the density is < 1.2 g/mL, the battery should then be recharged with a suitable charger.

19.2.2 Nickel-cadmium Accumulator

Due to its design, the nickel-cadmium accumulator (also: "Jungner accumulator") is/was often incorrectly referred to as a "rechargeable nickel battery". It was often used in portable electronic devices in particular. NiCd accumulators may no longer be placed on the German market since 2016 (Battery Act, BattG).

Unloading Process

Cathode: $NiO_2 + 2\,e^- + 2\,H_2O \rightarrow Ni(OH)_2 + 2\,OH^-$
Anode: $Cd + 2\,OH^- - 2\,e^- \rightarrow Cd(OH)_2$

Cell voltage: ~ 1.4 V

19.2.3 Nickel-metal Hydride Accumulator

Ni-MH accumulators are used wherever there is a high energy demand and high battery costs are to be avoided. They are also installed in electric cars. The high self-discharge of these accumulators made them unattractive for many years for devices with a desired long operating time (e.g. clocks, remote controls, fire alarms). Ni-MH batteries with low self-discharge have been on the market since 2006.

Unloading Process

Cathode: $NiO(OH) + e^- + H_2O \rightarrow Ni(OH)_2 + OH^-$
Anode: $MH + OH^- - e^- \rightarrow M + H_2O$

Cell voltage: ~ 1.3 V

19.2.4 Lithium-ion Accumulator

Li-ion accumulators have become indispensable in electrical devices such as smartphones, tablets, cordless screwdrivers, camcorders and many more. They are characterized by a high cell voltage and a high energy density. However, due to the strongly negative potential ($E^0(Li/Li^+) = -3.04$ V), aqueous electrolytes cannot be used.

The anode consists of lithium embedded in graphite, the cathode of a lithium metal oxide compound. Ion transport is ensured via anhydrous organic solvents (e.g. ethylene carbonate) and a membrane permeable only to Li^+ ions.

Unloading Process

Cathode: $CoO_2 + Li^+ + e^- \rightarrow CoLiO_2$
Anode: $Li_xC_n - x\,e^- \rightarrow x\,Li^+ + C_n$

Cell voltage: ~ 3.6 V.

Note If the internal electrodes of lithium-ion accumulators are short-circuited, e.g. due to mechanical damage, the battery may heat up considerably. In the worst case the battery can explode due to this "thermal runaway". Li-ion accumulators must therefore always be shipped in shock-proof packaging. In addition, they may only be packed in foil or placed in collection containers for used accumulators with taped contacts.

19.3 Fuel Cells

Unlike batteries and accumulators, fuel cells do not contain stored energy that is released as electrical energy. They are pure energy converters.

In some ways, fuel cells are similar to primary elements. The difference is that the substances needed for the reaction are constantly supplied "from outside".

In "classic power generation", the combustion of an energy source such as coal or gas is often used to generate steam. The steam then drives a turbine to generate electricity. Chemical energy is thus converted into thermal energy, which is then converted into electrical energy. In a fuel cell, on the other hand, electrical energy is generated directly from the chemical energy.

A fuel cell consists of an anode where the "fuel" (hydrogen, methanol, methane, etc.) is supplied. The cathode is supplied with air or oxygen. Both electrodes are separated from each other by a polymer membrane. Platinum or palladium are often used as catalysts at the electrodes.

Hydrogen Fuel Cell Reaction Process

Cathode: $O_2 + 2\,H_2O + 4\,e^- \rightarrow 4\,OH^-$
Anode: $H_2 + 2\,OH^- - 2\,e^- \rightarrow 2\,H_2O$

Cell voltage: ~ 1 V.

Fuel cells are used as propulsion systems for motor vehicles and are being tested for ships and submarines. They have been used in space travel since the 1960s.

19.4 Electrochemical Corrosion

▶ Corrosion is the destruction of metallic materials by chemical influences.

The term corrosion comes from the Latin *corrodere* = to gnaw away, to eat away – and corrosion spots often look the same. Especially the corrosion of iron and steels, the so-called rusting, plays an important role economically.

Corrosion phenomena are often electrochemical processes in which the metal is oxidized to metal ions, which in turn can react further in a variety of reactions, leading to a wide range of corrosion products.

In moist air, iron rusts strongly simplified according to $4\ Fe + 3\ O_2 + 2\ H_2O \rightarrow 4\ FeO(OH)$. If the water contains dissolved salts or acids, its conductivity increases and corrosion is accelerated. Chloride ions have a particularly corrosive effect, as they catalyse the redox process. Sea water and spray water after salt spreading in winter have a particularly strong effect.

Paint coatings ("rust protection paint"), plastic coatings or enamelling, but also metal coatings (chrome, nickel, tin, zinc, etc.) serve as corrosion protection. The latter show a different behaviour towards the metal to be protected, as a local element is formed.

19.4.1 Local Elements

If two metallic materials with different redox potentials come into contact and an electrolyte (e.g. rainwater) is present at the same time at the point of contact, then a so-called local element or corrosion element is formed.

This is nothing more than a very small, short-circuited galvanic element. The metal with the lower redox potential (the less noble metal) slowly dissolves here. Welds and rivets are particularly at risk, as two metals with different potentials come into direct contact with each other here.

If steel is coated with **zinc** (Fig. 19.1), the iron is protected from rusting. If the zinc layer is damaged, a local element is formed. Zinc, as the less noble element, forms the anode. The zinc dissolves, the iron still remains protected.

Since zinc may not be used for foodstuffs, the so-called tinplate can was originally provided with a **tin layer** (Fig. 19.2). However, if the tin layer is damaged, the base iron dissolves. As a result, the tin is destroyed. Today, genuine tinplate cans have become rare. In most cases, tin cans are coated with paint or plastic to protect the metal of the can from attack.

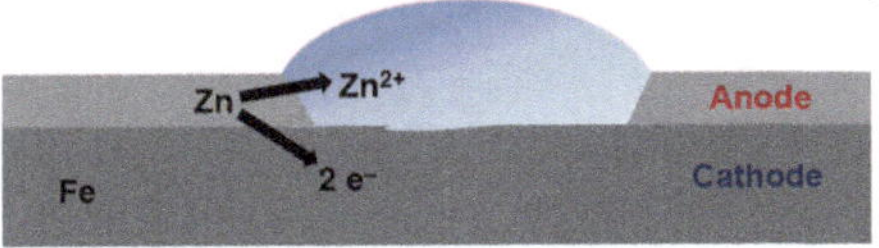

Fig. 19.1 Corrosion of zinc-plated steel

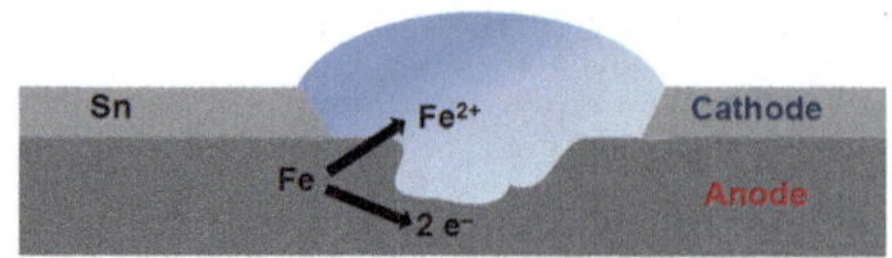

Fig. 19.2 Corrosion of tin-plated steel

19.4.2 Cathodic Corrosion Protection

For the protection of larger installations and objects, such as ships, bridges, pipelines, etc., local elements are intentionally produced. Metals with a strong negative potential (e.g. Mg or Zn) are used as anode. Since the material to be protected serves as the cathode, this is also called cathodic corrosion protection.

In the case of ship hulls, for example, zinc plates are applied to the steel hull. Pipelines and water pipes are usually protected with magnesium anodes. The effect is always the same: the more noble metal is protected because the less noble metal corrodes; it acts as an anode, is oxidized to the metal ion and dissolves. One therefore also speaks of **sacrificial anodes**.

20 Electrolysis

If an electric current is passed through a substance or its solution and chemical changes take place, this is commonly referred to as electrolysis. Electrolysis is more or less the reverse of the processes that occur when a battery or accumulator is discharged. It can also be said that the charging of an accumulator represents electrolysis.

▶ Electrolysis is the totality of all chemical changes in a substance when a direct electric current flows through it.

The chemical changes that occur during electrolysis are oxidation, reduction, decomposition. The term "a substance" does not exclusively mean a pure substance. It can also be heterogeneous or homogeneous mixtures (such as solutions).

Direct current must be used in electrolysis. With alternating current, the positive and negative poles constantly alternate back and forth, in Germany at 50 Hertz (Hz). This corresponds to a pole reversal every 20 ms. Electrode processes could not take place at this alternating frequency.

20.1 Electrode Processes

In order for a reaction to take place at the electrodes at all, conductivity must be ensured by free-moving ions. This is the case in solutions or in melts. An attempt to carry out electrolysis on a common salt crystal is doomed to failure, as the ions cannot move away from their lattice sites. Liquefied HCl gas (anhydrous!) or sugar water cannot be electrolyzed either, since there are no ions present that could carry the current. Both are electrically non-conductive.

If, on the other hand, free-moving ions or ion-like ("ionogenic") compounds are present, they align themselves in the electric field and migrate to the oppositely charged electrode:

T. Schmiermund, *The Chemistry Knowledge for Firefighters*,
https://doi.org/10.1007/978-3-662-64423-2_20

- Opposite charges attract each other.

Once the particles have reached the respective poles, the actual electrolysis process can take place.

20.1.1 Electrophoresis

The electrodes immersed in the solution or melt form an electric field within the liquid. This leads to the migration of the ions to the corresponding oppositely charged pole. This migration is called electrophoresis.

- A negatively charged particle moves to the positive pole, a positively charged particle to the negative pole.

Electrophoresis is mainly used as an analytical method in medicine and biology, e.g. for DNA analysis.

20.1.2 Electrolysis

If the voltage between the electrodes is high enough, electrolysis takes place. Cations can then draw electrons from the negative pole (cathode) and discharge themselves (= reduction process). Similarly, anions can also discharge at the positive pole (anode) by releasing their excess charge (= oxidation process).

$$\text{Cation} \rightarrow \text{Reduction at the cathode e.g.}: Cu^{2+} + 2\,e^- \rightarrow Cu$$
$$\text{Anion} \rightarrow \text{Oxidation at the anode e.g.}: 2\,Cl^- - 2\,e^- \rightarrow Cl_2$$

20.1.3 Decomposition Voltage

For the actual electrolysis, a certain minimum voltage is required between the electrodes, the decomposition voltage. If the voltage is lower than the decomposition voltage required for this process, then no electrolysis processes take place.

Theoretically, the decomposition voltage results from the difference of the normal potentials of the voltage series.

Example

What is the decomposition voltage (U_Z) for the electrolysis of a copper (II) chloride solution? E^0 (Cu/Cu^{2+}) = 0.34 V; $E°(Cl_2/2Cr)$ = 1.36 V.

$$U_Z = E^0(Cl_2/2Cl^-) - E^0(Cu/Cu^{2+}) = 1.36\ V - 0.34\ V = 1.02\ V$$

At voltages <1 V, no metal deposition will be observed. ◀

Due to the so-called **overvoltage,** anomalous increases in the decomposition voltage often occur in practice. In particular, when gases are generated during electrolysis, the gas bubbles often cover the electrode. This limits the effective electrode surface and a higher voltage is required to carry out the electrolysis.

20.2 Electrolysis of Aqueous Solutions

During the electrolysis of aqueous solutions, H_3O^+ and OH^- ions are present in addition to the ions of the dissolved salt. These ions originate from the autoprotolysis of the water or were intentionally added to the electrolyte solution by acid or alkali addition.

If the salt is composed of a very base metal and an oxygen-containing acid residue, components of the water (H_2 and O_2) may be deposited on the electrodes. This is due to the fact that the ions are always deposited as the element whose discharge requires the lowest voltage.

▶ If an electrolyte solution contains several cations and anions, the ions whose discharge requires the lowest voltage are deposited.

20.2.1 Dischargeability Series

According to the voltage series of metals, only copper and more noble metals should be deposited from aqueous solution. However, experiments show that metals less noble than copper can also be deposited. The reason for this is the overvoltage of the hydrogen: a higher voltage is required to discharge the H_3O^+ ions than would be expected according to the voltage series. The hydrogen moves "down" in the voltage series, so to speak, so that there are still other metals below the hydrogen. This series is called the dischargeability series.

From this dischargeability series (Fig. 20.1), it can be read which ions are deposited from aqueous solution containing several ions.

▶ From a solution with several ions, the cation and the anion which is furthest down the dischargeability series is deposited.

▶ If only non-precipitable ions are present in a solution, hydrogen and oxygen are precipitated.

20.2.2 Faraday's Law

To deposit one mole of silver from a solution containing Ag^+ ions, 1 mole of electrons is required. To separate one mole of copper from a solution containing

Cations	Precipitated substances	Anions	Precipitated substances	
K^+ Ca^{2+} Na^+ Mg^{2+} Al^{3+}		SO_4^{2-} NO_3^- CO_3^{2-} F^-		**Not** separable from aqueous solution
H_3O^+ Zn^{2+} Fe^{2+} Cd^{2+} Ni^{2+} Sn^{2+} Pb^{2+} Cu^{2+} Ag^+ Hg^{2+} Pt^{2+} Au^{3+}	**H_2** Zn Fe Cd Ni Sn Pb Cu Ag Hg Pt Au	**OH^-** Cl^- Br^- I^-	**O_2** Cl_2 Br_2 I_2	Can be separated from aqueous solution

Fig. 20.1 Dischargeability series

Cu^{2+} ions, 2 moles of electrons are required. If we determine the amount of electricity (Q) required to separate 1 mole of a substance from its ions with charge number 1, this amount of electricity is 96,485 coulombs (C). The following is true: 1 coulomb = 1 ampere · second (1 C = 1 As). This number, 96,500 C mol^{-1}, is also called Faraday's constant *(F)*. It can also be derived from the elementary electric charge of the electron (see Sect. 5.4):

$$F = e_0 \cdot N_A = 1.602177 \cdot 10^{-19} \text{A s} \cdot 6.022\,045 \cdot 10^{23} 1/\text{mol}$$
$$\approx 96\,500 \text{ A s mol}^{-1}$$

For the deposited amount of substance then results the Faraday's law named after its discoverer Faraday:[1]

$$n = \frac{Q}{z \cdot F}$$

About the relations

- $m = n \cdot M$ (mass = amount of substance · molar mass)
- $Q = I \cdot t$ (amount of charge = current · time)

Faraday's law can be reformulated as follows

[1]Michael Faraday, 1791–1867, English naturalist

$$m = \frac{I \cdot t \cdot M}{z \cdot F}$$

It means:

m	Mass of the separated substance in grams (g)
I	Current in amperes (A)
t	Electrolysis time in seconds (s)
M	Molar mass in grams per mole (g mol^{-1})
z	Number of electrons transferred
F	Faraday constant, 96,500 A s mol^{-1}

To determine the time to deposit a given amount, rearrange the equation to:

$$t = \frac{m \cdot z \cdot F}{M \cdot I}$$

Using the density of the deposited metal and the surface of the workpiece, the time required to achieve a desired layer thickness can be calculated.

Example

1. Fused-salt electrolysis of aluminium is carried out at 42 kA. What mass of Al is deposited in 24 hours?

$$m = \frac{I \cdot t \cdot M}{z \cdot F} = \frac{42\,000 \text{ A} \cdot 86\,400 \text{ s} \cdot 27 \text{ g}}{3 \cdot 96\,500 \text{ A s} \cdot \text{mol}} = 338\,437 \text{ g} = 338.4 \text{ kg}$$

A good 338 kg of aluminium is produced.

2. A cube with an edge length of 6 cm is to be silver-plated. Electrolysis is to be carried out at a current of 2 A and the layer thickness is to be 20 μm. How long must the electrolysis take?

a. The cube surface area is:

$$6 \text{ cm} \times 6 \text{ cm} \cdot 6 \text{ sides} = 216 \text{ cm}^2$$

b. The silver volume is:

$$216 \text{ cm}^2 \cdot 20 \text{ μm} = 0.0216 \text{ m}^2 \cdot 20 \cdot 10^{-6} \text{ m} = 0.000\,000\,432 \text{ m}^3 = 0.432 \text{ cm}^3$$

c. From the density of silver (ϱ (Ag) = 10.49 g/cm^3), the mass is:

$$m = \varrho \cdot V = 10.49 \text{ g cm}^{-3} \cdot 0.432 \text{ cm}^3 = 4.5317 \text{ g}$$

d. When inserted into the formula, the time required for electrolysis is calculated as:

$$t = \frac{m \cdot z \cdot F}{M \cdot I} = \frac{4.5317 \text{ g} \cdot 1 \cdot 96\,500 \text{ A s} \cdot \text{mol}}{107.87 \text{ g} \cdot \text{mol} \cdot 2 \text{ A}} = 2027 \text{ s} = 33 \text{ min } 47 \text{ s}$$

The desired layer thickness is achieved in about 34 min. ◀

20.3 Applications of Electrolysis

Electrolysis has found wide application in technology. The most important processes will be briefly discussed.

▶ If there are companies in the fire brigade's response area that use electrolysis processes, a visit to the company should take place. This is the only way to get to know the local conditions and to agree on safety rules for use. Particularly the sometimes very high currents and magnetic fields generated by the current often harbour dangers that can easily be misjudged.

20.3.1 Chlorine-alkali Electrolysis

Chlorine and caustic soda are among the most important basic materials in industry. Both can be obtained from common salt via electrolysis. On a large scale, a distinction is made between the amalgam process (with mercury as the cathode), the diaphragm process and the membrane process. In the latter two, the only difference is the "separating device" between the anode and cathode chambers.

Chlor-alkali Electrolysis

Cathode: $2\ H_2O + 2\ e^- \rightarrow H_2 + 2\ OH^-$

Anode: $2\ Cl^- - 2\ e^- \rightarrow Cl_2 \uparrow$

In the amalgam process, on the other hand, the sodium produced dissolves in the mercury. The resulting alloy (sodium amalgam) is then brought into contact with water:

$$Na_2Hg_n + 2\,H_2O \rightarrow 2\,NaOH + H_2\uparrow + n\,Hg$$

The summary redox equation for all three variants of chlorine-alkali electrolysis is as follows:

$$2\,NaCl + 2\,H_2O \rightarrow 2\,NaOH + H_2\uparrow + Cl_2\uparrow$$

20.3.2 Fused-salt Electrolysis

The electrolysis of molten salts ("fused-salt electrolysis", "melt electrolysis") can be used in particular to obtain highly reactive metals whose redox potential is lower than that of hydrogen. In this process, the salt is heated above the melting point so that a melt is obtained which is as fluid as possible and which is then electrolyzed. This method is used for the production of Li, Na, K, Ca and Mg, among others.

Example Lithium Production
Cathode: $2\,Li^+ + 2e^- \rightarrow 2\,Li$
Anode: $2\,Cl^- - 2\,e^- \rightarrow Cl_2\uparrow$

20.3.3 Aluminium Extraction

The production of aluminium is the most common fused-salt electrolysis on a large scale. The mineral bauxite (aluminium oxide, Al_2O_3) is used as the raw material. Because of the high melting point, cryolite (= $Na_3[AlF_6]$) is added. This allows the melting temperature to be reduced from over 2000 °C to 960 °C. Carbon electrodes are used as anode, which are oxidized to CO_2 with the resulting oxygen.

Aluminium Production

Cathode: $4\,Al^{3+} + 12\,e^- \rightarrow 4\,Al$
Anode: $6\,O^{2-} - 12\,e^- \rightarrow 3\,O_2$
$3\,C + 3\,O_2 \rightarrow 3\,CO_2$

The aluminium separates in liquid form at the prevailing temperatures and is either cast in ingots or transported in special thermal containers as molten metal for further processing (UN No. 3257; heated liquid, n.o.s.). Electrolysis takes place at a voltage of ~3.5 V and at a current of about 300 000 A. The production of one tonne of aluminium thus requires an energy input of about 15 000 kilowatt hours (kWh). This is enough to supply a single-family house for more than two years. Recycling aluminium requires only about 5 % of the energy needed to produce it from bauxite.

20.3.4 Metal Refining

Electro-refining is used to purify metals. This is a process that has become particularly widespread in the production of pure copper.

Here, the impure copper is connected as the anode. Due to the electron withdrawal (oxidation), Cu^{2+} ions are formed. These are then reduced again to elemental copper at the cathode. Metals that are more noble than copper (e.g. Ag, Au) do not go into solution and can be recovered from the so-called "anode slime". Less noble metals (e.g. Fe, Zn) remain dissolved. The "electrolytic copper" has a purity of 99.99 %.

Copper Refining

Cathode: $Cu^{2+} + 2\ e^- \rightarrow Cu$

Anode: $Cu - 2\ e^- \rightarrow Cu^{2+}$

20.3.5 Anodising Process

Electrolytically oxidised aluminium (anodising) uses electrolysis to create an artificial aluminium oxide layer on the workpiece. This is very resistant and can also be easily coloured. Mobile phone and tablet housings, as well as decorative carabiners, window frames, automotive components and many more are protected against corrosion in this way.

Eloxal Process

Cathode: $6\ H_3O^+ + 6\ e^- \rightarrow 6\ H_2O + 3\ H_2$

Anode: $2\ Al - 6\ e^- \rightarrow 2\ Al^{3+}$

$$\mathbf{2\ Al^{3+} + 9\ H_2O \rightarrow Al_2O_3 + 6\ H_3O^+}$$

20.3.6 Electroplating

Electroplating is the process of depositing a thin layer of metal on a workpiece. The workpiece is connected as the negative pole (cathode) and suspended in a bath containing a solution of the corresponding metal salt. After applying the necessary decomposition voltage, the corresponding metal is deposited on the surface of the workpiece.

Electroplating can be used to decorate the workpiece (e.g. gold plating, silver plating, chrome plating), but it can also be used to protect the workpiece, as corrosion protection (e.g. hard chrome plating).

Non-conductive objects (porcelain, glass, plastic) must be provided with an appropriate conductive lacquer before electroplating.

▶ **Attention!** Special attention is required when used in electroplating operations. Many electroplating baths contain toxic cyanides. If strong acids get into cyanide-containing solutions, hydrogen cyanide (HCN "prussic acid") may be released.

20.4 Terms of Electrochemistry

The terms used in electrochemistry go back to Faraday, who first used them in an article published in 1834.

20.4.1 Electrolysis terms

Term	Greek word origin	Meaning
Electrical	*electron* = amber	In the usual sense "electric"
Electrolysis	*lytikos* = dissolvable	Release / "detach" with current
Electrode	*hodos* = way	Current/electron conductor
Anode	*anodos* = ascent, way to the top	Positively charged electrode / (+)-pole
Anion	*ana* = up and *ion* = wanderer	Negatively charged ion (A^-)
Cathode	*kathodos* = way back, way down	Negatively charged electrode / (−)-pole
Cation	*kata* = down and *ion* = wanderer	Positively charged ion (Z^+)

Faraday assumed that electricity flows "from above" (= from the anode) "downwards" (= to the cathode) – i.e. from positive to negative. This way of looking at it is today called the "technical direction of current". It is opposite to the "physical direction of current" (= the direction of movement of the electrons).

20.4.2 Comparison Cathode/Anode

It should not have escaped the attentive reader that in processes in a battery the anode is the negative pole, but in electrolysis the positive pole. This seems confusing at first glance. As a mnemonic aid, the

Vowel Rule

▶ Anode – electron withdrawal – **al**ways – **o**xidation – **no**ble partner

- **Or, to put it a little differently:**

▶
- Anode = oxidation (mnemonic: vowels "A" and "O")
- Cathode = reduction (mnemonic: consonants "C" and "R")

For a better overview, the terms anode and cathode are summarized in the following table under the application aspects.

	Cathode	Anode
Chemical processes	Reduction Addition of electrons	Oxidation Electron withdrawal
Battery operations	Consumes electrons (+)-pole	Generates electrons (−)-Pole
Electrolysis processes	Attracts cations (Z^+) to (−)-pole	Attracts anions (A^-) to (+)-pole

▶ If you keep in mind that electrolysis is the reversal of the redox processes during the discharging process of a battery, remember the polarity *for* only one of the processes *(recommendation: electrolysis process)*.

Part VIII

Radioactivity

Few topics in chemistry and physics are as emotionally charged as the topic of radioactivity. Radioactive (better and more correctly: ionizing) radiation cannot be perceived with the human senses. It cannot be smelled, tasted, felt or heard. And yet it is there and can cause damage.

The use of the "power of the atom", along with the invention of atomic and hydrogen bombs, led to an initially almost euphoric mood. People wanted to tame "the atom", "clean nuclear energy" was supposed to ensure the power supply. This too has changed. In Germany, all nuclear power plants are being shut down. Not least a consequence of disposal problems concerning the resulting highly radioactive waste and serious accidents in nuclear power plants; e.g.: Three Mile Island (USA, 1979-03), Chernobyl (then USSR, now Ukraine, 1986-04) and Fukushima (Japan, 2011-03).

The radiation emitted by radioactive substances and the damage caused by them are hardly comprehensible, since they generally do not cause any clear damage, as is the case, for example, with acid burns. Doctors disagree about the effects of the smallest amounts of radiation on health. This also contributes to the uncertainty of the population.

Natural radiation exposure takes a back seat to man-made radioactivity. Many messages in the media aim at "big headlines" and neglect a critical, scientifically based approach.

The "unknown", the "intangible" creates uncertainty, sometimes even fear. This is what makes the subject of "radioactivity" so - shall we say - *sensitive.*

And yet, radioactive substances and the ionising radiation they emit have a firm place in our highly technical world: in medical diagnostics, for level and thickness measurements, for material testing, for sterilising food and medical products and even as medicines.

And another advantage for the emergency forces must not be neglected:

► Nothing from the field of NBC hazardous substances can be measured and determined as well and as quickly as the ionizing radiation of radioactive substances.

21 Background Knowledge "Radiation"

Before we familiarize ourselves with the actual processes of radioactivity, let us first take a look at the concept of radiation. Here we want to distinguish between particle radiation and wave radiation. For electromagnetic wave radiation, the relationship between energy, frequency and wavelength will also be explained.

21.1 Waves and Wave Radiation

If you throw a stone onto a mirror-smooth water surface, the still water is disturbed. However, this disturbance does not remain at the place where the stone hit. It spreads out in the form of waves on the water surface. The energy that the stone has given to the water is transported further by the waves.

▶ Wave phenomena are "disturbances" in a medium or field.

Another Example Will Illustrate the Wave Nature

We tie one end of a rope to a post, take the other end in our hand and pull the rope taut. If we now move our hand up and down, we disturb the rest position of the rope. This "disturbance of the rest position" runs as a wave through the entire rope. Each point of the rope merely moves up and down, there is no transport of matter. The up and down movement of the rope makes the energy transport visible. ◀

▶ Waves are a form of energy transport in a medium, without matter transport. Here, the energy is transferred in the form of oscillations.

T. Schmiermund, *The Chemistry Knowledge for Firefighters*,
https://doi.org/10.1007/978-3-662-64423-2_21

21.1.1 Electromagnetic Waves

The "rope waves" mentioned above as an example are bound to a medium (rope, water, air) just like water waves or sound waves. Light and other electromagnetic waves are not bound to such a material medium, but to the electromagnetic field.

Electromagnetic Field

Magnetic fields can cause a current flow (principle of the dynamo). Analogously, currents can generate a magnetic field (principle of the electromagnet). This suggests that there is a direct connection here. These connections were expressed in 1864 by Maxwell[1] in his Maxwell equations. These equations describe the behaviour of electric and magnetic fields and their interactions. The actual electromagnetic field is described in this model as a continuously alternating magnetic and electric phenomenon.

Wave-Particle Duality

Newton[2] was of the opinion that visible light consisted of very small particles. Huygens[3] explained optical effects as wave phenomena. In 1867, Maxwell was able to explain the connection between the speed of light and electrodynamics, and finally in 1888 Hertz[4] succeeded in proving electromagnetic waves. In 1900, Planck showed that the transfer of energy between radiation and matter is only possible in the form of energy packets ("quanta"). The energy (E) is linked to the frequency (ν) via the so-called "Planck's quantum of action" (h): $E = h \cdot \nu$. In 1905, Einstein[5] was able to prove that the energy transfer in the photoelectric effect can be explained by means of light quanta according to $E = h \cdot \nu = h \cdot c/\lambda$. For this he received the Nobel Prize. From 1909 onwards, he advocated the development of a theory that recognised both wave and particle properties of light.

Today, wave-particle dualism is an established fact. One can imagine light – and other electromagnetic waves – as small wave packets. However, matter (particles) is also subject to wave phenomena (Sect. 5.2.4). Matter and radiation can therefore neither be fully described by the particle picture nor by the wave picture. Both models must be used side by side. Each individual experience is then best described by the respective partial model. Einstein said: *"Light occurs in quanta. Light of a frequency v occurs in quanta, all of which have the same energy h · ν. More intense light means more quanta (= photons) of the same energy. A different frequency (color) of light changes the energy of the photons."*

[1] James Clerk Maxwell, 1831–1879, Scottish physicist.

[2] Isaac Newton, 1643–1727, English naturalist.

[3] Christiaan Huygens, 1629–1695, Dutch physicist.

[4] Heinrich Hertz, 1857–1894, German physicist.

[5] Albert Einstein, 1879–1955, theoretical physicist, Nobel Prize in Physics in 1921.

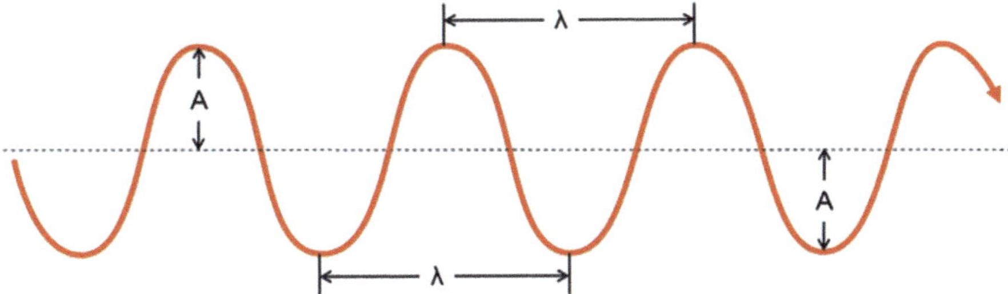

Fig. 21.1 Wavelength (λ) and amplitude (A) of a wave

Characterization of Electromagnetic Radiation

To characterize electromagnetic radiation are used:

- Wavelength (λ, Greek letter *lambda)*
  The distance between two consecutive points located at the same point on the wave (e.g. wave crest or wave trough). Formula: $\boldsymbol{\lambda = c/\nu}$ or $\boldsymbol{\lambda = c \cdot h/E}$.
- Amplitude (A)
  The distance from the wave crest to the baseline. The brightness (intensity) of the radiation is proportional to A^2.
- Propagation speed (c)
  The speed of light (in a vacuum) is $2.997\ 924\ 58 \cdot 10^8$ m s^{-1}. In denser substances, such as water or glass, it is somewhat lower. Usually it is sufficient to calculate with the value $c = 300\ 000$ km s^{-1}. Formula: $\boldsymbol{c = \lambda \cdot \nu}$ or $\boldsymbol{c = \lambda \cdot E/h}$.
- Frequency (ν, Greek letter *ny)*
- The number of waves passing a given location per second. The unit is the hertz (Hz); 1 Hz = 1 s^{-1}. Formula: $\boldsymbol{\nu = E/h}$ or $\boldsymbol{\nu = c/\lambda}$.
- Planck's quantum of action (h)
  Not all quanta (photons) contain the same energy. The short-wave, high-frequency X-rays are more energetic than the longer-wave heat rays. The proportionality factor is Planck's quantum of action or Planck's constant h. Its value is $6.626 \cdot 10^{-34}$ J s or $4.14 \cdot 10^{-15}$ eV s. Formula: $\boldsymbol{h = E \cdot \nu}$ or $\boldsymbol{h = \lambda \cdot E/c}$.
- Photon energy
  The energy of the light quanta (photons) can be derived from the wavelength ($\boldsymbol{E = h \cdot \nu}$) or from the frequency ($\boldsymbol{E = h \cdot c/\lambda}$).

(Compare also Fig. 21.1)

Example

1. What is the energy of light of wavelength 700 nm (= $700 \cdot 10^{-9}$ m)?

$$E = h \cdot \frac{c}{\lambda} = 6.626 \cdot 10^{-34}\ \mathrm{Js} \cdot \frac{3.0 \cdot 10^8\ \mathrm{m/s}}{700 \cdot 10^{-19}\ \mathrm{m}} = 2.84 \cdot 10^{-19}\ \mathrm{J}$$

2. What is the wavelength and frequency of light with an energy of $5.7 \cdot 10^{-19}$ J?

$$\nu = \frac{E}{h} = \frac{5.7 \cdot 10^{-19}\ \text{J}}{6.626 \cdot 10^{-34}\ \text{Js}} = 8.6 \cdot 10^{14}\ \text{s}^{-1.}$$

$$\lambda = \frac{c}{\nu} = \frac{3.0 \cdot 10^{8}\ \text{m/s}}{8.6 \cdot 10^{14}\ \text{s}} = 349 \cdot 10^{-9}\ \text{m} = 349\ \text{nm}$$

◀

▶ **Summary** Light **of** half the wavelength (= twice the frequency) has twice the energy.

21.1.2 Electromagnetic Spectrum

The bandwidth of electromagnetic waves ranges from radio waves to thermal radiation, visible light and UV radiation to high-energy X-rays. An energy can be assigned to each wavelength or frequency (Fig. 21.2).

If one wants to calculate the energy of "one mole" of photons, the energy has to be multiplied by the Avogadro constant ($N_A = 6.022 \cdot 10^{23}\ \text{mol}^{-1}$). This value is also called the **light equivalent** (E_{le}). Accordingly, it results in:

$$E_{le} = N_A \cdot h \cdot \nu$$

If these energies are calculated for visible light, Table 21.1 results.

Analogously, for example, for radio waves from the VHF range ($\lambda = 1$ m) a light equivalent of 0.12 J = 120 mJ or for gamma radiation ($\lambda = 10^{-10}$ m) of $1.2 \cdot 10^{9}$ J = 1.2 GJ results.

21.1.3 Energy Unit Electron Volt (eV)

Typical units of energy are the joule (J), the watt-second (W s), and the newton-meter (N m), noting that the watt (W) equals volts (V) times amperes (A). Therefore:

$$1\ \text{J} = 1\ \text{Nm} = 1\ \text{Ws} = 1\ \text{VAs}$$

In contrast, the electron volt (eV) is used as the atomic-physical unit of energy.

▶ An electron volt (eV) is the energy experienced by an electron when accelerated with 1 V (Fig. 21.3).

In classical screens, the energy of the electron beam that creates the image on the picture tube is about 20 keV (= 20 000 V).

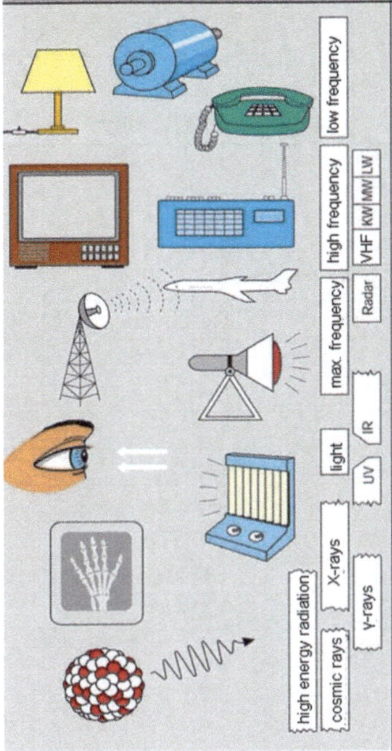

Type of radiation	Frequency in s−1	Wave length in m	Energy in eV	Energy in J
low frequency	$3 \cdot 10^{0}$	10^{8}	$1.24 \cdot 10^{-14}$	$1.99 \cdot 10^{-33}$
	$3 \cdot 10^{1}$	10^{7}	$1.24 \cdot 10^{-13}$	$1.99 \cdot 10^{-32}$
	$3 \cdot 10^{2}$	10^{6}	$1.24 \cdot 10^{-12}$	$1.99 \cdot 10^{-31}$
	$3 \cdot 10^{3}$	10^{5}	$1.24 \cdot 10^{-11}$	$1.99 \cdot 10^{-30}$
	$3 \cdot 10^{4}$	10^{4}	$1.24 \cdot 10^{-10}$	$1.99 \cdot 10^{-29}$
high frequency, LW	$3 \cdot 10^{5}$	10^{3}	$1.24 \cdot 10^{-9}$	$1.99 \cdot 10^{-28}$
MW	$3 \cdot 10^{6}$	10^{2}	$1.24 \cdot 10^{-8}$	$1.99 \cdot 10^{-27}$
KW	$3 \cdot 10^{7}$	10	$1.24 \cdot 10^{-7}$	$1.99 \cdot 10^{-26}$
VHF	$3 \cdot 10^{8}$	1	$1.24 \cdot 10^{-6}$	$1.99 \cdot 10^{-25}$
max. frequency, Radar	$3 \cdot 10^{9}$	10^{-1}	$1.24 \cdot 10^{-5}$	$1.99 \cdot 10^{-24}$
	$3 \cdot 10^{10}$	10^{-2}	$1.24 \cdot 10^{-4}$	$1.99 \cdot 10^{-23}$
	$3 \cdot 10^{11}$	10^{-3}	$1.24 \cdot 10^{-3}$	$1.99 \cdot 10^{-22}$
IR	$3 \cdot 10^{12}$	10^{-4}	$1.24 \cdot 10^{-2}$	$1.99 \cdot 10^{-21}$
	$3 \cdot 10^{13}$	10^{-5}	$1.24 \cdot 10^{-1}$	$1.99 \cdot 10^{-20}$
light	$3 \cdot 10^{14}$	10^{-6}	$1.24 \cdot 10^{0}$	$1.99 \cdot 10^{-19}$
UV	$3 \cdot 10^{15}$	10^{-7}	$1.24 \cdot 10^{1}$	$1.99 \cdot 10^{-18}$
	$3 \cdot 10^{16}$	10^{-8}	$1.24 \cdot 10^{2}$	$1.99 \cdot 10^{-17}$
X-rays	$3 \cdot 10^{17}$	10^{-9}	$1.24 \cdot 10^{3}$	$1.99 \cdot 10^{-16}$
high energy radiation, γ-rays	$3 \cdot 10^{18}$	10^{-10}	$1.24 \cdot 10^{4}$	$1.99 \cdot 10^{-15}$
	$3 \cdot 10^{19}$	10^{-11}	$1.24 \cdot 10^{5}$	$1.99 \cdot 10^{-14}$
cosmic rays	$3 \cdot 10^{20}$	10^{-12}	$1.24 \cdot 10^{6}$	$1.99 \cdot 10^{-13}$
	$3 \cdot 10^{21}$	10^{-13}	$1.24 \cdot 10^{7}$	$1.99 \cdot 10^{-12}$
	$3 \cdot 10^{22}$	10^{-14}	$1.24 \cdot 10^{8}$	$1.99 \cdot 10^{-11}$
	$3 \cdot 10^{23}$	10^{-15}	$1.24 \cdot 10^{9}$	$1.99 \cdot 10^{-10}$
	$3 \cdot 10^{24}$	10^{-16}	$1.24 \cdot 10^{10}$	$1.99 \cdot 10^{-9}$

Fig. 21.2 Frequency, wavelength and energy. (Source: Lit. [10-10], with kind permission of Kerntechnik Deutschland e.V. (KernD))

Table 21.1 Energy values of the light equivalents of visible light

Wavelength (in nm = 10^{-9} m)	Frequency (in THz = 10^{12} Hz)	Color	Light equivalent (in kJ/mol photons)
800	375	Infrared	150
750	351	Dark red	160
700	328	Red	171
650	305	Orange red	184
600	281	Yellow	199
550	258	Green	218
500	234	Blue-green	239
450	211	Blue	266
400	188	Purple	299
350	164	Ultraviolet	342

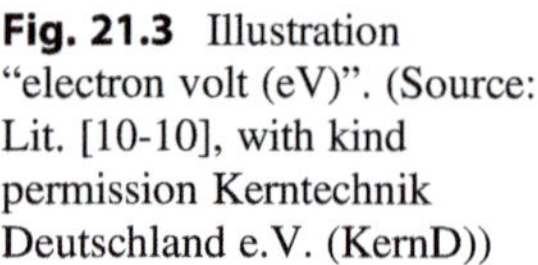

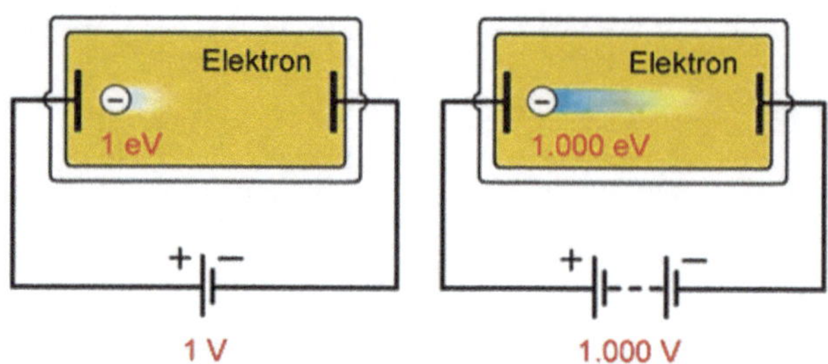

Fig. 21.3 Illustration "electron volt (eV)". (Source: Lit. [10-10], with kind permission Kerntechnik Deutschland e.V. (KernD))

Table 21.2 Comparison of rest mass and rest energy of different particles

Particle	Symbol	Rest mass (in u)	Rest energy (in MeV)
Photon	γ	0	0
Electron	e^- or β^-	0.000 548 573	0.511
Pion, neutral	π^0	0.144 93	135
Pion, loaded	π^+ resp. π^-	0.149 86	139.6
Proton	p	1.007 27	938.28
Neutron	n	1.008 66	939.57
Deuteron	^{2}H or d	2.013 54	1875.63
Alpha particle	^{4}He or α	4.001 48	3727.41

For conversions:

$$1\ \mathrm{J} = 6.242 \cdot 10^{18}\ \mathrm{eV} \qquad 1\ \mathrm{eV} = 1.602 \cdot 10^{-19}\ \mathrm{J}$$

This energy unit was determined because very small energy values occur during *individual* nuclear transformations.

21.1.4 Energy and Mass

Einstein's famous formula about the equivalence of energy and mass is $E = mc^2$. The energy of a particle is therefore its mass (m) multiplied by the square of the speed of light (c^2). Because of the large value of the speed of light ($c = 3 \cdot 10^8\ \mathrm{m\ s^{-1}}$), there is also a very high value for c^2: $9 \cdot 10^{16}\ \mathrm{m^2\ s^{-2}}$. Thus, even very small masses already have a high energy content.

Thus, every energy is also associated with a mass: $m = E/c^2$. Every change in energy (ΔE) is therefore associated with a change in mass (Δm) – and vice versa. It is therefore valid:

$$\Delta E = \Delta m c^2 \qquad \Delta m = \Delta E / c^2$$

It can now be calculated that the mass of $1.782\ 66 \cdot 10^{-36}$ kg can be assigned the energy of 1 eV. In nuclear physics, therefore, the masses of particles are also given in the unit electron volt and are referred to as rest mass (unit: eV/c^2) or rest energy (unit:

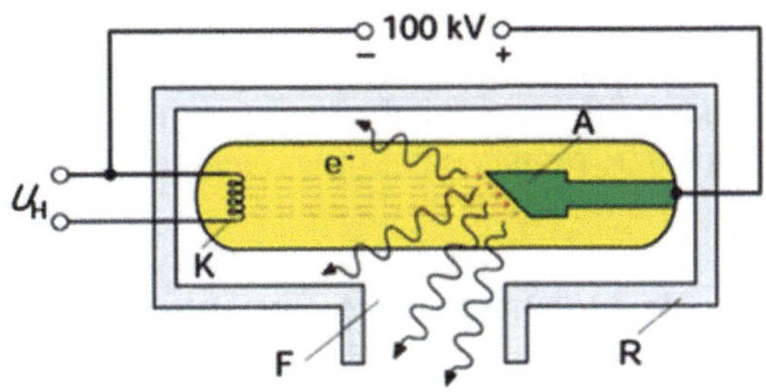

Fig. 21.4 X-ray device (schematic). U_H: Heating voltage; K: Cathode; A: Anode; e^-: Electrons; R: X-ray shielding; F: Window for the exit of the X-rays. (Source: Lit. [10–10], with kind permission Kerntechnik Deutschland e.V. (KernD))

eV). A comparison of the rest masses and rest energies of different particles can be found in Table 21.2.

21.1.5 X-Rays

In 1895, Röntgen[6] experimented with cathode rays. Here he noticed a fluorescence of crystals located near the cathode ray tube. Since the tube was covered with black paper, this should not have been the case. When he directed the radiation onto a photo plate to prove this, he succeeded in taking the first X-ray picture ever – the "fluoroscopy" of his wife's hand.

Röntgen referred to this "new type of rays" as "x-rays". This has survived in Anglo-American usage to this day *("x-rays"), whereas in German we speak of "Röntgen-rays". The* first applications in technology and medicine became possible as early as 1896.

X-rays are produced when fast-moving electrons hit an obstacle. The electrons generated in the cathode ray tube are very strongly accelerated by a high voltage of up to 50 000 V and hit the anode with high energy. Here the electrons are slowed down by the anode material. The energy is converted into electromagnetic radiation, the so-called bremsstrahlung (compare Fig. 21.4 and Sect. 25.3.3).

21.1.5.1 Properties of X-Rays

X-rays

- Can excite many substances to fluorescence or phosphorescence
- Blackens photographic plates/films
- Ionizes gases
- Can be used for crystal structure analysis
- Has a considerable penetrating capacity for matter

The penetrating power of X-rays in particular is exploited in medicine. The radiation is weakened differently when passing through different tissues (e.g. bones, healthy

[6] Wilhelm Conrad Röntgen, 1845–1923, German physicist, Nobel Prize in Physics in 1901.

and diseased body tissue). X-rays are therefore a valuable diagnostic tool (detection of bone fractures, stomach ulcers, etc.). In X-ray therapy (colloquially "irradiation"), tumor cells are irradiated and thus destroyed. In technology, X-rays are used in particular for material testing, e.g. for checking weld seams.

21.1.6 Luminescence, Phosphorescence and Fluorescence

The three terms luminescence, phosphorescence and fluorescence will be encountered again and again in the further course. Since there is often confusion between the terms, they will be briefly explained here.

21.1.6.1 Luminescence

Luminescence (from Latin *lumen* = light) is the generic term for the phenomena of fluorescence and phosphorescence. It refers to the emission of light without thermal side effects, i.e. without an increase in temperature. It can emanate from solids, liquids (mostly solutions) or gases. Luminescence occurs when atoms, ions or molecules are brought into an excited state and then return to their ground state by emitting radiation (mostly visible light, more rarely ultraviolet or infrared radiation).

The excited state can be caused by, for example:

- Irradiation with (visible) light (photoluminescence)
- UV light (UV luminescence)
- X-rays (X-ray luminescence)
- Electrons (cathodoluminescence)
- Ionising (radioactive) radiation (radioluminescence)
- Ions (Ionoluminescence)
- Chemical processes (chemiluminescence)
- Biochemical processes (bioluminescence; e.g. in fireflies)
- Mechanical influences (triboluminescence)
- Sound (sonoluminescence)

All luminescence phenomena occur at moderate temperatures and usually without noticeable temperature increases. One therefore also speaks of "cold light" or "cold luminescence". Natural phenomena, such as the northern lights or visible lightning, are based on luminescence. Technically, it is used, for example, in fluorescent tubes, screens or LEDs. The chemo-luminescence of luminol is used in forensics to detect occult (= hidden) traces of blood. Certain substances can literally store absorbed energy (e.g. through ionizing radiation) and then release this energy again in the form of visible radiation when heated. This is called thermoluminescence (see Sect. 28.1.6).

21.1.6.2 Phosphorescence

Phosphorescence describes a phenomenon of luminescence in which the afterglow lasts from a few seconds to weeks or even months. It occurs especially with solids.

Moderate increases in temperature shorten the time during which the object luminesces. Phosphorescent compounds are used, for example, in clock faces, escape route markings or firefighters' helmets. Here, the phosphorescent substances are added to the coating (paint or plastic). The term phosphorescence goes back to the element phosphorus. White phosphorus already oxidizes at room temperature in the air. In this case it glows weakly.

21.1.6.3 Fluorescence

The case of luminescence, where the glow occurs only during excitation (and until about 10^{-6} seconds after), is called fluorescence. Fluorescence was first observed in the mineral fluorite (CaF_2, "fluorspar"). Uranium compounds, rare earth salts, and various organic compounds, such as anthracene, naphthalene, or fluorescein show this phenomenon. In contrast to phosphorescence, fluorescence is independent of temperature.

21.2 Particle Radiation

Particle radiation always involves real particles that are "emitted" – just like the grains of sand leave the nozzle during sandblasting. Classic examples are: Electron radiation (e.g. in a picture tube), proton radiation, alpha radiation, neutron radiation or ion radiation. What they all have in common is that they are stopped by matter they encounter.

The energy of a particle radiation (also called corpuscular radiation from Latin *korpus* = body) depends on the mass of the respective particle and its velocity. The larger or faster the respective particle is, the greater is its energy. (Compare also Sect. 30.3.2)

21.3 Excursus: Ionizing Radiation

Radiation, whether particle or wave radiation, that is capable of generating ions is referred to as ionizing radiation. This is contrasted with non-ionising radiation. Since the typical binding energies of chemical bonds are in the range of about 3 eV – 7 eV, at least 3 eV are required to convert a molecule into an ion. Therefore, the boundary between ionizing and non-ionizing wave radiation has been set at approximately 3 eV.

Low-frequency radiation, especially thermal radiation (infrared) and microwave radiation, heat matter but are not able to create ions. The range of visible light covers the range from approx. 750 nm (red) to approximately 380 nm (violet). In the violet range, the first ionization effects already occur, since violet light with a wavelength of 380 nm already has a photon energy of 3.26 eV. However, only "particularly sensitive" molecules are attacked here.

The even shorter wavelength UV radiation, which is no longer visible to the human eye, of approx. 380 nm – 10 nm wavelength, already has a relatively high ionization potential. So-called "black light" (UV-A radiation, 380 nm – 315 nm) has a photon energy of approx. 3.3 eV – 4 eV, UV-B radiation (315 nm – 280 nm) approx. 4.0 eV – 4.4 eV and UV-C radiation (280 nm – 120 nm) of up to 12.4 eV.

▶ In **this context, please note:** Radioactive substances emit **ionizing radiation.** The radiation itself is not radioactive; only its generation is based on the phenomenon of radioactivity.

21.3.1 Labelling of Ionising and Non-ionising Radiation

Radiation sources which do not emit ionizing radiation, e.g. transmitters of radio and TV stations, must also be marked; correspondingly, of course, radiation sources which emit ionizing radiation must also be marked. According to DIN EN ISO 7010, the following warning signs are used:

Warnung vor nicht ionisierender Strahlung

Warnung vor radioaktiven Stoffen oder ionisierender Strahlung

If there are warning signs of non-ionising radiation in the area of roofs, towers, masts, ... during operations, protective measures must be taken. The most effective protection of the emergency forces against radiation is, of course, the shutdown of the system per se. If this is not possible, safety distances must be maintained depending on the transmitting power. These are indicated on an additional sign or can be requested from the operator.

Recommended:

- For mobile radio systems: 5 m
- For radio transmitters: 10 m
- For TV transmitters: 50 m

21.3.1.1 Application PID

With the photoionization detector (PID), a measuring device that uses ionizing radiation from the UV range is also available to the fire brigades on the NBC emergency response vehicle. With this device, air is sucked in and any hazardous substances contained are ionised by a UV lamp. The current flow of the ions generated in this way is measured. The built-in UV lamp has an ionization energy of 10.6 eV or operates with a wavelength of 117 nm. For the mode of operation see Fig. 21.5.

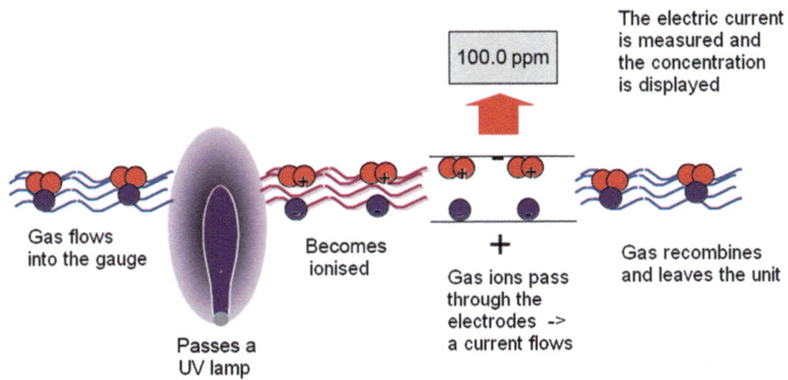

Fig. 21.5 Functional diagram PID

Table 21.3 Ionization energies of substances (examples)

Measurable with PID ionization energy <10.6 eV		Not measurable with PID ionization energy >10.6 eV	
Substance	Ionization energy	Substance	Ionization energy
Aniline	7.7 eV	Methanol	10.8 eV
Toluene	8.8 eV	Propane	11.1 eV
Benzene	9.2 eV	Ethine (acetylene)	11.4 eV
Iso-butene	9.2 eV	Chlorine	11.5 eV
Acetone	9.7 eV	Oxygen	12.1 eV
Hydrazine	9.9 eV	Sulphur dioxide	12.3 eV
Ammonia	10.2 eV	Hydrogen cyanide	13.8 eV
Hydrogen sulfide	10.4 eV	Carbon monoxide	14.0 eV
Ethanol	10.5 eV	Nitrogen	15.6 eV

The PID thus provides a non-specific sum signal of all gaseous substances whose ionization energy is <10.6 eV (compare Table 21.3). If it is a single hazardous substance, the concentration in the air can be determined down to 1 ppm – 2 ppm – provided the device is calibrated for this substance **and** the substance has an ionisation energy of less than 10.6 eV.

The German Federal Office of Civil Protection and Disaster Control (Bundesamt für Bevölkerungshilfe und Katastrophenschutz, BBK) provides a list of the 82 substances stored in the software of the PID of the NBC exploration vehicle at www.bbk.de. The list also contains the respective warning values (= alarm triggering by the device).

The latest developments in instrument technology combine the PID with an upstream gas chromatograph (GC). In the portable device, the ambient air to be analysed is first separated into its individual components by means of GC and then fed to the detector (the PID). This allows the simultaneous quantification of several substances.

22 History of Radioactivity

Radioactivity is the property of a number of atomic nuclei to spontaneously transform into other atomic nuclei. In this process, energy is emitted in the form of particles or electromagnetic radiation.

> ▶ Radioactivity is always a consequence of instabilities of the atomic nuclei. Radioactivity was not "looked for" as other phenomena were.

The discovery of natural radioactivity came about almost by chance in the course of other investigations. After its discovery, applications arose quickly. With the help of radiation from radioactive decay, it was possible, for example, to investigate "the inside of the atom" more closely (cf. Section 5.1.1).

22.1 Beginnings

22.1.1 Becquerel's Photoplate Experiment

At the end of the nineteenth century it was a well-known fact that cathode rays (cf. Section 5.1.2) produced a green fluorescence when the cathode ray tube was sealed with glass. In January 1896 Becquerel[1] heard a lecture on the "X-rays" of a Mr. Roentgen (compare Sect. 21.1.5), which are also cathode rays.

Becquerel now considered whether there might not be a connection between the fluorescence or phosphorescence of substances and the cathode rays. For this purpose, he took a strongly fluorescent compound (potassium uranyl sulphate), which he placed on a light-proof packed photographic plate and exposed this arrangement to sunlight in order to produce the fluorescence. The photoplate showed blackening where the compound rested on the photoplate. Becquerel initially felt

[1] Antoine Henri Becquerel, 1852–1908, French physicist, Nobel Prize in Physics 1903.

T. Schmiermund, *The Chemistry Knowledge for Firefighters*,
https://doi.org/10.1007/978-3-662-64423-2_22

vindicated. Fluorescence appearances seemed to be flanked by a kind of "accompanying radiation" similar to cathode rays (or X-rays).

On 02.26.1896 Becquerel prepared a new experiment which could not be finished at first because of the cloudy sky. As the experiment had to be stopped, Becquerel put the photo plate together with the uranium compound on it into the drawer. On 01 March the sun shone again and the experiment could be continued. Becquerel took a new photographic plate for the sake of comparability. However, he also developed the plate "from the drawer", which had only been exposed to sunlight for a short time.

To his astonishment, this showed a very distinct blackening – even though it had been in the dark. Further investigations revealed that the blackening of the photographic plates was independent of whether it was a fluorescent or phosphorescent compound – as long as it was a uranium compound. Becquerel therefore suspected that uranium emitted its own, hitherto unknown, radiation – "uranium radiation".

As early as March 1896, Becquerel was able to prove that this uranium radiation was capable of ionizing gases.

22.1.2 Discovery of New Elements

Inspired by Becquerel's investigations, Marie Curie[2], together with her husband Pierre[3], began to turn her attention to the study of "Becquerel rays". During their investigations, they discovered that pitchblende (a uranium-containing mineral) was much more "radioactive" than pure uranium compounds. They concluded that this mineral must contain other, also radioactive, elements. In 1898, they were able to extract 100 mg of radium chloride ($RaCl_2$) from 2 tons (!) of pitchblende. The element polonium postulated in 1898 could only be isolated in 1902.

22.1.3 Non-uniform Radiation

Around 1899/1900, Rutherford (alongside M. Curie and Becquerel) established that the radiation emitted by radioactive substances is not uniform. By deflection in the magnetic field, three types of radiation could be identified:

- positively charged helium atoms → α-radiation
- negatively charged electrons → β-radiation
- short-wave X-ray radiation → γ-radiation

[2]Marie Sklodowska Curie, 1867–1934, Polish-French chemist and physicist, Nobel Prize in Physics 1903, Nobel Prize in Chemistry 1911.

[3]Pierre Curie, 1859–1906, French physicist, Nobel Prize in Physics in 1903.

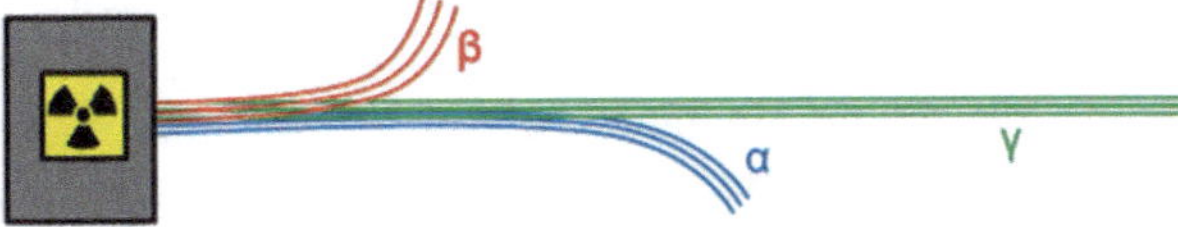

Fig. 22.1 Ionizing radiation in the magnetic field

The investigations further led to the assumption that the radiation owes its origin to the voluntary decay of the atomic nucleus.
As shown in Fig. 22.1, the low-mass β-radiation is deflected more easily than the high-mass α-radiation by a magnetic field. The γ-radiation is not affected.

22.1.4 Nuclear Conversion

Rutherford achieved the first nuclear transformation in 1919. He exposed nitrogen gas to α-radiation. In this process, the nitrogen was converted to oxygen and a free proton was also obtained.

22.1.5 Artificial Radioactivity

In 1934, the Joliot-Curie[4] couple succeeded for the first time in generating artificially induced radioactivity. By irradiating aluminium with a-particles, they were able to obtain a radioactive phosphorus isotope.

22.1.6 Discovery of Nuclear Fission

Already in 1934 Fermi[5] split uranium by means of neutrons. But he misinterpreted the results, as he was only looking for elements with a higher atomic mass than uranium. Irène Joliot-Curie also tried to produce heavier elements than uranium by neutron bombardment in 1937. She too misinterpreted the experimental results.

In 1938 Hahn,[6] together with his colleague Straßmann[7], repeated Fermi's experiments. But even Hahn and Straßmann were not able to interpret the results beyond doubt. At Hahn's request, L. Meitner[8] looked at the experimental results. In her exile in Sweden (Meitner was of Jewish descent and had therefore been forced to leave Nazi Germany), she succeeded in interpreting the results correctly at the

[4]Jean Frédéric Joliot-Curie, 1900–1958, French physicist, and Irène Joliot-Curie, 1897–1956, French physicist and chemist. Joint Nobel Prize in Chemistry 1935.

[5]Enrico Fermi, 1901–1954, Italian-American physicist, Nobel Prize in Physics in 1938.

[6]Otto Hahn, 1879–1968, German radiochemist, Nobel Prize in Chemistry in 1944.

[7]Friedrich "Fritz" Wilhelm Straßmann, 1902–1980, German chemist.

[8]Lise (Elise) Meitner, 1878–1968, Austrian nuclear physicist.

beginning of 1939, together with her nephew Frisch[9]: Hahn had split uranium. This finding was something completely new. Never before had larger fragments than protons or helium nuclei (α-particles) been split off from an atomic nucleus. Until Meitner's conclusions, the emission of larger particles was extremely unlikely according to the concepts of physics at the time. Meitner and Frisch were also able to estimate the energy that must have been released in the process: ~200 MeV (200 million eV). No other process known at that time was capable of releasing such huge amounts of energy.

22.2 Developments

22.2.1 The First Nuclear Reactor

Hahn's results and their interpretation by Meitner led to an intensive examination of nuclear fission. It quickly became clear that uranium could be split with relatively slow (so-called "thermal") neutrons. Fermi used natural uranium and employed graphite to slow down the neutrons. He was thus able to realize the first self-sustaining chain reaction in Chicago (USA) in December 1942. This "uranium machine" ran for 4½ min at a power of about ½ W and required no separate cooling. For the first time, man had released and controlled atomic energy.

22.2.2 Nuclear Weapons

The enormous amount of energy released during *fission* also led to the development of weapons. In 1942, the USA assumed that Nazi Germany had already begun developing "atomic bombs". To deal with this danger, the "Manhattan Project" was founded. Nazi Germany surrendered on 05.08.1945, the first atomic bomb, a plutonium bomb, was detonated on 07.16.1945 under the code name "Trinity" in New Mexico.

The US now decided to use nuclear weapons against Japan to force an end to the war with its surrender. Two bombs were dropped on Japanese cities:

- a uranium atomic bomb ("Little Boy") with an explosive force of 11 350 tons of TNT on 08.06.1945 on Hiroshima and
- a plutonium atomic bomb ("Fat Man") on Nagasaki on 08.09.1945 with an explosive force of approximately 22 000 tons of TNT

Further developments then led to the hydrogen bomb, which is based on the principle of nuclear *fusion*.

[9]Otto Robert Frisch, 1904–1979, Austrian physicist.

The largest hydrogen bomb ever detonated was the so-called "Tsar Bomb" (Russian name actually "Great Ivan"), which the USSR tested over the Arctic in October 1961. Its explosive power was 50 million tons of TNT.

23 Radioactivity: Terms and Notations

As already explained (Sect. 5.4), the atomic nucleus consists of protons and neutrons. Together these are called nucleons (A). The atomic nucleus now consists of Z protons (Z = atomic number) and $A - Z = N$ neutrons. Proton and neutron have nearly the same mass. However, the mass of the electron is only about one 1850th of the mass of the proton. Therefore, the mass of the electron can be neglected.

23.1 Notations of Nuclear Chemistry

An atom is defined by two numbers: Atomic number Z and mass number A. Certain notations have been established to represent the individual nuclides.

23.1.1 Atomic Number *Z*

► The atomic number Z is equal to the number of protons in the atomic nucleus.

Since the proton has the charge +1, this is also equal to the number of positive nuclear charges. In a neutral atom, the number of electrons in the atomic shell is also equal to the number of protons in the atomic nucleus.

► The atomic number Z indicates which element is involved.

T. Schmiermund, *The Chemistry Knowledge for Firefighters*,
https://doi.org/10.1007/978-3-662-64423-2_23

23.1.2 Mass Number *A*

▶ The mass number *A* is equal to the sum of the protons and neutrons in the atomic nucleus (*A* for "all").

Protons and neutrons are the components of the atomic nucleus, they are therefore also called nucleons (Latin *nucleus*). The mass number is practically equal to the atomic mass of the respective element, since both have the mass of ≈1 u and the small mass of the electrons ($^1/_{1850}$ u) can be neglected.

23.1.3 Nuclide Notation

$${}^{\text{Mass number}}_{\text{Proton number}}\text{Element symbol} \qquad {}^{A}_{Z}\text{Symbol}$$

A nuclide can only be described unambiguously with the mass number and the atomic number. For this purpose, a special notation has been agreed upon.

This notation is relatively easy to remember: The larger number is always at the top; or in the alphabet "A" above "Z". The atomic number can also be omitted, since it results automatically from the element symbol anyway. For easier calculation of the neutrons, however, it is recommended to include the atomic number.

Since each element is uniquely described by its atomic number, it is sufficient to write only the element symbol. To characterize the respective nuclide, simply add the mass number.

Symbol – mass number

Example

${}^{1}_{1}H$	${}^{2}_{1}H$	${}^{14}_{6}C$	${}^{37}_{17}Cl$	${}^{137}_{55}Cs$	${}^{234}_{92}U$
H-1	H-2	C-14	Cl-37	Cs-137	U-234

Analogously, the "building blocks" of atoms can also be represented in this notation.

${}^{1}_{1}p$	${}^{1}_{0}n$	${}^{0}_{-1}e$
p	n	e^-
Proton	Neutron	Electron

◀

23.2 Terms Relating to Nuclides and Isotopes

In connection with nuclides and isotopes, a number of terms exist which are explained below.

23.2.1 Number of Neutrons *N*

The number of neutrons (N) is the difference between the mass number and the atomic number. $N = A - Z$. It is usually not explicitly stated and must be calculated if necessary.

23.2.2 Nuclide

A nuclide is *one* type of atom that is clearly characterized by *one* mass number and *one* atomic number.

$^{59}_{27}Co$ (Co-59) and $^{60}_{27}Co$ (Co-60) are the same element (cobalt), but two different nuclides, which also have different radiochemical properties: Co-59 is stable, Co-60 is a β^-- and γ-emitter.

23.2.3 Isotopes

The number of protons in the nucleus is decisive for the chemical properties of an atom. These determine the structure of the electron shell and thus the chemical behaviour of an element. The term isotope (Greek isos = *equal* and *topos* = place) comes from the fact that all isotopes of an element are in the same place in the periodic table. The totality of the isotopes of an element used to be called the *pleiad* (after the "Pleiades" of Greek mythology).

► Isotopes are atoms with the same atomic number, regardless of the number of neutrons.

The elements occurring in nature are in most cases mixtures of several isotopes (**mixed elements**). Only 19 elements do not have a mixture of isotopes (**pure elements**).

Examples of isotopes and their composition, as well as the determination of the average atomic mass, can be found in Table 23.1.

The separation of the isotopes of an element is not possible with the means of chemistry, since all isotopes of an element behave chemically identically. Therefore, physical methods must be used. In analytics, a so-called mass spectrometer (MS) is used for this purpose. This separates the isotopes on the basis of their mass difference in a magnetic field and can thus identify them. Uranium isotopes are technically separated from uranium hexafluoride (UF_6) by means of gas centrifugation, among other methods. The more massive $^{238}UF_6$ molecules are preferentially driven to the outer wall and thus separated from $^{235}UF_6$.

Table 23.1 Isotopes and average atomic masses

Isotopes	Mass	Natural portion	Partial mass	Average atomic mass
$^{35}_{17}Cl$	35 u	75.77 %	26.52 u	35.49 u
$^{37}_{17}Cl$	37 u	24.23 %	8.97 u	
$^{24}_{12}Mg$	24 u	78.99 %	18.96 u	24.32 u
$^{25}_{12}Mg$	25 u	10.00 %	2.50 u	
$^{26}_{12}Mg$	26 u	11.01 %	2.86 u	
$^{12}_{6}C$	12 u	99.011 %	11.88 u	12.01 u
$^{13}_{6}C$	13 u	0.989 %	0.13 u	
$^{14}_{6}C$	14 u	10^{-10} %	0.0 u	

Isotope Rule

The Aston[1] isotope rule states that elements with odd atomic number never have more than two stable isotopes. Elements with even atomic number usually have more than two stable isotopes.

23.2.4 Pure Elements

19 elements (F, Na, Al, P, Sc, Mn, Co, As, Y, Nb, Rh, I, Cs, Pr, Tb, Ho, Tm, Au, and Bi) are naturally occurring elements composed of only one type of atom. They are called pure elements (or mononuclide elements).

What they all have in common is that they have both odd atomic numbers and odd mass numbers – and thus an even number of neutrons.

Bi, Th and Pu also belong to the pure elements. These three elements have even atomic numbers and are radioactive. Due to their long half-lives, however, they also occur naturally.

23.2.5 Mixed Elements

All elements that are not listed under the pure elements are mixed elements. This means that they occur in nature only as a mixture of different element isotopes. Here, up to ten natural isotopes of an element can occur. Among the naturally occurring non-radioactive elements (i.e. from hydrogen to lead), all mass numbers from 1 to 208 are represented – with the exception of the masses 5 and 8.

[1]Francis William Aston, 1877–1945, English chemist and physicist, Nobel Prize in Chemistry in 1922.

23.2.6 Isobares

Over 70 mass numbers occur several times in the naturally occurring elements. Atomic species with the same mass but different atomic numbers are called isobars (Greek *isos* = equal and *baros* = weight, heaviness).

Examples of isobars are: ${}^{36}_{16}S$ and ${}^{38}_{18}Ar$ or ${}^{64}_{28}Ni$ and ${}^{64}_{30}Zn$.

Isobar Rule

According to Mattauch's[2] isobar rule, atoms with odd mass numbers generally have only one stable atomic species. Nuclei with even mass numbers, on the other hand, can have several stable isobars. In this case, however, the atomic numbers (Z) must differ by at least two.

23.2.7 Isotones

Isotones are atomic nuclei with the same number of neutrons but different atomic numbers. The term is derived from the Greek *isos* = equal and *tonos* = rope. This means that the same number of neutrons binds any number of protons together "like a rope".

Example of three isotones with neutron number 32: ${}^{59}_{27}Co$, ${}^{60}_{28}Ni$, ${}^{61}_{29}Cu$

23.2.8 Core Isomers

In isomers (from Greek *isos* = equal and *meros* = proportion; i.e. "consisting of equal proportions") the numbers of nucleons are equal. However, they differ in their energy state. In addition to the ground state of the atomic nuclei, there exists a longer-lived (i.e. in this context >1 ns) so-called metastable state. To identify these metastable states, an "m" is added to the mass number, e.g.: Tc-99 m.

For a better overview, a summary is given in Table 23.2.

Overview Core Isomers

Table 23.2 Summary: Isotopes, isomers, isotones and isobars

Name	Proton number (Z)	Neutron number (N)	Mass number (A)	Examples
Isotopes	Equal	*Variously*	*Variously*	${}^{12}_{6}C$, ${}^{13}_{6}C$
Isomers	Equal	Equal	Equal	${}^{99}_{43}Tc$, ${}^{99m}_{43}Tc$
Isotones	*Variously*	Equal	*Variously*	${}^{60}_{28}Ni$, ${}^{61}_{29}Cu$
Isobars	*Variously*	*Variously*	Equal	${}^{64}_{24}Ni$, ${}^{64}_{30}Zn$

[2]Josef Mattauch, 1895–1976, German physicist.

The Atomic Nucleus 24

Radioactivity comes exclusively from the atomic nucleus. Here, the nucleus gives off energy in the form of particles or radiation. Before we turn to radioactive decay, we must first understand the structure of the atomic nucleus. We look at the two essential nuclear models, the forces that prevail in the atomic nucleus and the stability of the atomic nucleus.

24.1 Structure

From Sect. 5.3 we already know that the atomic nucleus is made up of nucleons (protons and neutrons). Let us now take a brief look at the two most important models of atomic nuclei.

24.1.1 Droplet Model

The droplet model compares the atomic nucleus with a drop of liquid. The nucleus consists of a *nuclear liquid,* which in turn is composed of protons and neutrons. Gamow[1] developed the basic idea in the early 1930s. Von Weizsäcker[2] and Bohr developed the model further in 1935/36. Meitner and Frisch in turn used this model to explain nuclear fission and to calculate the nuclear energy released.

This model is based on several facts of experience:

- The density of the nuclei is practically the same for all atoms – regardless of size. Just as is the case with drops of a liquid.

[1] Georgi Antonovich Gamov, 1904–1968, Russian-American physicist.

[2] Carl Friedrich von Weizsäcker, 1912–2007, German physicist, philosopher and peace researcher.

T. Schmiermund, *The Chemistry Knowledge for Firefighters*,
https://doi.org/10.1007/978-3-662-64423-2_24

- In a liquid, the cohesion of the particles is caused by the Van der Waals forces, which have only a small range. Mutual attractions only take place between neighbouring particles. It is also true for the strong nuclear force that it is only effective between neighbouring particles.
- In the case of water, for example, it does not matter whether a single water molecule is released from a drop, a bucket or a lake. The energy required for this is always the same. The situation is similar in the atomic nucleus. The energy required to separate a nucleon from the nuclear bond is on average about 8 MeV.

From the second fact follows that the cohesion of the nucleons in the atomic nucleus is similar to that of the water molecules in a drop of water. The energy that holds the (nuclear) drop together is also called "volume binding energy".
The repulsion of equal charges is used as a modification for the volume binding energy. The large range of the electrostatic force ensures that nuclei become more unstable the more protons they contain. According to the model, nuclei with more than 82 protons are unstable, i.e. radioactive.

24.1.2 Shell Model

The shell model of the atomic nucleus is similar to Bohr's atomic model. It is assumed that the nucleons in the atomic nucleus are arranged on very specific stable paths, also called shells – similar to the electrons in the atomic shell. As with electrons, these shells also correspond to very specific energy levels. However, the energies in the atomic nucleus are a million times higher than in the electron shell of the atom.

Similar to the electrons of the atomic shell, the nucleons can also be excited: They assume a state with higher energy (excited state). By radiating energy (γ-radiation) they can return to the ground state.

In a stable atomic nucleus, the various shells of this nucleus are fully occupied with nucleons. Neutrons cannot decay in these nuclei because the protons that are created cannot find any free space. The number of protons (Z) or neutrons (N) in the nucleus that form stable nuclides are: 2, 8, 20, 28, 40, 50, 82, 126. These numbers are also called magic numbers. Atomic nuclei with a magic number of protons or neutrons are also called magic nuclei. Because of their stability, atomic nuclei with magic numbers occur particularly frequently in nature and form many stable isotopes:

- $N = 20$: 5 stable atomic nuclei
- $N = 28$: 5 stable atomic nuclei
- $N = 50$: 5 stable atomic nuclei
- $N = 82$: 7 stable atomic nuclei
- $Z = 50$: 10 stable atomic nuclei

If both Z and N are one of the magic numbers, double magic nuclei are present. He-4, O-16 and Ca-40 stand out not only because of their stability, but also because of their frequency.

Shape of the Nucleus As a rule, atomic nuclei are represented as spherical structures. The particularly stable nuclei, such as He-4, O-16 or Pb-208, are spherical. Other nuclei, such as C-12 or Si-28, are discus-shaped, while still others (e.g. Ne-20, Mg-24) are more cigar-shaped.

24.2 Forces in the Core

Actually one should assume, atomic nuclei are most unstable. After all, there are lots of same (positive) charges at most narrow space – and charges of same are known to repel each other. However, experience shows, many atomic nuclei are most stable. Already elsewhere (Sect. 5.4) was pointed out, neutrons within nucleus are kind of 'glue' for protons.

24.2.1 Strong Nuclear Power

The actual effect of the "glue neutron" comes about through a constant interaction of the neutrons with the protons (Fig. 24.1). The so-called "strong nuclear force" is responsible for this. In the model developed by Yukawa[3] in 1935/37, the exchange of particles, so-called pi-mesons (pions for short), leads to a force that holds the nucleus together.

Three "force particles" are necessary for a complete explanation:

- an unloaded pion (π^0)
- a positively charged pion (π^+)
- a negatively charged pion (π^-)

One imagines the atomic nucleus in such a way that the nucleons there change their identity very often (approximately 10^{20} times per second) through the exchange of pions. Three different processes take place here:

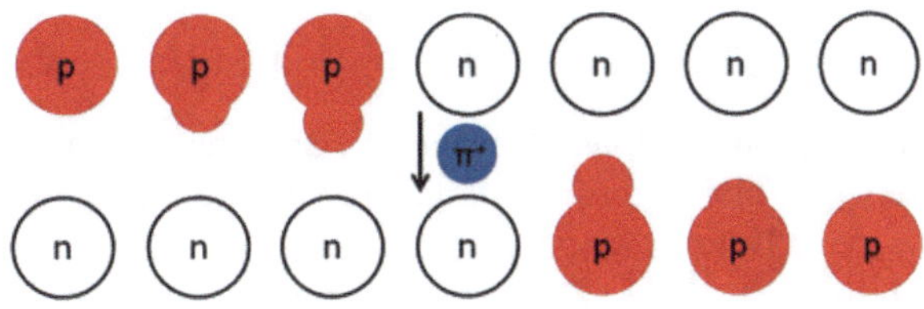

Fig. 24.1 Pion transfer

[3]Hideki Yukawa, 1907–1981, Japanese physicist, Nobel Prize in Physics in 1949.

a. A proton gives a positive pion to a neutron. The proton becomes a neutron by giving off the positive charge, the neutron, which takes up the charge, becomes a proton.

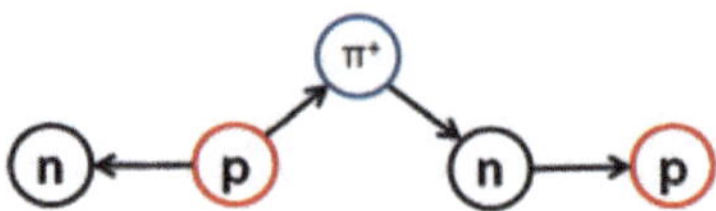

b. A neutron gives off a negative pion to a proton. A positive charge remains in the neutron: It becomes a proton. In the proton, which takes up the negative pion, the charge is "neutralized". It has become a neutron.

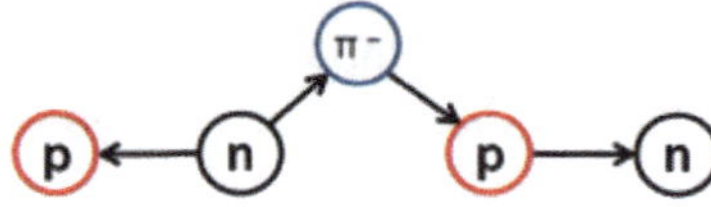

c. A proton exchanges an uncharged pion with another proton. Or a neutron exchanges such a particle with another neutron. In these cases, no "identity change" takes place.

The range of the pions is extremely small and is about the diameter of a nucleon: 10^{-15} m. This corresponds to the conclusions from the droplet model.

24.2.2 Weak Interaction

Within the atomic nucleus, the neutron is stable. However, a single neutron is unstable. It decays within 10–15 minutes according to the equation:

$$n \rightarrow p + e^{-} + \bar{\nu}_e$$

into a proton (p), an electron (e^{-}) and another particle – a so-called neutrino ($\bar{\nu}_e$). Responsible for this reaction is the so-called "weak interaction" (also "weak nuclear force"). It has a range of about 10^{-17} m, which is about $1/100$ the diameter of the proton (10^{-15} m). The weak interaction can be used to understand the emission of electrons from the atomic nucleus and various intermediate steps in nuclear fusion. Further information can be found in the literature [6-4] and [10-3].

24.2.2.1 Neutrino

Neutrinos are electrically uncharged particles with an (probably) extremely small mass. They travel at close to the speed of light and can pass through layers of lead several million kilometres thick without being damaged. Strictly speaking, several different types of neutrino exist. However, only two are of interest to us here: the electron anti-neutrino ($\bar{\nu}_e$) and the electron neutrino (ν_e).

The neutrino was initially a hypothetical particle to explain the energy distribution in β-decay. It was postulated by Pauli in 1933 and named neutrino ("small neutron") by Fermi. It was not detected until 1956.

24.2.2.2 Stability of the Proton

It is obvious to assume that also a single, unbound proton could decay due to the weak interaction. But this contradicts experience. The reason is relatively simple: the proton has a slightly smaller mass ($m_p = 1.007\ 27$ u) than the neutron ($m_n = 1.008\ 66$ u). If now a proton would change into a neutron, the "law of conservation of mass and energy" would be violated.

24.3 Stability of the Atomic Nucleus

Only nuclei with a "balanced" ratio of protons and neutrons are stable. But only if the nucleus as a whole does not become too large. If the ratio of protons to neutrons is shifted in favour of one of the two nuclear building blocks, then β-radiation is emitted. By this the nucleus tries to come back into an (energetic) equilibrium. If the nucleus becomes too large, it tends to emit α-particles or even to decay spontaneously.

► Of the 118 elements, about 3300 nuclides are known, of which about 250 are stable.

24.3.1 Nuclear Binding Energy

The mass of a hydrogen atom is exactly the sum of the mass of a proton and an electron. The masses of all (!) other atoms, however, are less than the sum of the masses of their constituents. This fact is called mass defect. This is a measure for the binding energy in the atomic nucleus itself.

According to Einstein's formula $E = mc^2$, mass and energy are related. The mass defect therefore also contains an energy difference. As the mass becomes smaller, the energy must also become smaller. The mass defect therefore represents the energy that is released during the formation of the respective atomic nucleus.

Example

The mass of the lithium nuclide Li-7 is 7.016 00 u. The mass of a proton is 1.007 276 u and that of a neutron 1.008 665 u. Li-7 has three protons and

4 neutrons in its nucleus. This results mathematical in a mass of 7.056 488 u. The mass defect is therefore 0.040 488 u – per Li-7 atom. This corresponds to a binding energy of about 38 MeV or $6.04 \cdot 10^{-15}$ kJ (still per atom). If we convert this to 1 mol of lithium, the energy is $3.64 \cdot 10^9$ kJ (= 3.64 Tera Joules). Figure 24.2 illustrates this once again.

This contrasts, for example, with the complete combustion of butane, which yields only 2857 kJ per mole of butane. Or to put it another way: to obtain the amount of energy released from protons and neutrons during the formation of 1 mol of lithium, one would have to burn over 71 tons of butane. ◀

It has proved advantageous to use the average binding energy per nucleon as a comparison. Plotted as a curve against the mass number, the result is a curve from which the relative stability of the respective atomic nucleus can be seen (Fig. 24.3).

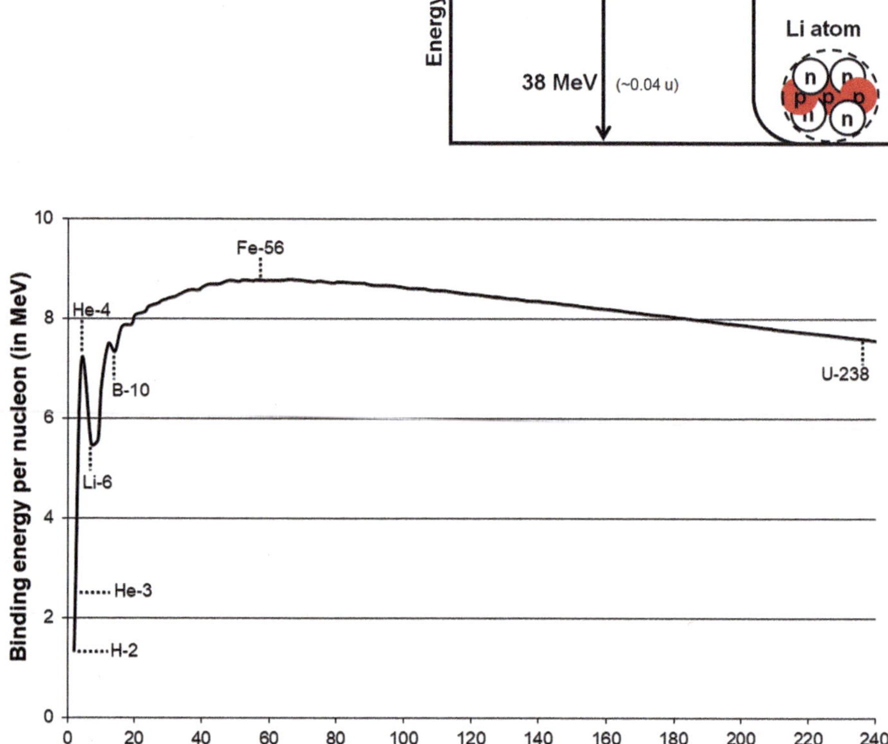

Fig. 24.2 Binding energy for lithium

Fig. 24.3 Binding energy per nucleon

The most stable nucleus is Fe-56. Nuclei with mass numbers 4 (He), 12 (C) and 16 (O) also prove to be quite stable. Light nuclei can be combined to form an intermediate mass nucleus (nuclear fusion). A heavy nucleus ($A > 100$) can be split into two nuclei of average mass (fission).

25 Radioactive Decay

Let us now turn to the original radioactivity. We want to take a closer look at the individual emitted "types of radiation" and their properties.

Radioactivity is caused by instabilities or unfavourable proton-neutron ratios in the atomic nucleus:

- An "excess" of energy causes the emission of γ-radiation.
- A nucleus that is "too big" can emit α-radiation or split spontaneously.
- If there are too many neutrons or protons in the nucleus, β-radiation is emitted.
- Neutron-deficient nuclei can capture an electron from the atomic shell (K capture).

Through certain nuclear processes, which in turn cause instabilities, free protons or free neutrons can also be produced.
In the individual considerations of this chapter, we will refer to radiation throughout as particles (gamma quantum, x-ray quantum, photon) and not as electromagnetic waves.

25.1 Representation of Nuclear Reactions

Similar to the chemical reaction equations, nuclear reactions can also be represented. In addition, there is a shortened notation as well as the possibility to represent nuclear transformations as an energy level diagram.

25.1.1 Nuclear Reaction Equation

The "normal" reaction equation for nuclear processes strongly resembles the notation known from chemical reactions.

T. Schmiermund, *The Chemistry Knowledge for Firefighters*,
https://doi.org/10.1007/978-3-662-64423-2_25

$$^{14}_{7}N + ^{4}_{2}He \rightarrow ^{17}_{8}O + ^{1}_{1}p$$

Nitrogen-14 reacts with an alpha particle to form oxygen-17 and a proton

Just as with the reaction equations in chemistry, care must be taken to ensure that the reaction equation is balanced. The sum of the number of nucleons (A) and the number of protons (Z) must be identical on both sides of the equation.

25.1.2 Abbreviated Nuclear Reaction Equation

Among nuclear chemists and physicists, a special shorthand notation has emerged. Here one writes down the initial nucleus and the final nucleus. In between, one puts (in brackets) both the "projectile" and the emitted particle. For the above reaction, the equation is then:

$$^{14}_{7}N\,(\alpha, p)\,^{17}_{8}O$$

N-14 is bombarded with alpha particles. O-17 is formed by the emission of a proton.

▶ **Note:** This form of the reaction equation requires a high level of attention and some practice to avoid mistakes.

25.1.3 Conversion Scheme

Sometimes it is more favourable or clearer to represent nuclear reactions schematically including the different energy levels. Especially if radiation of different kinds respectively different energy is emitted, this proves to be an advantage.

Rules for the Conversion Scheme

- Horizontal lines represent the respective energy level.
- Vertical lines indicate energy changes.
- Shifts to the left mean a decrease of the (positive) nuclear charge.
- Shifts to the right mean an increase in nuclear charge.
- An inserted "*m*" stands for "metastable".

Consider the Decay of Radium

Ra-226 emits predominantly (94.4 %) α-particles with an energy of 4.784 MeV. In addition, a small fraction (5.5 %) of α-particles with an energy of only 4.601 MeV is emitted. The "excess" energy still present in the nucleus is emitted in the form of γ-radiation, whose energy is 0.186 MeV.

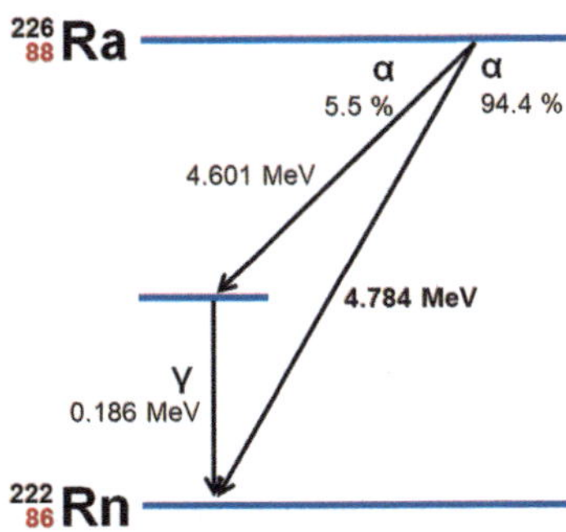

Fig. 25.1 Core transformation scheme (example). (After Lit. [10-10], modified)

As a classical nuclear reaction equation, two equations would have to be written here:

$$^{226}_{88}\text{Ra} \longrightarrow {}^{222}_{86}\text{Rn} + {}^{4}_{2}\text{He}\ (4.784\ \text{MeV})$$

$$^{226}_{88}\text{Ra} \longrightarrow {}^{222}_{86}\text{Rn}\ + {}^{4}_{2}\text{He}\ (4.601\ \text{MeV}) + \gamma\ (0.186\ \text{MeV})$$

This is much clearer in the form of the conversion scheme (Fig. 25.1). ◀

25.2 α-decay

The -decay leads to the generation of α-radiation. This consists of the nuclei of the noble gas helium. Notations for α-radiation are: α or He-4 or $^{4}_{2}$He respectively $^{4}_{2}\text{He}^{2+}$ (where the charges of the nuclei are usually not written). With binding energy of 28.4 MeV for the α-particle, the very high binding energy of 7.1 MeV per nucleon results. This explains the great stability of the helium nucleus and is the main reason why He-4 nuclei and not, for example, Li-6 nuclei (only about 5.5 MeV per nucleon) are emitted.

25.2.1 Origin of α-radiation

Heavy atomic nuclei ($Z > 83$, $A > 210$) are susceptible to α-decay. The strong nuclear force is no longer sufficient to compensate for the mutual repulsion of the protons. As a result, the sum of the masses (energies) of the decay products is often smaller than the mass (energy) of the initial nucleus. As a result, α-particles are emitted.

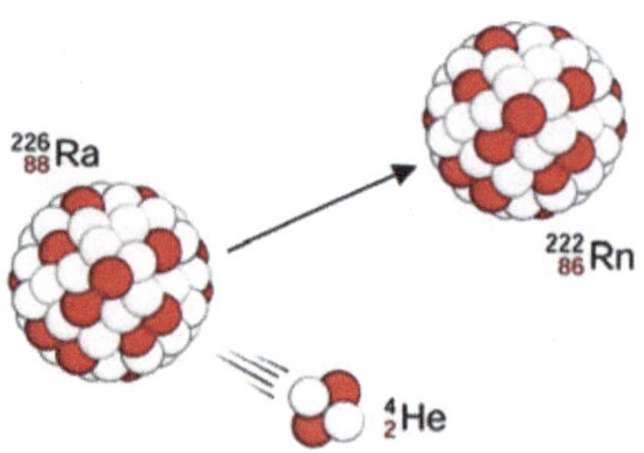

Fig. 25.2 Formation of a-radiation (schematic); radium-226 emits a helium nucleus and thus converts into radon-222: $^{226}_{88}\text{Ra} \longrightarrow {}^{222}_{86}\text{Rn} + {}^{4}_{2}\text{He}$. The mass number decreases by four, the atomic number by two. (Source: Lit. [10-10], with kind permission Kerntechnik Deutschland e. V. (before Deutsches Atomforum e.V.), Berlin)

Example: Decay of Th-232 to Ra-228

$$^{232}_{90}\text{Th} \rightarrow {}^{228}_{88}\text{Ra} + {}^{4}_{2}\text{He}$$

The mass of thorium is 232.038124 u. The sum of the masses of the two daughter nuclei is only 232.033 742 u (Ra-228: 228.031 139 u, He-4: 4.002 603 u) (Fig. 25.2).

This results in a mass difference of 0.004382 u – in favour of α-decay *per nuclear transformation*. This mass difference corresponds to an energy of $6.553 \cdot 10^{-19}$ Joules or 4.09 MeV. This does not seem much at first, but it refers to *a single* decay (compare also Fig. 24.2). ◀

▶ In the case of α-decay, the atomic number (Z) and the number of neutrons each decrease by two, i.e. the mass number (A) decreases by four: $^{A}_{Z}\text{X}(\alpha)^{A-4}_{Z-2}\text{Y}$.

25.2.2 Properties of α-radiation

Due to their charge ($\alpha = {}^{4}\text{He}^{2+}$), α-rays can be deflected in the magnetic or electric field (compare Fig. 22.1). They give off their energy essentially by ionizing atoms or molecules of the matter they encounter. In this process, the ions are produced at high density/proximity and the α-particles lose all their energy over a short distance. Their range in air is therefore only a few centimeters, in tissue less than a tenth of a millimeter (Table 25.1). In principle, α-rays can be completely shielded by just a sheet of paper or operational clothing (Fig. 25.12). Alpha particles easily pick up electrons from other particles and thus become helium gas, which evaporates.

Table 25.1 Range of α-particles of different energy

Particle energy in MeV	Range in Air	Range in Muscle tissue	Range in Aluminium
1	0.3 cm	4 μm	2 μm
3	1.6 cm	16 μm	11 μm
4	2.5 cm	31 μm	16 μm
6	4.6 cm	56 μm	30 μm
8	7.4 cm	91 μm	48 μm
10	10.6 cm	130 μm	67 μm

25.2.3 α-capture

For nuclei smaller than potassium (K-40), bombardment with α-particles leads to nuclear reactions called α-capture. Thus, Chadwick succeeded in producing and detecting free neutrons by bombarding Be-9 with α-particles.

$$^{9}_{4}\mathrm{Be} + ^{4}_{2}\mathrm{He} \rightarrow ^{13}_{6}\mathrm{C} \rightarrow ^{12}_{6}\mathrm{C} + ^{1}_{0}\mathrm{n} \; or \; ^{9}\mathrm{Be}\,(\alpha, \mathrm{n})\; ^{12}\mathrm{C}$$

25.3 β-decay

If the atomic nucleus contains too many neutrons or too many protons, then fast electrons are emitted from the nucleus. The weak nuclear force is no longer able to have a balancing effect on the nuclear particles. Protons and neutrons decay, releasing β-particles and neutrinos.

25.3.1 Origin of β^--radiation

In an atomic nucleus where there is a large excess of neutrons, these decay to emit electrons and neutrinos. One proton remains (Fig. 25.3). This increases the atomic number of the nucleus by one and decreases the mass number by one.

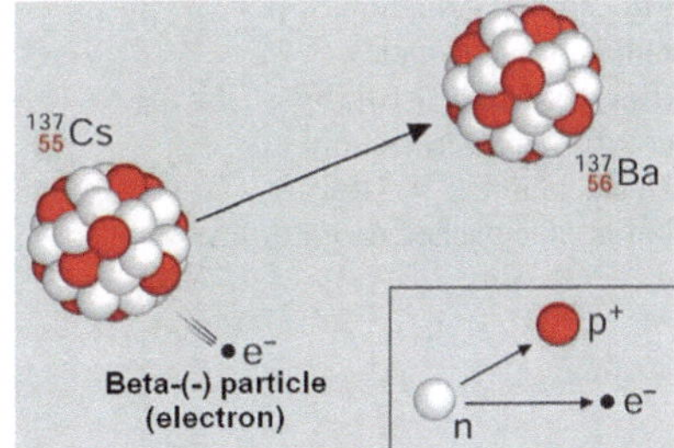

Fig. 25.3 Formation of β^--radiation (schematic). (Source: Lit. [10-10], by courtesy of Kerntechnik Deutschland e. V. (KernD, before: Deutsches Atomforum e. V.) Berlin)

$$^{40}_{19}\mathrm{K} \rightarrow \, ^{39}_{20}\mathrm{Ca} + \, ^{0}_{-1}\mathrm{e}^{-} + \bar{\nu}_{\mathrm{e}}$$

Beside the fast-flying electron, a particle called electron-anti-neutrino is released. This neutrino comes, like the electron, from the decayed neutron and thus from the atomic nucleus.

Typical β^--emitters include C-14, P-32, Ar-39, K-40, Co-60, Rb-87, Sr-89, Ra-228, or Es-256.

Similarly, a free neutron (one not bound in an atomic nucleus) also decays into a proton, an electron, and an electron-anti-neutrino:

$$^{1}_{0}\mathrm{n} \rightarrow \, ^{1}_{1}\mathrm{p} + \, ^{0}_{-1}\mathrm{e} + \bar{\nu}_{\mathrm{e}}$$

► In a β^--decay, the atomic number (Z) increases by one and the mass number stays equal: $^{A}_{Z}\mathrm{X}(\beta^-)^{A}_{Z+1}\mathrm{Y}$.

25.3.2 Origin of β⁺-radiation

The β^+-radiation is produced in a completely analogous way to the β^--radiation. In an atomic nucleus in which there is a large excess of protons, these decay under emitting of positrons and neutrinos. One neutron remains (Fig. 25.4). This reduces the atomic number of the nucleus by one, and the mass number also decreases by one.

$$^{22}_{11}\mathrm{Na} \rightarrow \, ^{21}_{10}\mathrm{Ne} + \, ^{0}_{1}\mathrm{e} + \nu_{\mathrm{e}}$$

In addition to the fast-flying electron, a particle known as the electron neutrino is released. This neutrino, like the positron, comes from the decayed proton.

The "elementary reaction" is, therefore: $^{1}_{1}\mathrm{p} \rightarrow \, ^{1}_{0}\mathrm{n} + \, ^{0}_{1}\mathrm{e} + \nu_{\mathrm{e}}$. However, this proton decay does not occur outside an atomic nucleus (as already explained in Sect. 24.2.2).

Positron radiation does not occur naturally. All β^+-emitters are artificially produced. Typical β^+-emitters are e.g. C-11, O-15, F-17, Na-22, Bi-203.

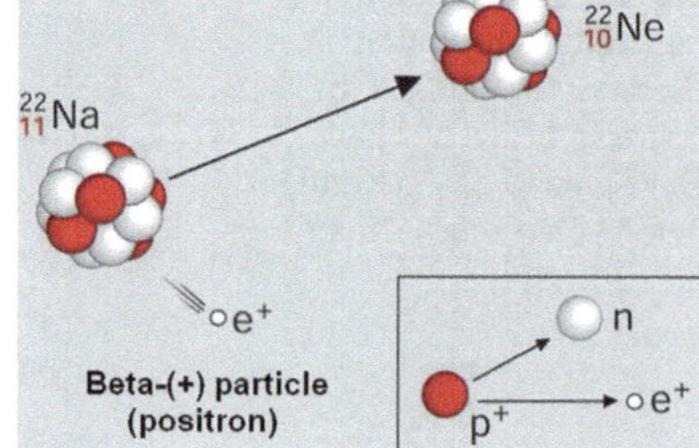

Fig. 25.4 Formation of β^+-radiation (schematic). (Source: Lit. [10-10], by courtesy of Kerntechnik Deutschland e. V. (KernD, before: Deutsches Atomforum e. V.) Berlin)

► In the case of a β^+-decay, the atomic number (Z) decreases by one, the mass number decreases by one: ${}^{A}_{Z}X(\beta^+)^{A-1}_{Z-1}Y$.

Positron (β^+) The positron is a particle that is identical to the electron in all its properties – except for its charge. The electron (e^-) is negatively charged, while the positron (e^+) is positively charged.

► The positron (e^+) is the antiparticle to the electron (e^-).

25.3.3 Properties of β-radiation

Originally, the term β-radiation referred purely to the β^--radiation of natural radioactive decay. Today, β^+-radiation is also included. Due to the charge, β-rays can be deflected in the magnetic or electric field.

When β-particles encounter matter, they lose their energy through a series of effects:

- ionizations,
- stimulation,
- generation of bremsstrahlung

Here, the β-particles essentially interact with the shell electrons of the matter they encounter (compare Fig. 25.6). The higher the density and the greater the atomic number of the shielding material, the stronger the interaction with the material. However, this also means that the generation of bremsstrahlung increases with an increasing atomic number of the shielding material (compare Fig. 25.7). This bremsstrahlung would then have to be shielded by additional layers of material. In practice, therefore, plastic shielding is usually used, the thickness of which is selected so that the ß-particles are completely absorbed. Only at higher particle energies is an additional shielding, usually made of lead, required to attenuate the bremsstrahlung (Fig. 25.5).
The maximum energy of the most commonly used radiation sources with β-emission is 1...2 MeV. Therefore, 4 mm aluminium is sufficient for complete shielding (compare Table 25.2; Fig. 25.12).

25.3.3.1 Bremsstrahlung

An accelerated electron enters the shell of an atom and is decelerated. The "lost" kinetic energy is released in the form of γ-quanta (Fig. 25.6). The generation of X-rays is also based on this principle (see Sect. 21.1.5).

As an analogy the braking process in a car may serve: The energy extracted from the wheel by the braking process is converted into thermal energy – and this, in turn, is emitted to the environment as thermal radiation.

Bremsstrahlung can also occur as a result of the photoelectric effect (Fig. 25.7; see Sect. 25.4.3). The electron knocked out of the atomic shell by the photoelectric

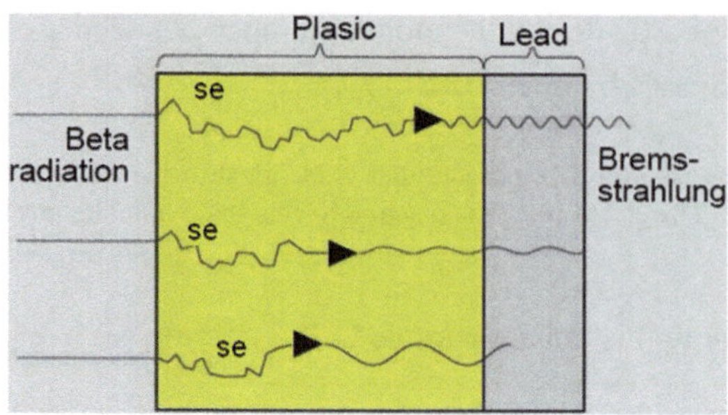

Fig. 25.5 Material combination for optimal shielding of β-radiation
The β-particles are deflected in their direction in the plastic (scattering effects, ST) and finally generate bremsstrahlung (upper beam path). This is weakened by the lead layer (middle beam path) or even completely absorbed at a lower energy of the β-radiation (lower beam path). (Source: Lit. [10-10], with kind permission of Kerntechnik Deutschland e. V. (KernD, before: Deutsches Atomforum e. V.) Berlin)

Table 25.2 Range of β-particles of different energy

Particle energy in MeV	Range in		
	Air	Muscle tissue	Aluminium
0.01	0.003 m	0.0025 mm	0.009 mm
0.1	0.1 m	0.16 mm	0.05 mm
0.5	1.2 m	1.87 mm	0.6 mm
1	3.1 m	4.75 mm	1.5 mm
2	7.1 m	11.1 mm	4.1 mm
5	19 m	27.8 mm	9.9 mm
10	39 m	60.8 mm	19.2 mm
20	78 m	123.0 mm	39.0 mm

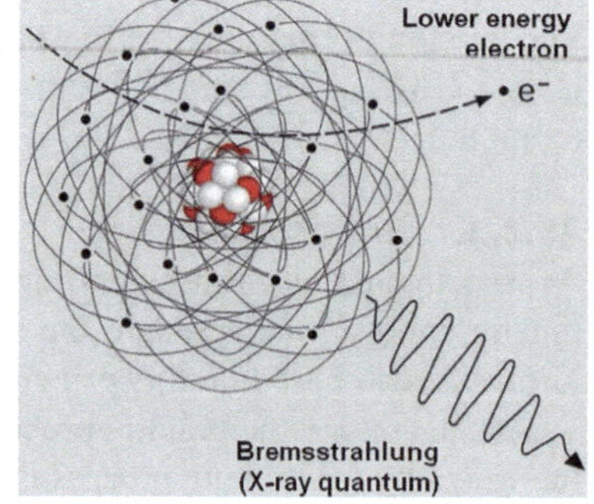

Fig. 25.6 Generation of bremsstrahlung (schematic). (Source: Lit. [10-10], by courtesy of Kerntechnik Deutschland e. V. (KernD, before: Deutsches Atomforum e. V.) Berlin)

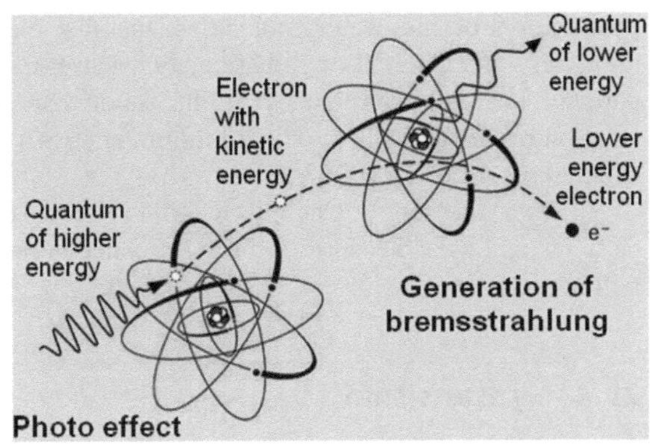

Fig. 25.7 Photoelectric effect with subsequent generation of bremsstrahlung (schematic). (Source: Lit. [10-10], with kind permission of Kerntechnik Deutschland e. V. (KernD, before: Deutsches Atomforum e. V.) Berlin)

effect is, in turn, decelerated in the atomic shell of a neighbouring atom, thus producing bremsstrahlung.

25.3.4 Nuclear Reaction "K-capture"

Related to the β-decay is the so-called "K-capture" or "electron capture". Related because here also a nucleon transformation (a proton becomes a neutron) occurs due to the weak nuclear force and under neutrino emission.

The nucleus of a neutron-poor nuclide captures an electron from the innermost electron shell (the K shell, compare Sect. 5.2.2). This transforms a proton into a neutron, and at the same time, an electron-neutrino is emitted.

$$ {}^{1}_{1}\mathrm{p} + {}^{0}_{-1}\mathrm{e} \rightarrow {}^{1}_{0}\mathrm{n} + \nu_{\mathrm{e}} $$

► During K-capture (electron capture), the atomic number decreases by one and the mass number stays equal.

25.3.5 Cherenkov Radiation

The radiation first observed by Cherenkov[1] in 1934 is a "blue glow" that occurs, for example, in high-energy radiation sources stored in water. Very fast particles, mostly electrons, trigger this phenomenon. The radiation occurs because the particles move at a higher speed in the medium than light itself can.

The speed of light in water is only $2.25 \cdot 10^8$ m s^{-1} and thus 75 % of the speed of light in vacuum ($3 \cdot 10^8$ m s^{-1}). If electrons now move through water at a speed of e.g. $2.5 \cdot 10^8$ m s^{-1}, their speed is higher than the speed of light in water. By

[1] Pavel Alexeyevich Cherenkov, 1904–1990, Soviet physicist, Nobel Prize in Physics in 1958.

polarization of the water molecules, the fast electrons produce electromagnetic radiation – the Cherenkov radiation. In a sense, this is the "supersonic cone" of the particle. Just as an airplane traveling faster than sound produces a bang, the fast electron produces a glow. The minimum energy a particle must possess to produce Cherenkov radiation is 263 keV.

The measurement of Cherenkov radiation is used for the detection of high-energy particles and as a measure of the instantaneous radioactivity in nuclear reactors or holding bassins.

25.4 γ-transition

By emitting high-energy radiation, an atomic nucleus passes from an excited state to the ground state. Gamma radiation occurs exclusively as accompanying radiation of other nuclear reactions.

25.4.1 Origin of γ-radiation

Following any nuclear transformation in which the subsequent nucleus remains in an excited state, the atomic nucleus releases its excess energy into the environment in the form of γ-radiation (Fig. 25.8). This is true for α-decay, β^-- and β^+-decay as well as for most nuclear reactions, including K-capture.

Since it is always a question of nuclear isomers (see Sect. 23.2.8) emitting their energy, the term "gamma decay" should not be used. The term "decay" indicates a change in atomic number (Z) and/or mass number (A). However, this does not apply to the gamma transition.

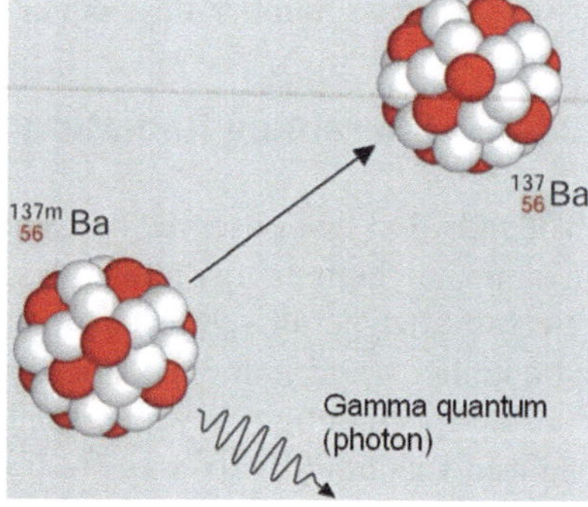

Fig. 25.8 Formation of γ-radiation (schematic). (Source: Lit. [10-10], by courtesy of Kerntechnik Deutschland e. V. (KernD, before: Deutsches Atomforum e. V.) Berlin)

▶ By emitting γ-radiation, the atomic nucleus does not undergo nuclear transformation. The mass number (A) and atomic number (Z) do not change.

25.4.2 Properties of γ-radiation

γ-radiation is electromagnetic wave radiation with a very small wavelength – and thus very high energy (Fig. 25.9). The wavelengths are in the range of 10^{-10} to 10^{-14} m. The energies of the radiation are up to 17.6 MeV per photon. For comparison: The energies of the photons emitted by electrons of the atomic shell are ~10 keV.

Due to its extremely short wavelength, **γ**-radiation is able to penetrate even thick layers of matter and can hardly be completely shielded (Fig. 25.12). Its range in air is a few hundred metres. Lead is more suitable than aluminum for attenuating the radiation since absorption is more probable at larger atomic numbers. In its effect, it is similar to X-rays but more energetic. X-rays have 1...100 keV, γ-rays between 100 keV and over 10 MeV (= 10 000 keV).

25.4.3 Interactions of γ-radiation

If γ-radiation hits matter, or rather if γ-radiation penetrates matter, a number of interactions occur. Depending on the energy of the photons, three different interactions occur:

- the photoelectric effect,
- the Compton effect,
- the formation or destruction of pairs.

All three effects overlap when γ-rays interact with matter. Depending on the energy of the radiation, one of the processes predominates. At particularly high photon energies, the nuclear photoelectric effect also occurs.

25.4.3.1 Photoelectric Effect

In the photoelectric effect, there is an interaction between the incident photon (gamma quantum) and the electrons of the atomic shell. The radiation knocks an

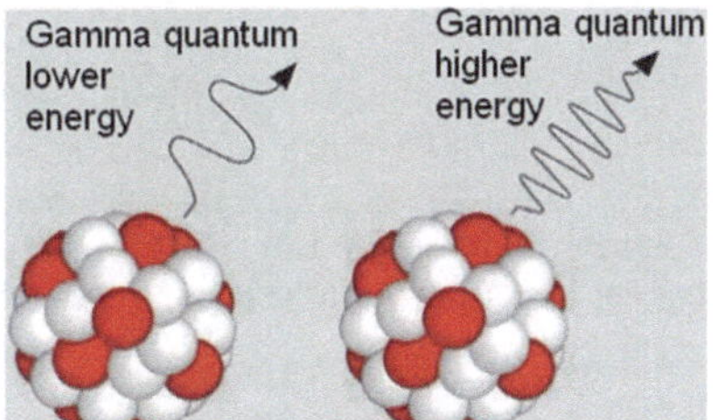

Fig. 25.9 γ-radiation: Wavelength and energy (schematic). (Source: Lit. [10-10], with kind permission of Kerntechnik Deutschland e. V. (KernD, before: Deutsches Atomforum e. V.) Berlin)

electron close to the nucleus out of the atomic shell (preferably out of the K-shell). This electron leaves the nucleus and is now called a photoelectron. The photoelectron can ionize or excite atoms in the environment (secondary radiation or delta radiation).

The ion produced by the photoelectric effect has an electron gap in the K-shell. From a shell further outside (e.g. the L-shell) an electron jumps into the K-shell to close this gap (= lowest possible energy for the electron). When changing from the L-shell to the K-shell, this electron again emits radiation energy. This is called X-ray fluorescence radiation (XRF radiation). The photoelectric effect occurs preferentially at small photon energies and large nuclear charge numbers.

25.4.3.2 Compton Effect

A photon transfers part of its energy to an electron of the outer electron shell. This electron is separated (Compton electron) and the atom is ionized. However, in contrast to the photoelectric effect, residual energy remains for the photon, so that it continues to move with lower energy (longer wavelength) on a modified path. The Compton electron can ionize further atoms in the environment until its energy is finally so far reduced that it disappears completely by the photoelectric effect. This phenomenon was first observed by Compton[2] in 1922.

25.4.3.3 Comparison of Photoelectric Effect and Compton effect

Photoelectric effect	Compton Effect
Small photon energy	Mean photon energy
Photon hits an electron close to the nucleus	Photon hits an outer electron
Photon transfers all its energy to the electron and disappears in the process	Photon transfers part of its energy to the electron, a photon of lower energy (longer wavelength) remains
Ionization of the atom: The electron is separated	Ionization of the atom: The electron is separated
The photoelectron can cause further interactions	The Compton electron can cause further interactions
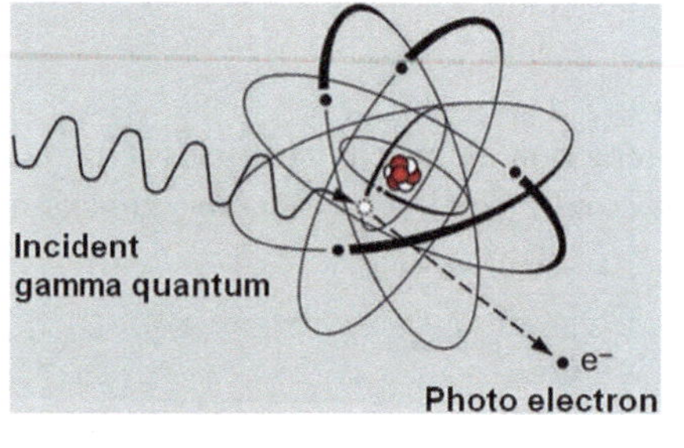	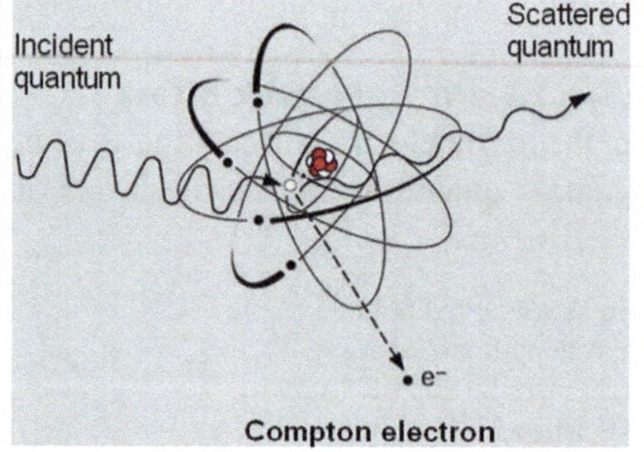

Source of illustrations: Lit. [10-10], with kind permission Kerntechnik Deutschland e. V. (KernD, before: Deutsches Atomforum e. V.) Berlin

[2] Arthur Holly Compton, 1892–1962, American physicist, Nobel Prize in Physics in 1927.

25.4.3.4 Pair Formation and Pair Destruction

Photons whose energy is greater than 1.022 MeV can trigger the effect of pair formation when they strike elements of higher atomic numbers. Here, the photon is converted into an electron (e^-) and a positron (e^+) in the electric field of the atomic nucleus. Particles are created from energy (Fig. 25.10).

$$\gamma \rightarrow e^- + e^+$$

The reverse case – particles become energy – usually takes place only a little later. If the positron meets an electron (e.g. from the atomic shell), both particles annihilate, producing two photons of 511 keV each (Fig. 25.11). This process is called pair annihilation or short annihilation. It is used, for example, in positron emission tomography (see Sect. 31.4).

$$e^- + e^+ \rightarrow 2\,\gamma$$

25.4.3.5 Nuclear Photoelectric Effect

At energies of 17 MeV and above, γ-radiation can trigger a nuclear reaction. This is called the nuclear photoelectric effect (or the "γ-induced nuclear reaction"). In most cases, neutrons are released. In light nuclei, protons and α-particles can also be ejected from the nucleus. The remaining nucleus is usually radioactive. The nuclear photoelectric effect plays only a minor role since the energies of γ-rays are usually much lower.

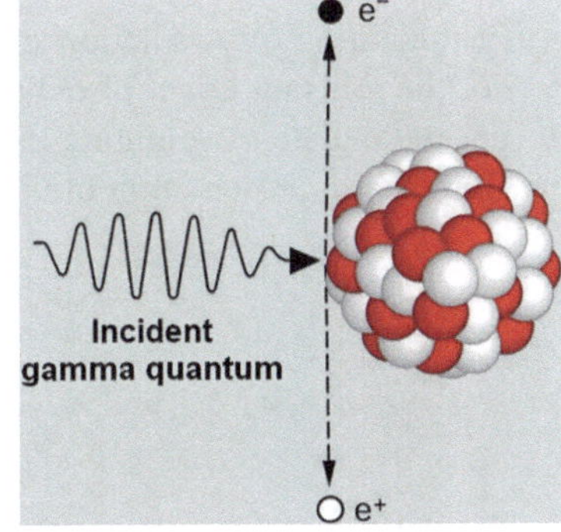

Fig. 25.10 Pair formation (schematic). (Source: Lit. [10-10], with kind permission Kerntechnik Deutschland e. V. (KernD, before: Deutsches Atomforum e. V.) Berlin)

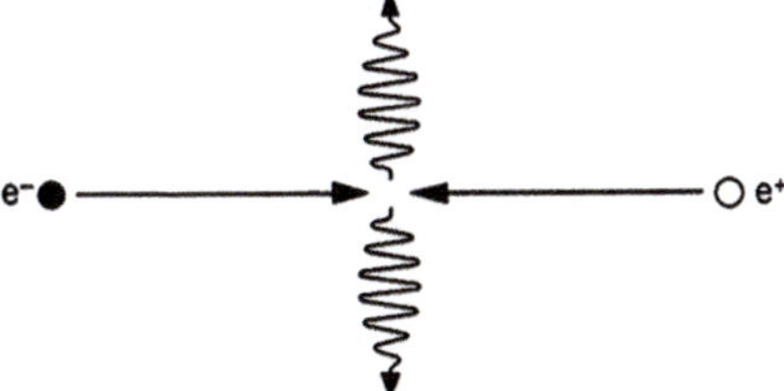

Fig. 25.11 Pair annihilation (schematic). (Source: Lit. [10-10], with kind permission Kerntechnik Deutschland e. V. (KernD, before: Deutsches Atomforum e. V.) Berlin)

25.5 Summary

Radiation type	α (alpha)	β (beta)	γ (gamma)
Image		β^- β^+	
Consists of:	He-core (^{4}He)	Electrons (e^-) or positrons (e^+)	Electromagnetic wave radiation
Loading	+2	−1	0
Number of nucleons	4	0	0
Penetration	Low	Medium	Large
Range in air:	Some cm	Several metres	Several hundred meters
Depth of penetration (tissue):	~50 μm	Several mm	Several tens of centimetres
Shielding through:	Sheet of paper Clothing	4 mm plexiglass 4 mm aluminium 15 sheet paper	*Thick layers of lead, steel or concrete only weaken*
Typical Energies:	4...6 MeV	~ 1...2 MeV	~ 1 MeV
Ionizing effect:	Highly ionizing	Ionizing	Ionizing
Radiation weighting factor w_R:	20	1	1

Notes on the overview:

- For shielding of γ-radiation see Sect. 34.2.
- For the ionising effect of radiation, see Sect. 30.3.2.
- For the radiation weighting factor, see Sect. 27.1.5 and Table 27.3.
- Graphical representation of the shielding in Fig. 25.12.

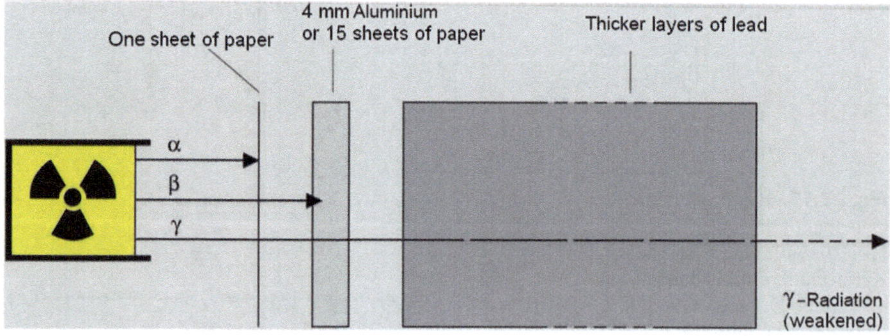

Fig. 25.12 Shielding of ionising radiation. (Source: Lit. [10-10], by courtesy of Kerntechnik Deutschland e. V. (KernD, before: Deutsches Atomforum e. V.) Berlin)

26 Nuclide Cards

Nuclide maps are overviews related to the periodic table, in which all nuclides and their decays or decay types are shown. The most common structure goes back to Segre.[1] The atomic numbers are plotted on the vertical axis, the neutron numbers on the horizontal axis.

26.1 Current Nuclide Maps

Paper-based nuclide maps are becoming less widespread. They are mostly only used as large wall charts for teaching purposes. Nowadays, interactive periodic tables with information on the individual nuclides can be found on the Internet, as can pure nuclide maps. Nevertheless, the basic structure usually remains the same.

26.1.1 Nuclide Overview

The following facts can be easily identified on the basis of the colored markings:

- In the diagonal, "$N = Z$" nuclei with the same number of protons and neutrons are plotted. Atoms with $Z > 20$ need more neutrons than protons in the nucleus to be stable.
- The black dots mark stable nuclides.
- The β^--decay (blue) occurs when the nucleus has an excess of neutrons.
- The β^+-decay (orange) occurs when the nucleus has an excess of protons.
- The α-decay occurs only in heavy atomic nuclei.
- Nuclei with a very large number of protons can emit them (red).

[1] Emilio Segrè, 1905–1989, Italian-American physicist, Nobel Prize in Physics in 1959.

T. Schmiermund, *The Chemistry Knowledge for Firefighters*,
https://doi.org/10.1007/978-3-662-64423-2_26

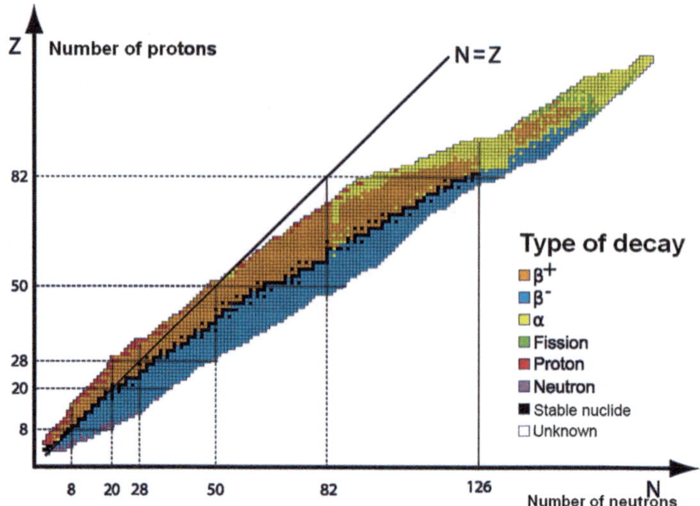

Fig. 26.1 Segrè's nuclide map (overview display). (Source: From isotope_table.svg: Table_isotopes_en.svg: Derivative work: Matt (discussion) Table_isotopes.svg: Napylkenobi derivative work: Sjlegg (discussion) derivative work: Matt (discussion) (isotopes_table.svg (different image)) [CC BY-SA 3.0 (http://creativecommons.org/licenses/by-sa/3.0) or GFDL (http://www.gnu.org/copyleft/fdl.html)], via Wikimedia Commons)

- Light nuclei with a large excess of neutrons can emit them (purple).
- Extremely heavy nuclei tend to spontaneously decay (fission, green).

Figure 26.1 shows a nuclide map according to Segrè with an indication of the radioactive decay type. Colored nuclides are unstable, black ones are stable. The diagonal line shows nuclides with equal numbers of protons and neutrons. Atomic nuclei with more than 20 protons need more neutrons than protons (neutron excess) to be stable.

26.1.2 Section of a Nuclide Map

The following is a detailed excerpt from a nuclide map:

Eu 144 β-	Eu 145 β-	Eu 146 β-	Eu 147 β- α	Eu 148 β-	Eu 149 ε	Eu 150 β-	Eu 151 stable
Sm 143 β-	Sm 144 stable	Sm 145 ε	Sm 146 α	Sm 147 α	Sm 148 α	Sm 149 α	Sm 150 stable
Pm 142 β-	Pm 143 β-	Pm 144 β-	Pm 145 ε α	Pm 146 β- β-	Pm 147 β-	Pm 148 β-	Pm 149 β-
Nd 141 β-	Nd 142 stable	Nd 143	Nd 144 α	Nd 145 stable	Nd 146 stable	Nd 147 β-	Nd 148 β-,β-
Pr 140 β-	Pr 141 stable	Pr 142 β- ε	Pr 143 β-	Pr 144 β-	Pr 145 β-	Pr 146 β-	Pr 147 β-
Ce 139 ε	Ce 140 stable	Ce 141 β-	Ce 142 β-,β-	Ce 143 β-	Ce 144 β-	Ce 145 β-	Ce 146 β-

Nuclide map (detail). In this nuclide map, the decays are both highlighted and indicated in the respective field (ε means electron capture, light blue fields). In other nuclide maps, the occurring energies and/or the half-lives are sometimes also indicated.

26.1.3 Decay types in the nuclide map

Depending on the type of decay, the nuclide following the decay is found at a certain subsequent position in the nuclide map. In the Segrè representation, the shift follows the pattern of this Fig.:

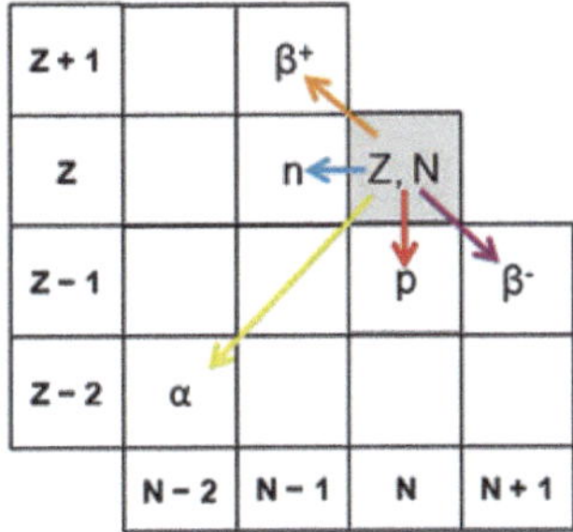

Decomposition Scheme. ß+-decay: **One up, one left (one up left) (*Z* + 1 & *N* – *1*).** Neutron emission (n): **One to the left (*N* – 1).** β−-decay: **One down to the right (Z – 1 & N + 1), (also applies to K-capture).** Proton emission (p): **One downwards (Z – 1).** α-decay: **Two down and two left (*Z* – 2 & *N* – 2).**

Example

If we apply this decay scheme to the above extract from the nuclide map, we can make the following observations, for example:

- Pm-144 converts to (stable) Sm-144 under β^{+}-emission.
- Nd-144 converts to (stable) Ce-140 under α-emission. ◄

27 Units of Measurement of the Radiation of Radioactive Substances

Due to the historical development, a multitude of units of measurement for radioactivity, ionizing radiation and their interactions has resulted. We will first look at the units commonly used today. Subsequently, the units which are no longer in use will be briefly explained.

27.1 Radiation Protection Units

Let us first consider the units commonly used in radiation protection today and their relationships to each other.

27.1.1 Impulse Rate (ips)

The impulse rate, also called count rate (cps, counts per second), is the number of particles measured through the radiation entrance window of a corresponding measuring device (e.g. contamination detection device) per second (ips = impulses per second). The measured pulse rate depends on the size (area) of the radiation entrance window, on the distance to the radiation source and on metrological conditions of the device used. It must not be equated with the activity (see Sect. 28.2.6 and Fig. 28.3).

▶ The impulse rate (ips) is an actual measured value that depends on the measuring conditions, the measuring device used and the radiation source.

27.1.2 Activity *A*

The activity of a radioactive source is the number of nuclear transformations per unit of time. Its unit is s^{-1} or Bq (Becquerel). If 5000 nuclear transformations take place per second, then this source has an activity of 5000 Bq or 5 kBq.

T. Schmiermund, *The Chemistry Knowledge for Firefighters*,
https://doi.org/10.1007/978-3-662-64423-2_27

The activity of a radiation source must not be confused with the pulse rate. The activity is primarily dependent on the respective radiation source and the amount of radioactive substance. Under non-laboratory conditions, a single counting tube will never measure the total activity of a source (see also Sect. 28.2.6).

▶ The activity *A* indicates the number of nuclear transformations per second.

Derived Units of Activity

Specific activity → activity per unit mass (Bq/g, Bq/kg, ...)
Activity concentration → Activity per unit volume (Bq/L, Bq/m^3, ...)
Area activity → activity per unit area (Bq/cm^2, Bq/m^2, ...)

Examples of Activities

Radio nuclide	spec. activity (in Bq/g)
I-131	$4.6 \cdot 10^{15}$
Fe-59	$1.8 \cdot 10^{15}$
H-3	$3.6 \cdot 10^{14}$
Co-60	$4.1 \cdot 10^{13}$

Radio nuclide	spec. activity (in Bq/g)
Sr-90	$5.3 \cdot 10^{12}$
C-14	$1.7 \cdot 10^{11}$
U (natural)	$2.5 \cdot 10^{4}$
K (natural)	$3.1 \cdot 10^{1}$

◀

As can be seen from the examples, the numerical values of the activities are usually rather large. Activity values in kBq (= 10^3 Bq) or MBq (= 10^6 Bq) are not unusual.

Rule of Thumb

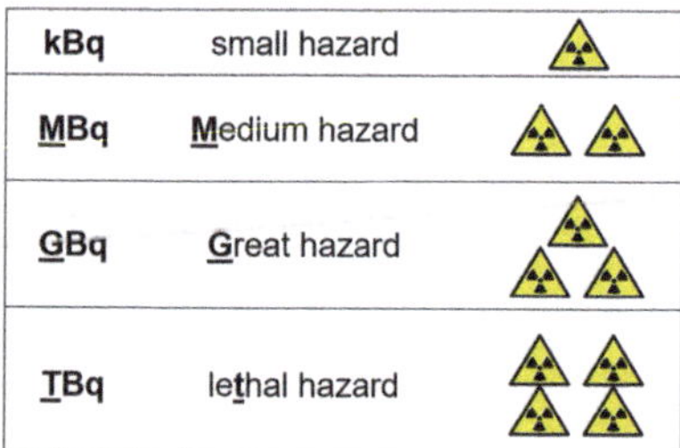

▶ Activity alone is not a direct measure of potential radiation damage. It depends on the type of emitter, the distance to the emitter and the period of exposure to radiation.

27.1.3 Half-Life $T_{1/2}$

Unlike chemical reactions, the reactions in the atomic nucleus cannot be influenced by temperature or pressure. However, it can be observed that the number of

Table 27.1 Half-lives (examples)

Nuclide	Half-life
U-238	$4.468 \cdot 10^9$ a
K-40	$1.277 \cdot 10^9$ a
Pu-239	$2.411 \cdot 10^4$ a

Nuclide	Half-life
Ni-65	2.52 h
Ir-197	5.8 m
Po-214	$1.643 \cdot 10^{-4}$ s

radioactive decays decreases with time. The time in which half of the radioactive substance has decayed is called the half-life. At the same time, the activity also decreases to half. The radioactive decays are different in different substances, therefore the half-life is characteristic for a decay.

▶ The half-life ($T_{1/2}$) of a radionuclide is the time in which half of the nuclei present have decayed.

The half-lives of the various nuclides span the range from billions of years to fractions of seconds (Table 27.1).

▶ The half-life is a constant characteristic of each nuclide that uniquely identifies the nuclide. It is independent of the history of the formation of the nuclide.

The mathematical derivation of the decay law or the resulting half-life is relatively complex. One of the reasons for this is that the time of the transformation of a *single* atomic nucleus cannot be determined or predicted exactly. Only the probability with which a certain portion of material decays can be given.

Introducing the nuclide-specific constant k, we get:

- For the quantity ratios between the initial quantity (N_O) and the quantity present at time t *existing quantity* (N), the logarithmic relationship:

$$\ln \frac{N_0}{N} = k \cdot t$$

- for the half-life:

$$T_{1/2} = 0.693\, k$$

Example Calculation

Co-60 has an half-life of $T_{1/2} = 5.27$ years. What mass of Co-60, starting from the original 10.0 mg, is still present after two years?

(a) Determination of the specific constant k:

$$k = \frac{0.693}{5.27\ \text{a}} = 0.1315\ \text{a}^{-1}$$

(b) Determination of $\ln(N_0/N)$:

$$\ln \frac{N_0}{N} = k \cdot t = 0.1315\ \text{a}^{-1} \cdot 2.0\ \text{a} = 0.263$$

(c) Determination of N_0/N: Calculator key [e^x] **or** [inv] [ln] = 1.301

(d) Calculation of the residual mass:

$$N = \frac{N_0}{1.301} = \frac{10.0\ \text{mg}}{1.301} = 7.69\ \text{mg}$$

◀

27.1.3.1 Relationship Between Activity and Half-Life

Nuclide	Spec. activity (in GBq/mg)	Half-life Years	Or:
I-135	129,900	$7.5 \cdot 10^{-4}$	6 h, 36 min
I-131	4600	0.022	8 d,7 h
Ca-45	660	0.446	162 d, 17 h
Co-60	41.8	5.27	5 a, 98 d, 12 h
Cs-137	3.2	30.2	30 a, 73 d
U-238	$12.4 \cdot 10^{-9}$	$4.47 \cdot 10^{9}$	$4{,}47 \cdot 10^{9}$ a

▶ If the half-life is very "short", the activity is very "high".

27.1.3.2 Illustration of the half-life

If the period of one half-life elapses, half of the respective radioactive substance has decayed. After two half-lives, only the ¼ original quantity is still present, and so on. Table 27.2 illustrates these relationships.

It may seem illogical at first that 500 g of an initial quantity of 1 kg of material decay in the course of the first half-life, but only 250 g in the course of the second half-life. Let us first realize that 1 kg of the material contains many more atoms than 1 mg. Since each atom has the same time probability of decaying, it follows that the number of atoms decaying in the 1 kg sample must also be many times higher than in the 1 mg sample.

Rule of Thumb

As can be seen, after 10 half-lives, the residual amount is reduced to about one-thousandth of the original amount. Therefore, the rule of thumb applies:

Table 27.2 Illustration of half-life

Number of half-lives	Factor	Nuclide residue	Graphical representation
0	$1/2^0 = 1/1$	100 %	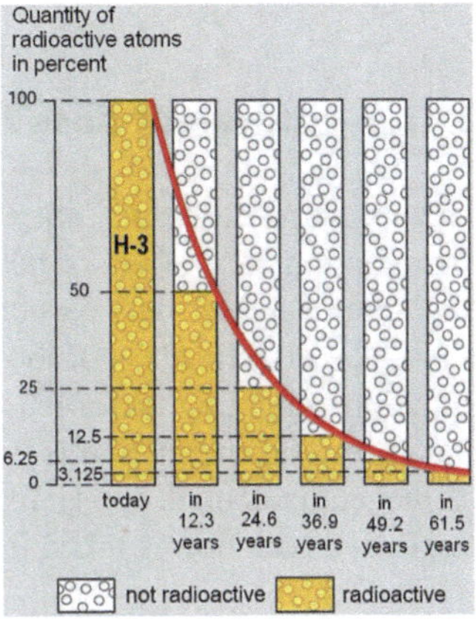
1	$1/2^1 = 1/2$	50 %	
2	$1/2^2 = 1/4$	25 %	
3	$1/2^3 = 1/8$	12.5 %	
4	$1/2^4 = 1/16$	6.25 %	
5	$1/2^5 = 1/32$	3.125 %	
6	$1/2^6 = 1/64$	1.56 %	
7	$1/2^7 = 1/128$	0.78 %	
8	$1/2^8 = 1/256$	0.39 %	
9	$1/2^9 = 1/512$	0.19 %	
10	$1/2^{10} = 1/1024$	0.098 %	
11	$1/2^{11} = 1/2048$	0.049 %	
12	$1/2^{12} = 1/4096$	0.024 %	

Source: Lit. [10-10], with kind permission Kerntechnik Deutschland e. V. (KernD; vormals Deutsches Atomforum e.V.), Berlin

► After the expiration of 10 half-lives, the activity of a particular nuclide can be considered negligible.

With this rule of thumb, keep in mind:

- Nuclides with a low activity have long half-lives. The period until ten half-lives have elapsed is therefore extremely long in some cases.
- It applies exclusively to the nuclide under consideration.
- If a likewise radioactive nuclide is created from the nuclide under consideration, then a "radioactive equilibrium" is formed in the course of these 10 half-lives. Depending on the half-life of the resulting nuclides, the activity of the resulting mixture may even increase.

27.1.4 Absorbed Dose *D*

Activity and half-life make no statements about the possible effect of the radiation emitted by the nuclide. For this purpose, the amount of energy, i.e. the dose ("quantity") absorbed by the irradiated mass, must first be determined. It is also said that the absorbed dose is the energy deposited (in the sense of "left behind") by the radiation.

► Absorbed dose (D) is the absorbed energy (E) per unit mass (m): $D = E/m$

The absorbed dose is a physical quantity that describes the absorbed portion of energy. Its unit is the Gray[1] (Gy). The following applies: 1 Gray = 1 Joule/Kilogram → 1 Gy = 1 J kg^{-1}.

27.1.5 Equivalent dose *H*

The absorbed dose does not provide any information about the biological effect of the radiation. The effect of heavy alpha particles on cells is much more devastating than that of lighter beta particles. In order to be able to make a statement about the biological effectiveness of the radiation emanating from radioactive substances, the radiation weighting factor w_R was introduced. This depends on the type of radiation and its kinetic energy and is used, among other things, for the simplified calculation of the equivalent dose. Here, the relative biological effectiveness (RBE) of the respective radiation is taken into account.

► The equivalent dose (H) is a measure of the biological effect of ionizing radiation.

It results from the absorbed dose (D) and the radiation weighting factor (wR): $H = D \cdot w\text{R}$. The unit of equivalent dose is the Sievert (Sv). The following applies: $1\ \text{Sv} = 1\ \text{Gy} \cdot w\text{R} = 1\ \text{J/kg}$.

27.1.5.1 Radiation Weighting Factor w_R

The values of the radiation weighting factor depend on the type and quality of the respective radiation (Table 27.3). They are specified in some regulations, e.g. the Radiation Protection Ordinance (Strahlenschutzverordnung - StrlSchV). (For the radiation weighting factor, see also Sect. 30.3.2).

Table 27.3 Radiation weighting factors (w_R) of various radiations

Radiation weighting factor w_R	Radiation
1	X-rays
1	γ-radiation
1	β-radiation
5	Proton Radiation
5	Neutrons (< 10 keV and > 20 MeV)
10	Neutrons (10...100 keV and 2...20 MeV)
20	Neutrons (100 keV to 2 MeV)
20	α-radiation
20	Heavy core fragments

[1] Louis Harold Gray, 1905–1965, British physicist and radiologist

Note
The radiation weighting factor is also referred to as the "weighting factor" ("q"). Until about 1990, the term "quality factor" ("Q") was in common use.

27.1.6 Effective Dose Equivalent H_{eff}

If only a partial irradiation has taken place and this is to be converted into a whole-body irradiation, this is referred to as the determination of the effective dose equivalent. For this purpose, the irradiated organs must be weighted with the so-called tissue weighting factor *w*T. This factor is the higher, the more sensitive the respective organ reacts to the radiation exposure.

The effective dose equivalent is calculated from the mean dose equivalent (HT) of the irradiated organs multiplied by the tissue weighting factor *w*T.

If several organs are affected, the sum is formed:

$$H_{\mathrm{eff}} = \sum_{\mathrm{T}=1}^{\mathrm{n}} w_T \cdot H_T$$

27.1.6.1 Tissue Weighting Factor w_T

The tissue weighting factors w_T are also specified e.g. in the Radiation Protection Ordinance (Strahlenschutzverordnung - StrlSchV) and depend on the sensitivity of the tissue or organ (Table 27.4).

It is assumed that a punctual irradiation of the body (organ irradiation) with an effective dose equivalent H_{eff} represents the same radiation risk as a uniform whole-body irradiation with $H = H_{\mathrm{eff}}$.

The unit of effective dose equivalent is the Sievert (Sv).

Table 27.4 Tissue weighting factors w_T for organs/tissues

Tissue weighting factor w_T	Organ/tissue
0.20	Gonads
0.12	Bone marrow
0.12	Colon
0.12	Lung
0.12	Stomach
0.05	Bubble
0.05	Chest
0.05	Liver
0.05	Esophagus
0.05	Thyroid
0.01	Skin
0.01	Bones
0.05	Other tissues/organs

27.1.7 Dose rate $\dot{D}$ and equivalent rate $\dot{H}$

Dose or equivalent dose make a statement about the amount of energy that is absorbed by a body. However, it almost goes without saying that the amount of radiation absorbed increases with the duration of the radiation exposure. The dose rate or the equivalent dose rate serve as the basis for this temporal dependence.

They are defined as:

$$\text{Dose rate} = \frac{\text{Absorbed dose}}{\text{Duration}} \qquad \dot{D} = \frac{D}{t}$$

$$\text{Equivalent dose rate} = \frac{\text{Equivalent dose}}{\text{Duration}} \qquad \dot{H} = \frac{H}{t}$$

The following are considered to be units

- *Dose* rate $\dot{D} \rightarrow$ Gray per hour $\rightarrow$ Gy/h
- Equivalent dose rate $\dot{H} \rightarrow$ Sievert per hour $\rightarrow$ Sv/h

Relationship Between Dose and Dose Rate

The dose corresponds to the odometer in the car, the dose rate to the speedometer display. If you drive a car for half an hour on the motorway at a speed of 120 km/h, you cover a distance of 60 km in this time. The same applies to the dose and the dose rate (Fig. 27.1).

If one is exposed to a radiation source with a dose rate (equivalent to tachometer) of 2 mSv/h over the period of half an hour, one has absorbed the dose (equivalent to kilometer) of 1 mSv.

27.1.8 Ion Dose J

The ion dose serves as a measure of the strength of the ionization capacity and thus as a measure of the quantity of ions generated. It is defined as the amount of charge generated per unit mass. Here, the amount of charge is given in coulombs. The mass usually refers to air, since radiation transmission generally takes place through the air.

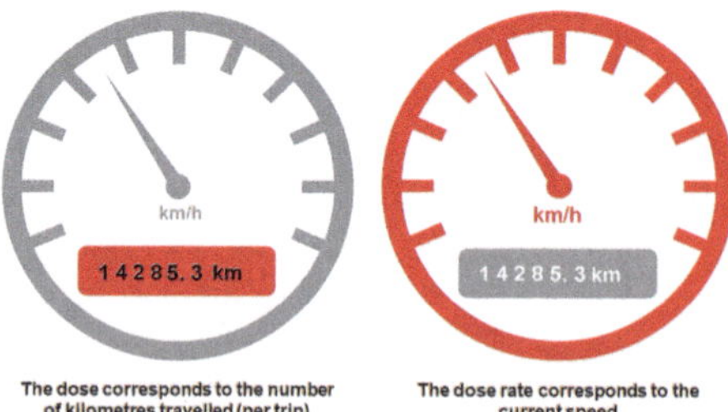

Fig. 27.1 Dose and dose rate

$$\text{Ion dose} = \frac{\text{Generated charge}}{\text{Mass } (\textit{usually air})} \qquad J = \frac{Q}{m}$$

The SI unit of the ion dose is therefore C/kg.

Various measuring devices, such as Geiger-Müller counter or rod dosimeters, determine the ion dose. From this, the absorbed dose can be estimated. A correction factor is necessary for this. This is as follows for

Air 35 Gy $(C/kg)^{-1}$

Soft tissue 37 Gy $(C/kg)^{-1}$

With an ion dose of 250 μC $kg^{-1} = 2.5 \cdot 10^{-4}$ C kg^{-1}, this results in a value of $8.75 \cdot 10^{-3}$ Gy = 8.75 mGy in air. For the ionisation density, see Sect. 30.3.2.

27.1.9 Gamma Dose Rate Constant Γ_H

For radiation sources, the dose rate constant Γ (= capital Greek letter gamma) can be used to determine the equivalent dose rate $\dot{H}$ for any distance from the radiation source if the activity is known.

$$\text{Equivalent dose rate} = \text{Dose rate constant} \cdot \frac{\text{Activity}}{\text{Distance}^2} \qquad \dot{H} = \Gamma_H \cdot \frac{A}{r^2}$$

The γ-dose rate constant (short: γ-DRC) is a nuclide-specific quantity and is often given in the unit [mSv m^2 h^{-1} GBq^{-1}] (see Table 27.5). For conversions applies:

$$1 \text{ mSv m}^2 \text{ h}^{-1} \text{ GBq}^{-1} = 10^{-3} \text{ mSv m}^2 \text{ h}^{-1} \text{ MBq}^{-1}$$
$$= 10^{-12} \text{ Sv m}^2 \text{ h}^{-1} \text{ GBq}^{-1}.$$

Example Calculation

Stand 30 min at a distance of 2 m from a Co-60 source with an activity of 5 GBq. What is the equivalent dose?

Table 27.5 γ-dose rate constants (examples)

Nuclide	γ-DRC (Γ_H) $\frac{\text{mSv m}^2}{\text{h GBq}}$
Co-60	0.348 75
Tc-95	0.106
Pb-201	0.104 67

Nuclide	γ-DRC (Γ_H) $\frac{\text{mSv m}^2}{\text{h GBq}}$
Cs-137	0.087 933
I-131	0.059 528
Kr-85	0.000 346

(a) Determination of the equivalent dose rate

$$\dot{H} = \Gamma_{\mathrm{H}} \cdot \frac{A}{r2} = 0.34875\,\frac{\mathrm{mSv\ m^2}}{\mathrm{h\ GBq}} \cdot \frac{5\ \mathrm{GBq}}{22\ \mathrm{m^2}} = 0.436\ \mathrm{mSv/h}$$

(b) Determination of the equivalent dose

$$0.436\ \mathrm{mSv/h} \cdot 0.5\ \mathrm{h} = 0.218\ \mathrm{mSv} = 218\ \mu\mathrm{Sv}$$

◀

Overview: Units of Measurement in Radiation Protection
An overview of the units of measurement in radiation protection is given in Table 27.6. The relationships between these units of measurement are shown in Fig. 27.2.

Table 27.6 Overview of units of measurement in radiation protection

Unit of measurement	Formula characters	Unit	Notes
Time	t	h (hour)	1 h = 60 min = 3600 s
Activity	A	Bq (Becquerel)	$1\ \mathrm{Bq} = 1\ \mathrm{s}^{-1}$
Absorbed dose	D	Gy (Gray)	$1\ \mathrm{Gy} = 1\ \mathrm{J\ kg}^{-1}$
Equivalent dose	H	Sv (Sievert)	$1\ \mathrm{Sv} = 1\ \mathrm{J\ kg}^{-1}$
Dose rate	$\dot{D}$	$\mathrm{Gy\ h}^{-1}$ (Gray per hour)	$1\ \mathrm{Gy\ h}^{-1} = 2.78 \cdot 10^{-4}\ \mathrm{W\ kg}^{-1}$
Equivalent dose rate	$\dot{H}$	$\mathrm{Sv\ h}^{-1}$ (Sievert per hour)	$1\ \mathrm{Sv/h} = 2.78 \cdot 10^{-4}\ \mathrm{W/kg}$
γ-dose rate constant	$\boldsymbol{\Gamma}_{\mathbf{H}}$	$\frac{\mathrm{mSv\ m^2}}{\mathrm{h\ GBq}}$	$1\ \frac{\mathrm{mSv\ m^2}}{\mathrm{h\ GBq}} = 10^{-12}\,\frac{\mathrm{Sv\ m^2}}{\mathrm{h\ Bq}}$

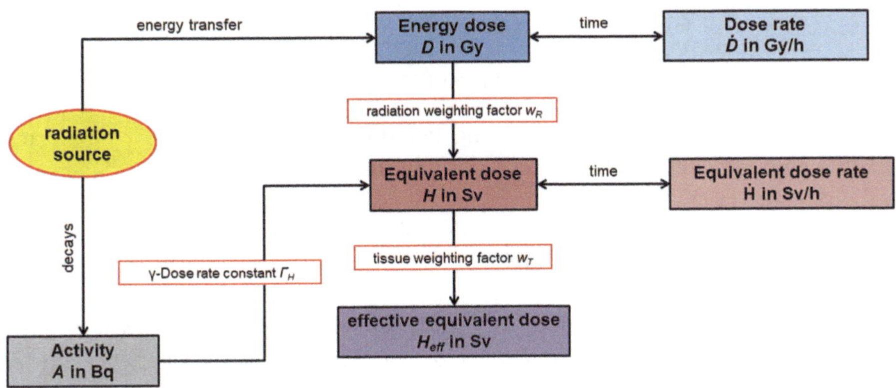

Fig. 27.2 Overview: Relationships between units of measurement in radiation protection

27.2 Other Units of Radiation Protection

The units listed below may no longer be used in Germany today. However, in order to enable the conversion of older German-language (and partly current American) literature, these units shall be briefly presented.

27.2.1 Activity in Curie

Formerly, the activity was expressed in the unit "Curie" ("Ci", sometimes also "C"). It originally referred to the number of decays that take place in one gram of radium per second.

The following conversions apply:

$$1\ \text{Curie (Ci)} = 3.7 \cdot 10^{10}\ \text{Bq} = 37\ \text{GBq}$$

$$1\ \text{Millicurie (mCi)} = 3.7 \cdot 10^{7}\ \text{Bq} = 37\ \text{MBq}$$

$$1\ \text{Microcurie (µCi)} = 3.7 \cdot 10^{4}\ \text{Bq} = 37\ \text{kBq}$$

$$1\ \text{Nanocurie (nCi)} = 37\ \text{Bq}$$

and

$$1\ \text{Becquerel (Bq)} = 2.7 \cdot 10^{-13}\ \text{Ci} = 27\ \text{pCi}$$

$$1\ \text{Kilobecquerel (kBq)} = 2.7 \cdot 10^{-10}\ \text{Ci} = 27\ \text{nCi}$$

$$1\ \text{Megabecquerel (MBq)} = 2.7 \cdot 10^{-7}\ \text{Ci} = 27\ \text{µCi}$$

$$1 \text{ Gigabecquerel (GBq)} = 2.7 \cdot 10^{-4} \text{ Ci} = 27 \text{ mCi}$$

$$1 \text{ Terabecquerel (TBq)} = 27 \text{ Ci}$$

For clarification: The amount of 27 mg radium-226, corresponding to ~35.5 mg radium chloride ($RaCl_2$), has an activity of 1 GBq.

27.2.2 Absorbed Dose in Rad

The former unit of *absorbed* dose was rad (from: *roentgen absorbed dose),* sometimes abbreviated as "rd". The relationships between rad and Gray apply:

- 1 rad = 10^{-2} Gy = 0.01 Gy
- 1 Gy = 10^{2} rad = 100 rad

27.2.3 Equivalent Dose in Rem

The former unit of equivalent dose was rem (from: *roentgen equivalent man).* The relationships between rem and Sievert apply:

- 1 rem = 10^{-2} Sv = 0.01 Sv
- 1 Sv = 10^{2} rem = 100 rem

27.2.4 Ion Dose in Roentgen

If one refers to a defined quantity of air (1 cm^3) and to the production of one electrostatic unit of charges by radiation, one arrives at the unit "Roentgen" ("R"). Between Roentgen and C/kg the relations are valid:

- $1 \text{ R} = 2.58 \cdot 10^{-4} \text{ C kg}^{-1}$
- $1 \text{ C kg}^{-1} = 3.876 \cdot 10^{3} \text{ R}$

Converted to the dose equivalent:

- In dry air: $1 \text{ R} \approx 8.77$ mGy
- In tissue: $1 \text{ R} \approx 9.6$ mGy

28 Measuring Instruments for Radiation Emitted by Radioactive Substances

This section provides an overview of the basic measuring principles of radiation measuring instruments. Furthermore, the different device groups are presented and instructions for their use are given.

28.1 Measuring Principles

The measuring devices for ionizing radiation used by the fire brigade are essentially based on four measuring principles:

- Ionization
- Photochemical reactions
- Scintillation
- Semiconductor Technology

28.1.1 Ionization Counter

One of the first devices for measuring ionizing radiation from radioactive substances was the Geiger-Müller counter, colloquially known as the "GM counter". It represents the prototype of an ionization-based measuring device for detecting the radiation emitted by radioactive substances.

The counter consists of a thin wire housed in a tube. The tube itself is usually filled with an inert gas under reduced pressure. A voltage of several 100 to over 1300 V is applied between the tube and the wire. Radiation from a radioactive substance ionizes the gas in the counter tube. The resulting gas ions or the released electrons are accelerated to the corresponding electrodes. The resulting electrical signal can be measured (Fig. 28.1). Depending on the voltage applied, the characteristics differ as shown below.

T. Schmiermund, *The Chemistry Knowledge for Firefighters*,
https://doi.org/10.1007/978-3-662-64423-2_28

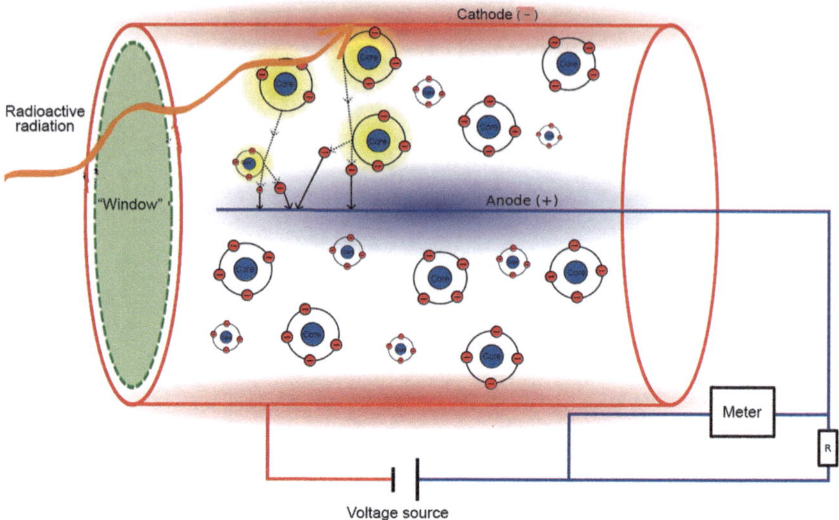

Fig. 28.1 Ionization counter tube (schematic). (Source: Geigerknueller (Own work) [CC BY-SA 3.0 (http://creativecommons.org/licenses/by-sa/3.0) or GFDL (http://www.gnu.org/copyleft/fdl.html)], via Wikimedia Commons)

Ionization Chamber

Ionization chambers have a working voltage of approximately 100. . .300 V (ionization range). All electrons released from the gas reach the anode. The measured current thus corresponds to the energy emitted by the radiation. Sensitive amplifiers are necessary for the measurement.

Proportional Counter

In the voltage range of approximately 300. . .900 V (proportional range), the counting tube operates as a so-called proportional counter. The free electrons created by the ionizing radiation knock out further electrons from the gas atoms and thus trigger electron avalanches. This effect, also called gas amplification, can yield 10^2 to 10^6 electrons. Proportional counters can distinguish between different types of radiation and measure the radiation energy.

Geiger-Müller Counter

By increasing the voltage to approx. 1100. . .1500 V (plateau range), even a single ionizing particle triggers an almost complete gas discharge. The triggered current pulses are so large that they can be made directly audible by means of a loudspeaker without further amplification.

28.1.2 Photochemical reaction

Becquerel already used the possibility of detecting the radiation of his radioactive uranium salts with the aid of photographic plates (compare Sect. 22.1.1). The ionizing radiation blackens the film. The amount of blackening is a measure of the absorbed dose. The darker the area in question, the higher the dose. This procedure is also called autoradiography. In addition to classic X-ray films (e.g. at the dentist), digital X-ray images can nowadays be produced with so-called X-ray cameras.

28.1.3 Scintillation Counter

As early as 1908, Rutherford and Geiger[1] discovered that a glass plate coated with zinc sulfide (ZnS) emits flashes of light when irradiated with particles (see Sect. 5. 3.1). This effect can be exploited for the detection and measurement of radiation emitted by radioactive substances. Substances that react with the emission of light when irradiated with radioactive substances are called scintillators.

Modern solid-state scintillation counters consist mainly of single crystals of inorganic substances, such as sodium iodide (Nal), which are mixed ("doped") with a tiny amount of thallium, for example. The ionizing radiation now excites an electron of the scintillator Nal, which enters an excited state and literally "wanders" through the crystal. If the electron now hits the "impurity" thallium, it falls back into the ground state and a photon is emitted. The photons are multiplied and counted by a so-called photomultiplier.

28.1.4 Semiconductor Counter

In simple terms, semiconductor meters are diodes. A DC voltage is applied to the diode in the reverse direction so that no current flows. When ionizing radiation is incident, free charge carriers are generated (similar to the ions in the Geiger counter), which migrate to the electrodes and can be measured as a current pulse.

28.1.5 Neutron Counter

For so-called neutron dosimeters, the neutrons must first be induced to produce charged particles in nuclear reactions. For this purpose, it is necessary to decelerate the neutrons beforehand with a moderator (often polyethylene, PE). The actual measurement can then be carried out, for example, using a counting tube filled with BF_3 as the counting gas. The helium nuclei produced according to the nuclear

[1] Johannes "Hans" Geiger, 1882–1945, German physicist.

reaction $^{10}B(n,\alpha)^{7}Li$ can thus be recorded and provide an indication of the neutron activity.

28.1.6 Further Detection and Measurement Options

The methods mentioned here are not relevant to firefighting. They are intended to show (without any claim to completeness) the different methods generally available for detecting and measuring the ionizing radiation of radioactive substances.

- Particle tracks can be made visible with cloud chambers, bubble chambers, and nuclear emulsions. The deflection in the magnetic or electric field and the range allow statements to be made about charge and mass.
- Phosphate glass dosemeters emit fluorescent radiation under UV light that is proportional to the absorbed dose. They can be evaluated repeatedly so that intermediate measurements are possible. However, they require expensive evaluation equipment.
- Thermoluminescence dosimeters (TLDs) emit light after excitation by ionizing radiation when heated. The amount of light is proportional to the irradiated absorbed dose. TLDs can only be evaluated once, but then used again.
- If no measured values are available after a radiation accident, the received radiation dose can be deduced, for example, by means of a neutron activation analysis of a hair sample ("accident dosimetry").

28.2 Measuring Instruments for Ionizing Radiation in Firefighting Operations

The devices commonly used by the fire brigade to measure the radiation emitted by radioactive substances can be divided into three groups:

- Devices for the measurement of personal dose
- Devices for measuring the dose rate
- Devices for the detection of contamination

Taking into account the permissible limit values, the devices are briefly addressed.

28.2.1 Overview of the Measuring and Detection Equipment

The following equipment is commonly used in firefighting operations:

To measure	Devices	Measurement in
Personal dose	Film dosimeter (badge meter)	mSv
	Dose warning device	mSv
Dose rate	Dose rate meters	μSv/h - mSv/h
	Dose rate warning devices	μSv/h
Contamination	Contamination detection equipment	ips

28.2.2 Film Dosimeter

This is an X-ray film that is blackened by the ionizing radiation of a radiation source. The film dosimeter must be worn by every (!) emergency worker in the danger zone. This also applies to exercises using radiation sources.

Notes on How to Wear the Film Dosimeter

- Under the protective clothing
- As close to the body as possible, in the upper chest area
- Labeling always points to the front
- Do not cover with other equipment
- Contamination must be safely prevented.

Structure and Function

The film dosimeter consists of light- and air-tight packed photographic film, which is shielded at certain points by metal plates (Fig. 28.2).

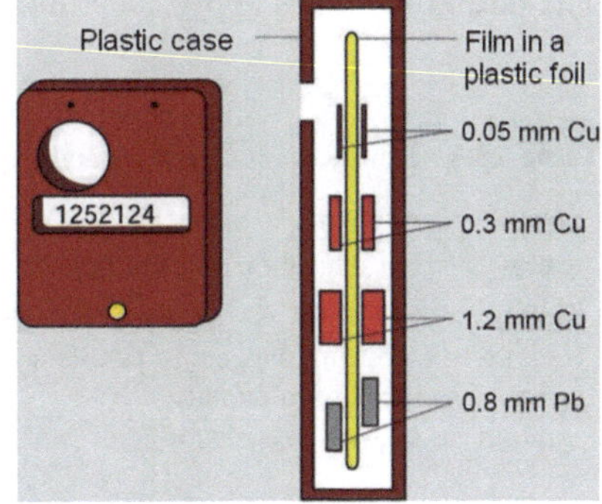

Fig. 28.2 Film dosimeter. (Source: Lit. [10-10], by courtesy of Kerntechnik Deutschland e. V. (KernD, before: Deutsches Atomforum, Berlin))

The copper filters are used to detect the energy strength of the absorbed radiation. The lead filters show the direction from which the radiation has impacted.

After the operation, the film dosimeters of the emergency forces are handed over to the responsible evaluation office. Based on the blackening of the film, this office determines the absorbed dose for each emergency worker.

► Only the (official) film dosimeter is used for the safe determination and documentation of radiation exposure.

28.2.3 Dose Warning Devices

Dose warning devices (DWD for short) are also worn on the body when entering the danger zone. They serve to protect the emergency forces by measuring the absorbed personal dose and emit an alarm signal when certain adjustable dose values are reached.

If the set dose threshold is exceeded, the responsible unit commander must be consulted.

► **Notes on the Dose Warning Device**

- Before the start of each **mission,** the warning threshold is set to 15 mSv.
- The devices must be adjustable to at least the three operational warning thresholds.
- The devices must be worn on the outside of the protective clothing to allow the warning threshold to be changed during the operation.
- The changeover to the next higher warning threshold is only to be carried out on the instruction of the responsible unit commander.
- For **exercises** involving the use of radiation sources, the warning threshold shall be set at 1 mSv.

Application Thresholds

The threshold values apply according to Table 28.1.

Table 28.1 Thresholds for exercise and deployment

Value	Activity	Period
1 mSv	For education and training	Per annum
15 mSv	Protection and salvage of property	Per use
100 mSv	Averting dangers to people and/or preventing a significant increase in damage	Per use and year
250 mSv	Saving lives	Per use and per life

28.2.4 Dose Rate Warning Devices

Dose rate warning devices (or dose rate monitors, short: DRM) are used to define the danger zone. The cut-off limit has been set at 25 µSv/h (Germany). When this dose rate is reached, the devices emit a warning tone, which is supported by light signals in some types.

Procedure

- Slowly approach the radiation source
- When triggering the device (25 µSv/h): Mark position
- Approach the source from the next side
- Target the source of the radiation

Explanatory notes to Table 28.1:

- The total absorbed dose equivalent during exercises per year shall not exceed one millisievert.
- If only material assets are affected, a maximum of 15 millisieverts per operation may be absorbed.
- To avert danger to people and or to prevent a significant increase in damage, a maximum of 100 mSv may be absorbed by the emergency worker either in one operation *or* in one year.
- In order to save human lives, 250 mSv should not be exceeded during operation.
- Once a lifetime dose of 250 mSv has been reached, the worker may no longer be deployed in the hazardous area. The maximum life dose for "radiation-exposed personnel", i.e. for persons who professionally handle radiation sources, on the other hand, is 400 mSv. The fire brigade is, therefore "on the safe side" here.

Once the danger zone has been defined in this way, emergency forces may only approach the radiation source from this boundary using personal dosimeters and suitable PPE.

28.2.5 Dose Rate Meters

Dose rate meters (DRM for short) are used for the protection of the emergency forces and the reconnaissance of the radiation situation. They are used for the search for the radiation source and the measurement of the (local) dose rate.

Notes

- Basically, work with the telescopic probe.
- Do not touch anything with the measuring probe. Danger of contamination!
- Proceed slowly. Consider the response time of the device.

- Thoroughly search.
- Mark detected radiation sources.

Please note: With some devices, it is possible to switch between the measured value of the telescopic probe and the measured value of the actual device in the display. This makes it easier to estimate the radiation exposure of the emergency forces. When proceeding further (e.g. when searching for another emitter), it must be ensured that the display has been switched back to the telescopic probe.

28.2.6 Contamination Detection Equipment

Contamination detection devices (abbreviated to CDD, also "contamination meter") are used to detect contamination of persons, animals, equipment, parts of buildings, vehicles, or the environment with radioactive substances. They are very sensitive measuring devices and also detect natural background radiation. In Germany, this amounts to approximately 10 cps (*counts per* second). Contamination detection devices are **not** able to measure the activity of a radiation source.

Proof of Contamination

- Keep the distance between the device and the surface as small as possible.
- Do not touch the surface. Danger of contamination!
- Guide slowly and without gaps over the surface.

A person, object, etc. is considered contaminated when the triple zero rate is exceeded.

Activity and Pulse Rate

The radiation emitted from all sides of the source is a measure of the number of nuclear decays and thus represents the activity of the source. It is given in Becquerel (Bq).

The contamination detection device is only able to detect a small part of the radiation (Fig. 28.3). The unit used is the pulse rate (cps, counts per second).

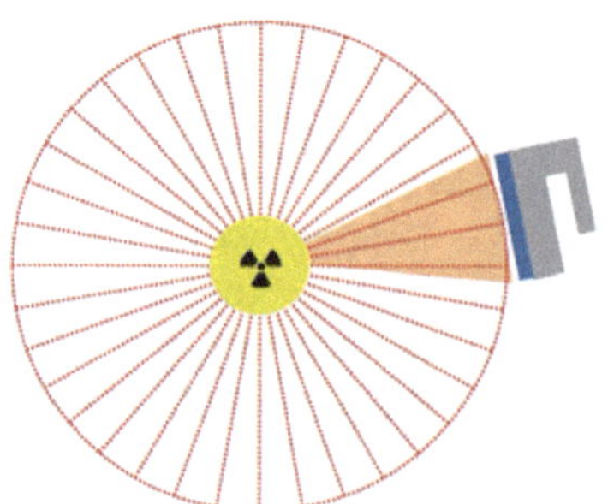

Fig. 28.3 Activity and pulse rate

28.2.7 NBR Probe

The radiological measurement component of the (German) NBC reconnaissance vehicle includes a so-called NBR probe (NBR = *natural background reduction).* By measuring the energies, this probe can distinguish between natural background radiation and artificial radiation sources. This makes it much easier to find radiation sources. The NBR probe used in this vehicle functions according to the principle of the scintillation counter.

28.2.8 Further Measuring/Detection Devices

The responsibility for incidents involving radioactive substances generally lies with the land environmental protection offices or the environmental protection departments of the responsible regional councils. Radiation protection specialists who come to the incident site may bring along extended measuring and detection equipment. This may also include devices for the identification of nuclides and devices for the detection of neutron radiation.

28.2.9 Supplement: Measured Variable H^*_{10}

Since August 1, 2011, the radiation measuring devices must be converted from the measurand H_x (photon equivalent dose) or $\dot{H}_x$ (photon equivalent dose rate) to the new measurand H^*_{10} (ambient dose equivalent) or $\dot{H}^*_{10}$ (ambient dose rate equivalent) (§ 117, StrlSchV).

Hereby mean

- H^*_{10}: The "biologically effective dose (H^*)" at 10 mm tissue depth
- $\dot{H}^*_{10}$: The "biologically effective dose rate ($\dot{H}^*$) at 10 mm tissue depth

This retrofitting requirement does *not* apply to fire department meters. Justification: The devices are not subject to calibration and proof of compliance with the limit values is provided by the official film dosimeters. The film dosimeters already record the measurandH^*_{10} .

However, fire departments can have this equipment retrofitted at their request.

To convert the values displayed by the fire brigade measuring instruments (in *Hx* or Ḣx) into the "new" measured quantity H^*_{10} or $\dot{H}^*_{10}$ is to be multiplied with the factors of Table 28.2 depending on the nuclides.

Table 28.2 Conversion H_x to H^*_{10}

Radiation field	Factor
X-rays: Generator voltage $\leq$ 50 or > 400 kV	1.0
Natural ambient radiation and gamma radiation with the exception of the nuclides listed below	1.0
Radiation field in the nuclear reactor	1.03
Radiation field after contamination in a reactor accident	1.06...1.1
Co-57, Ga-67, Se-75, Tc-99 m, Gd-153, Yb-169, Tm-170, Re-186, Ir-192, Hg-197, Au-199, Tl-201, Am-241	1.3
X-rays: Producer voltage 50...400 kV	1.3
Transmission device at the X-ray apparatus	up to 1.5

Summary

If the current measured value of the device (H_x) is multiplied by a factor of 1.3 in the case of an unclear situation, then one is on the "safe side" when calculating the duration of use or the equivalent dose. This assumes that any existing X-ray equipment has been switched- off.

Radiation Exposure 29

With regard to human exposure to radioactive substances, a fundamental distinction must be made between natural radioactivity and the additional radiation exposure caused by civilization.

29.1 Environmental Radioactivity

It is estimated that the Earth has a total natural activity of 10^{26} Bq. This is due to the decay of naturally occurring nuclides or decays triggered by natural processes.

29.1.1 Cosmic Radiation

From various sources in our Milky Way, the Earth is constantly hit by a stream of particles. This cosmic radiation consists mainly of hydrogen nuclei (protons) and helium nuclei (α-particles). Reactions with the atoms or atomic nuclei of the atmosphere produce a large number of other particles. The exposure of humans to this cosmic radiation amounts to approximately 0.3 mSv per year in Central Europe.

The dose varies strongly with altitude above sea level since the shielding effect of the air envelope decreases with increasing altitude. Thus, the exposure on the Zugspitze (highest mountain in Germany) is already 1.2 mSv/a and on Mount Everest even 20 mSv/a. The high altitude also has an effect on airplanes. For each hour of flight time, passenger planes at an altitude of 10. . .12 km add about 3 μSv of additional radiation exposure. Due to the shape of the Earth's magnetic field, cosmic radiation is stronger at the poles than at the equator.

T. Schmiermund, *The Chemistry Knowledge for Firefighters*,
https://doi.org/10.1007/978-3-662-64423-2_29

29.1.2 Terrestrial Radiation

The Earth's crust contains several naturally radioactive substances that also contribute to natural radiation exposure. The most important radionuclides are K-40, Ra-226, and Th-232, which are found in many building materials, such as concrete. The average exposure in Germany from terrestrial radiation is approximately 0.5 mSv/a.

The dose of this radiation shows strong regional variations depending on the predominant rock type of the area. For example, the annual dose around the Kaiserstuhl is about 1.5 mSv, in certain areas of the Black Forest up to 18 mSv.

29.1.3 Ingestion of Radionuclides

Ingestion is defined as incorporation through the mouth or the digestive tract. Through drinking water and food, humans absorb in particular K-40 and C-14, as well as smaller amounts of H-3, Po-210, Ra-226, and U-238. Through food intake, these substances accumulate in the body, so that humans themselves have an activity of about 7...9 kBq. The natural radiation exposure thus comes to approximately 0.4 mSv/a.

29.1.4 Radon Inhalation

From uranium-238, the uranium-radium decay series produces Rn-226 as an intermediate product, which escapes from the earth and is thus contained in the air we breathe. If the lung dose is converted to a whole-body dose, this results in an additional natural exposure of approximately 1.1 mSv/a. Thus, the inhalation of radon and its secondary products causes the largest part of the natural radiation exposure.

29.2 Civilisational Radiation Exposure

The technical and medical applications of radioactive substances lead to further exposure to radioactivity.

29.2.1 Medical Sources

The greatest exposure comes from diagnostic examinations, such as X-ray and computed tomography examinations. Here, the exposure from CT examinations is usually significantly greater than that from X-ray examinations (see Table 29.1).

Table 29.1 Effective dose per examination (examples)

CT-examination	Effective dose in mSv
Abdomen	30
Thorax	20
Spinal column	9
Head	2.5

X-ray examinations	Effective dose in mSv
Large bowel	20
Thorax	0.3
Lumbar column	2
Tooth	0.01

29.2.2 Technical Applications

Through research with and on radioactive substances, the operation of nuclear power plants, and the manifold use of radioactive substances in technology, radiation exposure of less than 0.02 mSv/a occurs.

29.2.3 Reactor Accidents and Nuclear Bomb Tests

The exposure from nuclear bomb tests, especially from the mid-1960s, and the reactor accidents of the more recent past (Chernobyl, 1986; Fukushima, 2011) amounts to only about 0.01 mSv/a each for Central Europe.

29.3 Total Load on Average

The following diagrams summarize the average radiation exposures for Germany:

Figure 29.1 shows the sources of natural radiation exposure. The main amount comes from the inhalation of radioactive radon and natural radiation of the soil.

Figure 29.2 shows that civil radiation exposure is almost exclusively due to medical applications (i.e. primarily X-rays and CT scans).

Figure 29.3 shows the total radiation exposure (mean values) for Germany, as the sum of the previous figures (Figs. 29.1 and 29.2).

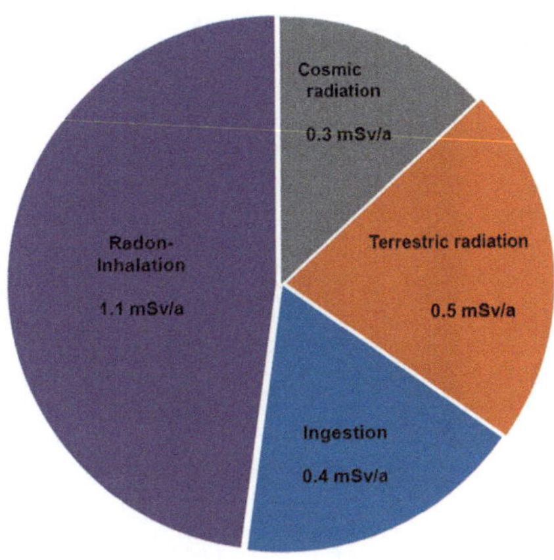

Fig. 29.1 Natural radiation exposure (total: 2.3 mSv/a)

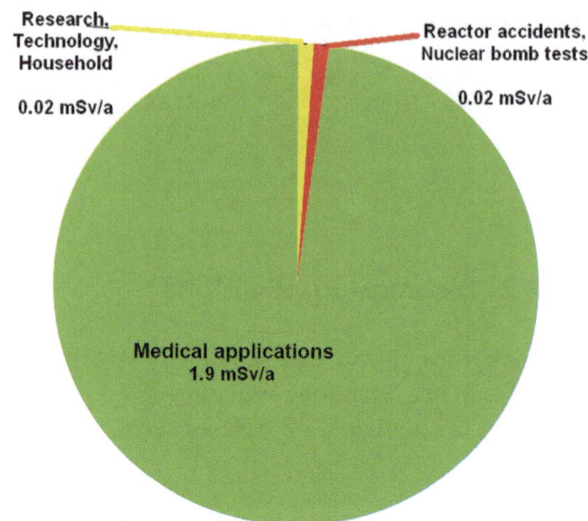

Fig. 29.2 Civilian radiation exposure (total: 1.94 mSv/a)

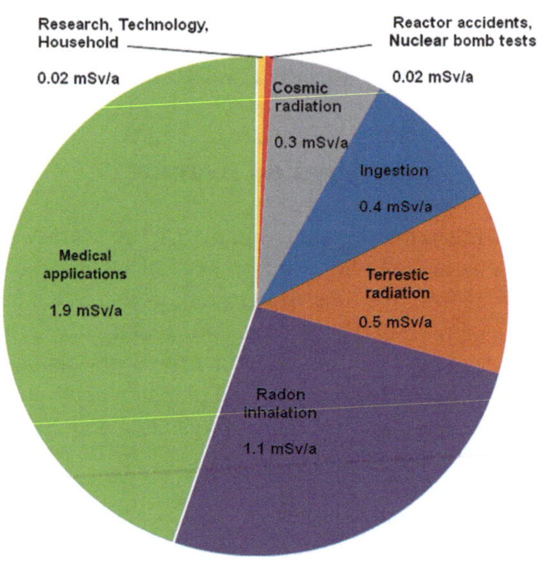

Fig. 29.3 Total radiation exposure (Germany) (total: 4.2 mSv/a)

30 Biological Effects of Ionizing Radiation

It is generally known that the ionising radiation emitted by radioactive substances can damage living cells. The biological (harmful) effect of radiation involves complex chains of biological processes at the molecular level, which are presented below in a simplified form.

30.1 Types of Radiation Damage

Most readers have probably had a sunburn at some point. This is also the effect of high-energy radiation on living tissue. In the case of sunburn, the so-called "UV-B radiation" with a wavelength of 280...320 nm is particularly "biologically effective".

A quick calculation (see Sect. 21.1) gives an energy of 4.14 eV for a *single* photon of UV-B radiation (300 nm). Sunburn, even when blistering occurs, is usually considered harmless. From a medical point of view, however, childhood sunburn is the most important risk factor for so-called "black skin cancer". Sunburn is therefore "not really" harmless.

30.1.1 Early Damage

Early damage occurs immediately after exposure to radiation. Their effect on the organism is proportional to the dose: at a whole-body dose of 250 mSv (see "Operational thresholds", Sect. 28.2.3), changes already occur in the blood count. At absorbed equivalent doses of approximately 750...1000 mSv, the so-called "radiation hangover" (headache, nausea) occurs. At even higher doses (over 1 Sv), the so-called "radiation sickness" occurs. At 4 Sv, the chance of survival is 50 % within 30 days of exposure. Doses above 7 Sv usually lead to death within a week.

T. Schmiermund, *The Chemistry Knowledge for Firefighters*,
https://doi.org/10.1007/978-3-662-64423-2_30

30.1.2 Late Damage

Typical late effects are cancers, some of which may not occur for decades. In the case of these late damages, the probability of occurrence of such damage is related to the dose – but not the severity of the disease itself. Calculated on the basis of an absorbed whole-body equivalent dose of 10 mSv, the total cancer risk is $5 \cdot 10^{-4}$, i.e. five people fall ill out of every 10 000 exposed to a radiation dose of 10 mSv. For a fatal accident (traffic accident, accidents at home or at work, etc.) the probability is also $5 \cdot 10^{-4}$.

30.1.3 Genetic Damage

If the germ cells are irradiated, mutations can be the result. This is not directly recognizable for the irradiated person. At the "genetically significant age" of humans (approximately up to the age of 35), about 140 gene mutations are produced by environmental factors – i.e. an average of 4 per year. Irradiation with 10 mSv causes two additional gene mutations. This corresponds to $^1/_{70}$ of the natural mutation rate. The additional risk for such heritable radiation damage (dose 10 mSv) is about $4 \cdot 10^{-5}$. I. e.: In 100 000 irradiated persons, hereditary damage is likely to occur in four persons in the first two subsequent generations.

▶ In order to minimise this risk during radiation protection operations, the command *may* decide to give preference to older personnel in the immediate danger zone.

30.2 Radiobiological Reaction Chain

The term "radiobiological reaction chain" is used to describe the processes which, starting from the physical effect of radiation, finally develops a biological effect via chemical and biochemical processes.

30.2.1 Time Course of the Physical-Biological Processes

The energy from the radiation acting on the body is absorbed within picoseconds (10^{-12} s). In the micro- to millisecond range ($10^{-6}...10^{-3}$ s), excited and/or ionized molecules are produced in the body tissue. Among other things, free radicals are generated in this process. At low equivalent doses, early damage occurs within minutes to hours, which is usually reversible. It may take days for mutations to occur. Late damage, on the other hand, usually only becomes noticeable after years.

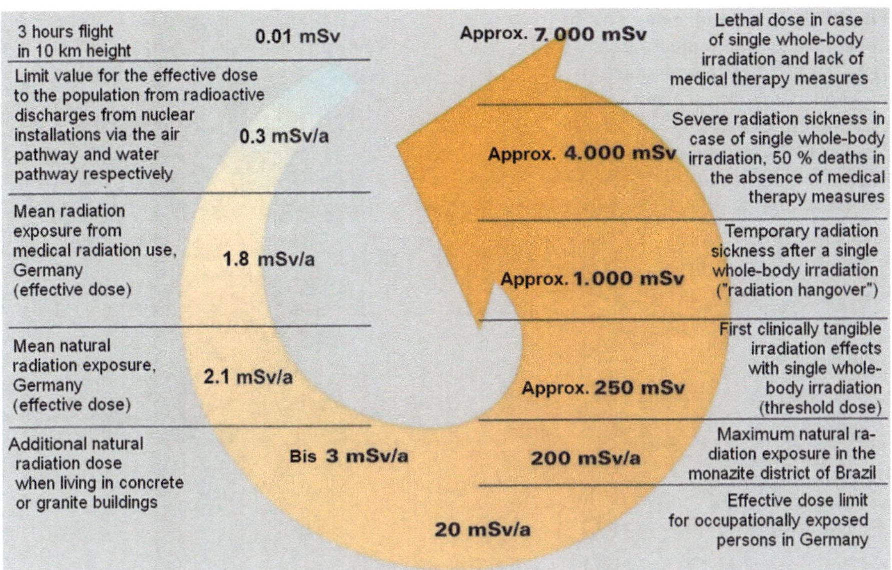

Fig. 30.1 Examples of equivalent doses and effects. (Source: Lit. [10-10], with kind permission of Kerntechnik Deutschland e. V. (KernD, before: Deutsches Atomforum e. V.) Berlin)
Note: The values from 250 mSv and above describe early somatic damage.

High doses cause biochemical changes, usually within a few hours, which also affect the shape (morphology) – and thus the functionality – of enzymes and genetic material within days. The death of individual cells, affected organs or resulted in the death of the individual (Fig. 30.1).

30.2.2 Processes at the Molecular Level

Through ionization and/or excitation, the atomic shells of the atoms of which the body cells are composed of are changed. In this process, molecules can be split if the bonding electrons are affected (Fig. 30.2). The resulting molecular fragments behave chemically or biochemically differently from the intact molecules. Also, the resulting new molecules can act as cytotoxins, e.g., H_2O_2 from H_2O.

Radiolysis

The splitting of chemical bonds by ionizing ("radioactive") radiation is called radiolysis (Greek: *radio* = beam, *lysis* = dissolution, separation). This produces ions or radicals, which in turn can react further.

When water, for example, is excited by γ-radiation, an excited water molecule (H_2O^*) is formed; when ionized, an H_2O^+-ion is formed.

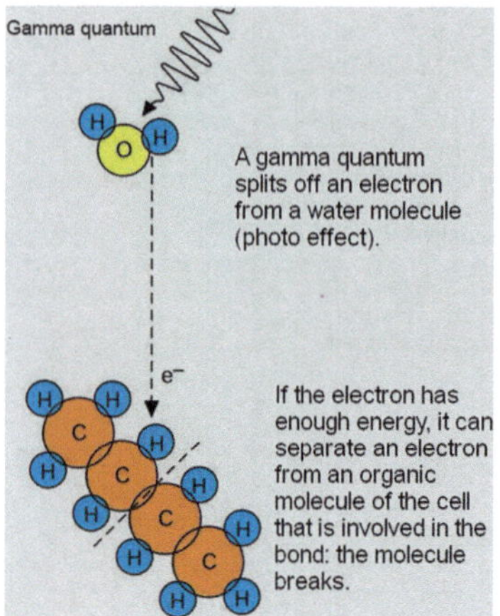

Fig. 30.2 Direct radiation effect (example). (Source: Lit. [10-10], with kind permission of Kerntechnik Deutschland e. V. (KernD, before: Deutsches Atomforum e. V.) Berlin)

- Excitation: $H_2O \xrightarrow{\gamma} H_2O^*$
- Ionization: $H_2O \xrightarrow{\gamma} H_2O^+ + e^-$

The H_2O^+ ion reacts very rapidly with water to form OH radicals (HO•):

$$H_2O^+ + H_2O \longrightarrow H_3O^+ + \bullet OH$$

The excited water molecules dissociate to form radicals:

$$H_2O^* \longrightarrow H\bullet + \bullet OH$$

The radicals eventually form the cell toxin hydrogen peroxide.

- $$2\ \bullet OH \longrightarrow H_2O_2$$

Irradiation of one litre of water with an absorbed dose of one Gray thus produces, for example, 0.073 µmol H_2O_2 and 0.28 µmol ·OH. Compare also Lit. [13-9].

30.2.3 Cellular processes

The primary physical effects (ionization, excitation) and the resulting secondary chemical effects (compare Fig. 30.3) can then lead to tertiary biological effects: The cell or rather parts of the cell (e.g. enzymes, DNA) are either no longer functional or show an altered biological behaviour.

As a rule, damaged cells can be recognised by the body's own immune system and rendered harmless. However, if this defence and repair system fails, e.g. because a large number of cells have been damaged, then so-called radiation damage occurs.

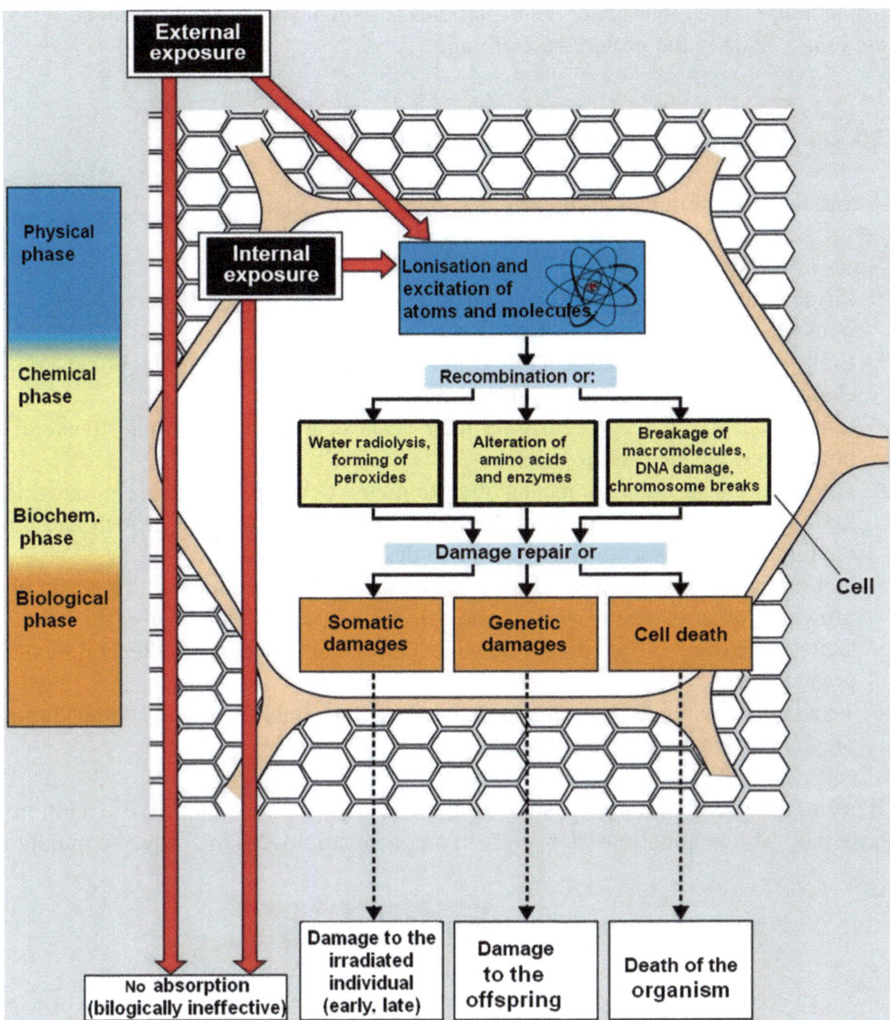

Fig. 30.3 Physical, chemical and biological processes in the cell. (Source: Lit. [10-10], with kind permission Kerntechnik Deutschland e. V. (KernD, before: Deutsches Atomforum e. V.) Berlin)

External and Internal Irradiation

After incorporation of radioactive substances by ingestion or inhalation, internal irradiation occurs. The nuclide has direct contact with the body cells. To assess the hazard to humans, the (main) decay mode of the incorporated emitter must be taken into account. In the case of nuclides that are only subject to the γ-transition, there is no great difference between internal and external irradiation – in terms of the radiation and the hazard to the emergency forces. For β-emitters, it depends primarily on the energy of the particles. However, if the nuclide is subject to α-decay, then internal irradiation causes massive damage (see Sect. 30.3.2). In the case of external irradiation, on the other hand, the α-particles are already completely retained by the clothing – usually the protective clothing.

30.2.4 Classification of Radiation Damage

A classification of radiation damage is shown in Fig. 30.4.

Notes and Explanations for Fig. 30.4

- Physical damage (also: somatic damage) occurs only in the irradiated individual himself.
- Early physical damage usually occurs after hours or days (e.g.: blood count changes, "radiation hangover", temporary sterility).
- Late physical damages occur only after years or even decades (e.g.: leukemia, bone, liver, bladder, kidney cancer)
- Late malignant damage means that tumours develop which form daughter tumours (metastases), have a high growth rate with destructive growth and that the tumour cells bear no resemblance to the original tissue.
- Not malignant proliferating are the colloquially "benign tumors". These do not grow into other tissues or vessels and spread less quickly than malignant tumors.
- Genetic damage is hereditary damage that usually occurs in the children or grandchildren.
- Teratogenic damage, i.e. damage caused to the embryo in the womb during pregnancy, was not included in the graph.

Reason: Pregnant firefighters are not allowed to work in radioactive operations. Extremely strict regulations also apply to pregnant employees in X-ray and radiation

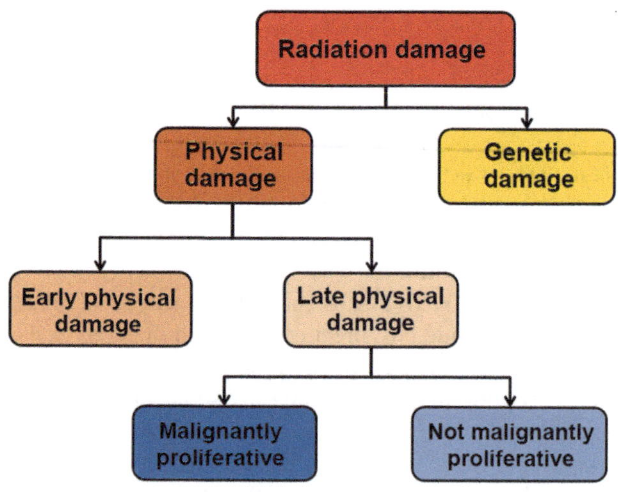

Fig. 30.4 Classification of radiation damage

laboratories and similar workplaces, so that reproductive toxicity caused by the ionising radiation emitted by radioactive substances can generally be ruled out here.

Regarding the Dose Dependence of Somatic Damage
The higher the dose,

- the more serious the consequences.
- the faster symptoms appear.
- the longer the radiation sickness lasts.
- the lower the chances of survival.

For example, 134 clean-up workers from the Chernobyl accident (1986) were treated for acute radiation syndrome (ARS). Of 41 individuals exposed to doses between 0.8 and 2.1 Gy, all survived. In the group that had received doses of 6.5 to 16 Gy (!) (21 cases), only one person survived. This corresponds to a mortality rate of 95 %.

30.3 Factor dependence of the radiation effect

For somatic damage in particular, various factors are essentially responsible for the radiation effect (Fig. 30.5).

30.3.1 Overview of Radiation Effect Factors

Explanations for Fig. 30.5.

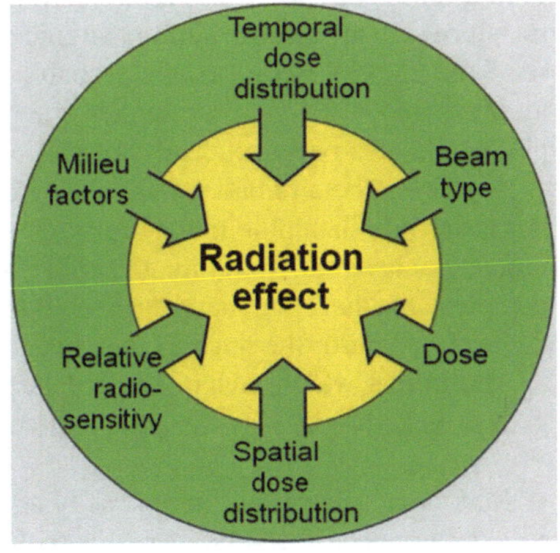

Fig. 30.5 Factors of radiation effects. (Source: Lit. [10-10], with kind permission of Kerntechnik Deutschland e. V. (KernD, before: Deutsches Atomforum e. V.) Berlin)

Radiation Type
The greater the ionisation density of the radiation in question, the greater the effect on biological systems (see Sect. 30.3.2). This is particularly important in the case of internal irradiation, i.e. after incorporation of radionuclides.

Dose
The radiation effect increases with increasing the dose.

Time Dose Distribution
The smaller the effect of a dose, the greater the time intervals between the partial doses.

Spatial Dose Distribution
The effect of the radiation depends on whether it is a whole-body, partial-body or organ dose. In the case of organ doses in particular, the internal and external radiation must also be considered separately.

Relative Radiosensitivity
The individual organs/tissues in the body have different sensitivities to the effects of ionising radiation (see Sect. 30.3.3).

Milieu Factors
Individual organs or the entire organism can be particularly sensitive to the effects of radiation due to dietary habits, medication, abuse of stimulants and other influences.

30.3.2 Ionization Density

X-rays and gamma rays penetrate deep into the tissue. Part of the energy is transferred to the electrons of the atomic shell (photoelectric effect, Compton effect, see Sect. 25.4.3). However, the probability of an interaction is relatively low. If the number of interactions were higher, the penetration depth of β-radiation, for example, would also be significantly lower.

Alpha and beta radiations are charged particles that can interact with the matter more strongly than pure radiation. When electrons interact with atoms, both usually remain unchanged. Therefore, the effect is similar to that of gamma radiation. Alpha particles, on the other hand, can collide with atomic nuclei and thereby change themselves and/or the nucleus they hit.

The rate at which a charged particle generates other charged particles along its path is related to its mass, charge, and velocity:

- Massive particles "hit" more easily and transfer more impact energy.
- A larger electric charge increases the probability of creating ion pairs.
- Slower particles have more time to interact with their environment.

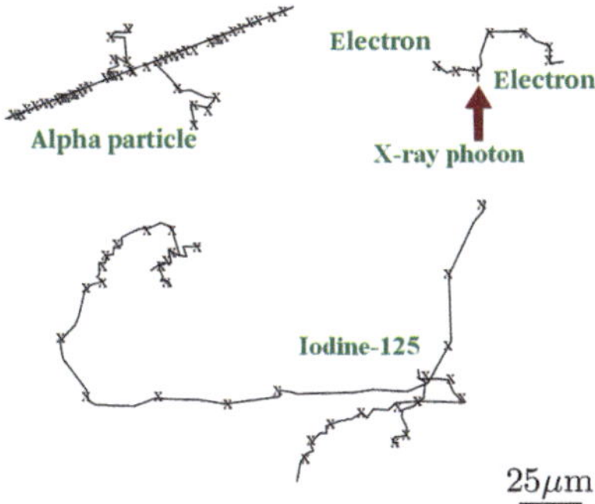

Fig. 30.6 Interaction of various radiations with water. The graph shows ionizations and excitations along the trajectories of particles in water: (**a**) for an α-particle at 5.4 MeV, (**b**) for the electrons released after absorption of a 1.5 keV X-ray photon, (**c**) that of the electrons released during the β^{+}-decay of I-125. Each "X" along the plotted path represents an interaction. (Source: Dirk Hünniger, CC BY-SA 3.0, https://commons.wikimedia.org/w/index.php?curid=1630953 (translated))

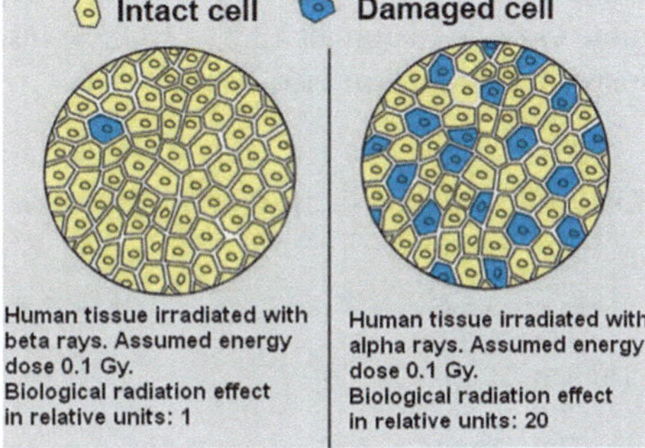

Fig. 30.7 Effect of α- and β-radiation on tissue. (Source: Lit. [10-10], with kind permission of Kerntechnik Deutschland e. V. (KernD, before: Deutsches Atomforum e. V.) Berlin)

One also speaks here of "loosely ionizing" (γ, β) and "densely ionizing" (α) radiation. The latter is biologically more effective because the effects (ionisation, excitation) caused by it are closer together. This means more damage in a smaller space and thus makes it more difficult for the body's repair mechanisms to work (See. Figure 30.6).

If we consider the effects on the cells of the body, we obtain the diagram of Fig. 30.7. It can be seen that the densely ionising α-radiation damages 20 times more cells than the loosely ionising β-radiation of the same absorbed dose. Here we arrive at the radiation weighting factor w_R, which has already been discussed in the consideration of the equivalent dose (compare Sect. 27.1.5).

30.3.3 Relative Radiation Sensitivity

The radiation sensitivity of individual organs/tissues of the body differs. In general, it can be stated that tissues with a high cell division rate react more sensitively to the effects of radiation. This is due to the fact that the radiation acts at the moment of cell division and thus delays or even blocks the division process. The ability of the affected cells to divide can also be completely lost. This severely disturbs the balance between cell loss and cell renewal.

Ranking of the Sensitivity of Human Tissue

Bone marrow > Gastrointestinal tract > Lungs > Mucous membranes > Gonads > Liver > Breast > Thyroid gland > Kidney > Skin > Bones.

30.3.4 Radiation Sensitivity of Animals

Different organisms are sensitive to radiation in different ways. The $LD_{50/30}$ is the lethal dose at which 50 % of the individuals die within 30 days. This dose is still quite close in mammals (3.5. . .10 Gy). Insects, bacteria and viruses are much more resistant to radiation (see Table 30.1).

Table 30.1 Radiation sensitivity of living organisms

	$LD_{50/30}$		$LD_{50/30}$
Goat	3.5 Gy	Bacterium E. coli	50 Gy
Pig	4.2 Gy	Snail	200 Gy
Human	4.5 Gy	Wasp	1 000 Gy
Dog	5 Gy	Amoeba	1 000 Gy
Rat	6 Gy	Tobacco mosaic virus	2 000 Gy
Bunny	8 Gy	Deinococcus radiodurans (Bacterium)	17 500 Gy
Hamster	10 Gy		

31 Use of Radioactive Substances

Radionuclides, or the radiation they emit, have many applications. Some important applications in science, technology, and medicine are briefly addressed.

Most people are probably not aware of the many ways in which radioactive substances are used in everyday life. The following list is not exhaustive and is only intended to provide a simplified overview.

Please note that neither β- nor γ-radiation is able to turn stable atomic nuclei into radioactive isotopes. Therefore, food or medical products do not become radioactive any more than you do during an X-ray or CT scan.

31.1 Application in the fire brigade

The IMS, which is mounted on some NBC vehicles, uses the ionizing effect of radiation emitted by radioactive substances to measure hazardous substances – both industrial chemicals and chemical warfare agents.

31.1.1 Ion Mobility Spectrometer (IMS)

Principle

Ions move in electric fields (compare electrolysis, Chap. 20). In the IMS, a radioactive source (Ni-63) generates ions that move in the electric field against the direction of the drift gas. Ions of different masses have different drift velocities, resulting in separation. These separated ions finally reach the detector.

How It Works

The radioactive nickel-63 emits β-radiation with a maximum energy of 67 keV. This is sufficient to ionize the gas molecules of the ambient air and to produce so-called reactant ions (RI, R^+/R^-). The ionization potential of N_2, for example, is only 15.6 eV.

T. Schmiermund, *The Chemistry Knowledge for Firefighters*,
https://doi.org/10.1007/978-3-662-64423-2_31

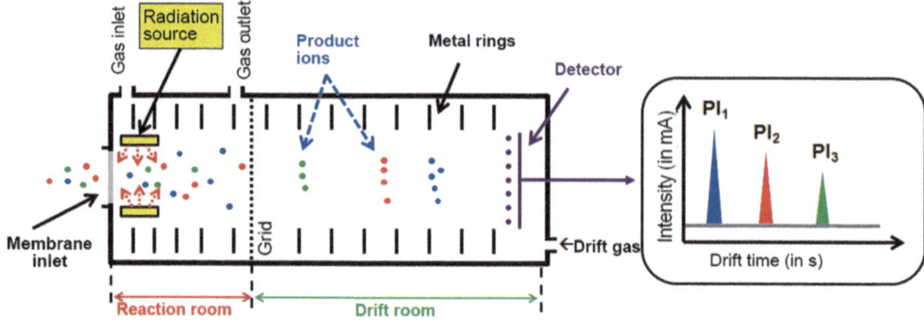

Fig. 31.1 IMS (principle sketch)

Table 31.1 Detection possibilities IMS

CWA mode	ITOX mode
Lewisite	Ammonia
Sarin	Hydrogen cyanide
S-Lost	Chlorine
Soman	Chlorinated hydrocarbons
N-Lost	Acetic acid
Tabun	Sulfur dioxide
VX	Toluene diisocyanate

$$\beta^-(67\ \text{keV}) \rightarrow R^+/R^-$$

These reactant ions react with the neutral sample molecules (P) and ionize them to so-called product ions (PI, P^+/P^-).

$$R^+/R^- + P \rightarrow P^+/P^- + R$$

Through the grid, which is alternately opened and closed, the formed PI is transferred as "packets" into the drift space. Due to the voltage gradient of the metal rings, the PI separates and can be detected at the detector (Fig. 31.1). The result is a time-dependent current signal. By comparing the spectrum obtained with a library stored in the instrument, the substances listed in Table 31.1 can be identified and quantified. The IMS operates in two different modes. The CWA mode for *chemical warfare agents* (compare Chap. 52) and the ITOX mode for some industrial chemicals.

31.2 Applications in Technology

Sterilization

Co-60 sources of high activity and energy dose (50...250 kGy) are used to sterilize medical devices (syringes, cannulas, catheters, etc.) but also implants and pharmaceuticals. Similarly, food packaging is also sterilized – and even imported children's toys from countries where a high microbial load is to be expected.

Thickness Measurement
For example, in the production of sheet metal in rolling mills, the rolls are controlled by measuring the thickness of the sheet metal, thus enabling a constant sheet thickness. The same applies to the production of plastic films. In road construction, the thickness of the asphalt layer is measured. In these cases, Co-60 is usually used.

Materials Testing
The quality of welds, but also changes in materials (corrosion, material fatigue) are checked, mostly with the aid of Cs-137 or Co-60 sources.

Isotope Batteries
Also called radionuclide batteries, they convert the thermal energy of spontaneous decay into electrical energy. Thus, the probes Viking 1 & 2, Voyager 1 & 2, and Galileo were equipped with Pu-238 isotope batteries. Today, the power of the batteries of the Voyager sondes (launch: 1977) is still a good 50 % of the initial power.

Tracer Methods (1)
Na-24 is often used for leak detection, e.g. in pipelines. The nuclide is injected into the pipeline and the activity is measured. The leak is located where the activity increases massively.

Ionization Smoke Detector
These smoke detectors use the radiation emitted by Am-241 and measure the ionization of the air. Because of the disposal problem, ionization smoke detectors are now only installed in special cases (Europe). The smoke alarms prescribed by the German states in private households are all-optical smoke alarms (see Sect. 11.3.3). Ionization smoke detectors are still widely used in the USA, Canada, and Australia.

Luminescent Light
Dials of clocks and certain instruments (e.g. in aircraft) are manufactured with a mixture of a radionuclide (Ra-226, Pm-147) and a luminescent substance (e.g. ZnS). The radiation stimulates the emission of luminescent light. In recent years, H-3 (tritium) has also been widely used.

Level and Density Measurement
Nuclides are also used to measure and control the levels in boilers and tanks. This method can be applied to both liquids and bulk solids and can also be retrofitted without interrupting production. The density of a medium in a pipeline can be measured continuously using nuclides and scintillation counters and is thus used for the continuous control of production processes. These methods are not only used in the chemical industry but also the food industry.

Nutrition

To reduce the germination capacity of microorganisms and for preservation, foodstuffs, such as dried spices are irradiated with doses of 0.1–10 kGy.

To control fruit flies and other pests, e.g. in seeds, doses of 0.1–0.2 kGy are used.

Other Applications

- For curing tire rubber and other plastics.
- Small amounts of radioactive material are used in photocopiers to prevent the paper from sticking together.
- Baggage scanners sometimes use radioactive sources.
- For decontamination of mail suspected of containing pathogens.
- For killing germs in wastewater and solid waste.

31.3 Applications in Science

Radiation-Induced Mutation

Irradiation can be used to produce mutations in (crop) plants. The plants with the highest yields and the greatest resistance to pests are further bred.

Tracer Methods (2)

Radioactive labeling of compounds can be used to elucidate reaction mechanisms and to investigate the mode of action of catalysts. First, a compound is prepared with a radionuclide (e.g. S-35, P-32, I-131, H-3, C-14) and then the reaction product that has taken up the nuclide is monitored.

This method is also well suited for the elucidation of biochemical processes. Other analytical methods are often hampered by the presence of thousands of other compounds and the extremely small concentrations.

Neutron Activation Analysis

The sample to be examined is bombarded with neutrons. The neutron capture in the nucleus produces radioactive nuclides. These can be identified and quantified. The method is particularly suitable for trace analysis and can detect over 50 elements.

Aging Using the C-14 Method

C-14 is constantly formed by the reaction of neutrons from cosmic rays with nitrogen in the atmosphere: N-14 (n, p) C-14, which decays back into N-14 by emitting β-radiation. There is an equilibrium between the C-14 that is formed and the C-14 that decays. C-14 is absorbed in the form of CO_2 by plants during photosynthesis and reaches animal organisms via the food chain. If a living organism dies, there is no "supply" of C-14, and the equilibrium shifts. By comparing the amounts of C-14 in living and dead tissue, the time at which the organism died can be determined, taking into account the half-life of C-14 (5730 years).

Geological Age Determination
The age of minerals and rocks can be determined, for example, by the isotope ratio of U-238 (half-life $4.468 \cdot 10^9$ years) and Pb-206 (stable). Other methods use the ratio of K-40 to Ar-40 or Sr-87 to Rb-87.

31.4 Applications in medicine

X-ray Examinations
Areas of the body are "X-rayed" and converted into an image.

Computed Tomography (CT)
Digital sectional images are reconstructed from a large number of X-ray images taken from different directions. This makes a three-dimensional representation possible.

Scintigraphy
The patient is administered a radiopharmaceutical that accumulates in the tissue to be examined. A gamma camera can then be used to take corresponding images. The most commonly examined tissues are the skeleton, thyroid gland, heart, liver, and kidney.

SPECT
*Single-**p**hoton **e**mission **c**omputed **t**omography* (SPECT) is used to examine heart muscles, bones, brain function and to detect tumors. Here, a nuclide is administered to the patient and the emitted γ-radiation is detected by means of gamma cameras. By using several such cameras rotating around the body, cross-sectional images can be obtained. This is a kind of "three-dimensional scintigraphy".

PET-CT
Positron **e**mission **t**omography uses nuclides that are subject to β^+-decay. The emitted positrons react with electrons present in the body and thus produce gamma radiation (pair annihilation, compare Sect. 25.4.3). This is registered in detector rings. In combination with CT, high resolutions can be achieved. Since it is primarily metabolic processes that are depicted, tumors can be detected particularly well.

Irradiation
The irradiation of disease foci, often tumors, with X-rays or gamma radiation has been used for many decades. Radiopharmaceuticals can also be used if an organ is to be specifically irradiated. More recently, the targeted destruction of tumors with ions accelerated to about 75 % of the speed of light has begun.

Radionuclides Frequently Used in Medicine
Radiation sources used for medical purposes.

Diagnosis		
Nuclide	Radiation	H-L
I-123	γ	8 d
In-111	γ	2.8 d
Tl-201	γ	73 h
Tc-99m	γ	6 h
F-18	β^+	110 m

Therapy		
Nuclide	Radiation	H-L
I-131	β^-	8 d
Y-90	β^-, γ	2.7 d
Er-169	β^-, γ	9.4 d
Re-186	β^-, γ	3.7 d
Sm-153	β^-, γ	1.9 d

32 Nuclear Reactions

Without the transformation of hydrogen into helium and all heavier elements in the course of the history of the formation of our universe, we would not be here today. The formation of all elements, with the exception of hydrogen, is exclusively due to nuclear reactions. Nuclear reactions also take place in our atmosphere, triggered by cosmic rays. Today, man is also capable of triggering such reactions.

32.1 Artificial Nuclear Transformations

The first nuclear transformation initiated by man was the reaction of nitrogen with α-radiation to oxygen observed by Rutherford in 1919: ^{14}N (α, p) ^{17}O. The first nuclide obtained in weighable quantities was P-30, produced after the reaction ^{27}Al (α, n) ^{30}P.

In addition to α-particles, neutrons, protons, deuterons (^{2}H-nuclei, d), and light atomic nuclei are used as "projectiles". A large number of nuclides can be produced in this way, as the following reaction equations show as examples.

^{75}As (α, n) ^{78}Br	^{9}Be (p, α) ^{6}Li	^{238}U (^{14}N, 6 n) ^{296}Es
^{106}Pd (α, p) ^{109}Ag	^{31}P (d, p) ^{32}P	^{239}Pu (α, n) ^{242}Cm
^{14}N (p, γ) ^{15}O	^{59}Co (n, γ) ^{60}Co	^{238}U (^{16}O, 4 n) ^{250}Fm

Nuclides produced by a large number of nuclear reactions are widely used in technology, science, and medicine (see Chap. 31).

32.2 Nuclear Fission

In nuclear fission, uranium-235 is bombarded with slow (called "thermal") neutrons (Fig. 32.1a). This neutron is absorbed into the nucleus, producing U-236. The nucleus begins to oscillate, deforms, and constricts like a drop of water (Fig. 32.1b

T. Schmiermund, *The Chemistry Knowledge for Firefighters*,
https://doi.org/10.1007/978-3-662-64423-2_32

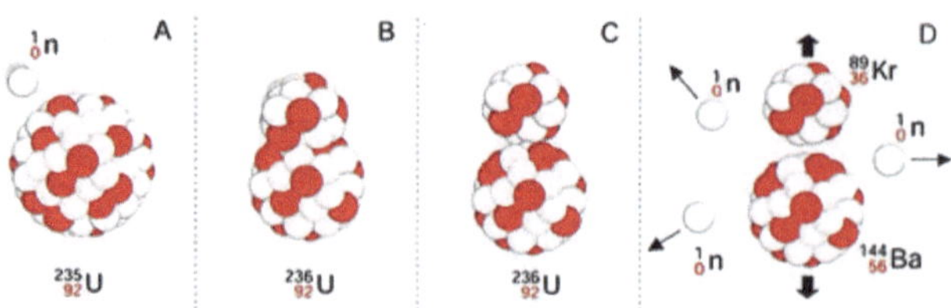

Fig. 32.1 Nuclear fission (schematic). (Source: Lit. [10-10], with kind permission Kerntechnik Deutschland e. V. (before Deutsches Atomforum e.V.), Berlin

and c). Eventually, the nucleus decays into several neutrons and two moderately heavy debris nuclei (Fig. 32.1d).

This nuclear fission releases the energy of 200 MeV (per fission!). The resulting neutrons can split further uranium nuclei. Depending on the conditions, this results in a controlled or uncontrolled chain reaction.

32.2.1 Controlled Nuclear Fission

When a uranium nucleus is split, two or three neutrons are released at a time. In nuclear power plants, excess neutrons are captured with an absorber material and thus removed from the chain reaction. In nuclear power plants, the number of neutrons released per nuclear fission is kept close to 1 (Fig. 32.2a).

32.2.2 Uncontrolled Nuclear Fission

If the number of neutrons produced per fission increases to more than one, or if no neutrons are absorbed, an avalanche of neutrons takes place. This triggers a chain reaction that can no longer be stopped (Fig. 32.2b).

32.2.3 Nuclear Power Plants

Controlled nuclear *fission* is carried out in nuclear reactors to generate electrical energy. In addition, nuclides are produced artificially and the resulting radiation is used for scientific purposes.

Natural uranium (U-nat), natural uranium enriched with U-235, Pu-233 or Pu-239 can be used as "fuel". Enriched uranium is the most commonly used. This fuel is put into zirconium alloy tubes in the form of UO_2 compacts ("pellets"). The "fuel rods" thus obtained are surrounded by a so-called "moderator" which slows down the neutrons produced. This is necessary because neutrons that are too fast do not fission with U-235. Water (H_2O), heavy water (D_2O), or graphite (C) serve as moderators (compare Sect. 22.2.1).

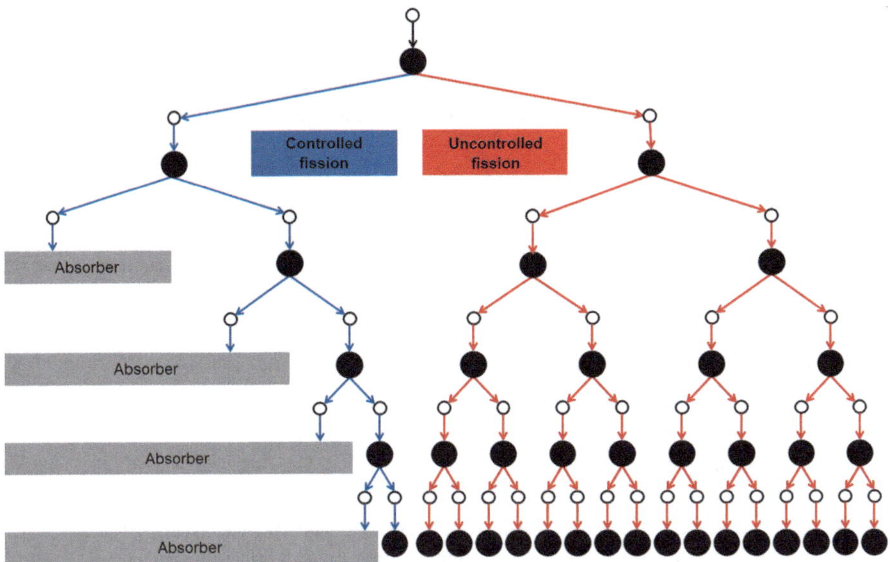

Fig. 32.2 **a** Controlled and **b** uncontrolled nuclear fission. The figure shows the effect of the absorber material in a nuclear power station (**a**) compared with the chain reaction in an atomic bomb (**b**). The circles represent the neutrons released which split the atoms *(black dots)*

The number of neutrons released must be kept close to one. This means that each individual nuclear fission should release only a single neutron on average in order to allow the reaction to continue. At a neutron reproduction factor < 1, the chain reaction comes to a standstill. If the reproduction factor rises to well above 1, the chain reaction becomes dangerously fast and the fuel rods can be damaged or the reactor *core* can even melt.

Control rods are used for control purposes. These are usually made of materials containing cadmium or boron, as these easily "capture" (absorb) neutrons. The control rods are pushed more or less far between the fuel rods, as required.

The heat generated during nuclear fission is dissipated via a cooling system, thus, generating steam outside the reactor system. This drives turbines and thus generates electrical energy.

Nuclear Energy Phase-out

Immediately after the accident at the Japanese nuclear power plant (NPP) Fukushima (March 2011), the German government decided to accelerate the phase-out of nuclear energy. The last German NPPs must be shut down by the end of 2022.

After the shutdown, the decommissioning of the power plants begins. This is where the nuclear part of the power plant is – dismantled. This is a complex task that takes ten years or more. The dismantling – up to the "greenfield"is possible without risk to the population, personnel, and the environment, as shown by the three plants that have already been completely dismantled.

32.3 Nuclear Fusion

Nuclear fusion is taking place inside our sun. Light atomic nuclei (H-1, H-2, H-3) fuse into helium under the high pressure inside the sun and at temperatures of several million degrees Celsius:

$$^{2}_{1}\mathrm{H} + {}^{2}_{1}\mathrm{H} \rightarrow {}^{4}_{2}\mathrm{He} \qquad \text{and} \qquad {}^{2}_{1}\mathrm{H} + {}^{3}_{1}\mathrm{H} \rightarrow {}^{4}_{2}\mathrm{He} + {}^{1}_{0}\mathrm{n}$$

Controlled nuclear fusion ("artificial sun") for energy production is not yet technically possible at present. For this, temperatures of 10^8 K must be generated in the reactor. Likewise, high magnetic fields are required to be able to control the plasma. Energy production from nuclear fusion has advantages over that from nuclear fission: Raw materials are available in practically unlimited quantities and much less radioactive material is produced.

32.4 Nuclear Weapons

A nuclear weapon is a device that obtains its explosive power from nuclear fission (atomic bomb) or a combination of nuclear fission and nuclear fusion (H-bomb). They are based on an uncontrolled nuclear chain reaction.

32.4.1 Atomic Bomb

If pure U-235 is used, a "uranium bomb" can be built. However, since only 0.7 % of U-235 is contained in natural uranium, complex isotope separation processes are necessary. If one has pure U-235, the decisive factor for triggering the chain reaction is that as few neutrons as possible escape. The minimum mass at which a "self-sustained" chain reaction occurs is called critical mass. For U-235 it is approximately 50 kg. The atomic bomb dropped on Hiroshima was based on this principle. During the complete fission of one kilogram of U-235, a mass loss of 1 g occurs. This corresponds to an energy release of $9 \cdot 10^{13}$ J. Analogously, an atomic bomb can also be built with Pu-239. Here, the critical mass is only about 6 kg (compare Sect. 22.2.2).

32.4.2 Hydrogen Bomb

Soon after the development of the atomic bombs ("fission bombs"), it was recognized that their explosive power could be significantly increased if fusion-capable isotopes were added. The hydrogen bomb ("H-bomb", "fusion bomb") uses the power of a "normal" atomic bomb to trigger a fusion process – similar to that in our sun.

32.4.3 Nuclear Crime

The term nuclear crime covers all illegal activities related to radioactive substances. In particular, illegal trade and fraudulent offers are to be mentioned here. In Germany, 200 to 250 cases become known per year.

The case of Russian intelligence agent A. V. Litvinenko also falls under nuclear crime. Litvinenko was poisoned with Po-210, a powerful α-emitter, in London in 2006 and died as a result of internal radiation.

32.4.4 Radiological Weapons

Radiological weapons" refers to devices used to disperse radioactive material. These are called *"radioactive dispersion devices* (RDD)", in common parlance *"dirty bomb"*.

All that is needed to make a "dirty bomb" is appropriate quantities of conventional explosives and "suitable" radionuclides. Numerous nuclides used in industry and medicine prove to be potentially suitable. However, the risk of acute radiation damage to the population of the affected area is considered to be low. The greater danger lies in the economic damage caused by the loss of local economic structures and the fact that the area may become uninhabitable for years. Terrorist groups could thus achieve one of their primary goals, namely to generate fear, panic, and uncertainty among the population.

33 Labelling of Radioactive Substances and Areas

With regard to labelling, a distinction must be made not only between transport regulations and regulations of chemicals, as is the case with "purely chemical" hazardous substances. In addition, regulations on the labelling of workplaces in which radioactive materials are handled must be observed.

33.1 Transport Labeling

In addition to the actual hazard labels, these also contain information on the radionuclide in question. This fundamentally distinguishes the labelling of radioactive substances from other hazardous substances.

33.1.1 Marking of Vehicles

Vehicles in which radioactive substances are carried must be marked with the adjacent danger label. Alternatively, in the case of tanks, the UN number of the radioactive material being carried may be indicated instead of the word "Radioactive".

T. Schmiermund, *The Chemistry Knowledge for Firefighters*,
https://doi.org/10.1007/978-3-662-64423-2_33

33.1.2 Marking of Packages

Packages (“cartons”) containing radioactive substances are divided into different categories depending on the dose rate on the outer surface and marked according to these categories. *These categories must not be confused with the hazard group.*

In addition to the hazard symbol, additional information is provided on these hazard labels:

Content*:*	Designation of the radionuclide
Activity*:*	Indication of the activity in Bq
Transport index *(TI):*	TI · 10 = highest dose rate in μSv/h at 1 m distance

Transport Index (TI)

Multiplying the transport index (short: TI) by a factor of 10 gives the maximum permissible dose rate in μSv/h which may prevail at a distance of one metre from the package. If the TI = 0 is indicated on a package of categories II or III, it is a pure α–/β-emitter which does not emit any accompanying y-radiation.

If a dose rate is measured for a **single** package which exceeds the permissible dose rate according to the TI, there is sufficient suspicion that this package has been damaged or that the shielding is no longer intact.

33.1.2.1 Fissile Material

If potentially fissile material is transported, the adjacent danger label is used. The criticality safety index (CSI) is entered in the “Criticality Safety Index” field.

Table 33.1 Transport labeling

Hazard label			
RADIOACTIVE I	RADIOACTIVE II	RADIOACTIVE III	RADIOACTIVE III
Category			
I – WHITE	II – YELLOW	III – YELLOW	III – YELLOW “Exclusive Use”[a]
Dose rate at the outer surface			
≤ 5 μSv/h	≤ 0.5 mSv/h	≤ 2 mSv/h	≤ 10 mSv/h
Dose rate at 1 m distance			
–	≤ 10 μSv/h	≤ 100 μSv/h	> 100 μSv/h
Transport index			
0 (≤ 0.05)	≤ 1	≤ 10	> 10

[a]“exclusive use” means direct journey from consignor to consignee

This indicates how far the packed material is from the critical mass. The maximum CSI is 50.

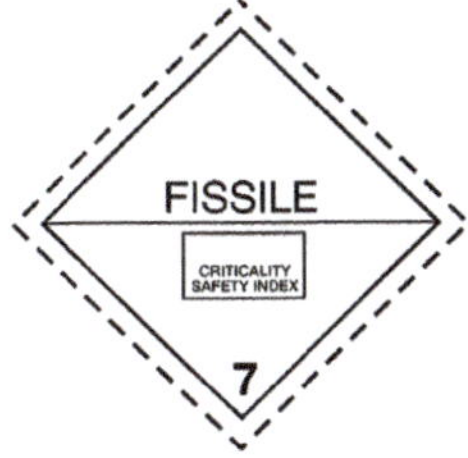

▶ Do not stack packages with fissile material or arrange them closely together.

33.1.3 Workplace Labeling

Work rooms in which radioactive substances are handled or in which equipment for the generation of ionising radiation is located must be labelled accordingly.

Examples of possible marking at workplaces (according to FwDV 500) are given in Fig. 33.1.

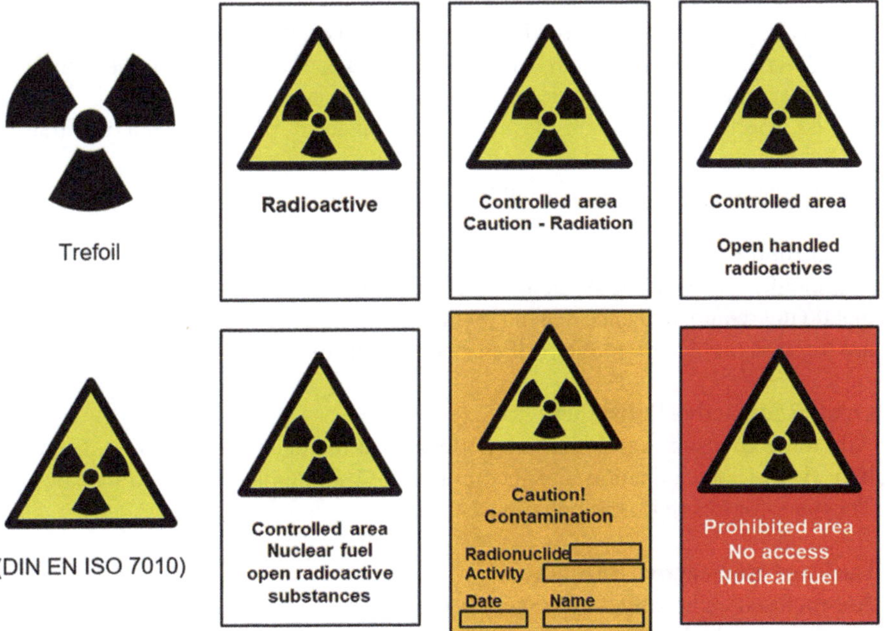

Fig. 33.1 Possible workplace marking (according to FwDV 500, modified)

Table 33.2 Classification into fire brigade hazard groups

Feuerwehr! Gefahrengruppe I	Feuerwehr! Gefahrengruppe II	Feuerwehr! Gefahrengruppe III
Areas with radioactive materials, if $\sum_A \leq 10^4$ of the EL	Areas with radioactive substances, if $\sum_A > 10^4$ but $\leq 10^7$	Areas with radioactive materials, if $\sum_A > 10^7$ of the EL
Areas with enclosed radioactive materials, if $\sum_A \leq 10^7$of the EL	$\leq 10^7$ of the EL	Areas where nuclear fuels are handled
Areas with radioactive materials, if $\sum_A \leq 10^7$ of the EL and in the type B container or in the type C container	Unless assigned to hazard group IA	Areas whose specific nature requires the presence of an expert in the event of an emergency

Legend: $\sum$: Total; A: Activity; EL: Exemption limit,Translation of the signs: "*Feuerwehr*": "Fire brigade!", "*Gefahrengruppe*": "Hazard group"

Table 33.3 Exemption limit and release according to StrlSchV (examples)

Radionuclide	Exemption limit in Bq	Release in Bq/g	Half-life
H-3	$1 \cdot 10^9$	$1 \cdot 10^3$	12,3 a
I-131	$1 \cdot 10^6$	2	8,0 d
Co-60	$1 \cdot 10^5$	$1 \cdot 10^{-1}$	5,3 a
Am-241	$1 \cdot 10^4$	$5 \cdot 10^{-2}$	432,6 a
Cs-137	$1 \cdot 10^4$	$1 \cdot 10^{-1}$	30,2 a

33.1.3.1 Classification into Fire Brigade Hazard Groups

The classification of the hazard groups IA, IIA and IIIA is based on the sum of the activities of the existing radiation sources and the exemption limits are e. g. specified in the German Radiation Protection Ordinance (Strahlenschutzverordnung, StrlSchV).

The classification is made according to Table 33.2.

Exemption Limit EL

In principle, the handling of radioactive substances is subject to licensing. Activities that do not require a licence depend on the type of radionuclide and its activity. They are listed in the Annex to the StrlSchV. Examples can be found in Table 33.3.

Open Radioactive Substances

All radioactive substances that are not enclosed radioactive substances. There is a high risk of contamination and incorporation. These emitters can be metabolized. They are, for example, radiopharmaceuticals.

Enclosed Radioactive Substances

An enclosed radioactive substance is defined as a radioactive substance that is encased or embedded in such a way that leakage of radioactive substance is safely

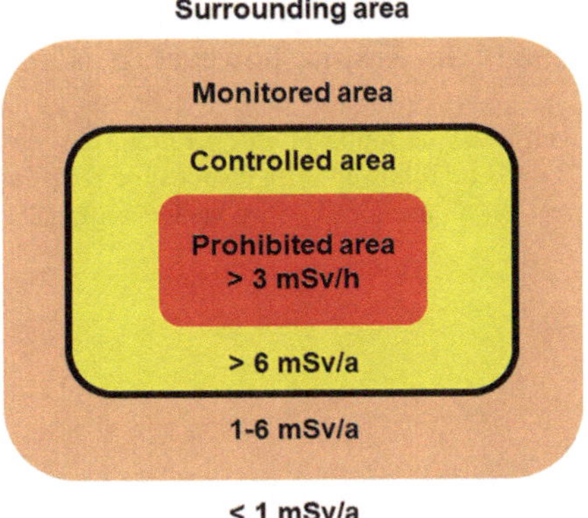

Fig. 33.2 Radiation protection areas according to § 36 StrlSchV. The environment outside the radiation protection areas must have a dose rate of < 1 mSv/a. The monitored area comprises dose rates of 1 to 6 mSv/a and does not belong to the controlled area. The controlled area has dose rates of > 6 mSv/a. Areas with more than 3 mSv/h shall be established as restricted areas. (Based on a 40-hour week with 50 working weeks per year)

prevented during normal exposure. Since the enclosure cannot be unscrewed or otherwise opened, it is a non-spreadable substance for which there is only a very low risk of contamination and incorporation.

33.1.3.2 Radiation Protection Areas

The German Radiation Protection Ordinance (Strahlenschutzverordnung, StrlSchV) also defines radiation protection areas in § 36 (Fig. 33.2).

33.1.3.3 IAEA Supplementary Marking

In 2007, the *International Atomic Energy Agency (IAEA),* together with the International Organization for Standardization (ISO), presented a new warning sign for ionizing radiation and recommended its introduction (Fig. 33.3). The starting point for the IAEA was that there had been massive injuries to people exposed to highly active radioactive sources who were unable to make sense of the "trefoil". This marking is intended as an additional marking directly on the actual radiation source – and is therefore not visible during normal operation. This symbol was defined in the ISO 21482 standard.

However, in the event that a device containing a highly radioactive source, such as Co-60 or Cs-137, is not disposed of properly or is so severely damaged by a catastrophic event that the actual radiation source is no longer in the device housing, this marking will become apparent.

Fig. 33.3 Additional label of the IAEA

► In Germany, this symbol has not yet been adopted in standards or accident prevention regulations. In Austria, however, it is standardised in ÖNORM ISO 21482.

Should this symbol nevertheless be found in an operation, the same procedure as for hazard group IIIA (see. Sect. 34.3) must be followed. The responsible radiation protection experts of the Ministry of the Environment or of the locally responsible regional council must be informed accordingly and consulted.

Protection Against Ionizing Radiation 34

Naturally, there is no "radiation protection suit" for gamma radiation. First and foremost, the protective clothing or the contamination suit protects against α-radiation. The greatest and best protection, however, is provided by the correct behavior during operations with radioactive materials. These behavioral measures are described and explained below.

34.1 Basic Behavior

As with all operations involving hazardous substances, the same applies to radiation protection operations:

- prevent incorporation,
- avoid contamination,
- contamination transfer must be prevented,
- exposure to energy (here: radiation energy) must be kept as low as possible.

34.1.1 Incorporation

The most effective measure to prevent incorporation in the hazardous area is the use of respiratory protective equipment (SCBA).

A danger that should not be underestimated is the risk of incorporation due to previous contamination transfer. Eating, drinking, and smoking in the danger zone should be prohibited as a matter of course. When setting up common areas, pay attention to the (main) wind direction and ensure that these areas are not entered with possibly contaminated clothing.

T. Schmiermund, *The Chemistry Knowledge for Firefighters*,
https://doi.org/10.1007/978-3-662-64423-2_34

34.1.2 Contamination/Contamination Transfer

Any contamination with hazardous substances must be avoided as far as possible. This applies both to the forces deployed and to the (measuring) equipment used.

In addition to the contamination protective clothing that must be removed at the boundary between the hazardous area and the containment area, simple measures can support contamination protection and/or prevent contamination transfer:

- Keep a sufficient number of plastic bags (e.g. tear-proof "garbage bags") ready for packing contaminated material, equipment, and clothing.
- Construction of a "trail" with paper, foil, etc. as a path for the measuring team. If necessary, contaminated soil does not need to be walked on in this way.
- Use of spray adhesive to bind dispersible radioactive substances or contaminated material.

34.2 Specific Protective Measures

The measures listed below also apply - at least in a figurative sense - to non-radiation protection missions. In radiation protection operations, these measures ensure that the dose absorbed by the emergency forces is as low as possible.

The important measures, besides the ALARA principle, is S-D-D-S:

- **S**witch off
- Keep large **D**istance
- Limit the **D**uration of stay
- Exploit **S**hielding

The aim of any radiation protection operation must be to keep the absorbed dose as low as possible. This is achieved by observing these protective measures.

34.2.1 ALARA Principle

The highest principle for avoiding unnecessary radiation exposure is the ALARA principle. ALARA stands for "**A**s ***L***ow ***A***s ***R***easonably **A**chievable" and means that radiation should be kept as low as reasonably achievable through technical and organizational measures. Accordingly, the aim should always be to fall below the limits specified in the radiation protection regulations.

This applies not only to the handling of radiation sources but also to their application. The ALARA principle must therefore also be maintained for technical, scientific, and medical applications (see Chap. 31) and forms the basis of the protective and behavioral measures listed below.

34.2.2 Switch Off

▶ X-ray units and other devices for generating ionizing radiation must be switched off.

It should be noted that residual radiation may still emanate from these devices for several minutes after they have been switched off. This applies in particular to devices for the generation of gamma radiation energy of > 20 MeV.

34.2.3 Distance

▶ The greater the distance to the radiation source, the lower the absorbed dose. The distance to a radiation source should therefore be kept as large as possible.

If a point source is used as a basis, then the inverse square law applies.

Inverse Square Law
The intensity of the radiation decreases with the square of the distance.
Meaning:

- 2-fold distance= $^1/_4$ of the radiation intensity at a 1-fold distance
- 3-fold distance = $^1/_9$ of the radiation intensity at a 1-fold distance
- 4-fold distance = $^1/_{16}$ of the radiation intensity at a 1-fold distance
- etc.

Figure 34.1 illustrates this situation.

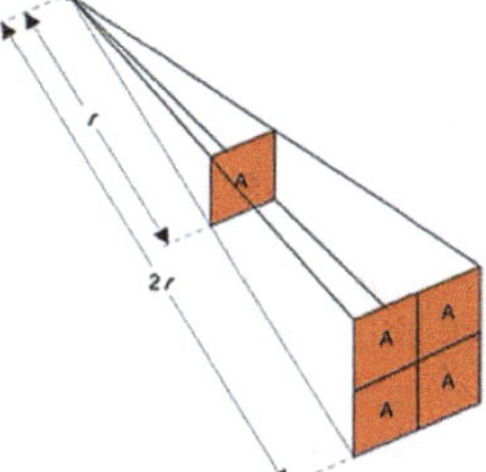

Fig. 34.1 Inverse square law (schematic). (Source: Lit. [10-10], with kind permission Kerntechnik Deutschland e. V. (KernD, before: Deutsches Atomforum e. V.) Berlin)

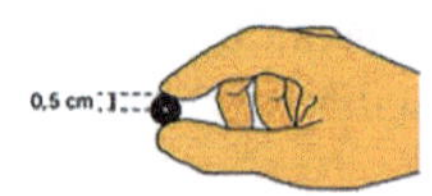

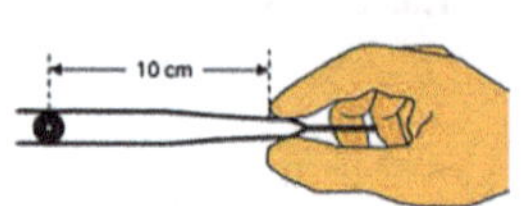

Fig. 34.2 Application of the inverse square law. (Source: Lit. [10-10], with kind permission Kerntechnik Deutschland e. V. (KernD, before: Deutsches Atomforum e. V.) Berlin)

Example 1
(for Fig. 34.2)

A small amount of radioactive substance is enclosed in a sphere with a diameter of 1.0 cm. If you hold it with your fingers, the distance to the radiation source is 0.5 cm. However, if tweezers are used, the distance increases to 10 cm. Due to the 20-fold distance, the intensity of the radiation is only $(^1/_{20})^2 = {}^1/_{400}$ of the original value.

Example 2
At a distance of 1.5 m from a radiation source, a dose rate of 735 μSv/h is measured. What is the dose rate at 4.5 m from the radiation source?

The formula is used for the calculation:

$$Dose\ rate_{\text{far}} = Dose\ rate_{\text{close}} \cdot \frac{(Distance_{\text{close}})^2}{(Distance_{\text{far}})^2} \Rightarrow \dot{H}_{\text{f}} = \dot{H}_{\text{c}} \cdot \frac{r_{\text{c}}^2}{r_{\text{f}}^2}$$

Substituting the numbers, you get:

$$\dot{H}_f = \dot{H}_c \cdot \frac{r_c^2}{r_f^2} = 735\ \frac{\mu Sv}{h} \cdot \frac{(1.5)^2\ m}{(4.5)^2\ m} = 735\ \frac{\mu Sv}{h} \cdot \frac{2.25}{20.25} = 735\ \frac{\mu Sv}{h} \cdot 0.111$$
$$= 81.7\ \frac{\mu Sv}{h}$$

The dose rate is now only about 82 μSv h^{-1}, which is just under 11 % of the initial value.

Conclusion

Dose rate measurements shall be made with a telescopic probe if possible. Remote grabs or other appropriate tools shall be used for source recovery or manipulation.

For practical purposes, it is important to remember that a radiation source may only be considered "point-like" if the smallest measuring distance to the source exceeds at least 5 times the diameter of the source. For smaller distances, the above calculation fails. In the case of area-shaped radiation sources, e.g. a spilled radioactive liquid, the intensity decreases less than indicated by the formula.

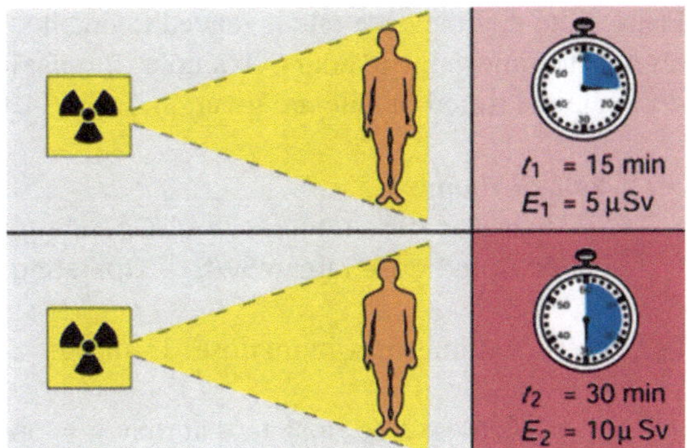

Fig. 34.3 Residence time (schematic). (Source: Lit. [10-10], with kind permission Kerntechnik Deutschland e. V. (KernD, before: Deutsches Atomforum e. V.) Berlin)

34.2.4 Duration of Stay (time factor)

- The time spent in the vicinity of a radiation source must be kept as short as possible.

- Short time = small dose.

The dose absorbed by a person is proportional to the irradiation time. This means that twice the dose is absorbed if the time is doubled (Fig. 34.3) and five times the dose is absorbed if the time is quintupled.

This time factor becomes more important as the strength of the radiation source increases (see also Table 34.1).

If the dose rate of a radiation source is known, the time until a certain dose is absorbed can be easily determined:

$$\text{Maximum time} = \frac{\text{Maximum dose}}{\text{Measured dose rate}}$$

Table 34.1 Maximum operating time (examples)

Measured dose rate	Maximum Dose = 15 mSv	Maximum Dose = 100 mSv	Maximum Dose = 250 mSv
25 µSv/h	600 h	4000 h	10,000 h
100 µSv/h	150 h	1000 h	2500 h
500 µSv/h	30 h	200 h	500 h
1000 µSv/h	15 h	100 h	250 h
10 mSv/h	90 min	600 min	1500 min
25 mSv/h	36 min	240 min	600 min
100 mSv/h	9 min	60 min	150 min
500 mSv/h	~ 2 min	12 min	30 min
1000 mSv/h	~ 1 min	6 min	15 min

Please note that the dose rate is related to one hour. The calculated time is therefore always the time value in hours. Examples for maximum operating times based on the dose rate measured on-site are given in Table 34.1.

Rule of Thumb

- Measured dose rate: μSv/h → Operating time: Hours
- Measured dose rate: mSv/h → Operating time: Minutes

Property protection (maximum dose: 15 mSv):

- 1000: Measured dose rate in mSv/h = operating time in min

Hazard prevention (maximum dose: 100 mSv):

- 6000: Measured dose rate in mSv/h = operating time in min

Mnemonic: The "μ" looks similar to an inverted and mirrored "h" and "m" like "min" (minute).

Calculation Example

What is the maximum operation time for salvaging property ($H_{max} = 15$ mSv) if a current dose rate of 185 μSv/h is measured?

$$t_{max} = \frac{H_{max}}{\dot{H}} = \frac{15\ \text{mSv}}{185\ \mu\text{Sv/h}} = \frac{15\,000\ \mu\text{Sv h}}{185\ \mu\text{Sv}} = 81.08\ \text{h} \approx 81\ \text{h}$$

34.2.5 Shielding/Covering

Exploiting opportunities for cover minimizes the dose rate and thus the personal dose. The more massive the material of the cover, the greater the effect.

The term "solid material" is to be understood as walls, concrete parts, earth walls, etc. These can be used as "natural" cover by the emergency forces. Any shielding, no matter how thin or thick, reduce the intensity of radiation and should therefore be used in any case.

Table 34.2 Examples of half-value layer (HVL)

Material	Material layer thickness in cm, at γ-quantum energy of					
	0.1 MeV	0.5 MeV	1 MeV	5 MeV	10 MeV	100 MeV
Water	4.15	7.18	32.7	76.6	105	133
Concrete	1.75	3.41	4.66	10.3	12.9	12.5
Iron	0.26	1.06	1.47	2.82	3.02	2.10
Lead	0.012	0.12	0.89	1.13	1.21	0.64

Source: Lit. [10-10]

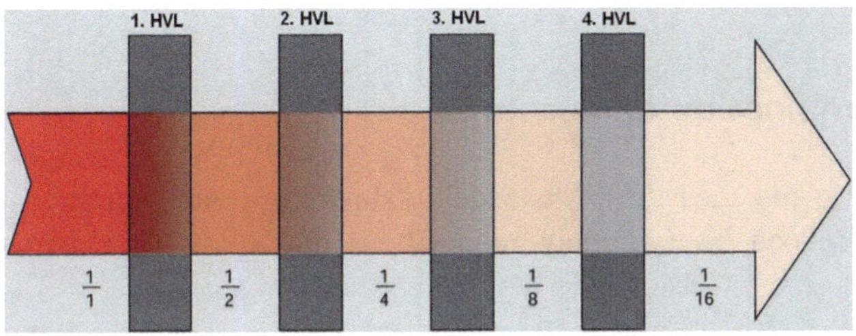

Fig. 34.4 Effect of half-value layers. (Source: Lit. [10-10], with kind permission of Kerntechnik Deutschland e. V. (KernD, before: Deutsches Atomforum e. V.) Berlin)

When recovering radioactive sources, care must be taken to ensure that the shielding effect of the salvage container is sufficient. This applies, in particular, if these containers have to be removed from the danger zone.

Half-Value Layer Thickness (HVLT)

The thickness of a certain substance that reduces the intensity of the radiation by half is called the half-value layer thickness. It depends on the material of the substance and the type and energy of the radiation Examples can be found in Table 34.2, Fig. 34.4 illustrates this relationship.

The intensity of the radiation decreases by half with each half-value layer (HVL). Meaning:

- 1 half-value layer = 1/2 of the radiation intensity without protection
- 2 half-value layers = 1/4 of the radiation intensity without protection
- 3 half-value layers = 1/8 of the radiation intensity without protection etc.

Calculation Example

The formula is used for the calculation:

$$\text{Dose rate}_{\text{HVL}} = \text{Dose rate}_0 \cdot \frac{1}{2^{\text{Number HVL}}}$$

At a distance of 1 m from a radiation source, a dose rate of 155 μSv/h is measured. The squadron takes up position behind a concrete wall, the thickness of which corresponds to three half-value layers. What is the dose rate behind the cover?

$$\dot{H}_{\mathrm{HVL}} = \dot{\mathrm{H}}_0 \cdot \frac{1}{2^{\mathrm{n}}} = 155\,\frac{\mu\mathrm{Sv}}{\mathrm{h}} \cdot \frac{1}{2^3} = 155\,\frac{\mu\mathrm{Sv}}{\mathrm{h}} \cdot \frac{1}{9} = 17.2\,\frac{\mu\mathrm{Sv}}{\mathrm{h}}$$

The dose rate is now only around 17 μSv/h, which is lower than the 25 μSv/h that marks the danger zone.

34.2.6 Combination of Measures

By combining the measures “distance”, “shielding” and “duration of stay”, the absorbed dose can be extremely reduced, as the following example calculation shows.

Example

Task

At a distance of one meter from a radiation source, a local dose rate of 685 μSv/h is measured. The squadron retreats behind a wall 6 m away from the source, the thickness of which corresponds to 1.5 half-value layers.

Questions:

(a) What dose would the squad have absorbed at 1 m from the source in 25 min?
(b) What is the dose rate due to the distance?
(c) What is the dose rate behind the wall?
(d) What dose does the squadron absorbs in 15 minutes behind cover?

Calculations:

(a) Calculation of the dose **without** protective measures

$$H = \frac{\dot{H}}{t} = 685\,\frac{\mu\mathrm{Sv}}{\mathrm{h}} \cdot \frac{25\ \mathrm{min}}{60\ \mathrm{min}} = 685\,\frac{\mu\mathrm{Sv}}{\mathrm{h}} \cdot 0.42 = 285.4\,\mu\mathrm{Sv}$$

(b) Calculation with distance law

$$\dot{H}_{\mathrm{f}} = \dot{H}_{\mathrm{c}} \cdot \frac{r_{\mathrm{c}}^2}{r_{\mathrm{f}}^2} = 685\,\frac{\mu\mathrm{Sv}}{\mathrm{h}} \cdot \frac{1^2\ \mathrm{m}}{6^2\ \mathrm{m}} = 685\,\frac{\mu\mathrm{Sv}}{\mathrm{h}} \cdot \frac{1}{36} = 685\,\frac{\mu\mathrm{Sv}}{\mathrm{h}} \cdot 0.028$$
$$= 19\,\frac{\mu\mathrm{Sv}}{\mathrm{h}}$$

(c) Calculation with additional shielding

$$\dot{H}_{\mathrm{HVL}} = \dot{\mathrm{H}}_0 \cdot \frac{1}{2^n} = 19\,\frac{\mu\mathrm{Sv}}{\mathrm{h}} \cdot \frac{1}{2^{1.5}} = 19\,\frac{\mu\mathrm{Sv}}{\mathrm{h}} \cdot \frac{1}{2.83} = 6.72\,\frac{\mu\mathrm{Sv}}{\mathrm{h}}$$

(d) Calculation of the dose **with** protective measures

$$H = \frac{\dot{H}}{t} = 6.72\,\frac{\mu\mathrm{Sv}}{\mathrm{h}} \cdot \frac{15\ \mathrm{min}}{60\ \mathrm{min}} = 6.72\,\frac{\mu\mathrm{Sv}}{\mathrm{h}} \cdot 0.25 = 1.68\,\mu\mathrm{Sv}$$

Result

Consistent implementation of protective measures reduced the dose to the troop to less than 1 % of the baseline observation. ◄

Figure 34.5 shows the protective effect of the number of half-value layers and the protective effect of increasing the distance. It can be seen clearly that two or four half-wave thicknesses produce approximately the same reduction in radiation exposure as doubling or quadrupling the distance. By combining both measures, the protective effect can be drastically increased (not shown graphically).

However, the figure also shows how important it is to keep a distance or to use shielding. Just look at the initial curve.

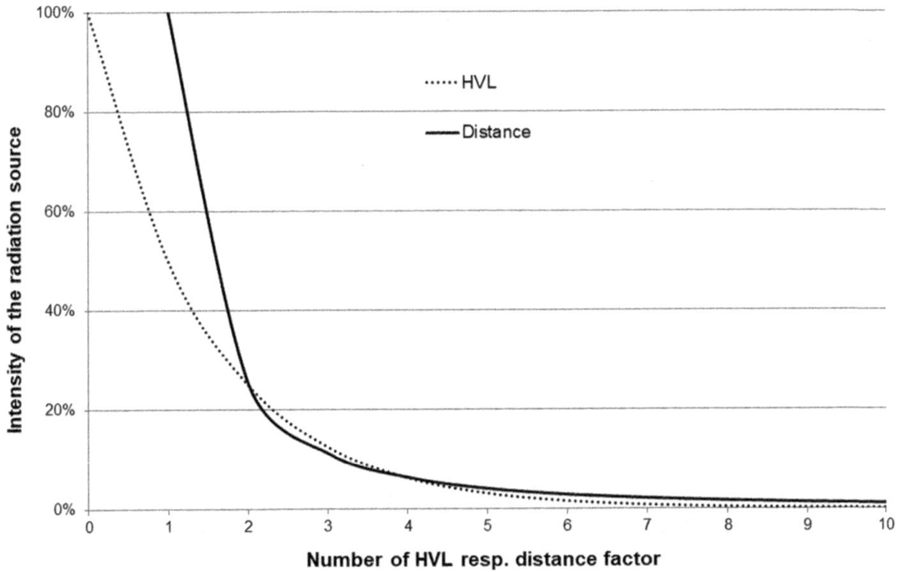

Fig. 34.5 Protective effect: distance and shielding

▶ **The "STICKS" marker scheme is proposed to combine the protective and behavioral measures explained in Sects. 34.1 and 34.2:**

S – Shielding
T – Time (limit)
I – Incorporation (prevention)
C – Contamination transfer (avoid)
K – Keep distance large
S – Switch off

34.3 Procedure According to Hazard Groups

Depending on the hazard group, certain operational-tactical measures are prescribed as a minimum standard in accordance with FwDV 500. These are to be adhered to without fail.

Feuerwehr! Gefahrengruppe I	Procedure *possible* without special equipment. Always: Breathing protection!
Feuerwehr! Gefahrengruppe II	With special surveillance (measuring and warning devices, documentation) With special equipment (protective clothing form 1 or 2) Prepare decontamination
Feuerwehr! Gefahrengruppe III	Like hazard group IIA, but: Additional expert adviser on site necessary

34.3.1 Special Operational Situations

- **Transport accidents are** initially to be treated as operations in hazard group IIA. For the rescue of people, the same procedure can be followed as for hazard group IA.
- Operations with a suspected **terrorist background** are in principle to be treated like operations of danger group IIIA.
- Areas of **hazard group IIIA** must not be entered under any circumstances without the presence of the responsible radiation protection officer or the radiation protection supervisor. *Not even for the purpose of rescuing people!*
- The care of **injured persons** by the rescue service has priority over radiation protection measures. The rescue service must be informed of any suspected contamination.
- If a dose of **≥50 mSv is** received or if **incorporation** is suspected, the affected emergency personnel must be presented to an authorized physician.

Part IX

Energy Conversion of Chemical Reactions

35 Energy

Energy (Greek *energeia* = acting force) is the "ability to do work". In other words, the energy of a system can produce an effect, which is then referred to as work. Work, in turn, is physically defined as a force acting on an object through a path. The law of conservation of energy applies to the sum of all energies in a closed system.

35.1 Law of Conservation of Energy

▶ Energy can neither be created out of nothing nor destroyed. It can only be transformed from one form of energy into one or more others.

This law of nature was recognized by von Helmholtz[1] and published in 1847. Figure 35.1 shows an example of how different forms of energy can be converted into each other.

35.2 Definitions

35.2.1 Systems

A system is a part of the universe. Namely, the part that you want to take a closer look at. It is advantageous if the system is separated from the environment by a boundary. We distinguish between open, closed, and isolated systems.

[1] Hermann Ludwig Ferdinand von Helmholtz, 1821–1894, German physicist and physiologist.

T. Schmiermund, *The Chemistry Knowledge for Firefighters*,
https://doi.org/10.1007/978-3-662-64423-2_35

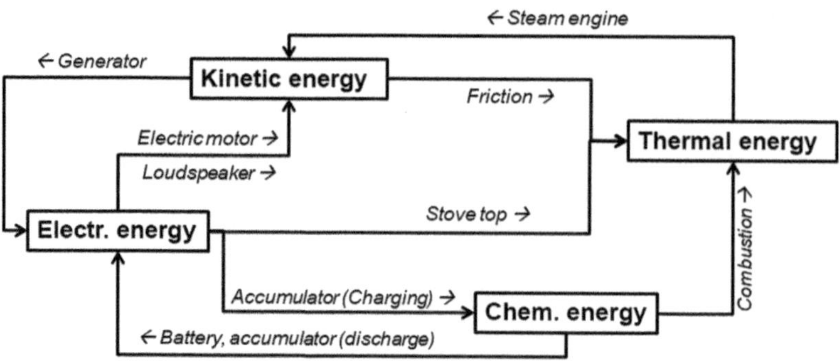

Fig. 35.1 Energies and their conversion into each other

Open System
In an open system, both an exchange of substances and an exchange of energy with the environment are possible. Examples are a campfire, any living thing, or a pot of boiling water.

Closed System
In closed systems, there is an exchange of energy but no exchange of substances with the environment. This is the case, for example, with the heating circuit of a central heating system or a sealed bottle.

Isolated System
In isolated systems, neither material nor energy exchange is possible, e.g. in an (ideal) thermos flask.

35.2.2 State Functions

Systems can be described by state functions (also state variables). A distinction is made here between:

Extensive properties - change with the size of the system. These include, for example, mass or volume. Extensive properties can be added.

Intensive properties - are independent of the system size. These include, for example, temperature, pressure, density, or refractive index. Intense properties cannot be added. Mixing water at 15 °C with water at 25 °C does not produce water at 40 °C.

State functions depend only on the initial and final state. Both the (reaction) path and the time required are irrelevant.

35.3 Energies

35.3.1 Energy Conversion of Chemical Reactions

The energy conversion in chemical reactions is many times lower than in nuclear reactions. For example, the energy released when a uranium atom is split is 200 MeV or $1.93 \cdot 10^{10}$ kJ per mole of uranium. In the formation of one mole of water, the energy of 284.5 kJ is released. This corresponds to about 3 eV per molecule.

The energy turnover in nuclear reactions is thus higher by a factor of $10^6 \ldots 10^8$ than in a classical chemical reaction. A consideration of the mass-energy equivalence can therefore be omitted.

35.3.2 Reaction Energy *E*

Chemical reactions can be thought of as molecules reacting with each other being "broken down" into groups of atoms or - in extreme cases - atoms. Subsequently, these atoms/atom groups reassemble in a new way. In the process of decomposition, energy must first be expended to separate the existing bonds. Energy is then released again when the bonds are "re-made". The reaction energy is essentially the difference between the energy of the products formed and the energy of the reactants used:

$$\Delta E = E_{\text{Products}} - E_{Educts}$$

35.3.3 Intrinsic Energy *U*

Every system has a certain amount of energy, which is called intrinsic energy (U). If energy is supplied to a system from the outside, its intrinsic energy increases. If the system releases energy to the environment, the intrinsic energy decreases. The intrinsic energy is composed of the thermal energy (Q) and the mechanical work (W) *of* the system under consideration:

$$\Delta U = \Delta Q + \Delta W$$

The work in a closed system is usually volume work. The supply of volume work, i.e. the decrease of the reaction volume, leads to an increase in intrinsic energy. Conversely, the performance of volume work (= increase in reaction volume) leads to a decrease in intrinsic energy.

This is already known to us from gases (compare Chap. 4). If gases are compressed, the temperature rises because the intrinsic energy increases. When gases flow out, volume work is done: The intrinsic energy decreases, the temperature decreases.

35.4 Enthalpy *H*

If a gas is formed in a chemical reaction, work must be done to create space for the gas against the external pressure of the atmosphere. Conversely, if gases condense into liquids, or if gases are consumed in the reaction, work is done externally on the reacting substances. By considering this volume work to be done, we arrive at the enthalpy (H). The volume work corresponds to the product of pressure (p) and volume change (ΔV):

$$\Delta H = \Delta U + p \cdot \Delta V$$

Enthalpy is best thought of as "corrected energy"; the energy conversion is corrected to atmospheric pressure.

▶ A distinction between enthalpy and intrinsic energy is usually only important for changes in the state of gases.

35.4.1 Energy of Formation and Enthalpy of Formation

Part of the intrinsic energy of a system comes from the energy stored in the chemical bonds and derived from the formation reaction of the substance. One can think of any chemical compound as being composed directly of the elements. In the case of reactions at constant volume, one speaks of the energy of formation ΔU_B^0, under the constant pressure of the enthalpy of formation ΔH_B^0. It is usually referred to 1 mol and standard conditions so that one also speaks of the standard molar enthalpy of formation.

▶ The standard molar enthalpy of formation of a compound ΔH_B^0 is the amount of heat that is absorbed or released when 1 mol of the compound is formed from the elements under standard conditions (1.013 bar, 25 °C = 298 K). The molar standard enthalpies of formation of the elements are set equal to zero here.

A selection of standard formation enthalpies can be found in Table A.8.

35.4.2 Reaction Enthalpy

If a reaction takes place directly from the elements, such as the combustion of hydrogen, the enthalpy of formation is equal to the enthalpy of reaction.

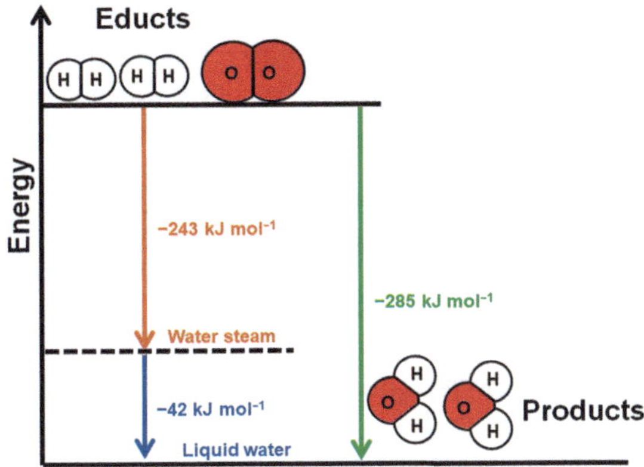

Fig. 35.2 Enthalpy of reaction of water

$$\begin{aligned} H_{2(g)} + \frac{1}{2}\, O_{2(g)} &\rightarrow H_2O_{(\ell)} \qquad -285\ \text{kJ} \\ \Delta H_B^0(H_2O) &= \Delta H_R^0(H_2O) = -285\ \text{kJ mol}^{-1} \end{aligned} \tag{35.1}$$

The negative sign results from the consideration of the system. The mixture of the gases hydrogen and oxygen lost energy when water vapor was formed. A further loss of energy has taken place during condensation to liquid water (Fig. 35.2).

Liquid water is 42 kJ mol^{-1} lower in energy than water vapor, or 285 kJ mol^{-1} more "stable" than a mixture of hydrogen and oxygen. We can solve reaction eq. G1. 35.1 into two parts as shown in Fig. 35.2:

$$\begin{array}{rcll} H_{2(g)} + \frac{1}{2} O_{2(g)} & \rightarrow & H_2O_{(g)} & -243\ \text{kJ mol}^{-1} \\ H_2O_{(g)} & \rightarrow & H_2O_{(\ell)} & -42\ \text{kJ mol}^{-1} \\ \hline H_{2(g)} + \frac{1}{2} O_{2(g)} & \rightarrow & H_2O_{(\ell)} & -285\ \text{kJ mol}^{-1} \end{array}$$

It is easy to see that the specification of the state of aggregation of the reactants and products is just as important as the specification of the formula conversion. If one mole of oxygen is brought to the reaction, 2 moles of water are formed. Therefore, one obtains

$$2\, H_{2(g)} + O_{2(g)} \rightarrow 2\, H_2O_{(\ell)} \qquad -570\ \text{kJ}$$

Exothermic and Endothermic Processes

If heat is released during a reaction, the energy content of the system under consideration is reduced. The enthalpy of reaction ΔH_R is negative. Such a reaction is called exothermic (Greek *exo* = outside and *thermos* = warm, hot).

If heat energy is consumed in a reaction, the process is called endothermic (Greek *endon* = inside). In order for the reaction to take place without a temperature change, heat energy must therefore be added. Since this increases the energy content of the system, the enthalpy of reaction ΔH_R is positive.

$$\text{Exothermic Reaktion}: \Delta H_R < 0\,(-\Delta H_R)$$
$$\text{Endothermic Reaktion}: \Delta H_R > 0\,(+\Delta H_R)$$

35.4.3 The Theorem of Hess

Based on a series of thermochemical investigations, Hess[2] came to the following conclusion in 1840:

► The enthalpy change ΔH of a process is the sum of the enthalpy changes of the individual steps. The result is independent of the reaction path. It depends only on the initial and final conditions.

$$\Delta H = \sum (H_{\text{Products}}) - \sum (H_{\text{Educts}})$$

This also means that the enthalpy of reaction - and thus also the enthalpy of formation - of substances can be determined by calculation. Therefore, not all enthalpies of formation or reaction have to be determined experimentally.

35.4.4 Born-Haber Cycle

The cycle process was found independently by M. Born[3] and F. Haber[4] in 1916. It is a direct consequence of the theorem of Hess and shows how unknown reaction or formation enthalpies can be calculated.

Example

Carbon monoxide, for example, cannot be produced directly from carbon and oxygen without carbon dioxide also being formed at the same time. Nevertheless, the enthalpy of the reaction can be determined. The reaction enthalpies of the combustion of carbon to CO_2 ($\Delta H = -393$ kJ mol^{-1}) and the combustion of CO to CO_2 can be determined experimentally ($\Delta H = -282$ kJ mol^{-1}).

The enthalpy of reaction of the reaction is calculated $C + \frac{1}{2} O_2 \rightarrow CO$ according to:

[2] Hermann Heinrich Hess (also Germain Henri Hess), 1802–1850, Swiss-Russian chemist.
[3] Max Born, 1882–1970, German mathematician and physicist, Nobel Prize in Physics in 1954.
[4] Fritz Haber, 1868–1934, German chemist, Nobel Prize in Chemistry 1918.

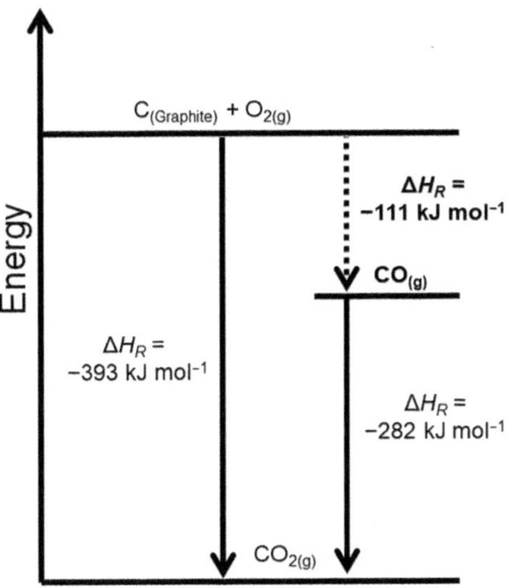

Fig. 35.3 Born-Haber cycle (example)

$$\begin{aligned}
\Delta H &= \sum(\Delta H_{\text{Products}}) - \sum(\Delta H_{\text{Educts}}) \\
\Delta H &= (\Delta H[CO_2]) - (\Delta H[CO] + \Delta H[C] + \Delta H[O_2]) \\
\Delta H &= (-393\ \text{kJ mol}^{-1}) - (-282\ \text{kJ mol}^{-1} + 0\ \text{kJ mol}^{-1} + 0\ \text{kJ mol}^{-1}) \\
\Delta H &= (-393\ \text{kJ mol}^{-1}) - (-282\ \text{kJ mol}^{-1}) \\
\Delta H &= (-111\ \text{kJ mol}^{-1})
\end{aligned}$$

written in reaction equations:

$$\begin{aligned}
C_{(s)} + O_{2(g)} &\rightarrow CO_{2(g)} \quad -393\ \text{kJ mol}^{-1} \\
CO + \frac{1}{2}O_{2(g)} &\rightarrow CO_{2(g)} \quad -282\ \text{kJ mol}^{-1} \\
\hline
C_{(s)} + \frac{1}{2}O_{2(g)} &\rightarrow CO_{(g)} \quad -111\ \text{kJ mol}^{-1}
\end{aligned}$$

Figure 35.3 shows the situation again graphically. ◀

35.5 Entropy *S*

Very many reactions that occur by themselves are exothermic, e.g. all combustion processes. This fact led to the conclusion in the nineteenth century that only exothermic (= heat energy releasing) reactions can proceed voluntarily. However, there are also endothermic (heat energy "consuming") reactions that proceed voluntarily. For example, a mixture of barium chloride-containing water of crystallization and ammonium thiocyanate can be used to "produce" a temperature of about −15 °C. Cooling is also observed when dissolving some salts (e.g. NH_4Cl) in water. Thus, the dissolving process is also endothermic. Nevertheless, these reactions occur spontaneously and voluntarily. If we look at the two processes at the level of the particles, it is noticeable that in both examples a solution arises from ordered crystal structures, in which the particles are present in a disordered manner.

$$2\,NH_4SCN_{(s)} + BaCl_2 \cdot 2\,H_2O_{(S)} \rightarrow 2\,NH^+_{4(aq)} + 2\,SCN^-_{(aq)} + 2\,H_2O$$
$$NH_4Cl \rightarrow NH^+_{4(aq)} + Cl^-_{(aq)}$$

Therefore, in addition to the enthalpy of reaction, there must be another quantity that is also responsible for the voluntariness of a reaction. This quantity is the entropy (Greek: *entropein* = *to* reverse). Entropy (formula symbol *S*) is actually what you can think of as "heat matter" in everyday life. It is what heats up the water in the pot; it is what the coffee loses when it gets cold in the cup.

If energy is supplied to a crystal, the particles vibrate around their rest position. The more energy is added, the stronger the vibration of the particles. If more energy is added, the crystal melts, and the particles can now also rotate around their axis. If energy continues to be added, the substance evaporates. The resulting gas particles can now move freely. The addition of energy (heating) has resulted in an increasingly disordered state, the order that the crystal initially had has decreased. The entropy, on the other hand, has increased. The following applies here: $\Delta S = \Delta Q/T$, where Q is the amount of heat supplied and T *is* the thermodynamic temperature (in K).

Generally speaking:

$$S_{(g)} > S_{(\ell)} > S_{(s)}$$

A gaseous system has a higher entropy (disorder) than a liquid, which has higher entropy than a solid.

One can also formulate:

- The smaller the entropy, the more ordered the system.
- The more disordered the system, the greater the entropy.

Therefore, entropy

- a measure of the state of order of a system,
- a measure of the non-reversibility of an operation,
- a measure of the "amount of restlessness" contained in a body.

Concept of Order

The concept of order in this sense is to be regarded as a statistical quantity. As illustrated in Fig. 35.4, a bottle of corn and peas can be mixed by shaking. The

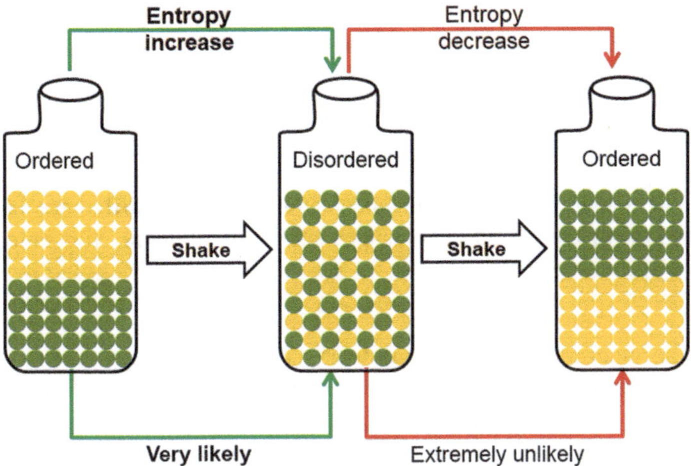

Fig. 35.4 Entropy and probability

entropy of the mixture is greater than that of the initial state. If segregation were to occur again by shaking the bottle again, the entropy should decrease because the state has greater order. This is theoretically possible, but extremely unlikely, so this segregation is not observed in nature. The number of possibilities of a homogeneous pea-corn mixture is many times greater than the few possibilities formed by a separated system.

Excursus: Entropy Examples

- The concept of entropy can be applied to the state of order of a room: Tidying up leads to a decrease in disorder, while not tidying up is a "voluntary" process that leads to an increase in entropy.
- Dust distributes itself approximately evenly on the furniture. This distribution (large entropy) is much more likely than the formation of a dust pile (small entropy) in the trash can.
- Entropy and the learning process: The acquisition of information, i.e. learning, leads to a decrease in disorder. Forgetting, on the other hand, which is a voluntary process, leads to an increase in entropy.

In the chemical-physical sense, entropy is generated because energy flows into a random movement of atoms or molecules that can no longer be reversed. This is the case with many voluntary processes, such as the diffusion of gases or the transfer of heat from a warm to a cold body.

It is interesting in this context that the entropy change also indicates the direction of the temporal sequence. All natural processes occur in such a way that the system under consideration always passes from a state of lower entropy (lower probability, higher-order) to a state of higher entropy (higher probability, lower order). The state of greater entropy is therefore always temporally after the state of smaller entropy.

35.5.1 Reaction Entropy

The entropy change in chemical-physical processes is the difference between the entropies of the products and the reactants and is called the reaction entropy ΔS_R.

$$\Delta S_R = \Delta S_{Products} - \Delta S_{Educts}$$

This can be determined - analogous to the reaction enthalpy (ΔH_R) - from the difference of the molar standard enthalpy of formation (numerical values compare Table A.8). However, the entropy change can already be estimated from the reaction equation:

- The same number of particles on both sides: ΔS_R is very small
- Increase in the number of particles: $\Delta S_R > 0$ ($= +\Delta S_R$)
- Decrease in the number of particles: $\Delta S_R < 0$ ($= -\Delta S_R$)

35.6 The Driving Force of Chemical Reactions

The voluntary course of a chemical reaction is influenced by two factors:

- Unit of the reaction enthalpy ΔH_R: kJ mol^{-1}
- Unit of the reaction entropy ΔS_R: J K^{-1} mol^{-1}

Both quantities are combined via the temperature in the Gibbs[5]-Helmholtz equation as *free* reaction *enthalpy* (*Gibbs energy*):

$$\Delta G_R = \Delta H_R - \Delta S_R \cdot T$$

Exergonic and Endergonic Processes

In every reaction that proceeds voluntarily, the free enthalpy of reaction decreases: $\Delta G_R > 0$. In this case, one speaks of an exergonic (from Greek *exo* = outside and *ergon* = work) reaction.

If $\Delta G_R < 0$, the reaction does not proceed voluntarily (in the direction given by the reaction equation) and is called an endergonic (from Greek *endo* = inside and *ergon* = work) reaction.

[5] Josiah Willard Gibbs, 1839–1903, American physicist.

35.6.1 Relationships ΔH_R, ΔS_R and ΔG_R

The course of the reaction can be estimated from the treated quantities of enthalpy (ΔH_R), entropy (ΔS_R) - including temperature - and the resulting free enthalpy (ΔG_R) of the reaction:

ΔH_R	ΔS_R	$\Delta G_R = \Delta H_R - \Delta S_R \cdot T$	Notes
< 0 (exothermic)	> 0	< 0	Exergonic High response probability
< 0 (exothermic)	< 0	< 0, T is small > 0, T is large	Exergonic At low temperature Endergonic At high temperature
> 0 (endothermic)	> 0	< 0, T is large > 0, T is small	Exergonic At high temperature Endergonic At low temperature
> 0 (endothermic)	< 0	> 0	Endergonic (at all temperatures)

▶ All exergonic reactions occur voluntarily. Endergonic reactions do not occur voluntarily.

It follows:

- if $\boldsymbol{\Delta G_R < 0}$ → voluntary response
- if $\boldsymbol{\Delta G_R = 0}$ → system is in equilibrium
- if $\boldsymbol{\Delta G_R > 0}$ → reaction does not proceed voluntarily

(Reaction, however, proceeds voluntarily in *reverse order*)

35.7 Transitions

Some reactions always occur immediately, such as neutralization. Other reactions could occur spontaneously but do not. For example, a mixture of hydrogen and oxygen can be kept for years without a reaction occurring. Most reactions require some energy to start, called activation energy. Usually, this is supplied in the form of heat. Sometimes, however, high-energy light is also sufficient to set a reaction in motion.

Depending on the reaction type, either the number of "favorable collisions" - and thus the formation of transition complexes - is increased, or a chain reaction is started.

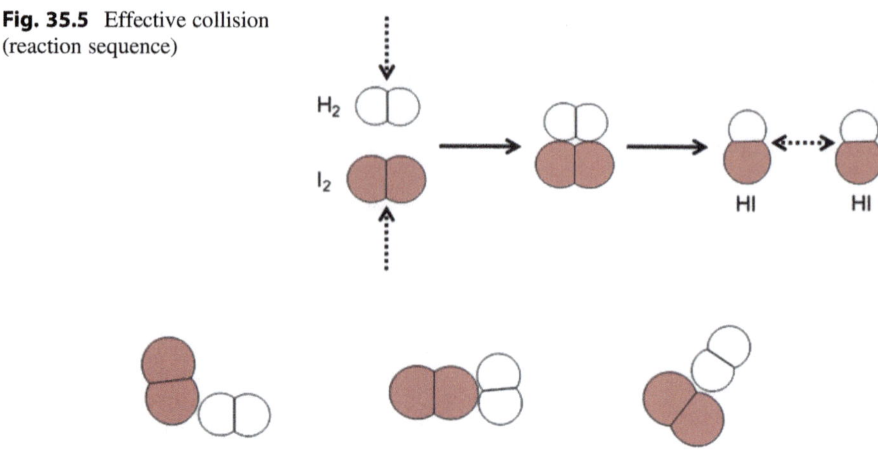

Fig. 35.5 Effective collision (reaction sequence)

Fig. 35.6 Non-effective collisions (reaction does not run)

35.7.1 Transition Complexes

For most reactions, so-called transition complexes are formed. These are formed when the particles collide "favorably". This is also referred to as effective collisions. Let us consider the reaction of iodine with hydrogen:

$$H_2 + I_2 \rightarrow 2\,HI$$

Figure 35.5 illustrates the sequence of an effective collision; Fig. 35.6 depicts several non-effective collisions in hydrogen iodide formation. The transition complex of the collision leading to the reaction is shown in the center of Fig. 35.5.

For a molecular collision to be effective, the molecules involved must collide in a suitable position relative to each other and the sum of the energies of motion must exceed a minimum value. Most collisions, therefore, do not lead to a reaction. Increasing the reaction temperature, however, leads to faster particle motion and increases the number of collisions - and thus also the number of collisions that lead to a reaction. This is also reflected in Van't Hoff's rule:

- An increase of the reaction temperature by 10 °C leads to a 2- to 3-fold increase in the reaction rate.

35.7.2 Activation Energy

If we consider the transition state from the energetic point of view, the transition complex consistently yields higher energy than the reaction products. Let us consider two different reactions:

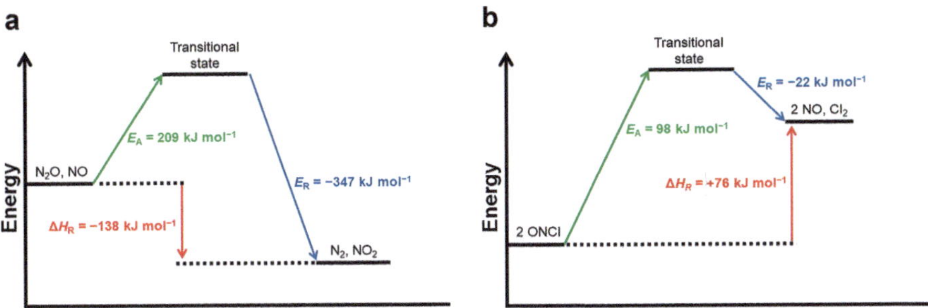

Fig. 35.7 Transition: (**a**) exothermic reaction, (**b**) endothermic reaction

$$N_2O + NO \rightarrow N_2 + NO_2 \qquad -138\ \text{kJ mol}^{-1}$$
$$2\ ONCl \rightarrow 2\ NO + Cl_2 \qquad +76\ \text{kJ mol}^{-1}$$

Figure 35.7a shows the energy diagram for the exothermic reaction of N_2O with NO. Also, for the endothermic reaction of the decomposition of nitrosyl chloride, as can be seen from Fig. 35.7b, the necessary activation energy must first be applied before the reaction starts.

36 Catalysis

It is estimated that 90 % of chemical products are produced using catalysts in at least one manufacturing step. In addition, many technical processes depend on catalysts, such as exhaust gas purification in motor vehicles.

First of all: No catalyst can make an "impossible" reaction "possible". This means that a reaction can and will only take place if it is possible at all - catalyst or not.

Every chemical reaction requires a certain amount of energy to get "going" at all. This energy is called activation energy and usually causes the formation of a transition state, which then reacts further to form the products. Once this amount of energy has been expended, exothermic processes, in particular, run "by themselves", so to speak. A catalyst now reduces this activation energy. This leads to the reaction already taking place at lower temperatures.

36.1 Introduction to Catalysis

The term catalysis (Greek *katalysis* = dissolution) was coined by Berzelius in 1836 and generally defined by Ostwald[1] in 1900. Most chemical products undergo at least one catalytic process during their manufacture. Furthermore, life processes are inconceivable without specific biological catalysts - in this case, called enzymes. Basically:

▶ No catalyst can make an impossible reaction possible.

In other words: Only if a reaction has a free enthalpy of reaction < 0 ($= -\Delta G_R$, compare Sect. 35.6) will it proceed, whether with or without a catalyst. However, a catalyst can "facilitate" this reaction.

[1]Friedrich Wilhelm Ostwald, 1853–1932, German-Baltic chemist, Nobel Prize in Chemistry in 1909.

T. Schmiermund, *The Chemistry Knowledge for Firefighters*,
https://doi.org/10.1007/978-3-662-64423-2_36

36.2 Influences of a Catalyst

36.2.1 Catalysis and Activation Energy

A certain amount of energy is required to ignite a hydrogen-oxygen mixture. If this starting energy is missing, such a mixture can be stored unchanged for decades at room temperature in a vessel. Only through ignition (flame, spark) does a reaction occur. This ignition energy causes a separation of the hydrogen molecules into hydrogen atoms, which in turn try to form a new bond: The chain reaction is started.

However, if a platinum catalyst is introduced into such a mixture, ignition occurs almost instantaneously (still at room temperature). This is due to the fact that the activation energy (i.e. the energy required to split the $H - H$ bond) is reduced by the catalyst to such an extent that the reaction already takes place at room temperature. The resulting free hydrogen atoms then set the chain reaction in motion, leaving the platinum unchanged (Fig. 36.2).

36.2.2 Catalysis and Reaction Rate

Hydrogen peroxide already decomposes into water and oxygen at room temperature:

$$H_2O_2 \rightarrow H_2O + O_2 \uparrow$$

The decomposition is accelerated by increasing the temperature or by light. Hydrogen peroxide should therefore be stored in a cool and dark place. If one now introduces a little manganese(IV) oxide ("manganese dioxide", MnO_2) into a solution of H_2O_2, the decay takes place extremely quickly. The MnO_2 acts as a catalyst for the decay and can subsequently be recovered unchanged:

$$H_2O_2 \overset{(MnO_2)}{\rightarrow} H_2O + O_2 \uparrow$$

(The catalyst is written in parentheses above the reaction arrow for identification.)

In many cases, the increase in the reaction rate is so strong that processes occur that would not occur at all under the given conditions without the intervention of a catalyst. This applies in particular to almost all chemical processes in living organisms, which can only take place at all with the help of "biocatalysts" (enzymes, ferments).

The two theorems of catalysis formulated by Ostwald can be seen clearly in the preceding examples:

▶

I. Catalysis is the acceleration of a slow chemical process by the presence of a foreign substance.

II. A catalyst is a substance which, without appearing in the final product of a chemical reaction, alters the rate of reaction.

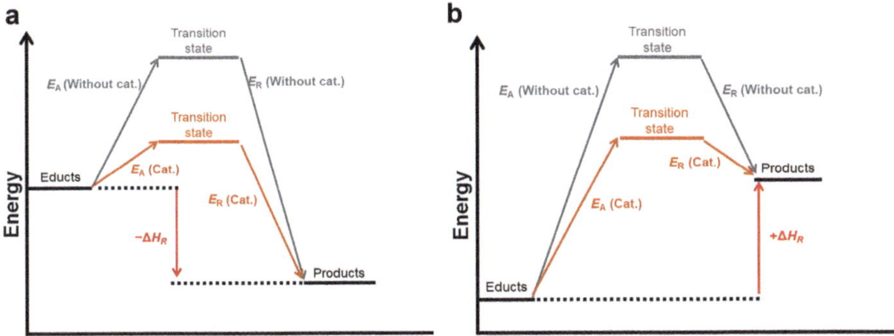

Fig. 36.1 (**a**) Exothermic reaction and catalyst, (**b**) Endothermic reaction and catalyst

36.2.3 Viewing the Energy Diagram

If we look at the curves of the individual energies of an exothermic (Fig. 36.1a) or endothermic reaction (Fig. 36.1b), we can see that only the activation energy is lowered by means of a catalyst. The reaction enthalpy (ΔH_R), and thus also the reaction entropy (ΔS_R) and the free reaction enthalpy (ΔG_R), remain unchanged.

Conclusions

If a catalyzed reaction results in a transition state that is at a lower energy level than in the same uncatalyzed reaction, this can only be explained by a modified reaction course. The transition state (the "activated complex") must have a different structure. This is the only way to explain the lower energy state.

At the same time, it follows that a catalyst only influences a very specific reaction process. Catalysts are therefore specific to a reaction and the resulting products.

- ▶ A reaction with a catalyst takes a different reaction course than without a catalyst. The catalyst opens up a new path for the course of the reaction, so to speak.

- ▶ The essential point here is that the catalyzed pathway requires lower activation energy. This results in a higher reaction rate for the overall reaction.

36.2.4 Catalysis and Reaction Course

A simple example of a changed reaction course due to the presence of the catalyst is the oxidation of thallium(I)-ions to thallium(III)-ions with cerium(IV)-ions catalyzed by manganese(II)-ions.

The reaction $2\ Ce^{4+} + Tl^{+} \rightarrow 2\ Ce^{3+} + Tl^{3+}$ proceeds extremely slowly. However, the addition of Mn^{2+}-ions greatly accelerates the reaction. In the uncatalyzed reaction, the three ions involved must collide simultaneously. This three-body mechanism rarely occurs. The addition of a manganese salt transforms the course of the reaction into a sequence of three two-body collisions:

$Mn^{2+} + Ce^{4+} \longrightarrow Mn^{3+} + Ce^{3+}$
$Mn^{3+} + Ce^{4+} \longrightarrow Mn^{4+} + Ce^{3+}$
$Mn^{4+} + Tl^{+} \longrightarrow Mn^{2+} + Tl^{3+}$
$2\,Ce^{4+} + Tl^{+} \overset{(Mn^{2+})}{\rightarrow} 2\,Ce^{3+} + Tl^{3+}$

The manganese ions take over the role of an electron transmitter. The difficult one-step reaction thus becomes a sequence of simple partial reactions.

Another example is the oxidation of ethylene to ethylene oxide on a silver catalyst.

$$H_2C{=}CH_2 + 4\,O_2 \longrightarrow 2\,CO_2 + 2\,H_2O$$

$$2\,H_2C{=}CH_2 + O_2 \overset{(Ag)}{\rightarrow} 2\,H_2C(O)CH_2$$

Total oxidation to carbon dioxide and water - the combustion reaction, so to speak - is the preferred reaction. The silver catalyst now ensures that the combustion reaction takes a back seat and the formation of the ethylene oxide is very strongly accelerated. The catalyst selectively accelerates one possible reaction here. Other reactions are not accelerated and thus fade into the background.

36.2.5 Catalyst Selectivity

To illustrate the selectivity of catalysts, reaction examples are given here with the same reactants, in this case, a mixture of carbon monoxide and hydrogen ("synthesis gas"):

Ni catalyst for methane synthesis: $CO + 3\,H_2 \overset{(Ni)}{\rightarrow} CH_4 + H_2O$

Fe/Co catalyst for gasoline production: $CO + H_2 \overset{(Fe/Co)}{\rightarrow} \text{Gasolin} + \text{Water}$

Rh catalyst for glycol production: $2\,CO + 3\,H_2 \overset{(Rh)}{\rightarrow} HO{-}C_2H_4{-}OH$

Cu/Cr/Zn catalyst for methanol production: $CO + 2\,H_2 \overset{(Cu/Cr/Zn)}{\rightarrow} CH_3OH$

Depending on the catalyst used, the different reaction possibilities are accelerated to different degrees. As a result, different reaction products are formed.

The specificity of enzymes is even more pronounced since they occur without any side reactions. An enzyme usually catalyzes only one reaction, which in turn leads to a single reaction product.

36.2.6 Promoters

Promoters (Latin *pro* = forward and *movere* = to move; meaning to move forward; in the broader sense "to multiply") are substances that are added to the actual catalyst. They are not catalytically active themselves but improve the effect of the catalyst. For example, Al_2O_3, K_2O, and CaO are added to the iron catalyst in ammonia synthesis. In this context, one also speaks of mixed catalysts.

36.3 Aggregate States of Catalysts

Catalysts can occur in all three states of aggregation. Not only solids or liquids but also dissolved substances and even gases can be used as catalysts. A distinction must be made here as to whether the reactants and the catalyst are in the same phase (so-called "homogeneous catalysis") or different phases ("heterogeneous catalysis").

36.3.1 Homogeneous Catalysis

In homogeneous catalysis, the feedstock and the catalyst are in the same phase. They are either present as gases or as a liquid or in solution. The decomposition of N_2O ("nitrous oxide") into the elements serves as an example of gaseous catalysis.

$$2\,N_2O \rightarrow 2\,N_2 + O_2 \qquad E_A = 240\,kJ\,mol^{-1}$$

$$2\,N_2O \overset{(Cl_2)}{\rightarrow} 2\,N_2 + O_2 \qquad E_A = 140\,kJ\,mol^{-1}$$

Another example of homogeneous catalysis in solution is the acid-catalyzed saponification of carboxylic acid esters. Although acetic acid is formed in addition to methanol during the reaction of methyl acetate with water - which already acts as a catalyst - the relative reaction rate increases sharply when catalytic amounts of other acids are used.

Reaction speed = 1

$$\mathrm{H_3C{-}COOCH_3 + H_2O \rightarrow H_3C{-}COOH + H_3C{-}OH}$$

Reaction speed: ~ 4

$$\mathrm{H_3C{-}COOCH_3 + H_2O \overset{(H{-}COOH)}{\rightarrow} H_3C{-}COOH + H_3C{-}OH}$$

Reaction speed: ~ 200

$$\mathrm{H_3C{-}COOCH_3 + H_2O \overset{(Cl_3C{-}COOH)}{\rightarrow} H_3C{-}COOH + H_3C{-}OH}$$

Reaction speed: ~ 300

$$H_3C{-}COOCH_3 + H_2O \overset{(HCl)}{\rightarrow} H_3C{-}COOH + H_3C{-}OH$$

The greatest difficulty with homogeneously catalyzed reactions is the recovery of the catalyst. By its very nature, it is difficult to remove the catalyst from the reaction mixture and return it to the process. In many technical processes, which are "actually" carried out as catalytically homogeneous processes, one, therefore, fixes the catalyst and thus achieves an easier separation of product and catalyst phase. This is achieved by chemically binding the catalyst, which is effective in the homogeneous phase, to a carrier material with as large a surface area as possible. Silica gel, aluminum oxide, zeolites, but also polypropylene or polystyrene are used for this purpose. These catalysts are called hybrid catalysts (from Latin *hybrida*).

Technical Processes and Products
By means of homogeneous catalysis, a large number of products are produced, e.g.:

- Butyraldehyde from propylene for plasticizers
- Acrylates from acetylene (ethine)
- Propionic acid from ethylene
- Butanol from propylene
- Acetic acid from methanol
- Production of polyethylene and polypropylene

36.3.2 Heterogeneous Catalysis

In heterogeneous catalysis, the reactants and the catalyst are in different phases. As a rule, the catalyst is solid, while the reaction takes place in the gas phase or solution. At the catalyst surface, chemical adsorption ("chemisorption") first takes place. Molecules are bound to the catalyst surface by forming chemical bonds. These bonds change the electron distribution of the adsorbed molecules. Bonds within the adsorbed molecule are weakened or broken. Subsequently, these molecules react further with the reaction partner to form the final product.

As mentioned in Sect. 36.2.1, a platinum catalyst introduced into an H_2/O_2-mixture initiates a spontaneous reaction. The Pt catalyst first loosens the H – H bond by attaching the H_2 molecule to the metal surface. Subsequently, the hydrogen atoms form bonds to the metal, loosening the H – H bond. Two hydrogen atoms (H radicals, H•) are formed, which then detach from the surface (Fig. 36.2). These hydrogen radicals are highly reactive and initiate the chain reaction between hydrogen and oxygen.

The decomposition of N_2O can be catalyzed heterogeneously with gold. The activation energy is only 120 kJ mol^{-1} (compare Sect. 36.3.1).

The effectiveness of a heterogeneous catalyst - and thus the speed of the reaction - is generally dependent on the surface area of the catalyst. The larger the specific surface, the better the catalytic effect. For this purpose, the actual catalyst is often

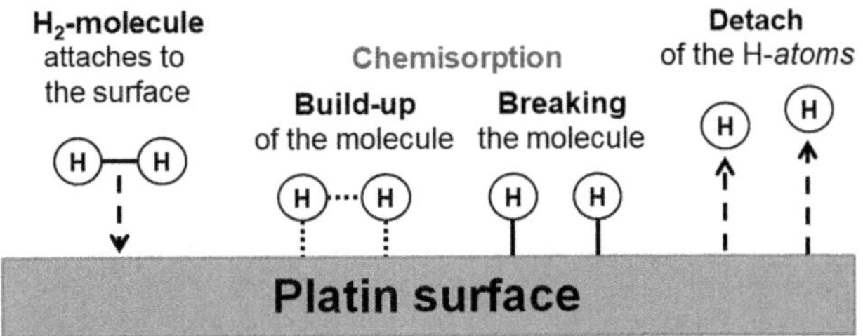

Fig. 36.2 Solid catalyst (example)

applied to carrier materials, such as Al_2O_3, SiO_2, TiO_2, activated carbon, etc., which have a large surface area.

Specific Surface

A short calculation example may explain the specific surface: A cube with an edge length of 1 cm is made of a material with a density of 6 g cm^{-3}. Since this cube has a surface area of $6 \cdot 1\ cm \times 1\ cm = 6\ cm^2$, the specific surface area is 1 cm^2/g. If we now divide the cube into 10 parts over each edge, we obtain $10 \cdot 10 \cdot 10 = 1000$ cubes, each with an edge length of 1 mm. The total surface area is now $1000 \cdot 6\ mm^2 = 6000\ mm^2 = 60\ cm^2$. The specific surface area thus increases to 10 cm^2/g.

Technical Processes and Products

Many technical processes are carried out using heterogeneous catalysts, e.g.:

- in the synthesis of sulphuric acid, the oxidation of SO_2 to SO_3 with V_2O_5,
- in nitric acid production, the oxidation of NH_3 to NO with Pt,
- for the addition of hydrogen ("hydrogenation") e.g. Ni, Pd or Pt are used,
- conversions of synthesis gas: see Sect. 36.2.5,
- in the polymerization of plastics, such as polypropylene,
- the introduction of alkyl groups into organic molecules with e.g. $AlCl_3$,
- the preparation of acrolein (for polyacrylates) from propene with bismuth molybdate and much more.

36.3.3 Catalysis in Everyday Life

We encounter a wide variety of plastic products in everyday life. The reaction of small starting molecules to form the actual plastic is usually controlled by catalysts to achieve optimum product properties. Another catalytic product that we frequently encounter is margarine. It is produced by the hydrogenation of unsaturated vegetable oils ("fat hardening").

However, it is probably the car exhaust catalytic converter that has made the term catalytic converter most familiar to the general public. The "cat" consists of a ceramic with many fine channels. The surface of the ceramic is enlarged by aluminum oxide (several 100 m^2/g) and the actual catalyst (Pt, Rh, Pd) is applied to it. During combustion in a motor vehicle engine, three main groups of pollutants are produced: Carbon monoxide, nitrogen oxides, and unburned gasoline components.

The three-way catalytic converter is capable of minimizing all three pollutant groups simultaneously.

Carbon monoxide: $2\,CO + O_2 \overset{(Kat)}{\rightarrow} 2\,CO_2$

Nitrogen oxides: $2\,CO + 2\,NO_2 \overset{(Kat)}{\rightarrow} 2\,CO_2 + N_2$

Gasoline residue: $C_mH_n + x\,O_2 \overset{(Kat)}{\rightarrow} m\,CO_2 + 0,5\,n\,H_2O$

To ensure that the optimum amount of air is always available, this is regulated by the so-called "lambda sensor" (λ sensor).

36.4 Inhibition

Inhibitors (Latin *inhibere* = "to prevent", "to stop"), also called "anti-catalysts", have exactly the opposite effect of a catalyst: They prevent a reaction or slow it down. In chemical engineering, the aim is to run reactions in such a way that they are easily controllable in terms of pressure and temperature. There are, therefore, also technical processes in which inhibitors are added to reduce the natural reaction rate.

In the context of firefighting, the inhibition effect is also intentionally exploited. The extinguishing success of extinguishing powder type BC is based on heterogeneous inhibition, that of halons on homogeneous inhibition.

36.4.1 Extinguishing Powder

The anticatalytic or inhibitory extinguishing effect is evident when extinguishing powder is used to fight flame fires. Complex chemical processes take place in the flame zone in which free radicals are involved as intermediates. These radicals hit the powder particles and are rendered ineffective there. In this phenomenon, known as the "wall effect", energy is extracted from the reactive intermediate stages of the combustion reaction. The radicals, which are particularly rich in energy and essential for the progress of the combustion reaction, lose part of their energy in contact with the cool powder particles. As a result, retrograde reactions, so-called recombinations, occur, which ultimately lead to the termination of the chain reaction.

36.4.2 Halons

The extinguishing effect of halons (abbreviation of halogenated *hydrocarbons*) is based on the principle of homogeneous inhibition. The halon splits off halogen radicals in the flame, which then reacts with the radicals formed from the fuel. The actual reaction chain breaks off, the fire is extinguished. This happens, for example, according to the following mechanism:

Radical formation of the halon: $CF_2ClBr \xrightarrow{\Delta T} \bullet CF_2\bullet + Cl\bullet + Br\bullet$

Radical formation of the fuel: $CH_4 \xrightarrow{\Delta T} H\bullet + CH_3\bullet$

Abort the combustion: $CH_3\bullet + Br\bullet \longrightarrow CH_3Br$

The classic halons have not been allowed to be used as extinguishing agents for several years. For new developments see Sect. 41.2.7, for extinguishing effect see Sect. 41.1.5.

36.5 Catalyst Poisons

Most catalysts are relatively sensitive. They can be rendered ineffective or unusable by so-called "catalyst poisons". A typical catalyst poison is hydrogen sulfide (H_2S, which smells like rotten eggs). Other substances, such as CO, COS (carbonyl sulfide), arsenic, lead, and mercury also act as catalyst poisons. They block the active centers of the catalysts by forming bonds with the catalyst.

The use of lead-free fuels, for example, stems from the fact that lead renders the motor vehicle catalytic converter ineffective. The lead tetraethyl ($Pb(C_2H_5)_4$) previously used for reasons of anti-knock properties had to be replaced by methyl tert-butyl ether (MTBE) and other compounds which also increase anti-knock properties.

Measuring devices of the fire brigade, especially the multi-gas measuring devices, often have electrochemical sensors (EC sensors). These sensors are also sensitive to certain substances that act as catalyst poisons in these sensors. The service life of the sensors (usually 2–5 years) is significantly reduced by contact with these substances. The operating instructions of the manufacturers provide further information on substances that have a damaging effect on the sensors.

Part X

Burning and Extinguishing

37 Fire and Blazes

Fire exerts a strong fascination on many people. Whether the glimmer of distant lights, the forest fire from a safe distance, the kerosene lamp, or the campfire – the flames catch the eye, attract the gaze. Even young children watch candle flames with great interest. Flames catch the eye, they twitch, blaze, glisten, they are calm, unsteady, swirling. Fire stands for warmth, comfort, and light, for peace and relaxation – think of the open fire – and, as a bonfire, for conviviality.

But fire also stands for annihilation, death, and destruction. Not only major fires (e.g. London 1666, Istanbul 1918, Hamburg 1943) but also "simple" room fires represent a catastrophe for those affected. Property and possessions are irretrievably lost. Often, important documents and mementos that have fallen victim to the flames represent a greater burden than the material damage incurred.

For historical reasons, the term fire is used both as a generic term and to identify a usefull fire. Useful fires are combustions that occur intentionally and in a controlled manner, often in so-called furnaces. They are used to produce or obtain light, heat, and other forms of energy, such as electricity. Damage fires or blazes, on the other hand, are combustion processes that are not or no longer under control. Neither the location nor the intensity of a damaging fire is intentional (Fig. 37.1).

37.1 Fire – An Oxidation Process

Combustion is an oxidation process; fire is the visible manifestation of this process. We have already encountered oxidations as redox reactions (see Chap. 16). Here, we had already defined:

> ▶ Oxidation is the reaction of a substance with oxygen. The substance reacts with oxygen, it is oxidized. The substance itself acts as a reducing agent.

The fact that a redox reaction does not necessarily have to be a reaction with oxygen has already been discussed. And not every reaction that produces the typical

T. Schmiermund, *The Chemistry Knowledge for Firefighters*,
https://doi.org/10.1007/978-3-662-64423-2_37

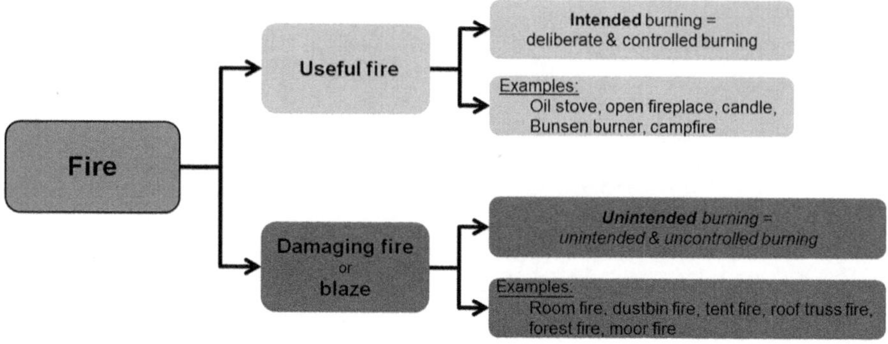

Fig. 37.1 Useful and damaging fires

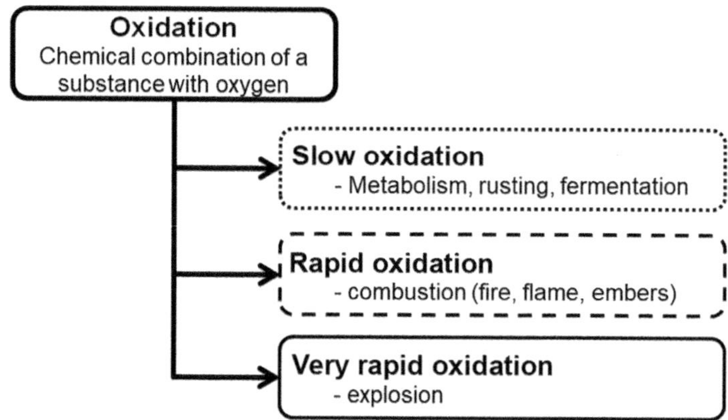

Fig. 37.2 Oxidations (overview)

combustion products CO_2 and H_2O is combustion. For example, the reaction of potassium permanganate ($KMnO_4$) with oxalic acid ($C_2O_4H_2$) in acidic solution:

$$2\,MnO_4^- + 5\,C_2O_4^{2-} + 16\,H^+ \rightarrow 2\,Mn^{2+} + 10\,CO_2 + 8\,H_2O$$

There is more to combustion, to fire: flames, smoke, heat, light.

We, therefore, first make a distinction according to the rate at which the oxidation reaction proceeds (Fig. 37.2).

37.1.1 Slow Oxidation

Typical slow oxidation processes that we encounter in everyday life are, for example, the rusting of iron or fermentation. The “burning” of nutrients in body cells for energy production is also such slow oxidation. These oxidation processes occur at room temperature or only a few degrees above. And: The reaction rate is significantly lower than in combustion.

However, if heat builds up (see Sect. 38.4.2), slow oxidation may turn into rapid oxidation, resulting in a fire. Typical examples are spontaneous combustion of lignite (oxidation process with atmospheric oxygen) or the ignition of hay triggered by bacteria.

37.1.2 Rapid Oxidation

The rapid combination of a combustible substance with oxygen, which takes place under the appearance of fire, is referred to as "burning" or combustion. Characteristic features are the formation of flames and/or embers, the high temperatures (up to over 3000 °C), the irreversibility of the process, and the release of light and thermal radiation.

► Combustion is a rapid, voluntary, and irreversible oxidation process. A combustible substance is reacted with oxygen (usually air) with the formation of flames and/or embers and perceptible heat development.

37.1.3 Very Rapid Oxidation

If combustion takes place extremely quickly, a pressure wave is also formed due to the speed of the reaction. This is an explosion.

38 The Process of Burning

Burning is a (natural) phenomenon that has been known since time immemorial. The use of fire has significantly influenced our development and our cultural history. Important steps in technology are also connected with fire.

If you look at the fire phenomena "flame" and "embers" a little more closely, you will see that gases and liquids burn exclusively with flame. In the case of liquids, such as petrol or diesel fuel, the actual liquid does not burn – only the (gaseous) vapors of a flammable liquid can burn. Solids, on the other hand, only burn with embers. This is most evident in metal fires.

But wood burns with flame *and* embers. This seems to contradict the above. If one heats wood chips under the exclusion of air, one can obtain wood tar and wood water as liquid phases and additionally combustible gases. What remains is charcoal. Similarly, coal gas (containing CH_4, H_2, CO, H_2S, NH_3, etc.), coal tar, and coke are obtained from hard coal. Both charcoal and coke burn as "completely degassed" compounds with embers only (Fig. 38.1).

38.1 Requirements for Burning

For a fire to burn at all, certain prerequisites are necessary. Two of the material prerequisites, which together are also referred to as the "combustible system", are conspicuous:

- Without a combustible substance, there is nothing that can burn at all. Without air (actually: Oxygen), no combustion occurs.

Although our environment consists of countless flammable substances, a fire only occurs in special cases. Consequently, another prerequisite must exist that at least sets the process in motion. This is the necessary (minimum) energy to initiate the combustion reaction.

T. Schmiermund, *The Chemistry Knowledge for Firefighters*,
https://doi.org/10.1007/978-3-662-64423-2_38

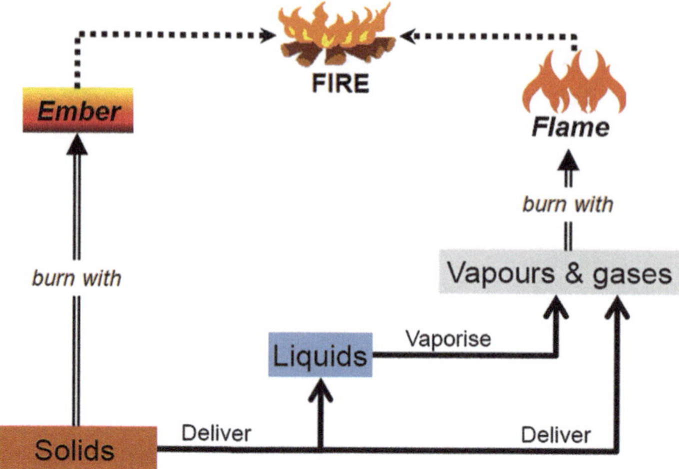

Fig. 38.1 Combustion vs. state of aggregation

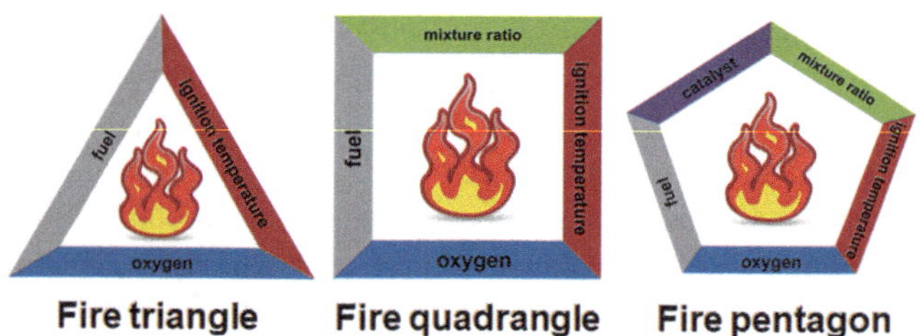

Fig. 38.2 Prerequisites for firing (illustrations I)

▶ A (suitable) ignition source is required.

These considerations lead to the graphical representation known as the **fire triangle** (Fig. 38.2).

In reality, however, the processes are more complicated. If we bear in mind that combustible material and oxygen must not interfere with each other, it quickly becomes clear that an excess of fuel inevitably displaces oxygen. If there is a strong imbalance between oxygen (air) and combustible material, no reaction will take place (see Sect. 3.3.4). The next requirement is therefore the mixing ratio.

▶ The ratio of combustible material to oxygen must be within certain limits.

By extending the fire triangle, one thus arrives at the **fire quadrangle** (Fig. 38.2).

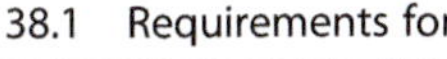

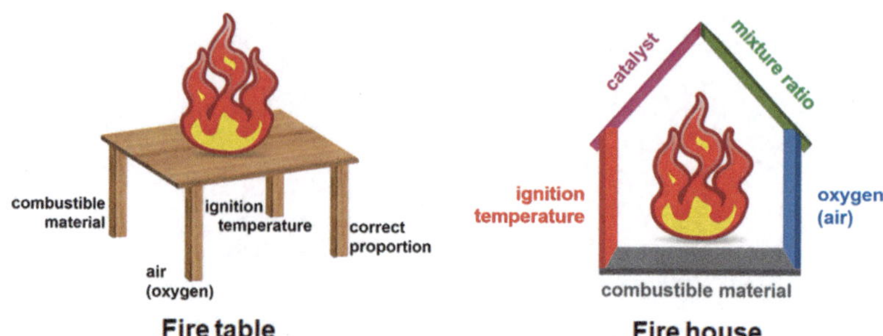

Fig. 38.3 Preconditions for burning (illustrations II). (© Felicitas "Cean" Schmiermund, after sketches by the author)

If we look at the chemical process of combustion in detail, we see that a catalyst (see Sect. 38.6) is always (!) required. The presence of catalytically acting intermediates also explains the fact that the inhibitory extinguishing effect is applicable at all. The fire quadrilateral thus expands into a **fire pentagon** (Fig. 38.2).

► No combustion without "combustion catalyst".

As an alternative to the fire square, the **fire table** can also be used to visualize the prerequisites. If one wishes to delve deeper into the course of combustion processes or deal specifically with the inhibitory extinguishing effect, for example, a representation as a **fire house** is suitable (Fig. 38.3).

38.2 Combustible Substance

Without a combustible substance, combustion cannot take place. But what makes a substance combustible? What conditions must be met in order for "combustibility" to occur?

Let us first consider the elements of the periodic table. We find three major groups into which we can classify the elements:

I. **Incombustible elements that do not form oxides**

The noble gases and some noble metals, such as gold or platinum, are not capable of forming oxides, or only under very special reaction conditions. Under combustion conditions, they do not react with oxygen.

II. **Incombustible elements forming oxides**

Halogens, nitrogen, and some metals (e.g. silver, mercury) are able to form oxides, but not under the conditions of a normal fire. The decay of the oxides and/or the low reaction rate prevails.

III. **Combustible, oxide-forming elements**

The non-metals hydrogen (H), carbon (C), sulfur (S), and phosphorus (P) as well as light metals, such as sodium (Na), potassium (K), magnesium (Mg), calcium (Ca), or aluminum (Al), form oxides in an exothermic reaction.

For combustible materials we can state:

- Combustible substances are combustible elements or are predominantly composed of them. Mixtures of combustible substances are also combustible. If non-combustible elements are involved in the composition of a combustible substance, the number, and type of chemical bond of the non-combustible elements are decisive for the combustibility of the substance.

38.2.1 Influence of the Molecular Structure on the Flammability

Most flammable substances consist mainly of carbon and hydrogen, so-called *hydrocarbons*. Therefore, we first consider the influence of the structure of some simple, very similar hydrocarbons on their flammability.

As shown in Table 38.1, boiling point and flash point decrease as the molecular structure becomes increasingly spherical (see also Sect. 8.2.3). At the same time, the ignition temperature rises. The latter is due to the poorer chemical attack properties – in terms of the combustion reaction – of the spherical molecules compared with the

Table 38.1 Comparison of flammability of hexane isomers

Name	Hexane	2-methylpentane	2,2-dimethylbutane
Sum formula	C_6H_{14}	C_6H_{14}	C_6H_{14}
Structural formula	$H_3C-CH_2-CH_2-CH_2-CH_2-CH_3$	$H_3C-CH(CH_3)-CH_2-CH_2-CH_3$	$H_3C-C(CH_3)_2-CH_2-CH_3$
Molar mass	86.2 g mol^{-1}	86.2 g mol^{-1}	86.2 g mol^{-1}
Boiling point	69 °C	60 °C	50 °C
Flash point	−22 °C	−32 °C	−48 °C
Ignition temperature	230 °C	300 °C	435 °C

Table 38.2 Comparison of flammability of ring-shaped compounds

Name	Cyclohexane	Cyclohexene	Benzene
Sum formula	C_6H_{12}	C_6H_{10}	C_6H_6
Structural formula	H_2C–CH_2 ring: H_2C, H_2C, CH_2, H_2C, CH_2, CH_2	H_2C, H_2C, CH, H_2C, CH, CH_2 (CH=CH double bond)	HC, CH, HC, CH, CH, CH (alternating double bonds)
Molar mass	84.2 g mol^{-1}	82.1 g mol^{-1}	78.1 g mol^{-1}
Boiling point	81 °C	83 °C	80 °C
Flash point	−18 °C	−17 °C	−11 °C
Ignition temperature	260 °C	265 °C	555 °C

Table 38.3 Flammability of halogen compounds of benzene

Name	Benzene	Fluorobenzene	Chlorobenzene	Bromobenzene
Sum formula	C_6H_6	C_6H_5F	$C_6H_5C_l$	C_6H_5Br
Molar mass	78.1 g mol^{-1}	96.1 g mol^{-1}	112.6 g mol^{-1}	157.0 g mol^{-1}
w % (X)	0 %	19.8 %	31.5 %	50.9 %
EN	H: 2.1	F: 4.0	Cl: 3.5	Br: 2.8
ΔEN X vs. C	−0.4	+1.5	+1.0	+0.3
Boiling point	80 °C	85 °C	132 °C	156 °C
Flash point	−11 °C	−15 °C	28 °C	51 °C
Ignition temperature	555 °C	630 °C	590 °C	565 °C

linear molecules. Thus, a single raw spaghetti can be broken more easily than a spiral noodle – although both have approximately the same mass.

If ring-shaped molecules are compared (Table 38.2), the differences in flash point and boiling point are, as expected, smaller. Benzene, however, is "out of the ordinary" with regard to its ignition temperature. This can be explained by the high stability of the ring system (see Sects. 7.3.1.4 and 44.4).

38.2.2 Influence of Non-combustible Elements on Combustibility

The influence of non-flammable elements on the flammability of a hydrocarbon can be easily illustrated using the example of halogen compounds (see Table 38.3). Thus, as expected, boiling point and flash point rise markedly with increasing mass fraction of halogen [*w*(X)]. The ignition temperatures are highest for the fluorine compound and almost back to the level of the pure hydrocarbon for the bromine compound. This is explained by the stronger bonding of the fluorine atom to the carbon atom compared to the bromine atom. The stronger bonds are in turn a direct consequence of the difference in the electronegativities (ΔEN) of the elements involved (compare Sects. 6.5.4 and 7.3.2).

The deliberate incorporation of non-flammable elements, especially halogens, finally leads to non-flammable substances which are or were even used as extinguishing agents, e.g. halons (see Sects. 36.4.2 and 41.2.7). Further examples are given in Tables A.15, A.16, A.17.

Oxygen itself also belongs to the incombustible elements in a broader sense. This can be seen in a comparison of various simple hydrogen compounds: Methane (CH_4) and ethane (C_2H_6) are combustible gases, as are hydrogen sulfide (H_2S) or hydrogen phosphide (PH_3). The compound of the flammable element hydrogen with oxygen (H_2O), on the other hand, is the most important extinguishing agent.

Summary

The influences of the molecular structure of flammable substances can be summarized as follows:

	Impact on	
Influencing variable	Flash point	Ignition temperature
Molar mass	Increase with an increase in molar mass	Decrease with an increase in molar mass
Carbon chain branching	Decrease with increasing branching	Increase with increasing branching
Double binds	Increase	Increase
Replacement of H by combustible elements	Increase or decrease	Increase or decrease
Replacement of H by incombustible elements	Increase (up to incombustibility or extinguishing efficiency)	Increase (up to incombustibility or extinguishing efficiency)

38.2.3 Classification of Flammable Substances I: Chemical-Physical

If we classify flammable substances according to scientific criteria (Fig. 38.4), we arrive at almost the same classification as we are already familiar with for substances in general (see Sect. 2.2).

Table 38.4 explains this classification using examples.

38.2.4 Classification of Flammable Substances II: Fire Classes

According to the European Standard EN 2, flammable substances are divided into fire classes (Table 38.5). This classification is necessary, among other things, to be able to select suitable extinguishing agents based on their fire behavior. The fire classes and the associated pictograms are also found on portable fire extinguishers

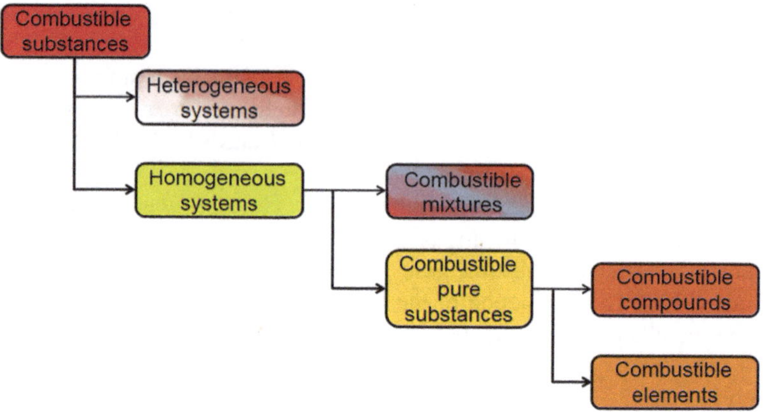

Fig. 38.4 Classification of flammable substances

Table 38.4 Combustible substances (examples)

Combustible ...	Examples
Heterogeneous systems	Wood, hay, straw, lignite, paper; but also: Cushion, sofa, curtain, Christmas tree and much more.
(homogeneous) mixtures	Synthesis gas (mixture of CO and H_2), gasoline, diesel, candle wax (hydrocarbon mixture)
Compounds	Octane, benzene, stearin, carbon monoxide (CO), hydrogen sulfide (H_2S), carbon disulfide (CS_2), phosphine (PH_3), plastics (PVC, PE, PP, PS, PET, PMMA, ...)
Elements	Hydrogen (H_2), carbon (C), phosphorus (P), Sulfur (S), sodium (Na), potassium (K), magnesium (mg), Aluminum (Al)

and thus enable even the layman to recognize whether the fire extinguisher is potentially suitable for the fire to be extinguished or not.

38.2.5 Fire Behavior of Building Materials

In order to classify the fire behavior of building materials, building material classes were defined according to DIN 4102, Part 1. This German standard (Deutsche Norm) has now been replaced by a European standard (EN 13501, Part 1). Table 38.6 gives an overview of the classification according to both standards.

In accordance with EN 13501–1, the classes for reaction to fire are given additional information on smoke development (code letter s for *smoke*) and on burning droplets (code letter d for *droplets*).

Table 38.5 Fire classes

Fire class	Fires of ...	Examples	Pictogram
A	Solids that produce embers	Wood, textiles, straw, paper, coal, plastics (thermosets)	A
B	Liquid substances or substances that become liquid	Gasoline, heating oil, wax, alcohols, tar, bitumen, plastics (thermoplastics)	B
C	Gases	Natural gas, acetylene, propane, hydrogen, carbon monoxide	C
D	Metals	Aluminum, magnesium, sodium	D
F	Edible oils and fats	Sunflower oil, frying fat, olive oil	F

Table 38.6 Building material classes (overview)

Building material class (DIN 4102–1)	Designation	Classes for reaction to fire (EN 13501–1)-NEW
A1	Non-combustible, without combustible components	A1
A2	Non-combustible, with combustible components	A2-s1,d0
B1	Of low flammability	A2 / B / C-s1, d0 / C-s1, d1 / C-s1, d2 / C-s2, d0 / C-s3, d0 / C-s3, d2
B2	Normally inflammable	D-s1, d0 / D-s1, d2 / D-s2; d0 / D-s2, d2 / D-s3, d0 / D-s2, d3 / E / E-d2
B3	Inflammable	F

Notes on the Building Material Classes

- Classes up to C (old: up to B1) are considered self-extinguishing. The fire goes out when the ignition source is removed.
- From class E (old: B2) the fire continues to burn even if the ignition source is removed.
- Highly flammable building materials (class F or old: B3) may only be installed in buildings as composite materials, which in turn may no longer be highly flammable.

- The building material classes must not be confused with the fire classes (Sect. 38.2.4) or the fire resistance classes (Sect. 38.2.6).

Notes to Table 38.5:

- Fire class B comprises liquid and liquid-becoming combustible substances. In addition to wax or tar, these also include some plastics (thermoplastics, see Chap. 49 and Table 49.4) which burn in a dripping manner. Restrict the use of water as an extinguishing agent (see Sect. 3.4.6).
- In the event of gas fires, always shut off the gas supply, otherwise, an explosive gas-air mixture may form.
- Never use water, foam, or CO_2 as extinguishing agents for fires involving Class D substances (see Sect. 3.4.6).
- Never use water on fat fires (fire class F) (see Sects. 3.4.6 and 41.2.6).

38.2.6 Fire Behavior of Building Components

In addition to building materials, so-called components are also used in the construction of buildings. The fire resistance classes are defined according to DIN 4102, Part 2. The associated European standard (EN 13501, Part 2) was adopted in the German Building Rules List in 2002.

According to DIN 4102–2, the fire resistance class consists of a code letter and the time in minutes of the guaranteed functional integrity of the component. Here mean:

F-	Fire resistance, general	**I-**	Installation duct (without functional integrity)
E-	Electrical installation with a support system	**K-**	Shut-off devise in ventilation ducts
F-	Walls, ceilings, stairs, beams	**L-**	Ventilation duct
F-	Fire-resistant glazing with heat protection	**R-**	Pipe bulkheading
		S-	Bulkhead, cable fire bulkhead
G-	Fire-resistant glazing without heat protection	**T-**	Doors, gates, flaps
		W-	Non-load-bearing exterior walls

About the functional maintenance and the designation by the building authorities, the following applies:

Fire resistance class	Designation	Function preservation via:
F30	Fire-retardant	30 min
F60	High fire-retardant	60 min
F90	Fire-resistant	90 min
F120	High fire-resistant	120 min
F180	Highly fire-resistant	180 min

EN 13501–2 sometimes combines several letters with the time in minutes. The following code letters are provided within the scope of this standard:

C-	Self-closing	**M-**	Mechanical strength
E-	Room closure	**P-**	Protection of electric cables
G-	Soot fire resistance	**R-**	Preservation of the load-bearing capacity
I-	Thermal insulation	**S-**	Smoke tightness
K-	Fire protection effect	**W-**	Limitation of thermal radiation

A wall that is designated as "F90" according to DIN 4102–2, for example, is thus given the classification "REI90" according to EN 13501–2.

38.3 Oxygen

Oxygen itself is *not* combustible. Without oxygen, however, combustion in the classical sense does not take place. It is the most widespread element on earth. Its total share in the mass of the earth's crust (including air and water) is 47.4 %. Even the human being consists of more than 60 mass-% of this element. The atmosphere of the earth contains $1.2 \cdot 10^{18}$ kg O_2 approximately. It, therefore, represents an almost unlimited source of oxygen. Through the metabolic processes of plants, algae, and cyanobacteria, oxygen is constantly released into the air from the water absorbed by them.

In combustion processes, oxygen can belong to the combustible system in various forms:

- as atmospheric oxygen
- as pure oxygen
- as bound oxygen

38.3.1 Free Oxygen of the Air

The oxygen in the ambient air is most frequently involved in combustion processes. The composition of dry air is shown in Table 38.7.

Table 38.7 Composition of dry air

Substance	Symbols	Volume fraction φ %	Mass fraction w %
Nitrogen	N_2	78.09 %	75.51 %
Oxygen	O_2	20.95 %	23.15 %
Argon	Ar	0.9325 %	1.286 %
Carbon dioxide	CO_2	0.03 %	0.04 %
Other noble gases	He, ne, Kr, Xe	0.021 %	0.014 %

Combustion in air proceeds at a moderate reaction rate. This is primarily due to the dilution effect of atmospheric nitrogen.

An increase in the reaction rate can be achieved by swirling the fuel with compressed air in a special combustion chamber. The higher oxygen pressure ensures faster combustion.

38.3.2 Pure Oxygen

Pure oxygen is usually found in compressed gas cylinders. These are marked differently – depending on the intended use – by EN 1089–3: In medicine a pure white compressed gas cylinder, in engineering a blue or grey compressed gas cylinder with a white shoulder. (However, only the mandatory dangerous goods label on the compressed gas cylinder is binding).

Pure oxygen is used, for example, in oxyacetylene welding (actually gas melting welding), flame-cutting (also gas cutting), or in so-called oxygen lances (also thermal lances). In chemical engineering, it is used, for example, for the production of synthesis gas (an H_2/CO mixture), ethylene oxide, vinyl chloride, terephthalic acid, and many other substances. It is also used in the steel industry for wastewater treatment and in the paper industry for pulp bleaching.

Even higher combustion rates than with compressed oxygen can be achieved with liquid oxygen. **L**iquid **ox**ygen (LOX) is therefore used as an oxidant in rocket engines. In the past, it was used in explosives ("oxyliquite").

► Handling liquid oxygen requires the utmost care. Wood, oils, and greases, but also magnesium and many other reducing agents can ignite spontaneously on contact with LOX and may even explode uncontrollably.

38.3.3 Bound Oxygen

A whole range of very oxygen-rich substances is used as oxidants or oxygen carriers. These include, among others

- Nitric acid (HNO_3) – used with hydrazine as rocket fuel
- Nitrates (e.g. $NaNO_3$, KNO_3) in black powder and pyrotechnic compositions
- Hydrogen peroxide (H_2O_2) in rocket propellants
- Peroxides (e.g. Na_2O_2, BaO_2) – are used as oxygen carriers for ignition mixtures, e.g. for the ignition of thermite mixtures.
- Nitrous oxide (N_2O, "laughing gas") – can be used in many cases as a substitute for O_2, e.g. in rocket technology or the automotive tuning sector. Carbon, sulfur, and phosphorus burn more vigorously in pure N_2O than in air (see also "Combustion temperature" in Sect. 38.3.6).

Fig. 38.5 Markings of oxidizing substances

All these substances have in common that they – like oxygen – are oxidizing. They must therefore be labeled accordingly (see Fig. 38.5).

38.3.4 Oxygen-Free Oxidizing Agents

The halogens are primarily known as oxygen-free oxidizing agents. Here, sometimes very violent reactions can occur. Chlorine, for example, reacts with phosphorus or turpentine, producing heat and flames. The mixture H_2/Cl_2 is called "chlorine detonating gas" due to its explosive reaction to hydrogen chloride – which can already be initiated by UV light.

Gaseous fluorine can, under certain circumstances, even "burn" the valve of the compressed gas cylinder as it escapes. It has also been suggested as an oxidizer in rocketry – using hydrazine as fuel.

38.3.5 Oxygen Measurement

Oxygen measuring devices ("ox-meters") are used to measure the oxygen concentration. These are intended to warn the emergency services of dangers due to oxygen enrichment or oxygen deficiency. The first interpretation of an increased oxygen concentration is relatively simple: Increased risk of fire and ignition. However, it should not be forgotten that a prolonged stay in an atmosphere strongly enriched with O_2 can also lead to oxygen poisoning (also *oxygen toxicity* or oxygen intoxication).

It is well known that a significantly reduced oxygen concentration in the ambient air leads to an oxygen deficiency (medical term: *hypoxia*), reduces the risk of fire, or even causes a fire to go out (For the physiological effect see Table A.12).

It is more difficult to assess an only slightly reduced oxygen concentration, e.g. due to noxious or asphyxiant gases mixed with the ambient air. If the expected 21 vol.-% O_2 is measured outside the danger zone, but 20.5 vol.-% O_2 is measured at the actual point of operation, the question arises "Insolating breathing apparatus (self-contained breathing apparatus, SCBA) or filter appliance?" This question can be answered with "SCBA!".

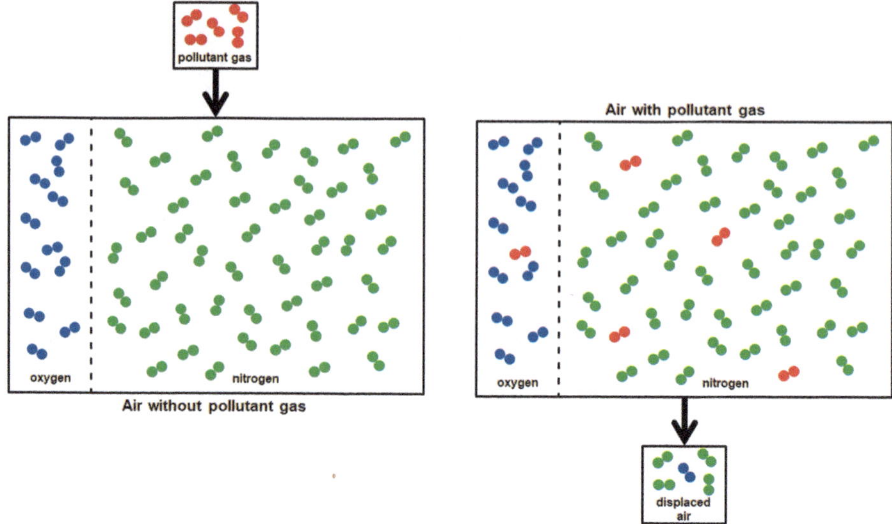

Fig. 38.6 Air displacement by other gases

Rough Calculation

Air consists of approximately $^1/_5$ (= 20 % by volume) oxygen and approximately $^4/_5$ (= 80 % by volume) nitrogen. If the oxygen concentration is reduced by 0.5 % by volume by a gas that displaces the air, then nitrogen is also displaced in addition to oxygen. However, the displaced N_2 volume corresponds to four times the displaced O_2 volume. The pollutant gas concentration is therefore 0.5 % by volume + 4 · 0.5 % by volume = 2.5 % by volume. ◄

Figure 38.6 illustrates this displacement effect.

Notes on Filter Devices

- Filter devices may only be used if sufficient atmospheric oxygen is available (usually ≥17 % by volume).
- Filter devices must not be used with unknown pollutants.
- Filter devices must not be used if the filter does not protect against the concentration of the respiratory toxin (e.g. ABEK2 filter: max. 0.5 % by volume of harmful gas).
- Gas filters may only be used if the respirator wearer can smell or taste the gases/vapors when the filter is breached.
- **All** the above conditions (etc.) must be fulfilled at the same time in order to be able to use a filter unit!

Further and more detailed instructions can be found in the latest version of Fire Brigade Service Regulation 7 "Respiratory Protection" (German Feuerwehrdienstvorschrift 7), the training documents of the Fire Brigade Schools, and the corresponding service instructions of the municipalities and districts.

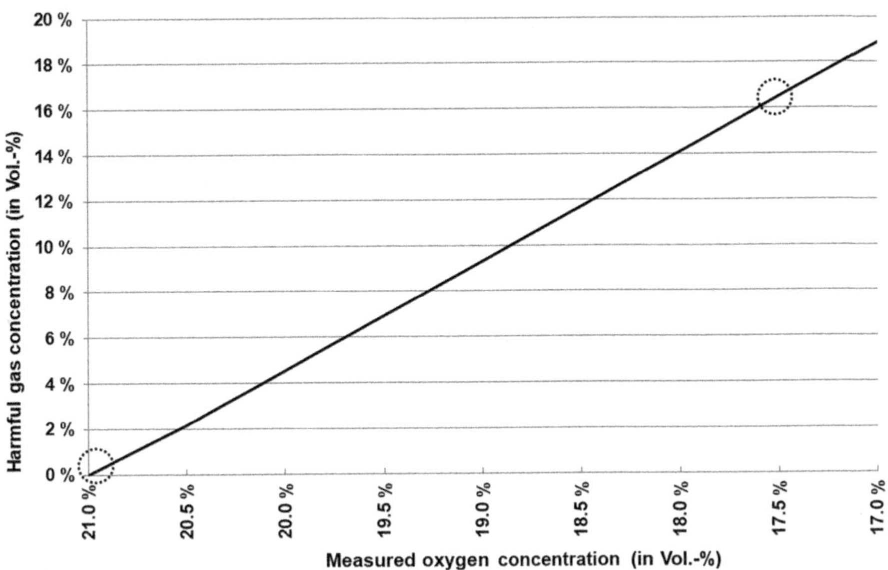

Fig. 38.7 Pollutant gas concentration versus oxygen concentration

The diagram in Fig. 38.7 again illustrates the relationship between measured oxygen concentration and real pollutant gas concentration. Two essential factors can be identified.
Recognize facts:

- With a measured O_2 concentration of 17.3 % by volume, the pollutant gas concentration is also 17.3 % by volume.
- To maintain a pollutant concentration of a maximum of 0.5 % by volume, which is still permissible for an ABEK2 filter, the measured oxygen concentration must not fall below the value of 20.85 % by volume.

38.3.6 Influence on the Course of Combustion

The O_2 concentration present in the ambient air influences the:

- Ignitability
- Combustion rate
- Combustion temperature

38.3.6.1 Ignitability

Examples of significantly increased flammability due to an increase in oxygen concentration include:

- The glowing splint test, a classic laboratory method for detecting oxygen: If a glowing, non-flaming wood splint is held in a vessel containing the gas to be tested, the glowing wood splint flames up in the presence of higher concentrations of oxygen and burns again.
- Even flame-retardant work clothing can be ignited by the flying sparks of an angle grinder if it has been mistakenly blown off with oxygen instead of compressed air.
- A thin steel wire burns in oxygen-enriched air like a sparkler, spraying sparks.
- Fittings of oxygen cylinders must not be oiled or greased, as oils and greases can ignite on contact with the oxygen even at room temperature.

38.3.6.2 Explosion Range

An increase in the oxygen concentration extends the explosion limits. The effect on the LEL is slight and on the UEL very marked (Table 38.8; see Sect. 3.3.4). The smaller change in the LEL is because there is an excess of oxygen in this range, i.e. the mixture is lean anyway. The LEL usually rises slightly in pure O_2 in order to be able to enter the ignitable range again at all. In a pure oxygen atmosphere, the LEL can assume values of up to over 90 vol.-% fuel, since the setting of a mixture that is too rich requires high fuel concentrations due to the high O_2 concentration.

38.3.6.3 Combustion Rate

Both the oxygen content of the ambient air and the content of fire-promoting substances in a corresponding mixture influence the speed of the combustion reaction. More available oxygen causes a faster reaction. For example, hydrogen in pure oxygen reaches a combustion speed of about 9 m s^{-1}. By comparison, gasoline vapors burn in the air at about 30 cm s^{-1}. A cigarette burns in an oxygen atmosphere in a few seconds with lively flame development, instead of slowly burning out as usual.

Mixtures of fire-promoting substances with corresponding fuels result in the optimum combustion rate in the most favorable ratio. This was achieved, for example, in the solid propellants of the space shuttle boosters by a mixture of 70% of the oxidant ammonium perchlorate (NH_4ClO_4) and 30 % fuel, consisting of aluminum and a polymer mixture as a binder.

38.3.6.4 Combustion Temperature

The oxygen concentration naturally influences the combustion temperature of the combustible material: The higher the O_2 concentration, the higher the combustion temperature. When burning in pure oxygen, temperature increases of over 900 °C are

Table 38.8 Comparison of explosion limits in air and explosion limits in oxygen (examples)

		Explosion limits in the air (% by volume)		Explosion limits in oxygen (% by volume)	
Substance	Formula	LEL	UEL	LEL	UEL
Dimethyl ether	$H_3C{-}O{-}CH_3$	2.0	27.0	3.9	61.0
Ethanol	$C_2H_5{-}OH$	3.1	20.0	4.0	93.0
Formaldehyde	H–CHO	4.0	57.0	4.0	93.0
Methane	CH_4	4.9	15.4	5.1	61.0

Table 38.9 Combustion temperatures in air and pure oxygen

Fuel gas	Combustion temperature In mid-air	In oxygen	ΔT
Propane (C_3H_8)	1925 °C	2850 °C	+925 K
Methane (CH_4)	1970 °C	2860 °C	+890 K
Hydrogen (H_2)	2130 °C	3080 °C	+950 K
Acetylene (C_2H_2)	2250 °C	3030 °C	+780 K

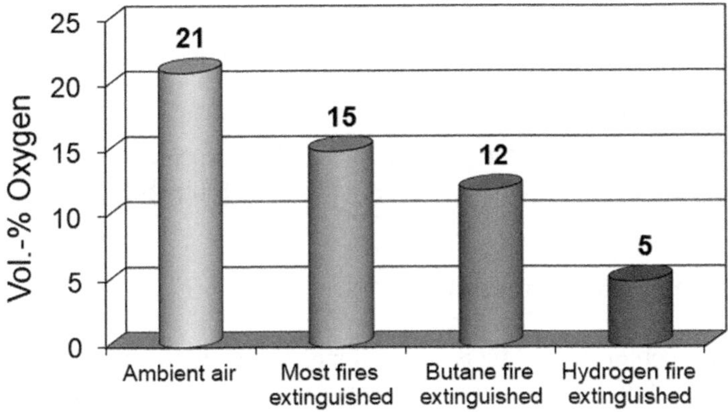

Fig. 38.8 Minimum oxygen concentration

possible. A cigar soaked in liquid oxygen (and then ignited) is even capable of "cutting" a steel plate several millimeters thick. Table 38.9 compares the combustion temperatures in air and oxygen.

Nitrous oxide (N_2O) decomposes endothermically into the elements at 600 °C according to the reaction eq. $2\ N_2O \rightarrow 2\ N_2 + O_2 + 164.2$ kJ. The resulting gas mixture thus consists of 33 vol.-% oxygen. The combustion of acetylene in N_2O therefore also shows an increased combustion temperature: 2800 °C.

38.3.7 Minimum Oxygen Concentration

If the oxygen content in the ambient air is reduced by inert gases, the flame temperature and reaction rate drop. Above a certain concentration, combustion comes to a standstill. The minimum oxygen concentration is the amount of oxygen that just enables the fire to continue burning. If the oxygen concentration falls below this level, the fire goes out or ignition is not possible. Most fires go out below an oxygen content of 15 vol.-% (Fig. 38.8).

38.3.8 Oxygen Index *OI*

The oxygen index (*OI*) indicates the ratio of the minimum proportion of oxygen in a nitrogen-oxygen mixture at which combustion just occurs. The oxygen index is a quantitative measure of the flammability of substances, which allows them to be divided into four flammability groups:

Oxygen index	Combustibility
0...0.20	Highly combustible
0.21...0.30	Normally combustible
0.31...0.50	Extinguishes by itself/combustible only when exposed to an ignition source
0.51...1.0	Highly self-extinguishing

An increase in (ambient) temperature lowers the oxygen index and thus leads to an increase in flammability. For example, values see Table A.28.

38.3.9 Combustion Equation

The complete combustion of fuels can be done classically by setting up the reaction equation and solving it as a redox equation. Here one finds that there is a general scheme – the combustion equation.

Organic substances burn after:

$$\begin{aligned} &C_aH_bO_cN_dX_eP_fS_g + x\ O_2 \\ &\rightarrow aCO_2 + gSO_2 + \frac{f}{4}P_4O_{10} + eHX + \frac{b-e}{2}H_2O + \frac{d}{2}N_2 \end{aligned}$$

Converted according to the oxygen demand x results in

$$x = a + g + \frac{5}{4}f + \frac{(b-e)}{4} - \frac{c}{2}$$

If we now write this equation by replacing the coefficients *a*-*g* with the element symbols, we obtain the form

$$xO_2 = \mathrm{C} + \mathrm{S} + \frac{5}{4}\mathrm{P} + \frac{(\mathrm{H}-\mathrm{X})}{4} - \frac{\mathrm{O}}{2}$$

This formula can be used to determine the oxygen demand (x) for a substance containing the elements C, H, O, S, N, X (= halogen) and P. For substances composed of fewer elements, the equation is simplified. An overview of the simplified formulae is given in Table 38.10.

These calculations give the number of oxygen molecules required for complete combustion. This oxygen demand value ($x\ O_2$) corresponds to the stoichiometric oxygen demand and can be used directly for other calculations, such as the oxygen

Table 38.10 Modifications of the combustion equation for different elemental compositions

Elements involved	Combustion equation
C, H, O, S, X, P, (N)	$xO_2 = C + S + \frac{5}{4}P + \frac{(H-X)}{4} - \frac{O}{2}$
C, H, O, S, X, (N)	$xO_2 = C + S + \frac{(H-X)}{4} - \frac{O}{2}$
C, H, O, S, (N)	$xO_2 = C + S + \frac{H}{4} - \frac{O}{2}$
C, H, O, X, (N)	$xO_2 = C + \frac{(H-X)}{4} - \frac{O}{2}$
C, H, X, (N)	$xO_2 = C + \frac{(H-X)}{4}$
C, H, O, (N)	$xO_2 = C + \frac{H}{4} - \frac{O}{2}$
C, H, (N)	$xO_2 = C + \frac{H}{4}$

demand for complete combustion (see Sect. 38.3.10) or the mathematical determination of explosion limits (Sect. 3.3.4).

Caution: With regard to the associated **chemical** reaction equations, it is important to note that:

- If the result is _.5, multiply the entire reaction equation by 2.
- If the result is _.25, multiply the entire reaction equation by 4.

38.3.10 Oxygen Demand or Air Demand

The amount of oxygen required for the complete combustion of a substance is – from a chemical point of view – initially derived from the combustion equation (see Sect. 38.3.9). In order to determine the mass-related or volume-related (only for gases!) theoretical oxygen demand, the molar mass of the substance must be converted.

The theoretical specific oxygen demand $V^0_{\text{s,m}}$ in m^3 O_2 per kg substance is given by:

$$V^0_{\text{s,m}} = \frac{32 \cdot x}{M}$$

The volume-related specific oxygen demand applies to the combustion of gases. It indicates how many m^3 of O_2 are required per m^3 of fuel gas and is equal to the oxygen demand of the combustion equation. The following applies: $V^0_{\text{s,V}} = x$.

To convert the mass-based oxygen demand to the volume-based oxygen demand, the following applies: $V^0_{\text{s,m}} = V^0_{\text{s,V}} \cdot 1.429$.

By taking into account the oxygen content of the air of 21 vol.-%, the theoretical air requirement can be easily determined by multiplying by 4.762 ($= {}^1/_{0.21}$) the theoretical oxygen requirement.

$$V^0_{\text{L,m}} = V^0_{\text{S,m}} \cdot 4.762 = V^0_{\text{S,m}} \cdot \frac{1}{0.21}$$

This gives the quantity of air in m^3 required for the complete combustion of 1 kg of substance (compare Table A.22). In reality, however, the mass-related air requirement $V_{L,m}$ is always somewhat greater than the theoretical value calculated here.

38.4 Ignition Energy

In our everyday lives, we move in an atmosphere consisting of 21% oxygen by volume. At the same time, we are surrounded by a myriad of flammable substances: Wood, paper, fuels, clothing, and decorative fabrics, plastics, and many more. Nevertheless, there is no fire everywhere all the time. Oxygen and combustible substances can be present next to each other for years without a combustion reaction starting. Only when a minimum amount of energy is supplied does the reaction start and then continue automatically. Systems of substances that react with each other under these conditions are also called inhibited systems. The reason for this inhibition of the reaction is the required activation energy (see Sect. 35.6).

Combustibility and Ignition Sources

The initiation of a noticeable oxidation reaction, i.e. the start of combustion, is called ignition. In this process, energy must be supplied to the combustible system – consisting of oxygen and combustible material. This energy supply can come from outside, so-called external ignition, or from the system itself, so-called self-ignition. In the case of self-ignition, systems are usually uninhibited or braked.

It seems to be a truism that a greater amount of energy introduced causes easier ignition of the flammable substance. Nevertheless, a rough classification can be made by comparing different ignition sources that provide different ignition energy.

- **Hardly combustible** substances require a strong ignition source, e.g. a gas burner.
- **Normally combustible** substances can be ignited with a match flame.
- **Highly combustible** substances can be ignited by sparks or cigarette embers.
- **Self-igniting** substances do not require an external ignition source.

38.4.1 External Ignition

In the case of external ignition, the energy required for ignition is supplied from the outside. The ignition source, flammable substance and oxygen must therefore coincide in space and time for ignition to occur. The main ignition sources are:

- open flames
- glow
- hot surfaces (e.g. exhaust system, halogen spotlights)
- electrical discharges (e.g. lightning, spark discharge, electric arc)

- mechanical sparks (e.g. grinding sparks, welding sparks, impact sparks)
- electromagnetic radiation (bundled visible light, thermal radiation)
- adiabatic compression (heat of compression, principle of the diesel engine)

Energy is transferred by heat conduction, heat flow, or heat radiation (see Sect. 3.1). The ignition temperature (ignition by hot surface, see Sect. 3.3.3) or the minimum ignition energy (ignition by electric sparks, see Sect. 38.4) serves as a measure of the ignition capability of a external-ignited system.

38.4.2 Spontaneous Ignition

In the case of spontaneous ignition (also called autoignition or self-ignition), a distinction must be made between uninhibited systems and inhibited systems. The uninhibited systems exhibit genuine spontaneous ignition. These include, for example, phosphine, PH_3, or organometallic compounds, such as diethylzinc, $(C_2H_5)_2Zn$. These substances ignite immediately on contact with atmospheric oxygen and must therefore be stored and handled in the absence of oxygen.

Self-ignition can also occur within seconds or a few minutes when oxygen-rich oxidizing agents come into contact with organic compounds. For example, pure nitric acid ignites sawdust, and chromium(VI) oxide ignites ethanol.

However, most spontaneous ignitions occur in braked systems. Oxidation processes lead to an increase in temperature. This accelerates the progress of oxidation and increases the reaction rate. If the resulting heat of reaction cannot be dissipated to the outside, ignition finally occurs. The thermal conductivity of the material has a significant influence on this process, which is known as **heat accumulation** (see Sect. 3.1.3). Aluminum powder, for example, is self-igniting because of its large surface area and its much poorer thermal conductivity. The same applies to lignite. The different thermal conductivities of the two materials can be seen in Table 38.11.

The spontaneous combustion of rags soaked in linseed oil, for example, or the microbially triggered self-heating of hay are well known.

38.4.3 Minimum Ignition Energy

The minimum ignition energy is the lowest energy with which an ignitable mixture can just be ignited (Table 38.12). The lower the minimum ignition energy, the

Table 38.11 Thermal conductivities of aluminum and lignite (comparison: Powder versus compact material)

	Thermal conductivity λ		
Substance	Compact piece	Powder	Factor
Aluminium	237 W m^{-1} K^{-1}	0.0745 W m^{-1} K^{-1}	3180
Lignite	0.20 W m^{-1} K^{-1}	0.022 W m^{-1} K^{-1}	9

Table 38.12 Minimum ignition energies in air at normal pressure (examples)

Substance	Minimum ignition energy (in mJ)	Substance	Minimum ignition energy (in mJ)
Carbon disulfide	0.009	Ethane	0.25
Hydrogen	0.019	Propane	0.26
Acetylene	0.019	Methane	0.28
Diethylether	0.19	Coal	40
Phosphorus, red	0.20	Wood flour	20 ... 60

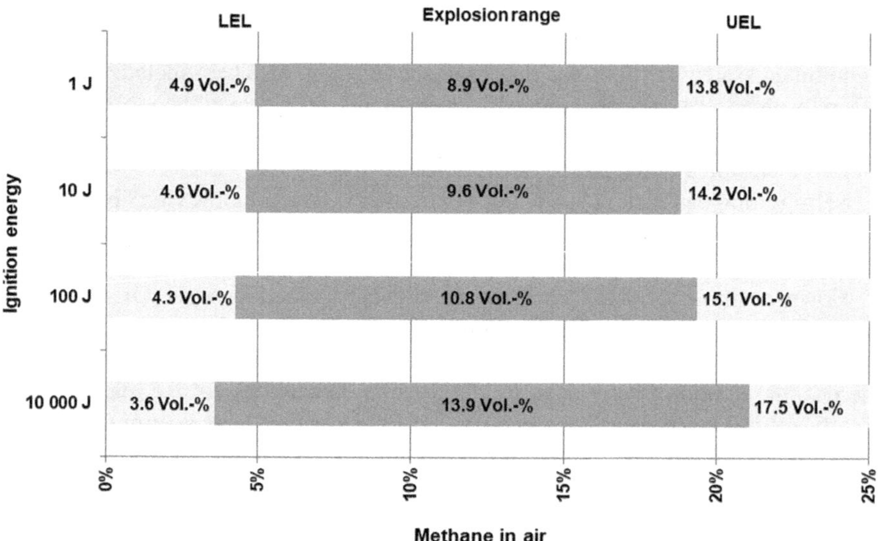

Fig. 38.9 Explosion ranges of methane as a function of ignition energy

greater the ignitability. The minimum ignition energy is usually given in units of J (joules) or mJ (millijoules $= 10^{-3}$ J). It is noticeably influenced by the oxygen concentration and the mixing ratio.

38.4.4 Influence of the Ignition Energy on the Explosion Range

With regard to the explosion ranges of flammable gases, studies exist on the relationship between the explosion limits (see Sect. 3.3.4) and the energy required to ignite the mixture. It has been found that higher ignition energies extend the explosion range. The expansion of the UEL is greater than that of the LEL (see Fig. 38.9).

38.5 Mixing Ratio

If we consider the mixing ratio between combustible material and air or oxygen, two guiding questions arise:

1. What influence does the mixing ratio have on the combustible system?
2. What influences the mixing ratio?

38.5.1 Influence of the Mixing Ratio on the Combustible System

As already explained, the explosion limits are influenced by the mixing ratio of flammable substances to air or oxygen. Below the LEL there is too little fuel (too much oxygen) and above the UEL too much fuel (too little oxygen) to ignite the mixture (see Sect. 3.3.4).

Within the explosion limits, it can be observed that the mixture reacts to varying degrees. The maximum explosion pressure is obtained slightly above the stoichiometric ratio of combustible to oxygen. All other mixtures within the explosion limits show a lower explosion pressure (Table 38.13).

This also applies analogously to combustible dusts. The dust cloud can only be ignited within a certain concentration range of dust in a dust/air mixture (see Sect. 39.1.3). The stoichiometric mixture composition of combustible dusts is generally in the order of 100...300 g of dust per cubic meter of air. In contrast to gases and vapors, the greatest explosive strength occurs with double to triple the quantity of dust.

38.5.2 Influence of Temperature on the Mixing Ratio

Most combustion takes place in the air. The concentration of available oxygen can therefore be considered practically constant. The amount of combustible substance is

Table 38.13 Volume concentration of fuel gas (butane, stoichiometric concentration = 3.1% by volume) against maximum explosion pressure

Fuel gas In mid-air	Maximum Explosion pressure
2 vol.-%	(< LEL)
3 vol.-%	6.1 bar
4 vol.-%	7.4 bar
6 vol.-%	5.9 bar
8 vol.-%	2.4 bar
10 vol.-%	(> UEL)

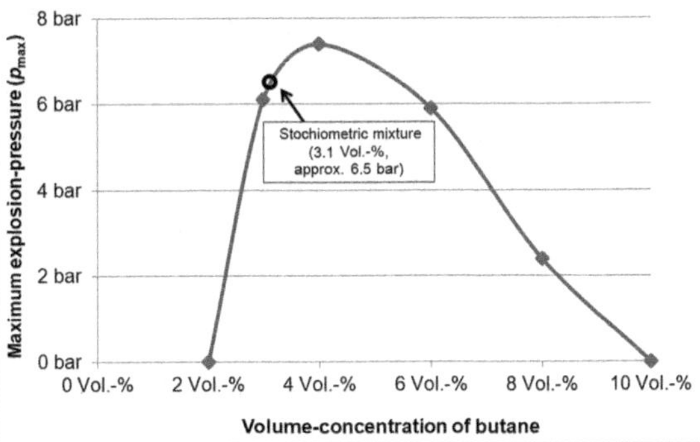

much easier to influence. The main variables influencing the mixing ratio are temperature and surface area. The instantaneous temperature of a flammable liquid influences the actual amount of vapors emitted by it. The flash point or the explosion points can be used to assess a possible danger from an explosive atmosphere.

Temperature influences the vapor pressure of a liquid. The higher the temperature, the higher the vapor pressure, i.e. the more vapor is released into the environment (see Sect. 3.3.2).

38.5.2.1 Flash Point/Inflammation Point

Below the flashpoint, so little vapor is released that the resulting mixture is too lean to be ignited. Above the flash point, an ignitable mixture is formed which can be ignited. At the inflammation point, which is usually only a few degrees above the flash point, the liquid supplies sufficient vapor due to its temperature to enable further burning (see also Sect. 3.3.3).

38.5.2.2 Lower and Upper Explosion Point (t_{ex})

The explosion points represent the limit temperatures of the ignition capability of gases or vapors. They are the temperatures at which vapor-saturated air in a closed space – e.g. in a closed container – reaches the lower or upper explosion limit (LEL or UEL, see Sect. 3.3.4) when mixed with air.

> **Caution:** The explosion points ($t_{ex,l}$ and $t_{ex,u}$; given in °C) must not be confused with the explosion limits (LEL and UEL, given in vol.-%).

The determination of the explosion points presupposes a dynamic equilibrium between the gas and liquid phases which has already been established (see Sect. 3. 2.4) and is therefore of primary importance for closed vessels that also contain liquid.

A temperature at least 10 °C below the lower explosion point is considered to be safe with regard to the formation of an explosive atmosphere. Assessing the upper explosion point is more difficult because it must be certain that the contents of the container have had sufficient time at the current temperature to reach this equilibrium, which is reached purely by diffusion processes (see Sect. 4.11). This will generally be difficult to achieve in use.

Examples of the explosion points of various gases and liquids can be found in Table A.21. In comparison, it can be seen that the lower explosion point is close to the flash point, usually a few (approximately 1–5) degrees below it.

> If the temperature of a liquid is more than 10 °C **below** the flash point of this liquid, **no** explosive atmosphere can form in the air.

The relationship between the lower and upper explosion points, LEL and UEL, and flash and fire points is illustrated in Fig. 38.10.

Solids

In the case of *degassable* solids – e.g. wood or coal – the temperature influences the amount of gas or vapor that is released into the environment per unit of time. The

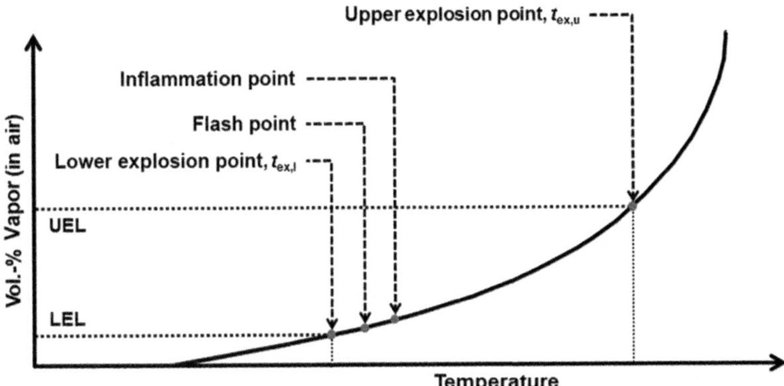

Fig. 38.10 Relationship between explosion points, explosion limits, flash point and inflammation point

higher the temperature, the faster gases/vapors are released from the solid (see Sect. 40.2.4).

38.5.3 Influence of the Surface on the Mixing Ratio

The surface area, i.e. the surface of attack of the combustible substance, also plays a major role in the resulting mixing ratio. The larger the surface area of a substance and the better it is mixed with air, the greater the resulting reaction rate. As an additional effect, smaller pieces are not only easier to heat, but also much more evenly and from all sides, which in turn favors the progress of the reaction.

In addition to the absolute surface area, the specific surface area is also given as a measurement in technology. The specific surface area is defined as the surface area of a substance in m^2 per gram of substance.

A wooden cube made of beech with a density $\varrho = 0.69$ g/cm^3 and an edge length of 10 cm has a surface area of $S_c = 600$ cm^2 $= 6$ dm^2. If this cube is divided into smaller cubes with an edge length of 1 mm, the result is a surface area of 600 dm^2 $= 6$ m^2. The surface area has increased a hundredfold.

The specific surface has increased from $8.7 \cdot 10^{-5}$ m^2/g (large cube) to $8.7 \cdot 10^{-3}$ m^2/g (small cube) – and thus also by a factor of 100.

The surface also plays a role with flammable liquids. Sprayed liquids have a significantly larger surface area than coherent liquid layers or droplets. This effect is even stronger with mists, as the droplets are even smaller here.

A glass dish with a diameter of 10 cm is filled ($\varrho = 0.8$ g/cm^3) 5 mm high with a flammable liquid. The surface area is 78.5 cm^2 and the volume is around 39 mL. If this volume is sprayed with an average droplet size of 200 μm, about 93,750 droplets with a total surface area of 11.8 dm^2 (corresponding to factor $= 15$) are obtained. If one creates a fog (droplet size 10 μm), $7.5 \cdot 10^{10}$ particles can be generated. The total

Table 38.14 Comparison of the surface area of a liquid to the droplet size

State	Droplet size	Surface	Specific surface	Factor
Liquid	–	78.5 cm^2	$2.5 \cdot 10^{-4}$ m^2/g	1
Sprayed	200 μm	1180 cm^2	$3.76 \cdot 10^{-3}$ m^2/g	15
Fogged	10 μm	235 600 cm^2	0.75 m^2/g	3000

surface area is 23.56 m^2 – and thus 3000 times in relation to the glass dish (see Table 38.14).

Another example are metals. A steel beam will not ignite. Steel wool, on the other hand, can already be ignited with a match, sparks, or the short-circuit current of a 9 V block. In the case of very fine iron dust (spec. Surface > 3.5 m^2/g), the iron behaves pyrophorically, i.e. it ignites by itself as soon as it comes into contact with air. Ignition stones made of the alloy "ceric iron" (composition: 30 % Ce, 70 % Fe) are based on the same principle. The particles produced during abrasion are so small that they ignite themselves in air and are thus capable of exciting other materials, e.g. lighter gas, to burn.

The so-called wick effect is also based on the principle of increasing the surface area. A wick enlarges the surface of the combustible material and thus ensures easier ignition. This is used for candles and kerosene lamps. This effect also occurs with clothing – including emergency clothing – that has been wetted with flammable liquids. For example, diesel fuel or pure acetic acid can easily ignite due to the increased surface area of the fabric.

38.5.3.1 Extinguishing Agent Surface

The influence of the surface of combustible materials is also reflected in the extinguishing agents. With a large, optimized surface area, an extinguishing agent develops the maximum extinguishing success. This is the case with extinguishing powder and water (spray jet).

The **optimum** droplet size of water is 0.2...0.4 mm. The formation of larger droplets (up to 1 mm), as well as the formation of smaller droplets (0.1...0.01 mm), should therefore be as low as possible. Fog nozzles fulfill this requirement better than multi-purpose jet nozzles. Irrespective of the nozzle used, thc following applies: Internal attack only with spray jet. Fog nozzles usually have a **"**grid position**"** which ensures the optimum spray cone for the internal attack.

Extinguishing powders for fire classes B and C must have the largest possible surface area in order to achieve a high extinguishing effect. However, if the powder particles are too small, they tend to agglomerate more strongly (**"**caking**"**, "balling") and have a shorter throwing distance. The particle size distribution of BC powders ranges from 2 μm to 200 μm (see Sect**.** 41.2.4).

38.5.4 Further Influencing Variables

The **density ratio** to the ambient air (see Sect. 3.3.5) also has an effect on the mixing ratio. The closer the gas/vapor density is to the density of the air, the easier it is for a uniform mixture to form. Thus, methane (CH_4, M = 16 g mol^{-1}) tends to rise and propane (C_3H_8; M = *44* g mol^{-1}) tends to fall. Ethene (C_2H_4, M = 28 g mol^{-1}), on the other hand, has about the same density as air and therefore easily forms homogeneous mixtures with it.

The density of a substance depends on the temperature. The Joule-Thomson effect (see Sect. 4.12.2) can lead to the so-called **heavy gas effect** in the case of compressed gases: The pressurized gas cools down so much during the expansion that it initially has a greater density than the ambient air and tends to collect near the ground.

Air streams and turbulences ensure better mixing. This can be caused by wind as well as by the use of fans. In addition, the evaporation rate of liquids can be significantly increased by air currents.

38.6 Combustion Catalyst

Carbon monoxide (CO) is a flammable and toxic gas. The combustion temperature in the air is approximately 2000 °C. The explosion range in air extends from 1.5 vol.-% (LEL) to 75 vol.-% (UEL). Nevertheless, it is not possible to ignite a really dry air/carbon monoxide mixtures, i.e. mixtures which are also free of traces of water vapor.

As studies on the chemistry of fires show, a catalyst, or rather a reaction carrier, is required for every fire. These are hydrogen radicals or hydroxyl radicals (H•, HO•), which are the first to significantly initiate the combustion reaction (see Sect. 40.1 ff.). Both the traces of water vapor that are normally always present and the radicals themselves can be described as triggering factors and thus as "catalysts of combustion".

39 Explosions

Explosions (from the Latin *explaudere* = to hit out, – to clap) are primarily very fast chemical reactions that trigger a bang due to a strong increase in the volume of the gases that are created and spread. Often the blast wave exerts a destructive influence on the surrounding area. Purely physical processes that take place with a strong increase in volume are also sometimes referred to as *explosions.*

39.1 Differentiation of Explosion Processes

Explosions can be distinguished on the basis of their cause, their reaction rate, or the type of exploding medium.

Explosions can be caused by chemical or physical processes. In terms of reaction speed, a distinction is made between relatively slow deflagrations (German: *Verpuffung*), faster deflagrations (German: *Deflagration*), and very fast detonations. With regard to the media, a distinction is made between gas explosions, dust explosions, and explosions of explosives.

39.1.1 Explosions with a Chemical or Physical Cause

The most important **physically induced explosions** are the steam explosion and the container explosion. **Steam explosions** occur when a cold liquid is added to a hot liquid whose temperature is greater than the boiling point of the cold liquid. The cold liquid now evaporates abruptly. The resulting vapor carries away the hot liquid. We have already encountered this effect in the so-called “fat explosion” (= FCI-steam explosion between oil and water; compare Sect. 3.4.6 and Fig. 3.18). The same effect occurs when a liquid, e.g. water, hits a hot molten metal. An attempt to extinguish a chimney fire with water would also result in a steam explosion (see Sect. 3.4.6 and Fig. 3.19).

T. Schmiermund, *The Chemistry Knowledge for Firefighters*,
https://doi.org/10.1007/978-3-662-64423-2_39

Container rupture or pressure vessel rupture refers to the rupture of a container due to a slowly building pressure that the container cannot withstand. Heating from the outside causes the temperature of the container contents to rise. This results in an ever-increasing pressure inside the container. Finally, the pressure becomes so great that the container can no longer withstand it and ruptures.

A pressure vessel explosion as a result of a fire can occur, for example, with pressurized gas cylinders (propane, acetylene, etc.). The BLEVE is also primarily the consequence of a vessel explosion (see Sect. 4.12.2).

Even moderate exposure to heat can cause a container to explode. For example, spray cans may explode if exposed to direct sunlight. If this heating takes place in a vehicle, e.g. in the case of a spray can on the parcel shelf in a car, the destruction of the window is usually the result.

► The sudden rupture of the vessel wall of a vessel under excess pressure is not called an explosion, but a **bursting.**

Furthermore, nuclear weapon explosions, volcanic eruptions, and meteorite impacts also count as physical explosions.

"Real" explosions, i.e. **chemically induced explosions**, occur due to a chemical reaction. In most cases, these are very rapid **combustion processes**, such as occur in gas or dust explosions, but also explosives. More rarely, **thermal explosions** are caused by chemical reactions. Such thermal explosions occur when the heat of reaction of a chemical reaction cannot be dissipated fast enough and thus the temperature of the reaction system rises. The higher temperature leads to an increase in the rate of reaction, which in turn causes a higher temperature. This, in combination with the thermal expansion of the system, eventually leads to an explosion. In chemistry, one also speaks in this context of "a reaction having gone through".

An overview of the distinction between physical and chemical explosion processes is given in Fig. 39.1.

Chemically and physically induced explosions can overlap:

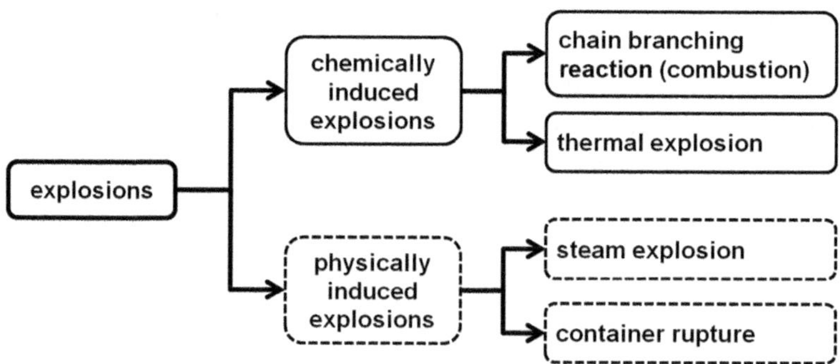

Fig. 39.1 Classification of explosion events according to their cause

- When an attempt is made to extinguish a metal fire, e.g. burning Mg, a vapor explosion occurs due to the sudden evaporation of water. However, part of the water is thermally dissociated, whereby the gases thus formed, H_2 and O_2, occupy a larger volume than the water vapor. At the same time, the hydrogen-oxygen mixture ignites.
- When water penetrates an aluminum melt ($\vartheta \geq 700$ °C), in addition to the steam explosion, there is also a reduction of the water in accordance with $2\ Al + 3\ H_2O \rightarrow Al_2O_3 + 3\ H_2\uparrow$. The released hydrogen can mix with the ambient air and thus trigger a secondary explosion.
- The situation is similar in the case of fat explosions or the explosion of a container with flammable contents: After the physical explosion, the flammable substance is swirled in the ambient air and ignites in the sense of a chemical explosion.

39.1.2 Reaction Speed for Explosions

Particularly in the case of chemically induced explosions in the sense of very fast oxidation reactions, distinctions are made with regard to the reaction rate. A distinction is made between deflagrations, deflagrations, and detonations (overview: Fig. 39.2).

39.1.2.1 Slow Deflagration

A slow deflagration is said to occur when the combustion reaction leads to an increase in temperature and volume, but not to a relevant increase in pressure. The speed at which the reaction front propagates can be between 0.01 and 1 m s^{-1} (up to 3.6 km h^{-1} or 2.24 mi/h). The former definition of a deflagration via the maximum occurring explosion pressure of <100 kPa (<1 bar) is no longer common.

Slow deflagrations frequently occur in the range of the lower explosion limit, i.e. when the mixture is very lean (low in fuel). Slow deflagrations may also originate

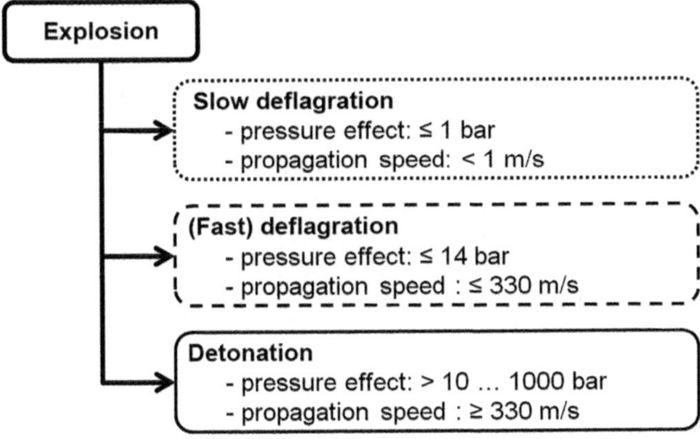

Fig. 39.2 Classification of explosion events

from non-dusty solids. A Christmas tree without roots and water will dry out within 4–6 days and **must** then **be** considered highly flammable. In addition, the needles and thin branches containing resin and oil form a very large surface area. Ignition by burning candles, table fireworks, etc. is then possible at any time without further ado. The burning usually takes place in the form of a deflagration. Depending on the size of the room and the stability of the components, the pressure shock of the deflagration can cause window panes to burst, doors to be pushed out of their frames or even walls to be displaced.

39.1.2.2 Deflagration

Deflagration (Latin *defraglare* = to burn off) is the term used to describe a rapid combustion process in which the resulting explosion pressure is caused only by the evolving and expanding combustion gases. This generally applies to gas and dust explosions. However, solid decompositions that occur with a release of heat and gas are also referred to as deflagrations. Examples of such solid deflagrations are:

- Black powder in firecrackers or rockets
- the decomposition of sodium azide (NaN_3) as a propellant for airbags
- Propellants in ammunition (a detonation would destroy the barrel of the weapon)

Much more frequently than solid deflagrations, hazards due to explosive atmospheres occur during operations. These are triggered by flammable gases or the vapors of flammable liquids (gas explosions) or by flammable dusts (dust explosions). The maximum pressures reached during such a deflagration are usually in the range of 0.7...1.4 MPa (7...14 bar). The reaction takes place purely via the propagation of the flame front and is therefore below the sonic velocity of the burning medium. A propagation velocity of up to 330 m s^{-1} (approximately 1190 km h^{-1} or 740 mi/h) is set as the limit. Thus, *back-draft* explosions with velocities of up to 20 m s^{-1} (72 km h^{-1} or 45 mi/h) also belong to the deflagrations.

39.1.2.3 Detonation

Detonation (Latin *detonare* = to thunder down) refers to the turnover of an explosive substance in which the reaction zone – in contrast to deflagration – spreads at supersonic speed. The pressure, energy, and density of the detonation zone increase to such extremely high values as a result of the compression shock which occurs that the chemical reaction is already completed within $10^{-7}...10^{-5}$ seconds. The detonation velocities for gas detonations are in the range of approximately 1500...3000 m s^{-1}, for solid and liquid explosives up to 9000 m s^{-1} (over 32 000 km h^{-1} = about 20 000 mi/h). Detonation pressures up to more than 100 MPa (>1000 bar) can occur. In pipelines, deflagrations can increase until detonation occurs.

39.1.3 Differentiation According to Explosive Medium

Depending on the substance system that explodes, a distinction is made between the space explosions "gas explosion" and "dust explosion" and explosions caused by explosive substances (overview: Fig. 39.3).

39.1.3.1 Gas Explosions

Explosions of flammable gases or the vapors or mists of flammable liquids mixed with air are probably the most common explosions. They are commonly referred to as gas explosions. The conditions under which an explosive mixture of flammable gas or vapor may occur have already been discussed (see Sect. 3.3). Gas explosions, like dust explosions, are space explosions: A mixture of flammable substance and gaseous oxidant (usually atmospheric oxygen) is initially formed in an available space. Accordingly, space explosions require the formation of explosive mixtures – and their subsequent ignition.

39.1.3.2 Dust Explosions

If combustible dusts are stirred up, dust explosions may occur. Compared with gas explosions, dust explosions are usually much more violent (see Sect. 39.2). Particular care must be taken during extinguishing operations where combustible dusts may be raised. In endangered areas, e.g. mills, carpenter's workshops, or furnaces that are operated with coal dust, extinguishing work may only be carried out with a spray jet. With a full jet, the dust may be whirled up, which may then lead to a dust explosion.

Whether a combustible dust is also capable of dust explosion depends primarily on the particle size distribution of the dust. The particle diameter for optimum ignition capability of dust is less than 40 μm. As a rule, dusts with a diameter of more than 400 μm cannot be made to explode. However, admixtures of 5 %–10 % fine dust (i.e. < 40 μm) in coarse dust (i.e. > 400 μm) can be sufficient to cause the coarse dust/fine dust mixture to explode.

Like gas/air mixtures, dust/air mixtures are subject to explosion limits. For most combustible dusts, LEL values of 20. . .60 g dust per m^3 air apply. The LEL is usually between 2 and 6 kg/m^3 and thus higher by a factor of 100.

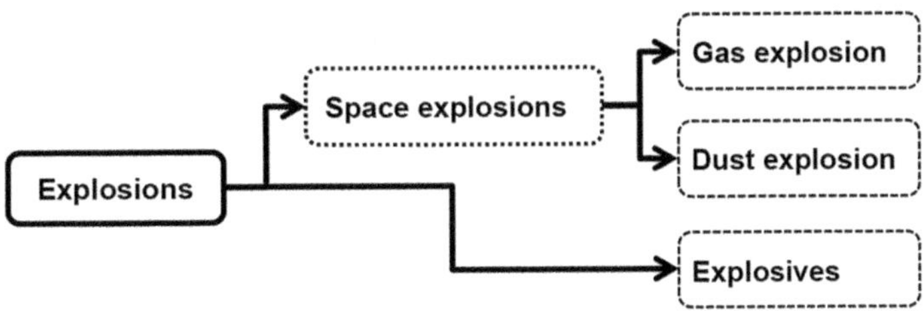

Fig. 39.3 Explosible media

Rule of Thumb
If you can no longer see your hand when your arm is outstretched (= visibility of $\leq$1 m) due to swirling dust, then there is a dust explosion hazard.

39.1.3.3 Explosive Substances

Explosives, together with primary explosives, low smoke powders, and pyrotechnic compositions, are classified as explosives. These, together with substances that are not expressly manufactured as explosives but which have such properties, are referred to as explosive substances. The (German) Explosives Act and its ordinances provide the relevant legal basis for these substances.

Explosive substances do not include the gases, vapors, mists, and dusts already mentioned, which only produce an explosive atmosphere when mixed with air or oxygen.

39.2 Explosion Indicators

Safety-related indicators which serve to assess the possible formation of an explosive atmosphere have already been presented (see Sect. 3.3). In this section, we shall consider those indicators which provide information on the course of a gas or dust explosion. Indicators used to assess explosives are **not** considered here.

Example values for the listed key figures can be found in Table A.19 and Table A.20.

39.2.1 Damage Due to Pressure Waves

To get an impression of the damage caused by blast pressure waves, we compare the pressure with which a blast wave *hits* an object (building, person) with the resulting damage, as shown in Table 39.1 (compare also Fig. 55.1).

Table 39.1 Damage caused by pressure waves

Maximum pressure of the impacting pressure wave	Damage
10 mbar	Glass damage to buildings
30 mbar	Minor damage to building structures
100 mbar	Severe damage to building Structures damage to roof structures
170 mbar	Violations of the eardrum Destruction of many buildings
600 mbar	Total destruction of common buildings
3 bar	Fatal rupture of the lung

39.2.2 Flame Propagation Velocity v_F

The speed at which the flame front moves into the still unburned, homogeneous gas-air mixture is the flame propagation speed. It is usually given in cm s^{-1} or m s^{-1}. It makes a statement about the speed at which the combustion reaction takes place in the gaseous state of the fuels.

39.2.3 Limiting Oxygen Concentration $\varphi_{O,min}$

The limiting oxygen concentration $\varphi_{O,min}$, also abbreviated to LOC, is the concentration of oxygen in a mixture with an inert gas at which an explosive mixture can just no longer be formed. It depends on both the flammable substance and the inert gas. The values for CO_2 as an inert gas, for example, are generally higher than the values for N_2.

In engineering, the LOC is applied to exclude explosive atmospheres from the outset by inerting. Systems are considered to be explosion-proof if they comply with the relationship

$$\varphi_{O,s} \leq 1.2 \cdot \varphi_{O,\,min} - 4.2 \text{Vol.}-\%$$

suffice. In general, when inerting with N_2 or CO_2, a system is considered explosion-proof if the oxygen content is < 5 vol-%, i.e.: $\varphi_{O,s} < 5$ vol.-%. This guideline value applies to all flammable substances; with the exception of H_2, CO, and the powders of Al, Mg, and Zr.

39.2.4 Maximum Explosion Pressure p_{max}

The ignition of an explosive mixture causes a pressure increase in the form of a pressure wave, the explosion pressure p_{ex}. If the pressure rise is observed over time, a pressure maximum is observed, the maximum explosion pressure $p_{ex,max}$, or p_{max} for short. Since the measurements are usually carried out at normal pressure, this is the maximum overpressure that occurs when the mixture is ignited.

The maximum explosion pressure depends, among other things, on the composition of the mixture. For gases and vapors, the maximum explosion pressure is in the range of the stoichiometric composition (see Sect. 38.5.1; Table 38.13). For dusts, two to three times the stoichiometric composition of the dust is required.

Stoichiometric Composition of a Vapor

The stoichiometric composition of a mixture of toluene ($C_6H_5{-}CH_3$) and the air is to be determined.

(a) Determination of the oxygen demand from the general combustion equation

$$C_aH_bO_c + xCO_2 \rightarrow aCO2 + \frac{b}{2}H_2O$$

$$\boldsymbol{x} = a + \frac{b}{4} - \frac{c}{2}$$

Toluene: $C_7H_8 \Rightarrow \boldsymbol{x} = a + \frac{b}{4} - \frac{c}{2} = 7 + \frac{8}{4} - \frac{0}{2} = 7 + 2 = \mathbf{9}$
or as a (chemical) reaction equation:

$$nC_6H_5{-}CH_3 + xO_2 \rightarrow aCO_2 + bH_2O$$

$$C_7H_8 + \mathbf{9}O_2 \rightarrow 7CO_2 + 4H_2O$$

(b) Determine the stoichiometric concentration (c_{st}) in air (see Sect. 3.3.4) with $\boldsymbol{x}$ = number of O_2 molecules (according to the combustion/reaction equation) according to:

$$c_{st} = \frac{100}{1 + \frac{\boldsymbol{x}}{0.21}}$$

$$c_{st} = \frac{100}{1 + \frac{\mathbf{9}}{0.21}} = \frac{100}{1 + 42.857} = \frac{100}{43.86} = 2.28$$

(c) The stoichiometric concentration of toluene (as vapor) in the air is 2.3 % by volume. ◀

39.2.5 Maximum Pressure Rise over Time *(dp/dt)*$_{max}$

When considering the time course of an explosion, the quotient of the pressure increase (*dp*) and the time required for this *(dt)* is a measure of the flame speed (reaction speed) and thus of the violence of an explosion. The faster the maximum explosion pressure is reached, the stronger the pressure surge. Consequently, the explosion is also more violent. The maximum pressure increase over time is given in bar s^{-1}. (Tab. A.19.)

39.2.6 Cubic Law: K_G and K_{St} Value

The cubic law describes the volume dependence of the pressure increase overtime of a space explosion: $(dp/dt)_{max} \cdot \sqrt[3]{V} = K$. It is a substance-specific quantity, which is specified for flammable gases or the vapors of flammable liquids as the K_G value and flammable dusts as the K_{St} value in the unit bar m s^{-1}. The higher the K_G or K_{St} value, the stronger the explosion.

Typical K_G values for flammable gases and the vapors of flammable liquids are in the range of 40...70 bar m s^{-1}. Propane, at 75 bar m s^{-1}, is in the upper limit. The exception is hydrogen: It is much higher at 550 bar m s^{-1} and thus produces explosions about 7½ times more violent than the liquid gas propane.

Many dust explosions are noticeably more violent than gas explosions. K_{St} values of various dusts are given as examples in Table A.20.

40 The Chemistry of Combustion

Combustion processes seem quite simple at first glance. If the molecular formula of the combustible substance is known, the reaction equation can be quickly established and the matter is settled. Carbon becomes CO_2, hydrogen becomes H_2O. And incomplete combustion, i.e. when there is too little oxygen, produces carbon monoxide.

The course of combustion reactions is nevertheless one of the most complex chemical reactions. The elucidation of the detailed chemical sequence of combustion processes is an important task for technology. In the context of "technical combustion", processes in combustion plants and combustion engines are optimized.

40.1 Chain Reactions

All combustion processes are chain reactions. Chain reactions are characterized by a sequence of constantly repeating identical individual reactions which propagate independently until there is no longer any possibility of a further reaction. Some chain reactions are self-contained (Fig. 40.1), others take place under chain branching (Fig. 40.2). The latter include combustion and the nuclear chain reaction (see Sect. 32.2).

40.1.1 Radical Chain Reactions

Radicals are atoms or molecules with an unpaired electron. This electron is represented in formulas with a dot (•). Radicals are – due to the unpaired electron – highly reactive particles.

T. Schmiermund, *The Chemistry Knowledge for Firefighters*,
https://doi.org/10.1007/978-3-662-64423-2_40

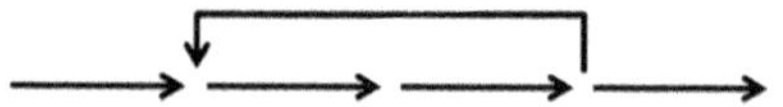

Fig. 40.1 Closed chain reaction

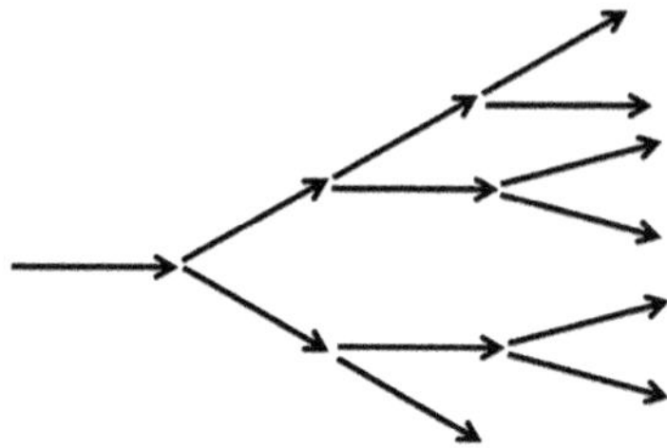

Fig. 40.2 Branched-chain reaction

> Radicals are unstable, rapidly reacting intermediates of a wide variety of chemical reactions. They are atoms or molecular fragments that have an unpaired electron.

40.1.1.1 Chain Start: Homolysis

The prerequisite for a radical chain reaction is the initial formation of radicals. Radicals are formed by homolysis (Greek *homoios* = similar, *lysis* = (dis)solution, separation) of suitable starting molecules. During homolysis or homolytic fission, an electron pair bond (EPB) is dissolved in such a way that each of the two resulting particles has an unpaired electron – from the original EPB.

The homolytic cleavage occurs preferentially at small differences of the electronegativity of the atoms to be separated. [*For larger EN differences, both electrons of the EPB remain with the more electronegative partner, which thus forms an anion. The other part becomes a cation. This process is called heterolysis.*] A not too high binding energy of the EPB concerned has a favorable effect. Relatively low dissociation energies – which are thus easy to bring to homolytic cleavage – possess, for example, the bonds: H–H, Cl–Cl, Br–Br, C–H, C–Cl, and C–Br.

Possible energy sources for the homolysis, which preferably takes place in the gas phase, are:

- Heat (thermolysis)
- Light (photolysis)
- Electrical discharges
- Ionizing radiation (radiolysis, see Sect. 30.2)
- Electron transfers in certain redox reactions

The decisive factor for all forms of energy is that the energy supplied is at least the dissociation energy of the respective EPB. (For examples of dissociation energies, see Table A.18).

Radicals can also be formed by substances that break down into radicals particularly easily. Such radical formers are used as initiators, e.g. for the production of plastics or in two-component synthetic resins.

40.1.1.2 Chain Propagation

When a radical, i.e. a free atom or a molecular group with an unpaired electron, reacts with a neutral molecule, a new radical is formed. The existing and newly formed radicals are also called chain carriers. They ensure the progress of the reaction. Radicals react in this way, for example, with

- unsaturated compounds under addition:

$$H_3C\bullet + H_2C{=}CH_2 \rightarrow H_3C{-}CH_2{-}CH_2\bullet$$

- other atoms of a molecule:

$$Cl\bullet + H_3C{-}CH_3 \rightarrow HCl + H_3C{-}CH_2\bullet$$

in each case, with the formation of further chain carriers. In some reaction steps, two or more of these chain carriers are formed. Here, a branching of the reaction chain takes place (compare Sect. 40.1.3).

40.1.1.3 Chain Termination

When chain carriers react with each other, these radicals disappear from the reaction system and thus break off the chain reaction. This process is also called recombination. When two radicals react with each other, recombination only succeeds if the excess energy produced by the formation of a new EPB can be dissipated. Very many recombinations are therefore trimolecular reactions. In a trimolecular reaction, the two recombining radicals must meet with an inhibitor molecule (M) or the vessel wall (wa). The inhibitor/wall absorbs the excess energy, thus reducing the total molecular energy to a value below the dissociation energy of the bonds involved. This results in exciting inhibitor molecules (M*) or a slightly more energetic – i.e. warmer – wall (wa*) (see also Fig. 41.2).

- Trimolecular reaction with inhibitor: $H_3C\bullet + H_3C\bullet + M \rightarrow H_3C{-}CH_3 + M^*$
- Trimolecular reaction with wall effect: $Cl\bullet + \bullet Cl + wa \rightarrow Cl_2 + wa^*$
- Bimolecular reaction: $H\bullet + \bullet Br \rightarrow H{-}Br$

40.1.1.4 Initiators and Inhibitors

Peroxides, such as dibenzoyl peroxide (DBPO), di-(2-ethyl-hexyl)-peroxidicarbonate (DEHPC) or H_2O_2, or azo compounds, such as azoisobutyronitrile (AIBN), usually serve as **initiators**, in particular, for starting polymerization reactions.

These compounds are used to synthesize plastics such as LD-PE (from ethylene), PVC (from vinyl chloride), PS (from styrene), PP (from propylene), and many others. Here, initiator doses of 0.02...0.1% are common.

Olefins, phenols, and aromatic amines are frequently used as **inhibitors.** They can be used, for example, to control chain growth in plastic production. However, gases, such as NO and O_2 are also used as inhibitors.

40.1.2 Unbranched Chain Reaction

A good example of a closed or unbranched chain reaction is the reaction of hydrogen with chlorine to form hydrogen chloride: $H_2 + Cl_2 \rightarrow 2\ HCl$. This reaction, also known as the chlorine detonating gas reaction, is easily triggered by UV light. Two chlorine radicals are formed for each chlorine molecule. These react with hydrogen molecules to form hydrogen chloride and hydrogen radicals. The latter reacts with chlorine molecules to form HCl and new chlorine radicals.

Chain start	$Cl_2 \xrightarrow{\text{UV Light}} Cl\bullet + Cl\bullet$
Chain propagation	$Cl\bullet + H_2 \rightarrow HCl + H\bullet$
	$H\bullet + Cl_2 \rightarrow HCl + Cl\bullet$
Chain termination	$H\bullet + \bullet Cl \rightarrow HCl$
	$H\bullet + \bullet H \rightarrow H_2$
	$Cl\bullet + \bullet Cl \rightarrow Cl_2$

The following figure illustrates that the chlorine oxyhydrogen reaction is a closed chain reaction.

$$Cl\bullet + H_2 \longrightarrow HCl + H\bullet$$
$$H\bullet + Cl_2 \longrightarrow HCl + Cl\bullet$$

40.1.3 Branched-Chain Reaction

In branched chain reactions, at least two new radicals are formed in individual partial reactions, which cause the reaction to branch at this point. The classic example is the combustion of hydrogen after the overall reaction: $2\ H_2 + O_2 \rightarrow 2\ H_2O$. For this simple reaction alone, up to 20 reaction steps can be considered important. Only a few of these will be considered.

Chain start is the thermal homolysis of a hydrogen molecule (R0 in the following overview) at approximately 600 °C. Here, two hydrogen atoms are formed, each of which starts a reaction chain. A hydrogen atom (H•) then reacts with an oxygen molecule to form a hydroxyl radical (•OH) and an oxygen atom (•O•), as shown in

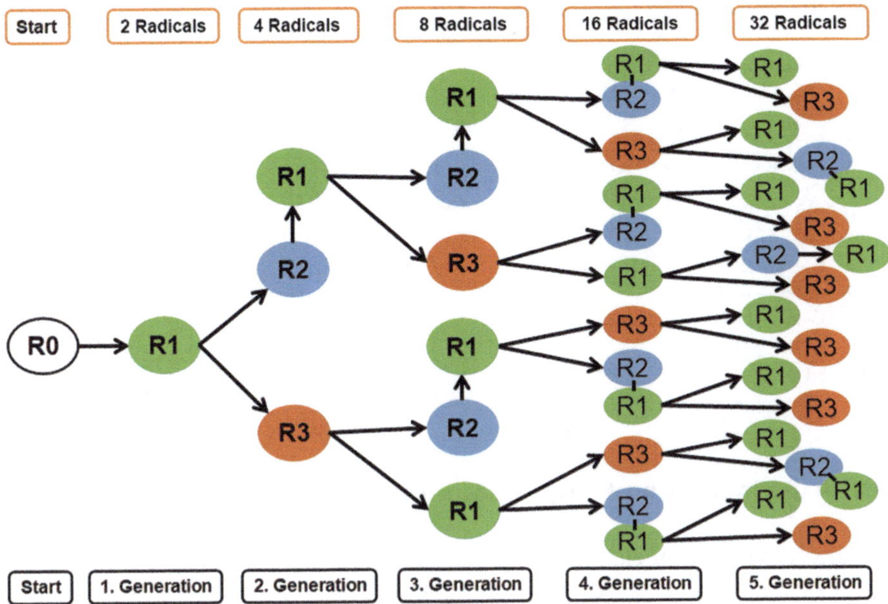

Fig. 40.3 Radical chain reaction of hydrogen combustion

reaction 1 (R1). The hydroxyl radical reacts with a hydrogen molecule to form the final product, water, leaving one hydrogen atom (Reaction 2). The oxygen atom formed in R1 reacts with a hydrogen molecule to form a new hydroxyl radical, leaving one hydrogen atom.

The substep R2 carries the reaction further, while the substeps R1 and R3 each lead to a chain branching. Figure 40.3 shows this situation schematically.

Chain start	$H_2 \xrightarrow{\text{Heat}} H\bullet + H\bullet$	**R0**: 2 radicals
	$H\bullet + O_2 \longrightarrow HO\bullet + \bullet O\bullet$	**R1**: branching (2 radicals)
Chain propagation	$HO\bullet + H_2 \longrightarrow H_2O + H\bullet$	**R2**: propagation (1 radical)
	$\bullet O\bullet + H_2 \longrightarrow HO\bullet + H\bullet$	**R3**: branching (2 radicals)

Other important steps in hydrogen combustion not considered for Fig. 40.3 are:

$$HO\bullet + \bullet O\bullet \longrightarrow H\bullet + O_2 \qquad H\bullet + HO\bullet \longrightarrow \bullet O\bullet + H_2$$
$$H\bullet + H_2O \longrightarrow HO\bullet + H_2 \qquad HO\bullet + HO\bullet \longrightarrow H_2O + \bullet O\bullet$$
$$H_2O + \bullet O\bullet \longrightarrow HO\bullet + HO\bullet \qquad HO\bullet + O_2 \longrightarrow HO_2\bullet$$

Table 40.1 Important reaction steps in burns (selection)

No.	Reaction	No.	Reaction
1	$CH_4 + HO\bullet \longrightarrow CH_3\bullet + H_2O$	13	$H\text{-}CHO + \bullet O\bullet \longrightarrow CHO\bullet + HO\bullet$
2	$CH_4 + H\bullet \longrightarrow H_2O + CH_3\bullet$	14	$CH_3\bullet + CH_3\bullet \longrightarrow C_2H_5\bullet + H\bullet$
3	$CH_3\bullet + O_2 \longrightarrow H\text{-}CHO + \bullet O\bullet$	15	$C_2H_5\bullet + \bullet O\bullet \longrightarrow C_2H_4O\bullet + H\bullet$
4	$H\text{-}CHO + HO\bullet \longrightarrow CHO\bullet + H_2O$	16	$C_2H_6 + \bullet O\bullet \longrightarrow C_2H_5\bullet + HO\bullet$
5	$CHO\bullet + \bullet O\bullet \longrightarrow CO + HO\bullet$	17	$CH_2\bullet + \bullet O\bullet \longrightarrow CO + H\bullet + H\bullet$
6	$CO + HO\bullet \longrightarrow H\bullet + CO_2$	18	$H\bullet + HO\bullet \longrightarrow H_2O$
7	$O_2 + HO\bullet \longrightarrow HO_2\bullet$	19	$CHO\bullet + HO\bullet \longrightarrow CO + H_2O$
8	$CH_3\bullet + CH_2\bullet \longrightarrow C_2H_4 + H\bullet$	20	$CH_3\bullet + CH_3\bullet \longrightarrow C_2H_6$
9	$C_2H_5\bullet \longrightarrow C_2H_4 + H\bullet$	21	$CHO\bullet + H\bullet \longrightarrow CO + H_2$
10	$C_2H_6 + HO\bullet \longrightarrow C_2H_5\bullet + H_2O$	22	$C_2H_5\bullet + CH_3\bullet \longrightarrow C_3H_8$
11	$CH_4 \longrightarrow H\bullet + CH_3\bullet$	23	$C_2H_5\bullet + C_2H_5\bullet \longrightarrow C_4H_{10}$
12	$CH_4 + \bullet O\bullet \longrightarrow HO\bullet + CH_3\bullet$	24	$C_2H_5\bullet + C_2H_5\bullet \longrightarrow C_2H_4 + C_2H_6$

40.1.4 Incineration of Organic Materials

For the combustion of such a simple organic compound as methane (CH_4), about 150 steps are relevant. In order to demonstrate the complexity, some of the important steps are shown as examples in Table 40.1.

Notes to Table 40.1

Reactions 1–10 are chain propagating; reactions 11–17 are chain branching; and reactions 18–24 result in chain terminations. **Reaction No. 6** is particularly noteworthy, representing **the** reaction of CO_2 formation. This also explains why a dry CO/O_2 mixture cannot be made to ignite (compare Sect. 38.6): Pure CO can only burn if a H• radical is present in some form from which an **•OH radical can** be formed. Only then can the reaction according to Eq. 6 take place at all. Without an **•OH radical** from any source, the carbon monoxide is not able to react to CO_2.

40.1.5 Course of Chain Reactions

The overall course of a radical chain reaction depends on the rate of radical formation and the rate of recombination of the radicals.

- If the chain breaks predominate, the reaction slows down and finally comes to a standstill. The **anticatalytic extinguishing effect** is based on the principle of forcing as many chain breaks as possible.
- If the number of radicals consumed by breaks and the number of newly formed radicals is in balance, a closed chain is present. If the resulting heat of the reaction is not dissipated quickly enough, this can lead to a **heat explosion** due to the heat accumulation that occurs.
- If two or more radicals are formed from a partial reaction, the speed of the branching is greater than that of the termination reaction. The very rapid

development of the chain avalanche is the consequence since each newly formed radical is capable of starting a new reaction chain. The avalanche-like increase in the number of chain carriers results in a **chain explosion**, which is usually accompanied by a heat explosion.

Addendum on Combustion Chemistry
The few reactions shown in Sects. 40.1.3 and 40.1.4 represent only a small section, an extremely abbreviated reaction mechanism, so to speak. Nevertheless, it can be seen that several parallel and competing reaction pathways are open to the overall system. In general, it can be said that three reaction types determine the reaction rate in combustion processes:

- The reaction initiates the consumption of the respective combustible substance. Mostly it is the homolytic splitting off of a hydrogen atom (H•) from a fuel molecule.
- The reaction of forming CO_2 from CO and a hydroxyl radical (**•OH**).
- The reactions of the hydrogen-oxygen system (Sect. 40.1.3).

The reactions specific to the fuel are important during ignition and in the initial phase. However, as soon as sufficient radicals are formed, they lose their influence on the reaction rate.

40.2 Fire Course

When it comes to the combustion of complex fuels, as we encounter them in our daily lives, a purely chemical view at the molecular level is not expedient. However, it has been found that almost all fires have the same course. This can be described in a fire progression curve. In addition, the propagation speed of fires and the process of pyrolysis are to be considered.

40.2.1 Fire Progression Curve

The framework conditions of each fire are different. These general conditions include, but are not limited to:

- Fire material (in type, quantity, and extent, . . .)
- Fire location (outdoors, indoors, size of the room, . . .)
- Air supply (no wind or wind, in a closed room, . . .)

The interplay of the above conditions, in turn, affects the fire temperature, the burning rate, the composition of the combustion products, and so on. The exact course must therefore inevitably differ from fire to fire.

► The following applies: no two fires are alike.

Despite a wide variety of conditions, the **course of the fire** is **in itself** uniform enough to be able to draw up a fire progression diagram, a so-called fire progression

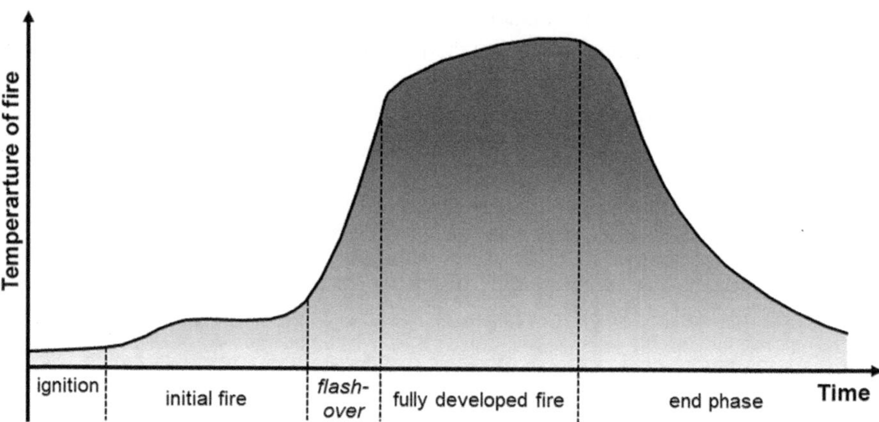

Fig. 40.4 Fire progression curve

curve (Fig. 40.4). The fire progression is divided into several phases, which differ in the fire temperature that occurs:

- **Ignition phase**: Supply of the minimum ignition energy by suitable ignition sources.
- **Initial fire phase**: There is a narrowly localized source of fire, an incipient fire. The fire temperature is relatively low but rises steadily. The environment is heated and pyrolysis of surrounding materials occurs. Smoldering fires are usually assigned to this fire phase.
- *Flash-over*: There is a sudden spread of fire to almost all combustible materials in a room. This phase is not reached in every fire, so it does not always occur.
- **Full fire phase**: When the fire is fully developed, the fire temperature and energy release reach their maximum. The development of smoke gas and heat is now at its highest. The danger of spreading to other parts of the building or the failure of supporting components is now at its greatest.
- **Final phase of the fire**: Temperatures drop, the rate of fire decreases – usually as a result of a lack of fuel.

40.2.2 Rate of Fire Spread

The distance covered by a fire in the course of fire propagation per unit of time is called the rate of fire spread and is given in the unit m min^{-1}. By its nature, it is not uniform and is therefore usually only determined as an approximate value. It depends, among other things, on

- the substance/object itself
- the distribution of flammable materials in the fire compartment (keyword: fire bridges)

- the air supply
- the intensity of heat generation or transmission

The rate of fire spread is:

- for flammable **gases**, high to very high (up to >200 m min^{-1}) and identical with the flame velocities.

Methane:	22.2 m min^{-1}	Ethylene:	46.8 m min^{-1}
Butane:	24.8 m min^{-1}	Acetylene (ethine):	101 m min^{-1}
Propane:	27.0 m min^{-1}	Hydrogen:	216 m min^{-1}

- for flammable **liquids**, strongly dependent on the temperature of the liquid and its flash point (approximately 1...50 m min^{-1})

n-Butanol:	4.8 m min^{-1}	Acetone:	19.0 m min^{-1}
Ethanol:	22.8 m min^{-1}	Toluene:	50.4 m min^{-1}

- for combustible **solids,** low (approximately 0.5...4 m min^{-1})

Residential buildings: 1.0...1.2 m min^{-1}	Wood, board stack, moisture 10%: 4.0 m min^{-1}
Paper (rolls): 0.27 m min^{-1}	Wood, board stacks, moisture content 30%: 1.2 m min^{-1}

40.2.3 Special Phenomena of the Course of Fire

In the course of a fire, especially in room and apartment fires, various phenomena occur. Despite their similarities (hot smoke layer and ignition of this), these phenomena differ from one another, but can also merge into one another. The countermeasures are therefore practically the same and belong to the basic tactics of internal attack: Room cooling and smoke gas cooling (compare Sect. 41.2.1).

40.2.3.1 Flash-Over

The abrupt leaping of the flames is the critical stage of every major (room) fire. Due to the high heat of the smoke layer (approximately 500...600 °C), this radiates a considerable amount of energy onto the combustible materials in the room. The heat radiation hitting the floor is equivalent to 20 times direct solar radiation – or more. Thermal processing, therefore, progresses rapidly, resulting in the almost simultaneous ignition of almost all combustible materials. As a rule, all the combustible material in the room burns after a *flash-over*.

A flash-over is influenced by many factors:

- Fire load (e.g. type, quantity, ignition behavior, calorific value, arrangement)
- Fire compartment (e.g. geometry, thermal properties of walls, ceiling, floor)
- Air supply (e.g. size of door/window openings)

The time until a fire jump occurs can therefore vary greatly. In fire tests, times ranging from less than one minute to over 30 minutes have been determined. On average, one can probably assume 8...12 minutes. The peak temperatures at the ceiling of the fire room after the flash-over amount to 900...1100 °C. If the oxygen supply is limited, flame tongues form at the boundary layer between smoke and air, analogous to *roll-over,* since an ignitable concentration ratio of pyrolysis gases occurs there.

If the fire does not jump after the development phase, the fire usually goes out. If there is a lack of oxygen, flue gases may accumulate and thus cause the danger of a back-draft.

40.2.3.2 Back-Draft

If flammable vapors from thermal processing cannot burn completely due to a lack of oxygen, a back-draft can occur. For example, the opening of a door or the destruction of a window provides the necessary oxygen supply. The temperature of the flue gas layer, open flames, or smoldering areas of the burnt material is possible ignition sources. The mixture, thus, created ignites and converts explosively. This results in a fireball and/or a massive flash fire. Windows and doors can be destroyed by the pressure wave.

After ignition, the flame front and pressure wave do not necessarily have to strike from the supply air opening. The reverse path (ignition at the supply air opening, flame front, and pressure wave strike into the room) is also possible if the pyrolysis gases are far above their ignition temperature. Under certain circumstances, pyrolysis gases may also accumulate in a room adjacent to the original fire room. A suitable ignition source can also trigger an explosion here (so-called "displaced back-draft"). Contrary to many assumptions:

► There is no sure sign of an imminent back-draft.

The following signs **may** indicate an imminent back-draft, without a law being able to be derived from this:

- A fire that has been undetected/undisturbed for a long time.
- Surge of fire smoke from door/window gaps.
- Hot door leaves and/or window panes.
- Air is sucked into the room after opening a door.

40.2.3.3 Roll-Over

The ignition of a hot layer of flue gas with a sufficient concentration of oxygen is called *roll-over,* (rarely *flame-over).* The hot pyrolysis gases – mixed with air –

collect under the ceiling. As soon as sufficient combustible products have accumulated in the flue gas layer, the mixture ignites.

If there is an insufficient supply of oxygen, the concentration of combustible flue gases in the flue gas layer increases above the upper ignition limit. Tongues of flame (dancing angels) then appear at the boundary layer between smoke and air. With an appropriate supply of fresh air (e.g. by opening a door or bursting a window), the jet of flame can even spread into the neighboring room and overtake the emergency services there.

40.2.4 Pyrolysis

Before they catch fire, many substances first undergo discoloration (browning, blackening), which is often accompanied by the formation of gaseous substances (degassing). This process is called **thermal decomposition** or **pyrolysis** (Greek *pyr* = fire, *lysis* = dissolution, separation); in fire brigade jargon, **thermal processing** is sometimes used instead of pyrolysis. Pyrolysis precedes combustion and takes place in every fire, especially in the initial stage and in the peripheral zones.

▶ Pyrolysis is the thermal decomposition of solid, liquid, or gaseous substances at a higher temperature without oxidation.

Depending on the temperature, a wide variety of reactions take place during the decomposition of organic substances, as shown in Table 40.2.

When the decomposition temperature of the combustibles is exceeded, a process known as **smoldering** begins, which generally takes place in the temperature range between 200 and 600 °C. Smoldering is generally characterized by the formation of smoke. This is often followed by **gleaming**. In this case, the solid material burns without flames appearing. In contrast to smoldering, light is emitted from the reaction zone during gleaming. With a sufficient supply of oxygen, a smoldering/gleaming fire can turn into an open fire, i.e. a fire with the appearance of flames.

Pyrolysis releases a wide range of different substances. For example, the carbonization of wood produces water (H_2O), methanol (CH_3OH), methane (CH_4), acetic

Table 40.2 Temperature-dependent decomposition processes of organic materials

Temperature	Operations
~100 °C	Drying, moisture (H_2O) escapes
200...300 °C	Elimination of H_2O, H_2S, CO_2, SO_2, and HCN from molecules and elimination of monomers from plastics
~350 °C	Separation of CH_4, C_2H_4, and homologous compounds
~500 °C	Crack processes: Formation of short-chain hydrocarbons
>600 °C	Crack processes: Formation of alkenes, benzenes, and polycyclic aromatic hydrocarbons

acid (CH_3COOH), formic acid ($HCOOH$), acetone ($(CH_3)_2CO$), formaldehyde ($HCHO$), carbon dioxide (CO_2), carbon monoxide (CO) and other substances. From the plastic polyamide, approximately 25 (main) smoldering products are obtained and during the pyrolysis of used tires, even more than 70 vaporizable substances are produced, which together account for more than 50% of the decomposition products.

40.3 Flue Gases

The terms flue gas, combustion gas (fire gas), and fire smoke are often interwoven, are sometimes used synonymously, and are not always clearly defined. Following the German Standard DIN 14011, which defines combustion gases (fire gases) as the gaseous components of the combustion products and smoke as the visible part of the gases and aerosols produced during combustion, we will first differentiate between the terms smoke and flue gas:

- **Smoke** is an aerosol of gases with very fine, solid, or liquid components.

Smoke is formed by thermal and/or chemical processes. An example of a purely chemical process is the formation of ammonium chloride smoke from the gas phase reaction of ammonia with hydrogen chloride: $NH_3 + HCl \rightarrow NH_4Cl$.

- **Flue gas** is a mixture (aerosol) of gaseous combustion products (fire gas), to which solid and liquid components (fire smoke) are admixed. It is produced as waste gas during the combustion or pyrolysis of fuels.

Flue gases, therefore, differ from "pure smoke" in the possible presence of liquid droplets and their direct formation during combustion processes. As exhaust gases from combustion, the liquid components are still in a heated, and thus gaseous, state.

A simple overview is given in Fig. 40.5.

Colour of the flue gas

The color of smoke gas for assessing the situation inside a fire object is usually overestimated. The influences that affect the color of the smoke are too varied: Fire stage, fire material, room volume, air inflow, and many more. Only basic, but **by no means** binding, statements can be made:

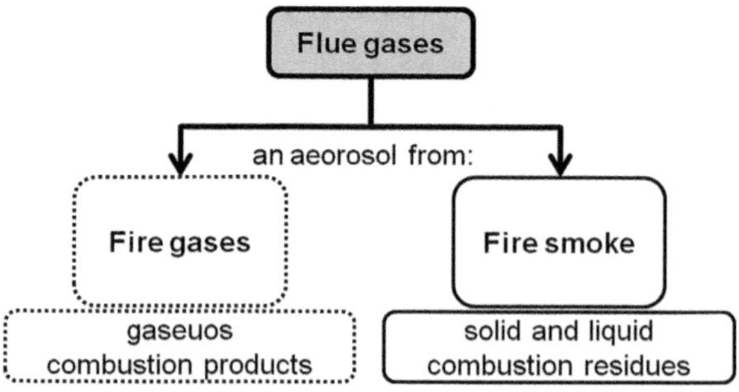

Fig. 40.5 Smoke gas: a mixture of fire gases and fire smoke

- White smoke indicates a high percentage of water vapor.
- Fresh pyrolysis gases often cause a yellowish to greenish color of the smoke.
- Black smoke indicates unburned particles, especially soot.
- Dark smoke is usually warmer than light smoke.

Smoke density

To assess the visual obstruction caused by flue gases, the smoke density is sometimes referred to. This refers to the quantity of solid and liquid particles in grams per cubic meter of smoke gas. An indication of the smoke density is only of importance as a comparative value between different objects that have caught fire.

Smoke spread

Smoke gases can easily spread in an uncontrolled manner and thus cause considerable damage in areas that are not directly affected by the fire. In addition, they may block the structural escape route stairwell, with corresponding consequences for persons still in the building. The prevention of an uncontrolled spread of smoke within the scope of defensive fire protection is therefore one of the tasks of the fire brigades. Suitable means are the appropriate use of ventilation equipment and mobile smoke barriers, but also the tactically correct procedure (see Lit. [7-6]).

Flue gas quantities

Incomplete combustion of organic products produces soot particles which cause the flame to glow. They are also responsible for the more or less strong smoke development when these substances burn in the air. Alkanes from hexane on burn with a sooty flame. The smoke development increases with increasing chain length. Branched and cyclic alkanes produce more smoke than straight-chain alkanes. Alkenes burn with more smoke, alkines with much more smoke than alkanes. Aromatic hydrocarbons also produce a lot of smoke. Oxygen atoms in the molecule, such as in alcohols or ketones, reduce smoke formation. Plastics such as polyethylene, polypropylene, polystyrene, or polyurethanes burn with the strong smoke formation and a strong sooty flame.

The experience of many fires shows that sometimes immense quantities of flue gas that are produced even before the transition to the full fire phase can make escape and rescue routes almost unusable. In order to dilute 1 m^3 of pure smoke gas so that visibility of 10 m remains in escape and rescue routes, approximately 1500 m^3 of clean air is required. If the visibility of 2.5 m is taken as a basis – which is most relevant for one's own home – the quantities of smoke gas shown in Fig. 40.6 are obtained.

Flue gases and respiratory protection

Flie gases continue to escape from so-called "cold fire grounds" for some time. The pollutants that continue to be released here (see Sect. 40.3.1) make it necessary in most cases to use respiratory protection during post-extinguishing work.

40.3.1 Fire Gases

▶ Fire gases are the gaseous components of the combustion products of a fire that are contained in the flue gas.

These gaseous components themselves are a highly complex mixture. They consist mainly of inert components of the ambient air (nitrogen, inert gases), unused oxygen, carbon dioxide, carbon monoxide, and water steam.

In addition, other gases and vapors are formed as a result of the fire event. The most important of these substances are:

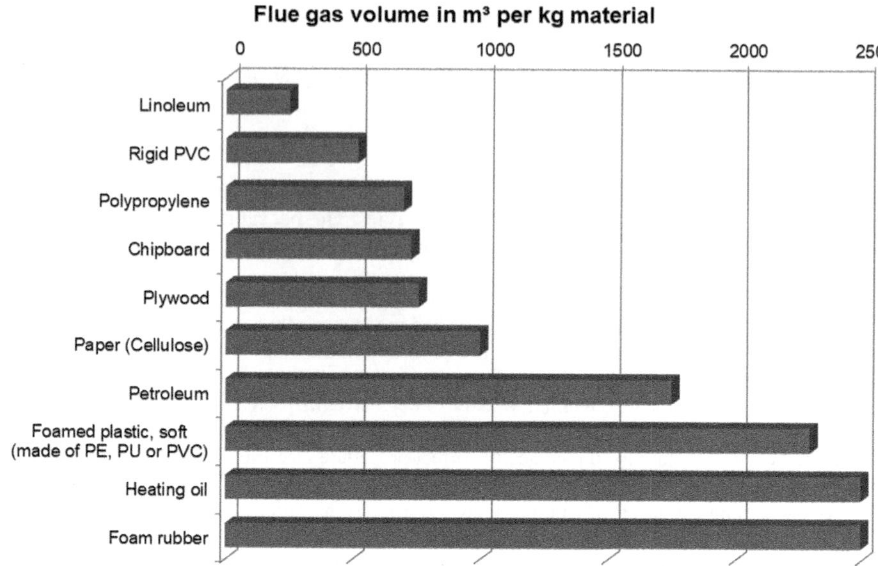

Fig. 40.6 Smoke development of various substances

- Short-chain alkanes and alkenes, e. g. methane (CH_4), ethane (C_2H_6), ethene (C_2H_4).
- Short-chain aldehydes, e.g. formaldehyde ($H{-}CHO$), acetaldehyde ($H_3C{-}CHO$).
- Short-chain organic acids, e.g. formic acid ($H{-}COOH$), acetic acid ($H_3C{-}COOH$).
- Short-chain alcohols, e.g. methanol ($H_3C{-}OH$), ethanol ($C_2H_5{-}OH$).
- For fire products containing organically bound nitrogen: Hydrogen cyanide (HCN), ammonia (NH_3) and nitrous gases (NO_x), amines (e.g. methylamine, $H_3C{-}NH_2$), and nitriles (e.g. methyl nitrile, $H_3C{-}CN$).
- For fire material containing organically bound sulfur: sulfur dioxide (SO_2), sulfur trioxide (SO_3), and hydrogen sulfide (H_2S).
- During combustion of organic substances containing chlorine: hydrogen chloride (HCl) and phosgene ($Cl{-}CO{-}Cl$).

In addition, there may be evaporated fuel residues or their pyrolysis products. Most of these substances are only contained in the ppm or per mille range in the fire gas, but may also increase to the percentage range.

Since the type and quantity of the fire gases formed depends on the factors of the respective fire, no statement can be made about the proportions of the individual substances. An *exact* statement on the composition can only be obtained through complex analyses of the actual fire gas. However, it can mainly be assumed that larger quantities of CO and CO_2 are involved, which are also responsible for the majority of fatalities.

▶ Fire gases are the cause of 80 % of fire deaths. Only 20 % of fire deaths are caused by fire.

40.3.1.1 Fire Gas Volume V_{FG}

The volume of combustion gas that results from the complete combustion of 1 kg of substance can be determined from the empirical formula and the oxygen coefficient x *of* the combustion equation (see Sect. 38.3.9). For the combustion equation

$$C_aH_bO_cS_g + x\ O_2 \rightarrow a\ CO_2 + \frac{b}{2}H_2O + g\ SO_2$$

valid

$$V^0_{FG} = \left(a + \frac{b}{2} + g - x + \frac{x}{0.21}\right) \cdot \frac{22.4}{M}$$

or written with the element symbols:

$$V^0_{FG} = \left(C + \frac{H}{2} + S - x\,O_2 + \frac{x\,O_2}{0.21}\right) \cdot \frac{22.4}{M}$$

This gives the theoretical fire gas volume V^0_{FG} in m^3 per kg of substance under standard conditions.

For the conversion to a real event, i.e. to an actual temperature at a given pressure, the following formula is used:

$$V_{FG} = \frac{3.71 \cdot V^0_{FG} \cdot T}{p}$$

Here, the temperature in Kelvin (K) and the pressure in hPa (= mbar) are to be used. If you want to use the pressure in the unit bar, the factor changes from 3.71 to 0.003 71.

Table 40.3 shows an example of how large the temperature influence is on the fire gas volume.

40.3.2 Fire Smoke

In addition to gases and vapors, solid combustion products are also produced, which are carried along as smoke by the fire gases. In addition, the vapors contained in the fire gases condense into tiny liquid droplets as they cool down. Both together form the fire smoke.

Table 40.3 Specific fire gas volumes of plastics at different fire gas temperatures

Material	Specific fire gas volume at different temperatures ($m^3\ kg^{-1}$)			
	20 °C	300 °C	600 °C	900 °C
Polyethylene (PE)	13.12	25.66	39.09	52.53
Polystyrene (PS)	11.46	22.41	34.15	45.88
Polyvinyl chloride (PVC)	5.15	10.08	15.35	20.64

▶ Fire smoke is the solid and liquid components contained in the flue gas. They are produced as a result of a fire or combustion.

Fire smoke consists mainly of fly ash, carbon in the form of soot and charcoal particles, and the substances adsorbed to these substances. The particle sizes of the constituents are 10^{-5} to 10^{-9} m (= 10 µm to 1 nm) and are thus partly respirable. By agglomeration, these smoke particles grow into larger structures; soot flakes are formed.

Metal oxides and carbonates form a significant proportion of the ash. Phosphorus pentoxide (P_2O_5) in fires of phosphorus-containing substances (e.g. certain pesticides) and other solid non-metal oxides are of secondary importance. Soot forms at temperatures between approximately 750 °C and 1700 °C from radicals such as C_3H_3•. Materials that contain a lot of carbon in the molecule in relation to hydrogen and oxygen, such as benzene, acetylene, and many plastics, burn in air with a strong formation of soot.

An important precursor in the formation of soot is the policyclic aromatic hydrocarbons (PAHs), which are formed in particular during rich combustion, i.e. with a high fuel or low oxygen content. Many of the PAHs, which also occur as pyrolysis products, are considered to be carcinogenic to humans. The lead substances are the toxic compounds benzo[a]anthracene, benzo[a]pyrene, and dibenzo[a,h]anthracene, which can be absorbed both via the respiratory tract and via the skin (see Fig. 40.7).

Polychlorinated dioxins and furans (especially dibenzofurans) are formed during the incineration of substances containing chlorine. The lead substance of these compounds is 2,3,7,8-tetrachloro-dibenzo-para-dioxin (TCDD, "Seveso dioxin"), one of the most toxic compounds known (see Fig. 40.7).

▶ **Caution:** PAHs, dioxins, and dibenzofurans are adsorbed on the soot particles and turn the soot from a fire into a toxic mixture. Clothing and equipment contaminated with soot must therefore be properly packaged and sent for cleaning/decontamination.

40.4 Energy Turnover During Fires

The amount of energy released in a fire event depends on many factors, such as the chemical energy stored in the burning materials, the influx of oxygen, the speed at which a material burns, the time that the fire has already lasted, and other influences.

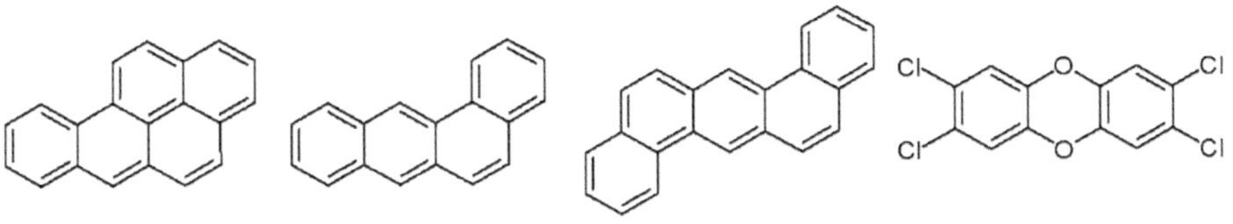

Benzo[a]pyren | Benzo[a]anthracen | Dibenzo[a,h]anthracen | 2,3,7,8-Tetrachlor-dibenzo-para-dioxin

Fig. 40.7 Polycyclic aromatic hydrocarbons and dioxin (lead substances)

If the basic variables are known, the energy that is freely available in the fire room can be estimated. From this, in turn, the amount of extinguishing agent required for successful fire fighting can be estimated.

40.4.1 Minimum Combustion Temperature

The minimum combustion temperature is the minimum temperature that must be reached for a substance to continue to burn automatically. It is lower than the combustion temperature, i.e. the temperature usually reached when a particular substance burns. Similarly, it is higher than the ignition temperature of the substance in question. This is due to the fact that when the ignition temperature is reached, there is not yet enough energy available to enable independent burning. Only when the minimum combustion temperature is reached is a further supply of energy from the outside no longer necessary.

To extinguish a fire using the "cooling" method (see Sect. 41.1.1), it is not necessary to cool the fire to temperatures below the ignition temperature. It is sufficient if the temperature falls below the minimum combustion temperature.

As can be seen from Table 40.4, the minimum combustion temperatures of many substances are generally several 100 °C above the respective ignition temperatures and the combustion temperatures (in the air) are even significantly higher.

40.4.2 Caloric Value (H_s) and Heating Value (H_i)

The calorific value H_s – obsolete: gross heat of combustion H_o – is a direct measure of the chemical energy contained in a substance. It can be calculated directly via the reaction equation of the combustion as standard enthalpy of combustion $\Delta_V H^0$. However, this is only possible without error for chemically well-defined compounds or mixtures of these with a known composition. For complex combustibles, such as wood, paper, or plastic mixtures, a calculation is only possible to a very limited extent. In technology, the calorific value is therefore determined experimentally.

The heating value H_i – obsolete net heat of combustion H_u – differs from the calorific value in that the water is still present in a gaseous state after combustion, i.e. as water vapor. The calorific value is therefore always higher than the heating

Table 40.4 Comparison of ignition, combustion, and minimum combustion temperatures (examples)

Substance	Ignition Temperature	Minimum Combustion Temperature	Combustion Temperature
Propane	460 °C	1200 °C	1925 °C
Petrol	260 °C	950 °C	2320 °C
Acetylene (ethyne)	305 °C	950 °C	2250 °C
Hydrogen	560 °C	630 °C	2130 °C
Carbon disulfide	95 °C	345 °C	2200 °C

value, as the energy released by the condensation of the water is not taken into account. The only exception is the combustion of CO to CO_2, in which no water is produced. For example, values, see Table A.23 and Table A.24.

Only the calorific value is relevant for the calculation of fire loads. The conversion from the calorific value is carried out according to

$$H_i = H_s - r_{H_2O} \cdot w_{H_2O}$$

where w_{H_2O} are the mass fraction of water and r_{H_2O} the specific heat of vaporization of water with 2.256 MJ kg^{-1}.If the caloric or heating value is related to a mass (1 kg) or a volume (1 m^3), this is also referred to as the specific caloric or heating value. The value is then given in the units MJ kg^{-1}, MJ m$^{-3,}$ or kWh kg^{-1}. The following conversion applies: 1 kWh = 3.6 MJ.

However, heating values cannot only be given as specific heating values. In some cases, heating values per unit are also listed in corresponding tables. Examples can be found in Table A.25.

40.4.3 Burning Velocity and Burning Rate

The **burning velocity** is the speed at which a substance burns perpendicular to its surface. This means that only the amount by which the height of the combustible material decreases is considered, so to speak. Consequently, the burning rate is usually given in the unit mm min^{-1}.

The **burning rate**, on the other hand, refers to the mass or volume of combustible material that burns per unit of time. It is therefore expressed, for example, in g s^{-1}; kg h^{-1}; L min^{-1}; m^3 h^{-1}. Consequently, the **burn-up** is the burnt mass of the fire load after a certain fire duration.

To be distinguished from this is the **specific burning rate** or **mass burning rate** (v_{ab}), which relates the burning mass or burning volume of the combustible substance per unit of time to the unit area of the fuel surface. Common units are kg m^{-2} s^{-1} or kg m^{-2} h^{-1} or m^3 m^{-2} h^{-1}.

Burning velocity and burning rate are dependent on, among other things:

- the substance/object itself,
- the degree of fragmentation (= ratio of surface area to mass or surface area to volume): The finer the fragmentation, the faster the combustion,
- the oxygen content of the ambient air: The more O_2, the faster the combustion.

40.4.4 Fire Load Density (*q*)

The amount of heat released in a fire per square meter of the floor area of the room is called the fire load density; the term fire load is also used synonymously. The calculation is based on the mass or number of combustible materials multiplied by the heating value and related to the room area. The unit is J m^{-2} or kWh m^{-2}.

$$q = \frac{\Sigma(m \cdot H_{\mathrm{i}})}{A}$$

40.4.5 Fire Load Density (I_{fi})

The fire intensity (I_{fi}) indicates the maximum amount of heat that can be released per m^2 of room area. It is determined from the specific burning rate (v_{ab}) and the heating value (H_{i}). If different substances are involved, the corresponding sum ($\sum$) must be formed.

$$I_{\mathrm{fi}} = \sum (\nu_{\mathrm{ab}} \cdot H_{\mathrm{i}})$$

In fire protection engineering, the fire intensity is usually called **area-related fire load density** (formula symbol $\dot{q}$).

40.4.6 Heat Release Rate (Q)

The *heat release rate (HRR)* indicates the total amount of heat released during the fire event under consideration. It can be determined from the fire load density and the fire area (A_{fi}).

$$\dot{Q} = I_{\mathrm{fi}} \cdot A_{\mathrm{fi}} \qquad \text{or} \qquad \dot{Q} = H_{\mathrm{i}} \cdot \nu_{\mathrm{ab}} \cdot A_{\mathrm{fi}}$$

The heat release rate in kW can also be estimated from the flame height. If the flame height h_{Fl} (measured in m) is not greater than 6 times the fire area ($h_{\mathrm{Fl}} < 6\, A_{\mathrm{fi}}$), the following applies:

$$\dot{Q} = (4.35 \cdot h_{\mathrm{Fl}})^{2.5}$$

The limit condition corresponds to a maximum flame height of 6 m per m^2 of fire area. With 10 m^2 this is already 60 m, with 25 m^2 an impressive 150 m flame height. The ratio should be undercut in almost all natural fires and the formula can therefore be used.

40.4.7 Extinguishing Water Demand

For a rough calculation of the extinguishing water demand, a full fire is assumed. In this "worst-case" scenario, the maximum he at release rate is used as a basis, i.e. it is assumed that the entire existing fire load is fully implemented. Naturally, this results in very large quantities of water, which are rarely required in practice. All specific features (type of storage, fire alarm systems, sprinklers, extinguishing water additives, etc.) are not taken into account here.

The calculated quantity of extinguishing water is significantly reduced if the fire load is not completely converted into thermal energy. This requires that the fire is detected at an early stage and that an effective extinguishing attack is carried out quickly. Similarly, the supply of extinguishing water must be ensured and the extinguishing agent must be applied optimally. Tactical errors or an insufficient water supply result in a higher extinguishing agent requirement.

An average value of 50% is assumed as the extinguishing efficiency. This can only be achieved under reasonably optimal conditions and correct application of the extinguishing water. "Errors" in the jet pipe routing reduce the extinguishing effectiveness and thus in turn lead to a higher extinguishing agent requirement.

To calculate the extinguishing water requirement (*EWR*), a heat binding capacity of 2.6 MJ L^{-1} is assumed. Dividing the heat release rate by this factor gives the quantity of extinguishing water required to absorb the total heat energy. To take into account the extinguishing efficiency, it is necessary to divide again by 50% = 0.5.

$$EWR_{50} = \frac{\dot{Q}}{2.6\ \mathrm{MJ/L} \cdot 0.5} = \frac{\dot{Q}}{1.3\ \mathrm{MJ/L}} = \frac{\dot{Q}}{2.6\ \mathrm{MJ/L}} \cdot 2$$

This results in the amount of extinguishing water in L s^{-1} that must be applied to dissipate the heat released and thus extinguish the fire. Multiplying by 60 gives the extinguishing water requirement in liters per minute.

40.4.8 Example Calculations

Example 1

In a large warehouse, a stack of wood with an extension of 8 m × 5 m is burning. The specific burning rate v_{ab} (wood) = 0.8 kg m^{-2} min^{-1} and the calorific value H_i (wood) = 14.5 MJ kg^{-1}. Calculate the firepower.

a. Conversion of the specific burning rate

$$v_{ab} = 0.8 \frac{\text{kg}}{\text{m}^2 \cdot \text{min}} \cdot 60 \frac{\text{min}}{\text{h}} = 48 \frac{\text{kg}}{\text{m}^2 \cdot \text{h}}$$

b. Conversion of the calorific value

$$H_i = 14.5 \ \frac{\text{MJ}}{\text{kg}} = \frac{14.5 \cdot 10^6 \text{ W s}}{\text{kg}} \cdot \frac{1 \text{ h}}{3\,600 \text{ s}} = 4.03 \ \frac{\text{kWh}}{\text{kg}}$$

c. Calculation of the fire load density

$$I_{fi} = v_{ab} \cdot H_i = 48 \frac{\text{kg}}{\text{m}^2\text{h}} \cdot 4.03 \frac{\text{kWh}}{\text{kg}} = 193.44 \frac{\text{kW}}{\text{m}^2}$$

d. Calculation of the heat release rate

$$\dot{Q} = I_{fi} \cdot H_i = 193.44 \frac{\text{kW}}{\text{m}^2} \cdot 40\text{m}^2 = 7737.6 \text{ kW} = 7.74 \text{ MW}$$

e. Solution
   The rate of heat release is $\approx$ 8 MW

Example 2
In a living room (6 m × 4 m) there is a coffee table (190 kWh), a sofa (three-seater, 500 kWh), two armchairs (175 kWh each), a dining table (250 kWh), four chairs (15 kWh each), a sideboard (350 kWh), a television (80 kWh) and a bookshelf (2.5 m × 2.0 m × 0.3 m = 1.5 m^3; about 600 kg total, 4.2 kWh kg^{-1}). The averaged specific burn rate was 0.6 kg m^{-2} min^{-1}. The estimated total mass was 900 kg.
Calculate the amount of extinguishing water at an efficiency of 50% or 30%.

a. Sum of the heating values
   The sum of the heating values of all items is 4300 kWh.
b. Calculation of the specific heating value

$$H_{\text{i}} = \frac{4300 \text{ kWh}}{900 \text{ kg}} = 4.78 \frac{\text{kWh}}{\text{kg}}$$

c. Calculation of the fire load density

$$I_{fi} = v_{ab} \cdot H_i = 36 \frac{\text{kg}}{\text{m}^2\ \text{h}} \cdot 4.78 \frac{\text{kWh}}{\text{kg}} = 172.1 \frac{\text{kW}}{\text{m}^2}$$

d. Calculation of the heat release rate

$$\dot{Q} = I_{\text{fi}} \cdot H_i = 172.1 \frac{\text{kW}}{\text{m}^2} \cdot 24\text{m}^2 = 4129.9\ \text{kW} = 4.13\ \text{MW}$$

e. Determination of the theoretical water flow rate for heat dissipation

$$EWR_{\text{th}} = \frac{\dot{Q}}{2,6\ \text{MJ/L}} = \frac{4.13\ \text{MW}}{2.6\ \text{MJ/L}} = \frac{4.13\ \text{MJ/s}}{2.6\ \text{MJ/L}} = 1.59 \frac{\text{L}}{\text{s}} = 95 \frac{\text{L}}{\text{min}}$$

f. Estimation of the practical water flow rate
For a 50 % extinguishing efficiency ("good branch guidance"), this results in an extinguishing water requirement of:

$$EWR_{50} = EWR_{\text{th}} \cdot 2 = 95\ \text{L}\ \text{min}^{-1} \cdot 2 = 190\ \text{L}\ \text{min}^{-1}$$

If the efficiency is only 30 % ("poor branch routing"), the extinguishing water requirement increases to:

$$EWR_{30} = EWR_{\text{th}} \cdot 3.33 = 95\ \text{L}\ \text{min}^{-1} \cdot 3.33 = 317\ \text{L}\ \text{min}^{-1}$$

g. Total water demand
Assuming that the extinguishing attack is finished after 30 minutes (maximum time for one squadron (team-of-two) under self-contained breathing apparatus), the total extinguishing water requirement for EWR_{50} can be calculated to 5700 L (5.7 m^3) and for EWR_{30} to 9510 L (9.5 m^3).

Example 3
A pile of rubbish is burning on an area of approx. 4 m × 5 m. On arrival at the scene of the fire, the flames are approximately 5.5...6 m high. Due to the conditions at the scene of the fire (difficult terrain, the danger of falling and stumbling, various types of fire debris, etc.), a very poor extinguishing efficiency of only 25 % must be expected.
Determine the type and number of branch pipes to be used.

a. Calculation of the heat release rate based on flame height

$$\dot{Q} = (4.35 \cdot h_{\mathrm{fl}})^{2.5} = (4.35 \cdot 6)^{2.5} = 3482.2\ \mathrm{kW} \approx 3.5\ \mathrm{MW}$$

b. Determination of the theoretical water flow rate for heat dissipation

$$EWR_{\mathrm{th}} = \frac{\dot{\mathrm{Q}}}{2,6\ \mathrm{MJ/L}} = \frac{3.48\ \mathrm{MW}}{2.6\ \mathrm{MJ/L}} = \frac{3.48\ \mathrm{MJ/s}}{2.6\ \mathrm{MJ/L}} = 1.34\ \frac{\mathrm{L}}{\mathrm{s}} = 80\ \frac{\mathrm{L}}{\mathrm{min}}$$

c. Estimation of the practical water flow rate

$$EWR_{25} = EWR_{\mathrm{th}} \cdot 4 = 80\ \mathrm{L\,min}^{-1} \cdot 4 = 320\ \mathrm{L\,min}^{-1}$$

d. Branch pipes to be used
The use of three branch pipes type C with a water flow of 100 L min^{-1} each is just as suitable as the use of one branch pipe type B with 400 L min^{-1}. Due to the higher flexibility, the overall lower water consumption throughout the operation and the lower probability of accidents, the use of the C-types is to be preferred.
[In reality, the conditions will probably be better than those on which the example calculation is based. It may be assumed that already 2 type C branch pipes are sufficient to extinguish the fire.]

Notes

I. The theoretical extinguishing water demand of example 2 correlates well with the rule of thumb for the **minimum extinguishing water demand** over the area of the fire room:

$$EWR_{\mathrm{th}} = A_{\mathrm{fi}} \cdot 4$$

II. The **efficiency** of the use of extinguishing water is assessed differently – depending on the author. The values range between 30 % and 50 %.
III. The assumed **energy absorption** of water varies – depending on the author – between 2.6 MJ L^{-1} (= heating from 18 °C to 100 °C and evaporation) and 3.6 MJ L^{-1} (= heating from 10 °C to 100 °C, evaporation, and heating of the steam to 600 °C). The example calculations here assume the worst case (steam does not heat above 100 °C).
IV. The example calculations are highly simplified assumptions and are intended only to illustrate the energy conversion. Correct calculations should be made by appropriately trained fire protection engineers.
V. Literature: [8-7], [12-5], [12-6], [12-7], [12-16]

Personnel Deployment Per Fire Area

The maximum areas that can be controlled in a full fire are (rule of thumb):

- Up to 4 m^2 per person (corresponding to 8 m^2 per squad)
- Up to 40 m^2 per firefighting group
- Up to 400 m^2 per fire brigade

41 Extinguishing

The German fire brigades extinguish around 200 000 fires every year. Water is used as the extinguishing agent in well over 90 % of these fires. The term "extinguishing" is defined according to the German DIN 14011 "Terms used in fire fighting":

- Extinguishing is the interruption of combustion.

41.1 Extinguishing Methods

Different extinguishing effects, also called extinguishing effects, occur with the different extinguishing agents. No single extinguishing effect occurs with any extinguishing agent. There is always a main extinguishing effect which is accompanied by secondary extinguishing effects.

In principle:

- Ember fires are cooled, flame fires are smothered.

DIN 14011 defines the extinguishing procedures as follows:

- **Cooling** is an extinguishing process in which the heat required to maintain combustion is removed from the burning substances by the extinguishing agent or by other measures.

- **Suffocation** is an extinguishing method in which combustion is prevented by changing the quantitative ratio between combustible material and oxygen.

The extinguishing method **suffocation** is divided into three subcategories:

- **Dilute**:
  Extinguishing effect by lowering the oxygen concentration.

T. Schmiermund, *The Chemistry Knowledge for Firefighters*,
https://doi.org/10.1007/978-3-662-64423-2_41

- **Reducing**:
  Extinguishing effect by throttling the further supply of combustible substances.

- **Separation:**
  Extinguishing effect by complete separation of oxygen and combustible material.

Reaction-inhibiting extinguishing effect.

- In the case of the reaction-inhibiting extinguishing effect, the catalyst of the combustion is acted upon in the combustion zone in such a way that the catalyst is no longer available for combustion.

41.1.1 Extinguishing by Cooling

Cooling is the main extinguishing effect of the extinguishing agent water. The extinguishing agent absorbs energy and heats up as a result. In the process, so much heat is extracted from the fire material that it is cooled below the ignition temperature. Further burning is then no longer possible. The higher the specific heat capacity (see Sect. 3.4.1) or the enthalpy of vaporisation (see Sect. 3.4.3), the more heat energy is extracted from the burning object and the less extinguishing agent is required, at least in theory. The extinguishing effect of cooling is also shown by extinguishing foam, whereby the effect is most pronounced with low expansion foam due to the high water content.

Containers with flammable liquids can be cooled with water (spray) or low expansion foam from the outside to below the flash point of the liquid. This method is clearly considered to be skimming, since it does not cool the reaction zone but primarily influences the ratio of flammable substance to oxygen (see Sect. 41.1.4).

Less conventional, but based on the same mechanism, is the addition of cold fuel or other heat-removing materials. For example, burning fat or oil in a pan or deep fryer can be cooled below the flash point by adding more oil or frying fat – provided there is little (!) burning material in the container.

Strictly speaking, in the event of a fire, there is no need to cool down to a temperature below the ignition temperature or even below the flash point. It is sufficient to reach a temperature below the minimum combustion temperature (see Sect. 40.4.1) to prevent further combustion. Pyrolysis is already restricted to such an extent that further combustion can no longer take place.

41.1.2 Extinguishing by Suffocation: Separation

Separating combustible material and (air) oxygen is one of the simplest measures of all. By shutting off the further oxygen supply, the fire inevitably comes to a

Table 41.1 Combustion temperatures of metals (examples)

Metal	Combustion Temperature
Aluminium	~ 2200 °C
Magnesium	~ 2400 °C – 2500 °C
Zirconium	~ 4600 °C

standstill. Incipient fires can often be effectively fought by using a fire blanket. It is important not to want to check the extinguishing success too early, as there is a risk of re-ignition.

Instead of a special extinguishing blanket, other tightly woven and not easily combustible fabrics can also be used. This means that fires involving people or clothing can also be extinguished by covering them with a jacket, for example.

In the case of fires involving flammable liquids, especially fat fires, it is better to use the appropriate lid, e.g. of the pan, instead of a fire blanket. The fabric of a blanket can burn through due to the high temperatures. In addition, grease can be absorbed as liquid or vapour, come back into contact with oxygen and re-ignite (wick effect, see Sects. 13.1.5 and 38.5.2).

With the extinguishing agents foam and glowing fire powder (ABC extinguishing powder) the separating effect occurs. Due to its density (see Sect. 3.4.6), foam is particularly suitable for fighting fires involving flammable liquids. It forms a layer on the surface of the burning liquid, prevents further oxygen ingress and at the same time prevents further evaporation of the liquid (so-called **cover effect**). Glowing fire powder has a relatively low melting point and thus forms a protective layer on the burning material, separating it from atmospheric oxygen. In both cases, an additional **insulating effect** occurs due to the low thermal conductivity of the layer.

Metal fires can practically only be extinguished by cutting. In addition to metal fire powder (PM, D extinguishing powder), dry sand, cement and other special extinguishing agents can be used (see Sect. 41.2.7). Other extinguishing agents, such as water or CO_2, dissociate due to the high combustion temperatures (see Table 41.1 and Sect. 3.4.6).

Removing objects that have already caught fire from the surrounding area is also a form of separation in a broader sense. Simply "throwing" the Advent wreath that caught fire out of the window prevents the fire from spreading to furniture and curtains. Outdoors, the burning material can usually be easily extinguished. Always provided that the burning material hits an area that is free of combustible material and people.

In a broader sense, the closing of shut-off valves (taps, valves, slide valves) is also a form of disconnection. In this way, the supply of fuel is cut off and the fire goes out. In many firing systems, the "emergency stop button" also cuts off the fuel supply.

41.1.3 Extinguishing by Suffocation: Dilute

Dilution refers to the objective of primarily reducing the oxygen content. This is usually done with inert gases, such as carbon dioxide, nitrogen, argon or inergen, and often in the context of stationary extinguishing systems. Water vapour, which is produced when water is used as an extinguishing agent, also has a diluting effect, albeit a small one. When diluting, the oxygen concentration of the ambient air (21% by volume) must be reduced to the extinguishing concentration of $\leq$ 15 % by volume, i.e. below the minimum oxygen concentration. To achieve this, the minimum extinguishing concentration must be around 29 vol.-%. Large quantities of the extinguishing agent are required to ensure this safely throughout the entire range of combustion. In addition to oxygen, combustible gases and vapours are also diluted (**displacing effect**).

41.1.4 Delete by Suffocation: Reducing

Reducing the amount of available combustible substance is shortly called reducing. In the case of flammable liquids, cooling below the flash or burning point can reduce the rate of evaporation to such an extent that the fire goes out for lack of fuel. Adding water to water-miscible flammable liquids raises the flash point and extinguishes the fire (see Sect. 3.3.3).

The blowing out of a candle flame is also a reducing. The reaction zone, i.e. the flame, is pushed away from the fuel and thus no longer receives a supply. The "blasting out" of burning oil and natural gas wells is to be considered analogously. With these blasting methods – depending on the explosive – in some cases the oxygen content is also reduced locally and thus the extinguishing effect is further enhanced by the dilution effect.

41.1.5 Reaction-Inhibiting Extinguishing Effect

To have a reaction-inhibiting effect, the extinguishing agent in question must intervene directly in the chemical process of combustion. In this case, the fire catalyst of the combustion reaction is rendered ineffective. This is therefore also referred to as the anticatalytic extinguishing effect or the inhibitory extinguishing effect. Depending on whether a gaseous or a liquid/solid extinguishing agent is used, a distinction is made between homogeneous inhibition and heterogeneous (= "inhomogeneous") inhibition.

41.1.5.1 Homogeneous Inhibition

In homogeneous inhibition, an extinguishing gas acts on the flames. This form of anticatalytic extinguishing effect occurs preferably with halogenated hydrocarbons, such as halons.

Put simply, the extinguishing gas molecules catch the reactive particles of the combustion reaction. The extinguishing gas molecules represent – in the imagination – multi-handed structures which catch and hold the reactive particles like balls and thus remove them from the combustion (Fig. 41.1).

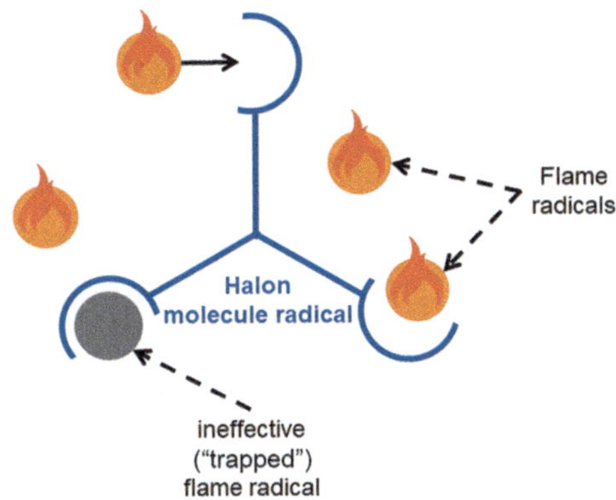

Fig. 41.1 Homogeneous inhibition (symbolic); © Felicitas "Cean" Schmiermund, after a sketch by the author

Chemical Consideration

Due to the formation of radicals from the extinguishing gas, the radicals of the combustion reaction – especially the hydroxyl radicals, HO• – are bound. This brings the reaction to a standstill: the flame goes out.

Bromotrifluoromethane ($CBrF_3$, Halon 1301), for example, is split by the flame temperature into a trifluoromethyl radical (F_3C^*) and a bromine radical (Br*). These two radicals react with the combustion radicals in the flame – e.g. with H_3C•, HO•, H• – to form more or less stable compounds and thus interrupt the radical chain reaction.

The strongest contribution to the extinguishing effect is made by the "destruction" of H• and HO• with new formation of Br• with intermediate formation of HBr:

$$\text{Br}\bullet + \text{H}\bullet \rightarrow \text{HBr}$$

$$\text{HBr} + \text{HO}\bullet \rightarrow \text{H}_2\text{O} + \text{Br}\bullet$$

The "effective" radicals in the flame zone, H• and HO•, are rendered "harmless", so to speak, by this reaction sequence (compare Sect. 40.1.3). This "rendering harmless" is sometimes also referred to as "flame poisoning".

$$\text{Br}\bullet, \text{F}_3\text{C}\bullet \begin{cases} + \text{HO}\bullet \longrightarrow \text{F}_3\text{C—OH}, \text{Br—OH} \\ + \text{H}\bullet \longrightarrow \text{CHF}_3, \text{H-Br} \\ + \text{H}_3\text{C}\bullet \longrightarrow \text{H}_3\text{C—CF}_3, \text{H}_3\text{C—Br} \end{cases}$$

41.1.5.2 Heterogeneous Inhibition/Wall Effect

In heterogeneous inhibition – in the case of extinguishing powder – solid particles act on the flame. In simple terms, the particles necessary for maintaining the combustion reaction and transferring the reaction energy lose their energy through impact with the relatively large powder particle. The reaction comes to a standstill (Fig. 41.2).

The process can be compared to throwing a dry sponge onto a blackboard: The reactive particle (*sponge*) hits the powder particle (*the blackboard*), loses its

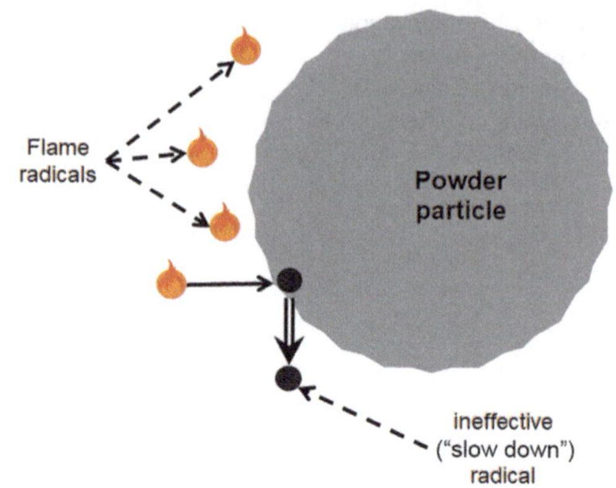

Fig. 41.2 Heterogeneous inhibition (symbolic); © Felicitas "Cean" Schmiermund, after a sketch by the author

(*motion*) energy and is no longer available (*falls to the floor*). The powder particle (*the blackboard*) is unimpressed by this.

Chemical Consideration
If we take a closer look at the reaction sequence, we see that this is the wall effect already mentioned (see Sect. 40.1.1): The chain carriers of the combustion reaction, e.g. hydrogen radicals (H•), hydroxyl radicals (HO•) or organic radicals (R•), hit the surface of the powder particles [P]. There they react with each other and release their now excess energy to the powder particle. The powder particle absorbs the energy and is excited or heated [P*]. The reaction then proceeds, for example, according to:

$$\mathrm{H\bullet} + \mathrm{HO\bullet} + [\mathrm{P}] \rightarrow \mathrm{HO{-}H} + [\mathrm{P}^*]$$

$$\mathrm{R\bullet} + \mathrm{H\bullet} + [\mathrm{P}] \rightarrow \mathrm{R{-}H} + [\mathrm{P}^*]$$

41.2 Extinguishing Agent

First of all, there is no universal extinguishing agent that can extinguish all fires equally well. The flammable substances and their combustion phenomena are far too different.

According to the definition of DIN 14011:

▶ Extinguishing agents are solid, liquid or gaseous substances suitable for extinguishing burning substances.

Let us now take a closer look at the common extinguishing agents, as well as some others.

41.2.1 Water

Water has always been the most important extinguishing agent. It is easy to transport, non-toxic, chemically neutral, odourless, tasteless, and is available to us (Middle Europe, North America) in almost unlimited quantities. Its extinguishing effect is based on the cooling of the material to be extinguished, so that it is primarily suitable for class A fires. The fire brigade uses water either as a full jet or as a spray jet:

	Advantages	Disadvantages
Full jet	• Long throw • Good mechanical effect (impact force, penetration depth)	• Bad cooling effect • Danger of water damage • Safety distances required for electrical installations (Table A.29) • Danger of dust explosion due to whirling up of dusts
Spray jet	• Good cooling effect • Lower risk of water damage • Smaller safety distances for electrical installations	• Shorter throw (Table A.29) • Lower impact • Danger of scalding by water steam, especially in closed rooms

41.2.1.1 Application Limits

The following hazards **may** occur when using water as an extinguishing agent:

- Whirling up of dusts → dust explosion (Sect. 39.1.3)
- Burning overflow → risk of spreading (Sect. 3.3.3)
- Chemical contact
  - Heat release (e.g. with: conc. H_2SO_4, CaO)
  - Release of flammable gases (e.g. with: Na, CaC_2)
- Fat/oil fires → FCI-steam explosion between oil and water (Sect. 3.4.6)
- Chimney fires → steam explosion, danger of collapse (Sect. 3.4.6)
- Metal fires → thermal dissociation, formation of oxyhydrogen gas (Sect. 3.4.6)
- Swellable materials → danger of collapse due to volume expansion
- Absorbent materials → danger of collapse due to mass increase

41.2.1.2 Room Cooling/Temperature Check/Flue Gas Cooling

Room cooling is the discharge of water into the hot flue gas layer (spray jet or "grid position" in the case of using a fog nozzle branch pipe) before entering the fire room. It serves to reduce the thermal energy in the room. For this purpose, the door is opened, water is discharged, the door is closed and thus the water is given time to act. Only then is the fire room entered.

Temperature check refers to the release of water into the flue gas layer during the operation in the fire area. If the water evaporates completely, flue gas cooling must be initiated. If water drips back, the operation can be continued.

Flue gas cooling serves to prevent flue gas ignition (*roll-over*). During fire-fighting measures, the flue gas layer is cooled if necessary (see also Sect. 40.2.3).

41.2.2 Foam

The extinguishing effect of the extinguishing agent foam is based on suffocation by separation. To a small extent, foam also has a cooling effect due to the water it contains. Extinguishing foam, more precisely: air foam, is freshly produced in foam branch pipes (FBP), light foam generators, portable fire extinguishers and in permanently installed extinguishing systems as required. For this purpose, a foam solution is intensively mixed with air.

As the foam has a significantly lower density than the water (compare Table 3.18) used to produce it, foam is suitable for fighting fires involving burning liquids. Similarly, puddles of flammable liquids that have not yet caught fire can be covered with a foam carpet and thus prevented from igniting (cover effect).

When medium or low expansion foam is introduced into rooms, ducts, plant components, etc., the foam displaces air and flammable gases/vapours (displacing effect).

> **Note: There** is an acute risk of suffocation for persons in rooms flooded with foam!

41.2.2.1 Admixing Rate

When creating a foam solution using a inline inductor, the admixture rate (AR) can be adjusted on most models. This is the percentage proportioning of the foam compound to the extinguishing water flow. A proportioning rate of 3% means a mixture of 3 L foam compound and 97 L water.

41.2.2.2 Foaming Ratio

The foaming ratio (FR) indicates the factor by which the volume of the foam solution is increased by the admixture of air.

$$\text{Foaming ratio} = \frac{\text{Foam volume}}{\text{Flow volume}} \Rightarrow \text{FR} = \frac{800\ \text{L}}{100\ \text{L}} = 8$$

The foaming rate cannot be set separately. It is fixed by the design of the currently used foam generating unit. Typical foaming rates are listed in Table 41.2.

Table 41.2 Foaming ratios (FR) of the three foam types

Foam Type	FR Area	Typical FR
Low expansion foam (LX)	4...20	15
Medium expansion foam (MX)	20...200	75
High expansion foam (HX)	200...1000	300

Heavy and medium expansion foam can be produced with foam branch pipes. These automatically suck in the air required for foam expansion. Special generators are required for low expansion foam (LX), as the required amount of air must be supplied by a fan.

41.2.2.3 Foam Destroying Influences

Extinguishing foam consists of a water-foam-skin in which air is enclosed. This "skin" is relatively sensitive to various influences. Foam can be destroyed by the following influences:

- Other extinguishing agents (extinguishing powder in particular has a foam-destroying effect).
- Chemicals (Many substances such as alcohols have a foam-destroying effect).
- Mechanical influences (wind, rough surfaces, emergency forces, . . .)
- Heat (radiant heat or heated surfaces)

In addition, there is the so-called water half-time, WHT. During this time, 50 % of the foam decomposes by itself, i.e. without external influences, and the water flows out. For low expansion foam the WHT is ~20 min.
If flue gas is sucked in with the foam jet pipe instead of fresh air during a foam jet pipe operation, no foam or only very poor foam is obtained due to the foam-destroying influences of the flue gases. In the model, the solid particles of the fire smoke literally cut the soap skin of the foam bubbles.

41.2.2.4 Foam Application Notes

- Provide the entire foam compound supply of all fire trucks at the inline inductor and open all canisters to avoid foam compound supply shortages. Shortages may result in firefighting failure.
- Do not direct the foam jet onto the fire material until the foam has the desired consistency. Only a well-formed foam achieves a good extinguishing effect.
- Suck in only clean air with the foam branch pipe. If the air is unclean/contaminated with fire gases, it may not be possible to produce foam at all.
- Only use foam in de-energised high-voltage systems. The foam produced at the scene of the incident is usually electrically conductive to high voltage. This can possibly lead to a hazard for the emergency personnel. For low voltage the conductivity is negligible (see also Sect. 50.4.2).
- Observe the danger of suffocation when flooding rooms. Emergency personnel wearing respiratory protection must also leave these areas, as the foam can lead to the failure of important parts of the respiratory protection equipment.
- Clean equipment thoroughly after use. Foam compound residues tends to stick to everything as it dries.
- Prevent discharge of foam into open waters. Foam compounds are harmful to aquatic organisms.
- For Class A fires, the foam can be delivered directly onto the fire.

- When extinguishing fires of liquids (fire classes B and F), the foam must be applied indirectly in order to develop an optimum extinguishing effect. In this case, the foam is applied over an auxiliary surface—e.g. a container wall or the surface of the ground—and thus away from the burning area. From here, the foam then flows onto the surface of the burning liquid.
- For fires involving polar, flammable liquids, an alcohol-resistant foaming agent is required for optimum extinguishing success. This also applies to carburetor fuels mixed with bio-ethanol (in Germany named E10 and E85).

▶ Fist values foam concentrate requirement

- Low expansion foam (LX): 2 L per m^2 of fire area
- Medium expansion foam (MX): 0.5 L per m^2 of fire area

41.2.2.5 Environmental Protection and Foam Using

- Extinguishing exercises with foam should only be carried out if
  - the foam solution can be collected and thus disposed of in an orderly manner,
  - the exercise is designed in such a way that a positive exercise effect is also ensured.
- Demonstrations with foam (e.g. "open day" at the fire station) should not be carried out as a matter of principle—even with so-called exercice/practice foam agent.
- Foam compounds must always be stored in such a way that even in the event of container leakage, no contamination of the soil, water, groundwater or sewage system can occur (e.g. catch basins in the storage area).
- In the event of an emergency, it must be checked where the foam/foam solution resulting from the extinguishing measures remains. The civil engineering office, the lower water authority, the sewage treatment plant, the water management office and other authorities, e.g. residents with standing surface waters (fish ponds) or industrial companies (extraction of water for work processes), must be informed.
- Extinguishing water retention is essential to avoid impacts on the environment, particularly on water bodies such as streams, rivers, lakes and ponds.

41.2.2.6 Diffusion of Flammable Vapours

Especially in the case of non-polar, lipophilic (Greek *lipos* = fat, i.e. "fat-loving") flammable liquids such as petrol, diffusion of the vapours through the closed foam cover often occurs. These vapours, which now escape from the foam layer, can ignite on hot surfaces, for example, and thus cause so-called "ghost flames". Due to their heat radiation, these cause the extinguishing foam to disintegrate more quickly. Polar liquids – in contrast to non-polar liquids – mix with the water contained in the foam, so that the formation of ghost flames does not generally occur with the vapours of polar liquids.

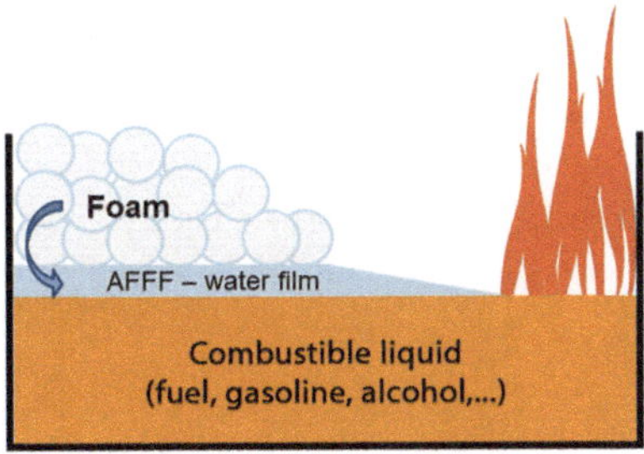

Fig. 41.3 Mode of operation AFFF

41.2.3 Special Foams

Multigrade foams are commonly used by fire brigades. Synthetic multigrade foams serve as foaming agents and can be used for fire classes A and B and for the production of both heavy and medium foams as well as light foams. Protein foams, fluoroprotein foams and film forming fluorinated protein foams (FFFP – *Film Forming Foam Protein) can* only be used for the production of low expansion foam. In addition, there are two other important foams with special foaming agents.

41.2.3.1 Water Film Forming Foams (AFFF)

AFFF foams (AFFF = *Aqueous Film Forming Foam)* contain special fluorinated surfactants which form a water-containing layer which in turn is lipophobic (= fat-repellent). This thin, water-containing film is impermeable to vapour and remains on the surface of the flammable liquid – hence the name *light water* (Fig. 41.3). Depending on the conditions of use, this aqueous film can remain effective for a long time beyond the decay of the actual foam layer.

Notes

- AFFF foaming agents can be used to produce heavy and medium expansion foams for fighting fires in fire classes A and B.
- Instead of the term "lipophobic" for the property of the aqueous surfactant film, the term "oleophobic" (= oil-repellent), which originates from textile finishing, is sometimes used.
- Alcohol resistant variants are designated AFFF-AR *(**A**lcohol **R**esistant)* or AFFF-ATC *(**A**lcohol **T**ype **C**oncentrate).*
- AFFF foams should – if possible – only be used for extinguishing liquid fires. The fluorosurfactants are not biodegradable (sewage treatment plant) and accumulate in the environment.

41.2.3.2 Compressed Air Foam System (CAFS)

The compressed air foam (CAF) is already fed as foam into the hose by the supply of compressed air to the foam solution and then delivered to the source of the fire through a full jet pipe (Fig. 41.4). This is therefore a foam generation system installed in the vehicle as a whole: *Compressed Air Foam System.* Only so-called "*Class-A-Foam*" foam compound is used for foam generation and is thus – as the name already indicates – intended for fighting Class A fires. The foaming number is usually FR = 7.

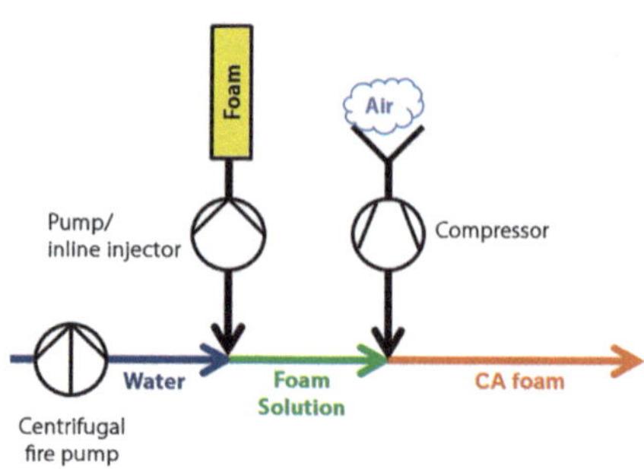

Fig. 41.4 CAFS principle

Advantages

- Significantly less water damage compared to water as an extinguishing agent
- No environmental pollution due to the foam agent, as there is no uncontrolled run-off of the extinguishing agent due to its economical use.
- Short foam bursts possible (interval extinguishing method).
- Foam-filled hoses weigh only about ½ to $^{1}/_{5}$ opposite hoses filled with water.
- No foam concentrate containers are to be carried.
- A high throw of the foam.
- The white foam, which is clearly visible in the fire room, enables a good visual control of the already extinguished or exposed areas.

Disadvantages and Other Comments

- No time advantages in setting up the attack line.
- Intermittent extinguishing results in 5 to 9 times the foam concentrate consumption compared to the set point.
- No cooling of the fire smoke possible.
- Indirect extinguishing is only possible by applying the CAF to walls and ceilings of the fire room.
- Hoses must be laid in large radii and without kinks.
- The extinguishing of solid fires is based on the cooling effect. Due to the enclosed air per unit volume, foam has a lower cooling effect than water.
- Hoses are more sensitive to heat. The hose burns through more easily because the foam has a significantly lower heat capacity than water.

41.2.4 Extinguishing Powder

Basically, there are three different types of extinguishing powder that are used to fight different fires:

ABC extinguishing powder	(for glowing fires and others)	For fire classes A, B, C
BC extinguishing powder	(for burning liquids and gases)	For fire classes B, C
D extinguishing powder	(for burning metals)	For fire class D

Table 41.3 Powder types: Composition (extinguishing components) and extinguishing effects

Type	Composition (Example)	Erase Effect
ABC extinguishing powder	$(NH_4)H_2PO_4$, with $(NH_4)_2HPO_4$ and $(NH_4)_2SO_4$	Ember fires: Separate
		Flame fires: Inhibition
BC extinguishing powder	$NaHCO_3$ or $KHCO_3$ or K_2SO_4	Inhibition
D extinguishing powder	NaCl	Separate

The D extinguishing powder melts together on the surface of the burning metals to form a layer which prevents further oxygen access and thus smothers the fire (extinguishing effect: suffocation by separation). The ABC extinguishing powder has a similar effect when used against solid, ember-forming substances: at temperatures above 70 °C, the powder begins to melt and thus separates the burning material from the oxygen.

The extinguishing effect of BC extinguishing powder or ABC extinguishing powder is completely different when fighting flame fires. Here, the anticatalytic or inhibitory extinguishing effect occurs. Complex chemical processes take place in the flame zone, in which free radicals are involved as intermediates. These radicals hit the powder particles and are rendered ineffective there. This phenomenon, known as the "wall effect" (see Sects. 40.1.1 and 41.1.5), removes energy from the reactive intermediates in the combustion reaction. The particularly energetic radicals, which are essential for the progress of the combustion reaction, lose some of their energy on contact with the cool powder particles. As a result, retrograde reactions, so-called recombinations, occur, which finally lead to the termination of the chain reaction. Table 41.3 gives an overview of possible chemical compositions (extinguishing components) and the extinguishing effects of the different powder types.

41.2.4.1 Particle Sizes and Surfaces

For an optimum effect, the majority of the powder particles must of course be very small and free-flowing. Due to the extinguishing effects, it inevitably follows that D powder types must be relatively coarse and BC powder types very fine. Smaller particles mean that a larger surface area is available for inhibition; coarser particles are more suitable for covering the fire material. Typical values for the different powder types are listed in Table 41.4.

41.2.5 Carbon dioxide CO_2

The extinguishing effect of CO_2 is based practically only on the suffocating effect. Since CO_2 has a density 1.5 times higher than air, it spreads from the ground over the fire and prevents further air access. To achieve a sufficient suffocating effect, a residual oxygen concentration of $\leq 15\%$ by volume should be aimed for. For this purpose, a CO_2 concentration of approximately 30 vol.-% is required. Carbon dioxide is applicable for fire classes B and C, especially in areas that must not be contaminated. As it evaporates quickly in the open air, it is only suitable for firefighting in closed rooms (for the physiological effect of CO_2, see Table A.11).

Table 41.4 Comparison of extinguishing powder types

	ABC Extinguishing Powder	BC Extinguishing Powder	D Extinguishing Powder
Extinguishing agent density	~ 1.8 g cm^{-3}	~ 2.5 g cm^{-3}	~ 2.1 g cm^{-3}
Apparent density	~ 850...900 g L^{-1}	~ 950...1000 g L^{-1}	~ 1000...1100 g L^{-1}
Apparent density of powder	~ 1250 g L^{-1}	~ 1500 gL^{-1}	~ 1500 g L^{-1}
Average grain size (D_{50})	~ 40 μm	~ 30 μm	~ 70 μm
Grain size distribution:			
< 40 μm	~ 70 %	~ 80 %	~ 35 %
40–63 μm	~ 15 %	~ 10 %	~ 30 %
63–125 μm	~ 10 %	~ 5 %	~ 30 %
125–250 μm	~ 5 %	~ 5 %	~ 10 %
> 250 μm	up to 1 %	up to 1 %	up to 3 %
Specific surface	~ 4200 cm^2 g^{-1}	~ 4200 cm^2 g^{-1}	~ 2200 cm^2 g^{-1}
Mass of extinguishing powder for 1 m^2 surface	2.4 g	2.4 g	4.5 g

41.2.5.1 Application Limits

- Danger of suffocation in confined spaces (firefighters: self-contained breathing apparatus).
- No effect outdoors (rapid volatilization).
- Danger of backfiring.
- No effect with highly ember-building substances.
- Thermal decomposition in metal fires (compare Sect. 3.4.6 and Fig. 3.20)
- For personal fires: Risk of frostbite (CO_2 outlet temperature: –78 °C).
- Danger of electrostatic charge.

41.2.6 Extinguishing Agents for Burning Fat

Fire hazards in kitchens – both in the household and in the catering trade – must not be underestimated. Oil and fat fires in particular pose a great danger. If a pan is left unattended on the heated stove, then in an unfavourable case a period of less than 5 minutes is already sufficient to ignite the fat.

Attempts to extinguish with water or foam can lead to a "fat explosion" (= FCI-steam explosion between oil and water), CO_2 is almost ineffective and powder causes a corresponding loss of use in large kitchens due to the subsequent cleaning work required. Even fire blankets are only suitable to a limited extent (wicking effect).

The solution is portable fire extinguishers for fire class F (see Sect. 41.2.6), which are mandatory for commercial kitchens. The liquid extinguishing agent contained in the extinguisher is suitable to fight fires of fire classes A, B and F (German approval). Therefore, this extinguisher is the almost ideal standard extinguisher for household use. The extinguishing agent is **not** suitable for fires of organic acids (e.g. acetic acid) and for metal fires.

Mode of Action for Fat Fires

The extinguishing agent is a highly concentrated solution of the alkali salts of carboxylic acids with small amounts of auxiliary substances. The water content is usually less than 50%. From a chemical point of view, edible fats and oils are esters of long-chain fatty acids with glycerine. As in soap manufacture, these fatty acids are split off by the extinguishing agent and converted into soap (see Sect. 50.2.2). The process is therefore known as **saponification.** The resulting soap foams up and forms a barrier layer on the oil or fat. At the same time, the extinguishing agent gently cools the burning liquid.

41.2.7 Other Extinguishing Agents

In addition to the "standard" extinguishing agents described above, there are a number of special and auxiliary extinguishing agents. A few will be briefly addressed here.

41.2.7.1 Sand, Cement, Grey Cast Iron Filings, Common Salt

Dry sand, cement and grey cast iron chips (these consist of cast iron and are produced during the manufacture and finishing of cast components) can be used as extinguishing agents to fight metal fires. *Dry* common salt is also suitable for this purpose (do not use damp road salt!). Like metal fire extinguishing powder, these substances separate the fire material from the ambient air.

41.2.7.2 Hollow Glass Granulate

Glass consists predominantly of silicon dioxide (sand, SiO_2), so hollow glass granulate has the properties of SiO_2 in it: it has high chemical and thermal resistance, a high melting point, low thermal conductivity and is an electrical insulator. Due to the special manufacturing method of the granulate, which is marketed as PyroBubbles®, it has a density of only ~ 0.25 g mL^{-1} and is therefore also buoyant on flammable liquids, such as gasoline. It is approved as an extinguishing agent for fire classes A, B, D and F, whereby within fire class B alcohol fires can also be fought.

The possibility of preventive use in structural fire protection, as well as active firefighting by feeding the granulate from storage containers, the recyclability, the easy absorption of the extinguishing agent after firefighting and special applications, such as the use as a filling agent for transport containers of used lithium-ion batteries (compare Sect. 19.2.4), or a special container filled with it for the transport of Bengal fire (so-called "Bengalos") make this new extinguishing agent highly interesting.

41.2.7.3 Halon Substitutes

In the Montreal Protocol (which came into force in January 1989), the signatory states undertook to take measures to protect the ozone layer. Based on this, Germany created the CFC-Halon Ban Ordinance in 1991, which in turn was replaced by the

Chemicals Ozone Layer Ordinance in 2006. Ozone (O_3) influences the temperature of the stratosphere and shields high-energy UV radiation. Protecting the ozone layer is therefore essential. Unfortunately, in the early 1980s it was found that chlorofluorocarbons (CFCs) cause lasting damage to this ozone layer. As a result, halons, which had been used successfully as extinguishing agents for many decades, were also banned.

The advantages of halons were obvious. They are less toxic than CO_2, leave no residue and can be used in electrical installations because they are non-conductive. In contrast to other extinguishing gases (CO_2, N_2, argon), which are based on the effect of oxygen displacement, much less extinguishing agent was required. The addition of 4 %–6 % by volume of halon to the ambient air was already sufficient to be able to report "Fire out!". Even if the O_2 content was still close to 20 vol.-% due to additional oxygen supply. With the extinguishing gases mentioned, or their mixture, the "Inergen®" (8 % CO_2, 40 % Ar, 52 % N_2), the oxygen concentration of the ambient air must be reduced to below 15 % in order to achieve extinguishing success.

However, in certain areas, e.g. in aviation for fighting engine fires, the halons proved to be almost indispensable. In particular, industry has sought substitutes for the formerly very common halons bromochlorodifluoromethane (Halon 1211, CF_2ClBr) and bromotrifluoromethane (Halon 1301, CF_3Br). These "new halons" (Table 41.5), which have no ozone-depleting effect, are, for example:

Table 41.5 Halon substitutes (examples)

Trade Name	Trigon® 300	FM-200®	Novec® 1230
Other designations	Trifluoromethane; "Fluoroform"; HFC-23	1,1,1,2,3,3-Heptafluoropropane; R-227ea	1,1,1,2,4,5,5-Nonafluoro-4-(trifluoromethyl)-3-pentanone
Structural formula	F–C(F)(F)–H	F–C(F)(F)–C(H)(F)–C(F)(F)–F	F–C(F)(F)–C(F)(F)–C(=O)–C(F)(CF₃)–CF₃
Sum formula	CHF_3	C_3HF_7	$C_6F_{12}O$
Molar mass	70 g mol^{-1}	170 g mol^{-1}	316 g mol^{-1}
Physical state (20 °C)	Gaseous	Gaseous	Liquid
Boiling point	−82,2 °C	−17,3 °C	49 °C
Vapour pressure (20 °C)	41.8 bar	3.99 bar	0.325 bar
Vapour density vs. air	2.4	5.9	10.9
Further use	Refrigerant	Refrigerant	–

- Trifluoromethane (Halon 13, CHF_3, Trigon®)
- Pentafluoroethane (Halon 25, $F_3C{-}CHF_2$)
- Heptafluoropropane (Halon 37, $F_3C{-}CHF{-}CF_3$, FM-200®)
- Hexafluoropropane (Halon 36, $F_3C{-}CH_2{-}CF_3$)
- Perfluoroethyl isopropyl ketone ($F_3C{-}CF_2{-}C(O){-}CF(CF_3)_2$, Novec 1230®)

However, the disadvantage is that the effect of these substances as greenhouse gases is greater than that of CO_2. Only the latter compound has the same greenhouse potential as carbon dioxide and is not harmful to the ozone layer. However, since the substance has a boiling point of 49 °C, it must be atomized in order to be used effectively as an extinguishing agent.
For the extinguishing effect of halons, see Sects. 36.4.2 and 41.1.5.

42 Flame Retardants

Plastics have become indispensable in today's highly technical world. The wide range of applications they offer as materials has led to their steadily increasing use in all areas of life. However, plastics are largely based on hydrocarbons and are therefore – with very few exceptions – well flammable. For safety reasons, however, various materials such as computers and accessories, mobile phones, televisions, circuit breakers, cable insulation, aircraft seats and clothing must not catch fire easily.

Various regulations and standards require that building products, just like the materials used in the electrical and electronics industry, must keep the risk of fire starting or spreading as low as possible. To make this possible, plastics are provided with flame retardants.

On the other hand, steel girders easily lose their strength in case of fire. To prevent steel girder structures from collapsing in the event of a fire, they must also be specially equipped.

42.1 Flame Retardant

Flame retardants (FR) are intended to reduce the flammability of plastics, slow down the spread of fire or – ideally – cause self-extinguishing. In most cases, the FRs are already added to the plastics during production. For some FRs, just a few percent are sufficient, while for others up to 60 % are added to the plastic.

A distinction is made between halogenated flame retardants (HFR) and halogen-free flame retardants (HFFR). The latter can be further subdivided into inorganic, nitrogen-containing and phosphorus-containing flame retardants.

Flame retardants are not a substance class of their own. They are a variety of different organic and inorganic substances which are used alone or in mixtures with each other. For structural formulae of FRs see Table A.26.

T. Schmiermund, *The Chemistry Knowledge for Firefighters*,
https://doi.org/10.1007/978-3-662-64423-2_42

42.1.1 Halogenated Flame Retardants

The polychlorinated and especially the polybrominated compounds, which were frequently used in the past, have proven to be persistent, i.e. difficult to degrade in nature, hazardous to water and bioaccumulative, i.e. they accumulate in organisms. The use of polybrominated biphenyls (PBBs) and polybrominated diphenyl ethers (PBDEs) was banned in 2004 for electrical and electronic ("E&E") applications in Europe. The use of hexabromocyclododecane (HBCD) was massively restricted in 2015 by the European REACH regulation, which classified the compound as "of very high concern". However, most halogenated flame retardants, such as tetra-bromo-bisphenol A (TBBA), are still permitted.

A European directive (keyword: "WEEE") stipulates that brominated FRs must be removed from electrical and electronic components before recycling or recovery. In the case of fires in engine compartments of modern passenger cars, it takes at least five to ten minutes for the flames to reach the passenger compartment. For the occupants, therefore, the greater danger lies not in the flames themselves but in the fumes that are produced. The use of halogen-free flame retardants prevents the formation of toxic substances such as halogenated dioxins and corrosive substances such as HCl and HBr.

42.1.1.1 Application Examples

TBBA is mainly used for epoxy resins and ABS plastics. Brominated polystyrene is used in polybutylene terephthalate (PBT), polyethylene terephthalate (PET) and polyamides (PA). Tris-(2-chloro-*iso*-propyl)-phosphate (TCPP) is used for polyurethane (PU) foams. Dibromo-neopentyl-glycol (DBNPG) is found in polyester resins, tribromo-neopentyl-alcohol (TBNPA) in polyols and pentabromo-benzyl-acrylate (PBBA) in nylon and polycarbonate (PC).

42.1.1.2 Mode of Operation

The effect of halogenated flame retardants is similar to that of halons. The fire causes the formation of halogen radicals (Cl•, Br•) which interfere with the radical chain reaction of combustion (cf. Sect. 41.1.5). In the field of flame retardant technology, the term "*flame poisoning*" is often used instead of "inhibitory extinguishing effect".

42.1.2 Inorganic Flame Retardants

The most commonly used FSMs in terms of quantity include aluminium (tri) hydroxide $Al(OH)_3$ (abbr.: ATH) and magnesium (di) hydroxide $Mg(OH)_2$ (abbr.: MDH). However, in order to achieve the desired effects, these substances must be added to the plastic in quantities of up to 60 % or more. Antimony trioxide (Sb_2O_3) is often used together with halogenated FRs to enhance their effect.

42.1.2.1 Application Examples

ATH is mainly found in polyethylene (PE), polyolefin-based thermoplastic elastomers (TPE-O), ethylene vinyl acetate (EVA), but also in latex-based carpet backing coatings. MDH is mainly used in plastics such as polypropylene (PP) and polyamide (PA 6).

42.1.2.2 Mode of Operation

Due to the strongly endothermic reaction during the formation of the oxides from the metal hydroxides used as FRs, *heat* is extracted from the reaction system (*heat sink) and* pyrolysis is thus reduced. As a result, fewer pyrolysis gases are available for combustion. Ideally, the flame goes out due to the resulting lack of fuel and the energy consumed. At the same time, water is released as steam and dilutes the combustion gases. The radical concentration is lowered and the reaction rate decreases.

$$2\ Al(OH)_3 \rightarrow Al_2O_3 + 3\ H_2O_{(g)} \qquad + 1075\ kJkg^{-1}$$

Furthermore, the high application concentration means that less combustible material is available overall.

42.1.3 Flame Retardants with Organic Nitrogen

Instead of the urea used in the past, melamine, melamine polyphosphates (MPP) and the melamine salt of cyanuric acid (melamine cyanurate, MC) are now used. These compounds consume energy during their decomposition, which they extract from the reaction zone. In the process, elemental nitrogen and carbon dioxide are released, which dilute the gas phase. In intumescent coatings (see Sect. 42.2), they thus provide a significant increase in the volume of the layer.

More recent developments are based on the further development of compounds originally used as light stabilizers. These substances, known as sterically hindered amines, prevent embrittlement and cracking caused by the effects of UV light on the plastic (HALS – hindered amine light stabilizers). By making suitable changes to the molecular structure, it is also possible to produce FRs on this basis that have an inhibitory effect in the gas phase.

42.1.4 Phosphorus as a Flame Retardant

At first glance, it may seem absurd to use a combustible element as a flame retardant. But the P_2O_5 produced during the combustion of the phosphorus forms polyphosphoric acids with hydroxyl radicals (•OH) and water produced during combustion, which in turn are highly hygroscopic and lead to greater charring of the plastic.

42.1.5 Flame Retardants Based on Inorganic Phosphorus Compounds

Probably the most frequently occurring inorganic phosphorus-containing FRs are ammonium polyphosphates (APP). They form polyphosphoric acids in the heat with ammonia splitting off, which lead to greater charring.

42.1.6 Flame Retardants Based on Organic Phosphorus Compounds

The organic phosphorus-containing FRs may be halogenated (TCPP, see Sect. 42.1.1) or halogen-free. Of the halogen-free FRs, metal salts of alkyl phosphinates, such as aluminium tris(diethylphosphinate), (DEPAL) and phosphonic acid derivatives are important compounds. Their action is not based solely on that of the inorganic phosphorus-containing FRs. By releasing PO radicals into the flame zone, they act in a similar way to halogenated FRs (see Sect. 42.1.1). The most important radical reactions are:

$$PO\bullet + H\bullet \rightarrow HPO \qquad HPO\bullet + H\bullet \rightarrow H_2 + PO\bullet$$

$$PO\bullet + HO\bullet \rightarrow HPO_2 \qquad HPO\bullet + HO\bullet \rightarrow PO\bullet + H_2O$$

42.1.7 Clothing with Flame Protection

Some synthetic fibres are already difficult to ignite "by themselves". Such inherently flame-retardant fabrics are, for example, the aramid fibres poly-(para-phenylene-terephthalamide) (PPTA, e.g. Kevlar©) and poly-(meta-phenylene-iso-phthalamide) (PMPI, e.g. Nomex©) as well as fabrics made of polybenzimidazole (PBI). If other fibres are to be used for specific reasons, such as special colouring, wearing comfort or breathability, they must be specially finished. Protective and work clothing for the metal and petrochemical industries, but also for the military, police and fire brigade can also be made from viscose fibres if the fibre from which the fabric is made has a flame-retardant finish. To ensure protection against fire, flash fire, flying sparks and heat radiation as well as against electric arcs and drops of molten metals, sulphur-containing dioxaphosphorinanes (DOP-S) are used, for example. Similarly, seat covers, mattresses, etc. can also be made from these flame-retardant viscose fibers.

Summary: Effect of flame retardants

- Halogen and P-containing flame retardants have an inhibitory effect.
- The reaction zone is diluted by the formation of CO_2, N_2 and H_2O.
- Certain flame retardants extract a lot of energy from the decomposition zone.
- The strengthening of the carbon layer causes the formation of a thermal protective layer (compare: Fig. 42.1).
- An overview of different structures of flame retardants can be found in Table A.26.

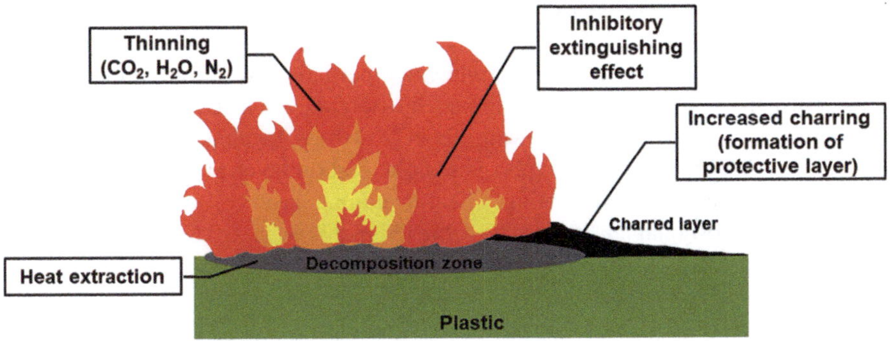

Fig. 42.1 Effect of flame retardants

42.2 Intumescent Coatings

Steel has only about half its original strength and load-bearing capacity when it reaches a temperature of 500 °C. In the event of a fire, steel beams can reach this temperature in less than ten minutes. In the event of a fire, steel beams can reach this temperature in less than ten minutes. To prevent steel girder structures from collapsing, it makes sense to protect steel girders from the occurrence of such temperatures. In the case of fires in cable shafts or other wall and ceiling penetrations, a fire can spread further via this route. Sealing these penetrations can prevent the spread of fire. For this purpose, intumescent (Latin *intumescere* = to swell) substances are used, i.e. substances or mixtures that greatly increase their volume in the event of fire.

42.2.1 Applications

In the application, a distinction must be made between the prevention of fire spread and the protection of components. To prevent the spread of fire, cable insulation or pipes in which electrical cables and pipelines are laid are provided with flame retardants which cause foaming. In this case, the plastic used serves as a "carbon donor". Special putties, with which penetrations in walls and ceilings are provided, are also used as so-called fire protection putties for bulkheads.

To prevent excessive temperature on steel components, they are provided with an appropriate coating that forms an insulating layer. Here it is possible to maintain the strength of steel beams for a period of one hour or more. These coatings can also be applied to doors, wall cladding and roof substructures, thus protecting wooden components from the effects of fire.

The effect of intumescent systems is similar, regardless of the application or the components to be protected. The only difference is the source of the carbon layer to be foamed. In the case of paints, this must be integrated in the system, whereas in the case of plastics, the polymer itself is usually the "carbon donor".

42.2.2 Mode of Operation

Heat causes pyrolysis of the plastics, which are partially carbonized in the process. This thin carbon layer can be reinforced by polyols, such as pentaerythritol. In combination with a "thermal blowing agent", such as melamine or urea, and an acid donor, e.g. ammonium polyphosphate, the carbon layer grows into a voluminous foam. In the case of coatings, heat-insulating layers 10 to 100 times the thickness of the original coating can be achieved in this way. A 0.5 mm thick coating thus easily becomes 10 cm of insulating layer.

Chemism

Ammonium polyphosphate (APP) decomposes at temperatures above 250 °C. In addition to ammonia, condensed, longer-chain phosphoric acids are produced:

$$(NH_4PO_3)_n \rightarrow (HPO_3)_n + n\, NH_3$$

These react with the polyol to form phosphate esters, which in turn split off CO_2 and leave behind a phosphate-containing carbon layer that is difficult to ignite. Due to the simultaneous presence of gas-producing "blowing agents", such as melamine, this carbon layer foams strongly:

$$C_3N_3(NH_2)_3 + 3\, O_2 \rightarrow 2\, NH_3 \uparrow + 2\, N_2 \uparrow + 3\, CO_2 \uparrow$$

Summary: Effect of Intumescent Coatings

Structural components, such as steel beams, doors or walls, are provided with a coating that foams up when exposed to fire. The layer formed in this way acts as insulation and protects the components from increased heat exposure (Fig. 42.2).

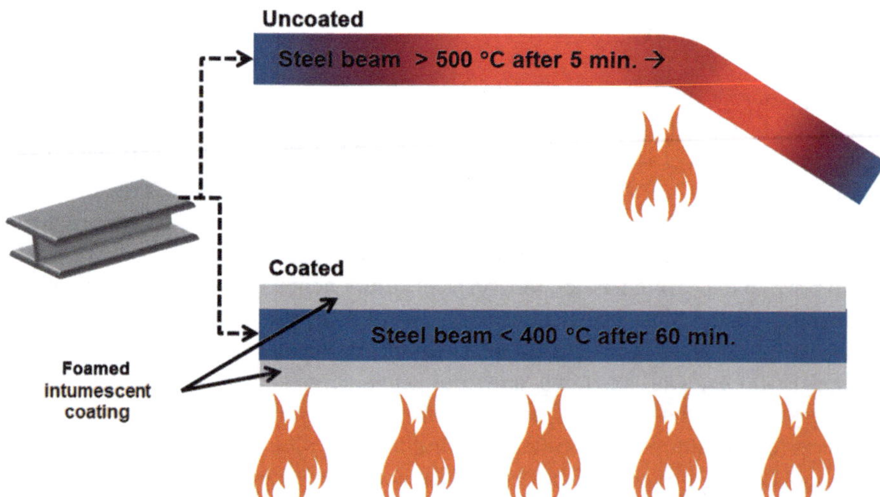

Fig. 42.2 Effect of intumescent coating (schematic)

Part XI

Organic Chemistry

Indispensable Organic 43

So-called "organic chemistry" is the part of chemistry that has the most far-reaching and visible impact on our lives. Millions of people are cured (medicines), clothed (synthetic fibres, dyes) and fed (pesticides) without realising the importance of this branch of chemistry. In our daily lives we encounter another multitude of their products: Plastics of all kinds, detergents and cleaners, cosmetics, composites, paints and coatings, fragrances in perfumes, dyes in displays, LEDs in lamps, and many more. Motor fuels (gasoline, diesel, liquefied petroleum gas, kerosene) and heating fuels (natural gas, heating oil) are also included in this area, as are natural substances, which are mostly obtained from plants but are also produced synthetically to a large extent.

Organic chemistry ("OC") is in a way the "science of molecules". It has created new materials, copied substances from the animal and plant kingdoms, or even modified some of them, with the aim of making them "better". There are stable and reactive compounds, harmful and helpful, toxic and healing – but there is only one kind of chemistry. It takes place in our bodies, e.g. as metabolism or as brain function; it takes place in plants and bacteria – and in the flasks and reactors of the chemical industry.

Naturally, only an extremely rough overview of some important substance classes can be given here. More detailed explanations of synthesis techniques, reactions and reaction mechanisms and other substance classes must be omitted. The interested reader is referred to the literature in the appendix.

43.1 Introduction

Let us first deal with what makes organic chemistry so special compared to inorganic chemistry, and the special position of the element carbon.

T. Schmiermund, *The Chemistry Knowledge for Firefighters*,
https://doi.org/10.1007/978-3-662-64423-2_43

43.1.1 Demarcation Organic ↔ Inorganic Chemistry

Bergmann[1] proposed as early as around 1780 to separate the "chemistry of the living" (= organic chemistry) from the "chemistry of minerals" (= inorganic chemistry). Berzelius adopted this idea and disseminated it in natural science from 1807 onwards. According to this idea, "organic" substances could only be produced by living organisms (plants, animals) with the help of a "life force" (Latin: *vis vitae).* This doctrine persisted until about 1850, although Wöhler[2] succeeded as early as 1828 in producing the organic compound urea by heating the ammonium salt of cyanic acid (H—N=C=O) – i.e. from an inorganic substance:

$$NH_4^+ \ NCO^- \xrightarrow{\Delta T} H_2N\text{–}CO\text{–}NH_2$$

Other substances, such as nitrobenzene ($C_6H_5-NO_2$) or chloroform ($CHCl_3$), do not occur in living beings and therefore cannot be easily obtained from them. So, according to the original "life force" notion, they are inorganic substances. And ammonia could be obtained from inorganic salts (sal ammoniac, NH_4Cl) as well as from urine. The dichotomy had lost its original meaning over time.

However, the division into "inorganic chemistry" and "organic chemistry" has proven itself due to the nature of the substances and has rightly been retained. However, a really sharp dividing line cannot be drawn. Carbonic acid (H_2CO_3 or $(HO)_2CO$) is inorganic, urea ($(NH_2)_2CO$) is organic. Inorganic CO_2 can be added to organic molecules under certain conditions to form organic acids. Carbon monoxide (inorganic) is used with alkenes and hydrogen to produce aldehydes (organic). And organic ethine (C_2H_2) can be obtained from inorganic calcium carbide (CaC_2). For delimitation we stick to the definition introduced by Gmelin[3] in 1848, which we extend only slightly:

▶ Organic chemistry is the chemistry of carbon compounds, especially compounds of carbon containing C—C and C—H bonds.

Carbon compounds that are not counted as organic chemistry are: Carbon itself (element), the oxides (CO, CO_2), carbonic acid (H_2CO_3) and car-bonates $\left(CO_3^{2-}\right)$, hydrogen cyanide (HCN) and its salts, cyanic acid (HCNO), its isomers and associated salts, and carbides (Me_xC_y).

[1] Torbern Olof Bergman, 1735–1784, Swedish chemist and mineralogist.

[2] Friedrich Wöhler, 1800–1882, German chemist.

[3] Leopold Gmelin, 1788–1853, German chemist.

43.1.2 Why Carbon?

The approximately 80 000 to 100 000 inorganic compounds are currently compared with more than 130 000 000 organic compounds [Lit. 13-10]. But what of all things enables carbon to form such a large number of compounds?

As an element of main group IV (cf. Chap. 6), carbon has four outer electrons. The tendency to form ions has almost completely disappeared, the tendency to form stable covalent bonds has reached a maximum. Carbon and the silicon below it in the PTE should therefore already be capable of building stable molecules due to their position in the PTE. The question arises as to why carbon appears to be preferred over silicon. There are several reasons for this:

1. The character of the compounds formed is different. Silicon forms predominantly macromolecular silicates. Carbon forms predominantly low-molecular substances; especially in combination with hydrogen.
2. Si is in the third period of the PSE and therefore already possesses the possibility of shifting electrons into a lower-energy d orbital. Silicon therefore reacts particularly easily with nucleophilic, electron-rich agents that can fill these d orbitals with their lone electron pairs.
3. The oxidation of Si–H to Si–O compounds releases more energy than the reaction of C–H to C–O compounds. As a result, silanes (Si–H compounds) are extremely sensitive to atmospheric oxygen and moisture. In the case of carbon, on the other hand, the C–H bond is stable.

For these reasons, only carbon – especially when combined with hydrogen, which occupies free bonds – remains capable of building such a huge number of compounds. Carbon atoms can form chains, branched chains and rings. Carbon skeletons with 10 000 and more C-atoms are thus possible. In addition, organic molecules – with few exceptions – always contain hydrogen (H) in the molecule. All other elements involved in the structure of an organic molecule are called heteroatoms. Besides the halogens (F, Cl, Br, I; general: X), oxygen (O), nitrogen (N), sulfur (S) and – especially in biochemistry – phosphorus (P) are involved in the structure of organic molecules.

► Elements frequently involved in the structure of organic molecules: CHONS.

43.1.3 Basic Structure of Organic Compounds

Carbon has four bonding possibilities. All bonds that are not used by another carbon atom are available for hydrogen or for heteroatoms. And it is precisely these heteroatoms or the so-called "functional groups" formed from them that make up organic chemistry. Organic molecules consist of a "backbone", a chain of carbon atoms, which is occupied by functional groups in one or more places. This results in a few simple rules:

1. Since the C–C and C–H bonds are relatively inert, many reactions take place at the functional group or in its immediate vicinity.
2. Each functional group always reacts in the same way. The subdivision of organic substances into compound classes of different chemical character is therefore based solely on the functional groups contained in the molecule.
3. In addition to single bonds (C–C), the carbon skeletons also contain double bonds (C=C) and triple bonds (C≡C) between the C atoms, which in themselves increase reactivity. They can therefore also be regarded as functional groups.
4. Hydrocarbon residues often – in contrast to the functional groups – do not participate in reactions. These residues "migrate" as a radical from one part of the molecule to another. In general formulas, they are usually represented as "R" for "radical" or "rest".
5. If there are several radicals in the molecule, different radicals are marked with dashes (R', R") or with indices (R_1, R_2, ...). In some formulas, R can also stand for hydrogen, in which case the legend for the formula illustration must be observed.

43.1.4 Functional Groups

Functional groups – sometimes also referred to as characteristic groups – are atoms or groups of atoms that can replace a hydrogen atom on the carbon chain. A functional group defines a class of compounds and gives them characteristic chemical and physical properties. The most important functional groups are briefly described in Chaps. 44, 45, 46, 47 and 48.

Hydrocarbons 44

Hydrocarbons (HCs), i.e. compounds composed only of carbon and hydrogen, are the simplest organic compounds. They can occur as linear or branched chain molecules, as closed ring structures or as so-called aromatic compounds. The non-aromatic compounds are called aliphatic hydrocarbons or aliphatics for short. An overview is given in Fig. 44.1.

44.1 Aliphatic Hydrocarbons

Aliphatics occur as chain or ring structures. The bonds between the carbon atoms can consist only of single bonds (C–C, alkanes) or contain double bonds (C=C, alkenes) or triple bonds (C≡C, alkines). Ring-shaped aliphatics are analogously called cycloalkanes, cycloalkenes or cycloalkines.

44.1.1 Alkanes

The alk**an**es (formerly: paraffins, from Latin *parum affinis* = not very reactive) are the hydrocarbons with the simplest structure. The simplest alkane, the parent compound, is methane, CH_4. By adding a CH_2 group the next alkane is formed: ethane, C_2H_6. Adding another CH_2 group gives propane, C_3H_8. Such a successive series is called a homologous series. At the ends of the carbon chain, each C atom is surrounded by three hydrogen atoms, all other C atoms carry two hydrogens each. The general molecular formula for alkanes is C_nH_{2n+2}. Since all possible bonds of the carbon are completely saturated by C–C or C–H bonds, they are also called saturated compounds.

The names of the first four alkanes (methane, ethane, propane, butane) are historical. From the fifth carbon atom on, the compounds are named with Greek numerals: Pentane *(penta* = five), Hexane *(hexa* = six), Heptane *(hepta* = seven)

T. Schmiermund, *The Chemistry Knowledge for Firefighters*,
https://doi.org/10.1007/978-3-662-64423-2_44

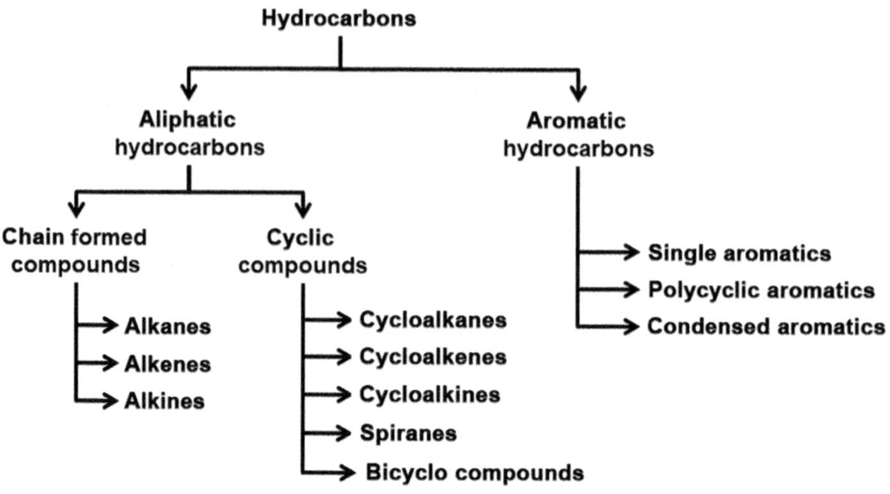

Fig. 44.1 Overview hydrocarbons

etc. If a terminal H atom is removed, a molecular residue remains, which in the case of alkanes is called an alk**yl** rest.

44.1.1.1 Formula Representation

The transition from the structural formula to the skeletal formula can be clearly seen in the example of alkanes.

Name	Structural Formula (with H Atoms)	Structural Formula (Without H Atoms)	Carbon Skeleton	Skeleton Formula
Methane	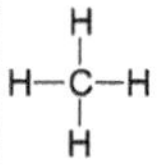		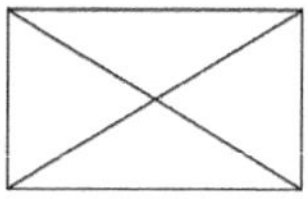	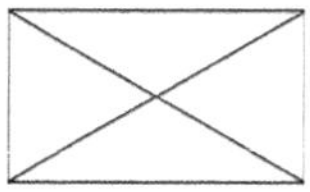
Ethane	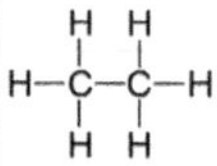	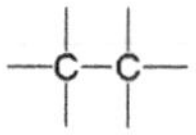	C–C	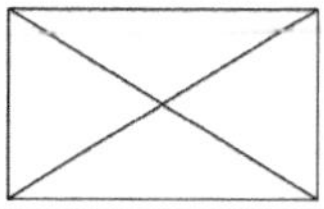
Propane	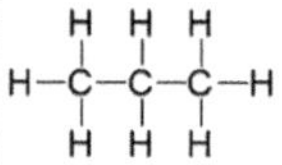	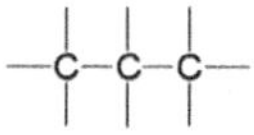	C–C–C	
Butane	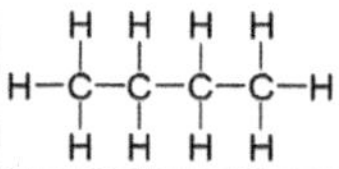	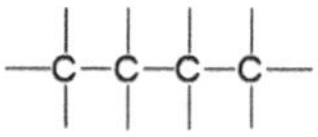	C–C–C–C	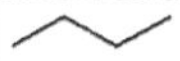

In the following, the different forms of presentation are chosen depending on the facts of the case.

44.1.1.2 Branched Chains

Starting with butane, it is possible to form branched structures in addition to the linear chain. To distinguish them, the straight-chain compounds are given an "*n*" for "normal". The branched chains get additions like "*iso*" from "isomer" or "*neo*" for "new". However, this naming system is confusing in the long run, so that a specific naming system, the nomenclature (Sect. 44.1.1.3), has been agreed upon.

Alkan	Sum Formula	Formula	Alkyl Radical	Formula
Methane	CH_4	| —C— |	Methyl- (*Me*)	CH_3-
Ethan	C_2H_6	C – C	Ethyl- (Et)	C_2H_5-
Propane	C_3H_8	C – C – C	*n*-Propyl- (*n*-Pr)	*n*-C_3H_7-
n-butane	C_4H_{10}	C – C – C – C	*n*-Butyl- (*n*-Bu)	*n*-C_4H_9-
iso-butane	C_4H_{10}	C \ C–C / C	*iso*-Butyl- (*i*-Bu)	*i*-C_4H_9-
n-pentane	C_5H_{12}	C – C – C – C – C	*n*-Pentyl-	*n*-C_sH_{11}-
iso-pentane	C_5H_{12}	C \ C–C–C / C	*iso*-Pentyl-	*i*-C_5H_{11}-
neo-pentane	C_5H_{12}	C | C–C–C | C	*-unusual-*	*-unusual-*

Hexane (C_6H_{14}) has 5 position isomers, heptane (C_7H_{16}) 9 and octane (C_8H_{18}) already 18 isomers. For dodecane ($C_{12}H_{26}$) there are already 355.

44.1.1.3 Nomenclature of the Alkanes

The term nomenclature (lat. = list of names) includes the totality of names and naming rules in chemistry. For each group of substances, the rules are expanded somewhat, so that in the end you have a clear set of rules for naming the various substances.

From pentane onwards, the names of the alkanes are formed from the Greek number words according to the number of carbon atoms and are given the suffix **-an.** Branched alkanes are given the name of the longest, unbranched carbon chain as the "parent name", whose C atoms are numbered consecutively. Alkyl radicals are given in alphabetical order, indicating the number of the C atom to which they are attached. If there are several identical alkyl radicals, this is indicated by a preceding Greek numeral. The numbers of the C atoms should always be as low as possible.

Examples

3-Ethyl-hexane

3-Ethyl-2-methyl-pentane

2,2,4-Trimethyl-pentane

44.1.1.4 Examples, Occurrence, Properties, Use

The first four alkanes are gases, pentane (C_5H_{12}) to heptadecane ($C_{17}H_{36}$) are liquids, alkanes with even longer chains are solids. The liquid alkanes have a petrol-like odour, all others are odourless. They are not water miscible (*hydrophobic* = water repellent) and flammable. The liquid alkanes have a lower density than water and therefore float.

Methane is the main component of natural gas, mine gas, swamp gas and biogas. Like ethane, propane and butane, it is used as a raw material in the chemical industry and as a heating gas. Hexane, heptane and octane isomers are essential components of gasoline ("petrol"). As a measure for the knock resistance, the so-called octane number (ON), *n*-heptane (ON = 0) and 2,2,4-trimethyl-pentane ("*iso*-octane", ON = 100) are used. Similarly, the cetane number (CN) is given for the ignitability of diesel fuels. The comparative substances here are *n*-hexadecane ($C_{16}H_{34}$, CN = 100) and α-methyl-naphthalene ($C_{11}H_{10}$, CN = 0). Paraffin oils are mixtures of the alkanes from C_{12} to C_{16}. Paraffin wax consists of mixtures of solid alkanes from C_{22} to C_{40}, with hard paraffin wax containing more longer, higher melting chains than soft paraffin wax.

Alkanes are relatively inert. The most important reactions, apart from combustion, are halogenation, sulfochlorination and nitration.

44.1.2 Alkenes

Alkenes, also called olefins (French *gaz olefiant* = oil-forming gas), have a double bond between two C atoms. Their general molecular formula is C_nH_{2n}. The (final) syllable **-en** is used to indicate the double bond. The position of the double bond is indicated by the number of the C atom at which it is present. If there is more than one double bond in the longest, name-giving chain, the identifier "en" is preceded by the corresponding Greek numeral.

Alken	Sum Formula	Formula	Other Notations	Alkyl Radical
Ethene	C_2H_4	$H_2C{=}CH_2$	“Ethylene”	Ethenyl, “Vinyl”
Propene	C_3H_6	$CH_3{-}CH{=}CH_2$	“Propylene”	Propenyl-
1-butene	C_4H_S	$H_2C{=}CH{-}CH_2{-}CH_3$	Buta-1-ene	1-Butenyl
2-butene	C_4H_8	$CH_3{-}CH{=}CH{-}CH_3$	Buta-2-ene	2-Butenyl
1,3-butadiene	C_4H_8	$H_2C{=}CH{-}CH{=}CH_2$	Butadiene	1,3-Butadienyl-
1-pentene	C5H_{10}	$CH_3{-}CH_2{-}CH_2{-}CH{=}CH_2$	Penta-1-en	1-Pentenyl
2-pentene	C5H_{10}	$CH_3{-}CH_2{-}CH{=}CH{-}CH_3$	Penta-2-en	2-Pentenyl
1,2,4-pentatriene	C_5H_6	$H_2C{=}CH{-}CH{=}C{=}CH_2$	Penta-1,2,4-triene	1,2,4-Pentatrienyl-

44.1.2.1 Cis-Trans Isomerism

In the alkanes, the C atoms can rotate freely around the axis of the single bond. This is not possible with the double bond of the alkenes. It behaves as if in the case of the alkanes a lath is attached to a board with *one* nail (= one bond), in the case of the alkenes the lath is attached to a board with *two* nails (= two bonds). In the first case the batten can be turned freely, in the second case it is fixed in its position.

This circumstance leads to a structural isomerism, which yields distinguishable, so-called stereoisomers (same molecule, but different spatial arrangement). If the substituents are on the same side of the molecule, we speak of *cis*-compounds (lat. *cis*= on this side); if they are on opposite sides, we speak of *trans*-compounds (lat. *trans* = across).

Examples

Substance	Structural Formula	Melting Point	Boiling Point
cis-1,2-Dichloro-ethene	H H C=C Cl Cl	−80 °C	60 °C
trans-1,2-Dichloro-ethene	H Cl C=C Cl H	−50 °C	48 °C
cis-But-2-ene	H H C=C H_3C CH_3	−139 °C	3,7 °C
trans-But-2-ene	H CH_3 C=C H_3C H	−106 °C	1 °C
cis-Butenedioic acid, *cis*-Ethylene-1,2-dicarboxylic acid, “maleic acid”	H H C=C HOOC COOH	141 °C	–

(continued)

Substance	Structural Formula	Melting Point	Boiling Point
trans-butenedioic acid *trans*-ethylene-1,2-dicarboxylic acid "fumaric acid	H(COOH)C=C(H)HOOC — H, COOH / C=C / HOOC, H	299 °C (decomposition)	–

44.1.2.2 (*Z*)-(*E*)-Isomerism

In the presence of different substituents, the naming according to the *cis-trans*-rule proves to be no longer sufficient. It is necessary to name according to a new rule: The (Z)-(*E*)-rule. Here the *(Z)* stands for "together" and the (*E*) for "opposite". This is considered from the substituent that has the highest priority. The priority is determined -simplified- according to the number of bonds (triple before double before single) and then according to the atomic number (higher atomic number goes first). In the case of alkyl residues, one moves one place further out, if necessary, until the priorities of all groups are determined.

Examples

Substance	Structural Formula (Z-Compound)	Structural Formula (*E*-Compound)	Substance
(Z)-But-2-ene	H, H / C=C / H_3C, CH_3	H, CH_3 / C=C / H_3C, H	(*E*)-But-2-ene
(Z)-1-Chloro-2-methyl-butene	Cl, C_2H_5 / C=C / H, CH_3	Cl, CH_3 / C=C / H, C_2H_5	(*E*)-1-chloro-2-methyl-butene
(Z)-1-Brom-2-chlor-1-iod-ethene	Cl, I / C=C / H, Br	Cl, Br / C=C / H, I	(*E*)-1-Brom-2-chlor-1-iod-ethene

44.1.2.3 Polyene

In the case of longer chains, alkenes with several double bonds may also be present. With two double bonds one speaks of dienes, with three double bonds of trienes, generally of polyenes.

The position of the double bonds to each other influences the reactions of the polyenes and the resulting products. A distinction is therefore made between:

- **accumulated double bonds:** several (at least two) double bonds without being separated from each other by single bonds (lat. *cummulare* = to accumulate)
- **conjugated double bonds:** an alternating sequence of single and double bonds (lat. *koniugare* = to connect, meaning: belonging together)
- **isolated double bonds:** Double bonds that are separated from each other by more than one single bond (lat. *isolare* = to make an island, to separate).

Examples

Double Binds	Formula	Name	*Comment*
Cumulated	$CH_2{=}C{=}CH{-}CH_2{-}CH_3$	1,2-Pentadiene (Penta-1,2-diene)	*Sequential numbering*
Conjugated	$CH_2{=}CH{-}CH{=}CH{-}CH_3$	1,3-Pentadiene (Penta-1,3-diene)	*Only even or odd digits*
Isolated	$CH_2{=}CH{-}CH_2{-}CH{=}CH_2$	1,4-Pentadiene (Penta-1.4-diene)	*Digit spacing: at least 3*

44.1.2.4 Examples, Production, Use, Reactions

The lower four alkenes (C_2 to C_4) are gases, from C_5 liquids and greater C_{15} solids. The most important alkenes are probably ethene ("ethylene"), propene ("propylene") and the butenes. These substances are produced, for example, in appreciable quantities during the thermal cracking of crude benzenes.

Isobutene (2-Methylprop-1-ene) is used for the production of *iso*-octane. Otherwise, plastics are mainly produced from the lower alkenes (cf. Chap. 49).

The double bond reacts more readily than the single bond. In this way, epoxides can be obtained by the action of peroxo acids and other oxidizing agents. Via ozonides as an intermediate stage, aldehydes and ketones are obtained. Typical for alkenes are so-called addition reactions. This means that suitable groups are "added" to the molecule. The double bond is opened, so that at the end there is a C−C bond instead of a C=C bond. Important addition reactions of alkenes are listed in Table 44.1.

By the way, overripe apples, bananas and tomatoes give off ethene. This has the effect that fruits around them ripen faster. Ethene is a true plant hormone and is used by plants to regulate the ripeness of their fruit.

44.1.3 Alkine

In addition to the alkanes and the alkenes, there is another group of compounds that also form a homologous series: the alkynes. Characteristic for this class of compounds is the C-C triple bond (C≡C), their general molecular formula is C_nH_{2n-2}. After the trivial name of the simplest alkyne, they are also called "acetylenes".

Table 44.1 Important addition reactions of alkenes

Addition of:	R−CH=CH−R becomes	Emerging connection class
Halogen	R−CHCl−CHCl−R	Dihaloalkanes
Hydrogen halide (HCl, HBr)	$R{-}CH_2{-}CHCl{-}R$	Halogenalkanes
Conc. Sulphuric acid (H_2SO_4)	$R{-}CH_2{-}CH(OSO_3H){-}R$	Alkyl hydrogen sulfonates
Hypochlorous acid (HClO)	R−CHCl−CH(OH)−R	Chloroalkanols

Alkin	Sum Formula	Structural Formula	Other Notations	Alkyl Radical
Ethine	C_2H_2	$HC{\equiv}CH$	"Acetylene"	Ethinyl-
Propin	C_3H_4	$H_3C{-}C{\equiv}CH$	"Methyl acetylene"	Propinyl-
But-1-in	C_4H_6	$HC{\equiv}C{-}CH_2{-}CH_3$	"Ethyl acetylene"	1-Butinyl-
But-2-in	C_4H_6	$H_3C{-}C{\equiv}C{-}CH_3$	"Dimethylacetylene"	2-Butinyl-

Alkines are also subject to *cis-trans or* (*Z*)-(*E*) isomerism. Poly-ines, i.e. compounds with several $C{\equiv}C$ groupings, are also known.

44.1.3.1 Examples, Production, Use, Reactions

The most important alkyne is ethine ("acetylene"), which is used in large quantities both in technical organic chemistry and as a welding gas. It can be produced from methane, e.g. by incomplete combustion in O_2, or by pyrolysis in an electric arc.

Pure ethine is a narcotic, etheric smelling gas. The "typical" smell of the technical gas comes from impurities, mostly PH_3. Ethine dissolves noticeably in water, even better in propanone (acetone): 1 L dissolves 15 L ethine at normal pressure. Liquefied ethine decomposes explosively on impact or heating, as it is a strongly endothermic compound:

$$C_2H_2 \rightarrow 2\,C + H_2 \qquad \Delta H_R = -226\ \text{kJ mol}^{-1}$$

In order to transport ethine safely, the pressurized gas cylinders are filled with a porous material such as diatomaceous earth or pumice stone, which is then soaked in acetone. The acetone then serves as a solvent for the ethine: at 12 bar, 300 L = 325 g = 12.5 mol ethine dissolve per liter of acetone. *Other alkynes are of no importance in industrial applications.*

Like alkenes, alkines can also undergo addition reactions. These take place via the intermediate stage of the corresponding alkene, which can be produced in this way if necessary. In addition, the combination of three or four ethyne molecules to form a ring system and polymerization are interesting. Important reactions are shown in the following overview.

Addition Reactions (Examples)

Reaction, Level 1	Reaction, Level 2
$H - C = C - H + Cl_2 \rightarrow ClHC = CHCl$	$ClHC = CHCl + Cl_2 \rightarrow Cl_2HC - CHCl_2$
$H - C = C - H + HBr \rightarrow H_2C = CHBr$	$H_2C = CHBr + HBr \rightarrow CH_3C - CHBr_2$

Cyclizations

Trimerization	Tetramerization
$3\,H - C = C - H \rightarrow C_6H_6$	$4\,H - C = C - H \rightarrow C_8H_8$

Benzene (C_6H_6) is required in large quantities for petrol and for the production of benzene derivatives.

Polymerization

$$n\,\mathrm{H{-}C{\equiv}C{-}H} \rightarrow -[\mathrm{CH{=}CH{-}CH{=}CH{-}CH{=}CH}-]_n$$

The resulting plastic "polyacetylene" is an interesting compound. On the one hand, it is a polyene with continuously conjugated double bonds, which has an electrically insulating effect. On the other hand, by doping with sodium, it is possible to produce an electrically conductive plastic whose conductivity at low temperatures is of the order of mercury.

> **Note** The correct name "ethine" is not very common outside chemistry. Therefore, the compound is also usually referred to here by its trivial name "acetylene".

44.2 Excursus: Bonding Ratios at the C Atom

The addition reactions of the alkenes and alkynes indicate that the double or triple bonds must be of a different nature than the single bonds. Measurements of bond lengths and bond energies also show differences.

Type	C–H bond length	C–C bond length	C–C binding energy
Alkane –C–C–	110 pm	154 pm	346 kJ mol^{-1}
Alkene –C=C–	108 pm	134 pm	602 kJ mol^{-1}
Alkyne –C≡C–	106 pm	121 pm	836 kJ mol^{-1}

These measurement results suggest that there are differences in the single bond versus the additional bonds of the double or triple bond.

44.2.1 Electron Structure of the C Atom

Carbon is in group 14 or the IV. main group of the periodic table. Its electron structure is [Ne]$2s^2 2p^2$. Since the electrons of the s-orbitals do not have the same energy as the electrons of the p-orbitals, it would initially be assumed that there are two somewhat stronger and two less stronger C–H bonds in the methane molecule. This is not the case, as measurements show. Therefore, there must be a mechanism that makes the four bonds in the methane molecule "equal". For a better understanding we use the box notation (compare Sect. 6.4.2).

44.2.2 sp^3 Hybridization

The four outer electrons of the carbon "mix" before the carbon forms bonds. This "mixing process" is called hybridization. In the simplest case, four energetically exactly identical orbitals are created from two times two energetically different orbitals.

First, an excited state (b) is formed from the ground state (a). For this purpose, one electron of the s-orbital is transferred to the p-orbital. Afterwards, four equal hybrid orbitals (c) are formed from these 3 p-orbital electrons and the remaining electron of the s-orbital, which are then called sp^3 hybrid orbitals and lie energetically between the energy level of the s- and the p-orbital.

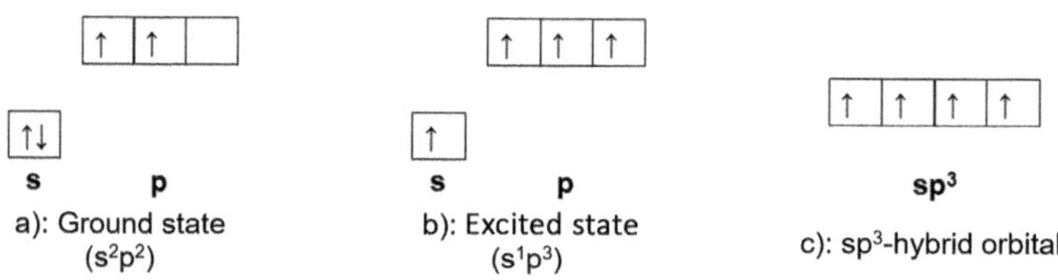

The four identical hybrid orbitals form four equivalent bonds, called sigma bonds *(σ-bonds)*.

44.2.3 sp^2 Hybridization

Analogous to the sp^3 hybridization, the excited state (s^1 p^3) is formed from the ground state (s^2p^2). But now the electron of the s-orbital mixes with only two electrons of the p-orbital. One unpaired electron remains in the p orbital.

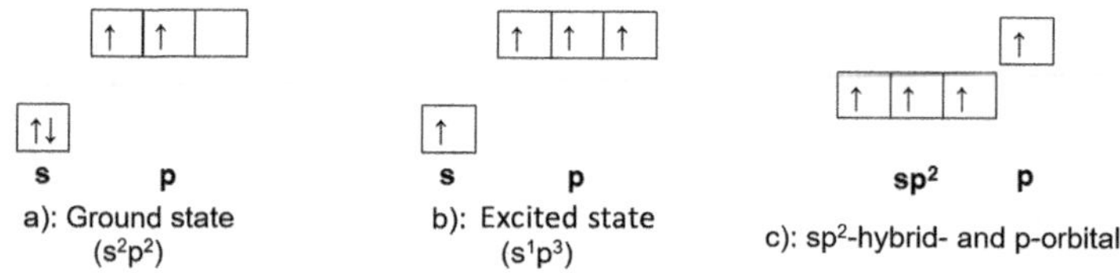

The three electrons of the sp^2 -hybrid orbitals form three equivalent single bonds (σ-bonds). The remaining p-electron forms a so-called π-bond (pi-bond) with the remaining p-electron of another carbon atom. This is the second, quasi the "reactive part" of the double bond. A double bond is composed of a σ-bond and a π-bond.

44.2.4 sp-Hybridization

The sp. hybridization is analogous to the other two hybridizations, except that now two electrons remain in the p orbital.

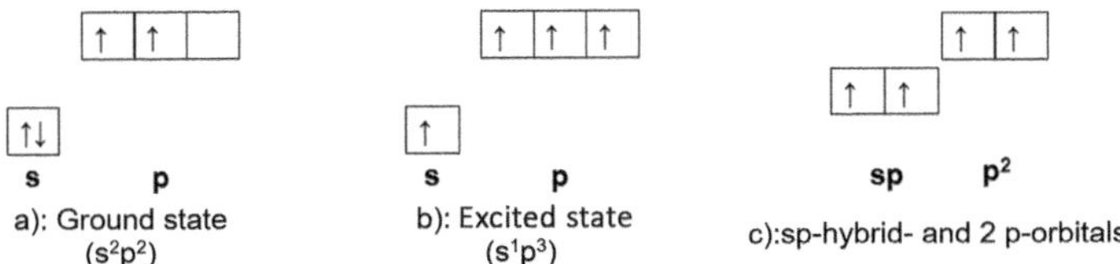

The two electrons of the sp. hybrid orbitals form two equivalent single bonds (sp hybrid orbitals). The two electrons in the p orbital each form a π-bond. The triple bond thus consists of one σ- and two π-bonds.

44.3 Annular Aliphatic Hydrocarbons

Alkanes, alkenes and alkines are able to form ring-shaped compounds, so-called carbocycles, to "bite themselves in the tail", so to speak. Compounds with several ring-shaped structures are also known. Depending on the number of ring structures, these are called bicyclic or tricyclic hydrocarbons.

44.3.1 Cyclic Hydrocarbons

To distinguish them from the chain-shaped compounds, the syllable "cyclo-" is prefixed to the ring-shaped alkanes / alkenes / alkines, the final syllables -an / -en / -in remain.

44.3.2 Bicyclic and Polycyclic Compounds

In bicyclic or polycyclic compounds, several ring systems share one or more carbon atoms. In spiro compounds or spiranes (lat. *spira* = pretzel), the rings are connected by a common C atom. The name is given as the total carbon system and the number of members per ring is given in square brackets. If the rings share two C atoms, they are called fused rings. In the name, in addition to the designation "bicyclo-", the number of C atoms forming a separate bridge is given as the third digit.

A polycyclic system called Adaman-tan is particularly interesting, since it represents a section of the structure of the diamond lattice (compare Sect. 8.3.2).

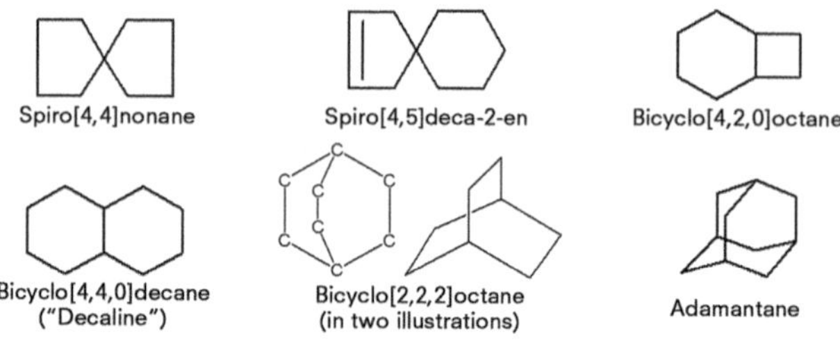

44.4 Aromatic Hydrocarbons

Examining the formal "cyclohexa-1,3,5-triene", we find that the C–H bond distance is 108 pm, the same as an alkene. The C–C bond distance is 139 pm for all C–C bonds, which is intermediate between the distance for alkane (154 pm) and alkene (134 pm); although two different bond lengths would be expected. At the same time, the molecule proves to be rather inert and does not undergo various reactions typical of double bonds, such as the addition of Br_2 (shaking with aqueous bromine solution).

In the case of hydrogenation – i.e. the "addition" of hydrogen – the energy required is outside the expected range: In the hydrogenation of cyclohexene to cyclohexane, 120 kJ mol^{-1} are released. The hydrogenation of cyclohexa-1,3-diene to cyclohexane releases about 240 kJ mol^{-1}; corresponding to $2 \cdot 120$ kJ mol^{-1}. For the hydrogenation of "cyclohexa-1,3,5-triene" to cyclohexane, about 360 kJ mol^{-1} are expected. In fact, however, only 208 kJ mol^{-1} are released. This means that the molecule is about 150 kJ mol^{-1} more stable than expected (Fig. 44.2).

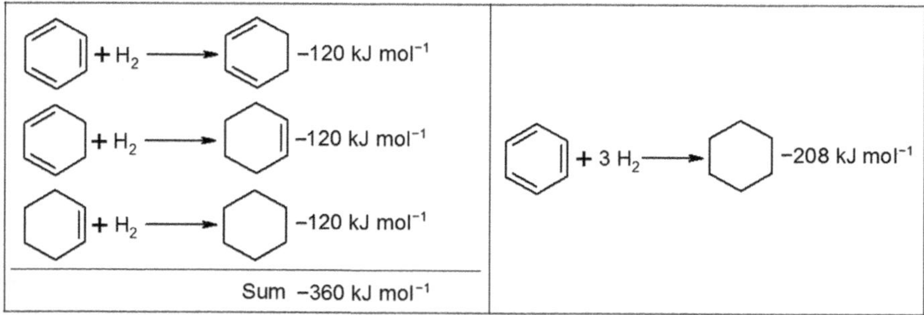

Fig. 44.2 Comparison of the reaction energy during hydrogenation: (**a**) the calculated value for multi-stage hydrogenation, (**b**) the experimentally determined value

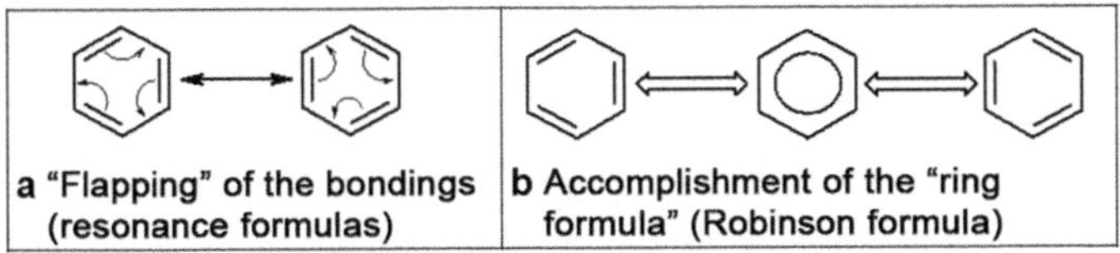

Fig. 44.3 Benzene formulae

The chemical behaviour of this molecule can therefore *not* be explained by an alternation of single and double bonds in the ring system. Rather, it is as if there are 1½ bonds between the C atoms. There must therefore be a mechanism that stabilizes the molecule.

This is indeed the case. The π-bond, as a component of the double bond, flaps, as it were, from one C atom to the next. And it does so constantly and across all six C atoms, as shown in Fig. 44.3a. Thus, two mesomeric boundary formulas give rise to the "intermediate formula" shown in Fig. 44.3b.

Both the left and right structures in Fig. 44.3b represent the boundary formulas. The middle structure is closest to the real molecule. Following Lit. [4-1] (p. 363, German edition) the following comparison may be allowed: *A traveller of the Middle Ages explains a rhinoceros as a mixture of a dragon and a unicorn to those who stayed at home. The right structure is therefore a dragon, the left one a unicorn. Both animals are mythical creatures that do not exist. The "mixture" of both animals, the rhinoceros (i.e. the middle structure), on the other hand, is real.* Nevertheless, we will keep the representation with the alternating single and double bonds, because it has several advantages -despite: It remains a *"fable formula"*. The cyclohexa-1,3,5-triene ("dragon") is just as non-existent as the cyclohexa-2,4,6-triene ("unicorn").

The ring-shaped structure of the compound better known as benzene was recognized by Kekulé[1] in 1865. The boundary structures shown here are also known as the "Kekulé benzene formulae". The structure with the π-electron ring is also called the Robinson formula.

Compounds which, like benzene, undergo stabilization via π-electron pair bonds are called **aromatics** in reference to the aromatic odor of benzene. The theoretical interpretation of **aromaticity** was not made until 1931 by Hückel[2] .

Aromaticity and Hückel Rule

A compound is considered aromatic if the number of π-electrons follows the Hückel rule ($4n + 2$). For $n = 1$, this gives the 6 π-electrons of benzene. Further requirements for an aromatic molecule are:

- There must be a ring-shaped carbon skeleton.
- The ring must be flat ("planar").

[1]Friedrich August Kekule von Stradonitz, 1829–1896, German chemist.

[2]Erich Armand Arthur Joseph Hückel, 1896–1980, German chemist and physicist.

- Conjugated double bonds must be present.
- The number of double bonds must be odd.

If all conditions apply, the molecule is more stable than a chain-shaped molecule or a cyclic molecule with a different arrangement of the double bonds. The reason for this stability is the ring-shaped closed system of delocalized π-electrons, which has become the hallmark of "aromatic compounds".
The Hückel rule also applies to ring systems with heteroatoms (e.g. N, O, S) in the molecule, but in the narrower sense it is only valid for individual rings. In the case of polycyclic aromatic hydrocarbons, for example, there are ring systems that satisfy the rule as well as those that do not.

44.4.1 Simple Aromatic Compounds

The most common aromatic building block is benzene, which must be correctly called benzene because of the double bonds formally present. Many compounds are derived from it, which are important starting materials for pharmaceuticals, pigments, dyes, plastics, and many more. The C_6H_5 group is called the phenyl radical and is sometimes abbreviated to "Ph".

Name(s)	Structure	Notes
Benzene		UN No. 1114, solvent, important precursor for other aromatics
Toluene (methylbenzene)	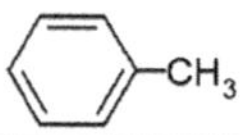 $-CH_3$	UN No. 1294, solvent, important precursor, TNT production
Nitrobenzene	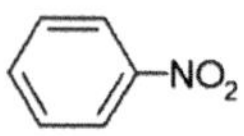 $-NO_2$	UN No. 1662, Oxidizing agent, precursor for aniline, etc.
Aniline (aminobenzene)	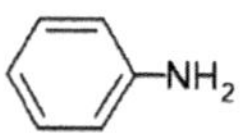 $-NH_2$	UN No. 1547, precursor for pharmaceuticals, pigments, dyes, etc.
Styrene (vinylbenzene)	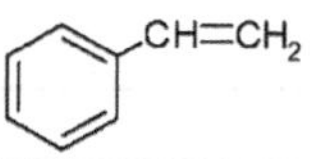 $CH{=}CH_2$	UN No. 2055, manufacture of polystyrene (PS)
Chlorobenzene	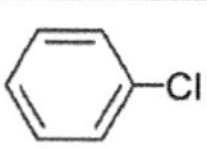 -Cl	UN No. 1134, solvent, precursor
1,2-dichlorobenzene	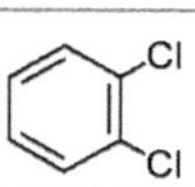 Cl, Cl	UN No. 1591, solvent, precursor

44.4.1.1 Naming of Simple Aromatic Compounds

If two or more hydrogens of benzene are replaced by other atoms or groups of atoms, different molecules result. In order to enable a clear naming, the C-atoms of the ring system are numbered consecutively. Special names have become established for certain constellations of disubstituted benzene derivatives:

- 1,2-position: *ortho* (abbreviation "*o*-") from Greek *ortho* = correct, straight
- 1.3-position: *meta* (abbreviation "*m*-") from Greek *meta* = between
- 1,4-position: *para* (abbreviation "*p*-") from Greek *para* = beside, along

While retaining the trivial names of individual compounds, this gives rise to various naming possibilities, as the examples show.

Examples "Naming of Aromatics"

Structure	Designations
H_2N–C₆H₄–Cl	1-Amino-4-chloro-benzene 4-Chloroaniline / *para*-Chloroaniline
H_2N–C₆H₄–NH_2	*para*-Phenylenediamine 4-Aminoaniline 1,4-Diaminobenzene
O_2N, NH_2	1-Amino-3-nitro-benzene 3-Nitroaniline / *meta*-nitroaniline
Cl, Cl	1,2-Dichlorobenzene *ortho*-Dichlorobenzene / o-DCB
NH_2, CH_3	1-Amino-2-methyl-benzene 2-Amino-toluene / *ortho*-Amino-toluene 2-Methyl-aniline / *ortho*-Methyl-aniline
Cl, Br, Br	1.3-Dibromo-2-chloro-benzene 2.3-Dibromo-1-chloro-benzene
CH_3, O_2N, NO_2, NO_2	1-Methyl-2,4,6-Trinitro-benzene 2,4,6-Trinitro-toluene TNT

44.4.2 Polynuclear Aromatic Compounds

If aromatic ring systems are connected to each other by single bonds or by aliphatic (chain-like) C atoms, they are referred to as polynuclear aromatic compounds. Typical examples are biphenyl, diphenylethane or the triphenylmethane used for dye synthesis.

Examples

Biphenyl Diphenylethane Triphenylmethane

44.4.3 Condensed Aromatic Compounds

Fused aromatic compounds are aromatic systems in which two or more rings are linked to each other via several common C atoms. These include in particular the polycyclic aromatic hydrocarbons (PAHs). The simplest representative of the PAHs is naphthalene, which consists of two fused benzene rings. Anthracene and phenanthrene consist of three aromatic rings fused together. With an odd number of C=C bonds, they satisfy the Hückel rule. Pyrene and fluoranthene, so-called because it is highly fluorescent, have an even number of C=C bonds, so are antiaromatic compounds. The [18]annulene ($C_{18}H_{18}$), on the other hand, is an aromatic compound.

Examples

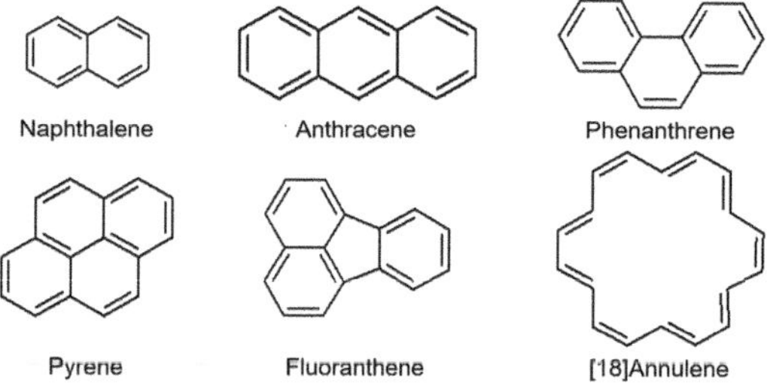

45 Organic Halogen Compounds

A large number of organic compounds of industrial importance are halogenated hydrocarbons. As end products, the chlorine derivatives, as intermediates, especially in laboratory syntheses, the bromine derivatives are of greater importance. Instead of the specific halogen, "X" is sometimes written in overviews to indicate that it is any halogen.

The halogen compounds include many technically important solvents, such as methylene chloride (dichloromethane, CH_2Cl_2, UN No. 1593), chloroform (trichloromethane, $CHCl_3$, UN No. 1888), ethylene chloride (1,2-Dichloroethane, $ClCH_2{-}CH_2Cl$, UN No. 1184) and others.

The plastic precursors tetrafluoroethylene ($F_2C{=}CF_2$, UN No. 1081) and vinyl chloride ($H_2C{=}CHCl$, UN No. 1086), both of which are highly flammable, are of great industrial importance.

45.1 The "Dirty Dozen" of Halogen Compounds

Some of the organic halogen compounds are virtually non-degradable in the environment and accumulate in body tissue, among other places. The use or production of these persistent organic pollutants (POPs) was banned or severely restricted by the Stockholm Convention (2001). As of 2009, further substances were added to the list of the Stockholm Convention.

The substances originally included in the Stockholm Convention are:

- the insecticides aldrin, chlordane, dieldrin, dichlorodiphenyltrichloroethane (DDT), endrin, heptachlor, mirex and toxaphene
- the cereal fungicide hexachlorobenzene (HCB)
- polychlorinated biphenyls (PCBs) as industrial chemicals
- polychlorinated dibenzodioxins (PCDD) and dibenzofurans (PCDF), which are formed as combustion by-products, among other things

T. Schmiermund, *The Chemistry Knowledge for Firefighters*,
https://doi.org/10.1007/978-3-662-64423-2_45

Among others, the following have been added in the meantime:

- γ-Hexachlorocyclohexane (lindane, insecticide), hexabromobiphenyl (flame retardant), perfluorooctanesulfonic acid (used, inter alia, for AAAF foaming agents) and endosulfan (thiodane, insecticide).

46 Organic Oxygen Compounds

Oxygen forms a large number of important compound classes in organic substances (Fig. 46.1). The functional groups formed with oxygen are polar and therefore cause solubility/miscibility with polar liquids, such as water. The associated carbon chain is nonpolar, thus causing miscibility with nonpolar solvents such as hexane or benzene. The longer the C-chain becomes, the worse the solubility in polar liquids. Substances that dissolve in both polar and nonpolar substances are called amphiphilic (Greek *ampho* = both and *philos* = loving) because they have both lipophilic and hydrophilic properties. The addition of amphiphilic substances to non-polar substances gives them a solubility for polar substances that depends on the concentration.

46.1 Alcohols

Alcohols, actually alkanols, can be thought of as water molecules in which a hydrogen atom has been replaced by a hydrocarbon radical. Their general formula is therefore R–OH. The hydroxy group –OH is one of the most important functional groups and gives alcohols their characteristic chemical and physical properties.

The name is given after the longest chain to which the –OH group is attached, with the suffix -ol. A number is used to indicate the C atom on which the hydroxyl group is located. This C-atom should receive the lowest possible number.

Alcohols react with water to form alkenes or ethers (see Sect. 46.4), with acids to form esters (see Sect. 46.6) and with oxidizing agents to form aldehydes (Sect. 46.2) or ketones (Sect. 46.3).

Examples of technically important alcohols are given in Table 46.1.

T. Schmiermund, *The Chemistry Knowledge for Firefighters*,
https://doi.org/10.1007/978-3-662-64423-2_46

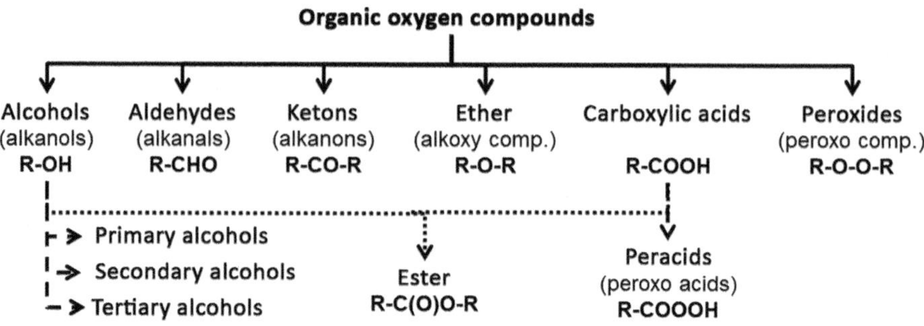

Fig. 46.1 Overview of organic oxygen compounds

Table 46.1 Technically important alkanols (alcohols)

Name	More Designations	Formula	Notes
Methanol	Methyl alcohol, wood alcohol	CH_3OH	Toxic, flammable; important solvent, precursor for formaldehyde and formic acid; UN No. 1230
Ethanol	Ethyl alcohol, (drinking) alcohol	C_2H_5OH	In alcoholic beverages, flammable, important solvent, additive in fuels (E10, E85), precursor for the synthesis of other substances; UN No. 1170
Propan-1-ol	*n*-Propanol, Propyl alcohol	C_3H_7OH	Flammable, solvent, for synthesis of other substances; UN No. 1274
Propan-2-ol	*iso*-Propanol, Isopropyl alcohol	$CH_3CH(OH)CH_3$	Flammable, solvent, in antifreeze, de-icers and glass cleaners, precursor for the synthesis of other substances; UN No 1219
2-Methyl-propanol	*iso*-Butanol, Isobutyl alcohol	$(CH_3)_2C_2H_3OH$	Flammable, solvent, precursor for the synthesis of other substances; UN Nr. 1212
Butan-1-ol	*n*-Butanol, *n*-Butyl alcohol	$C_4H_{10}OH$	Flammable, solvent, precursor for the synthesis of other substances; UN Nr. 1120
Butan-2-ol	*sec.* Butanol	$C_2H_5CH(OH)CH_3$	Flammable, solvent, precursor for the synthesis of other substances; UN Nr. 1120
Ethane-1,2-diol	Ethylene glycol, Glycol	$C_2H_4(OH)_2$	Harmful to health, tastes sweet, in antifreeze and de-icers, precursor for the synthesis of other substances
Propane-1,2,3-triol	Glycerin, Glycerol	$C_3H_5(OH)_3$	Tastes sweet, humectant in cosmetics, precursor for the synthesis of other substances, building block in almost all animal fats and vegetable oils
Propane-1,2-diol	1,2-Propylene glycol	$CH_3C_2H_5(OH)_2$	Precursor

46.1.1 Primary, Secondary, Tertiary Alcohols

Depending on the number of carbon atoms bonded to the C atom carrying the OH group, a distinction is made between primary, secondary and tertiary alcohols. This distinction is important because primary (*prim.*), secondary (*sec.*) and tertiary (*tert.*) alcohols behave chemically differently – especially with regard to their oxidation behaviour.

In primary alcohols, only a maximum of one carbon atom is directly bonded to the carbon atom carrying the OH group. In the case of secondary alcohols, two C atoms are bonded to the C atom carrying the OH group, and in the case of tertiary alcohols, three C atoms are bonded accordingly.

$H_3C{-}CH_2{-}CH_2{-}CH_2{-}OH$

1-Butanol
(Primary alcohol)

$H_3C{-}CH_2{-}CH(OH){-}CH_3$

2-Butanol
(Secondary alcohol)

$H_3C{-}C(CH_3)(OH){-}CH_3$

2-Methyl-2-propanol
(Tertiary alcohol)

46.1.1.1 Oxidation of Primary Alcohols

Primary alcohols can be oxidized under *mild* conditions to alkanals (aldehydes), which are then further oxidized to carboxylic acids. The oxygen written in square brackets means "oxygen from any oxidizing agent". As an example, the oxidation of ethanol to ethanal ("acetaldehyde") and further to ethanoic acid ("acetic acid").

$$H_3C-CH_2-OH \overset{+[O]/-H_2O}{\rightarrow} H_3C-CHO \overset{+[O]}{\rightarrow} H_3C-COOH$$

46.1.1.2 Oxidation of Secondary Alcohols

Alkanones (ketones) are formed during the oxidation of secondary alcohols. For example, propan-2-ol becomes propanone, better known as "acetone":

$$H_3C-CH(OH)-CH_3 \overset{+[O]/-H_2O}{\rightarrow} H_3C-C(O)-CH_3$$

46.1.1.3 Oxidation of Tertiary Alcohols

During the reaction of tertiary alcohols with oxygen-donating oxidizing agents, they are always oxidized to CO_2 and H_2O. The conversion of the alcoholic OH group into another oxygen-containing functional group is not possible.

$$(H_3C)_3C-OH \overset{[O]}{\rightarrow} 4\,CO_2 + 5\,H_2O$$

46.1.2 Polyhydric Alcohols

The number of OH groups determines the so-called "valence" of an alcohol. Monohydric alcohols have one, bihydric alcohols have two, trihydric alcohols have three OH-groups and so on. Here, the Erlenmeyer[1] rule applies, which states that two OH groups on a C atom are not possible.

Typical examples of technically important polyhydric alcohols are, for example, ethane-1,2-diol (also called glycol or ethylene glycol) and propane-1,2,3-triol (= glycerol). Other polyhydric alcohols are also important intermediates in organic syntheses.

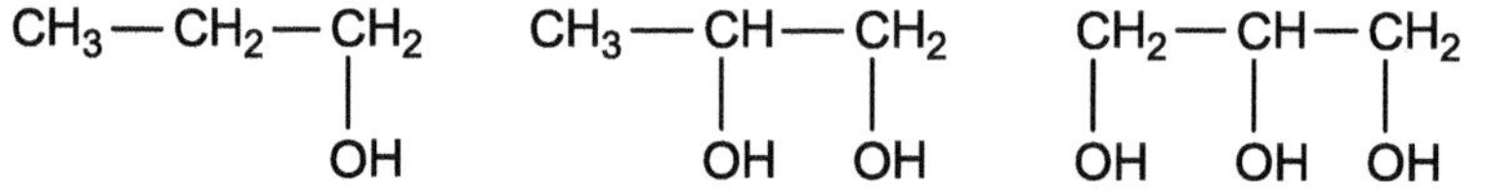

$CH_3—CH_2—CH_2(OH)$ $CH_3—CH(OH)—CH_2(OH)$ $CH_2(OH)—CH(OH)—CH_2(OH)$

Propan-1-ol (einwertiger Alkohol)

Propan-1,2-diol (zweiwertiger Alkohol)

Propan-1,2,3-triol *(„Glycerin")* (dreiwertiger Alkohol)

46.1.3 Phenols

If one or more OH-groups are directly bound to the benzene ring, these substances are called phenols. To a certain extent, they can be regarded as tertiary alcohols of aromatic compounds. Depending on the number of OH groups, they are also subdivided into monovalent, divalent and trivalent phenols. Phenols are more acidic than alcohols and show a different chemical behavior than "typical" alcohols. They are therefore usually treated as a separate group.

The simplest representative is the phenol ("hydroxybenzene") which was formerly used as disinfectant ("carbolic acid"). Also the 3-methyl-phenol (*m*-cresol) has an antiseptic effect and is used in disinfecting liquid soaps *("Lysol")*.

Phenol and its derivatives are used in the manufacture of pharmaceuticals and plastics (phenoplastics or "Bakelites"), 1,2,3-trihydroxybenzene ("pyrogallol") is used in photographic developers.

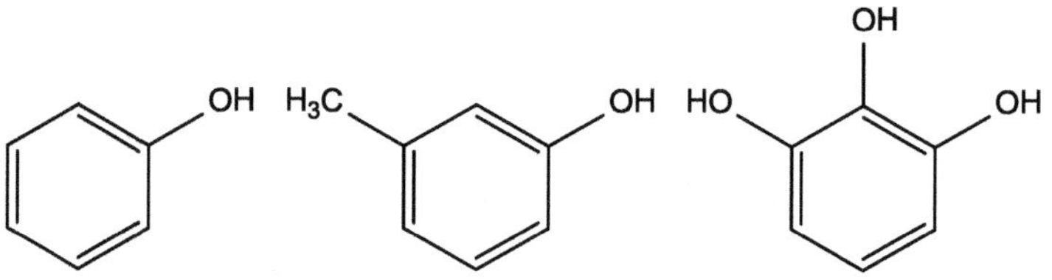

OH H_3C OH HO OH OH

Phenol

3-Methyl-phenol (*m*-Kresol)

1,2,3-Trihydroxybenzen *(Pyrogallol)*

[1] Emil Erlenmeyer, 1825–1909, German chemist.

2,4-Dichlorophenoxyacetic acid (2,4-D) is a weedkiller (herbicide), as is 2,4,5-trichlorophenoxyacetic acid (2,4,5-T). The latter is *teratogenic,* i.e. it causes malformations in embryos. As "Agent Orange", a 1:1 mixture of the n-butyl esters of 2,4-D and 2,4,5-T was used in the Vietnam War as a defoliant and to destroy crops. It was later found to contain up to 14 ppm of the highly toxic 2,3,7,8-TCDD (see Sect. 40.3.2) as an impurity.

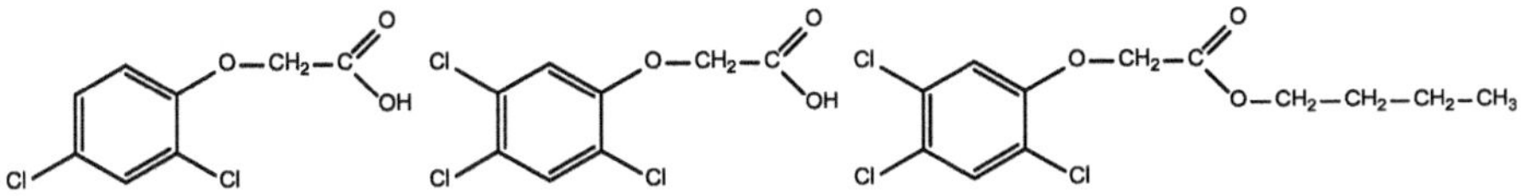

2,4-D 2,4,5-T 2,4,5-T-n-butylester (Bestandteil von „Agent Orange")

46.2 Aldehydes

The name "aldehyde" is derived from the term ***alcohol dehydrogenatus*** (dehydrated alcohol = alcohol with fewer hydrogens). The IUPAC designation is alkanals, the compounds accordingly end in -al. The –CHO group is characteristic. The nomenclature is analogous to that of alcohols.

Unlike alcohols, a carbon atom is part of the functional group. The C=O group, called "carbonyl", and an H atom located at this C atom together form the aldehyde function. They are found as natural substances in almonds (benzaldehyde), vanilla (vanillin) or oranges (citral). Examples are given in Table 46.2.

46.3 Ketones

Ketones are characterized by the carbonyl group (C=O). Both free bonds are connected with C-atoms. The IUPAC designation is alkanone, the compounds accordingly end in **-one** (for the structure see Table A.27). In nature, ketones are found in many plants as fragrances and aroma compounds, e.g. in rose petals, raspberries and jasmine. Examples can be found in Table 46.3.

46.4 Ether

The alkoxyalkanes or ethers can be understood as derivatives of water, where both hydrogen atoms of the water molecule have been replaced by alkyl groups. Thus, the oxygen atom connects two carbon chains with each other, so to speak. A distinction is made between symmetrical ($R_1 = R_2$) and asymmetrical ($R_1 \neq R_2$) ethers. In

Table 46.2 Examples of alkanals (aldehydes)

Name	Other Designations	Formula	Notes
Methanal	Formaldehyde	$H{-}CHO$	toxic, disinfectant, basic material for resins, adhesives, plastics etc.; UN No. 1198 (solution)
Ethanal	Acetaldehyde	$CH_3{-}CHO$	for the production of acetic acid, raw material for plastics, medicines, etc.; UN No. 1089
Propanal	Propionaldehyde	$C_2H_5{-}CHO$	Basic material for medicines, plastics, perfumes and pesticides; UN No. 1275
Benzaldehyde	Formylbenzene	$(C_6H_5){-}CHO$	for the manufacture of medicaments, perfumes and aromatic substances, dyestuffs; UN No. 1990
Propenal	Acrolein	$H_2C{=}CH{-}CHO$	lachrymatory liquid, precursor for plastics and medicines; UN No. 1092

Table 46.3 Important alkanones (ketones)

Name	Other Designations	Formula	Notes
Propanone	Acetone	$H_3C{-}CO{-}CH_3$	important solvent, plastic precursor; UN No. 1090
Butan-2-one	Methyl ethyl ketone (MEK)	$H_3C{-}CO{-}C_2H_5$	Solvent; UN No. 1193
Butane-2,3-dione	Diacetyl	$H_3C{-}CO{-}CO{-}CH_3$	Butter flavouring; UN No. 2346
Pentane-2,4-dione	Acetylacetone	$H_3CC(O)CH_2C(O)CH_3$	Precursor for medicinal products; UN No. 2310
Cyclohexanone	–	$C_6H_{10}O$	For the production of plastics and as solvent; UN No. 1915
Phenylethanone	Acetophenone	$C_6H_5{-}CO{-}CH_3$	For the production of medicines, flavourings and dyestuffs
Diphenylmethanone	Benzophenone	$C_6H_5{-}CO{-}C_6H_5$	as UV absorber in sun creams, lacquers and plastics; UN No. 3077

addition, there are cyclic ethers in which the O atom forms a member of the ring system. Of particular importance are the three-membered epoxides.

Some of the ethers are water-miscible and dissolve well in alcohols. Because they are unable to form hydrogen bonds, they have much lower boiling points than alcohols of the same empirical formula (see Table 46.4).

Table 46.4 Comparison of boiling points of alcohols and ethers with the same molar mass

	Alcohol		Ether	
Sum formula	Name	Boiling point	Name	Boiling point
C_2H_6O	Ethanol	78.4 °C	Dimethyl ether	−24.9 °C
$C_4H_{10}O$	butan-1-ol	117.5 °C	Diethyl ether	34.6 °C

Table 46.5 Technically important ethers

Name	Other Designations	Formula	Notes
Methoxymethane	Dimethyl ether	$H_3C{-}O{-}CH_3$	Flammable gas, propellant in aerosol cans; UN No. 1033
Ethoxyethane	Diethyl ether, ether	$C_2H_5{-}O{-}C_2H_5$	Important solvent and extraction agent; UN No. 1155
2-methoxy-2-methylpropane	Methyl *tert.*-butyl ether (MTBE)	$(CH_3)_3C{-}O{-}CH_3$	Solvent, as anti-knock agent in fuels; UN No. 2398
1,2-epoxyethane	Ethylene oxide, oxirane	$(CH_2)_2O$	For the manufacture of ethylene glycol, crown ethers, plastics and waxes; UN No. 1040/1041
1,4-epoxybutane	Tetrahydrofurane (THF)	$(CH_2)_4O$	Solvent, for the manufacture of plastics; UN No. 2056
1,4-dioxane	–	$(CH_2)_4O_2$	Solvent; UN No. 1165

▶ **Caution:** If ethers are stored in the air for a longer period of time, nonvolatile peroxides are formed under the influence of light, which accumulate when the ether is distilled off and can lead to dangerous explosions.

Some important ethers are listed in Table 46.5.

46.4.1 Symmetrical and Asymmetrical Ether

The molecules of symmetrical ethers carry the same alkyl group ($R = R'$) on both sides of the O atom. They can be prepared, for example, by reacting the corresponding alcohol with sulfuric acid.

$$2\ C_2H_5{-}OH \xrightarrow{H_2SO_4,\ 140°\ C} C_2H_5{-}O{-}C_2H_5 + H_2O$$

Unsymmetrical ethers can be obtained, for example, from the alkali salts of alcohols with alkyl halides.

$$\text{R-O}^-\ \text{Na}^+ + \text{R}'\text{-X} \rightarrow \text{R-O-R}' + \text{Na}^+\text{X}^-$$

The *tert.*-butyl methyl ether is obtained from *iso*-butene and methanol.

$$(CH_3)_2C{=}CH_2 + CH_3{-}OH \overset{H_2SO_4}{\rightarrow} (CH_3)_3C{-}O{-}CH_3 + H_2O$$

46.4.2 Cyclic Ethers

Cyclic ethers are also called epoxides. The most important are ethylene oxide (oxirane or 1,2-epoxyethane), tetrahydrofuran (THF, 1,4-epoxy-butane) and the diether 1,4-dioxane.

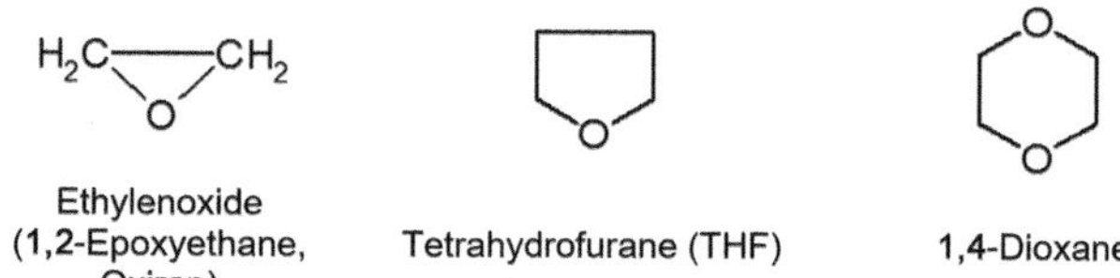

46.4.2.1 Crown Ether

Cyclic polyethers also include the so-called crown ethers. The name "crown" is derived from the jagged structure of the molecules, in which the O atoms act like the beaded tips of a crown. A separate nomenclature has evolved for these ethers. The ring structure of the compound 1,4,7,10-tetraoxacyclododecane, which is made up of 8 C and 4 O atoms, is called 12-crown-4 for short. The number before "crown" indicates the number of all ring atoms, the number after indicates the number of oxygen bridges.

Due to their molecular structure, crown ethers have an outer, hydrophobic sphere and an inner, hydrophilic sphere. Depending on the ring size, crown ethers can therefore selectively complex metal ions. 12-crown-4 dissolves Li ions, 15-crown-5 Na ions and 18-crown-6 K ions. With the aid of crown ethers, inorganic salts such as $KMnO_4$ can be dissolved in non-polar substances such as benzene.

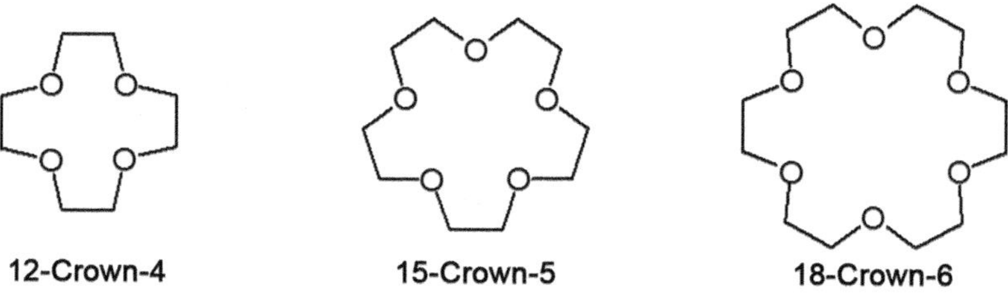

46.5 Carboxylic Acids

Organic acids or carboxylic acids are familiar to us from everyday life. We know formic acid from ants and nettles, with acetic acid – even if diluted – we season salads and sauces. Citric acid, malic acid, tartaric acid, oxalic acid and others contribute significantly to the taste of fruits and plants. And longer-chain acids, so-called fatty acids, are a component of animally and vegetational fats and oils.

Characteristic for carboxylic acid molecules is the carboxy group –COOH, which is formally composed of a carbonyl group with attached hydroxy group. The acidity, the acidic effect, is based on the electron attraction that the double-bonded oxygen atom exerts on the H atom of the OH group. This loosens the H–O bond, so to speak, and the H atom can leave the functional group as an H^+ ion.

46.5.1 Monocarboxylic Acids

Monocarboxylic acids contain only one COOH group in the molecule. The short-chain acids ($C_1...C_3$) are pungent-smelling liquids, the medium-chain acids ($C_4...C_9$) have a rancid/sweaty odor, and the longer-chain acids ($\geq C_{10}$) are white, odorless, waxy solids. Important monocarboxylic acids are listed in Table 46.6.

Table 46.6 Technically important monocarboxylic acids

Name	Other designations	Formula	Notes
Methanoic acid	Formic acid	H–COOH	Salts: Formates, decalcifiers, preservatives, for many syntheses, with conc. H_2SO_4: development of CO; UN No. 1779
Ethanoic acid	Acetic acid	H_3C–COOH	Salts: Acetates, biolog. Production by acetic acid bacteria, for many syntheses, table vinegar (5...8%), pure acid solidifies at 16 °C (glacial acetic acid); UN No. 2789
Propanoic acid	Propionic acid	C_2H_5–COOH	For the manufacture of esters; UN No. 1848
n-Butanoic acid	Butyric Acid	C_3H_7–COOH	Rancid odour, for the manufacture of perfumes and aromatic compositions (e.g. as esters); UN No. 2820
Propenic acid	Acrylic acid	CH_2=CH–COOH	Pungent odour, for the manufacture of polyacrylic acid; UN No. 2218
Methylpropenoic acid	Methacrylic acid, methylacrylic acid	CH_2=C(CH_3)–COOH	For the manufacture of esters for the production of plastics; UN No. 2531

46.5.2 Halogenated Carboxylic Acids

Halogen atoms can occur in two places in a carboxylic acid molecule. Either they replace H atoms of the carbon chain or the OH function of the carboxy group. In the latter case, one speaks of acid halides, which are characterized by a high reactivity with other substances. The acid chlorides react under HCl evolution, the acid bromides under HBr release. Acetic acid chloride ($CH_3{-}COCl$, also acetyl chloride or ethanoyl chloride), for example, is a frequently used reagent for the introduction of the acetyl group ($CH_3CO{-}$).

If, on the other hand, the H atoms of the carbon chain are replaced by halogen atoms, this primarily affects the acid strength (compare Sect. 14.5.3). In particular, if the halogen is located at the C atom directly adjacent to the COOH group, the so-called α-C atom.

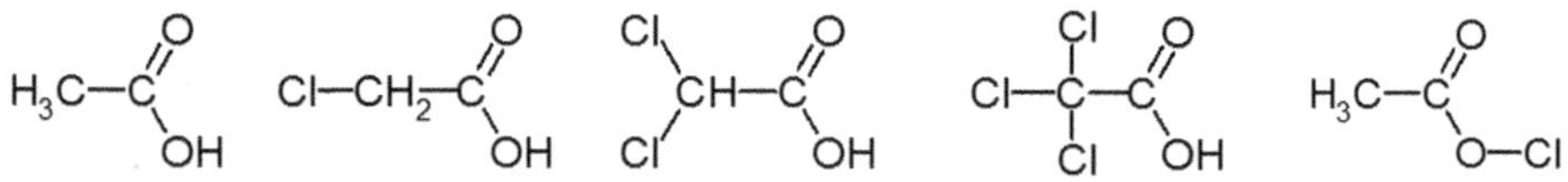

Acetic acid $\mathrm{p}K_s = 4.8$ — Chloroacetic acid $\mathrm{p}K_s = 2.9$ — Dichloroacetic acid $\mathrm{p}K_s = 1.3$ — Trichloroacetic acid $\mathrm{p}K_s = 0.7$ — Acetylchloride

46.5.3 Dicarboxylic Acids

Dicarboxylic acids contain two COOH groups in the molecule. The simplest representative of this class, ethanedioic acid (oxalic acid), is found in clover and rhubarb, among others. Propane-1,3-dicarboxylic acid (malonic acid) and hexane-1,6-diacid (adipic acid) are used for a variety of technical syntheses, such as drugs and plastics.

As a example of a tricarboxylic acid is 3-carboxy-3-hydroxy-pentanedioic acid, which is found mainly in pineapples, lemons, oranges, cranberries, Johannesberries and strawberries – and is better known as citric acid.

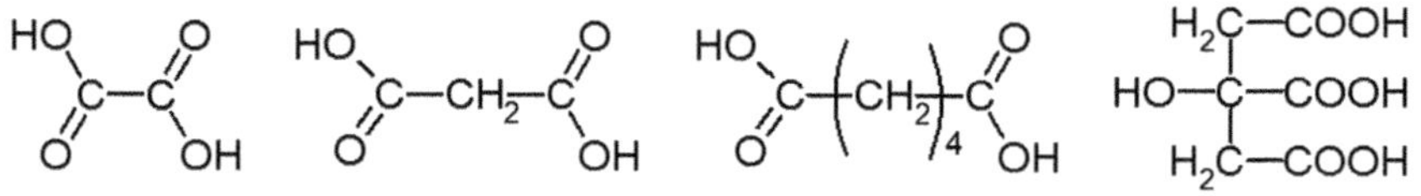

Oxalic acid — Malonic acid — Adipic acid — Citric acid

46.5.4 Carboxylic Acid Anhydrides

The most important representatives of the carboxylic acid anhydrides are the acetic anhydride and the phthalic anhydride. A carboxylic acid anhydride molecule is formed from two carboxylic acid molecules by the splitting off of a water molecule from two COOH groups each.

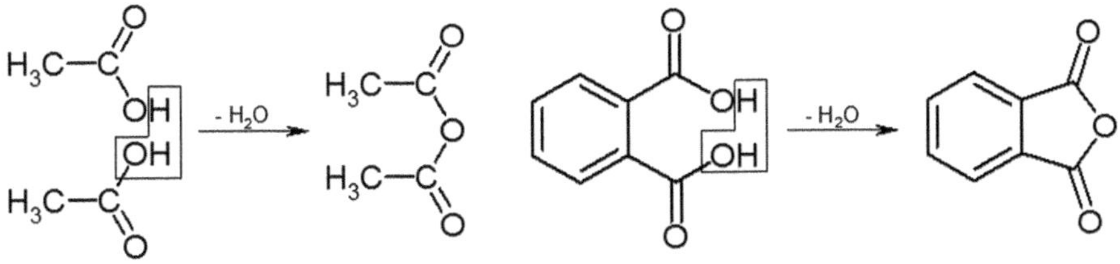

Acetic acid → Acetic anhydride Phthalic acid → Phthalic anhydride

Acetic anhydride (UN No. 1715) is used in the manufacture of medicines, dyes and acetyl cellulose. Phthalic acid and phthalic anhydride (UN No. 2214) are used in the manufacture of plasticizers for PVC and of dyestuffs.

46.6 Ester

One of the most interesting groups is that of the so-called carboxylic acid esters. They are named by adding the alkyl of the alcohol to the name of the acid and ending with the word ester, e.g. acetic acid butyl ester. Another variant is to designate the ester as the alkyl salt of the acid, e.g. butyl acetate.

In the simplest case, esters are formed by the reaction of a carboxylic acid with an alcohol. Here, the OH group of the carboxylic acid function (–COOH) and the hydrogen atom of the alcoholic OH group form a molecule of water as a by-product. The organic residues then form the ester together. As an example, we consider the synthesis – the esterification – of ethyl acetate from acetic acid and ethanol:

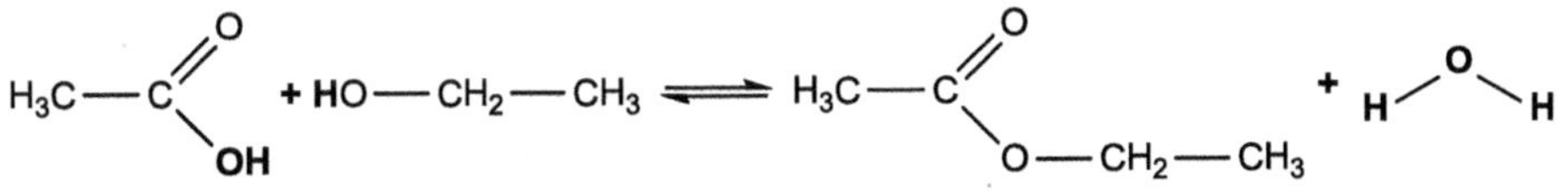

$$H_3C-COOH + HO-CH_2-CH_3 \rightleftharpoons H_3C-CO-O-CH_2-CH_3 + H_2O$$

The equilibrium arrow indicates that a dynamic equilibrium is established in the reaction. This means that at a certain point the number of ester molecules formed from alcohol and acid molecules is identical to the number of ester molecules that are (re)split by water into acid and alcohol molecules. This reverse reaction is called saponification. The position of the reaction equilibrium, i.e. the amount of ester in the reaction mixture, depends on the starting materials. In technical processes, water *or* esters are distilled off in order to achieve as complete a conversion as possible.

Esters can also be produced from other esters – by so-called transesterification -, from acid anhydrides, from acid halides and in other ways.

The interesting thing about esters is their wide range of uses.

- Esters of unsaturated alcohols or unsaturated acids are important precursors for plastics, e.g. vinyl acetate ($H_2C{=}CH{-}O{-}CO{-}CH_3$) or methyl methacrylate ($H_2C{=}C(CH_3){-}CO{-}O{-}CH_3$). When the dicarboxylic acid terephthalic acid is esterified with ethanediol, a polyester is formed which is used as a plastic: PET (polyethylene terephthalate).
- Some esters are important solvents, e.g. ethyl acetate ($CH_3{-}CO{-}O{-}C_2H_5$) or butyl acetate ($CH_3{-}CO{-}O{-}C_4H_9$).
- Large quantities of esters are produced for fragrances and flavours, e.g. acetic acid butyl ester (apple flavour), acetic acid iso-pentylester (banana flavour), butyric acid ethyl ester (pineapple flavour) and many more.
- The esters of carboxylic acids and alcohols with chain lengths of C $\geq$ 16 find a variety of applications as waxes.

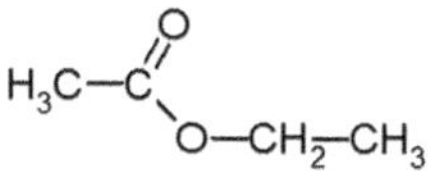

Ethyl acetate 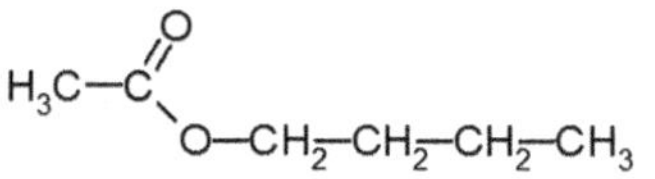 | Butyl acetate 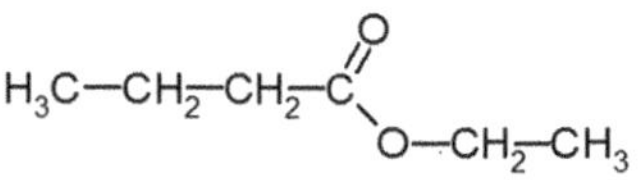 | Ethyl butyrate

46.7 Peroxides

Organic peroxides can be thought of as derivatives of hydrogen peroxide (H−O−O−H) in which one or both hydrogen atoms have been replaced by alkyl radicals. Like H_2O_2, organic peroxides are not particularly stable compounds. They tend, especially at elevated temperatures, to decompose, sometimes violently.

The dissociation energy of the O−O bond is about 200 kJ mol^{-1} and is thus only about half that of C−C, C−H or C−O bonds. This explains the easy decomposition. With regard to storage, it is therefore imperative that the maximum storage temperature be observed in order to counteract decay – and thus a loss of peroxide. In addition, the SAD temperature (SADT: self-accelerating decomposition temperature) should be observed. If this temperature is exceeded, more heat is generated by the decomposition than can be dissipated to the outside. Under certain circumstances, this can lead to a heat accumulation with subsequent explosion.

Examples

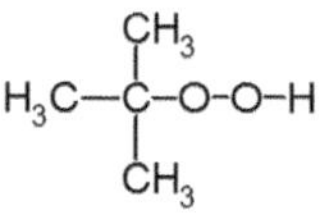

tert-Butyl hydroperoxide 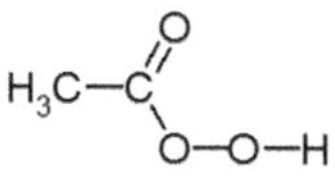 | Peracetic acid 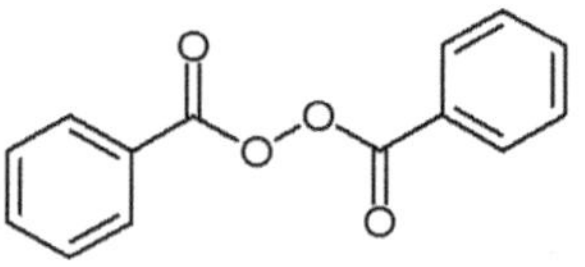 | Benzoyl peroxide 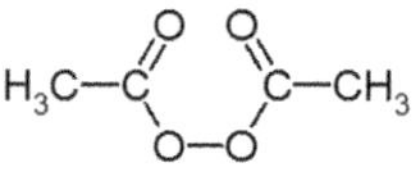 | Diacetyl peroxide

Some peroxides are used as starters for radical chain reactions, especially in polymerizations. Benzoyl peroxide, for example, decomposes into two benzoic acid radicals, which then start the reaction.

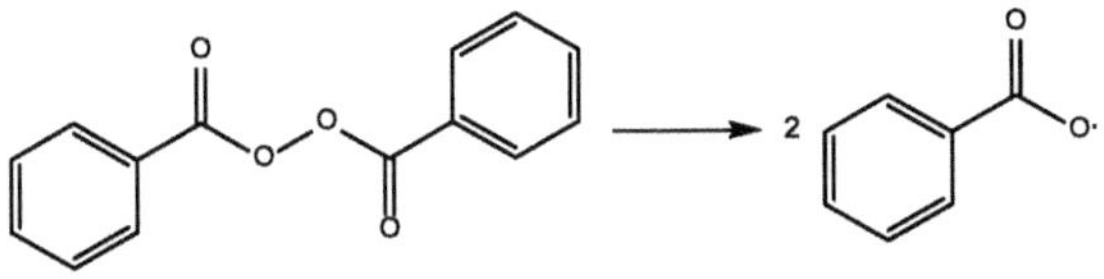

46.7.1 Peroxycarboxylic Acids

In peroxycarboxylic acid molecules, another O atom is "inserted" between the C atom of the acid function and the OH group. The hydrogen atom can now no longer dissociate as H^+, so that there is also no acid effect. Similar to hydrogen peroxide ($H_2O_2 = H{-}O{-}O{-}H$), the C(O)–O–O–H group also tends to release oxygen and therefore has a disinfecting effect. The decomposition under oxygen release is accelerated by temperature increase or catalytically acting impurities. The substances may only be stored in the original container with a pressure-relief cap. Removed substance must not be poured back. Higher concentrated solutions may decompose explosively.

Peroxoacetic acid (Peracetic acid, PAA), the most important representative of the class, is used as an almost universal disinfectant. The low pH of the solution is a result of the equilibrium between CH_3COOH and CH_3COOOH (see also Sect. 13.2.6).

Organic Nitrogen Compounds

47

The next large class of functional groups is based on the element nitrogen. In contrast to the partly acidic oxygen compounds, organic nitrogen compounds are basic in many cases and form salts with acids. An overview is given in Fig. 47.1.

47.1 Amines

Amines are the most important organic compounds that show a distinct basicity. They dye litmus paper blue and can form salts with acids. Formally, they are derivatives of ammonia, in the molecule of which one or more hydrogen atoms have been replaced by organic radicals.

47.1.1 Primary, Secondary, Tertiary Amines

Depending on the number of alkyl groups attached to the amine nitrogen, a distinction is made between primary, secondary and tertiary amines, with the general formulae $R{-}NH_2$ (primary), R_2NH (secondary) and R_3N (tertiary).

Primary amine	Secondary amine	Tertiary amine
$H_3C{-}NH_2$ Methylamine	$H_3C{-}NH{-}CH_2{-}CH_3$ Ethyl-methyl-amine	$(H_3C)_2N{-}C_2H_5$ Ethyl-dimethyl-amine
$C_6H_5{-}NH_2$ Aniline	$C_6H_5{-}NH{-}C_2H_5$ N-Ethylaniline	$C_6H_5{-}N(CH_3)_2$ N,N-Dimethylaniline

T. Schmiermund, *The Chemistry Knowledge for Firefighters*,
https://doi.org/10.1007/978-3-662-64423-2_47

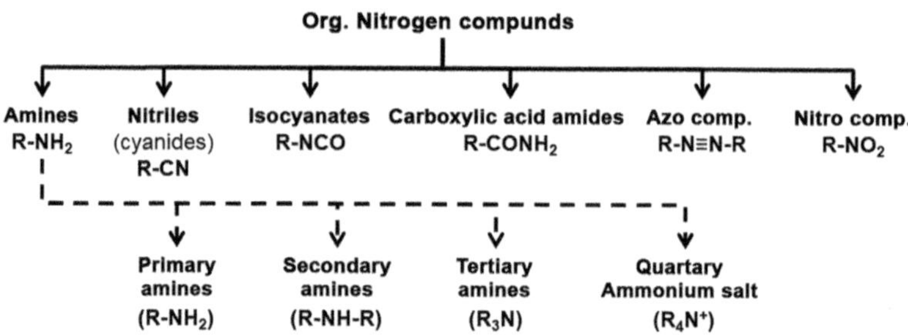

Fig. 47.1 Overview of organic nitrogen compounds

The addition of an H^+ ion to the free electron pair of the nitrogen atom results in amine salts, which are usually water-soluble.

$$R{-}NH_2 + HCl \rightarrow R{-}NH_3^+ + Cl^-$$
$$R_2{-}NH + HCl \rightarrow R_2{-}NH_2^+ + Cl^-$$
$$R_3{-}N + HCl \rightarrow R_3{-}NH^+ + Cl^-$$

The primary diamine 1,6-diaminohexane ($H_2N{-}C_6H_{12}{-}NH_2$, UN No. 2280) is an important plastics precursor. Aniline ($C_6H_5{-}NH_2$, UN No. 1547) and its derivatives are used for dyes, pigments and medicines.

47.1.2 Quaternary Ammonium Salts

Through protonation, the ammonia molecule (NH_3) becomes the ammonium ion (NH_4^+). If all four H atoms are replaced by alkyl radicals, so-called quaternary ammonium compounds (QAC) are obtained. These are water-soluble salts which are used as surfactants and in fabric softeners due to their surface-active properties. As they also have a disinfecting effect, they are contained in many hand and surface disinfectants and in anti-algae agents.

Typical compounds are benzalkonium chloride, cetylalkonium chloride and cetylpyridinium chloride. The latter is also used as a medicinal substance in lozenges to relieve throat complaints.

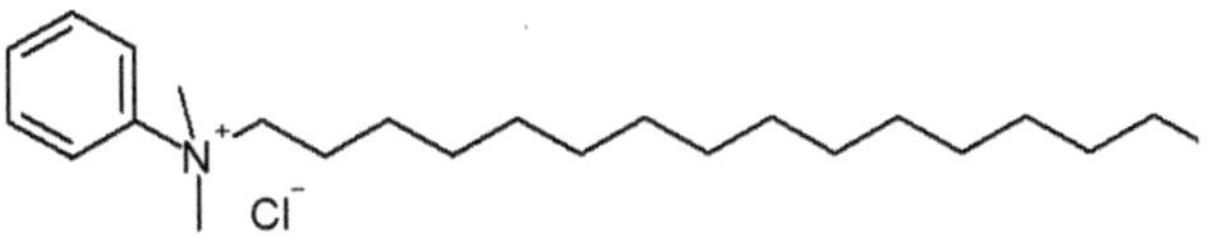

Cetylalkoniumchloride

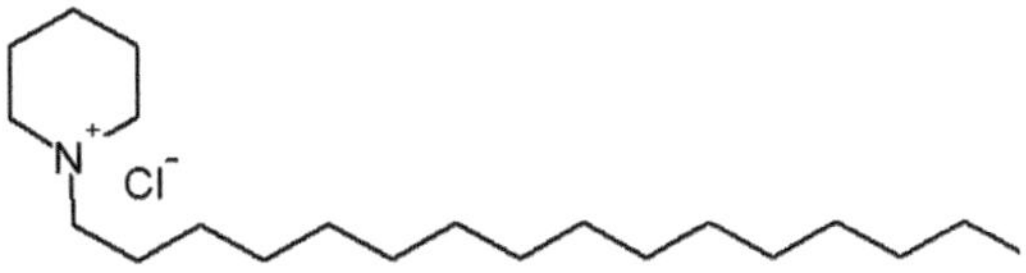

Cetylpyridiniumchloride

47.1.3 Alkaloids

Alkaloids are organic amines of plant origin with basic character, complicated structure and high physiological effect. Many alkaloids are secondary or tertiary amines whose nitrogen is in a ring structure. The bitter taste of the alkaloids protects the plant from predators. Many of the substances themselves have antispasmodic, blood pressure-increasing, stimulating, but also psychogenic effects in humans; some are strong poisons.

Typical representatives are nicotine (UN No. 1654), which – in addition to nicotine sulphate (UN No. 3445) – is used as a plant protection agent, cocaine, codeine, heroin, lysergic acid, caffeine and atropine, which is used, among other things, in ophthalmology. The poison strychnine was used against rodents. The carrier of the pepper flavour, piperine, is also an alkaloid.

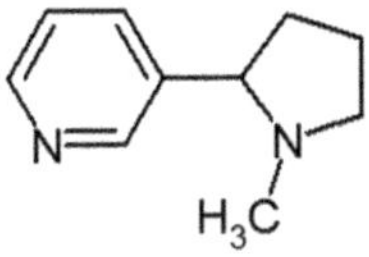

Nicotine 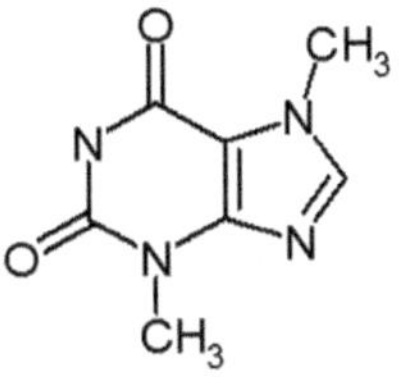 Caffeine 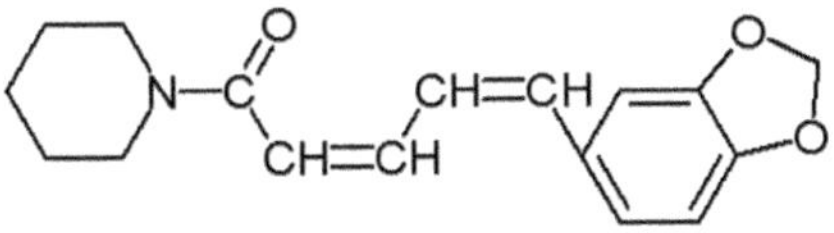 Piperine

47.1.4 Amino Acids

Amino acids are substances whose molecules contain both the carboxyl group $-COOH$ and the amino group $-NH_2$. Amino acids have their greatest importance as building blocks of proteins. Proteins are macromolecular compounds with molar masses of 10 000 to 100 000 u. In the molecules of so-called polypeptides, which are made up of many amino acids, the amino acid molecules are linked to each other via the peptide bond –CO–NH. The –COOH group of one amino acid molecule reacts with the $-NH_2$ group of the other amino acid molecule to form the acid amide molecule.

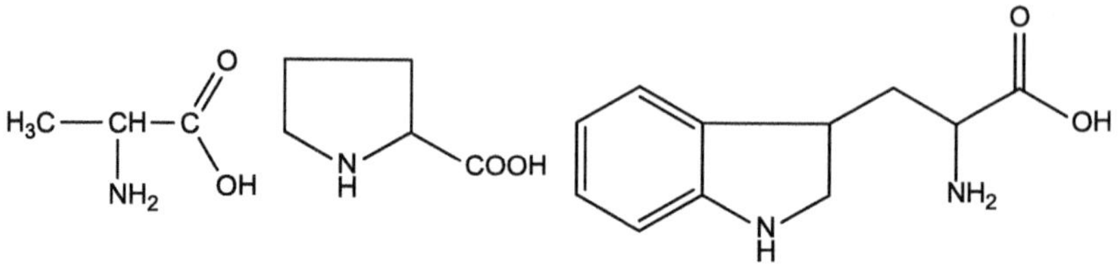

Alanin Prolin Tryptophan

47.2 Nitro Compounds

Nitro compounds containing the nitro group $-NO_2$ can be produced directly from alkanes or aryls by so-called nitration with nitric acid. They are usually colorless, pleasant-smelling liquids or odorless solids. The lower primary nitroalkanes are often used as solvents, e.g. for polystyrene or acetyl cellulose.

They often serve as starting materials for the production of amines and are therefore important intermediates in the production of dyes and drugs. For example, aniline is easily obtained from nitrobenzene, which is obtained by nitration of benzene.

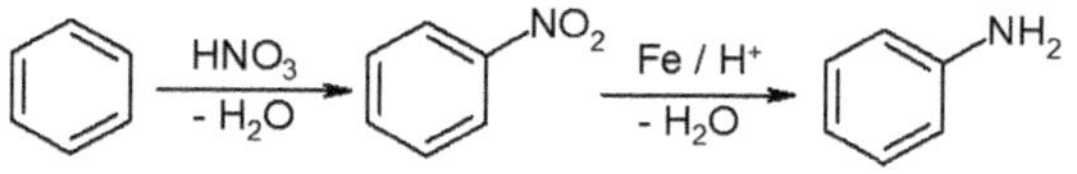

$$C_6H_6 \xrightarrow[-H_2O]{HNO_3} C_6H_5NO_2 \xrightarrow[-H_2O]{Fe/H^+} C_6H_5NH_2$$

Higher nitrated compounds – i.e. those with several nitro groups – are sometimes used as explosives. Known examples are 2,4,6-trinitrotoluene (TNT) and 1,3,5-trinitro-[1,3,5]-triazinane (hexogen).

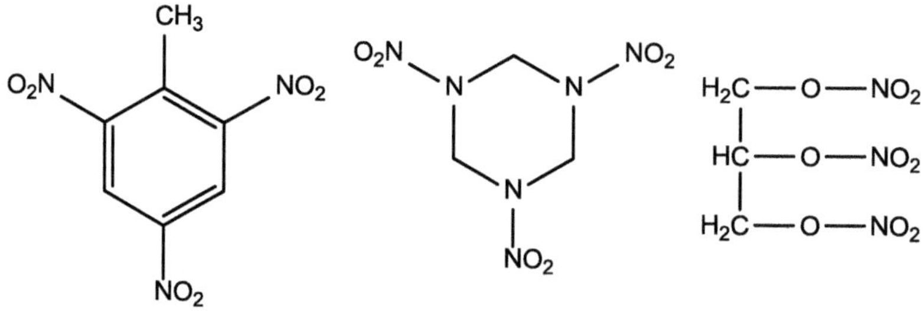

TNT Hexogen Nitroglycerin

47.2.1 Nitric Acid Ester

The esters of the inorganic nitric acid with the organic alcohols are also assigned to the nitro compounds, since their molecules also contain the $-NO_2$ grouping. They are formed analogously to the organic esters by reaction of the acid with an alcohol. In contrast to the "real" nitro compounds, the nitro group is not directly linked to a C atom, but is attached to the molecule via an O-bridge.

An important representative is glycerol trinitrate, better known as nitroglycerin. This compound explodes on impact or shock, producing only gaseous decomposition products. By absorbing it in diatomaceous earth, the safe-to-handle dynamite is formed. In medicine, glycerol trinitrate is used as a vasodilator, especially for heart problems.

47.3 Nitrile

Nitriles contain the so-called cyano group –CN in their molecules. Their names are usually derived from the acid formed during their hydrolysis and bear the suffix -nitrile. Acids whose molecules contain the –CN group are usually named cyano compounds, such as cyanoacetic acid, which can also be called malonic acid mononitrile.

The reaction of nitriles with acids *or* alkalis yields a carboxylic acid while retaining the C-atom of the cyano group in the molecule. Due to this extension of the carbon chain, nitriles are important intermediates in organic chemistry.

Typical examples are: Acetonitrile (CH_3-CN, UN No. 1648) is an important solvent, methacrylonitrile ($CH_2=C(CH_3)-CN$, UN No. 3079) and acrylonitrile ($CH_2=CH-CN$, UN No. 1093) are plastic precursors, phthalodinitrile is used for pharmaceuticals and dyes.

H_3C-CN Acetonitrile

$H_2C=C(CH_3)-CN$ Methacrylnitrile

Phthalodinitrile

$NC-CH_2-C(=O)OH$ Cyanoacetic acid

47.4 Isocyanates

Isocyanates contain in their molecules the $-C=N=O$ – group known as isocyanate. They are all highly reactive compounds that are used as hardeners for coatings, PU foams and for polyurethane production. By varying the alkanediol ("base component") and the diisocyanate ("crosslinker", "hardener") used, it is possible to produce hard to soft polyurethanes which we encounter in everyday life in shoe soles, insulation, car seats and much more.

Typical products are toluene-2,6-diisocyanate (TDI, UN No. 2078), diphenyl-methane-2,4′-diisocyanate (methylene diphenyl diisocyanate, 2,4-MDI) and 1,6-hexa-methylene diisocyanate (HDI, UN No. 2281).

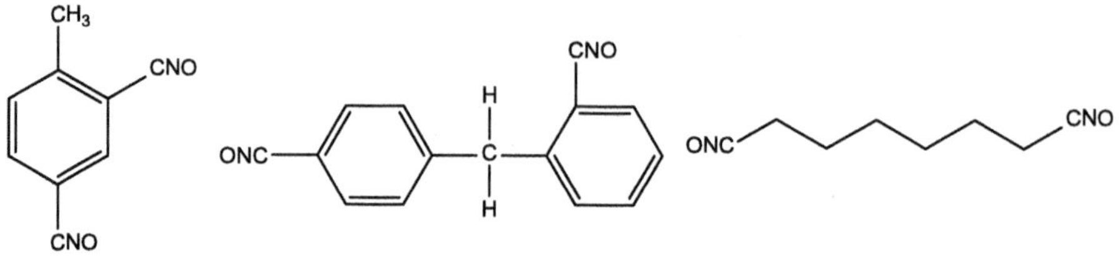

TDI 2,4-MDI HDI

47.5 Carboxylic Acid Amides

If the –OH group of the carboxylic acids is replaced by the $-NH_2$ group, carboxylic acid amides are obtained. Alkylated amides, such as *N*-methyl-formamide (UN No. 1213), *N,N′*-dimethyl-formamide (DMF, UN No. 2265) or *N,N′*-dimethyl-acetamide (DMA) are important, water-miscible solvents for various processes and applications.

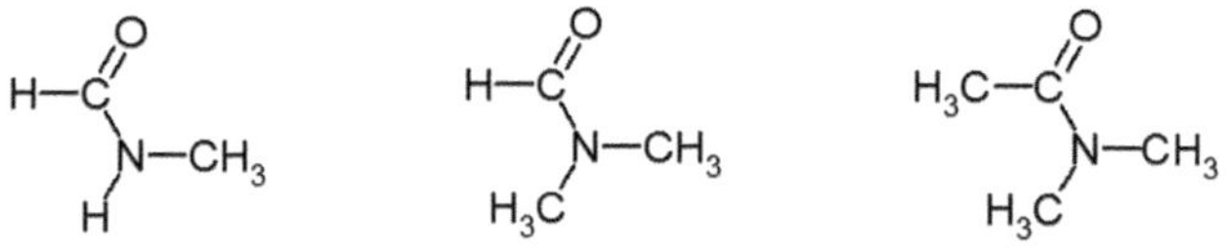

N-Methyl-formamide *N′N′*-Dimethyl-formamide *N′N′*-Dimethyl-acetamide

Other important acid amides are acrylamide (UN No. 2074), which like ε-caprolactam is used in the manufacture of plastics, and urea (carbonic acid diamide). Urea is an important fertilizer and starting material for melamine resins. Humans excrete about 20 g of urea daily as an end product of metabolism. (For acid amide binding in proteins, see Sect. 47.1.4).

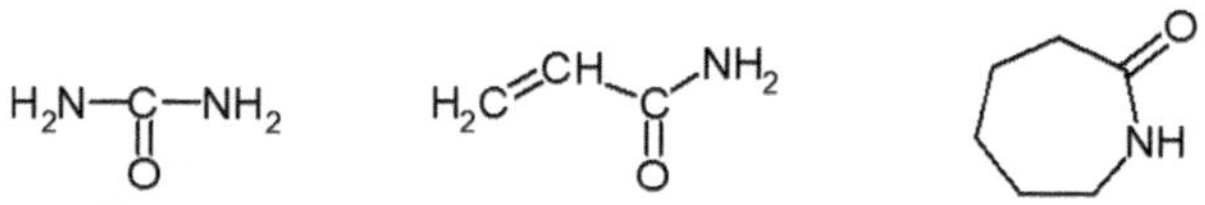

Urea Acrylamide ε-caprolactame

47.6 Azo Compounds

Aromatic azo compounds are coloured. Many dyes and pigments therefore contain the azo group (–N=N–). For example, the indicator methyl orange (see Sect. 14.6.1) is an azo dye.

Another important representative of this class of compounds is azo-bis (isobutyronitrile), or AIBN (UN No. 3234). It is a crystalline solid which readily decomposes into nitrogen, nitrous gases and HCN and must therefore be stored in a cool place. Under controlled conditions, AIBN decomposes into N_2 and two isobutyronitrile radicals. These radicals serve as starters for radical polymerizations.

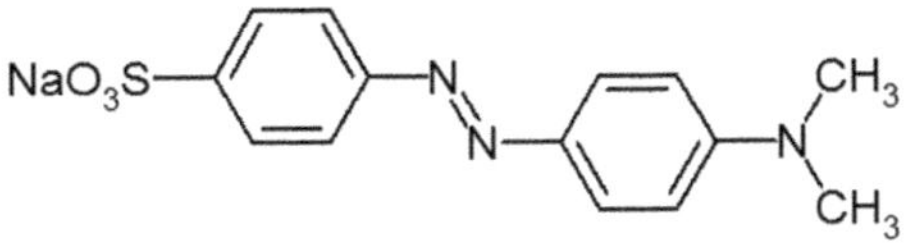

Methyl orange

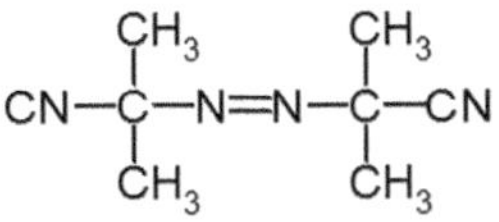

AIBN

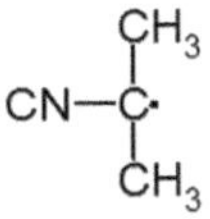

Isobutyronitrile radical

48 Organic Sulphur Compounds

Organic sulfur compounds are frequently found in nature: in the amino acids cysteine, taurine and methionine, in vitamin B1, in the anticoagulant heparin, but also, for example, in garlic and in the glandular secretions of skunks. An overview is given in Fig. 48.1.

48.1 Thiols

Just as in the case of alcohols one hydrogen atom of the water molecule can be thought of as being replaced by an organic radical, so in the case of thiols one hydrogen atom of the hydrogen sulphide molecule (H_2S = H–SH) is formally replaced by an organic radical (R). This results in the general formula R–SH. Thiols - formerly called mercaptans - occur in broccoli, onions, garlic, asparagus etc. and are released besides H_2S during decay processes of organic materials.

Due to the lower electronegativity of the sulfur atom, the molecules of thiols form significantly weaker hydrogen bonds than the analogous alcohols. The boiling points of thiols are therefore lower than those of the corresponding alcohols: C_2H_5–OH: bp = 78 °C; C_2H_5–SH: bp = 37 °C.

The low-molecular thiols in particular have a nauseating odor even in the smallest concentrations and are therefore added to the actually odorless natural gas as odorants ("odor generators"). Methylthiol (CH_3–SH, odor threshold: 0.002 ppm = 2 ppt) or ethylthiol (C_2H_5–SH, odor threshold: 1 ppt) are used, for example.

The preparation is carried out, for example, by the reaction of halogen alkanes with sodium hydrogen sulphide:

$$C_3H_7Br + NaSH \rightarrow C_3H_7SH + NaBr$$

They are used as starting materials for other S-containing compounds such as sulfones, sulfonic acids and thioethers. They are also used in the production of medicines, pesticides and sulphur-containing amino acids.

T. Schmiermund, *The Chemistry Knowledge for Firefighters*,
https://doi.org/10.1007/978-3-662-64423-2_48

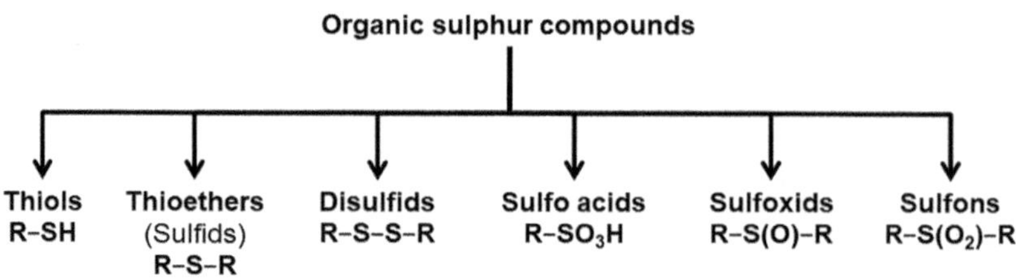

Fig. 48.1 Overview of organic sulphur compounds

$H_3C{-}CH_2{-}SH$ Ethylthiole

$H_2C{=}CH{-}CH_2{-}SH$ 2-Propen-1-thiole

$HO{-}C(=O){-}CH(NH_2){-}CH_2{-}SH$ Cysteine

48.2 Thioethers and Disulphides

If both hydrogen atoms of the H_2S molecule are replaced by organic radicals, the thioethers $R{-}S{-}R'$ are obtained. The naming follows the schemes alkyl-thio-hydrocarbon or alkyl-alkyl-sulfide. The cyclic thioether tetrahydrothiophene (THT) is also used as an odorant (odor threshold value: 1 ppt). Disulfides contain two interconnected S atoms in their molecules as a functional group: $R{-}S{-}S{-}R'$. The substances are named according to the scheme alkyl-alkyl-disulfide.

$H_3C{-}S{-}CH_3$ Dimethylsulfide (Methylthiomethane)

$H_3C{-}S{-}CH_2{-}CH_2{-}CH_3$ Methylpropylsulfide (Methylthiopropane)

$H_3C{-}S{-}S{-}CH_2{-}CH_3$ Methylethyldisulfide

48.3 Sulphonic Acids

Sulphonic acids or sulphonic acids can be regarded as derivatives of sulphuric acid in whose molecule a hydrogen atom has been replaced by an alkyl radical. The characteristic group is therefore the $-SO_3H$ group. For naming purposes, the designation -sulfonic acid is added to the alkyl radical. The free acids are often crystalline compounds. The salts, especially the Na and K salts, are usually water soluble and are called sulfonates. Linear alkyl sulfonates or their Na salts are used as detergent substances. Other sulfonic acids are dyes or indicators or are used for ion exchangers.

The sulfonic acid chlorides ($-SO_2Cl$) can be converted into the sulfonic acids by reaction with water or into sulfonamides by reaction with ammonia or with amines. Some sulfonamides possess antibacterial activity and are therefore important drugs. The sulfonamides derived from the aniline are called anilides.

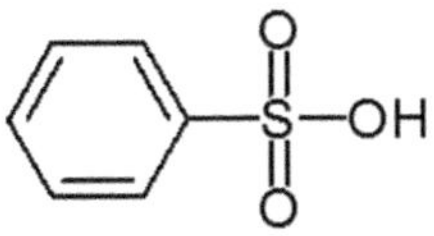

Benzene sulphonic acid

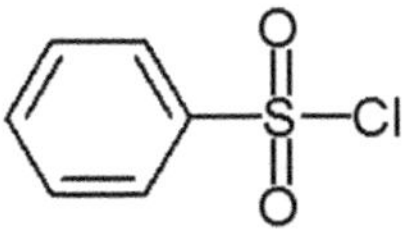

Benzene sulphonyl chloride

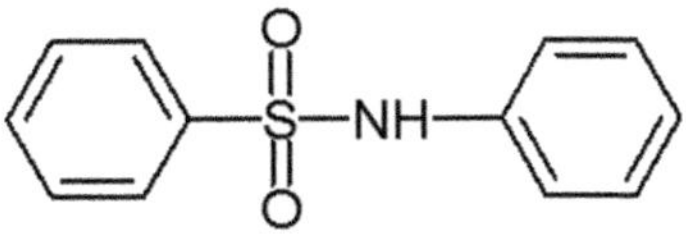

Benzene sulphonanilide

48.4 Sulphoxides and Sulphones

Sulphoxides (general: $R_2S{=}O$) and sulphones ($R{-}SO_2{-}R'$) are other important classes of compounds. Typical compounds are dimethyl sulfoxide (DMSO) and sulfolan. DMSO is used as a solvent, for the separation of aromatic compounds from gasoline and as a *carrier* for medicinal substances, as it can easily penetrate cell walls. Substances dissolved in DMSO can thus easily enter the body. DMSO has a characteristic, onion-like odor. In higher concentrations in the air, it produces a metallic taste on the tongue.

Sulfolane is an important high-boiling solvent (Kp = 285 °C) used, among other things, to remove H_2S and CO_2 from natural gas and to remove aromatic components from liquid hydrocarbons.

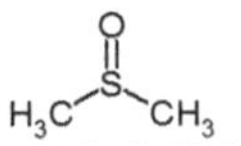

Dimethylsulfoxide (DMSO)

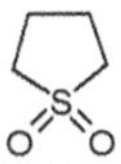

Sulfolane

49 Plastics

Plastics, also known as polymers (Greek *poly* = much, *meros* = parts) or plaste or colloquially plastic, is the name given to a whole range of artificially produced substances consisting of very large molecules, so-called macromolecules (Greek *makro* = large, huge). Unlike most natural materials, many plastics can be melted and shaped. Through the type of starting molecules ("monomers"), the manufacturing conditions and through additives, the chemical and physical properties can be varied over a wide range and tailored to specific applications.

The first plastics were chemically modified natural materials. For example, a horn-like plastic made from the casein of milk or the flammable celluloid made from nitrated cellulose. After a boom in plastics based on petroleum products, the trend today is to develop plastics based on renewable raw materials, which above all should be biodegradable.

Around 12 million tonnes of plastics were consumed in Germany in 2015. Figure 49.1 provides an overview of the areas of application. It is impossible to imagine our daily lives without plastics. Just take a look around ...

49.1 Classification According to Thermal Properties

On the basis of their physical, in particular thermal properties, a distinction is made between three large groups of polymers:

a. Thermoplastics (Tpl)
b. Thermosets (Tst)
c. Elastomers (Elm)

Thermoplastics and thermosets are sometimes also referred to under the generic term "plastics", elastomers also as "elastomers". In principle, it is also possible to produce foamed plastics from each of these groups. Because of their special properties, they are sometimes listed in the literature as a separate group. An overview is given in

T. Schmiermund, *The Chemistry Knowledge for Firefighters*,
https://doi.org/10.1007/978-3-662-64423-2_49

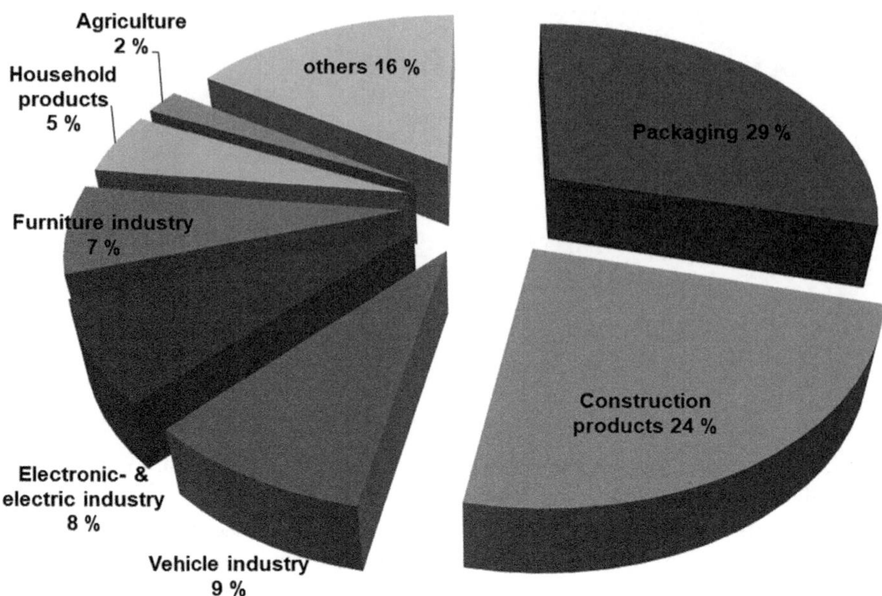

Fig. 49.1 Applications of plastics

Table 49.1. For the abbreviations of the individual plastics, see the overview in Table 49.2.

49.1.1 Thermoplastics

Thermoplastics (Greek *thermos* = warm, hot, *plassein* = to form, shape) soften when heated, up to the point of flowability. They harden again on cooling. If thermoplastics are not heated beyond their decomposition point, they can be softened as often as desired and thus reshaped.

This is made possible by a linear or linear-branched structure of the macromolecules, whereby the individual molecular chains are only present next to each other, i.e. are not cross-linked with each other.

Important thermoplastics are e.g.: PE, PP, PVC, PA, PS, PMMA, PTFE, PVA.

▶ Thermoplastics are plastics that can be changed in shape by heating.

49.1.2 Thermosets

Thermosets, which should actually be called "thermodure" (lat. *durus* = hard), are plastics that harden during shaping and can no longer be plastically deformed by heating. However, most thermosets can be easily processed with tools: Drilling, milling, sawing etc. are possible.

Table 49.1 Basic properties of plastics; comparison of thermoplastics, thermosets and elastomers

Features	Thermoplastics	Thermosets	Elastomers
Mechanical strength	Soft, semi-hard or hard – depending on type	Hard, brittle	Elastic, rubbery
Behaviour during heating	Softening	No change	Slight softening
Behaviour during cooling	Solidification	Less elastic	No change
Chemical structure	Linear chains, partly with side chains	Close intermolecular bonds	Extensive intermolecular bonding
Cross-linking of the macromolecules	No cross-linking	Strong networking	Low networking
Structure model			
Application	Plastics, fibres	Pressing and spraying compounds, adhesives	Rubber, gum

In thermosets, the molecules form a three-dimensional, solid lattice. This leads to a very rigid structure, which is mechanically and thermally very stable. When the decomposition temperature is exceeded, the material decomposes, and breaks down under strong mechanical impact.

Important thermosets are e.g.: EP, PF, UP. Alongside a few thermoplastics, they form the basis for CFRP and GFRP materials.

▶ Thermosets are hard, non-meltable plastics.

49.1.3 Elastomers

Elastomers (Greek *elastos* = stretchable, bendable) or elastics are macromolecular substances that exhibit rubber-like ("rubber-elastic") behavior. A change in shape due to external forces is completely reversed once the force has ceased to act.

A wide-meshed cross-linking of the polymer strands enables elastic stretching or compression. At the end of the force application, the microstructure returns to its initial state, as this is the lowest-energy state in relation to the crosslinked molecular chains.

Table 49.2 Overview: important plastics (sorted by abbreviation)

Abbreviation	Plastic Name	Sort	Formation	Application/Use
ABS	Acrylic-butadiene-styrene copolymer	Tpl	Pme	Housings for telephones and household appliances, for 3D printers
ASA	Acrylic ester styrene Acrylonitrile copolymer	Tpl	Pme	Coffee maker, microwave, electric kettle
BR	Butadiene rubber	Elm	Pme	Car tyres
CA	Cellulose acetate	Tpl	ssP	Rayon, cigarette filter
CN	Cellulose nitrate	Tpl	ssP	Films ("celluloid"), table tennis balls (until 2014), for nitro varnishes, membranes, filters
CR	Chloroprene rubber	Elm	Pme	Hoses, cable sheathing, diving and surfing clothing
EP	Epoxy resins	Tst	Pad	Construction adhesives, casting resins, corrosion protection, for glass reinforced (GRP) and carbon fibre reinforced (CFRP) plastics
PA	Polyamide	Tpl	Pad	Equipment parts, moulded parts, coatings, synthetic fibres (Nylon©, Perlon©)
PAN	Polyacrylonitrile	Tpl	Pme	Synthetic fibre
PBI	Polybenzimidazole	Tpl	Pco	Fire protection clothing, membranes, temperature-resistant coatings
PC	Polycarbonate	Tpl	Pco	Moulded parts, safety discs, CD & DVD
PE (HD-/ LD-)	Polyethene (polyethylene)	Tpl	Pme	Foils, packaging material, sheets, pipes, moulded parts, household appliances
PET	Polyethylene terephthalate	Tpl	Pco	Beverage bottles, fibres (so-called "fleece")
PF	Phenol-formaldehyde condensates	Tst	Pco	Moulded parts, appliance parts, laminates, electro-technical articles
PLA	Polylactic acid	Tpl	Pme, ssP	For 3D printers, is well biodegradable
PMMA	Polymethyl methacrylate	Tpl	Pme	Glazing, sanitary equipment, lamp bodies, household appliances
POM	Polyoxymethylene	Tpl	Pco	Precision and gear parts, fuel pumps, snap connections, lighters
PP	Polypropene (Polypropylene)	Tpl	Pme	Foils, packaging material, sheets, pipes, moulded parts, household appliances,

(continued)

Table 49.2 (continued)

Abbreviation	Plastic Name	Sort	Formation	Application/Use
				transport containers, beverage crates
PS	Polystyrene	Tpl	Pme	Household articles, containers, packaging material
PTFE	Polytetrafluoroethene (Polytetrafluoroethylene)	Tpl	Pme	Coatings, hoses, seals, filter bags, membranes
PU	Polyurethane	Tpl	Pad	Foams (hard, soft, integral skin), profiles, mouldings
PVC (-H/-W)	Polyvinyl chloride	Tpl	Pme	Foils, packaging, tubes, profiles, cable sheathing, flooring, window profiles
SI	Silicones	Tst	Pco	Hoses, lubricants
UP	Unsaturated polyester resins	Tst	Pco	Glass reinforced plastics (GRP), fillers, casting resins, paint binders

Abbreviations: *Tst* thermoset, *Elm* elastomer, *Pad* polyaddition, *Pco* polycondensation, *Pme* polymerisation, *Tpl* thermoplastic, *ssP* semi-synthetic plastic

▶ Elastomers are rubber-like plastics.

49.2 Classification According to Education Mechanisms

Depending on the type of chemical reaction that leads to a particular polymer, a distinction is made between the following reactions

a. Polymerization
b. Polycondensation
c. Polyaddition

The degree of polymerization indicates how many monomeric units on average have come together to form a macromolecule. In the following polymer formulas, "n" stands for the degree of polymerization. Regardless of the reaction type, the following applies:

▶ Chemical reactions link small basic molecules (the "monomer") to form the macromolecules of the plastic (the "polymer").

49.2.1 Polymerization

Polymerisation (Pme) is the chemical linking of small molecules – the monomers – to form a macromolecule – the polymer – by reaction between multiple bonds or the opening of ring-shaped compounds to form longer chains. Many polymerizations take place as radical chain reactions, such as the polymerization of ethylene (ethene) to polyethylene (polyethene) (PE):

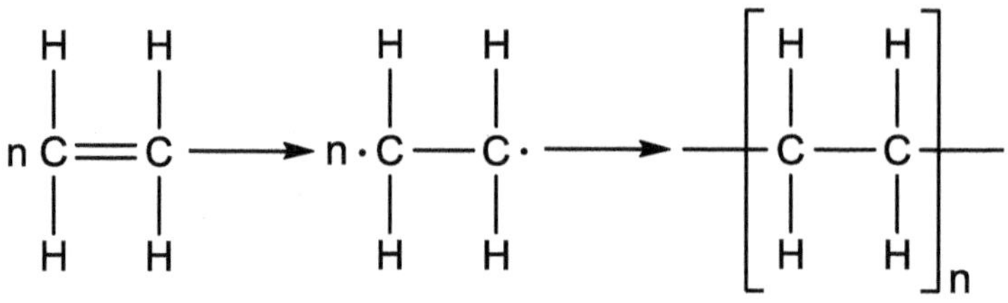

$$n\,CH_2{=}CH_2 \longrightarrow n\cdot CH_2{-}CH_2\cdot \longrightarrow [{-}CH_2{-}CH_2{-}]_n$$

The synthesis of polyamide 6 (PA 6; see Sect. 49.4.3) from ε-caprolactam serves as an example of a polymerization with prior ring opening.

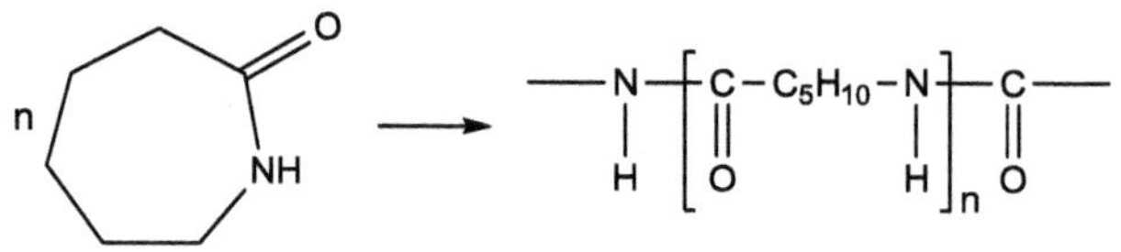

$$n\ \text{ε-caprolactam (ring: C=O, NH)} \longrightarrow {-}NH{-}[{-}CO{-}C_5H_{10}{-}NH{-}]_n{-}CO{-}$$

49.2.2 Polycondensation

If the formation of a macromolecule takes place with the elimination of water molecules or other small molecules, this is called polycondensation (Pco). As an example, the reaction of a dicarboxylic acid with a dialcohol to form a polyester is shown.

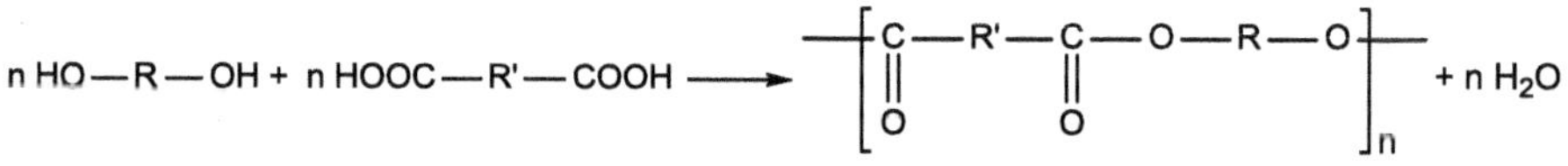

$$n\,HO{-}R{-}OH + n\,HOOC{-}R'{-}COOH \longrightarrow [{-}CO{-}R'{-}CO{-}O{-}R{-}O{-}]_n + n\,H_2O$$

Water splitting takes place according to the following scheme:

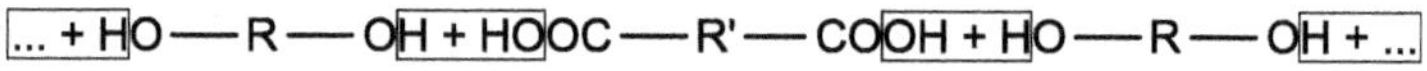

$$\boxed{...+H}O{-}R{-}O\boxed{H+HO}OC{-}R'{-}CO\boxed{OH+H}O{-}R{-}O\boxed{H+...}$$

49.2.3 Polyaddition

Polyaddition (Pad) is the linking of smaller molecules to form a macromolecule by reacting the functional groups with each other without splitting off a smaller molecule. Thus, polyaddition of butane-1,4-diol with hexane-1,6-diisocyanate produces a polyurethane according to the following scheme:

$$\mathrm{n\,HO{-}C_2H_4{-}OH + n\,O{=}C{=}N{-}C_6H_{12}{-}N{=}C{=}O \longrightarrow \left[O{-}C_6H_{12}{-}O{-}\underset{\underset{O}{\|}}{C}{-}\underset{\underset{H}{|}}{N}{-}C_6H_{12}{-}\underset{\underset{H}{|}}{N}{-}\underset{\underset{O}{\|}}{C} \right]_n}$$

49.3 Fully Synthetic/Partially Synthetic Plastics

Fully synthetic plastics are polymers obtained purely from so-called "petroleum and petroleum derivatives". Semi-synthetic plastics (ssP) were originally the reaction products of high-molecular natural substances. These primarily include the plastics cellulose acetate (acetylcellulose, CA), cellulose nitrate (celluloid, CN) and the viscose fibres, which are obtained from cellulose. The cellulose required comes from wood or cotton waste. A semi-synthetic plastic based on casein (milk protein) no longer has any economic significance today.

In the wake of the ever-increasing quantities of plastic waste, research has been conducted since the 1990s into biodegradable plastics, i.e. plastics that can be degraded by composting. This is, so to speak, the "second generation" of semi-synthetic plastics. Polysaccharides (starch, cellulose), succinic acid or lactic acid (PLA) are used for this purpose. Polyhydroxybutyric acid (PHB) is also used. Only a few petroleum-based products, such as certain polycaprolactones and polyvinyl alcohols, are also biodegradable as fully synthetic plastics.

49.4 Important Plastics

This section provides a rough overview of the large number of different plastics. Table 49.2 provides an overview, and some important polymers will be considered in more detail.

49.4.1 Polyolefins: Polyethylene (Polyethene, PE) and Polypropylene (Polypropene, PP)

PE is produced by polymerisation of ethene ("ethylene"). PE produced by the high pressure process (150...300 MPa, 150...320 °C) has a density of ~0.92 g cm^{-3} and is called LD-PE (low density). The product produced by the low pressure process (1.5 MPa, cat.) has a density of ~0.96 g cm^{-3} and is called HD-PE (high *density*).

PP is produced by catalytic low-pressure polymerisation of propene ("propylene"). The catalysts used have a significant influence on the properties of the end product. The density is ~0.90 g cm^{-3}.

$H_2C{=}CH_2$ $\left[CH_2{-}CH_2\right]_n$ $H_2C{=}CH{-}CH_3$ $\left[CH_2{-}CH(CH_3)\right]_n$

Ethen PE Propen PP

49.4.2 Polyvinyl Chloride (PVC) and Polytetrafluoroethylene (Polytetrafluoroethene, PTFE)

For the production of PVC, chloroethene ("vinyl chloride") is polymerised. Rigid PVC (*PVC-U;* u = *unplasticized*) is obtained without the addition of plasticizers, while *plasticized* PVC (*PVC-P;* p = *plasticized)* is obtained with plasticizers, which can account for up to 40 %. The plasticizers used are primarily diesters of phthalic acid. Rigid PVC has a density of approx. 1.4 g cm^{-3}, flexible PVC (PVC-P) between 1.20 and 1.35 g cm^{-3}. There is also a special type of post-chlorinated PVC (PVC-C), which has a higher continuous service temperature.

The second important halogen-containing plastic is PTFE, better known as Teflon©. It is produced by polymerization of tetrafluoroethene. The density is relatively high at ~2.2 g cm^{-3}, the chemical and thermal resistance extremely good. In addition to its use as a coating material for frying pans, for example, it is an important material for seals and is used as a lining for apparatus in chemical engineering, as an electrical insulator and as a dry lubricant.

$H_2C{=}CHCl$ $\left[CH_2{-}CHCl\right]_n$ $F_2C{=}CF_2$ $\left[CF_2{-}CF_2\right]_n$

Chlorethen („Vinylchlorid") PVC Tetrafluorethen PTFE

49.4.3 Polyamides (PA)

Polyamides (PA) can be prepared by ring-opening polymerization of lactams (see Sect. 49.2.1) or by polycondensation of diamines with dicarboxylic acids (see Sect. 49.2.2). Their characteristic feature is the acid amide group (–CO–NH–).

Polyamides made from lactams are given a number indicating the number of carbon atoms in the monomer. PA 6 is made from a monomer with 6 carbon atoms (= ε-caprolactam), PA 12 from laurinlactam (12 carbon atoms).

In the case of condensed polyamides, the number of C atoms of the diamine and that of the dicarboxylic acid are indicated and separated by a dot. PA 6.6 thus consists of the monomers hexamethylenediamine (6 C atoms) and adipic acid (6 C atoms). PA 6.9 consists of hexamethylenediamine and azelaic acid (9 C atoms).

Polyamides can also be used to produce fibers and spun into clothing or utility textiles, such as firefighter's cord. Known trade names are, for example, nylon, Perlon® or Cordura®.

If the main chain of the polyamide does not contain chain molecules but aromatic groups (colloquially: "benzene rings"), fibres made from it are called aramids. The most important types are poly(*p-phenylene terephthalamide*) (PPTA, Kevlar®) and poly(*m-phenylene isophthalamide*) (PMPI, Nomex®).

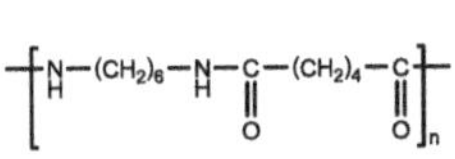

PA 6.6 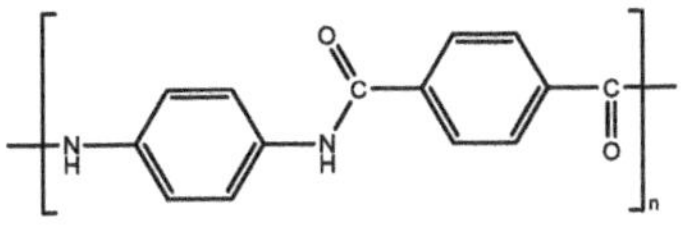 PPTA 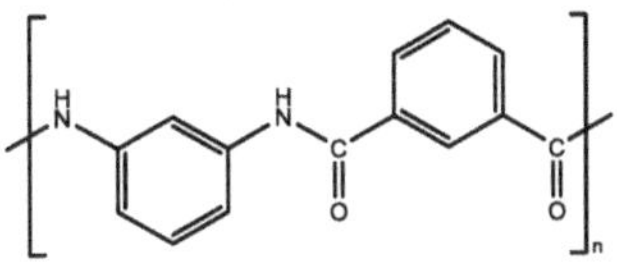 PMPI

49.4.4 Polyurethanes (PUR/PU)

The urethane group (–NH–CO–O–) characterizes the polyurethanes. They are produced by polyaddition of diisocyanates with diols. They are used as casting and adhesive resins, as fibrous materials and, in the majority of cases, as foams. Due to the manufacturing conditions, both soft and hard foams can be produced, which are widely used in the furniture and automotive industries, among others.

49.4.5 Formaldehyde Plastics

Formaldehyde (H–CHO) can be polymerized with itself (polyoxymethylene, POM) or serve as a crosslinking agent in polycondensation with other substances. For example, phenolics (PF) are formed with phenols, melamine resins (MF) with melamine, or urea resins (UF) with urea. POM is a thermoplastic, melamine resins are thermosets and phenoplastics can be reacted with "hardeners" (e.g. urotropin) to form three-dimensionally crosslinked thermosets.

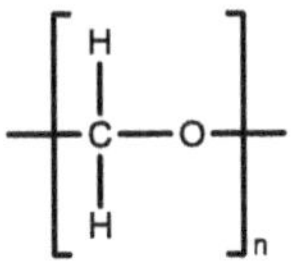

POM 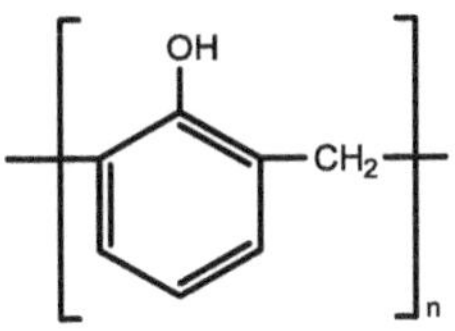 PF 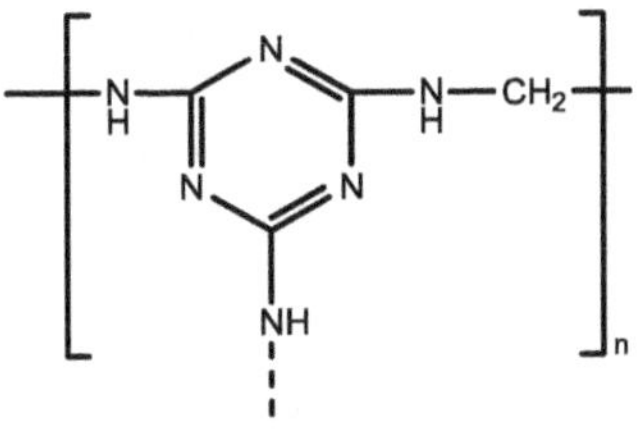 MF

49.4.6 "Organic glasses": PMMA & PC

Polymethacrylic acid methacrylate (PMMA) is produced by polymerisation of methacrylic acid methyl ester and is also known as "acrylic glass" (trade name e.g. Plexiglas®). It is resistant up to approximately 70 °C and is often used as a glass substitute.

Polycarbonates (PC) are formally polyesters of carbonic acid ($[-R-O-C(O)-O-]_n$) and are produced by polycondensation of diols with phosgene. The diol used is, for example, bisphenol A (trade name e.g. Makrolon®). Polycarbonates burn less easily than PMMA and are also a glass substitute.

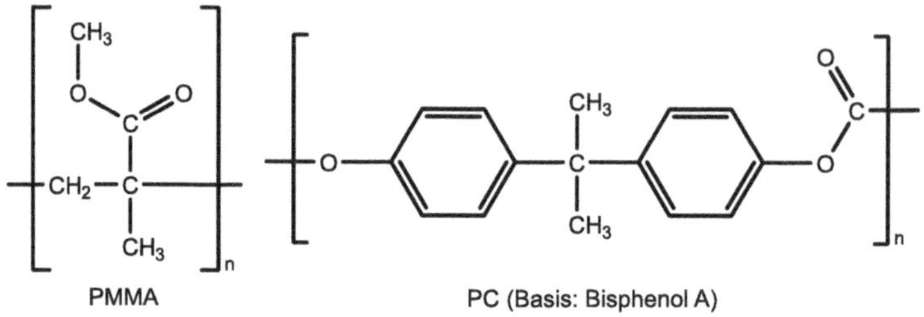

PMMA PC (Basis: Bisphenol A)

49.4.7 Polystyrene (PS)

Polystyrene is produced by polymerisation of ethenylbenzene ("styrene" or also "vinylbenzene"). It is used as a packaging material (e.g. yoghurt pots), in the electrical industry and many other applications. Application. As a foamed plastic (colloquially: Styropor©) it is an important packaging and thermal insulation material.

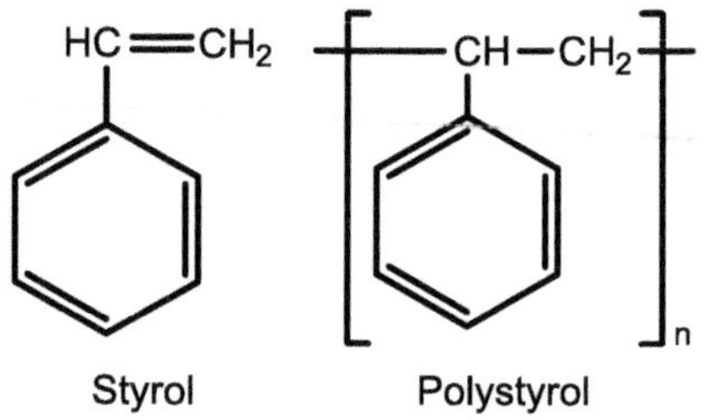

Styrol Polystyrol

49.4.8 Other polymers

By far not all plastics can be dealt with. Finally, three more polymers should be mentioned which are of some importance.

Polyethylene terephthalate (PET)
Besides polycarbonates, PET is one of the most important polyesters. In addition to its use as food packaging, the PET fibre ("fleece fibres") is of great importance.

Polybenzimidazole (PBI)
Due to its resistance to high temperatures, PBI has been used as an outer material for firefighters' protective clothing and hoods since the mid-1980s.

Silicones (SI)
Silicones, chemically poly(organo)siloxanes, is the group name for a whole series of synthetic polymers in which Si atoms are linked via O atoms to form chains, rings or cross-linked structures.

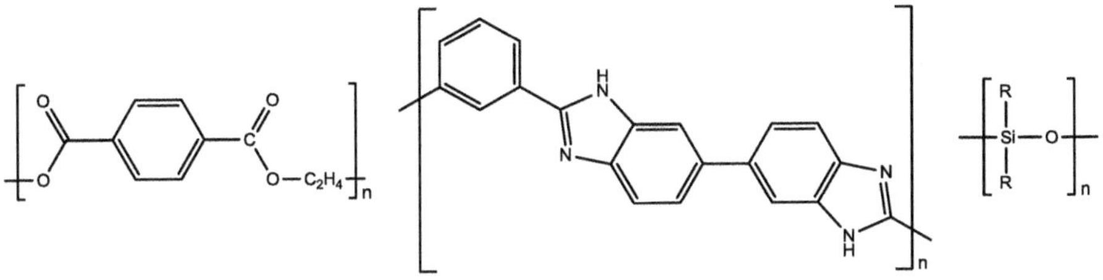

PET PBI SI (Bsp.)

Silicones exhibit a spectrum of properties that differs significantly from other polymers. Silicone oils and greases, but also silicone elastomers can be produced. SI elastomers in particular have been increasingly used in households in recent years (e.g. baking or ice cube moulds) and have been processed into chemical-resistant hoses for many years.

49.5 Fire Behaviour of Plastics

The essential components of all organic substances – and thus also of plastics – are the elements carbon and hydrogen. For this reason, most organic substances can either burn or influence the course of a fire (see Sect. 38.2). Depending on the type of plastic and admixtures, such as flame retardants (Sect. 42.1), the flammability can vary considerably.

The fire behaviour of plastics is not a purely material-specific property and, as with other combustible solids, also depends on the shape, surface properties and distribution.

Low ignition temperatures facilitate the process of ignition and promote the progression of a fire. High combustion heats (heating values, H_i) favour a faster fire progression. The heat generated by the fire heats up areas not yet affected, thermal decomposition begins. This may result in a flash-over (Sect. 40.2.3).

49.5.1 Heating values in comparison

The calorific values of many plastics that are processed in large quantities are above 30 MJ kg^{-1}. This corresponds roughly to the heating value of coal, which – depending on the type – is 27...33.5 MJ kg^{-1}. The calorific values of PE, PP and PS even reach the order of magnitude of fuel oil or petrol, as can be seen from Fig. 49.2.

For a real fire we can therefore be formulated – simplified:

- Plastics are "like solid fuel oil", that is, the amounts of heat released when plastics are burned are similar to those of heating oil.

For better clarification, the calorific value of household items with a high plastic content can be converted, for example, into the volume of petrol that provides this calorific value. Values are obtained as shown in Table 49.3 as an example.

49.5.2 Fire Behaviour When Different Substances Come Together

Pure fires of a single plastic are extremely rare. In most cases, a wide variety of plastics are stored together with other materials, such as paper, cardboard packaging or (wooden) pallets. Plastic containers in the form of sacks, bottles, crates, canisters or drums are used to store a wide variety of substances at the same time. The storage of flammable liquids in plastic containers naturally poses a particular hazard.

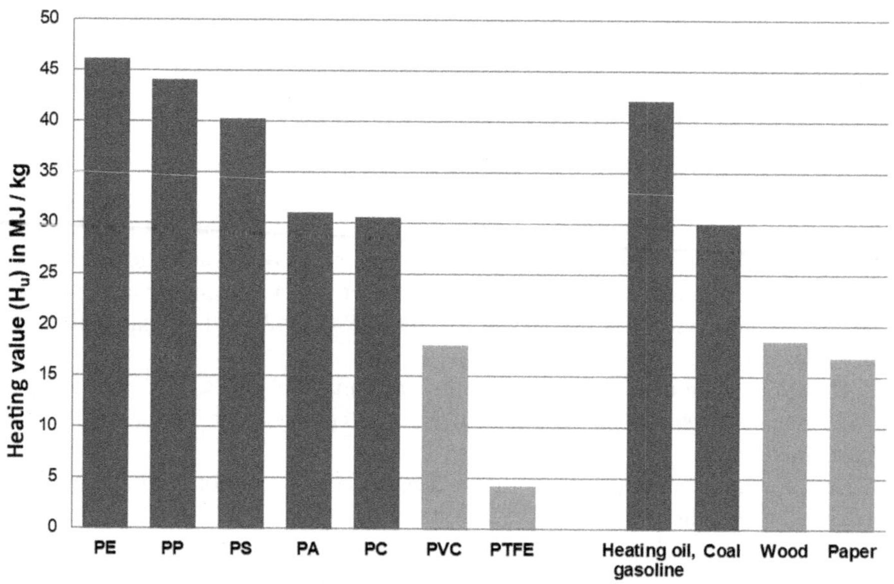

Fig. 49.2 Comparison: Heating values of plastics against heating materials

Table 49.3 Petrol equivalents of household goods

Examples	Mass of Plastic Per Unit (average)	Gasoline Equivalent
LCD TV	8.4 kg	6.1 L
Tube TV	2.8 kg	2.3 L
Vacuum cleaner, iron, electric kettle	2.1 kg	1.3 L
PC keyboards, printers	1.4 kg	0.9 L
Microwaves, air conditioners	1.3 kg	0.8 L
DVD players, webcams, cameras	0.9 kg	0.6 L

If combustible materials are present which provide a so-called supporting fire in the event of a fire, even flame-retardant plastics with a low heating value, such as PVC or PTFE, can burn and, in addition, contribute to an increase in damage due to the corrosive gases produced. The HCl evolution (calculated to 20 °C gas temperature) in rigid PVC is as a function of ambient temperature:

- from 140 °C: Start of HCl evolution
- from 180 °C: noticeable HCl development
- from 210 °C: PVC begins to melt
- from 300 °C: Separation of 85 % of the chlorine as HCl: approx. 330 L $HCl_{(g)}$ per kg PVC
- from 400 °C: Separation of 92 % of the chlorine as HCl: approx. 350 L $HCl_{(g)}$ per kg PVC
- from 530 °C: Separation of 99 % of the chlorine as HCl: approx. 380 L $HCl_{(g)}$ per kg PVC

49.5.3 Burning Behaviour of Pure Plastics

Plastics that have been treated with flame retardants or plasticizers, for example, can exhibit completely different burning behavior than pure polymers. Therefore, only basic statements on the pure polymers can be made here.

Thermosets exhibit similar combustion behaviour to wood or coal and therefore do not require detailed consideration. The flue gases of thermosets containing melamine or urea are slightly alkaline due to NH_3 admixtures.

Most elastomers burn with a strong soot and a luminous flame. Exceptions are halogenated elastomers and silicones, which are difficult to ignite.

The thermoplastics show in part very different burning behaviour, which is shown by way of example in Table 49.4 for a few plastics.

Table 49.4 Burning behaviour of some plastics (examples)

Smoke	Flame	Drips Burning	Melt	Burns Without Ignition Source	Smell	Possible Polymers
None/ little	Yellow, blue core	Yes	Clear	Yes	Candle	PE, PP
None/ little	Blue	Yes	Light charring	Yes	Pungent (formaldehyde)	POM
None/ little	Yellow-orange	No	Foams, chars	Yes	Mawkish	PMMA
White/ little	Yellow	Hardly	Froths	Badly	Burnt hair	PA
White	Yellow, blue core	Yes	Charring	Yes	Mawkish	PET
Sooty	Gleaming	Yes (sooty)	Foams, chars	Yes	Unpleasant (isocyanate)	PU
Sooty	Yellow-orange	Yes	Charring	Yes	Sugary	PS, ABS
Sooty	Yellow-orange	No	Charring	Goes out slowly	Phenolically	PC
Sooty	Yellow-orange	No	Charring	No	Pungent (HCl)	PVC

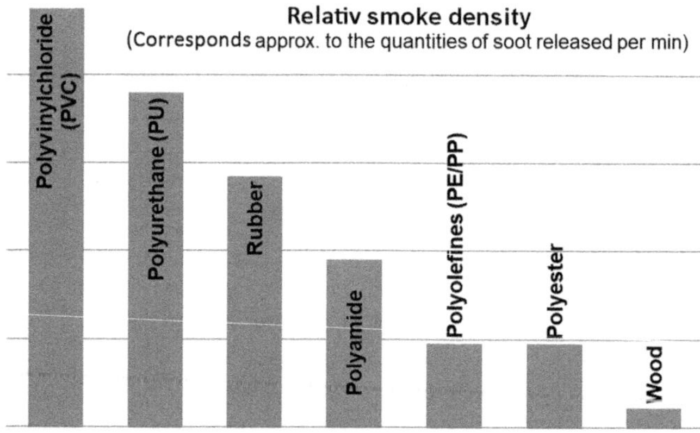

Fig. 49.3 Relative smoke density in plastic fires

49.5.4 Smoke Generation from Plastics

Many plastics burn with noticeable soot formation / strong (fire) smoke development (compare Sect. 40.3). In some plastic fires, e.g. when burning POM, almost no soot is produced. Other plastics, e.g. PS or PU, burn with strong soot formation and thus with high smoke emission.

Soot formation – and thus smoke development – depends on many factors, including:

- the type and quantity of burning plastics
- from the plastic surface (foamed plastics in particular have a greater tendency to form soot)
- the finishing of plastics with additives, flame retardants, plasticizers, etc.
- the melting, decomposition and ignition temperatures of the respective plastics involved
- of other substances involved in the fire
- from oxygen access (see Sect. 38.3)
- from the fire phase (compare Sect. 40.2.1)

Figure 49.3 shows the relative smoke densities of various plastic fires compared to wood. The relative smoke density here correlates approximately with the quantities of soot released per minute.

Surfactants

50

Surfactants or tensids (lat. *tensio* = tension) are substances that reduce the surface tension. This can be demonstrated in a simple experiment: With a fork and due caution, one can "place" a paper clip on the surface of water contained in a glass. If you add a drop of "dishwashing liquid", the paper clip will sink. The dishwashing liquid has greatly reduced the surface tension of the water.

50.1 Structure and Surfactant Groups

Characteristic for the structure of surfactants is at least one hydrophilic (Greek *hydro* = water, *philos* = loving; meaning "water-soluble") and one hydrophobic (Greek *phobos* = fear; meaning "water-repellent") functional group.

The hydrophilic moieties are mostly polar functional groups, such as the carboxylic acid group ($-COO^- \ H^+$), the sulphonic acid group ($-SO_3^- \ H^+$), the sulphate group ($-OSO_3^- \ H^+$), the quaternary ammonium group ($R_4N^+ \ X^-$) or polyethylene oxide groups ($-O-(CH_2-CH_2O)_n-H$) with n = 5–10. The hydrophobic groups used are mainly the long-chain aliphatic (chain length: C_{11} to C_{21}) or aliphatic-aromatic radicals (chain length on the aromatic: C_8 to C_{16}).

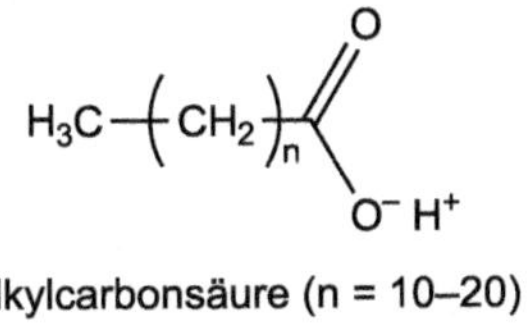

$H_3C-(CH_2)_n-COO^- \ H^+$

Alkylcarbonsäure (n = 10–20)

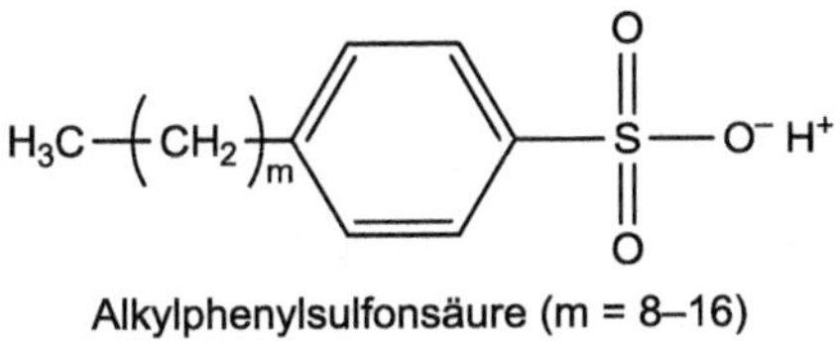

$H_3C-(CH_2)_m-C_6H_4-SO_2-O^- \ H^+$

Alkylphenylsulfonsäure (m = 8–16)

In the symbolic representation of surfactants, the hydrophilic part is shown as a circle, the hydrophobic part as a bar (Fig. 50.1). While aliphatic acids or sulphonic acids with a short hydrocarbon chain are readily soluble in water, the solubility in

T. Schmiermund, *The Chemistry Knowledge for Firefighters*,
https://doi.org/10.1007/978-3-662-64423-2_50

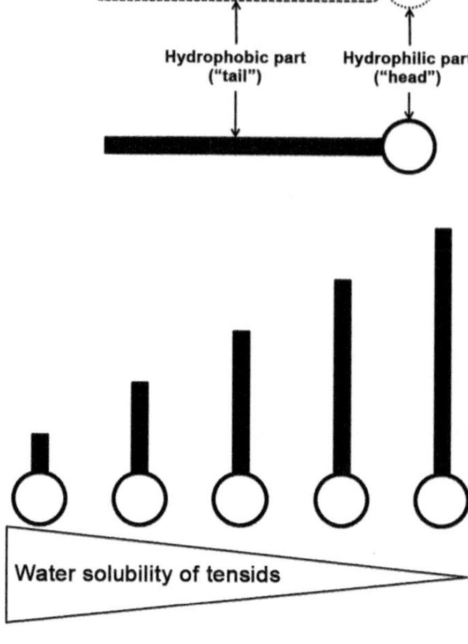

Fig. 50.1 Surfactants, schematic

Fig. 50.2 Solubility of surfactants as a function of the length of the carbon chain

water decreases again with increasing chain length of the hydrophobic group (Fig. 50.2).

50.2 Excursus: Fats and Oils

Fats, oils and tallow are important substances in the animal and plant kingdoms. They serve organisms as building materials and as energy stores. These three types of fat differ in their melting ranges. While oils are liquid at room temperature, tallows melt between 30 °C and 45 °C, fats at about 35 °C–40 °C. Chemically, they essentially consist of the glycerol esters – also called glycerides – of unbranched, higher carboxylic acids with an even C-number. Because of their origin, these carboxylic acids (C_{12}–C_{20}) are also called fatty acids. The higher the proportion of unsaturated fatty acids (oleic, linoleic and linolenic acids), the lower the melting range of the respective fat.

Naturally occurring fats, such as butter, lard, coconut fat, peanut oil, olive oil or linseed oil, are not pure substances but mixtures of different fat molecules. In each fat molecule, glycerol is esterified either with three identical fatty acids or with two or three different fatty acids. Vegetable oils contain the so-called essential fatty acids linoleic and linolenic acid, which cannot be produced by the human body itself. The most important fatty acids are listed in Table 50.1.

Table 50.1 Important fatty acids

Name	Formula	Molar Mass (in g mol^{-1})	Melting Point (in °C)
Lauric acid	$C_{11}H_{23}-COOH$	200.32	44
Myristic acid	$C_{13}H_{27}-COOH$	228.38	54
Palmitic acid	$C_{15}H_{31}-COOH$	256.43	63
Stearic acid	$C_{17}H_{35}-COOH$	284.48	68
Zoomaric acid	$C_6H_{13}-CH{=}CH-C_7H_{14}-COOH$	254.41	1
Oleic acid	$C_8H_{17}-CH{=}CH-C_7H_{14}-COOH$	282.46	16
Linoleic acid	$C_4H_9-(CH_2-CH{=}CH)_2-C_7H_{14}-COOH$	280.45	−7
Linolenic acid	$H_3C-(CH_2-CH{=}CH)_3-C_7H_{14}-COOH$	278.44	−11

Table 50.2 Smoke points of oils

Oil	Smoke point	Oil	Smoke Point
Butter	~ 175 °C	Rapeseed oil, refined	220 °C
Peanut oil, refined	230 °C	Sunflower oil, unrefined	107 °C
Peanut oil, unrefined	130 °C	Grape seed oil, unrefined	130 °C
Olive oil, refined	> 220 °C	Walnut oil, unrefined	160 °C

Smoke Point of Oils

The smoke point is the temperature above which an oil produces smoke when heated. The higher the smoke point, the more suitable the oil is for frying and deep-frying. It has been shown (Table 50.2) that oils which "only" smoke – but do not yet burn – generally already exhibit temperatures of over 100 °C. The introduction of water for cooling purposes may therefore lead to dangerous steam explosions (see Sect. 3.1.6 and Sect. 39.1.1).

50.2.1 Fat Hardening

Fat hardening is the term used to describe the process by which margarine has been produced since 1902. Vegetable or animal fats or oils containing unsaturated fatty acids are converted into solid, saturated fats by catalytic hydrogenation. For example, liquid unsaturated oleic acid is converted into solid stearic acid. The desired consistency can be largely controlled by suitable process control.

50.2.2 Saponification

Technically and economically important is the alkaline hydrolysis of glycerides, the so-called saponification. As early as the 3rd millennium B.C., the Sumerians produced soap from plant ash and oil. The process of saponification is, so to speak, the reverse of esterification and provides the salts of the fatty acids in addition to glycerol as an alcohol component. Saponification is shown as an example in Fig. 50.3.

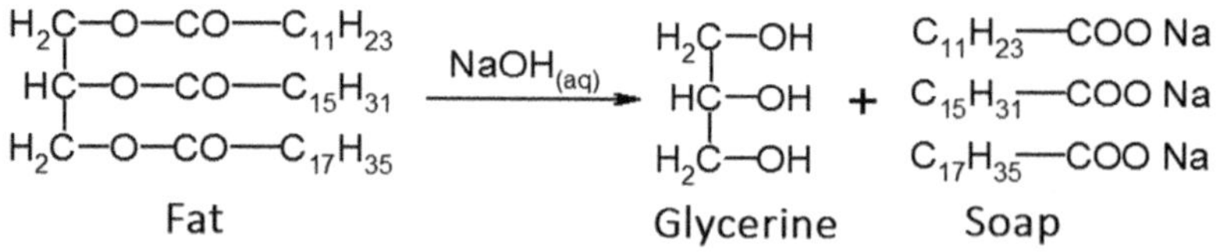

$$\begin{array}{l} H_2C{-}O{-}CO{-}C_{11}H_{23} \\ HC{-}O{-}CO{-}C_{15}H_{31} \\ H_2C{-}O{-}CO{-}C_{17}H_{35} \end{array} \xrightarrow{NaOH_{(aq)}} \begin{array}{l} H_2C{-}OH \\ HC{-}OH \\ H_2C{-}OH \end{array} + \begin{array}{l} C_{11}H_{23}{-}COO\,Na \\ C_{15}H_{31}{-}COO\,Na \\ C_{17}H_{35}{-}COO\,Na \end{array}$$

Fat Glycerine Soap

Fig. 50.3 Saponification

The acids themselves can be obtained by acidification if necessary. As a rule, however, the fatty acid salts, the so-called soaps – are of interest. The sodium salts form solid soaps, which are also called curd soap. The potassium salts are pasty and are sold as so-called soft soap. With hard water, i.e. water containing relatively high amounts of calcium and magnesium ions, the Ca and Mg salts of the fatty acids, which are extremely difficult to dissolve, form the so-called lime soaps:

$$2\,C_{17}H_{35}{-}COO^-\,Na^+ + Ca^{2+} \rightarrow 2\,Na^+ + (C_{17}H_{35}{-}COO^-)_2Ca^{2+}\downarrow$$

The actual soap is thus removed from the solution, the cleaning effect decreases. This is also noticeable, among other things, in a lower foam formation. Some of the lime soaps adhere to the fabric, making it "hard" and also leading to a grey film on the laundry. In addition, soap solutions react alkaline: the fatty acids are weak acids, the alkali ion is a strong base (compare Sect. 14.4). In households and industry today, other washing-active substances are used instead of soaps, and these are much more effective.

50.3 Mode of Action of the Surfactants

The effect of soaps and detergents during washing is based on chemical-physical processes which, taken together, bring about cleaning. Some dirt particles are water-soluble, such as sugar, inorganic salts or some proteins; others, such as grease or soot, are insoluble in water. The water-insoluble impurities are now dispersed (more precisely: suspended or emulsified) in the washing solution by the surfactants, thus bringing about a cleaning effect. The reduction of the surface tension and the formation of micelles play an important role here.

50.3.1 Surface Tension

A phenomenon known as surface tension occurs at the boundary between gases (usually air) and liquids. It occurs because the molecules inside the liquid interact with their neighbouring molecules. At the surface, on the other hand, the molecules of the liquid only experience inward forces on one side (compare Fig. 3.7). Due to the greater potential energy of the particles at the surface, physical work must be done to increase the surface area of the liquid. The liquid surface therefore resembles

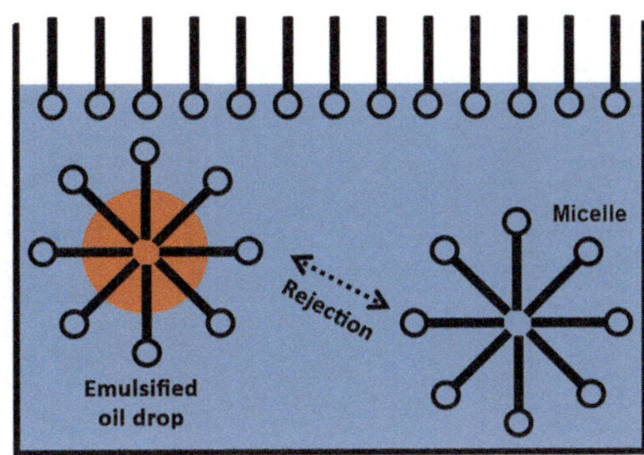

Fig. 50.4 Surfactant effects

a thin film in its behaviour. The force required is expressed as energy (in Joule) per area (in square metres) or as force (in Newton) per length (in metres). In former times the unit dyn cm^{-1} was common. The following applies: $1\ J\ m^{-2} = 1\ Nm^{-1} = 10^3$ dyn cm^{-1}.

Surfactants accumulate at the interfaces of water towards other substances with the molecules oriented vertically. The surfactant head is oriented towards the water, while the tail turns away from the water (Fig. 50.4).

This reduces the forces between the water molecules at the interface – the surface tension decreases and non-wettable substances are coated by the surfactants and can be washed away. At the same time, the water can thus penetrate solids more easily.

50.3.2 Micellation

If the entire surface layer is provided with surfactant molecules, so-called micelles (lat. *mica* = small lumps) are formed. Here the surfactant molecules arrange themselves with their hydrophobic molecule parts directed inwards to form spherical structures – the micelles. Hydrophobic substances such as carbon black or oil are enclosed in these micelles and thus distributed in the water (Fig. 50.4). Because the hydrophilic part of the surfactant molecule points away from the dirt particle, two effects occur:

a. The dirt remains stably distributed in the suds and does not stick again to e.g. clothing.
b. The micelles repel each other, so that the emulsified or suspended particles cannot form larger structures.

50.3.3 Foaming

The reduction of the surface tension by the aligned surfactant layer causes another effect: If air is introduced into a surfactant solution, a new interface is formed at

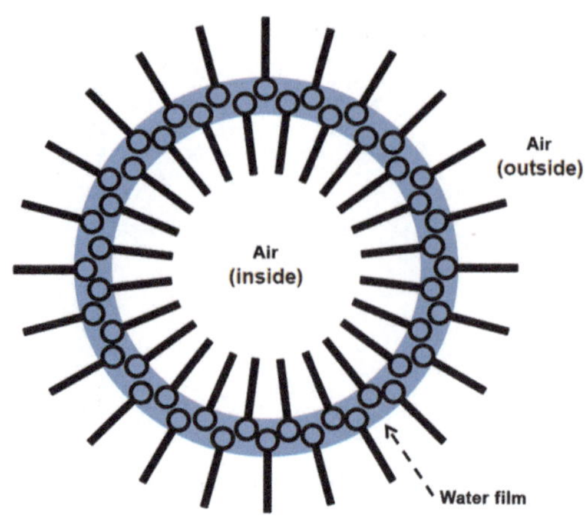

Fig. 50.5 Foam bubble

which the surfactant molecules can align themselves again. The air bubbles are thus partially stabilized (Fig. 50.5). Foam is thus a dispersion of air in a surfactant-containing solution.

► In foam, a liquid layer is stabilized by the surfactant molecules.

50.4 Surfactants in Fire Fighting

Surfactants used by fire brigades are usually the so-called foaming agents. These are used in low concentrations (0.1 %–0.5 %) as wetting agents and in higher concentrations (3...6 %) as foam formers to produce extinguishing foam. (For information on foam extinguishing agents, see Sect. 41.2.2).

50.4.1 Surfactants as Wetting Agents

Adding small quantities of foaming agent to the extinguishing water significantly reduces its surface tension. Pure water has a surface tension of $72.5 \cdot 10^{-3}$ N m^{-1} at 20 °C; the addition of surfactant reduces this to $20...30 \cdot 10^{-3}$ N m^{-1}. The extinguishing water thus penetrates more easily into the fire material and also wets materials that have caught fire and tend to be water-repellent, such as straw or lignite. The resulting increase in extinguishing efficiency when using mains water for solid fires is undisputed.

50.4.2 Surfactants and Electrical Conductivity

The electrical conductance is the reciprocal of the electrical resistance (unit = Ohm, Ω) and is given in the unit Siemens (S). Accordingly, the following applies: 1 S = 1/Ω.

The electrical conductivity is the reciprocal of the specific electrical resistance, its unit is S m^{-1}. For measurements of the conductivity, the converted unit S cm^{-1} is usually used in chemical analysis, thus obtaining more manageable figures and avoiding conversion-related calculation errors.

Concerns regarding the increased electrical conductivity of mains water or foam due to the addition of surfactants are unnecessary at voltages in the household range (230...400 V). According to the German Drinking Water Ordinance, drinking water may have a conductivity of up to 2 mS cm^{-1}. In most cases, the values are below 1 mS cm^{-1}. The foam additive causes increases in conductivity in the order of magnitude of about 0.1...0.5 mS cm^{-1}. This means that the conductivity of mains water is generally still below the limit of the Drinking Water Ordinance.

Drinking water itself has a significantly higher conductivity than extinguishing foam. Due to the lower electrical conductivity (= higher electrical resistance) of extinguishing foam, lower leakage currents and thus also a lower risk to persons – are to be expected than when using water. It can therefore be stated:

▶ The increase in conductivity due to surfactants or surfactant solutions is of no practical significance in firefighting operations at low voltages.

50.4.3 Fluorine-Containing vs. Fluorine-Free Foam Agents

The first surfactants used to produce extinguishing foam were free of halogens. The search for foaming agents with better resistance to alcohols and ketones led to perfluorinated surfactants. Besides fluorinated protein-based foaming agents (fluorinated protein foam, FPF), perfluorooctanoic acid (PFOA*)*, perfluorooctane sulfonic acid (PFOS) and similar perfluorotensids (PFT) became the "agents of choice". In particular, the fact that the hydrophobic "surfactant tail" is also lipophobic at the same time, resulting in the formation of an aqueous film on the surface of the burnt material (cf. Section 41.2.3), has contributed to the widespread use of PFTs.

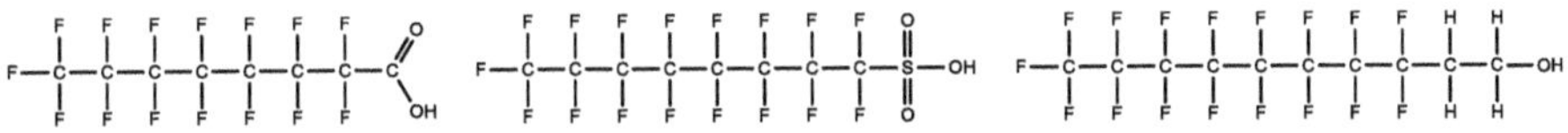

PFOA PFOS FTOH (8 : 2)

Organic fluorine compounds are even rarer in the natural environment than organic chlorine compounds. Perfluorinated compounds are exclusively of human origin. Therefore, there are no bacteria or fungi that are able to decompose perfluorinated compounds quickly and easily. As a result, perfluorinated and polyfluorinated chemicals (PFCs) accumulate in the environment – they are said to be persistent. Since June 2011 (in Germany), foam agents containing PFOS are no longer allowed to be in fire brigade stocks. So-called fluorotelomer alcohols

(FTOH), which in turn still represent polyfluorinated compounds, are currently acting as a substitute for PFOS.

Investigations of the IdF Heyrothsberge (literature [13-11]) show that the extinguishing capacity of fluorine-free foam agents is equal to that of fluorine-containing foam agents in fighting fires of polar, foam-destroying liquids.

- The biodegradable substitutes are just as suitable for fire fighting with foam as the non-degradable fluorosurfactants.

Part XII
CBRNE Hazards

The fire brigade is familiar with hazards associated with Hazmat (= hazardous materials) and NBC hazards (NBC = nuclear, biological, chemical) or Hazmat-ABC operations (Hazmat-ABC = hazardous materials: atomic (nuclear), biological, chemical), and the abbreviations are sufficiently well known. To specify the "nuclear" hazards, the abbreviation CBRN from the Anglo-American linguistic usage for chemical, biological, *radiological, nuclear* has become established. Nuclear here stands for the special case of the detonation of a nuclear weapon. Radiological stands for all other incidents involving sealed or open radiation sources. In more recent times, incidents involving *explosives have* also been explicitly included. *This* finally leads to the abbreviation CBRNE.

As examples of chemical hazards, poisons (Chap. 51) and then chemical warfare agents (Chap. 52) will be considered first. Subsequently, B-hazards are considered in Chap. 53 (Biological agents) and Chap. 54 (Biological warfare agents). Finally, there is a brief outline of explosives (Chap. 55).

Radiological and nuclear hazards have already been dealt with in detail:

- Chap. 29: Radiation exposure
- Chap. 30: Biological effects of ionizing radiation
- Chap. 32: Nuclear reactions
- Chap. 34: Protection against ionizing radiation

Poisons

51

Poisons are substances which, even in relatively small quantities, cause dysfunctions in a living body which may even lead to the death of the living being. The danger of a substance acting as a poison depends to a large extent on the dose. Paracelcus already expressed this in the sixteenth century:

> *"What is that is not poison? all things are poison / and nothing is without poison / Alone the dose makes that a thing is not poison. "*

This dose-effect-relationship is not only found in poisons, but in many substances, as long as the body is able to absorb them. Taken in quantities of 3...5 g per kg body weight, even the essential table salt has a toxic effect. And the smallest doses of the "poisonous" digitalis glycosides have a life-sustaining effect as heart medication; higher doses are nevertheless lethal.

In general, it can be said that highly concentrated poisons are more dangerous than diluted forms – even with the same absolute amount of poison. Fine powder is more harmful than a single, large poison cube. Dissolved substances act faster than solids.

51.1 General Information on Poisons

Poisons are naturally occurring or artificially produced inorganic or organic compounds. The effect of poisons is a consequence of biochemical interactions of these substances with endogenous substances. It is irrelevant whether the toxin is of plant-animal-mineral origin or has been synthetically produced.

> The same molecule has the same toxicological effect regardless of the route by which it was produced.

T. Schmiermund, *The Chemistry Knowledge for Firefighters*,
https://doi.org/10.1007/978-3-662-64423-2_51

Table 51.1 Poison groups (examples)

Designation	Examples
Corrosives	Nitric acid, potassium hydroxide solution
Blood toxins	HCN, CO, nitrobenzene, aniline
Respiratory toxins	Chlorine, phosgene
Gastrointestinal toxins	Hg, Pb salts
Liver toxins	Tetrachloromethane, many fungal toxins
Kidney toxins	Glycol, oxalic acid
Heart toxins	Aconitine, digitalis, toad poisons
Nerve agents	H_2S, CS_2, strychnine, cocaine

51.1.1 Poisons and Poison Effects

Toxicology (Greek *toxon* = arrow poison), i.e. the study of poisons, differentiates according to the type or site of damage as shown in Table 51.1. Other groups of poisons are, for example, enzyme, skin, nose or eye poisons.

According to the application, a distinction is made between, for example: Arrow poisons, fish poisons, pesticides, combat agents, insecticides, herbicides and many more.

The toxin effect depends – apart from the dose – also on age, sex, the application or intake of the toxin and the sensitivity of the organism to this toxin. The effects may be reversible, so that complete recovery occurs, or irreversible, so that long-term damage remains, e.g. scars after chemical burns.

Some poisons, such as skin or corrosive poisons, have a predominantly local effect and primarily damage the affected areas. Other poisons are absorbed via the gastrointestinal tract, through the skin, the mucous membranes or the lungs and are distributed throughout the entire body before a toxic effect occurs. This toxic effect then unfolds as a so-called **systemic** effect.

Ingestion of small doses over longer periods of time may lead to **chronic**, i.e. slowly progressing, poisoning. Typical examples are poisoning by heavy metals or pesticides, in case of regular direct contact with these substances. In contrast to this are **acute**, i.e. rapid, severe poisonings, the severity of which is usually strongly dependent on the dose. The extent of the poisonous effect is often directly proportional to the amount ingested. However, some poisons accumulate in the body. In this case, there is no linear relationship between the strength of the effect and the ingested dose. In principle, therefore, there is no difference in the effect of toxins and drugs.

In addition to **exogenous** (i.e. coming from outside) toxins, there are also **endogenous** (i.e. produced in the body) toxins. These are metabolic products that arise in the organism itself and lead to health problems if they are not excreted or detoxified in the body.

51.1.2 Therapy of Poisoning

The mechanisms of action are as different as the poisons themselves. If it is not known what caused the poisoning, the therapy is symptomatic, i.e. the treatment is based on the symptoms that occur. If, on the other hand, the toxin is known, an antidote can be administered for some toxins.

▶ Attention: In case of suspected poisoning with regard to first aid measures, it is essential to observe personal protection (gloves, respirator, . . .)!

Remove contaminated clothing. Flush skin/eyes with water (collect water!?). After ingestion, rinse out mouth of the affected person.

▶ Note: Do not induce vomiting in case of: Drowsiness of consciousness, convulsions, poisoning with acids, alkalis, solvents or surfactants!

51.2 Labelling and Classification of Toxic Substances

The classifications according to the German Chemicals Act into "very toxic" (T^+), "toxic" (T) and "harmful" (Xn; also "less toxic") have in the meantime been replaced by the **G**lobally **H**armonised **S**ystem of Classification and **Labelling** of Chemicals (GHS) based on the CLP Regulation (**C**lassification, **L**abelling, **P**ackaging).

51.2.1 GHS Classification

According to the GHS system, substances are considered to be acutely toxic if they show a harmful effect in a single dose or within 24 hours in several doses by oral or dermal administration or by inhalation over 4 hours. An overview of the GHS classifications is given in Table 51.2.

51.2.2 Transport Law

According to ADR, toxic substances are classified in hazard class 6.1 "toxic substances". This applies to substances of which it is known, or may be assumed on the basis of investigations, that even relatively small quantities may cause damage to health or death by inhalation, skin contact or ingestion after a single or brief exposure. The classifications according to ADR can be found in Table 51.3. The danger label and the warning plates are shown in Fig. 51.1.

Table 51.2 GHS classification "acute toxicity"

Category	Cat. 1	Cat. 2	Cat. 3	Cat. 4	
Symbol					
Signal word	Danger	Danger	Danger	Warning	
LD_{50} (oral)	≤ 5	$> 5\ldots50$	$> 50\ldots300$	$> 300\ldots2000$	mg kg^{-1} body weight
LD_{50} (dermal)	≤ 50	$> 50\ldots200$	$> 200\ldots1000$	$> 1000\ldots2000$	mg kg^{-1} body weight
LC_{50} (inhalation): Gases	≤ 0.1	$> 0.1\ldots0.5$	$> 0.5\ldots2.5$	$> 2.5\ldots20$	mL L^{-1} air
Vapours	≤ 0.5	$> 0.5\ldots2$	$> 2\ldots10$	$> 10\ldots20$	mg L^{-1} air (4 h)
Dusts, mists	≤ 0.05	$> 0.05\ldots0.5$	$> 0.5\ldots1$	$> 1\ldots5$	mg L^{-1} air (4 h)
Danger notice	Very toxic danger to life	Very toxic danger to life	Toxic	Unhealthy	

Table 51.3 Classification of toxic substances according to ADR

Danger notice	Hazard number	LD_{50} (oral) mg kg^{-1}	LD_{50} (dermal) mg kg^{-1}	LC_{50} (inhalation) mg L^{-1}	Packing group
Very poisonous	66	≤ 5	≤ 50	≤ 0.2	**I**
Toxic	60	$> 5 \leq 50$	$> 50 \leq 200$	$> 0.2 \leq 2$	**II**
Slightly toxic	60	$> 50 \leq 300$	$> 200 \leq 1000$	$> 2 \leq 4$	**III**

Fig. 51.1 Transport labelling of toxic substances (hazardous goods labels and warning signs)

51.2.3 Classification into Fire Brigade Hazard Groups

According to German FwDV 500, C-hazardous substances are classified into hazard groups IC, IIC and IIIC. The following applies here:

- Hazard group IC
  Risks that are likely to be countered by standard fire brigade (fire engine) is to be encountered.
- Hazard group IIC
  Risks which are likely to require additional special equipment.
- Hazard group IIIC
  Risks that are likely to be controllable only with special equipment **and** external specialist advice.

In addition to the quantity of hazardous substances, the packaging group (cf. Table 51.3) is used as an aid to classification into the fire brigade hazard groups. Transport accidents are generally treated as hazard group IIC. If very toxic substances (hazard number 66, packing group I) are involved, it is recommended to proceed as for hazard group IIIC.

51.3 Operational Measures

51.3.1 Human Life in Danger

Quote from FwDV 500:

▶ For rescue operations in hazard groups IIC and IIIC, the rescue personnel must be equipped with at least insulating equipment and Form 1 body protection.

51.3.2 Body Protection

Protective clothing appropriate to the situation must be worn, and this applies particularly to toxic substances, if contact with a hazardous substance cannot be completely ruled out.

The FwDV 500 considers as (at least) suitable:

- Hazard group IIC Body protection Form 1
- Hazard group IIIC Body protection Form 2 or 3

As a general rule, the following applies to all C operations:

▶ If there is insufficient information available, the teams proceeding with the investigation must be equipped with Form 3 body protection.

This is particularly true when it is already clear that toxic substances are or may be involved and the exact location has yet to be explored.

Table 51.4 Measures group 6 according to FwDV 500 (modified)

Dangers	• Substance itself and vapours / dusts / mists are toxic. • Risk of poisoning with – Inhalation – Incorporation – Contamination • Fire gases can also be toxic.
Special measures	• Breathing and body protection • Prevent spreading, collect substance, seal leakage • Secure sewers, deeper spaces and water bodies • Use measuring instruments or indicators
Additional notes	• Avoid skin contact with free substances at all costs • In case of contact or suspected contamination: – Initiate decon measures immediately – Initiate medical examination immediately

51.3.3 Action Group 6 "Toxic Substances"

Toxic substances and objects containing toxic substances are assigned to MG 6 (FwDV 500, MG = *Maßnahmengruppe* → action group) (Table 51.4).

52 Chemical Warfare Agents

The term "chemical warfare agents" brings to mind the horrors of World War I, when large quantities of these substances were used. But it also brings to mind the danger posed by terrorism (e.g. the sarin attack in Tokyo in 1995). However, the danger does not only emanate from "real" warfare agents, i.e. those produced only for this purpose. Potentially, "normal" chemical products such as chlorine, hydrocyanic acid or phosgene, so-called "dual use" substances, are also suitable. In addition, there are a large number of substances that are suitable for localised attacks: sabotage poisons.

52.1 Introduction

52.1.1 Definition

Chemical warfare agents are substances used in the military that affect the body in such a way that the soldier's ability to fight is greatly reduced or death occurs. This goal can be achieved in open combat or covertly, through sabotage.

52.1.2 Early Bans on Chemical Weapons

The history of "chemical substances" to weaken the enemy is practically as old as the history of war itself. As early as 3000 years ago, "pitch and sulphur" were used in warfare. In general, sulfur was very popular for centuries as an additive to incendiary projectiles, because the SO_2 produced during combustion has a strong irritant effect and sulfur itself burns well. In the thirteenth century, the use of arsenic compounds in incendiary devices was suggested as a way to poison opponents with arsenic oxide fumes. Irritants and corrosives were also used, such as caustic lime, which was commonly used in the Middle Ages. In modern times, other irritating, foul-smelling and poisonous substances were considered in the context of warfare.

T. Schmiermund, *The Chemistry Knowledge for Firefighters*,
https://doi.org/10.1007/978-3-662-64423-2_52

The first bans on the use of toxic substances date back to the seventeenth century and prohibited the use of poisons, e.g. to poison wells or foodstuffs. This was followed in 1899 and 1907 by the Hague Peace Conferences. Thus, Annex IV to the Hague Convention – the so-called **Hague Land Warfare Regulations** – stipulated in 1907:

- Article 22
  "The belligerents have no unlimited right in the choice of means to injure the enemy"
- Article 23
  "Except for prohibitions established by special contract, it is prohibited by name:
  a. The use of poisons or poisoned weapons
  b. The use of weapons, projectiles and substances capable of causing unnecessary suffering
  c. The use of projectiles whose sole purpose is to disperse poisonous or asphyxiating gases. The fragmentation effect must always exceed the poisonous effect."

Despite this international agreement, which was ratified by all European states, the history of large-scale use of warfare agents began during the First World War. On 22 April 1915, 146 tons of chlorine gas were released by German troops against French soldiers north of Ypres. In the further course of the war, a multitude of poisonous, irritating and suffocating substances were used.
In the **Geneva Protocol** of June 1925, which was also ratified by the German Reich, the signatory states declared to respect the prohibition of the use in war of asphyxiating, poisonous or similar gases, as well as all similar liquids, substances and methods. Likewise, they agreed to extend this prohibition to bacteriological warfare agents and to induce other states to accede to this protocol.

Regardless of this, the search for "weapons-grade chemicals" continued between the two world wars and led to the discovery of the so-called nerve gases, the further development of which was the subject of military research during and also after World War II. With the **Brussels Treaty** in 1954, the German government committed itself not to produce the nuclear, chemical and biological weapons listed in the treaty.

52.1.3 Chemical Weapons Convention (CWC)

The "Convention on the Prohibition of the Development, Production, Stockpiling and Use of Chemical Weapons and on their Destruction", in short the **Chemical Weapons Convention** (CWC) entered into force on 29 April 1997. Monitoring is the responsibility of the *Organisation of the Prohibition of Chemical Weapons* (***OPCW***), which carries out regular inspections in the chemical industry.

In Germany, the **Federal Office of Economics and Export Control (Bundesamt für Wirtschaft und Ausfuhrkontrolle,** BAFA) is responsible for collecting reporting data for the OPCW, monitoring inspections and controlling the import and export of substances listed in the CWC.

On behalf of the OPCW, residual stocks of ammunition containing chemical warfare agents, some of which date back to World War I, are being destroyed in a controlled manner. As a rule, the ordnance is first x-rayed (content: solid or liquid? cracks in/on the container?) and the chemical signature of the ordnance is determined by means of a neutron activation analysis. The projectile is then deep-frozen and the warhead is separated from the explosive container using a remote-controlled saw. The explosive and the ordnance are then disposed of separately in an orderly manner, usually by incineration in special facilities.

52.1.4 War Weapons Control Act

The "Gesetz zu Artikel 26 Absatz 2 des Grundgesetzes (Gesetz über die Kontrolle von Kriegsw affen)" (Act on the Control of War **Weapons),** abbreviated **KrWaffKontrG,** also **KWKG, lists** in the Annex (Kriegswaffenliste, KWL) in Part A, Section III, the chemical warfare agents and precursors whose production and proliferation is restricted or prohibited under the Act (see also Sect. 54.1.2).

52.1.5 "Dual Use" → Substances & equipment

These are substances produced on a large scale and consumed by industry in large quantities. They can (also) be used as direct precursors for chemical weapons or are toxic bulk chemicals (Schedules 2 and 3 of the CWC). Some of these substances have already been used militarily as warfare agents, such as chlorine or phosgene, and could be used today as "substitute warfare agents".

Plants and plant components that are needed for the production of substance XY can of course also be used for the production of other substances. In particular, plants in which insecticides are produced on the basis of phosphoric acid esters can relatively easily be converted to the production of warfare agents (and back again). Pre- and intermediate products as well as certain components for the construction of chemical plants are therefore subject to control by BAFA or the OPCW. The legal bases include the KrWaffKontrG, the Foreign Trade and Payments Ordinance (Außenhandelswirtschaftsverordnung, AWV) and the EC Dual-Use Regulation No. 428/2009.

The possible precursors of chemical warfare agents include a whole range of organic and inorganic substances. A few examples may illustrate the problem of dual use:

Dimethylamine [$(CH_3)_2NH$]	Phosphorus oxychloride ($POCl_3$)
Hydrogen fluoride (HF)	Triethanolamine [$(HO{-}C_2H_4{-})_3N$]
Phosphorus trichloride (PCl_3)	Sodium cyanide (NaCN)
Thionyl chloride ($SOCl_2$)	Phosphorus pentachloride (PCl_5)
Sulphur dichloride (SCl_2)	

52.1.6 Classification of Chemical Warfare Agents

Usually, C warfare agents are divided into groups according to their (main) effect:

- Irritants (white cross, blue cross warfare agents)
- Pulmonary agents (green cross warfare agents)
- Blood warfare agents
- Skin warfare agents (yellow cross warfare agents)
- Nerve agents
- psychotoxic agents

These are usually so-called tactical agents that have a direct effect (Fig. 52.1). They cause injuries and pain, can lead to death, cause confusion, create fear, and force the opponent to adapt his tactics to the effect of the agent. In addition, there is the group of strategic warfare agents, which is not unimportant for military command.
Strategic – or indirect – warfare agents are combat agents designed to disable the enemy in a variety of ways. These include phytotoxic (plant-damaging) substances such as herbicides or defoliants. These are intended to cause crop failure, for example, or to take away cover opportunities by destroying vegetation. The destruction or rendering useless of the enemy's drinking water and food supplies is also one

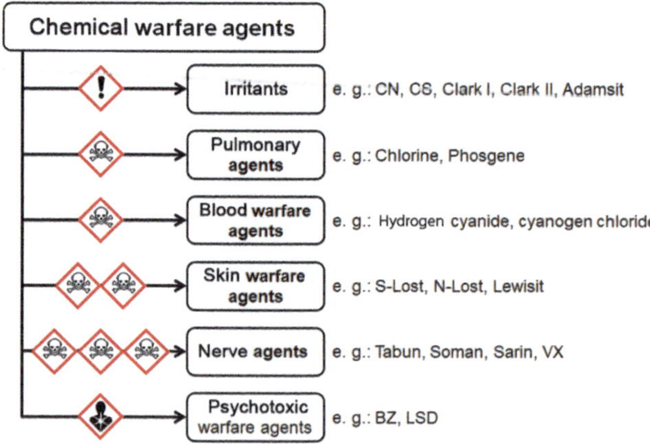

Fig. 52.1 Overview of warfare agents (with examples)

of the strategic measures. In addition, some of these warfare agents, as well as other substances, can be used as sabotage poisons, e.g. against groups of people or public institutions.

Notes

- The ammunition of the German troops (WW I) containing combat agents was marked with a colored cross to identify the combat agent group. This gives rise to the purely historical designations "...kreuzkampfstoff" ("...cross warfare agent").
- The letter codes, which are indicated in the following, e.g. GA or CN, were introduced by the US Army and represent common abbreviations.
- *The measures of the respective subsections "Protective measures and decontamination" given for the groups of warfare agents (Sects. 52.3, 52.4, 52.5, 52.6, 52.7 and 52.8) are general instructions. No obligations can be derived from them. Specific measures for the respective (warfare) agent are to be taken from the relevant special literature.*

52.2 Physical, Chemical and Toxicological Properties of Warfare Agents

In order for a toxic substance to be considered as a military warfare agent, it must possess a number of properties, both of a physico-chemical and toxicological nature, which will be briefly explained here.

52.2.1 Melting and Boiling Point

A large difference between melting point and boiling point is desirable for warfare agents to allow year-round use. For "winter use", the substance should have a melting point well below 0 °C.

52.2.2 Vapour Pressure p_D

For all substances, including warfare agents, a higher vapour pressure causes faster evaporation and thus a higher concentration in the ambient air. The temperature dependence of the vapour pressure must be taken into account if necessary (see Sect. 3.2.4).

In the relevant literature, the vapour pressure is often given in the unit "mm Hg" (= millimetre mercury column = Torr). To convert to mbar, multiply by the factor 1.333.

52.2.3 Saturation Concentration c_S

The saturation concentration, often called "volatility" in the older literature, represents the concentration of a warfare agent, given in mg L^{-1} or mg m^{-3}, which forms in a closed vessel as a result of the vapour pressure. The saturation concentration is reached when as many molecules pass into the gas phase as migrate back into the liquid (solid) phase.

The saturation concentration depends on the molar mass (M in g mol^{-1}), on the temperature and on the vapour pressure of the substance (p_D in mbar) at this temperature. It can be approximated according to the formula

$$c_S = \frac{M \cdot p_D \cdot 12}{T} \quad (52.1)$$

where the temperature, as in all calculations with gases and vapours, must be given in Kelvin. The formula Eq. 52.1 gives the saturation concentration in the unit mg L^{-1}.

In the open air, the saturation concentration cannot usually be reached, since warfare agent vapours are constantly being carried away by air currents and other processes.

52.2.4 Sedentariness

Settling time indicates how long the agent is able to remain in the atmosphere in open terrain. It is not a constant specific to the agent, but depends on a variety of factors. Influencing factors include:

- Hydrolysis rate of the substance
- Air/ground temperature
- Soil composition
- Vapour pressure of the substance
- Air/soil humidity
- Ground wind speed

Depending on the agent, sedentariness can vary over long periods of time (Table 52.1). Example values for the influence of soil temperature on the sedentariness of two selected warfare agents are given in Table 52.2.
Mostly, however, warfare agents are divided into sedentary and non-sedentary (volatile) warfare agents solely on the basis of their boiling point.

- Warfare agents with boiling points > 150 °C are considered to be settled
- Warfare agents with boiling points < 130 °C are considered volatile.

Table 52.1 Sedentariness of various chemical warfare agents

Combatant	Sedentariness
VX	5...10 d
S-Lost	3...5 d
Soman	10...15 h
Sarin	1...4 h
Hydrogen cyanide	5...10 min

Table 52.2 Influence of soil temperature on the sedentariness of two warfare agents (VX – sedentary, sarin – volatile)

Combatant	−10 °C	0 °C	10 °C	20 °C	30 °C	40 °C
VX	70 d	25 d	18 d	10 d	5 d	2 d
Sarin	2 d	1 d	10 h	4 h	2 h	1 h

- For warfare agents with boiling points 130...150 °C, the ambient conditions are decisive for the assessment of settling.

52.2.5 Latency

The latency period is the period of time between the effect of a poison on the organism and the appearance of the first symptoms of poisoning. The latency period depends, among other things, on the respective substance, the type of agent, the concentration and the type of absorption.

▶ With inhalation, the latency period is shorter than with oral or percutaneous intake. The higher the concentration, the shorter the latency.

52.2.6 Resistances

Combat agents should be as insensitive as possible to environmental conditions, but also to decontamination tests. At the same time, they should not decompose when used in bombs and grenades due to the temperatures generated by the explosion. **Thermal resistance** is given for most warfare agents.

With regard to environmental conditions, the greatest role is played by the **resistance to hydrolysis.** The lower the reaction with water (humidity, dew, body fluids after penetration into the organism), the more resistant it is.

To make decontamination more difficult, a high **chemical resistance** is desirable. Most warfare agents are hydrolyzable in an alkaline environment and/or are destroyed by chlorine-containing oxidants such as sodium hypochlorite or calcium hypochlorite (chlorinated lime). In contrast, N-lost, for example, must be rendered harmless with an alcoholic sodium sulfide solution.

52.2.7 Mean Lethal Concentration-Time Product *(Lct_{50})*

Especially in the case of toxins that are absorbed via the respiratory tract, the specification of the toxin concentration (c, specified in mg L^{-1} or mg m^{-3}) in the ambient air is of great importance. The time spent in air of a given concentration of toxin (t in min) determines the amount of toxin that enters the body. The determined concentration multiplied by the time at which 50 % of the test animals die is called the median lethal concentration-time product, or Lct_{50} for short. It is expressed in mg min m^{-3} or mg min L^{-1}.

Thus, an Lct_{50} *of* 100 mg min L^{-1} means that at a concentration of 100 mg L^{-1} after 1 min or at a concentration of 5 mg L^{-1} after 20 min, there is a 50 % probability of death.

▶ The lower the Lct_{50}, the more toxic the substance.

52.2.8 Combat Incapacitating Concentration Time Product (Ict_{50})

With many poisons and warfare agents, bodily reactions set in even before a lethal dose has been ingested. These physical reactions can lead to the affected person no longer being able to perform the tasks assigned to him. In analogy to Lct_{50}, the concentration-time product causing *incapacitation* is referred to as Ict_{50} (from incapacitate) if 50 % of the persons affected are no longer able to perform their tasks.

▶ The incapacitating concentration-time product Ict_{50} is always lower than the mean lethal concentration-time product Lct_{50}.

52.2.9 Mean Lethal Dose (LD_{50})

The lethal dose is the amount of a toxin that causes fatal poisoning after ingestion into the organism. The dose at which 50 % of the test animals die is called the median lethal dose (MLD) and is designated LD_{50}. It is expressed in mg per kg body mass (mg kg^{-1}). The MLD is an important comparative value for the assessment of toxic substances.

The MLD depends, among other things, on the route of absorption into the body. Some substances are absorbed via the skin (= percutaneously), others via the respiratory tract (= inhalation) or enter the body via the mouth (= orally). The MLD can differ greatly depending on the route of absorption, so it is useful to specify the route of absorption.

52.2.10 Threshold Value and Tolerance Limit

The **threshold value** is the concentration of a warfare agent that triggers a noticeable irritation or causes a typical symptom when it affects the organism. In the case of eye irritants, for example, it is the onset of lacrimation, in the case of skin irritants it may be itching, and in the case of nerve agents it may be the onset of pupil constriction.

The **tolerance limit** is the concentration that can just be tolerated by a person without permanent damage. In particular, the efficacy of irritants can be compared well, especially since a temporary inability to act already exists when the tolerance limit is reached.

Both values are given in mg L^{-1} or mg m^{-3} and are usually based on an exposure time of one minute.

52.2.11 Poison Strength *pLD*

Strey has introduced the so-called **toxic strength** as a comparative value (cf. Lit. [9–24]). To make it easier to compare the toxicity, or more precisely the LD_{50} *values,* of different toxic substances, the LD_{50} *values* are logarithmized and undergo a change of sign. The arithmetic operation here corresponds to the conversion of the H^+ concentration to the pH value (cf. Sect. 14.1.1). The following therefore applies:

$$\mathrm{p}LD = -\log LD_{50}$$

Just as the H^+ concentration is always given in mol L^{-1} for the pH value, the LD_{50} *value in* connection with the pLD value is always given in the unit kg kg^{-1} body weight. The higher the p*LD* value, the greater the toxicity. The p*LD* values can be compared directly with each other, which is otherwise difficult in the case of substances with widely differing toxicity. Examples are given in Table A.30, a comparative scale is shown in Fig. A.4.

Note that an increase of 1 in the p*LD* value means a tenfold increase in toxicity.

52.2.12 Detection of C Warfare Agents

Chemical warfare agents can be detected by detection and measuring devices. These are primarily detection paper, detection powder, test tubes and ion mobility spectrometers.

Detection powder can be used for a rough examination of suspicious surfaces. It reacts to many different substances with red discoloration and is therefore very non-specific.

CWA detection-paper is designed for the detection of liquid chemical agents on surfaces. Skin warfare agents (e.g. Lost, Lewisite) cause a red discoloration, nerve agents (Sarin, Soman, Tabun) an ochre yellow to ochre brown discoloration and nerve agents of the type "VX" a green discoloration (compare Fig. 52.2). CWA detection-paper reacts positively with a wide variety of chemicals and pesticides (e.g. insecticides based on organic phosphoric acid esters).

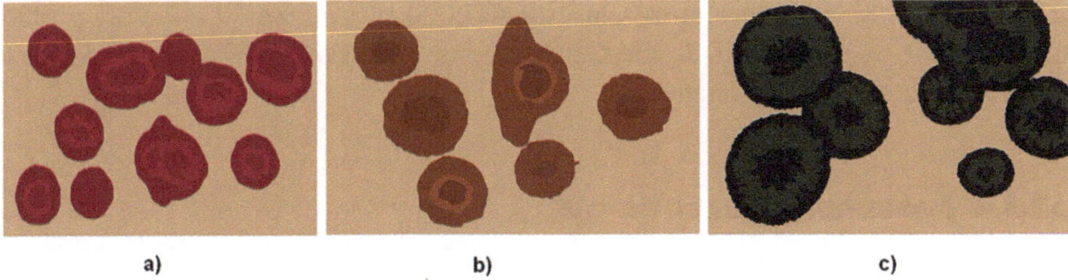

Fig. 52.2 Color change of CWA detection-paper; © drawing by Felicitas "Cean" Schmiermund based on information provided by the author

> Caution: The color reaction of the CWA detection-paper is **not** limited to CWA alone.

If a positive result is obtained with the tracing paper, a more precise detection must be carried out with **test tubes.** Manufacturers offer both single test tubes for the common warfare agents and simultaneous tests.

The **ion mobility spectrometer** (IMS, see Sect. 31.1.1) can detect extremely small quantities of skin or nerve agents. Due to the high sensitivity of the device, false-positive readings can also occur here.

For the unequivocal detection and identification of CWA, sampling with subsequent **laboratory analysis is** necessary. The early involvement of an Analytical Task Force (ATF) can significantly shorten the time frame required for this.

52.3 Irritants

Irritants are intended to cause temporary incapacitation by triggering certain physiological effects. Eye irritants and nasopharyngeal irritants are used, whereby the transitions are fluid. These irritants are fast-acting substances with a short duration of action. They are usually applied in aerosol form, as they are mostly solids. In contrast to "real" warfare agents, there is a large difference between the threshold value and the lethal concentration for these substances. In confined spaces or shafts, fatalities may nevertheless occur. An overview is given in Table 52.3.

52.3.1 Eye Irritants

A typical representative of the eye irritants is chloroacetophenone (CN). It is a colourless to yellow solid with a melting point of 59 °C, which is thermally quite stable and dissolves in organic solvents. Its solubility allows it to be used in atomizers. After atomizing the CN solution, the solvent evaporates and tiny CN crystals remain in the air as an aerosol. The substance can also be evaporated thermally (boiling point 245 °C), which also produces an aerosol. CN is also used, in addition to CS, by police forces for riot control.

The *ortho*-chloro-benzylidene-malodinitrile, better known as "CS" was introduced in the 1960s as an irritant in the US Army and was used, among other things, in the Vietnam War. In the German color code (World War I), eye irritants were white cross warfare agents.

52.3.2 Nasopharyngeal Irritants

Typical are the organo-arsenic compounds diphenylarsine chloride (Clark I, DA) and diphenylarsine cyanide (Clark II, DC), which were used as "mask breakers" in World War I, as the aerosolized solid particles penetrated the gas mask filters. In the

Table 52.3 Overview of irritants

Designation	Formula	Threshold value (in mg L^{-1})	Tolerance limit (in mg L^{-1})	Lct_{50} (in mg min L^{-1})
CN	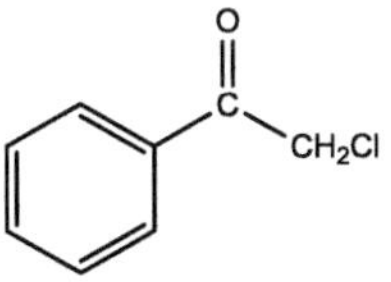	~ 0.00 04	~ 0.005	11
CS	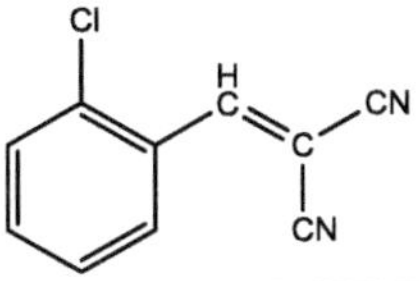	–	–	61
Clark I (DA)	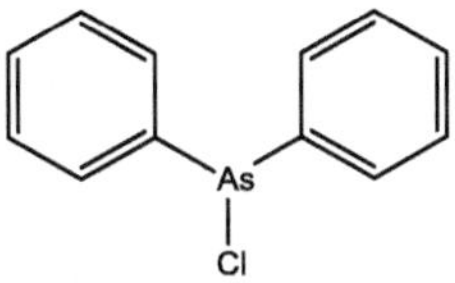	~ 0.0001	~ 0.001	15
Clark II (DC)	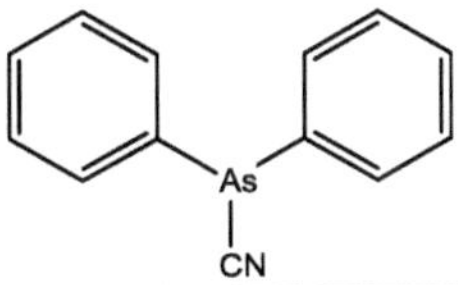	~ 0.000 01	~ 0.0005	10
Adamsite (DM)	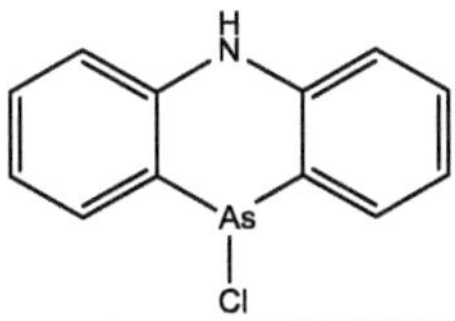	~ 0.0001	~ 0.001	12

Vietnam War, the even more effective phenarsazine chloride (Adamsite, DM) was used. DA and DC in particular are likely to have only historical significance, as much more effective HEPA filters are available today. Adamsite also induces vomiting and is therefore also assigned to the group of *vomiting agents* in English-speaking countries.

In the German color code (World War I), the nasopharyngeal irritants were the so-called Blue Cross warfare agents.

Pelargonic acid morpholide is said to have an even stronger irritant effect than the substances mentioned above, with a lower toxicity than CS. The so-called "pepper spray" should also not remain unmentioned. The active agent is capsaicin {pronounced cap-sa-i-cin}. It is extracted from the resin oil of peppers or chillies (plant genus: *Capsicum*) and is therefore also called *Oleum Capsicum* (OC). (*Unfortunately, no reliable data on effective concentrations could be determined for CS and OC.*)

The German name 'pepper spray' goes back to a mistranslation from English. Here *pepper* can mean both 'pepper' and 'paprika'. In Germany, the spray is approved for private use exclusively as an animal repellent. It is used by the police of many countries as a substitute for CS.

52.3.3 Protective Measures and Decontamination

Good protection by the NBC protective mask. Remove contaminated clothing. Decontamination with soapy water or, in the case of arsenic compounds, with alkaline solutions or chlorinated lime.

52.3.4 Irritant Industrial Chemicals

A large number of substances used in the chemical industry, whether in the laboratory or industrially, are also strong irritants. As examples his mentioned at this point:

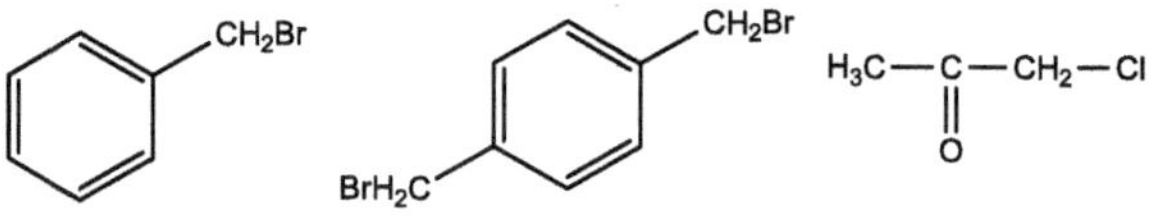

CH_2Br CH_2Br $H_3C-C(=O)-CH_2-Cl$

BrH_2C

Benzylbromid *para*-Xylylbromid Chloraceton

52.4 Pulmonary Warfare Agents

Pulmonary warfare agents are substances whose main effect is to damage the lungs; they usually lead to pulmonary oedema. At the appropriate concentration in the air they have a lethal effect. The best known pulmonary warfare agents are the dual use substances chlorine (Cl_2) and phosgene ($COCl_2$), both of which were used in World War I. Although they are no longer of military significance as warfare agents today, it must nevertheless be taken into account that they are used in large quantities in industry and can therefore escape in an uncontrolled manner in the event of incidents.

As gases, they are relatively quickly diluted by the air outdoors and thus belong to the group of volatile warfare agents. In closed rooms, however, life-threatening concentrations can easily occur. For example, 1000 ppm (0.1 % by volume) of chlorine over a period of 10 minutes is lethal, while 15...20 ppm of phosgene over the same period causes pulmonary oedema.

Examples of substances used as pulmonary warfare agents in World War I can be found in Table 52.4.

Table 52.4 Substances harmful to the lungs (examples)

Substance	Chlorine (CL)	Phosgene (CG)
Formula	Cl_2	$COCl_2$
Boiling point	−34 °C	7.6 °C
UN no.	1017	1076
Smell	Typical (swimming pool)	Rotten hay
Latency	4...6 h	4...6 h
Lct_{50}	–	3.2 mg min L^{-1}
Odour threshold	0.2 ppm	0.4 ppm
AGW	1.5 mg m^{-3}	0.4 mg m^{-3}

Table 52.5 Blood toxins (examples)

Substance	Hydrogen cyanide	Chlorcyan
Formula	H−CN	Cl−CN
boiling point	26 °C	13 °C
UN no.	1051	1589
Smell	Bitter almond	"unbearable"
Odour threshold	0.8 ppm	1 ppm
AGW	11 mg m^{-3}	–

52.4.1 Protective Measures and Decontamination

If inhalation is suspected, maintain absolute physical rest. Seek medical treatment and supervision immediately. The NBC protective mask prevents inhalation. Remove contaminated clothing. Decontamination outdoors not necessary. Ensure good ventilation. Respiration only with equipment.

52.5 Blood Warfare Agents

Blood warfare agents are blood poisons that act on the red blood cells and thus disrupt the transport of oxygen to the cells. The most important representatives of this group are also dual-use substances: hydrocyanic acid (HCN) and cyanogen chloride (ClCN). Data on these two blood poisons, which were already used in World War I, can be found in Table 52.5.

▶ Warning: Concentrations of > 1 vol.-% HCN are to be regarded as hazardous to life even for short periods (< 5 min) and when using self-contained breathing apparatus, due to the strong skin absorption of the substance. Strong skin absorption is also assumed for cyanogen chloride.

52.5.1 Protective Measures and Decontamination

Protection against inhalation is only provided by a specially impregnated respiratory filter or self-contained breathing apparatus. Decontamination outdoors is generally

Table 52.6 Overview of skin warfare agents (examples)

Designation	S-Lost (H/HD)	N-Lost (HN-3)	Lewisite (L)
Formula	$S(CH_2-CH_2-Cl)_2$	$N(CH_2-CH_2-Cl)_3$	$Cl-CH{=}CH-AsCl_2$
Melting point	14 °C	−4 °C	−18 °C
Boiling point	217 °C	230 °C (dec.)	190 °C (dec.)
Vapour pressure	~ 0.09 mbar	~ 0.009 mbar	~ 0.5 mbar
Latency	2. . .6 h	6. . .12 h	few min
LD_{50} (percutaneous)	~ 60 mg kg^{-1}	~ 20 mg kg^{-1}	~ 20 mg kg^{-1}
Lct_{50}	~ 1.5 mg min L^{-1}	~ 1.5 mg min L^{-1}	~ 1.3 mg min L^{-1}
Ict_{50}	~ 0.2 mg min L^{-1}	~ 0.2 mg min L^{-1}	~ 0.3 mg min L^{-1}

not necessary due to the high volatility. Ensure good ventilation. Respiration only with equipment.

52.6 Skin Warfare Agents

Skin warfare agents destroy the skin, causing wounds that heal only slowly. These warfare agents were used in large numbers during World War I. They are all sedentary substances that are well suited for terrain poisoning. These substances affect the entire organism not only as pure contact poisons, but also as eye and inhalation poisons (so-called systemic poisonous effects). Table 52.6 provides an overview.

52.6.1 Type "S-Lost"

The sulfur lots belong chemically to the group of halogenated thioethers. The best-known representative is probably the 2,2′-dichlorodiethyl sulfide, also known as yperite, sulfur perite, lost, H, HD or mustard *gas*.

The multitude of designations goes back to historical circumstances:

- During World War I, the German chemists **Lo**mmel and **St**einkopf developed a large-scale synthesis (→ Lost, German name).
- Near the French town of Ypres, the agent was first used at the front in July 1917 (→ Yperit, Belgian city).
- The substance smells of mustard or horseradish (→ *mustard gas*, British term).

S-Lost is insoluble in water, but dissolves well in organic solvents. The good lipid solubility causes rapid absorption – and thus an effect on all organs – regardless of the route of absorption. Swelling and reddening of the skin occurs after 2 . . . 6 hours. The tissue is severely destroyed (medical term: necrosis). Purulent wounds form, which take weeks to months to heal.
More recently, S-Lost was used by the terrorist militia "Islamic State" against Syrian and Iraqi Kurds respectively in the summer of 2015.

52.6.2 Type "N-Lost

Nitrogen losts or nitrogen yperites are aliphatic tertiary amines whose aliphatic C-chain has been halogenated at the end. Tris-(2-chloroethyl)-amine (abbreviation: HN-3) is the most potent substance of this group. The entire group was discovered between the two world wars and industrially produced during World War II.

All N-losts, which have the general formula: $R{-}N(C_2H_5Cl)_2$, are highly resistant to hydrolysis and form ammonium salts that are readily soluble in water with acids. These properties make them potentially suitable as sabotage poisons.

52.6.3 Type "Lewisite"

Halogenated aliphatic arsines are called Lewisites after their discoverer, the American chemist Lewis. Initially, great importance was attached in particular to 2-chloroethylene-dichloroarsine and it was called "death dew" in the press. Its action, however, is comparable to that of S-lostes. It is also more sensitive to hydrolysis, so that its importance as a military warfare agent rapidly declined. The solubility in water is low ($\sim 0.5\ g\ L^{-1}$), in organic solvents it is well soluble. It penetrates quickly into leather, rubber, wood, cardboard and textiles.

52.6.4 Protective Measures and Decontamination

The NBC protective mask prevents inhalation poisoning. Contamination protection is provided by the NBC self-help kit (protective tarpaulin, overshoes, gloves) or – for a short time – a liquid-tight contamination protection suit. Change contaminated clothing immediately. Pay attention to self-protection.

Dab toxins from the skin, do not rub! Pay attention to self-protection! Soap and up to 10 % $NaHCO_3$ solution are suitable for washing off. Rinse eyes with 3 % $NaHCO_3$ solution.

Decontamination with sodium hypochlorite solution, chloramine T or chlorinated lime possible. For N-Lost preferably use sodium sulfide in ethanolic solution.

52.7 Nerve Agents

The nerve agents, sometimes also referred to as "nerve gases", are lethal warfare agents that exert an effect on the central nervous system (CNS) and cause death in relatively low concentrations or doses. Typical representatives are sarin, soman, tabun and VX, which are among the most dangerous warfare agents of all.

They are all organic phosphorus compounds (organophosphates), which are to be understood as phosphoric acid esters (phosphates) or phosphonic acid esters (phosphonates). Phosphonates are characterized by a direct C–P bond.

This group of warfare agents was only discovered after World War I during the search for insecticides. It was discovered during the search for insecticides, but some of them were already produced on a large scale during World War II – even though they were not used.

The structural formulae of the nerve agents dealt with in Sects. 52.7.2, 52.7.3, 52.7.4 and 52.7.5 are given in Table 52.7, and physico-chemical and toxicological data are summarised in Table 52.8.

52.7.1 Mode of Action

All nerve agents share a common mechanism of action: inhibition of the enzyme acetylchlolinesterase (AChE).

Between the nerve end and the muscle, the stimulus is transmitted by the transmitter substance acetylcholine. It is used to control both voluntary muscles (striated muscles, skeletal muscles) and involuntary muscles (smooth muscles; eye, heart, bronchial tubes, stomach, intestines, bladder and blood vessels). In order to

Table 52.7 Formula overview of nerve agents

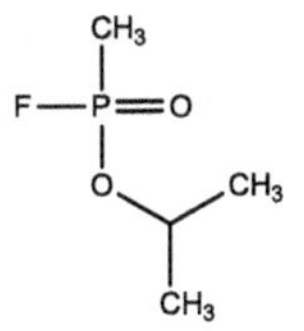 Sarin	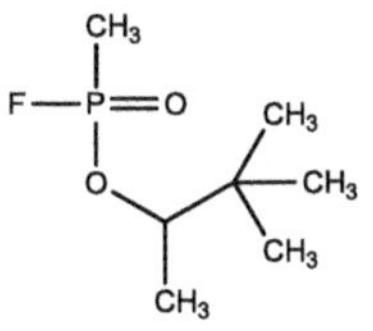 Soman	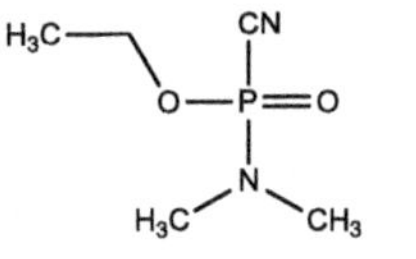 Tabun	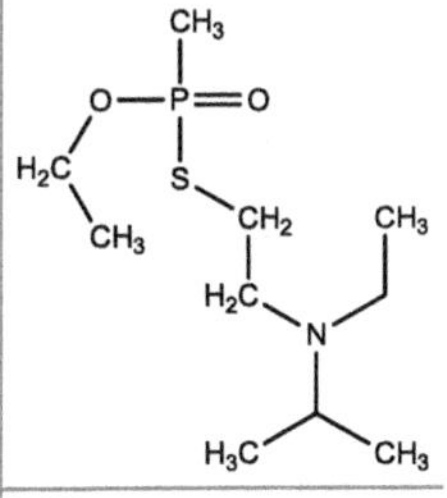 VX

Table 52.8 Nerve agents (data)

Designation	Sarin (GB)	Soman (GD)	Tabun GA.	VX
Molar mass (in g mol^{-1})	140.1	182.2	162.1	267.4
Melting point (in °C)	−56	−42	−50	−39
Boiling point (in °C)	147	167	246	298
Density (in g mL^{-1})	1.089	1.022	1.073	1.008
Solubility in water (at 25 °C)	∞	1.5 %	12 %	3 %
Vapour pressure (at 25 °C in mbar)	~ 4	~ 0.4	~ 0.009	~ 0.0009
c_S (at 25 °C, in mg L^{-1})	16.8	3.1	0.6	3.0
Lct_{50} (in mg min L^{-1})	0.07. . .0.1	0.04. . .0.07	0.2. . .0.4	~ 0.015
Ict_{50} (in mg min L^{-1})	0.02. . .0.06	~ 0.025	0.02. . .0.1	~ 0.005
LD_{50} (percutaneous) (in mg kg^{-1})	25	15	20	0.2
LD_{50} (oral) (in mg kg^{-1})	0.14	0.14	0.6	0.06
pLD_{50} (oral)	6.8	7	6.2	8.2

ensure a regulated transmission of stimuli, the "messenger substance" (acetylcholine) must be broken down again in the shortest possible time at the receptor of the muscle after it has been released by the nerve end. If this is not the case, there are sometimes serious physiological effects.

The degradation of acetylcholine into choline and acetic acid is carried out by the enzyme AChE. Each molecule of AChE can cleave approx. 300 molecules of acetylcholine per millisecond (!).

$$[H_3C{-}N^+(CH_3)_2{-}CH_2{-}CH_2{-}O{-}C(=O){-}CH_3]\,HO^- \xrightarrow[H_2O]{AChE} [H_3C{-}N^+(CH_3)_2{-}CH_2{-}CH_2{-}OH]\,HO^- + HO{-}C(=O){-}CH_3$$

Acetyl choline — Choline — Acetic acid

The main effect of the nerve agents is now the **inhibition** of the enzyme AChE and other cholinesterases. Due to structural similarities with acetylcholine, the antagonist attaches itself to the enzyme and thus blocks it in the long term. As a result, acetylcholine can no longer be broken down and accumulates.

Here, not only the effect on the smooth or striated muscles must be distinguished. In addition, the effect on the two different ACh receptors (nicotinic or muscarinic) must be considered.

AChE inhibitors have an excitatory effect on the nicotinic receptors of the striated or smooth muscles at low concentrations (e.g. constriction of the pupils) and a paralysing effect at high concentrations. At the muscarinic smooth muscle receptors, AChE inhibitors at low concentrations cause nausea, headache, and dizziness, among other effects. In contrast, at high concentrations, they cause dyspnea, hypotension, paralysis, muscle twitching, and convulsions. The effects on muscarinic receptors are more severe and may eventually lead to death by respiratory paralysis.

> Caution: Some insecticides, such as parathion ("E 605"), which is no longer permitted in Europe, are also AChE inhibitors.

52.7.2 Tabun

Tabun was discovered as early as 1936 by G. Schrader during the search for new insecticides and was industrially produced for the German Wehrmacht from 1942. The agent was not used during the war. It was given its name because it was far too toxic as an insecticide – and thus "taboo" for use as an insecticide. Chemically it is dimethylphosphoramido-cyanide acid ethyl ester, the code name of the Wehrmacht was "Trilon 83" (T 83). The low vapour pressure and the low volatility make Tabun a sessile warfare agent that causes inhalation poisoning for many hours at 20 °C.

Tabun was used by Saddam Hussein in the Iran-Iraq war of 1983–1988 and against Kurdish separatists (March 1988).

52.7.3 Sarin

Methylfluorophosphonic acid isopropyl ester belongs to the phophonates and, like tabun, was discovered in the course of insecticide research. The substance was first given the code name "Trilon 144" (T 144), later "Trilon 46" (T 46). The name Sarin, in common use today, results from the names of the people involved in the discovery and the development of a large-scale manufacturing process: **S**chrader, **A**mbros, **Ri**tter and von der **Lin**de.

It has a higher vapor pressure than tabun, making it about 20 times more volatile than the latter.

Sarin may have been used by Saddam Hussein against the Kurdish minority in Iraq (1988). The Japanese Aum sect used it in two terrorist attacks (in Matsumoto in 1994 and Tokyo in 1995) that killed 20 people and injured hundreds. Sarin was used several times in the Syrian civil war, in April 2013, August 2013 and April 2017.

52.7.4 Soman

Soman, 1,2,2-trimethylpropylfluorophosphonic acid isopropyl ester, is a development of the Austrian-German chemist Kuhn and was produced in 1944 for test purposes.

It is a relatively thermally stable liquid. No reliable data are available on its use after 1945.

52.7.5 VX

Chemically, VX is *O*-Ethyl-S-2-di*iso*propylamino-ethyl-methyl-phosphono-thiolate. The substance, along with several derivatives, was also discovered around 1955 in a laboratory for the development of insecticides. Due to its very low vapour pressure and its low volatility, VX is an extremely sedentary agent, which also shows a particularly strong toxicity.

In February 2017, Kim Jong-nam, the half-brother of North Korean strongman Kim Jong-un, was killed in an assassination attempt at Kuala Lumpur airport using VX, according to officials.

52.7.6 Protective Measures and Decontamination

The NBC protective mask prevents inhalation poisoning by nerve agents. Protection against contamination is provided by the specially developed protective clothing from the NBC self-help kit (protective tarpaulin, overshoes, gloves). Contaminated clothing must be removed immediately (!). Pay attention to self-protection!

Contaminated skin must be washed immediately with water or soapy water. Wash/rinse hair and eyes with 3 % $NaHCO_3$ solution. Decontamination is possible with sodium hypochlorite solution, chloramine T or chlorinated lime. In case of symptoms of poisoning self-injection of atropine.

52.7.7 Other Nerve Agents

The substances already used in World War I (mustard gas, N-lost, phosgene, etc.) are referred to as “first generation” of CWA. The “second generation” consists of the phosphoric acid esters developed in the 1930s and 1940s (Sarin, Soman, Tabun). The “third generation” is the series of V warfare agents (VX, VR, etc.) and the Tammelin esters developed in the 1950s. The latest development, the “fourth generation” of warfare agents, are the compounds of the Novichok series.

For the sake of completeness, the Tammelin esters (chemical: phosphorylcholines) and the *Novichok* agents should be mentioned here.

52.7.7.1 Tammelin’s Ester

The Tammelin esters (phosphorylcholines) are named after Lars-Erik Tammelin (Sweden), who discovered this class of compounds in the 1950s. The most important representatives are the alkyl-fluorophosphorylcholines, which differ from the group of V-fighting agents primarily in that they have an oxygen atom in the same place in the molecule instead of sulfur.

Substances in which the "X" in the general formula (cf. Table 52.9) represents fluorine are particularly suitable as warfare agents. If alkyl or alkoxy groups are inserted at this position, the toxicity decreases.

52.7.7.2 Novichok Group

These compounds are probably the latest developments in the field of nerve agents. They were developed in the USSR in the 1970/80s. Little is known about the exact chemical structures. However, the most probable information is provided by the Russian chemist Mirsajanow, according to which they are phosphonamidates. Here, a – N=C grouping is directly bound to the phosphorus atom. The camouflage designations of the three most probable types are A-230, A-232 and A-234. As a binary warfare agent, A-232 was made available for military use as "Novichok-5" (compare Table 52.9).

No reliable data exist on the chemical-physical or toxicological properties. According to the British government, a substance from the Novichok group was used in March 2018 to carry out an attack on the Russian defector S. Skripal. This led to a certain degree of notoriety for this group of warfare agents.

52.7.7.3 Comparison of the Structures

Table 52.9 shows the general formulas of the V warfare agents ("V-WA"), the Tammelin esters ("Tam-E") and the Novichok compounds ("Nvc"). Next to them are the structures of the warfare agent A-232 and VR ("Russian VX").

52.8 Psychotoxic Warfare Agents

These substances are said to cause mental disorders (psychoses) in normal people. Only a few of these compounds, also known as hallucinogens, are suitable for military use. Lysergic acid diethylamide (LSD) and 3-quinuclidine benzilate (BZ), which must be used as solid aerosol, have the strongest effects. For BZ, a concentration of $\leq$ 0.4 mg per litre of breath is sufficient to produce an effect that "incapacitates". For LSD, this value is even much lower, at 0.03 mg L^{-1}. The symptoms (mental disturbances, hallucinations) may last for several hours.

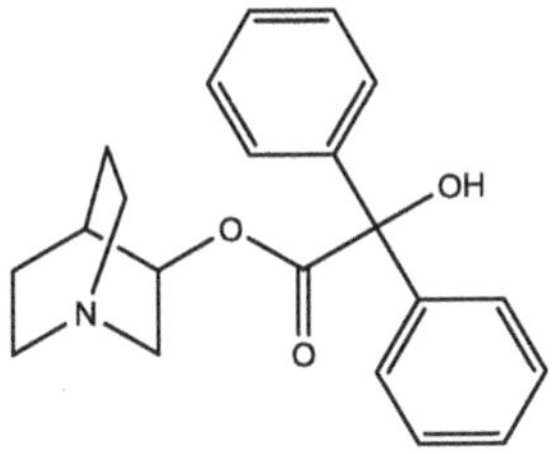

BZ

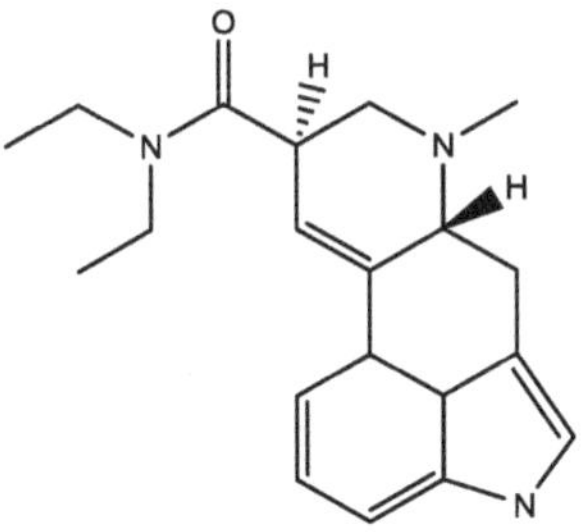

LSD

Table 52.9 Nerve agents – structural formula comparisons (Abbreviations: **V-KS**: V-warfare agents, **Tam-E**: Tammelin's ester; **Nvc**: Novichok)

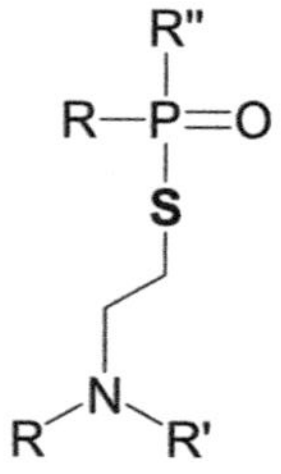 General formula "V-WA"	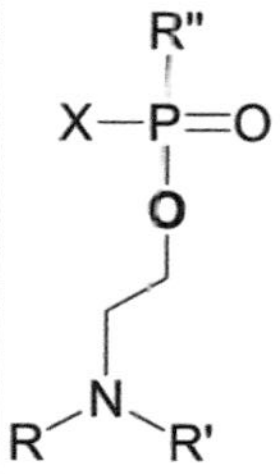 General formula "Tam-E"	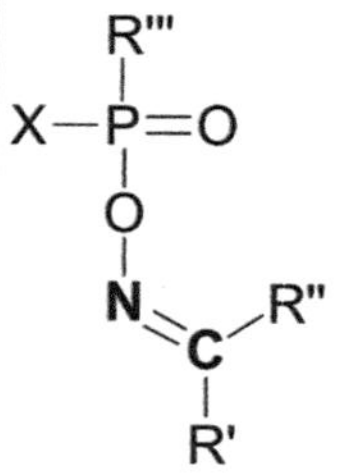 General formula "Nvc"	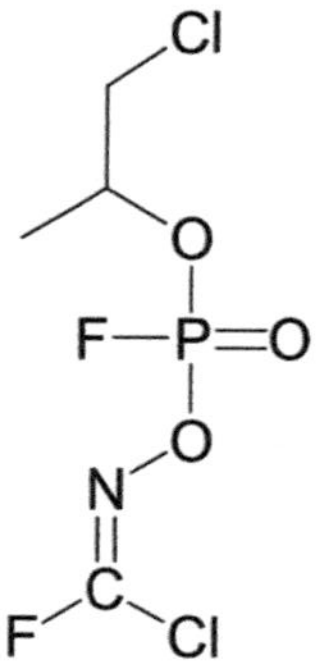 A-232 (Novichok-5)	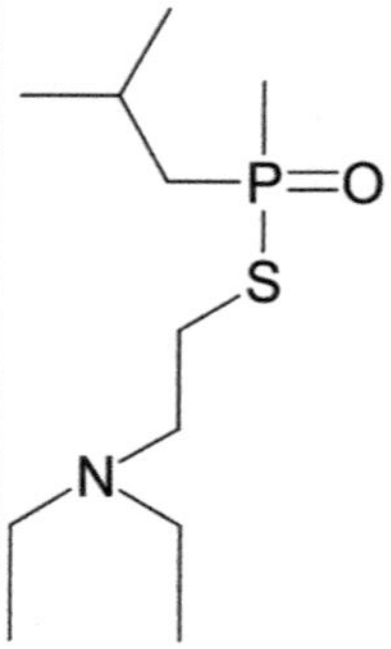 VR ("Russian VX")

52.8.1 Protective Measures and Decontamination

The NBC protective mask with HEPA filter prevents inhalation. Alternatively use P3 filter. Remove contaminated clothing. Wash off with soap and water.

52.9 Sabotage Poisons

Sabotage poisons are primarily intended as large-scale poisonings to render the enemy's drinking water supplies unusable or to decimate livestock. The enemy's use of these poisons can also have a major impact on the pharmaceutical, cosmetic and food industries. Furthermore, they may be used against small groups of people or individuals in key positions to massively disrupt leadership and direction. In principle, a vast number of different substances can be considered as sabotage poisons, which, however, must "ideally" combine a number of properties. The main criteria are:

- High toxicity, i.e. lethal effect of smallest quantities
- Odourless and tasteless
- High resistance to water (hydrolysis resistance)
- Delayed impact
- difficult analytical detection of the toxin
- a toxic effect which makes the use of antidotes difficult or impossible

Looking at the poisons that have been applied for such purposes, it can be seen that they are based on "practical considerations" rather than the above-mentioned "ideal requirements". These practical considerations are primary:

- Easy availability
- Simple manufacturing conditions
- Possibility of camouflage as "industrial accident" or "mistake"

52.9.1 Examples of Sabotage Poisons

In addition to the warfare agents, some of which are quite suitable as sabotage poisons, many other compounds come into consideration, whereby the list is to be seen as exemplary and by no means complete or conclusive.

Plant Ingredients
Many plants produce potent poisons or hallucinogens that could be used as sabotage poisons, e.g. strychnine, aconitine, atropine or mescaline.

Animal toxins
Among the poisons produced by animals, the poison of the Japanese puffer fish (tetrodotoxin), the poison of some species of mussels (saxitoxin) and the poison

of the poison dart frog (batrachotoxin) should be mentioned as possible poisons for acts of sabotage, in addition to various snake poisons.

Synthetic compounds

Besides insecticides based on organic phosphoric acid esters, arsenic oxide (As_2O_3), salts of arsenic acid (arsenates) and arsenious acid (arsenites), potassium and sodium cyanide (KCN, NaCN) as well as mercury, lead or thallium salts, and many others, can be used as sabotage poisons.

52.9.2 Targets

Possible targets for the use of BWA and CWA are large cities, large gatherings of people, artificially ventilated systems such as subways, and possibly individuals.

If one considers the probabilities with which terrorists *would* deliberately cause contamination, one arrives at the following list:

Probability of contamination of

Staple foods & beverages:approx.	45 %
Other dishes:	approx. 13 %
Drinking water:	approx. 12 %
Personal effects:	approx. 4 %
Projectiles:	approx. 3 %
Gases, vapours, mists:	approx. 15 %.

It is noted that BWA and CWA – analogous to other poisons – would preferably be used in the sense of being used as sabotage poisons.

52.10 Strategic Ordnance

In particular, phytotoxic (= plant-damaging) agents are of strategic military importance. These substances cause damage to plants through growth abnormalities or defoliation or destroy them completely. They also occupy a special position among military poisons, since they are or were used in large quantities as weed killers. Attacks on crops can thus cause massive crop failures, which in turn lead to hunger and demoralization of the population.

Among the most important phytotoxic substances are the esters of 2,4-dichlorophenoxyacetic acid (2,4-D) and 2,4,5-trichlorophenoxyacetic acid (2,4,5-T), which were used as “Agent Orange” by the US armed forces in the Vietnam War (see also Sect. 45.1).

The substances listed under sabotage poisons are suitable for the strategic poisoning of drinking water and food supplies.

52.11 Binary CWA

Binary CWA are chemical warfare agents in which two or more relatively non-toxic precursors are separated in the projectile. When fired, the components mix and, in the course of time until impact in the target area, produce the actual warfare agent due to a chemical reaction taking place in the projectile.

Example: Reaction of methylphosphonic acid difluoride with *iso*-propanol to sarin:

$$(H_3C)_2HC{-}OH + F{-}P(=O)(F){-}CH_3 \xrightarrow[-\,HF]{(Kat)} (H_3C)_2HC{-}O{-}P(=O)(F){-}CH_3$$

The advantages are the lower toxicity of the chemicals to be filled, the better storage stability, which prevents the actual warfare agent from decomposing, and generally less corrosion of the projectile casings.

52.12 Final Remark C-Weapons

The use of chemical warfare agents seems more likely than the use of biological warfare agents because of the better controllability of their spread. However, no one is currently anticipating the military use of chemical warfare agents in Europe. The potential threat posed by groups with extortionist or terrorist intentions, on the other hand, cannot be dismissed out of hand. Globally, the Middle East is viewed with great concern, as there have already been several uses of chemical weapons here and further uses – including against the civilian population – cannot be ruled out.

Terrorist attacks using chemical weapons cannot be completely ruled out and are at the same time more likely than the terrorist use of biological weapons. At the same time, there is a high level of sensitivity, and in some cases fear, among the general public with regard to chemical warfare agents.

It should also not be forgotten that Germany still has many tons of ammunition from the World Wars that are filled with CWA and whose exact location is not always known. Particularly in the vicinity of the former production and storage sites, buried munitions or munitions sunk in surface waters can lead to leaks and thus to corresponding operational situations.

In principle, it is advisable, at least for the NBC platoons, to keep a supply of decontamination/detoxification agents on hand. In the meantime, the industry also offers universal agents that can be used for a large number of warfare agents and are very effective.

Biological Substances 53

Biological agents or biological agents are natural and genetically modified (micro) organisms (GMOs) – bacteria, fungi, algae, viruses – as well as cell cultures and parasites. They can trigger infections, allergies or toxic effects.

53.1 Introduction

Biotechnology encompasses many centuries-old processes such as cheese production, beer brewing or the breeding of (useful) plants and animals. In more recent times, this also includes tissue cultivation in cell cultures, as used, for example, for the synthesis of skin for burn victims.

This must be distinguished from genetic engineering, which deals with the targeted modification of genetic information in organisms.

53.1.1 Fields of Application of Biotechnology

In livestock farming, breeding/cross-breeding is used in an attempt to make (farm) animals resistant to various diseases or to increase their yields (milk, meat, etc.). In the case of plants, the aim is to achieve resistance to fungi, viruses and other plant pests, as well as better adaptation to environmental factors (temperature, water availability, amount of sunlight, etc.), both by breeding/cross-breeding and by genetic engineering methods. Genetically modified plants with resistance to herbicides are already being marketed.

For human medicine, vaccines and pharmaceuticals can be obtained from genetically modified microorganisms. In addition, enzymes, e.g. for detergents, flavourings and aromas, but also rennet for cheese production are obtained. Research is also being carried out into means of combating cancer. The treatment of wastewater or the decontamination of contaminated soil (e.g. from oil spills) is also possible using biotechnical methods.

T. Schmiermund, *The Chemistry Knowledge for Firefighters*,
https://doi.org/10.1007/978-3-662-64423-2_53

53.2 Classification of Biological Agents

According to the "Ordinance on Safety and Health Protection during Work with Biological Agents" (Biostoffverordnung – BioStoffV), biological agents are defined as microorganisms, cell cultures and endoparasites, including their genetically modified forms, which can endanger humans through infections, transmissible diseases, toxin formation, sensitising effects or other effects which are harmful to health. Biological agents include bacteria, viruses and fungi.

53.2.1 Bacteria

Bacteria (Greek: *bakteria* = rod) are single-celled microorganisms without a true cell nucleus. Their size is usually 1...10 μm. So far, more than 3000 species have been described. Most bacteria can be traced back to basic forms: Sphere (cocci), straight rods (bacterium), curved cylinders (vibrios) and helical cylinders (spirilla). Cell clusters (staphylococci), cell packages (sarcines) or cell threads (streptococci) are formed by sticking together after cell division. Some bacteria form permanent forms, so-called spores, which are able to survive adverse environmental conditions: Dehydration, UV radiation, extreme temperatures, too high or too low oxygen content/pH value, nutrient deficiency, etc. Spores are often resistant to common household disinfectants and can survive in the environment for months, sometimes decades. Spore-forming agents are, for example, the pathogens of anthrax *(Bacillus anthracis),* tetanus *(Clostridium tetani)* or food poisoning *(Clostridium botulinum).*

Bacteria are found in all habitats on earth, most commonly in the soil. One gram of soil can contain up to 5 billion bacteria. Most bacteria are harmless or even indispensable for humans, animals or plants. For example, the intestinal bacteria of mammals or the nodule bacteria of certain plants are necessary for survival.

The harmful effect of some bacteria that are dangerous to humans is based on the excretion of toxins, so-called exotoxins. These include, for example, *Yersinia pestis* (plague) or *Clostridium tetani* (tetanus).

Examples of Diseases Caused by Human Pathogenic Bacteria

Cholera	Diphtheria	Whooping Cough
Leprosy	Meningitis	Anthrax
Salmonellosis	Plague	Tuberculosis

◀

53.2.2 Viruses

Viruses are cell parasites 20...400 nm (0.02...0.4 μm) in diameter. Unlike bacteria, viruses do not have their own metabolism. They require living cells for their reproduction. The virus enters the cell and then causes it to produce viral DNA and viral proteins. This eventually produces new viruses that are shed from the infected cell or cause the cell to die and decay, thus releasing them. Virus particles outside cells are called virions.

Examples of Human Pathogenic *Viruses* or Viral Diseases

Ebola virus	*Dengue fever*	*Mumps*
Flu virus	*Hepatitis*	*Polio*
HI virus (AIDS)	Lassa fever	Pseudocroup
Marburg virus	*Measles*	*Rabies*

◄

53.2.3 Mushrooms

Fungi occupy a special position between animals and plants. They are a very heterogeneous group of living organisms, ranging from clearly visible life forms with large fruiting bodies (e.g. edible mushrooms) to microorganisms (e.g. yeast fungi).

The mushrooms themselves and/or their metabolic products can be useful or harmful to us. Edible mushrooms such as champignons or chanterelles are suitable for consumption. Some fungi are used to refine/ferment foods, giving them a characteristic aroma (e.g. blue cheese) or producing certain desirable ingredients (e.g. alcohol). Still others provide important medicines, such as penicillin. In addition, there are fungi that produce toxins, so-called mycotoxins (fungal toxins), which can have a strongly damaging effect.

53.2.4 Rickettsia

Rickettsiae are very small (0.1...4 μm), mostly spherical organisms. They differ from classical bacteria in that they require a living host cell (similar to viruses). The best-known disease caused by rickettsiae is spotted fever (also: typhus).

53.2.5 Prions

The pathogens that most recently became known as such are the prions (derived from *proteinaceous infectious particle).* They were only discovered in 1997 and, as proteins, belong to the subviral particles, i.e. they are much smaller than viruses. However, they are not living organisms, but rather misfolded proteins that are able to transfer their conformation to other proteins. Infection routes are contaminated food and smear infections. Prion diseases are, for example, Creutzfeldt-Jakob disease in humans or *bovine spongiform encephalopathy* (BSE) in cattle.

53.3 Classification of Biological Agents - Criteria

Biological agents pose a wide variety of hazards with different effects. The most important effects of biological agents are:

- Risk of infection
- Risk of poisoning
- Triggering allergies
- Environmental hazard
- Inheritance damage
- Disgust (smell!)

These effects give rise to a wide range of criteria which are necessary to classify biological agents into risk categories. Important criteria are briefly described in Sects. 52.3.1 to 53.3.10.

53.3.1 Infection Rate/Transmissibility

The transmissibility or infection rate indicates how easily a pathogen can be transmitted from one organism to another. The transmissibility depends, among other things, on the transmission route. For example, the infection rate for airborne pathogens is usually higher than for pure contact infections with body fluids, for example. The crossing of the species barrier also plays a role. Some pathogens can be transmitted from animal to human. Others are only transmissible from human to human or from animal to animal.

53.3.2 Infectivity

Infectivity (from Latin *inficere* = to put in, to infect) is a measure of the ability of a microorganism to infect another organism. The number of microorganisms required to cause the disease is usually given as a "measure". The lower the number of microorganisms, the higher the infectivity. The number of microorganisms that can trigger the disease is to be understood here as the **infectious dose** (**ID**).

Examples of Infectious Doses

Cholera	10^8	Smallpox	100...1000
Salmonellosis	$10^3...10^6$	Lung plague	100...500
Anthrax	10^4	Ebola fever	1...10

◄

53.3.3 Incubation Time

The incubation period is the symptom-free period between infection by a pathogen and the outbreak of the infectious disease. It can range from a few hours (e.g. anthrax or influenza) to several years (e.g. leprosy or AIDS).

53.3.4 Lethality

Lethality indicates the percentage of the total number of people who die from the disease after infection. A distinction is sometimes made between the number of individuals who died with or without treatment. In contrast, for toxins, the lethal dose at which, for example, 50 % (LD_{50}) or 95 % (LD_{95}) of those poisoned die is often reported. Lethality must not be confused with mortality (= deaths per inhabitant).

53.3.5 Morbidity

Morbidity (from Latin *morbidus* = making ill) refers to the frequency of illness. It is the number of people who fall ill within a certain period of time (e.g. one week) in relation to the population (e.g. 10 000 or 100 000 inhabitants).

53.3.6 Mortality

Mortality is the number of deaths in relation to the population (e.g. 10 000 or 100 000 inhabitants). It must not be confused with lethality (= deaths in relation to the number of patients).

53.3.7 Pathogenicity

Pathogenicity (from Greek *pathos* = suffering, pain, disease and *genos* = to form, cause) is the ability of a microorganism to cause a disease in another organism. Human pathogens are those that are capable of causing the respective disease in humans.

53.3.8 Stability/Tenacity

Stability against environmental influences (temperature, humidity, sunlight, etc.), but also against chemicals such as disinfectants, essentially determines the possibilities of rendering a pathogen harmless. If necessary, predictions can be made about the time a pathogen remains virulent under certain environmental conditions. Storage and transport stability provide information on sampling and shipment.

Tenacity (Latin *tenacitas* = to hold on) describes the ability of a microorganism to survive even under less than optimal environmental conditions. In particular, it usually indicates the period of time over which this microorganism can survive outside its usual habitat.

53.3.9 Toxicity

Toxicity is a measure of the toxicity of a substance. The higher the toxicity, the more severe the damage caused. At the same time, as toxicity increases, the dose that causes symptoms of poisoning decreases.

53.3.10 Virulence

Virulence (from Latin *virulentus* = poisonous) describes the infectious power of a pathogen. This means that the more easily a pathogen penetrates an organism and is then able to multiply there until the outbreak of the disease, the more virulent it is.

53.4 Classification of Biological Agents – Risk Groups

Four risk groups can be derived from the classification criteria for biological agents. Table 53.1 shows the criteria and risk groups based on a compilation by the World Health Organization (WHO).

These four risk groups are translated into four *biological safety levels* (BSL) or safety levels via the Biological Substances Ordinance (BioStoffV) and the Genetic Engineering Safety Ordinance (GenTSV) (Table 53.2). These protection/safety levels define the construction and equipment of laboratories and production facilities as well as the handling of the corresponding biological agents, hygiene measures, and much more. Safety level 4 has the highest requirements. Since the safety levels build on each other, all regulations of the lower protection levels always automatically apply to the higher levels as well.

Worldwide, only 34 BSL-4 laboratories are officially known, four of which are located in Germany:

- Berlin, Robert Koch Institute (RKI)
- Hamburg, Bernhard Nocht Institute for Tropical Medicine (BNITM)
- Riems Island, Institute for New and Novel Animal Pathogens (INT) of the Friedrich-Löffler-Institute (FLI)
- Marburg, Institute for Virology of the Phillips University Marburg

The classification of individual biological agents into risk groups can be found, for example, in the "Technical Rules for Biological Agents (TRBA)":

Table 53.1 Risk group classification of biological agents (according to WHO, modified)

Classification of biological agents into risk groups	Examples (bacteria, fungi, viruses, diseases)
Risk group 1 • No diseases • Human (animal) unsuitable as "growth medium • Pathogenic effect lost • No diseases documented (according to experience)	Baker's yeast, brewer's yeast, lactic acid bacteria, polio vaccine viruses
Risk group 2 • Harmless diseases • Rarely severe diseases despite human contact • Special transmission pathways necessary for infection • High infectious dose • Good possibilities of prophylaxis (e.g. effective vaccination) • Well treatable • High immunity among the population	Salmonella, Legionella, Hepatitis, Noro virus, measles and flu virus, Pertussis, tetanus *Staphylococcus aureus, clostridium botulinum*
Risk group 3 • Serious diseases, transmissible through the air • Severe diseases with difficult/long-term therapy options • Low infectious dose • Rapid disease progression • Fatal diseases with specific infection routes	*RGr3:* Anthrax, tuberculosis, plague, yellow fever
Risk group 3** • As risk group 3, but infection via the air is generally *not* possible	*RGr3**:* Hepatitis B, HIV (AIDS), rabies, dysentery, typhoid fever, EHEC
Risk group 4 • Fatal diseases • After infection no therapy possible • High contagiousness • Rapid transmission through air, insects or direct skin/mucous membrane contact with infected persons	Viruses only: Lassa, Crimean-Congo, Marburg, Ebola, smallpox

Note: Not all of the above criteria need to be met at the same time for classification into the risk groups

Table 53.2 Implementation of the risk groups in the BioStoffV and the GenTSV

	BioStoffV			GenTSV
Risk group	Protection level	Lab	Production	Security level
1	S 1	L 1	P 1	S 1
2	S 2	L 2	P 2	S 2
3	S 3	L 3	P 3	S 3
4	S 4	L 4	P 4	S 4

- TRBA 460 – Classification of fungi
- TRBA 462 – Classification of viruses
- TRBA 466 – Classification of bacteria/prokaryotes

53.5 Labelling and Classification of Biological Agents

In principle, of course, facilities, rooms and transport containers containing B-hazardous substances or contaminated with B-hazardous substances must be marked in some form.

53.5.1 Transport Law

According to ADR/GGVSEB, biological substances are classified in hazard class 6.2 "infectious substances". However, this only applies to biological substances in risk groups 2, 3 and 4, so that the number of UN numbers remains manageable (Table 53.3).

For **UN 2814,** a list of biological agents falling under category A is deposited in the ADR. Note that this UN No. contains the agents with the highest hazard potential and that transports are often carried out in cars or vans.

The danger labels shown in Fig. 53.1 are used as dangerous goods labels on packages.

Table 53.3 UN numbers of hazard class 6.2 according to ADR/GGVSEB

UN no.	Kemler number	Designation	Notes
2814	606	Infectious substance for humans, category A	Can cause a life-threatening or fatal disease in humans or animals
2900	606	Infectious substance for animals, category A	May cause a life-threatening or fatal disease in animals
3291	606	Clinical/medical waste	If there is a low probability of the presence of infectious substances.
3373	606	Biological substances, category B	Infectious substance not meeting the conditions for category A
3245	**90**	Genetically modified organisms, not in risk group 1	Hazard class 9 (!), neither toxic nor infectious organisms

Fig. 53.1 Dangerous goods labelling of packages with infectious goods

Fig. 53.2 Labelling of biological substances

53.5.2 Workplace Labeling

Facilities, rooms and transport containers containing or contaminated with B-hazardous substances must be marked with a specified sign and the addition "Biohazard". Marking with the warning sign "Biohazard" according to DIN EN ISO 7010 is only present in areas with safety/protection level 3 and 4. For safety level 1 this warning sign is not to be found, for safety level 2 only partially.

In addition, there may be further references to safety levels S 1 to S 4, L 1 to L 4, P 1 to P 4 or labels such as "Genetic engineering work area" or similar.

The bio hazard symbol, the warning symbol according to DIN EN ISO 7010 and two example labels can be found in Fig. 53.2.

Genetic engineering facilities, i.e. laboratories and animal houses, are permanently marked with the sign "BIO I", "BIO II" or "BIO III" in accordance with state fire protection regulations.

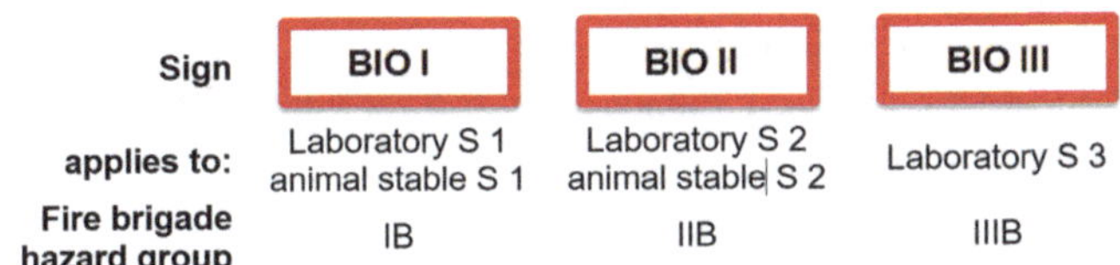

53.5.3 Classification into Fire Brigade Hazard Groups

According to FwDV 500, B hazardous substances are divided into hazard groups IB, IIB and IIIB. Table 53.4 compares the risk groups with the hazard groups for organic operations of FwDV 500.

Table 53.4 Hazard groups for B-operation (according to FwDV 500, modified)

Risk group (protection levels, safety levels)	Fire brigade danger group (FwDV 500)	Health risk	Risk of infection
1	**IB**	Unlikely	Without risk of infection
2	**IIB**	Low	With Risk of infection
3**		Moderate (not over the air)	
3	**IIIB**	Moderate to high	
4		High	

53.6 Operational Measures

The tasks of the fire brigade in a bio incident essentially include: preventing the spread of the incident, rescuing people, cordoning off and all measures to support the relevant competent authority within the framework of administrative assistance.

53.6.1 Scenarios for the Release of Biological Agents

It is relatively easy for emergency personnel to come into contact with biological agents. Important or frequent scenarios in which contact can occur are, for example:

- Operations in the field of waste water/flooding
- Salvage of corpses/disposal of animal carcasses
- Transport of infected persons/animals
- Transport accidents involving dangerous goods of Class 6.2
- Releases or fires in stationary equipment, such as:
  - Hospitals
  - biological laboratories
  - Production plants (e.g. for vaccines, insulin, etc.)
  - Collection point for contaminated dressing material etc.
  - Waste disposal systems
  - Composting plants/biogas plants
- Use of biological weapons (terrorist attacks)

53.6.2 Principles of Use

In accordance with FwDV 500, various principles for B operations are listed here in bullet form:

- Any unnecessary spread of biohazardous substances or any contamination of humans and the environment must be avoided.
- Airlocks may only be deactivated after consultation with expert personnel. Hoses may not be laid through airlocks. If an operation site was entered via a sluice, it may also only be exited via this route.
- In facilities with B-hazardous substances, doors/windows may only be opened if absolutely necessary. They must be closed again immediately after entering/inspecting the room (prevention of spreading).
- Closed containers, packaging, cabinets with unknown contents or B hazardous substances must not be opened.

53.6.3 Human Life in Danger

For rescue operations in hazard groups IIB and IIIB, the rescue team must be equipped with at least insulating equipment and Form 1 body protection.

Quote from FwDV 500:

> Areas of hazard group IIIB in which biological substances of safety or protection level or risk group 4 are handled may not be entered under any circumstances – not even to save human lives – without the presence of the responsible permit holder in accordance with the Infection Protection Act or a competent person named within the framework of an action agreement concluded between the operator and the fire brigade.

53.6.4 Body Protection

► It is not possible to determine contamination or incorporation with B hazardous substances "on site" by means of measurement.

During all operations, all persons, equipment and other objects are therefore considered to be contaminated until they have been disinfected and cleaned accordingly or a competent person can safely rule out contamination with B-hazardous substances.

Protective clothing appropriate to the situation must be worn if contact with a hazardous substance cannot be completely ruled out.

The FwDV 500 considers as (at least) suitable:

- Hazard group IIB Body protection Form 1
- Hazard group IIIB Body protection Form 2 or 3

54 Biological Agents

Even if it seems that so-called biological weapons are an invention of modern times, this impression is deceptive. Poisons of plant or animal origin have a long history, some of which were also used in warlike conflicts. As early as the fourth century BC, herbs that gave off poisonous fumes were used. In the fourteenth century, plague corpses were used to poison drinking water. In the fifteenth century, the Spanish distributed smallpox-infected clothing to the Indians, and four hundred years later, the Indians of North America also received smallpox-infected blankets from the "whites".

The first large-scale test of modern times took place on Gruinard Island in 1942 *(Operation Vegetarian).* Researchers at the Porton Down Chemical and Biological Weapons Centre (UK), founded in 1914, were commissioned by the War Office to release anthrax to be tested on sheep. After the island was declared a restricted area because of the contamination, decontamination began in 1986. It was not until April 1990 that Gruinard Island was declared habitable again by the British Ministry of Defence.

54.1 Introduction

Biological warfare agents include bacteria, viruses, fungi and – for historical reasons – toxins from plant, animal and microbial sources. The latter, as so-called ABOs *(Agents of Biological Origin), are* by definition classified as B warfare agents rather than C warfare agents. Biological *warfare* (*BW*) *agents* are used with the intention of harming, sickening or even killing humans, animals or plants.

There are about 200 toxins and pathogens potentially suitable for use in bioweapons. The American CDC (Centers of Disease Control and Prevention) has divided the potentially bioterrorist-relevant pathogens and toxins into three categories and compiled a list of twelve agents that have become known as the "*dirty dozen*". These are the biological warfare agents most likely to be used militarily.

T. Schmiermund, *The Chemistry Knowledge for Firefighters*,
https://doi.org/10.1007/978-3-662-64423-2_54

54.1.1 Biological Weapons Convention

The Hague Regulations on Land Warfare (1899 and 1907 respectively) and the Geneva Protocol (1925) had already – more or less clearly – prohibited the (military) use of biological agents and toxins (cf. Sect. 52.1.2).

The Convention on the Prohibition of the Development, Production and Stockpiling of Bacteriological (Biological) and Toxin Weapons and on their Destruction (Biological *and Toxin Weapons* Convention, *BTWC*), which entered into force in 1975, obliges states under no circumstances to produce weapons based on microorganisms, other biological substances and toxins, or to develop weapons systems suitable for this purpose. Unlike the Chemical Weapons Convention (see Sect. 52.1.3), however, there is no monitoring body that regularly inspects states.

54.1.2 War Weapons Control Act

The "Gesetz zu Artikel 26 Absatz 2 des Grundgesetzes (Gesetz über die Kontrolle von Kriegswaffen)" (Act on the Control of War **Weapons),** abbreviated **KrWaffKontrG,** also **KWKG,** lists in the Annex (Kriegswaffenliste, KWL) in Part A, Section II, the biological warfare agents whose production and proliferation is restricted or prohibited under the Act (see also Sect. 52.1.4).

54.1.3 "Dual Use"

In the field of bioweapons, primarily plants and plant components for the cultivation of bacteria and viruses fall under the "dual use" concept. These components are subject to control by BAFA. The legal bases include the War Weapons Control Act (KWKG), the Foreign Trade and Payments Ordinance (AWV) and the EC Dual Use Regulation No. 428/2009 (see also Sect. 52.1.5).

54.1.4 Detection of Biological Warfare Agents

Is a white powder flour, powdered sugar, baking powder, a classic chemical or anthrax spores? If the substance cannot be identified beyond doubt by the fire brigade – and this is likely to be the case in most cases – "the problem only begins".

- Could it be a biohazard?
- Where do we put the powder? Who takes it?
- Who releases the assignment site?
- Who does the exact laboratory analysis? (Biological substance?, Chemical?)

At present, the only option is to request the Analytical Task Force or to clarify, in cooperation with the police, whether there are any suspicions of an attack. The sample can then be transferred, for example, to the State Criminal Police Office or a specialist authority designated by the police. The laboratories commissioned in this way need several hours to several days to be able to make a definitive statement. Research, industry and the military are working together to develop rapid tests that can detect pathogens and toxins within a few minutes. The first rapid tests have already been tested and can detect ricin, botulinum toxin, SEB, anthrax, tularaemia and plague.

54.1.5 Risks from B Attacks

Fire brigade operations in approved and labelled bio-laboratories are rare, but – in contrast to terrorist attacks involving biological substances – can be planned relatively well (e.g. by means of object-related operational plans).

In the context of armed conflicts, specialists consider the use of bacteria or viruses as bioweapons extremely unlikely. "Diseases do not stop at borders."

However, the terrorist attacks of the last 20 years show that there seems to be no limit to the imagination of terrorists. It is therefore impossible to develop operational concepts to deal with terrorist attacks on the basis of specific scenarios. All that can be done is to develop framework plans that can be applied in the event of suspicion.

Attacks with B-hazardous substances have not been perpetrated in recent years. This is probably primarily due to the fact that the use of conventional explosives has an immediate effect. The complicated preparation and no less complicated execution as well as the incubation periods, uncertainties in further spread (infection) and impact (illness, downtime, death) make bioterrorist attacks less attractive for terrorist groups.

Even if bio-terrorism thus represents a rather hypothetical danger, many people are nevertheless afraid of it. Likewise, such an attack cannot be completely ruled out. The high subjective evaluation and the potential possibility require to deal with it. To this end, practicable standard measures must be defined that can be applied to all dangerous situations and that are usually

- justifiable (towards: population, emergency forces, politics, ...)
- correct (technically, factually, medically, legally ...) and
- applicable (resources, tactical and strategic possibilities, cooperation of relief and intervention forces, ...) are.

The emergency forces are/would be overburdened with having to train or apply special operational measures on a case-by-case basis.

Further information on the topic of bioterror can be found on the homepage of the German Federal Office of Civil Protection and Disaster Assistance BBK (www.bbk.bund.de).

54.2 Classification of Pathogens and Toxins

The American CDC *(Centers for Disease Control and Prevention)* has classified bacteria, viruses and toxins that may potentially be of bioterrorist relevance into three groups (Table 54.1):

- Category A includes pathogens and toxins with the highest hazard potential.
- Category B includes pathogens and toxins with a high hazard potential
- Category C includes, inter alia, agents that could be potential biological weapons of the future.

This results in a list that has become known as the "*dirty dozen*" (Table 54.2).

Table 54.1 Categories of bioterrorism-relevant pathogens

	Category A	Category B	Category C
Hazard potential	Particularly high	High	Low
Distribution	Extremely light	Light	Light
Risk of infection	High	Medium	Potentially high
Morbidity	High	High	Potentially high
Mortality	High	Low	Potentially high
Examples of bacteria-induced Diseases	Anthrax, plague, tularaemia	Melioidosis, brucellosis, glanders, spotted fever, Q fever	–
Examples of diseases caused by viruses	Smallpox, VHF	VEE, EEE, WEE	Hanta, Nipah, yellow fever
Examples toxins	Botulinum toxin	Ricin, SEB	–
Notes	–	Additionally, pathogens that contaminate food and drinking water (e.g. salmonella, cholera bacteria)	–

Source: CDC, modified

Table 54.2 The "dirty dozen" of biological warfare agents

Disease (bacterium)	Viral diseases	Toxins
Anthrax *(B. anthracis),* Plague *(Y. pestis),* Tularaemia *(F. tularensis),* Brucellosis (Brucella spp.), Q fever *(C. burnetii),* Glanders *(B. mallei)*	Smallpox, VEE / EEE / WEE, VHF (Marburg, Ebola)	Botulinum toxin, Ricin, SEB

Sections 54.3 and 54.4 describe some pathogens and toxins in more detail. Further information can be obtained, for example, from the homepage of the German Robert Koch Institute (www.rki.de).

54.3 Important Bacteria and Viruses

An overview of the most important bacteria and viruses that can be used as biological agents is given in Table 54.3. The most important toxins are listed in Table 54.6.

54.3.1 Anthrax

The USA and the former USSR processed anthrax spores for use as BWA in the 1950s/1960s. Iraq is reported to have had weapons containing this pathogen in 1995. In September and October 2001, a total of seven letters were mailed in the United States containing anthrax spores. Eleven people contracted pulmonary anthrax, five of whom died.

The pathogen, *Bacillus anthracis, is* a spore-forming rod-shaped bacterium. It is found in cattle, sheep, goats and horses. Transmission to humans can occur through contact with e.g. hair, meat, blood or excreta of infected animals. The actual infection occurs through scratches, via wounds, the consumption of insufficiently cooked meat, by mosquitoes or the inhalation of spores. The spores are very stable and remain active for many years. Worldwide there are about 2000 cases of anthrax per year. In Germany, the last case of cutaneous anthrax was reported in 1994. Cases of injection anthrax (see below) last occurred in 2012.

Table 54.3 Overview of bacteria and viruses

Diseases caused by bacteria/ viruses	Human-to-human transmission	Infectious dose (organisms)	Incubation time	Lethality
Brucellosis	No	10...100	5...60 d	< 5%
Cholera	Rare	10...500	4 h...5 d	Untreated high
Anthrax (inhalation anthrax)	No	8000...50 000 (spores)	1...6 d	High
Plague (pneumonic plague)	High	100...500	2...3 d	High
Smallpox	High	10...100	7...17 d	High to moderate
Q fever	Rare	1...10	10...40 d	Low
Glanders	Low	Low	10...14 d	> 50%
Tularaemia	No	10...50	2...10 d	Moderate
VEE	Low	10...100	2...6 d	Low
VHF (Ebola, Marburg)	Moderate	1...10	4...21 d	High

Source: USAMRIID, modified

Anthrax manifests itself in humans in various forms: cutaneous anthrax, intestinal anthrax, pulmonary anthrax and injection anthrax. Regardless of the form, anthrax is not transmissible from person to person.

Cutaneous Anthrax

Cutaneous anthrax accounts for about 95 % of all infections and is caused by contact of infected material with small skin lesions. It can be treated well with antibiotics. If left untreated, fatal poisoning occurs in up to 20 % of cases due to the exotoxin secreted by the bacteria. The LD_{50} for anthrax toxin is given as $< 200\ \mu g\ kg^{-1}$ (mouse, i.v.).

Intestinal Anthrax

Intestinal anthrax usually occurs through the consumption of insufficiently cooked meat from infected animals. If left untreated, the infection is usually fatal.

Pulmonary Anthrax

Pulmonary anthrax is caused by the inhalation of spores. The disease usually progresses very rapidly and leads to death if left untreated. In case of deliberate aerosol application, pulmonary anthrax would be the predominant mode of progression.

Injection Anthrax

Injection anthrax was first described in heroin users in Germany and the United Kingdom in 2009/2010. Heroin contaminated with anthrax bacteria is assumed to be the source of infection. In 2012, cases occurred again among heroin users in Germany, some of which were fatal.

54.3.2 Plague

Plague is a notifiable bacterial disease (pathogen: *Yersinia pestis*), which makes quarantine of the infected person mandatory. Transmission usually occurs through fleas, which ingest the bacterium from rats.

The term "plague" is capable of causing fear and panic. The disease is one of the most significant in history. The plague epidemic in Europe (1348–1351) claimed 20...30 million victims, the epidemic in China and India (1833) claimed 12 million victims. Both the Japanese (during World War II) and Soviet militaries saw the plague as a viable BWA, as the pathogen is relatively easy to replicate and causes a severe, very often fatal, clinical picture. Depending on the route of infection, different forms of the disease occur.

Bubonic Plague

If the bacteria penetrate the skin through small injuries or bites (fleas), they spread along the lymph vessels and multiply in the nearest lymph node, which swells and turns black. If the plague bump empties outwards after a few days, it can heal. Untreated, the lethality is 50...60 %, medically treated still 5...15 %.

Pneumonic Plague

If plague pathogens are inhaled, pneumonic plague may occur. Transmission from person to person is possible through droplet infection. Death usually occurs after a few days. Untreated, the lethality rate is 100 %. In the case of deliberate aerosol spread, pneumonic plague would be the predominant form of progression.

Plague Sepsis

If a plague bump empties into the body or the pathogens enter the bloodstream by other means, internal organs are colonised. This clinical picture has a lethality of almost 100 % and, if left untreated, usually leads to death within a few hours. Even with clinical treatment, the lethality is still up to 50 %. The LD_{50} of the pure toxin excreted by the bacteria is given as $< 10\ \mu g\ kg^{-1}$ (mouse, i.v.).

54.3.3 Tularaemia

The causative agent of tularemia or rabbit plague, *Francisella tularensis, is* found throughout the northern hemisphere and in over 150 mammalian species. Transmission to humans occurs via skin/mucous membrane contact with body fluids from infected animals. Transmission by ticks and mosquitoes plays a role in both animal-to-animal and human-to-human transmission.

A use as a biological warfare agent was considered by the USA, other countries are suspected to have at least researched this pathogen as a BWA. If the bacteria are inhaled, a lethality of 35 % (untreated) is assumed.

54.3.4 Brucellosis

The infections caused by the six species of the species *Brucellae are* among the most important animal diseases of all. Four of these species are dangerous to humans, although mortality is low. It is about 5 % of untreated cases. Brucellosis was part of the American bioweapons program until the destruction of the BWA arsenal in 1969.

54.3.5 Q-Fever

The causative agent of Q fever – also known as *query fever* or Queensland fever – *Coxiella burnetii*, belongs to the rickettsiae and is found in cattle, sheep, goats, dogs, cats and birds. Infected animals usually do not become ill, but excrete large amounts of the pathogen with their body fluids – including milk. Humans usually become infected by inhaling aerosols containing the pathogen. The pathogen is very resistant to heat and dehydration – and therefore at least interesting as a biocontaminant. Sick people suffer from fever, headache and muscle pain, flu-like symptoms and pneumonia.

However, an outbreak of Q fever in the Netherlands in 2007–2009 and the knowledge gained from it make its use as a bioweapon unlikely.

54.3.6 Glanders & Melioidosis

Glanders, also called malleus, is a dangerous animal disease that occurs in horses, donkeys and mules in Southeast Asia, South America and Africa and is caused by the bacterium *Burkholderia mallei.* The incubation period is 1–5 days for skin infections and 10–14 days for lung infections. Infection occurs during close contact with infected animals.

Closely related to the glanders pathogen is the pathogen of melioidosis, which is caused by the bacterium *Burkholderia pseudomallei.*

Both glanders and melioidosis can cause acute pneumonia or a rapidly progressing sepsis (Greek *sepsis* = putrefaction; colloquially: blood poisoning). The pathogens are quite stable against environmental influences, the lethality is over 50 %, the infection dose is low – therefore both the USA and the USSR are said to have experimented with these bacteria as bioweapons.

54.3.7 Cholera

Cholera causes severe diarrhoea. The pathogen is the bacterium *Vibrio cholerae,* which secretes a toxin. The bacteria are mainly ingested through contaminated drinking water contaminated with faeces, less frequently through contaminated food. Cholera can be treated well with antibiotics. However, if left untreated, the mortality rate is about 60 %. Death occurs through fluid loss and kidney failure. The LD_{50} of pure cholera toxin is less than 250 μg kg^{-1} (mouse, iv).

Cholera is less suitable as a military bioweapon due to the low survivability of the bacteria in the environment and the possibility of vaccinating oneself against cholera. However, a bioterrorist use cannot be ruled out, as the bacterium occurs in India, South America and Africa and is thus relatively readily available.

54.3.8 Smallpox

Smallpox has been considered eradicated from the wild since 1980 through a comprehensive WHO vaccination programme. With the permission of the WHO, smallpox pathogens of the *Orthopox Variola major* type still exist at the CDC (Atlanta, USA) and at the State Institute of Virology and Biotechnology ("Vektor", Koltsovo, RUS). Smallpox was an early focus of bioweapons research. For unvaccinated – and thus immunized – persons, the probability of infection is almost 100 %, and the lethality is up to 40 %. Infection occurs through the air and the virus is relatively easy to produce.

In Germany (West) all persons were vaccinated against smallpox until 1976, in Austria vaccination was compulsory until 1981, for the US military even until 1989. Persons born later are no longer vaccinated. This makes – at least in theory – smallpox (again) interesting as a possible bioweapon. In particular, the psychological effect in the event of a threat of a smallpox attack by terrorist groups should not be underestimated. Nevertheless, the probability of a real threat from smallpox may be considered extremely low.

54.3.9 Equine Encephalitis (VEE, WEE, EEE)

In equine encephalitis (encephalitis = inflammation of the brain), which is triggered by viruses of the alphavirus genus, three different types are distinguished:

1. Venezuelan equine encephalitis (VEE) Occurrence: Northern South America, Central America, Venezuela, Colombia
2. Western equine encephalitis (WEE) Occurrence: West coast of the USA, western Canada, Mexico, Central and South America.
3. Eastern equine encephalitis (EEE) Occurrence: East Coast USA and Canada (esp. New England, Florida, Gulf Coast).

Transmission usually occurs through mosquitoes of various species. There is currently no generally available vaccination. Table 54.4 compares the three types with regard to incubation period, infection rate and lethality.

54.3.10 Viral Haemorrhagic Fever (VHF)

Some viruses cause severe infections, which are characterized by the fact that in the course of the disease there are disturbances in blood clotting. As a result, bleeding occurs internally into hollow organs and body cavities and externally through body orifices, such as pores in the skin. VHFs are among the infections with a high mortality rate.

Table 54.5 compares, among other things, the incubation periods and lethality of viral haemorrhagic fevers.

Table 54.4 Equine encephalitis; comparison of the three types VEE, WEE, EEE

	Incubation time	Encephalitis at approx.	Lethality
VEE	1...6 d	1 % (total)	~ 10 %
WEE	5...10 d	0.1 % (adults) – 2 % (children)	3...7 %
EEE	4...10 d	2.5 % (adults) – 6 % (children)	50...70 %

Table 54.5 Overview of viral haemorrhagic fevers (VHF)

VHF	Incubation time	Lethality	Infection route
Ebola/Marburg fever	2...21 d	Up to > 90 %	Contact with body fluids
Dengue fever	3...14 d	Up to 30 %	Mosquitoes
Lassa fever	6...21 d	Up to 20 %	Contact with feces, urine, Blood of infected rats
Crimean-Congo fever	1...12 d	Up to 50 %	Ticks

54.4 Important Toxins

Table 54.6 provides an overview of the most important toxins that can be used as biological warfare agents.

54.4.1 Botulinum Toxin (BTX)

The spore-forming bacterium *Clostridium botulinum* secretes extremely potent nerve-damaging toxins (neurotoxins), types A, B, E and F of which are dangerous to humans. The associated disease is called botulism and is also known as food or meat poisoning. In particular, inadequately preserved meat and sausage (lat. *Botulus* = sausage) can lead to poisoning if the food is not cooked sufficiently before consumption. Untreated, the mortality rate is 90 %.

Botulinum toxin A is the most toxic substance known to man. It is about 15 000 times more toxic than VX and 100 000 times more toxic than sarin. The lethal dose by oral ingestion is estimated to be 70...100 ng (= 0.07...0.1 μg), by inhalation 150...300 ng and by injection 3...10 ng. A comparison of the LD_{50} values of biogenic and synthetic poisons determined in the laboratory is given in Fig. 54.1 (cf. also Fig. A.4).

The drug "Botox" is highly diluted and purified BTX. It is used, among other things, for the treatment of hyperfunction of the sweat glands, headaches and muscle pain and in cosmetic medicine.

54.4.2 Ricin

Ricin is a polypeptide with a molar mass of approximatey 66 000 g mol^{-1} and is found in the seeds of the castor bean plant ("miracle tree"). When taken orally, as few as 6 seeds in a child or 20 seeds in an adult are said to have a lethal effect, although this also depends, among other things, on the intensity of chewing.

The LD_{50} of the toxin is reported to be 0.1...1.0 μg kg^{-1} (= 100...1000 ng kg^{-1}).

Table 54.6 Overview of toxins

Toxin	Lethality	Latency	LD_{50} (in μg kg^{-1})	pLD
Botulinum toxin	High	12...36 h	< 0.01	13.5
SEB	< 1 %	1...6 h	High	–
Ricin	High	18...24 h	< 5	7.7
T2 mycotoxin	Moderate	2...4 h	< 150	> 3.8
Saxitoxin	Moderate	< 1...4 h	< 50	6.6

Source: USAMRIID, modified

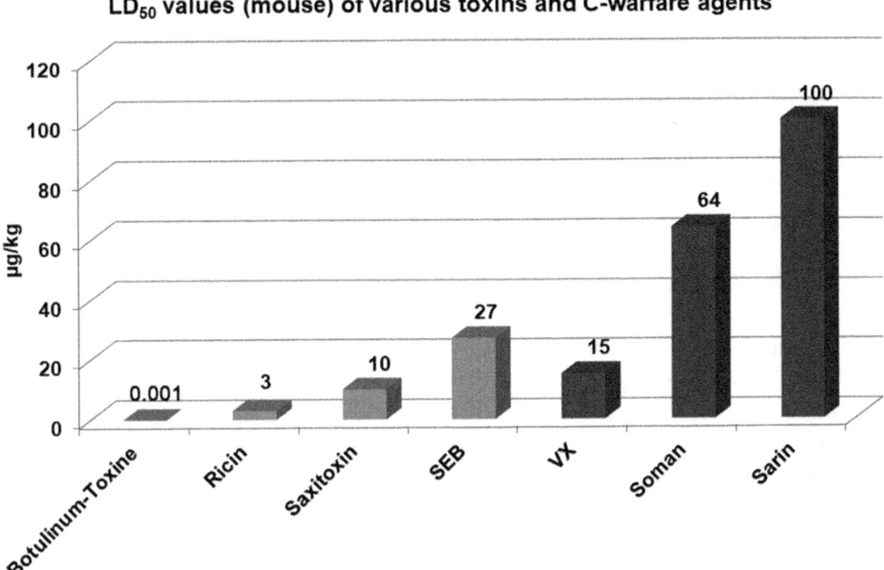

Fig. 54.1 Toxicity comparisons. *Light grey* biogenic toxins, *dark grey* chemical warfare agents

54.4.3 SEB

Among the bacteria of the genus *Staphylococcus* with about 30 species, there are individual representatives that secrete toxins, the effects of which are often summarized under the term "food poisoning". It is a group of closely related proteins with a molar mass of about 35 000 with currently 10 described subtypes. The subtype B toxin produced by *Staphylococcus aureus,* among others – commonly referred to as staphylococcal enterotoxin B (SEB) – is probably the most dangerous of these toxins. The ingestion of 20. . .25 µg SEB (equivalent to approx. 3 $\mu g\ kg^{-1}$) causes severe convulsions and vomiting in humans, sometimes lasting several days [13-17]. The lowest dose that induces emesis is reported to be 0.5 $\mu g\ kg^{-1}$ [9-3]. Reliable data on lethal concentrations in humans could not be determined by the author.

54.4.4 T2 Mycotoxin

T2-mycotoxin belongs to the group of trichothecenes, which comprise more than 50 substances and which are secreted e.g. by molds of the genera *Fusarium* or *Trichoderma* (widespread grain pests). The very environmentally stable toxins are the only mycotoxins that are skin-active and cause vesicle formation after a short exposure time.

The toxins can be absorbed orally, by inhalation or through the skin. Extensive poisoning of humans and animals by feeding mouldy grain or eating bread made from contaminated flour has been reported. The lethal dose for humans is estimated to be 7. . .10 mg (about 75. . .125 μg kg^{-1}).

Mycotoxins are said to have been sprayed from aircraft in *yellow* rain incidents in Laos (1975–81), Cambodia (1979–81) and Afghanistan (1979–81). Other sources state that this was supposed to have been bee droppings.

54.4.5 Saxitoxin

The alkaloid saxitoxin is primarily produced by unicellular algae, so-called dinoflagellates. These in turn serve as food for various *shellfish.* The toxin accumulates in the mussels and thus repeatedly leads to poisoning of humans.

The lethal dose to humans is estimated to be 0.2. . .0.5 mg (equivalent to 2.5. . .6.5 μg kg^{-1}) when ingested orally. The Lct_{50} is reported to be 5 mg min m^{-3}.

Explosives 55

In addition to the hazards that can be caused by the handling and transport of explosives, so-called "incendiary explosive devices (IEDs)" also fall within the scope of explosives.

55.1 Explosives Hazards

Depending on the subgroup in which the explosive substances and articles are classified, different hazardous situations arise. These range from unusually strong fire to an explosion of the entire explosive present. The subclasses and the resulting hazards are described in Table 55.1. For definitions of explosion/detonation, see Sect. 39.1.

Figure 55.1 shows examples of the pressures that can occur at what distance from a source of explosion and the resulting damage.

It follows:

▶ In the event of a fire, even relatively small quantities of explosives can pose a life-threatening hazard to the emergency services.

55.2 Labelling of Explosive Substances and Goods

The labelling regulations of the ADR / GGVSEB apply:

- **Explosive substances** (= solid or liquid substances or mixtures of substances which, as a result of a chemical reaction, may evolve gases of such temperature, pressure and velocity that they may cause destruction in the environment).
- **Pyrotechnic compositions** (= substances or mixtures of substances intended to produce an effect in the form of heat, light, sound, gas, mist or smoke, or a

T. Schmiermund, *The Chemistry Knowledge for Firefighters*,
https://doi.org/10.1007/978-3-662-64423-2_55

Table 55.1 Description of the subclasses of Class 1. (Source: ADR, modified)

Subclass	Description
1.1	Substances and articles capable of mass explosion. That is, almost the entire charge/explosive will explode at the same time. Main hazard: Pressure effect
1.2	Substances and articles which present a risk of forming fragments, explosive fragments and thrown fragments, but which do not have a mass explosion hazard. *This means that each individual piece explodes "on its own" – but often in rapid succession.* Main hazard: Splinter effect
1.3	Substances and articles which have a fire hazard and which present either a low hazard from blast waves or a low hazard from fragments, explosive fragments, thrown fragments or both, but which do not have a mass explosion hazard, the combustion of which gives rise to considerable radiant heat, or which burn in succession so as to produce a low hazard from blast waves or fragments, explosive fragments, thrown fragments or both. Main hazard: Fire
1.4	Substances and articles with a low explosion hazard. The effects are limited to the package. Main hazard: Fire
1.5	Very insensitive mass-explosive substances. Minimum requirement in the test: Must not explode in an external fire test.
1.6	Extremely insensitive non-mass explosive substances. I.e. substances for which there is a negligible probability (under normal conditions) of accidental ignition or propagation of an explosion.

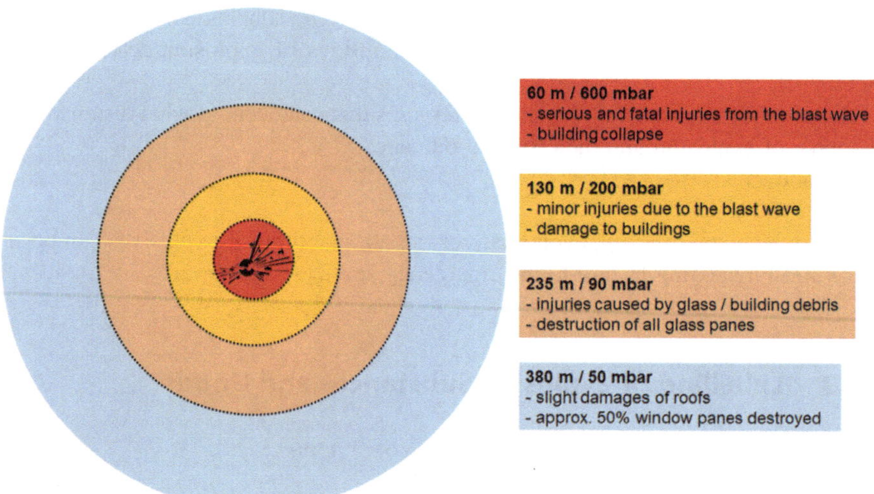

Fig. 55.1 Hazard radii due to pressure waves with 4 t mass of explosives

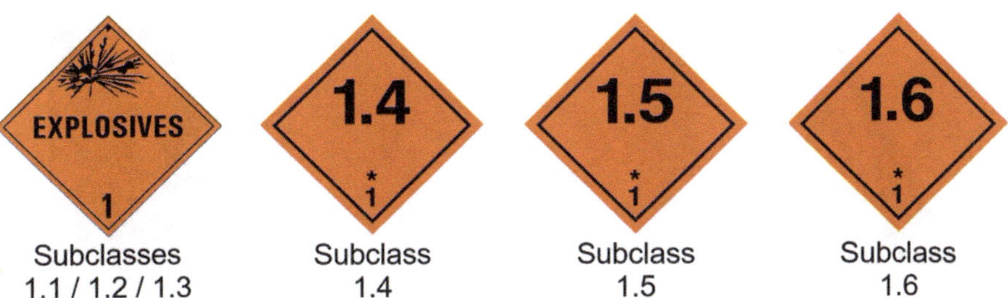

Fig. 55.2 Labeling of the hazard class 1 according to ADR

combination thereof, as a result of a non-detonating, self-sustaining, exothermic chemical reaction).

- **Articles containing explosives** (= articles containing one or more explosive substances or pyrotechnic compositions).
- All other substances and articles manufactured with a view to producing an explosive or pyrotechnic effect.

55.2.1 Transport Law

According to the ADR / GGVSEB, explosive goods are classified on the basis of the subclasses **and** the compatibility groups into which the substances or articles are classified.

The identification letter of the compatibility group is indicated for subclasses 1.4 to 1.6 instead of the asterisk (*) on the dangerous goods label (Fig. 55.2). The compatibility groups are not discussed in detail here. A list can be found, for example, in ADR (Chap. 2.2.1.1.6; ref. [9-7]).

55.2.2 Workplace Marking

Facilities in which explosives are handled, storage rooms and containers containing explosives must also be labelled accordingly. This is done according to DIN EN ISO 7010 or according to the GHS labelling. Labelling in accordance with the (German) Chemicals Act (ChemG) is outdated and should actually no longer be found anywhere (Fig. 55.3).

The GHS marking is not only found on fireworks or so-called "Bengalos", but also, for example, on the propellants or triggering devices of airbags, which are themselves considered "articles containing explosives".

Fig. 55.3 Marking of explosive substances and mixtures or of the areas

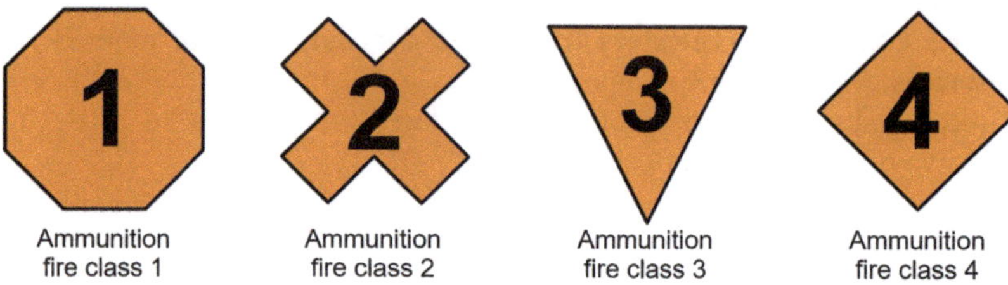

Fig. 55.4 NATO Ammunition Fire Classes

55.2.3 NATO Ammunition Fire Classes

For the storage and keeping of ammunition in NATO military facilities, the marking according to ammunition fire classes is applied (Fig. 55.4). Ammunition fire classes 1 to 4 *correspond to* subclasses 1.1 to 1.4 of the transport law.

It should be noted that the marking according to the ammunition fire classes is **not** a transport marking. Military transports are generally marked in accordance with ADR/GGVSEB (cf. Sect. 55.2.1).

55.3 Operational Measures

If it is known or becomes known after arrival at the scene of the fire that it involves large quantities of explosives or military ammunition, the danger zone must be extended to at least 300 metres and the cordon zone to 1000 m, provided there is sufficient cover.

55.3.1 Human Life in Danger

Quote from FwDV 500:

> Hazard Group IIIC areas, which are military installations containing munitions or chemical warfare agents, shall not be entered under any circumstances, **including**

Table 55.2 **Measure group 1** according to FwDV 500 (modified)

Dangers	• Explosion/detonation hazard • Main hazards subclass – 1.1: Pressure – 1.2: Splitter – 1.3: Fire • Fire gases can be toxic.
Special Measures	• In case of fire: Initiate evacuation of all bystanders from the danger area. • Working from cover (e.g. water cannon) • Use as few personnel as possible in the danger zone
Additional Notes	• Adjust barriers – Danger zone: 500 m (except for division 1.4) – Barrier area: 1000 m (except subdivision 1.4) • If necessary, observe ammunition fire classes

for the purpose of saving human life, without the presence of a responsible and competent military personnel.

This statement should be sufficient per se. Of course, this statement also applies to civilian facilities where explosives are handled: Without a competent person on site, the self-protection of the emergency forces takes precedence.

55.3.2 Group of Measures "1 Explosive substances and articles"

If the type and properties of the hazardous substances are known, they can be assigned to so-called measure groups (MG) – in accordance with FwDV 500. In the MG the dangers, special measures and additional instructions are listed. Explosive substances and objects are assigned to MG 1, compare Table 55.2.

55.3.3 Operation Sequence

Figure 55.5 shows a general flow chart for incidents involving explosives, but this should be understood only as a "rough framework" which, moreover, relates primarily to transport accidents.

In the last days of each year, however, there is also the possibility of finding relatively large quantities of fireworks in the stores of large discounters. It should be borne in mind here that their central warehouses receive the relevant deliveries some time in advance, so that they can then be distributed to the individual stores on schedule.

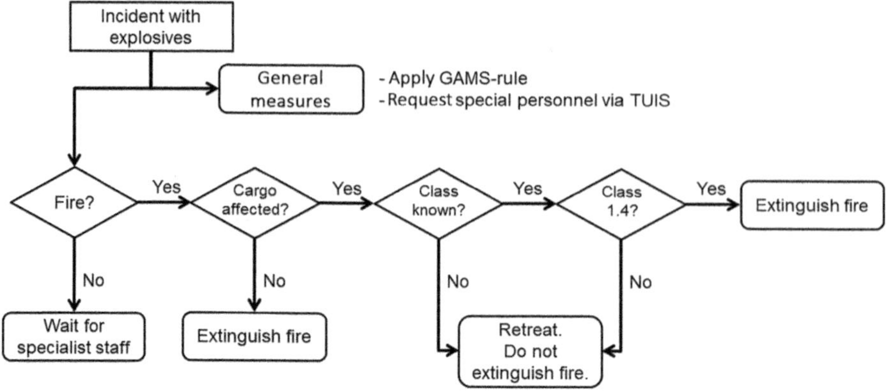

Fig. 55.5 General flowchart "Incident involving explosives"

55.3.4 Suspicion of Terrorist Attacks

If there is a suspicion of a terrorist act, special operational plans are generally used. This takes into account, among other things, the fact that after a primary attack there is a risk of a second attack in which the rescue and law enforcement forces are to be weakened. In particular, "incendiary explosive devices" (IEDs) in the form of so-called "booby traps", in addition to actively remotely detonated explosive devices, may pose a high risk to the emergency services.

As a general rule, all unusual observations, findings and suspicions must be reported in accordance *with the reporting channels.* The command organisation must be adhered to, cover opportunities must be used, objects must not be touched or moved.

▶ Further information can be found in the "Handlungsempfehlung zur Eigensicherung für Einsatzkräfte der Katastrophenschutz-und Hilfsorganisationen bei einem Einsatz nach einem Anschlag" (Recommendation for action on self-protection for emergency services and relief organizations during an operation after an attack) (in short: HEIKAT), which is available for download from the Federal Office of Civil Protection and Disaster Assistance (www.bbk.bund.de).

▶ If there is a suspicion of a terrorist attack, service instructions, alarm and deployment plans as well as orders received must be strictly adhered to.

Afterword

No matter what your reason for purchasing this book: I hope you enjoy it and find it useful.

But this work which you hold here in your hands is not "set in stone". Possibly one or the other topic "came too short" for you. Or other parts are simply "too long" for you - and possibly not relevant for you or your (location) training.

Are you missing something? You have suggestions? You have discovered an error? You wish more additional material to the book?

Basically, if you think something needs improvement, share it. Don't be afraid to express your wishes.

Just send an email with your question/comment/suggestion to: **Fw_Chemie@aol.com**.

All e-mails will be read and considered for a new edition if necessary.

T. Schmiermund

Frankfurt am Main, June 2018

T. Schmiermund, *The Chemistry Knowledge for Firefighters*,
https://doi.org/10.1007/978-3-662-64423-2

Tables and Figures

Table A.1 SI units

Name	Formula symbol	Base unit	Unit symbol
Length	ℓ	Meter	m
Time	t	Second	s
Mass	m	Kilogram	kg
Amount of substance	n	Mol	mol
Electr. current	I	Ampere	A
Light intensity	I_v	Candela	Cd
Thermodynamic temperature	T	Kelvin	K

Table A.2 Physical and mathematical constants

Size	Symbol	Numeric value	Unit
Atomic mass constant	u	$1.660\ 5655 \cdot 10^{-27}$	kg
Avogadro constant	N_A	$6.022\ 045 \cdot 10^{23}$	mol^{-1}
Bohr radius	a_0	$5.291\ 7706 \cdot 10^{-9}$	m
Elementary charge	e	$1.602\ 176\ 565 \cdot 10^{-19}$	C
Electron radius	r_e	$2.817\ 938 \cdot 10^{-15}$	m
Electron, rest mass	m_e	$9.109\ 382\ 91 \cdot 10^{-31}$	kg
Faraday constant	F	96.484 56	$C\ mol^{-1}$
Gas constant, universal	R	8.314 41 0.083 1441	$J\ mol^{-1}\ K^{-1}$ $L\ bar\ mol^{-1}\ K^{-1}$
Speed of light	c	$2.997\ 924\ 58 \cdot 10^{8}$	$m\ s^{-1}$
Mass ratio proton : electron	m_p/m_e	1836.152	–
Molar standard volume	$V_{m,0}$	22.413 83	$dm^3\ mol^{-1}$
Standard pressure	p_n	$1.013{,}250 \cdot 10^{5}$ 1.013,25	Pa Bar
Standard temperature	T_n	273,15	K
Pi	π	3.141 592 65	–
Proton, rest mass	m_p	$1.672\ 621\ 777 \cdot 10^{-27}$	kg

T. Schmiermund, *The Chemistry Knowledge for Firefighters*,
https://doi.org/10.1007/978-3-662-64423-2

Table A.3. Attachments for multiples and parts

Symbol	Name	Value 1	Value 2	Meaning
Y	Yotta...	10^{24}	1 000 000 000 000 000 000 000 000	Septillion
Z	Zetta...	10^{21}	1 000 000 000 000 000 000 000	Sixillion
E	Exa...	10^{18}	1 000 000 000 000 000 000	Quintillion
P	Peta...	10^{15}	1 000 000 000 000 000	Quadrillion
T	Tera...	10^{12}	1 000 000 000 000	Trillion
G	Giga...	10^{9}	1 000 000 000	Billion
M	Mega...	10^{6}	1 000 000	Million
k	Kilo...	10^{3}	1 000	Thousand
h	hecto...	10^{2}	100	One hundred
da	deca...	10^{1}	10	Ten
–	–	$\mathbf{10^{0}}$	**1**	**One**
d	dezi...	10^{-1}	0.1	Tenth
c	centi...	10^{-2}	0.01	Hundredths
m	milli...	10^{-3}	0.001	Thousandth
μ	micro	10^{-6}	0.000 001	Millionth
n	nano	10^{-9}	0.000 000 001	Billionth
P	pico...	10^{-12}	0.000 000 000 001	Trillionth
f	femto	10^{-15}	0.000 000 000 000 001	Quadrillionth
a	atto	10^{-18}	0.000 000 000 000 000 001	Quintillionth
z	zepto...	10^{-21}	0.000 000 000 000 000 000 001	Sixillionth
y	yokto...	10^{-24}	0.000 000 000 000 000 000 000 001	Septillionth

Prefixes are the **decimal** prefixes defined for use in the International System of Units (SI). They are based on powers of ten with integer exponents. A distinction is made between the name and the symbol of the prefix. The symbols are internationally uniform.
Characters for parts of a unit are written as lowercase letters, while most characters for multiples of a unit are written as uppercase letters.
Exceptions are - for historical reasons - the characters “da”, “h” and “k”.

Rules:

When combining with a unit, a new, inseparable unit character is created (also at line break!). One writes without a space.
Quantities without a unit of measurement cannot receive a prefix.
Only one prefix per unit may be used (μμm is incorrect; pm is correct).
No prefixes are used for minute (min), hour (h) and day (d).
For “kilogram” no other prefixes; use “grams” instead.
“m” stands for “meter” as well as for “milli”. Therefore: “Meter” at the end, “Milli” at the beginning: mN = Milli-Newton; Nm = Newtonmeter (They are prefixes!)
When exponentiating, the exponent for the prefix is: $1\ cm^2 = (10^{-2}\ m)^2 = 10^{-4}\ m^2$.

Table A.4 Derived quantities

Size	Unit	Unit symbol	Conversions
Length	Light year	ly	1 ly = $9.461 \cdot 10^{12}$ km
Area	Square meter	m^2	–
Volume	Cubic meters	m^3	–
	Litres	L	1 L = 10^{-3} m^3 = 1 dm^3
Volume flow	–	m^3/s	1 m^3/s = 3600 m^3/h
Mass	Gram	g	1 g = 10^{-3} kg
	Ton	t	1 t = 10^3 kg = 10^6 g
	Atomic mass unit	u	1 u = $1.660\ 5655 \cdot 10^{-27}$ kg
Mass flow	–	kg/s	1 kg/s = 60 kg/min = 3.6 t/h
Density	–	kg/m^3	1 kg/m^3 = 1 g/L = 1000 g/mL
Time	Year	a	365 d = $3.1536 \cdot 10^7$ s
	Day	d	24 h = 1440 min = 86 400 s
	Hour	h	60 min = 3600 s
	Minute	min	1 min = 60 s
Frequency	Hertz	Hz	1 Hz = 1/s
Force, weight force	Newton	N	1 N = 1 kg m/s^2
Pressure	Pascal	Pa	1 Pa = 1 N/m^2 = 1 kg/s^2 m
	Bar	bar	1 bar = 10^5 Pa = 10^3 hPa = 10^3 mbar
	Torr	*Torr*	*1 Torr = 1.333,224 mbar*
Work, energy, heat quantity	Joule	J	1 J = 1Nm = 1Ws 1 MJ = 0.277 778 kWh
	Kilowatt hour	kWh	1 kWh = 3.6 MJ
	Electronvolt	eV	1 eV = $160.218\ 92 \cdot 10^{-21}$ J
	Calorie	cal	1 cal = 4.1868 J = $1.163 \cdot 10^{-3}$ W h
Calorific/heating value, specific	–	MJ/m^3	1 MJ/m^3 = 0.277 778 kW h/m^3
	–	MJ/kg	1 MJ/kg = 0.277 778 kW h/kg
Heat capacity	–	J/K	1 J/K = 1 m^2 kg/(s^2 K)
Energy, specific	–	J/kg	1 J/kg = 1 m^2/s^2
Energy, molar	–	J/mol	1 J/mol = 1W s/mol
Heat capacity, molar	–	J/(mol K)	1 J/(mol K) = 1m^2 kg/(s^2 K mol)
Power	Watt	W	1 W = 1 J/s = 1 N m/s = 1 VA
	Voltampere	VA	1 VA = 1 W
Thermal conductivity	–	W/(m K)	1 W/(m K) $\approx$ 0.860 kcal/(m h °C)
Heat flux density	–	W/m^2	–
Electr. voltage	Volt	V	1 V = 1W/A
Electr. resistance	Ohm	Ω	1 Ω = 1V/A = 1/S = 1 W/A^2
Electr. conductance	Siemens	S	1 S = 1A/V = 1/Ω = 1W/V^2
Electr. Charge	Coulomb	C	1 C = 1As
	Ampere-hour	Ah	1 A h = 3600 A s = 3600 C
Activity (radioactivity)	Becquerel	Bq	1 Bq = 1/s

(continued)

Table A.4 (continued)

Size	Unit	Unit symbol	Conversions
Absorbed dose	Gray	Gy	1 Gy = 1 J/kg = 1 W s/kg
Equivalent dose	Sievert	Sv	1 Sv = 1 J/kg = 1W s/kg
Absorbed dose power	–	Gy/s	1 Gy/s = 1 W/kg
Equivalent dose power	–	Sv/s	1 Sv/s = 1 W/kg
Ion dose	–	C/kg	1 C/kg = 1 A s/kg

Table A.5 Greek numeral words

Numeric value	Intent	Numeric value	Intent
1	Mono	7	Hepta...
2	Di...	8	Octa...
3	Tri...	9	Nona...
4	Tetra	10	Deca...
5	Penta...	11	Undeca...
6	Hexa...	12	Dodeca...

Table A.6 Greek characters

Characters	Value	Meaning
α	alpha	- First - Helium core (radioactivity)
β	beta	- Second - Electron (radioactivity) - Mass concentration
γ	gamma	- Third - X-rays
Γ	Gamma	Gamma dose rate constant
δ	delta	Small difference, shift
Δ	Delta	Calculated difference
ϑ	theta	Temperature in °C
λ	lambda	Wavelength
μ	my	Prefix 10^{-6}
ν	ny	- Stoichiometric factor - Frequency
π	pi	- Circle number (3.14...) - Binding type
ϱ	rho	Density
σ	sigma	- Volume concentration - Binding type
Σ	Sigma	Total
φ	phi	Volume fraction
ω	omega	Last
Ω	Omega	electrical resistor

Table A.7 Electronegativity and ionic character

Δ EN	Ion character	Δ EN	Ion character	Diagram representation
0.0	0 %	1.7	51 %	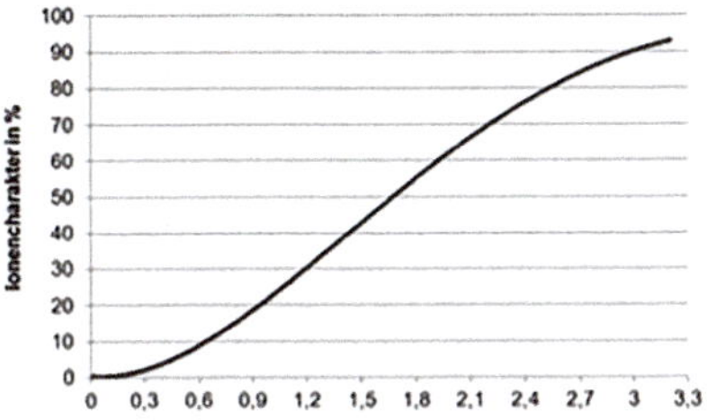 Δ EN as % ionic character
0.1	0.5 %	1.8	55 %	
0.2	1 %	1.9	59 %	
0.3	2 %	2.0	63 %	
0.4	4 %	2.1	67 %	
0.5	6 %	2.2	70 %	
0.5	9 %	2.3	74 %	
0.7	12 %	2.4	76 %	
0.8	15 %	2.5	78 %	
0.9	19 %	2.6	82 %	
1.0	22 %	2.7	84 %	
1.1	26 %	2.8	86 %	
1.2	30 %	2.9	88 %	
1.3	34 %	3.0	89 %	
1.4	39 %	3.1	91 %	
1.5	43 %	3.2	93 %	
1.6	47 %	3.3	94 %	

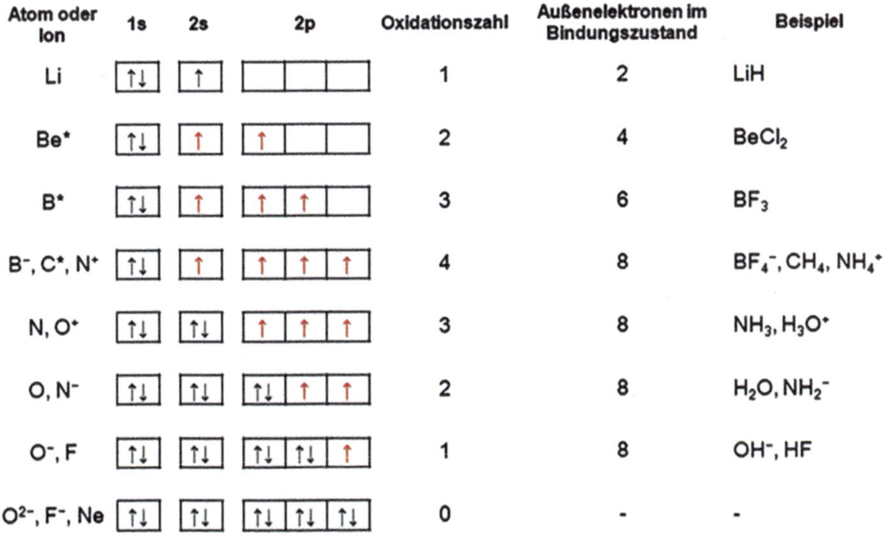

Atom oder Ion	1s	2s	2p			Oxidationszahl	Außenelektronen im Bindungszustand	Beispiel
Li	↑↓	↑				1	2	LiH
Be*	↑↓	↑	↑			2	4	$BeCl_2$
B*	↑↓	↑	↑	↑		3	6	BF_3
B^-, C^*, N^+	↑↓	↑	↑	↑	↑	4	8	BF_4^-, CH_4, NH_4^+
N, O^+	↑↓	↑↓	↑	↑	↑	3	8	NH_3, H_3O^+
O, N^-	↑↓	↑↓	↑↓	↑	↑	2	8	H_2O, NH_2^-
O^-, F	↑↓	↑↓	↑↓	↑↓	↑	1	8	OH^-, HF
O^{2-}, F^-, Ne	↑↓	↑↓	↑↓	↑↓	↑↓	0	-	-

Fig. A.1 Oxidation number and electron configuration of the elements of the second period (* means "excited state")

Atom oder Ion	3s	3p	3d	Oxidationszahl	Außenelektronen im Bindungszustand	Beispiel
Na	[↑]	[ ][ ][ ]	[ ][ ][ ][ ][ ]	1	2	NaH
Mg*	[↑]	[↑][ ][ ]	[ ][ ][ ][ ][ ]	2	4	MgO
Al*	[↑]	[↑][↑][ ]	[ ][ ][ ][ ][ ]	3	6	$AlCl_3$
Si*	[↑]	[↑][↑][↑]	[ ][ ][ ][ ][ ]	4	8	$SiCl_4$
P	[↑↓]	[↑][↑][↑]	[ ][ ][ ][ ][ ]	3	6	PH_3
P*	[↑]	[↑][↑][↑]	[↑][ ][ ][ ][ ]	5	10	PF_5
S	[↑↓]	[↑↓][↑][↑]	[ ][ ][ ][ ][ ]	2	4	H_2S
S*	[↑↓]	[↑][↑][↑]	[↑][ ][ ][ ][ ]	4	8	SF_4, H_2SO_3
S**, Si^{2-}, P^-	[↑]	[↑][↑][↑]	[↑][↑][ ][ ][ ]	6	12	SF_6, H_2SO_4
Cl	[↑↓]	[↑↓][↑↓][↑]	[ ][ ][ ][ ][ ]	1	2	HCl, HClO
Cl*	[↑↓]	[↑↓][↑][↑]	[↑][ ][ ][ ][ ]	3	6	ClF_3, $HClO_2$
Cl**	[↑↓]	[↑][↑][↑]	[↑][↑][ ][ ][ ]	5	10	$HClO_3$
Cl***	[↑]	[↑][↑][↑]	[↑][↑][↑][ ][ ]	7	14	$HClO_4$
S^{2-}, Cl^-, Ar	[↑↓]	[↑↓][↑↓][↑↓]	[ ][ ][ ][ ][ ]	0	–	–

Fig. A.2 Oxidation number and electron configuration of the elements of the third period (* means "excited state")

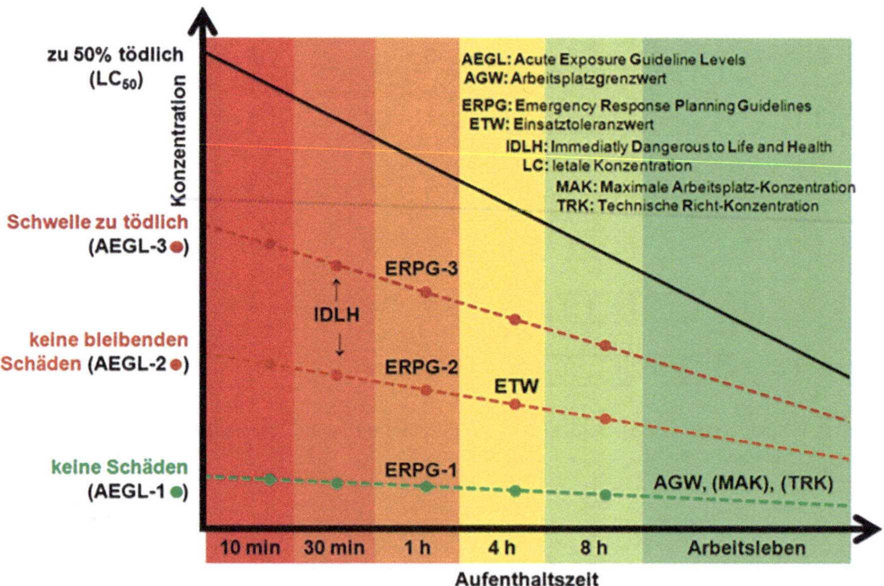

Fig. A.3 Overview of toxicological limit values, © Feuerwehrkoordination Schweiz (2014), Bern, modified

Table A.8 Thermodynamic data of selected compounds

Substance		ΔH_B^0 in kJ mol^{-1}	ΔG_B^0 in kJ mol^{-1}	S_O in J mol^{-1} K^{-1}
$\mathbf{Ag_{(s)}}$	Silver	0.0	0.0	42.7
$\mathbf{AgCl_{(s)}}$	Silver chloride	−127.0	−109.7	96.1
$\mathbf{Al_{(s)}}$	Aluminium	0.0	0,0	28.3
$\mathbf{Al_2O_{3(s)}}$	Aluminium oxide	−1669.8	−1576.4	51.0
$\mathbf{C_{(s)}}$	Carbon (graphite)	0.0	0,0	5.69
$\mathbf{CH_{4(g)}}$	Methane	−74.9	−59.8	186.2
$\mathbf{C_2H_{2(g)}}$	Ethine	+226.7	+209.2	200.8
$\mathbf{C_2H_{4(g)}}$	Ethene	+52.3	+68.1	219.5
$\mathbf{C_2H_{6(g)}}$	Ethane	−84.7	−32.9	229.5
$\mathbf{C_6H_{6(\ell)}}$	Benzene	+49	+ 129.7	159.8
$\mathbf{H_3C{-}COOH_{(\ell)}}$	Acetic acid	−487.0	−392.5	159.8
$\mathbf{H_3C{-}OH_{(\ell)}}$	Methanol	−238.6	−166.2	126.8
$\mathbf{C_2H_5{-}OH_{(\ell)}}$	Ethanol	−277.6	−174.8	160.7
$\mathbf{CO_{(g)}}$	Carbon monoxide	−110.5	−137.3	197.9
$\mathbf{CO_{2(g)}}$	Carbon dioxide	−393.5	−394.4	213.6
$\mathbf{Ca_{(s)}}$	Calcium	0.0	0.0	41.6
$\mathbf{CaO_{(s)}}$	Calcium oxide	−635.5	−604.2	39.8
$\mathbf{CaCl_{2(s)}}$	Calcium chloride	−795.0	−750.2	113.8
$\mathbf{CaCO_{3(s)}}$	Calcium carbonate	−1206.9	−1128.8	92.9
$\mathbf{Ca(OH)_{2(s)}}$	Calcium hydroxide	−986.6	−896.8	76.1
$\mathbf{Cl_{2(g)}}$	Chlorine	0.0	0.0	223.0
$\mathbf{Cu_{(s)}}$	Copper	0.0	0.0	33.3
$\mathbf{CuO_{(s)}}$	copper(II) oxide	−155.2	−127.2	43.5
$\mathbf{H_{2(g)}}$	Hydrogen	0.0	0.0	130.6
$\mathbf{HBr_{(g)}}$	Hydrogen bromide	−36.2	−53.2	198.5
$\mathbf{HCl_{(g)}}$	Hydrogen chloride	−92.3	−95.3	186.7
$\mathbf{H_2O_{(g)}}$	Water (steam)	−241.8	−228.6	188.7
$\mathbf{H_2O_{(\ell)}}$	Water	−285.9	−237.2	69.96
$\mathbf{Hg_{(\ell)}}$	Mercury	0.0	0.0	77.4
$\mathbf{HgO_{(s)}}$	Mercury oxide	−90.7	−58.5	72.0
$\mathbf{N_{2\,(g)}}$	Nitrogen	0.0	0.0	191.5
$\mathbf{NH_{3(g)}}$	Ammonia	−46.2	−16.7	192.5
$\mathbf{NH_4Cl_{(s)}}$	Ammonium chloride	−315.4	−203.9	94.6
$\mathbf{NO_{2(g)}}$	Nitrogen dioxide	+33.8	+51.8	240.5
$\mathbf{Na_{(s)}}$	Sodium	0.0	0.0	51.0
$\mathbf{NaCl_{(s)}}$	Sodium chloride	−411.0	−384.1	72.4
$\mathbf{NaOH_{(s)}}$	Sodium hydroxide	−426.7	−377.1	52.3
$\mathbf{O_{2(g)}}$	Oxygen	0.0	0.0	205.0
$\mathbf{S_{(s)}}$	Sulphur	0.0	0.0	31.9
$\mathbf{SO_{2(g)}}$	Sulphur dioxide	−296.9	−300.4	248.5
$\mathbf{SO_{3(g)}}$	Sulphur trioxide	−395.2	−370.4	256.2

ΔH_B^0: standard molar enthalpy of formation (at 1.013 bar, 298 K)
ΔG_B^0: free standard enthalpy of formation (for 1 mol, at 1.013 bar, 298 K)
S^0: standard molar entropy (at 1.013 bar, 298 K, per mole)

Table A.9 Hazard classes - with memory aid

Hazard No.	Hazard classes	Meaning	Hazard label	Danger mnemonic
1	1.1 1.2 1.3 1.4	Explosion hazardous	1	Initial: Extend "1" to E: 1≡ → E = Explosion
2	2	Gases	2	**Two** = "**whoosh**" → Noise when a gas escapes
3	3	Flammable liquid		Think of the 3 flammable liquids used in vehicles: gasoline, diesel, kerosine
4	4.1	Combustible solid		Red beams → glowing wooden beams = combustible, solid material
	4.2	Self-igniting substance		If the (red) substance comes into contact with air (white), it ignites.
	4.3	Substance reacts dangerously with water		Blue area = water (lake) → Substance in the lake = fire
5	5.1	supporting combustion	5.1	Symbol "O under flames": **O** = **o**xygen (oxygen supports the combustion)

(continued)

Table A.9 (continued)

Hazard No.	Hazard classes	Meaning	Hazard label	Danger mnemonic
	5.2	Organic peroxide	5.2	
6	6.1	Toxic		Comparison of "6" and mirrored "P" → Poison
	6.2	Contagious-dangerous		
7	7.1 7.2 7.3	Radioactive substances	RADIOACTIVE 7	Comparison of "7" and "r" (mirrored): r = rays (emitting)
8	8	Caustic		The "8" consists of four compounded "c": Caustic, caustic, caustic, caustic
9	9	Other dangerous substances		NINE: No Indication, No Explanation

Table A.10 Physiological effect of elevated CO_2 concentrations

CO_2 concentration	Impact
≥ 1 % by volume	Drowsiness, slight increase in respiratory rate
≥ 3 % by volume	Mild anesthesia, increased heart rate, decreased hearing
≥ 5 % by volume	Dizziness, confusion, headache, shortness of breath, sweating
≥ 8 % by volume	Fainting, blurred vision, tremors, sweating; fatal in 1–4 hours.
≥ 12 % by volume	Sudden unconsciousness, fatal in 30–60 min
≥ 20 % by volume	Sudden unconsciousness, fatal in minutes
≥ 30 % by volume	Sudden unconsciousness, fatal in seconds

Table A.11 Indoor air quality according to DIN EN 13779

CO_2 concentration		Indoor air quality
< 800 ppm	(<0.08 % by volume)	good
800–1000 ppm	(0,08–0,1 % by volume)	medium
1000–1400 ppm	(0,1–0,14 % by volume)	moderate
> 1400 ppm	(> 0,14 % by volume)	low

Table A.12 Physiological effect in case of O_2 deficiency

O_2 concentration	Impact
21–17 % by volume	No effects
17–15 % by volume	Shortness of breath
15–13 % by volume	Increased hyperventilation
13–8 % by volume	Disorders of the central nervous system
< 8 % by volume	Unconsciousness
< 3 % by volume	Rapid suffocation

Table A.13 Physiological effect at increased CO concentration

CO concentration	Impact
≥ 200 ppm	Headache after 23 h
≥ 300 ppm	Signs of CO poisoning after 2–3 h
≥ 500 ppm	Hallucinations within 0.5–2 h
≥ 1000 ppm	Motor disturbances, fatal in 1–2 h
≥ 1500 ppm	Fatal within 1 h
≥ 3000 ppm	Lethal after 30 min
≥ 8000 ppm	Sudden asphyxia

Table A.14 Explanatory toxicological values

AEGL-1	*Acute Exposure Guideline Levels, Level 1:* Threshold to discomfort, slight impairment possible, but no permanent or serious damage. (DIRECTIVE 96/82/EC)
AEGL-2	*Acute Exposure Guideline Levels, Level 2:* Threshold for restriction, from here on permanent or long-term health damage possible. (Directive 96/82/EC) 4-hour value as a substitute for ETW.
AEGL-3	*Acute Exposure Guideline Levels, Level 3:* Threshold to lethal effect, at concentrations AEGL-3 or greater a life-threatening situation or even death is imminent. (DIRECTIVE 96/82/EC)
AETL-1	*Acute Exposure Threshold Level, Level 1:* The maximum concentration in air to which the general population may be exposed up to a specified exposure time without more than mild and reversible adverse health effects.
AETL-2	*Acute Exposure Threshold Level, Level 2:* The maximum airborne concentration to which the general population may be exposed for up to a specified exposure time without causing irreversible or other serious adverse health effects, including symptoms that interfere with escape.
AETL-3b	*Acute Exposure Threshold Level, Level 3b:* The concentration in air to which the general population is predicted to be able to be exposed for a specified exposure time without suffering life-threatening health effects or death.
AETL-3a	*Acute Exposure Threshold Level, Level 3a:* The concentration in air for which it is predicted that a certain proportion of the general population (e.g. : 1, 5 and 50 %) will die after defined exposure times.
AGW	*Arbeitsplatzgrenzwert (Occupational exposure limit;* replaced MAK, BAT and TRK values in 2005) Concentration at which health is not adversely affected during 8 hours of daily exposure and a 40-hour week.
BAT	*Biologischer Arbeitsplatztoleranzwert (Biological workplace tolerance value):* maximum permissible quantity of a biological substance at which health is not impaired in the case of 8-hour daily exposure and a 40-hour week. (TRGS 900) Outdated - replaced by AGW since 2005.
ERPG-1	*Emergency Response Planning Guidelines, Level 1:* Mild, temporary health effect possible. (USA)
ERPG-2	*Emergency Response Planning Guidelines, Level 2*: temporary eye/respiratory irritation possible, but no permanent or serious damage. (USA)
ERPG-3	*Emergency Response Planning Guidelines, Level 3*: health impairment possible, but no life-threatening effects. (USA)
ETW	*Einsatztoleranzwert (Operational tolerance values for fire brigades;* according to German vfdb-guideline 10/01): Below the ETW there is generally no health hazard to be feared. Applies to temporary stay (4 hours without respiratory protection). They can be replaced by the AEGL-2 (4h) values.
IDLH	*Immediately Dangerous to Life and Health:* Specified concentration must not cause irreversible damage within 30 minutes. (USA)
LC_{50}	**Lethal concentration** at which 50 % of the test animals die. Indication in mg/L for 4 hours inhalation time (hazardous substances legislation).
LD_{50}	**Lethal dose** at which 50 % of the test animals die. Indication in mg/(kg body weight of test animal).
MAK	*Maximale Arbeitsplatzkonzentration (Maximum workplace concentration):* maximum permissible concentration of a pollutant at which health is not impaired in the case of 8-hour daily exposure and a 40-hour week. (TRGS 900) Outdated - replaced by AGW since 2005.

(continued)

Table A.14 (continued)

TEEL	*Temporary Emergency Exposure Limits:* similar to the ERPG value (USA), but used less frequently.
TRK	*Technische Richtkonzentration (Technical guideline concentration):* according to the state of the art, the concentration that can be achieved. Serves for risk minimization for substances without MAK value (or similar). (TRGS 102) Obsolete - AB replaced by AGW since 2005.

Rank order of use if multiple values available: ERPG-2 (TEEL-2;STH) → IDLH → AGW

Table A.15 Flammability and molecular structure: comparison of methane and the four chloromethane derivatives

Name	Methane	Chloromethane	Dichloromethane	Trichloromethane	Tetrachloro ethane
Sum formula	CH_4	CH_3Cl	CH_2Cl_2	$CHCl_3$	CCl_4
Structure	H H—C—H H	H H—C—Cl H	Cl H—C—Cl H	Cl H—C—Cl Cl	Cl Cl—C—Cl Cl
Molar mass	16 g mol^{-1}	50.5 g mol^{-1}	84.9 g mol^{-1}	119.4 g mol^{-1}	153.8 g mol^{-1}
w%(Cl)	0 %	70.2 %	83.5 %	89.1 %	92.2 %
Boiling point	−161 °C	−24 °C	40 °C	61° C	77 °C
Flash point	−188 °C	< −30 °C	hard. combust.	incombustible	Ex-agent
Ignition temperature	595 °C	625 °C	605 °C	–	–

hard. combust. = hardly combustible, Ex-agent = extinguishing agent, but: no longer permitted

Table A.16 Flammability and molecular structure: comparison of ethane with four chlorine derivatives

Name	Ethane	Chloroethane	1,1-dichloro ethane	1,2-dichloro ethane	1,1,2-trichloro-ethane
Sum formula	C_2H_6	C_2H_5Cl	$C_2H_4Cl_2$	$C_2H_4Cl_2$	$C_2H_3Cl_3$
Structure	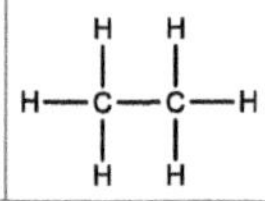 H–C(H)(H)–C(H)(H)–H	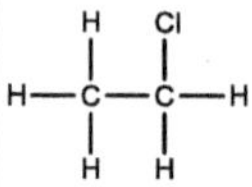 H–C(H)(H)–C(Cl)(H)–H	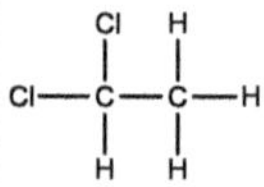 Cl–C(Cl)(H)–C(H)(H)–H	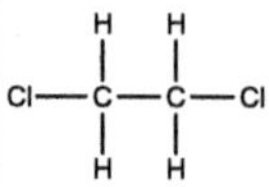 Cl–C(H)(H)–C(H)(H)–Cl	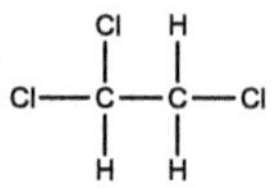 Cl–C(Cl)(H)–C(H)(H)–Cl
Molar mass	30.1 g mol^{-1}	64.5 g mol^{-1}	99.0 g mol^{-1}	99.0 g mol^{-1}	133.4 g mol^{-1}
w%(Cl)	0 %	55.0 %	71.6 %	71.6 %	79.7 %
Boiling point	−89 °C	13 °C	57 °C	84 °C	114 °C
Flash point	−135 °C	−50 °C	−10 °C	13 °C	hard. combust.
Ignition temp tur	515 °C	510 °C	440 °C	440 °C	460 °C

hard. combust. = hardly combustible

Table A.17 Flammability and molecular structure: comparison of organic nitrogen compounds

Name	Benzen	Aniline	Pyridine
Formula	C_6H_6	$C_6H_5NH_2$	C_5H_5N
Structure	 (benzene ring)	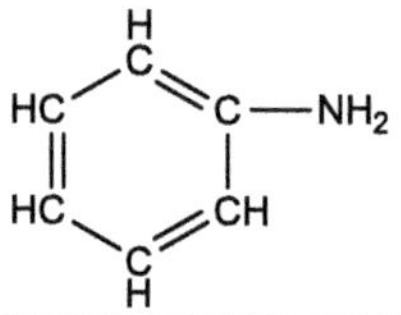 HC, HC, CH, C–NH_2 ring	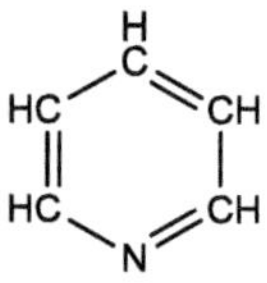 HC, CH, N ring
Molar mass	78.1 g mol^{-1}	93.1 g mol^{-1}	79.1 g mol^{-1}
w% (N)	0 %	19.8 %	31.5 %
Boiling point	80 °C	85 °C	132 °C
Flash point	−11 °C	−15 °C	28° C
Ignition temperature	555 °C	630 °C	590 °C

Table A.18 Bond energies

Bond	Bond energy	Bond	Bond energy
H–H	436 kJ mol^{-1}	C–H	414 kJ mol^{-1}
F–F	158 kJ mol^{-1}	C–F	489 kJ mol^{-1}
Cl–Cl	244 kJ mol^{-1}	C–Cl	339 kJ mol^{-1}
Br–Br	193 kJ mol^{-1}	C–Br	285 kJ mol^{-1}
C–C	346 kJ mol^{-1}	O–H	464 kJ mol^{-1}
O=O	498 kJ mol^{-1}	S–F	368 kJ mol^{-1}
N≡N	945 kJ mol^{-1}	N–H	391 kJ mol^{-1}
H–Cl	431 kJ mol^{-1}	H–Br	366 kJ mol^{-1}

Table A.19 Explosion characteristics of flammable liquids and gases

Substance	Flame speed (in m s^{-1})	$\varphi_{O,min}$ (in N_2) (in % by volume)	p_{max} (in bar)	$(dp/dt)_{max}$ (in bar s^{-1})
Acetone	0.32	11.9	8.7	313
Benzene	0.41	11.0	8.8	291
Butane	0.40	12.0	8.5	158
Diethyl ether	0.49	10.7	9.0	206
Di-*iso*-propoylether	0.45	n. i. a.	7.0	115
Ethane	0.40	11.0	6.9	171
Ethanol	0.55	10.5	7.3	171
Ethene	0.78	10.0	8.7	583
Ethine	1.68	6.5	10.1	824
Hexane	0.39	11.9	8.5	233
Carbon monoxide	0.21	4.7	8.2	n. i. a.
iso-Propanol	0.41	8.7	6.9	132
Methanol	0.87	10.0	7.2	239
Propane	0.45	11.5	8.4	220
Toluene	0.39	12.5	6.6	164
Hydrogen	3.6	4.3	8.3	n. i. a.

n. i. a. = no information available

Table A.20 Explosion characteristics of combustible dusts

Substance	pmax (in bar)	kst (In bar m s^{-1})
Activated carbon	5.8...6.5	~ 12.5
Aluminium	4.8...6.5	20...148
Lignite	6.3...10	42...71.5
Iron	5.7...6.5	61...111
Fishmeal	~ 6.5	~ 90
Grass seed	3.5...5.1	3.0...8.7
Charcoal	2.8...5.7	2.2...13.4
Potato starch	8.2...8.6	91...116
Concentrated feed	7.2...8.1	14...78
Magnesium	7.0...17.5	12...508
Cornstarch	9.3...100.4	85...200
Melamine resin	5.4...10.5	88...190
Methylcellulose	~ 6.5	~ 51
Powdered milk	5.8...9.8	28...140
Paper dust	~ 6.5	~ 27
Polyethene (HD-PE)	5.1...6.0	16.4...48.3
Polyethylene (LD-PE)	5.5...8.8	13...157
Polyvinyl chloride (PVC)	1.2...9.6	20.1...168
Wheat flour	6.7...8.8	21...70
Pulp	9.3...10	~ 9

The exact values depend strongly on the respective product (particle size distribution, moisture, pre-treatment, etc.). The values are therefore only guide values.

Table A.21 Explosion points, flash points and boiling points of flammable liquids and gases (examples)

Substance	Formula	Boiling point (in °C)	Flash point (in °C)	$t_{ex,u}$ (in °C)	$t_{ex,o}$ (in °C)
Acetone	$CH_3{-}CO{-}CH_3$	56	−21	−20	6
Benzene	C_6H_6	80	−11	−14	13
Butane	C_4H_{10}	−0.5	–	−74	−50
Diethyl ether	$(C_2H_5)_2O$	35.6	−40	−45	13
Acetic acid	$CH_3{-}COOH$	118	34	35	76
Ethan	C_2H_6	−88.6	–	−136	−118
Ethanol	$C_2H_5{-}OH$	78,5	12	11	41
Ethene	C_2H_4	−104	–	−154	−128
Ethyne	C_2H_2	−83	–	−127	−87
Hexane	C_6H_{14}	69	−22	−26	4
Methane	CH_4	−161.6	–	−188	−180
Methanol	$CH_3{-}OH$	65	11	7	39
Pentane	C_5H_{12}	36	< -40	−51	−23
Propane	C_3H_8	−42	–	−102	−83
iso-Propanol	$CH_3{-}CH(OH){-}CH_3$	82	12	8	37
n-Propanol	$C_3H_7{-}OH$	98	15	20	53
Propene	C_3H_6	−48	–	−108	−87
Tetrahydrofuran	C_4H_8O	67	−17	−20	10
Toluene	$C_6H_5{-}CH_3$	111	5	0	30
para-Xylene	$C_6H_4{-}(CH_3)_2$	138	25	24	55

The boiling point, flash point (but not for gases) and the lower ($t_{ex,u}$) or upper ($t_{ex,o}$) explosion point of selected substances are given as examples.

Table A.22 Specific oxygen/air demand and specific fire gas volume of selected substances

Substance	Spec. O_2 demand (in $m^3\ kg^{-1}$)	Spec. air requirement (in $m^3\ kg^{-1}$)	Spec. fire gas volume (in $m^3\ kg^{-1}$)
Acetone	1.5	7.4	8.1
Acetylene	2.2	10.3	10.7
Benzene	2.2	10.3	107
Carbon Monoxide	0.4	1.9	2.3
Magnesium	0.5	2.2	1.8
Methane	2.8	13.3	14.7
Methanol	1.1	5.0	6.0
Polyethylene (PE)	2.4	11.42	12.23
Polystyrene (PS)	2.15	10.25	10.68
Polyvinyl chloride (PVC)	0.9	4.26	4.8
Propane	2.5	1.1	13.1
Sulphur	0.7	3.3	3.3

Note: Data under standard conditions

Table A.23 Specific calorific values of various gases

Substance	MJ m^{-3}	kWh m^{-3}	MJ kg^{-1}	kWh kg^{-1}
Ammonia	14.2	3.9	18.6	5.2
n-butane	123.81	34.39	45.72	12.70
iso-butane	122.91	34.14	45.57	12.66
Ethane	64.35	17.87	47.49	13.19
Ethene	59.46	16.52	47.15	13.10
Ethine	56.49	15.69	48.22	13.40
Carbon monoxide	12.63	3.51	10.10	2.81
Methane	35.88	9.97	50.01	13.89
Propane	93.22	25.89	46.35	12.88
Propene	87.58	24.33	45.78	12.72
Hydrogen	107.83	29.95	119.97	33.33

Note: The specific volumetric calorific value is given per m^3 of gas (in standard condition).

Table A.24 Specific calorific values of various solids and liquids

Substance	MJ kg^{-1}	kWh kg^{-1}	MJ m^{-3}	kWh m^{-3}
Acetaldehyde	24.52	6.81	19 209	5336
Acetone	28.4	7.89	22 453	6237
Acetonitrile	29.16	8.1	22 835	6343
Waste paper	13.4	3.7	–	–
Aniline	34.7	9.64	35 221	9783
Cotton	17.4	4.8	–	–
Petrol	42.8	11.9	30 816	8560
Benzene	40.6	11.28	35 679	9911
Bitumen roofing felt, sanded	16.8	4.7	–	–
Books	16.8	4.7	–	–
1,2-dichlorobenzene	18.84	5.23	24 452	6792
Diesel fuel	41.8	11.61	35 530	9869
Diethyl ether	33.9	9.42	24 197	6721
1,4-dioxane	26.68	7.41	25 576	7660
Ethanol (90 %)	22.6	6.3	–	–
Ethanol (96 %)	26.9	7.47	21 348	5930
Fat	39.9	11.1	–	–
Cereal	16.8	4.7	–	–
rubber, tire pile	27.7	7.7	–	–
Wood, Beech	17.8	4.9	–	–
Wood, Oak	19.8	5.5	–	–
Wood, spruce	20.0	5.6	–	–
Candles	46.9	13.0	–	–
Flour	15.9	4.4	–	–
Methanol	19.9	5.53	15 824	4396
iso-octane	44.6	12.39	30 859	8572
Olive oil	41.9	11.64	38 380	10 661

(continued)

Table A.24 (continued)

Substance	MJ kg^{-1}	kWh kg^{-1}	MJ m^{-3}	kWh m^{-3}
PE	43.4	12.1	–	–
Phosphorus	48.6	13.5	–	–
PMMA	27.2	7.6	–	–
PP	43.0	11.9	–	–
iso-Propanol	30.0	8.33	23 553	6543
PUR	29.2	8.3	–	–
PVC, hard	23.0	6.4	–	–
PVC, soft	26.0	7.2	–	–
Sawdust	16.8	4.7	–	–
Chocolate	23.5	6.5	–	–
Carbon disulfide	13.8	3.83	17 533	4870
Plywood	19.3	5.4	–	–
Styrofoam®	38.0	10.6	–	–
ortho-xylene	41.25	11.46	36 345	10 096
meta-xylene	42.99	11.94	37 212	10 337
para-xylene	42.95	11.93	36 980	10 272
Sugar	17.1	4.8	–	–

Note: The specific heating value by volume is given per m^3 of liquid.

Table A.25 Heating values of various items per piece

Subject	MJ piece^{-1}	kWh piece^{-1}
Filing cabinet	1172.3	325.6
Ball, leather	6.2	1.7
Bed, complete	1088.6	302.4
Bookshelf, per m^2 area	840.0	233.3
Double bunk bed	1188.0	330.0
Bucket, plastic	16.7	4.6
Spring mattress	171.6	47.7
Wooden pallet (Euro-P.)	418.7	116.3
Cot	251.2	69.8
Stretcher	90.3	25.1
Mattress (average)	188.4	52.3
Piano	4060.0	1127.8
Duvet	46.0	12.8
Armchair, upholstered	267.1	74.2
Sofa, 2-seater	1256.0	348.9
Sofa, 3-seater	1884.0	523.3
Stacking chair, wood/steel	42.0	11.7
Wallpaper table	322.1	89.5
Door, inside	411.1	114.2
Wool blanket	23.0	6.4

Table A.26 Structures of flame retardants (examples)

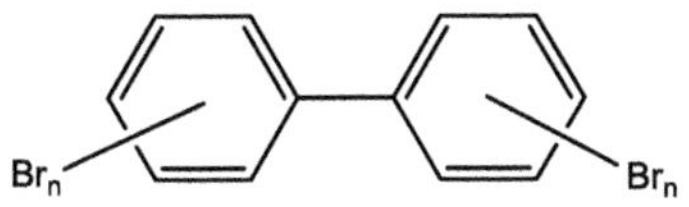 PBB	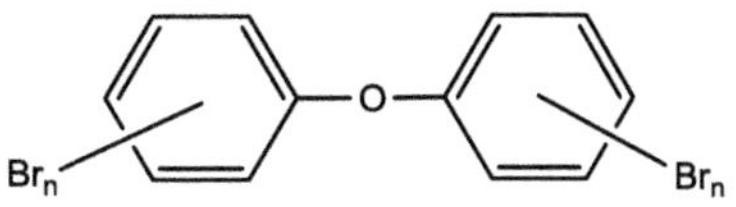 PBDE
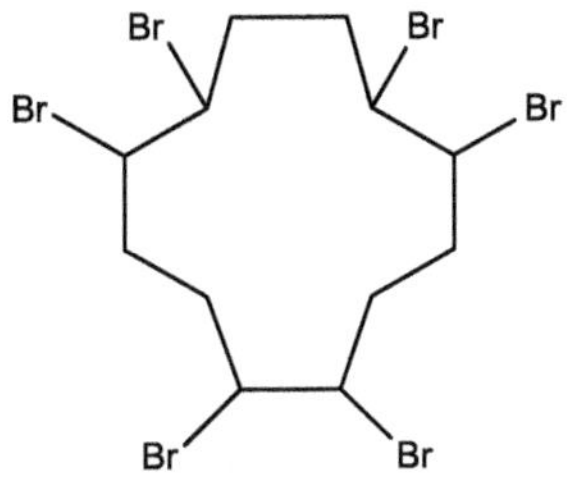 HBCD	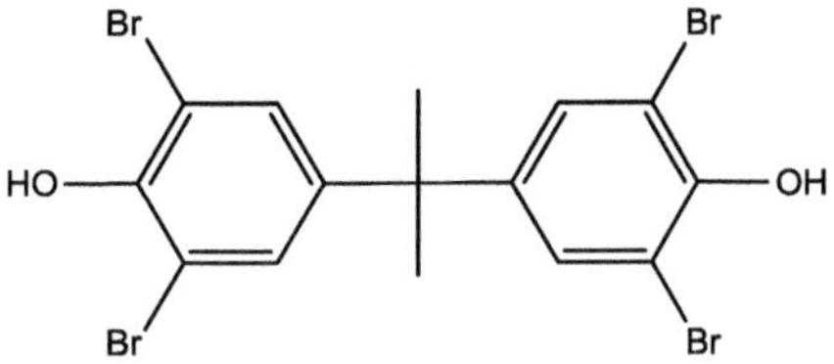 TBBA
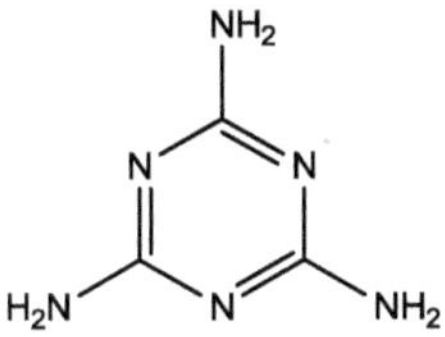 Melamin	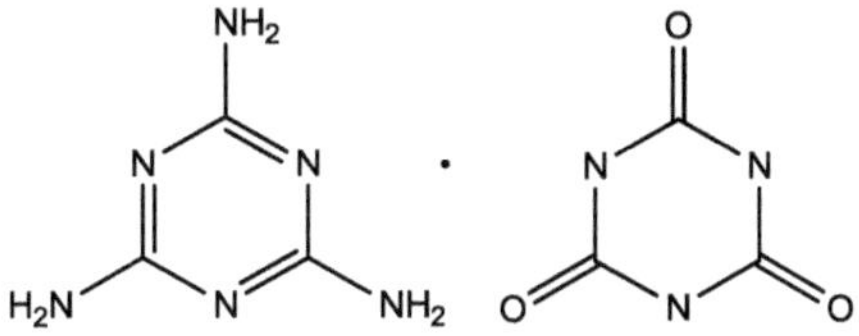 Melamin-cyanurat (MC)
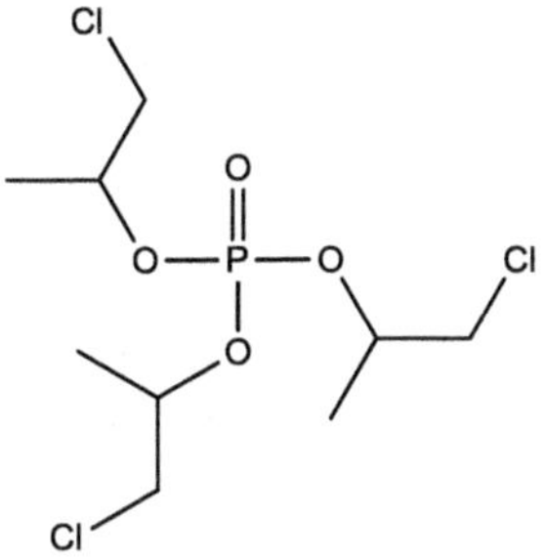 TCPP	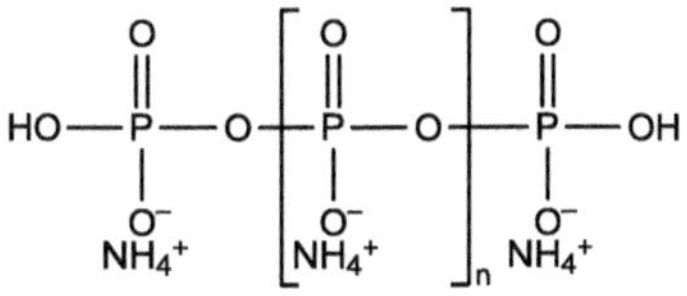 APP
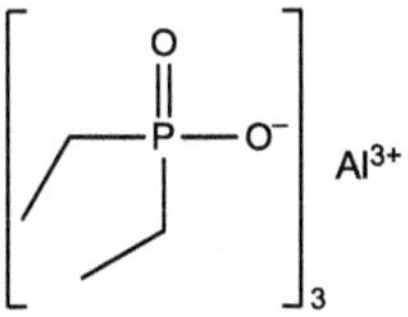 DEPAL	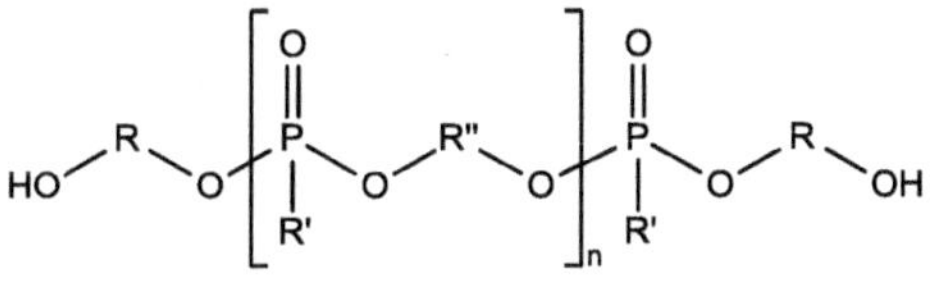 Phosphonsäure-Derivat (Beispiel)
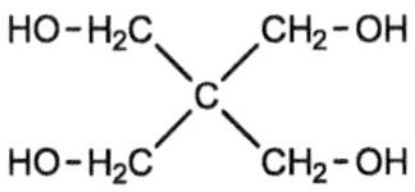 Pentaerythritol	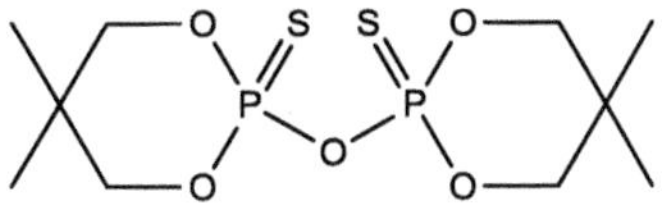 DOP-S

Table A.27 Functional groups

Designation	Structure element	See:
Alcohol	R–OH	Section 46.1
Aldehyde	R–C(=O)–H	Section 46.2
Ketone	R–C(=O)–R	Section 46.3
Ether	R–O–R	Section 46.4
Carboxylic acid	R–C(=O)–OH	Section 46.5
Ester	R–C(=O)–O–R	Section 46.6
Percarboxylic acids	R–C(=O)–O–OH	Section 46.5
Peroxides	R–O–O–R	Section 46.7
Amine	R–NH_2	Section 47.1
quaternary ammonium salts	R–N^+(R′)(R″)–R‴	Section 47.1
Nitro compounds	R–NO_2	Section 47.2
Nitrile	R–CN	Section 47.3
Isocyanates	R–N=C=O	Section 47.4
Carboxylic acid amides	R–C(=O)–NH_2	Section 47.5
Azo compounds	R–N=N–R′	Section 47.6
Thiols	R–SH	Section 48.1
Thioether	R–S–R′	Section 48.2
Disulfide	R–S–S–R′	Section 48.2
Sulphonic acids	R–S(=O)(=O)–OH	Section 48.3
Sulfoxides	R(R′)S=O	Section 48.4
Sulfone	R–S(=O)(=O)–R′	Section 48.4

Table A.28 Oxygen index of selected substances

Substance	Oxygen index	Substance	Oxygen index
Hydrogen	0.05	**Cotton**	~ 0.19
Carbon Monoxide	0.07	**Ethan**	0.19
Methanol	0.11	**Rubber**	0.19
Petrol	~ 0.13	**Paper**	0.20
Methane	0.14	**Acetic acid**	0.204
Nitrobenzene	0.14	**Pine**	0.21
Ethanol	0.15	**Polyamide**	0.24
Aniline	0.16	**Wool**	~ 0.25
Benzene	0.16	**Ammonia**	0.27
Pentane	0.16	**PVC**	0.41
Polystyrene	0.18	**PTFE**	0.95

Table A.29 Jet pipe distances according to VDE 0132, throw distances and flow rates

German jet pipe type		Flow rate	Throw distance, approx.	fire nozzle distances near electrical installations	
				Voltage up to 1000 V	Voltage above 1000 V
CM-9	Spray jet	100 L min^{-1}	5 m	1 m	5 m
CM-9	Full jet	100 L min^{-1}	15 m	5 m	10 m
CM-12	*Spray jet*	*200 L* min^{-1}	–	–	–
CM-12	*Full jet*	*200 L* min^{-1}	*20 m*	–	–
BM-16	Spray jet	400 L min^{-1}	6,5 m	4 m	8 m
BM-16	Full jet	400 L min^{-1}	25 m	8 m	11 m
BM-22	*Spray jet*	*800 L* min^{-1}	–	*8,5 m*	*12,5 m*
BM-22	*Full jet*	*800 L* min^{-1}	*30 m*	*12,5 m*	*15,5 m*

Table A.30 *LD_{50}values* (rat, oral intake) and *pLD* values of various toxins according to Lit. [9-24]

Substance	*LD_{50}* (in mg kg^{-1})	*pLD*
Acetaldehyde	1900	2.7
Acetylsalicylic acid (ASS)	200	3.7
Aniline	250	3.6
Benzo[a]pyrene	250	3.6
Benzene	50	4.3
Botulinum toxin	0.000 000 03 (i. v.)	13.5
Capsaicin	47	4.3
Caffeine	192	3.7
DDT	113	4
Dioxin (2,3,7,8-TCDD)	0.02	7.7
Ethanol	10 300	2
Potassium cyanide	5	5.3
Sodium chloride	3000	2.5
Parathion *(E 605)*	2	5.7
Ricin	0.02	7.7
Sarin	0.15	6.8
Soman	0.1	7
Tabun	0.6	6.2
VX	0.007 (i. v.)	8.2
Water	100 000	1

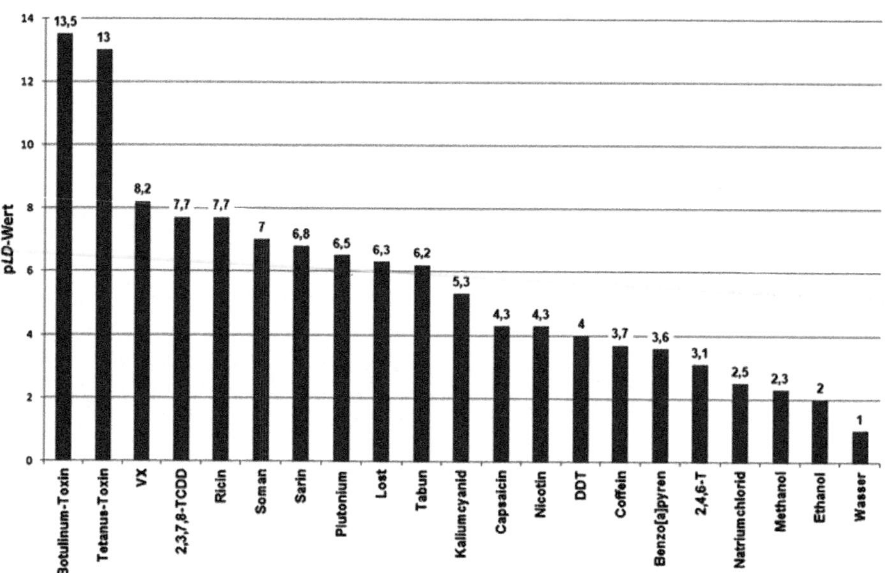

Fig. A.4 Comparison of p*LD* values (rat, oral) of different substances (values according to lit. [9-24])

Bibliography

The literature used and further literature on the individual sub-areas

1. General Literature

[1-1] Feil, S.; Resag, J.; Riebe, K. (2017) *Faszinierende Chemie*, Springer Verlag, Heidelberg

[1-2] Jackson, T. (2016), *Die Elemente – 100 Meilensteine der Chemie*, Librero IBP, Kerkdriel (NL)

[1-3] Fischer E. P. (2014) *Unzerstörbar – Die Energie und ihre Geschichte*, Springer Verlag, Heidelberg

[1-4] Quadbeck-Seeger, H.-J. (1997) *Chemie-Rekorde*, Wiley-VCH, Weinheim

[1-5] Rossotti, H. (1993) *Feuer*, Spektrum Akademischer Verlag, Heidelberg

[1-6] Welsch; Schwab; Liebmann (2013) *Materie: Erde, Wasser, Luft und Feuer*; Springer Verlag, Heidelberg

2. Chemistry, Total

[2-1] Arndt; Halberstadt (1972) *Grundzüge der Chemie*, Diesterweg Verlag, Frankfurt am Main

[2-2] Christen, H. R. (1974) *Einführung in die Chemie*, Diesterweg Salle, Frankfurt am Main

[2-3] Christen H. R. (1974) *Struktur, Stoff, Reaktion*, Diesterweg Salle, Frankfurt am Main

[2-4] Dickerson; Gray; Haight (1978) *Prinzipien der Chemie*; Walter de Gruyter Verlag, Berlin // english Edition: *Chemical Principles*, 2nd ed., W. A. Benjamin, Menlo Park

[2-5] Gärtner, u. a. (2004) *Chemie verständlich*, Weltbild, Augsburg

[2-6] Gärtner, u. a. (o. J.) *Das große Buch der Chemie*, Sonderausgabe

[2-7] Habermaier (2010) *Chemie – Grundwissen für die Feuerwehr*; Die Roten Hefte 59, Kohlhammer Verlag, Stuttgart

T. Schmiermund, *The Chemistry Knowledge for Firefighters*,
https://doi.org/10.1007/978-3-662-64423-2

[2-8] Hauptmann, S. (1996), *Starthilfe Chemie*, BG Teubner Verlag, Stuttgart
[2-9] Blecker, J. (o. J.) *Chemie für Jedermann*, Weltbild, Augsburg
[2-10] Kuballa M.; Kranz J (2017) *Pocket Teacher Chemie*, 5. Auflage, Dudenverlag, Berlin
[2-11] Kurzweil, P. (2015) *Chemie – Grundlagen, Aufbauwissen, Anwendungen und Experimente*, Springer Verlag, Heidelberg
[2-12] Mortimer, C. (1996) *Chemie – Das Basiswissen der Chemie*, Thieme-Verlag, Stuttgart // english edition: *Chemistry*, Wadsworth Publishing, Belmont
[2-13] Wolber, D. (2014) *Chemie ist gar nicht so schwer …*, Books-on-Demand, Norderstedt

3. General and Inorganic Chemistry

[3-1] Beyer, L (1993) *Grundkurs Anorganische Chemie*, Johann Ambrosius Barth Verlag, Leipzig
[3-2] Christen; Meyer (1997) *Grundlagen der allgemeinen und anorganischen Chemie*, Salle + Sauerländer, Frankfurt am Main
[3-3] Cotton, F. A; Wilkinson, G.; Gaus, P. L. (1990) Grundlagen der Anorganischen Chemie; VCH, Weinheim; english edition: *Basic Inorganic Chemistry*, 2nd edition, John Wiley & Sons
[3-4] Greenwood, N.N.; Earnshaw, A. (1988) *Chemie der Elemente*, VCH, Weinheim; english edition: *Chemistry oft he Elements*, Pergamon Press, Oxford
[3-5] Hofmann; Rüdorf (1973) *Anorganische Chemie*, Vieweg Verlag, Braun schweig
[3-6] Hollemann; Wiberg (1985) *Lehrbuch der Anorganischen Chemie*; Walter de Gruyter Verlag, Berlin
[3-7] Kolditz, L. (1993) *Anorganikum*, Johann Ambrosius Barth Verlag, Leipzig
[3-8] Kuhn, N.; Klapötke, Th. M. (2014) *Allgemeine und Anorganische Chemie – Eine Einführung*, Springer Verlag, Heidelberg
[3-9] Latscha; Klein (1996) *Anorganische Chemie – Chemie Basiswissen I*, Springer Verlag, Heidelberg
[3-10] Latscha; Mutz (2011) *Chemie der Elemente – Chemie Basiswissen IV*, Springer Verlag, Heidelberg
[3-11] Pscheidl. H. (1992) *Grundkurs Allgemeine Chemie*, Johann Ambrosius Barth Verlag, Leipzig
[3-12] Riedel (1999) *Anorganische Chemie*, Walter de Gruyter Verlag, Berlin
[3-13] Wawra; Dolznig; Müllner (2009) *Chemie verstehen*; Facultas Verlag, Wien
[3-14] Westermann; Näser; Brandes (1988) *Anorganische Chemie*; VEB Deutscher Verlag für Grundstoffindustrie, Leipzig

4. Organic Chemistry

[4-1] Allinger; Cava; De Jong, Johnson; Lebel; Stevens (1980) *Organische Chemie*, de Gruyter, Berlin; english edition: *Organic Chemistry*, 2nd edition, Worth Publishers, New York

[4-2] Beyer; Walter (1988) *Lehrbuch der organischen Chemie*, S. Hirzel Verlag, Stuttgart

[4-3] Clayden J.; Greeves N.; Warren S. (2013) *Organische Chemie*. 2. Auflage, Springer-Verlag, Heidelberg; english edition: *Organic Chemistry*, 2nd edition, Oxford University Press

[4-4] Latscha; Klein (1997) *Organische Chemie – Chemie Basiswissen II*, Springer Verlag, Heidelberg

[4-5] Neubauer, D. (2014) *Kekulés Träume – Eine andere Einführung in die Organische Chemie*, Springer Verlag, Heidelberg

5. Physical Chemistry

[5-1] Atkins, P. W. (1996) *Physikalische Chemie*, VCH, Weinheim; english edition: *Physical Chemistry*, 5th edition, Oxford University Press, Oxford

[5-2] Hug; Reiser (1999) *Physikalische Chemie*, Europa Lehrmittel Verlag, Haan-Gruiten

[5-3] Moore, W.J. (1990) *Grundlagen der Physikalischen Chemie*, Walter de Gruyter, Berlin; english edition: *Basic Physical Chemistry*, Prentice-Hall, Englewood Cliffs

[5-4] Näser; Regen; Brandes (1990) *Physikalische Chemie*, VEB Verlag für Grundstoffindustrie, Leipzig

6. Physics Textbooks

[6-1] Ballif; Dibble (1987) *Anschauliche Physik*, Walter de Gruyter, Berlin; english edition: *Conceptual Physics*, John Wiley & Sons, New York

[6-2] Grehn; Krause (1998) *Metzler Physik*, Schroedel Verlag, Hannover

[6-3] Heywang; Nücke; Timm (1992) *Physik für Techniker, Handwerk und Technik*, Hamburg

[6-4] Höfling, O. (1990) *Physik*; Dümmlers Verlag, Bonn

[6-5] Tipler, Paul A. (1998) Physik; Spektrum Akademischer Verlag, Heidelberg, english edition: *Physics for Scientists and Engineers*; 3d edition, Worth Publishers, New York

[6-6] Wahl, R. (1971) *Physik für Schule und Beruf*, Girardet-Verlag, Essen

7. General Firefighting Literature

[7-1] Feuerwehrdienstvorschrift Atemschutz – FwDV 7 (2005)

[7-2] Feuerwehrdienstvorschrift Einheiten im ABC-Einsatz – FwDV 500 (2012)

[7-3] Knorr, K.-H. (2000) *Die Gefahren der Einsatzstelle*, Kohlhammer Verlag, Stuttgart

[7-4] Rönnfeldt; König (2010) *Messtechnik im Feuerwehreinsatz*, Kohlhammer Verlag, Stuttgart

[7-5] Rönnfeldt, J. (2003) *Feuerwehr-Handbuch der Organisation, Technik und Ausbildung*, Kohlhammer Verlag, Stuttgart

[7-6] Pulm, M. (2006) *Falsche Taktik – große Schäden*, Kohlhammer Verlag, Stuttgart

8. Encyclopedias, Tables and Reference Works

[8-1] Autorenkollektiv (1999) *Lexikon der Chemie in drei Bänden*, Spektrum Akademischer Verlag, Heidelberg

[8-2] Bierwerth (1997) *Tabellenbuch Chemietechnik*; Europa Lehrmittel Verlag Haan-Gruiten

[8-3] Brandes, E.; Möller, W. (2003) *Sicherheitstechnische Kenngrößen, Band 1 – Brennbare Flüssigkeiten und Gase*, Wirtschaftsverlag NW, Bremerhaven

[8-4] Dembeck, H. (1981) *Chemie-ABC für Feuerwehr- und* Sicherheitsfachkräfte, Kohlhammer Verlag, Stuttgart

[8-5] Dembeck, H. (1997) *Gefahren beim Umgang mit Chemikalien*, Kohlhammer Verlag, Stuttgart

[8-6] Emsley, J. (1993) *Die Elemente*, Walter de Gruyter, Berlin; english edition: *The Elements*, 2nd edition, Claredon Press, Oxford

[8-7] Hähnel, Erich (1977) *Brandschutz Formeln und Tabellen*, Staatsverlag der DDR, Berlin

[8-8] Hähnel, Erich (1990) *Fachlexikon Brandschutz Explosionsschutz*, Rudolf Haufe Verlag, Berlin

[8-9] Kaltofen, u. a. (1998) *Tabellenbuch Chemie*; Verlag Harri Deutsch; Frankfurt am Main

[8-10] Küster; Thiel (1985) *Rechentafeln für die chemische Analytik*, Walter de Gruyter Verlag, Berlin

[8-11] Latscha, Schilling, Klein (1990) *Chemie-Datensammlung*, Springer Verlag, Heidelberg

[8-12] Nabert; Schön (1978) *Sicherheitstechnische Kennzahlen brennbarer Gase und Dämpfe*, Deutscher Eichverlag, Braunschweig

[8-13] Portz, H. (2005) *Brand- und Explosionsschutz von A-Z*, Vieweg-Verlag, Wiesbaden

[8-14] Prendke; Hrsg.: Schröder (2001) *Lexikon der Feuerwehr*, Kohlhammer Verlag, Stuttgart

[8-15] Regen, u. a. (1987) *Chemisch-technische Stoffwerte*, Verlag Harri Deutsch; Frankfurt am Main

[8-16] Steinleitner, H.-D.; u. a. (1979) *Tabellenbuch brennbarer und gefährlicher Stoffe*, Staatsverlag der Deutschen Demokratischen Republik, Berlin

[8-17] Steinleitner, H.-D.; u. a. (1989) *Brandschutz- und Sicherheitstechnische Kennwerte gefährlicher Stoffe*, Verlag Harri Deutsch, Frankfurt am Main

[8-18] SUVA (2010) *Sicherheitstechnische Kenngrößen von Flüssigkeiten und Gasen*, pdf, source: www.suva.ch; last access 2017-07-16

[8-19] Synowietz, C.; Schäfer, K. (1984) *Chemiker-Kalender*, Springer Verlag, Heidelberg

[8-20] Wußnig; Dietrich; Purkert; Tutzke (1992) *Fachlexikon Forscher und Erfinder*, Verlag Harri Deutsch, Frankfurt am Main

9. Hazardous Substances / Dangerous Goods / Warfare Agents / Toxicology

[9-1] CEFIC (2008) *ERI-Cards*, Kohlhammer Verlag, Stuttgart

[9-2] Fürnsinn, G. (2001) *Der biologisch-chemische Katastrophenfall – Ein Handbuch für* Einsatzkräfte, Springer Verlag, Wien

[9-3] Franke, S; Koehler, K; Zaddach, H (1994) *Chemie der Kampfstoffe – Teil I*, Dr. Koehler GmbH, Munster

[9-4] Gartz, J. (2003) *Chemische Kampfstoffe*, Grüne Kraft, Löhrbach

[9-5] Geissmann, F.; Ludescher, U. (2006) *Chemiewehr für Einsatzkräfte*, Simowa Verlag, Bern

[9-6] Klimmek; Szinicz; Weger (1983) *Chemische Gifte und Kampfstoffe - Wirkung und* Therapie, Hippokrates Verlag, Stuttgart

[9-7] Krautwurst, M. (2017) *ADR 2017*, Verkehrsverlag Fischer, Düsseldorf

[9-8] Kurzweil, P. (2013) *Toxikologie und Gefahrstoffe – Gifte Wirkungen* Arbeitssicherheit, Europa Lehrmittel Verlag, Haan-Gruiten

[9-9] Lanz; Rossetti (1980) *Katastrophenmedizin*, Ferdinand Enke Verlag, Stuttgart

[9-10] Martinetz, D. (1996) *Vom Giftpfeil zum Chemiewaffenverbot*, Verlag Harri Deutsch, Frankfurt am Main

[9-11] Matyba, St. (o. J.) *Biowaffen – Biologische Waffen im Kontext moderner Gen*-und *Biotechnologien*, Books-on-Demand, Norderstedt

[9-12] Meinholz, H.; Förtsch, G. (2010) *Handbuch für Gefahrstoffbeauftragte*, Vieweg + Teubner, Wiesbaden

[9-13] Möse, J. R. (2002) *Biowaffen – Erreger Erkrankung Diagnose Therapie Vorbeugung*, Kneipp Verlag, Leoben (A)

[9-14] Müller, N. (2011) *GHS – Das neue Chemikalienrecht*, Ecomed Verlag, Landsberg/Lech

[9-15] Müller, N.; Arenz, Th. (2011) *Sichere Lagerung gefährlicher Stoffe*, Ecomed Verlag, Landsberg/Lech

[9-16] Müller, U. (1943) *Die chemische Waffe*, Verlag Chemie, Berlin

[9-17] Robert-Koch-Institut (2011) *Steckbriefe seltener und importierter Infektionskrankheiten*, Robert-Koch-Institut, Berlin

[9-18] Rodewald, G.; Heuschen R. (2000) *Gefährliche Stoffe und Güter*, Kohlhammer Verlag, Stuttgart

[9-19] Roth; Weller (1990) *Chemie-Brände*, Ecomed Verlag, Landsberg/Lech

[9-20] Rump, A. F. E. (1999) *Gift-, Chemie-und Brandunfälle*, Schattauer Verlag, Stuttgart

[9-21] Schäfer, A. Th. (2002) *Bioterrorismus und biologische Waffen – Gefahrenpotential Gefahrenabwehr*, Verlag Dr. Köster, Berlin

[9-22] Schäfer, A. Th. (2003) *Lexikon biologischer und chemischer Kampfstoffe*, Verlag Dr. Köster, Berlin

[9-23] Schnedlitz, M. (2008) *Chemische Kampfstoffe: Geschichte, Entwicklung und Einsatz*, Grin Verlag, Norderstedt

[9-24] Strey, K. (2015) *Die Welt der Gifte*, 2. Auflage, Lehmanns Media, Berlin

[9-25] Widetscheck, O. (2012) *Der große Gefahrgut-Helfer*, Leopold Stocker Verlag, Graz

10. Radioactivity, Radiation Protection

[10-1] Grupen, C. (2000) *Grundkurs Strahlenschutz*, Vieweg Verlag, Braunschweig

[10-2] Hinsch, Hermann (2010) *Radioaktivität – Aberglaube und Wissenschaft*, Books on Demand, Norderstedt

[10-3] Höfling, O.; Waloscheck, P. (1984) *Die Welt der kleinsten Teilchen*, Rowohlt Verlag, Reinbek

[10-4] Kiefer, H; Koelzer, W. (1987) *Strahlen und Strahlenschutz*, Springer Verlag, Heidelberg

[10-5] Koelzer, W. (2014) *Lexikon zur Kernenergie*, Karlsruher Institut für Technologie (KIT) - KIT Scientific Publishing, Internet: www.ksp.KIT.edu; (pdf) Publikation: https://doi.org/10.5445/KSP/1000042198; last access 2017-12-16

[10-6] Krieger, H. (2009) *Grundlagen der Strahlungsphysik und des Strahlenschutzes*, Vieweg + Teubner, Wiesbaden

[10-7] Povh; Rith; Scholz; Zetsche (2009) *Teilchen und Kerne*, Springer Verlag, Heidelberg

[10-8] Rhodes, R. (1990) *Die Atombombe – oder die Geschichte des 8. Schöpfungstages*, Verlag Volk und Welt, Berlin; english edition: *The Making oft he Atomic Bomb*

[10-9] Volkmer, M. (2006) *Basiswissen Kernenergie*, Deutsches Atomforum e. V. (vormals: Informationskreis Kernenergie), Berlin

[10-10] Volkmer, M. (2007) *Radioaktivität und Strahlenschutz*, Deutsches Atomforum e. V. (vormals: Informationskreis Kernenergie), Berlin

[10-11] Volkmer, M. (2012) *Radioaktivität und Strahlenschutz*, Deutsches Atomforum e. V. (DAtF), Berlin / as pdf: www.kernenergie.de; last access 2017-07-16

[10-12] Volkmer, M. (2013) *Basiswissen Kernenergie*, Deutsches Atomforum e. V. (DAtF), Berlin / as pdf: www.kernenergie.de; last access 2017-07-16

11. Chemical Reaction and Chemical Engineering

[11-1] Autorenkollektiv (1990) *Technische Anorganische Chemie*, Deutscher Verlag für Grundstoffindustrie, Leipzig

[11-2] Bartholome u. a. (1972–84) *Ullmanns Encyklopädie der technischen Chemie in 24 Bänden*, Verlag Chemie, Weinheim; newer english edition: *Ullmann's Encyclopedia of Technical Chemistry*

[11-3] Behr; Agar; Jörissen; Vorholt (2016) *Einführung in die Technische Chemie*, Springer Verlag, Heidelberg

[11-4] Hopp, V.; Hennig, I. (1986) *Chemie-Kompendium*, Kaiserlei Verlag, Dreieich

[11-5] Semjonow, N. N. (1961) *Einige Probleme der chemischen Kinetik und Reaktionsfähigkeit*, Akademie Verlag, Berlin

12. Burning and Extinguishing

[12-1] Bartknecht, W. (1980) *Explosionen – Ablauf und Schutzmaßnahmen*, Springer Verlag, Heidelberg

[12-2] Beythien, R. (1960) *Die Löschmittel und ihre Anwendung – Kleine Fachbücherei der Feuerwehr Heft 6*, Verlag des Ministeriums des Inneren der DDR, Berlin

[12-3] Cicha, J. (2004) *Die Ermittlung von Brandursachen*, Boorberg Verlag, Dresden

[12-4] Demidow, P. G.; Sauschew, W. S. (1980) *Verbrennung und Eigenschaften* brennbarer *Stoffe*, Staatsverlag der Deutschen Demokratischen Republik, Berlin

[12-5] Grimwood, P.; Desmet, K. (2003) *Taktische Brandbekämpfung – Ein umfassendes Handbuch zur Innenangriffs- und Realbrandausbildung*, english edition: *Tactical Firefighting – a comprehensive guide to compartment firefighting & live fire training (CFBT).*

[12-6] Grimwood, P. (2002) *Flashover & Strahlrohtechniken*, english edition: *Flashover and nozzle techniques*

[12-7] Grimwood, P.; Barnett, C. (2005) *Fire-Fighting Flow-Rate*, www. firetactics. com; last access 2017-12-17

[12-8] Hamberger (1995) *Sicherheitstechnische Kennzahlen brennbarer Stoffe*; Die Roten Hefte 5, Kohlhammer Verlag, Stuttgart

[12-9] Kaufhold; Klingsohr (1996) *Verbrennen und Löschen*; Die Roten Hefte 1, Kohlhammer Verlag, Stuttgart

[12-10] Klingsohr (1996) *Brennbare Flüssigkeiten und Gase*; Die Roten Hefte 59, Kohlhammer Verlag, Stuttgart

[12-11] Menzel, H. (1924) *Die Theorie der Verbrennung*, Theodor Steinkopff Verlag, Dresden

[12-12] Minkoff, G. J.; Tipper, C. F. H. (1962) *Chemistry of Combustion Reactions*, Butterworth, London (GB)

[12-13] Purt, G. A. (1969) *Einführung in die Brandlehre*, Rentsch Verlag, Stuttgart

[12-14] Rempe, A. (1997) *Feuerlöschmittel – Eigenschaften Wirkung Anwendung*, Kohlhammer Verlag, Stuttgart

[12-15] Rodewald, G. (1998) *Brandlehre*, Kohlhammer Verlag, Stuttgart

[12-16] Särdqvist, S. (1996) *An Engineering Approach to Fire-Fighting Tactics*, Department of Fire Safety Engineering, Report 1014, Lund

[12-17] Scheichl, L (1958) *Brandlehre und chemischer Brandschutz*, Alfred Hüthig Verlag, Heidelberg

[12-18] Schreiber; Porst (1972) *Löschmittel – chemisch-physikalische Vorgänge beim Verbrennen und beim Löschen*, Staatsverlag der Deutschen Demokratischen Republik, Berlin

[12-19] Sommerer, J. (2011) *Kinetische Untersuchungen von Reaktionen kurzlebiger Intermediate im Zündfunken und bei der Verbrennung*, Karlsruher Institut für Technologie (KIT) – KIT Scientific Publishing, (pdf) Internet: www.ksp.KIT.edu; Publikation: https://doi.org/10.5445/KSP/1000024875; letzter Zugriff am 22.12.2017

[12-20] Warnatz, J.; Maas, U.; Dibble, R. W. (2001), *Verbrennung*, Springer-Verlag, Heidelberg; english edition: *Combustion*, Springer, Heidelberg

13. Magazines and Articles

[13-1] „How *geckos defy gravity*“; www.highered.mcgraw-hill.com, (pdf), last access 2017-05-28

[13-2] „*Sofortmaßnahmen bei Mineralölunfällen: Liste der geprüften Ölbindemittel*“, veröffentlicht vom Verband der Hersteller geprüfter Öl- und Chemikalienbindemittel GÖC e. V., www.goec-ev.de/Downloads (laztest version); last access 2017-05-10

[13-3] Falbe; Bahrmann (1981) *Homogene Katalyse in der Technik*, in: Chemie in unserer Zeit, 15, S. 37–45

[13-4] Schüth, F. (2006) *Heterogene Katalyse*, in: Chemie in unserer Zeit, 40, S. 92–103 Wiley-VCH, Weinheim (https://doi.org/10.1002/ciuz.200600374

[13-5] Spektrum der Wissenschaft, Ausgabe 8/1996 zum Thema „*Explosivstoffe*“

[13-6] Geo Kompakt, Nr. 31 (2012) *Wie uns Chemie die Welt erklärt*, Gruner + Jahr, Hamburg

[13-7] Der Spiegel – Geschichte (2015) *Die Bombe – Das Zeitalter der nuklearen Bedrohung*, Spiegel, Hamburg

[13-8] Brickmann, Klöffler, Raab (1978) *Atomorbitale*, in: Chemie in unserer Zeit, 12. Jahrgang, S. 23–26

[13-9] Rink, H. (1971), *Die Radiolyse des Wassers* in: Chemie in unserer Zeit, 5. Jahrgang, S. 90–95

[13-10] www.cas.org/about/cas-content, last access:2017-12-06

[13-11] Keutel, K.; Koch, M. (2016) *Untersuchung fluortensidfreier Löschmittel und geeigneter Löschverfahren zur Bekämpfung von Bränden häufig verwendeter polarer (d. h. schaumzerstörender) Flüssigkeiten*, Reihe: Brandschutzforschung der Bundesländer – Bericht 187, FA-Nr.: 86 (3/2014) IdF, ISSN 170-0060, Heyrothsberge

[13-12] *Rauch- und Wärmeabzugsgeräte* (o. J.) Heft 1 der Informationen des Fachverbandes Tageslicht und Rauchschutz e. V., Detmold

[13-13] Volkmann, P (1986) *Automatische Brandlüftung (RWA) – wirksame Maßnahme des Vorbeugenden Brandschutzes*, in: Vorbeugender Brandschutz, Ausgabe 1/86

[13-14] Rasbash, D. J.; Drysdale, D. D. (1982) *Fundamentals of Smoke Production*, in: Fire Safety Journal, 5 (1982), 77–86

[13-15] Rasbash, D. J.; Phillips, R. P. (1978) *Quantification of Smoke Produced at Fires*, in: Fire and Materials, Vol. 2, No. 3, (1978), 102–109

[13-16] Millich, D. (2003) *Entrauchungskonzepte im modernen Brandschutz*; Sonderdruck aus: db – Deutsche Bauzeitung, Ausgabe 10/03

[13-17] Russmann H (2003) *Toxine – Biogene Gifte und potenzielle Kampfstoffe* in: Bundesgesundheitsbl – Gesundheitsforsch – Gesundheitsschutz 2003, 46:989–996 / https://doi.org/10.1007/s00103-0716-0

[13-18] Merkblatt M 005 „*Fluorwasserstoff, Flusssäure und anorganische Fluoride*" (DGUV Information 213-071)

Index

A
A-232, 641, 642
Absolute zero, 21, 64–65
Absorbed dose, 361–362, 364–366, 368, 371, 372, 374, 375, 386, 391, 412, 413, 418, 682
α-capture, *see* Alpha capture
Accumulator
  lead-acid battery, 290
  nickel-cadmium, 291
  nickel-metal hydride, 291
Acetate buffer, 256
Acetic acid, 40, 49, 149, 175, 179, 187, 191, 205, 206, 227, 228, 238, 239, 242, 253, 255, 259, 260, 394, 441, 442, 478, 504, 565, 568, 571–574, 638, 685, 694, 699
Acetylchlolinesterase, 637
Acetylchlolinesterase inhibition, 637
Acetylcholine, 637, 638
Acetylenes, *see* Alkines
Acid
  corrosion effect, 219
  corrosivity, 217, 221
  measures after release, 260
  neutralization (*see* Neutralization)
  strength of, 215, 217, 237, 239, 243, 244
Acid-base theory
  Arrhenius, 205–206
  Arrhenius, limits, 208
  Brønsted & Lowry, limits, 212
  Brønsted u. Lowry, 209
  Brønsted u. Lowry, comparison with Arrhenius, 211
  Lewis, 212
  Lewis, limits, 213
Acid constant, 240
Acid, definition
  to Arrhenius, 206
  to Brønsted u. Lowry, 209
  to Lewis, 212
Acid halide, 572, 573
Acids
  important, 210, 217, 221–228, 582
Acids, organic, *see* Carboxylic acids
Acid strength, 238, 242, 243, 274
Acrylic glass, 598
Actinoids, 120, 127
Action quantum, Planck's, 83, 308, 309
Activation energy, 433–435, 437–439, 442, 472
Activity (radioactive material)
  derived units, 358
  examples, 358
  and half-life, 358–361
Acute Exposure Guideline Levels (AEGL), 187
Acute Exposure Threshold Level (AETL), 689
Acute radiation syndrome (ARS), 389
Adamsite, 631
Addition reaction
  alkenes, 551–553
  alkines, 552, 553
Admixing rate, 522
Aerosol, 10, 199, 221, 502, 569, 630, 641, 664, 665
AFFF foam, 525
Age determination, geological, 397
Agent Orange, 567, 644
Agents, biological, *see* Biological agents
Agents of Biological Origin (ABO), 659
AGW, 633, 689, 690
Air, composition, 463
Air liquefaction, 70
ALARA, 412–413
ALARA principle, 412–413
Alchemy, 80, 126, 167
Alcohols
  multivalent, 566

T. Schmiermund, *The Chemistry Knowledge for Firefighters*,
https://doi.org/10.1007/978-3-662-64423-2

Alcohols (*cont.*)
primary, 565
secondary, 565
tertiary, 565, 566
Aldehydes, 172, 504, 542, 551, 563, 565, 567, 568, 698
Alkali metals, 113–116, 181, 205, 219, 230, 266
Alkaline earth metals, 112, 115, 116, 181, 196, 199, 219
Alkaloids, 579, 670
Alkanes
examples, 546, 585
nomenclature, 547–548
Alkenes
examples, 549, 551, 552
Alkines
examples, 552
Allotropy, 153, 155
Alloys
amalgams, 11
substitution alloy, 11
Alpha capture, 343
Alpha decay, 341, 342, 348, 353, 355
Alpha radiation
ionization density, 390–391
properties, 339
Aluminium
formation, by electrolysis, 301
formation, electricity consumption, 299
Ambient dose equivalent, 377
Ambient dose equivalent rate, 377
Amines
primary, 577–578
Amino acids, 117, 579–580, 585
Ammonia solution, 194, 230, 242, 253, 254, 259
Ammonium buffer, 256
Amontons, G., 62, 64
Amount of substance, 65, 66, 98, 99, 161, 167, 183, 184, 298, 679
Amphiphilic, 563
Amplitude, 309
Analysis, 12, 13, 159, 160, 184, 189, 197, 247, 296, 313, 372, 396, 611, 623, 630, 660
Analytical Task Force, 660
Analyze, 245–248
Annihilation, *see* Pair annihilation
Anode, 276, 284, 288–294, 296, 300–304, 313, 370
Anomaly of the water, 150, 153
Anthrax, 648, 650, 651, 653, 660–664
Antidote, 617, 643
Antimony, 117, 118, 121, 126, 534
A-operations
hazard groups, 408, 420
human rescue, 420
operational thresholds, 374
and pregnancy, 388
special operational situations, 420
terror suspicion, 420
transport accident, 420
Application thresholds, A-operations, 374
α-radiation, *see* Alpha radiation; Beta radiation; Gamma radiation
Aramids, 597
Argon, 102, 109, 110, 119, 139, 463, 501, 517, 530, 631
Aromaticity, 557
Aromatics
condensed, 560
nomenclature, 558
polycyclic, 501, 507, 560
Arrhenius, S., 205–209, 211, 215, 236–238
Arsenic, 110, 117, 118, 121, 445, 621, 630, 632, 644
Astat, 118
Aston, F.W., 328
Atom
diameter, 82, 84, 92, 96, 97
Atomic bomb, 322, 401, 402
Atomic bond, 142, 148
Atomic lattice, 153–156
Atomic mass, absolute, 97
Atomic mass, relative, 81, 97–98
Atomic mass units, u, 97
Atomic models
Bohr, 83–84, 91, 105, 332
Bohr-Sommerfeld, 84–86
Dalton, 80–81, 91, 126
orbital model, 84, 86–91
quantum numbers, 85–89
Rutherford, 82, 84
spherical model, 80–81
Thomson, 81–82
Atomic nucleus
cohesion, 94–95, 332
density of the, 95
diameter, 82, 84, 92, 96
droplet model, 331–332
shape, 91, 333
shell model, 332–333
stability, 331, 332, 335–337
structure, 90–94, 331–333
Atomic radius, 110–112, 115, 120

Atomic shell, 79–100, 112, 325, 332, 339, 345, 347, 349–351, 385, 390
Atomic volume curve, 110, 111
Attraction, electrostatic, 139, 148
Attraction, magnetic, 95
Autoradiography, 371
Avogadro, A.
  sentence of, 65
Avogadro constant, 99, 100, 310, 679
Avogadro number, 98
Azo compounds, 493, 583, 698

B
Back-draft, 484, 500
  speed, 484
Bacteria
  human pathogens (examples), 648
  incubation time, 663
  infectious dose, 663
  lethality, 663
BAFA, 623, 660
Balloon model, 89
Barium, 101, 116, 121
Base constant, 5, 240–241
Base, definition
  to Arrhenius, 206–207
  to Brønsted and Lowry, 209–212
  to Lewis, 212–213
Base strength, 212, 238, 242, 243
BAT, 689
Battery
  alkaline manganese cell, 289
  Daniell element, 288
  historical, 288–290
  Leclanché element, 288–289
  zinc-carbon battery, 289
  zinc mercury battery, 290
Becquerel (unit), 357, 366, 681
Becquerel, A.H., 319–320, 367, 371, 376
Bengalos, 528, 529, 673
Benzilate (BZ), 641
Bergmann, T.O., 542
Beryllium, 109, 110, 116, 121
Berzelius, J.J.
  element symbols, 126
Beta decay, 335, 343–348, 355
Beta radiation
  emergence β^+, 344–345
  emergence β^-, 343–344
  ionization density, 390–391
  properties, 345–347
Binary weapons, 645
Binder, 200, 220, 260, 468, 593
Binding, 32, 57, 131, 137–144, 146, 149–151, 153, 156, 196, 220, 315, 332, 335–337, 341, 442, 464, 492, 502, 510, 553, 582, 682
Binding, chemical
  atomic bond, 142, 148
  covalent, 142, 144, 215
  dative, 130
  and dipole-dipole interaction, 137, 147–148
  electron pair bond, 137, 142–146, 151, 154, 155, 192
  electrostatic, 138, 139
  electro-valent, 139
  Hedgehog model, 137
  homeopolar, 142
  hydrogen bond, 137, 148–151
  ionic bond, 137–146
  ionogenic, 139
  main valence bond, 137
  metal binding, 137–138, 146
  nonpolar, 142
  polarized electron pair bond, 142–144
  secondary valence bond, 137
  strong bond, 137–147
  transitions, 144–146
  unitarian, 142
  valence bond, 137
  Van der Waals forces, 137, 142, 146, 148, 151–156
  weak bond, 137, 147–153
Binding energy
  C–C (alkanes), 553
  C–C (alkenes), 553
  C–C (alkynes), 553
  C–C bond, 553
  by mass defect, 335, 336
  per nucleon, 336, 341
Binding length
  C–C (alkanes), 553
  C–C (alkenes), 553
  C–C (alkynes), 553
  C–C (aromatic), 556
  C–C bond, 553
  C–H (alkanes), 553
  C–H (alkenes), 553
  C–H (alkynes), 553
  C–H (aromatic), 556
  C–H bond, 553
Biochemicals
  categories of the CDC, 662
  classification, 654
  classification according to ADR, 654

Biochemicals (*cont.*)
fire service hazard groups, 655–656
labeling, 654–656
release, scenarios, 656
workplace labeling, 655
Bioengineering, 647
Biological agents
classification, 654–656
criteria for classification, 650–652
mushrooms, 649
rickettsia, 649
risk groups, 652–654
stability, 651
toxicity, 652
viruses, 648–649
See also Biochemicals
Biological warfare, 659–662, 665, 668
Biological Weapons Convention, 660
Bio safety level (BSL), 652
Biotechnology, application areas, 647
Bio-terrorism, 661, 662
Bioweapons, 659–661, 665, 666
Bismuth, 117, 118, 121, 443
Black, J., 59
BLEVE, 72–73, 482
Blood warfare agents
decontamination, 633–634
examples, 633
protective measures, 633–634
Blue Cross, 624, 631
Blue glowing, 347
Body protection
B-operations, 656
C-operations, 619
Bohr, N.
atomic model, 83–84, 105, 332
postulates, 83–84
Bohr-Sommerfeld atomic model, 84–86
Boiling point, 4, 12, 14, 15, 21, 30, 31, 34, 36–38, 47, 49, 55, 71, 72, 110, 119, 138, 139, 142, 146, 148–150, 152–154, 457, 458, 481, 530, 531, 549, 550, 568, 569, 585, 625–627, 630, 633, 634, 638, 691, 692, 694
gases, 31, 32, 34, 36, 37, 71
pressure dependence, 31, 34, 37
Boil-over, 55
B-operations
body protection, 657
human life in danger, 657
operational measures, 656–657
principles of use, 656–657
Born-Haber cycle, 428–429
Born, M., 428
Boron, 11, 102, 109, 116–117, 121, 126, 129, 130, 154, 401
Botulinum toxin (BTX), 661, 662, 668, 700
Box notation, *see* Orbital diagram
Boyle, R., 61, 80
Breathing air, *see* Gas, real
Bremsstrahlung, 313, 345–347
Brightness, 309
Bromine, 101, 118, 149, 458, 519, 556, 561
Brønsted, J.N., 209–212, 237, 238, 265
Brown, R.
molecular motion, 68, 99
Brucellosis, 662, 663, 665
Brussels Treaty, 622
Bubonic plague, *see* Plague
Buffer
acetate buffer, 256
acetic acid/acetate, 255
buffer equation, 255
important buffer systems, 256
pH value calculation, 255–256
Buffer solution
calculation pH value, 255
functionality, 255
Building material class, 460–462
Burning
requirements, 453–456
Burning behaviour
building materials, 460–462
components, 462–463
plastics, 601–602
PVC, 602
Burning behaviour (plastics), 602
Burning rate, 497, 508–511
Bursting, 482, 501

C

C-14, 326, 344, 380, 396
Caesium (Cs), 113–115, 121, 328
CAF, *see* Compressed air foam (CAF)
Calcium (Ca), 13, 101, 109, 110, 112, 114, 116, 121, 127, 270, 288, 301, 457, 608, 685
Calculations, note to, 232
Calorie, 49, 681
Calorific value, 500, 507, 508, 510, 511, 600, 681
Calorific value, specific, 695
Capsaicin, 631, 700
Car battery, *see* Lead accumulator

Carbon, 11, 13, 91, 97–99, 109, 117, 126, 130, 135, 154, 155, 161, 247, 263–265, 289, 301, 428, 443, 457–460, 464, 491, 506, 536–538, 541–545, 547, 553–555, 557, 563, 565, 567, 572, 581, 592, 596, 599, 606, 609, 685, 693, 694
Carbonate buffer, 256
Carbon dioxide (CO_2), 12, 13, 19, 49, 50, 53, 56–57, 59, 67, 70, 71, 74, 147, 154, 161, 164, 166, 263, 264, 428, 440, 463, 502, 503, 517, 527–528, 531, 535, 685
Carbonyl group, 567, 571
Carboxy group, 571, 572
Carboxylic acid amides, 582, 698
Carboxylic acid anhydrides, 572–573
Carboxylic acid esters, *see* Ester
Carboxylic acids, 172, 227, 441, 529, 565, 571–574, 581–582, 606, 698
  anhydrides, 572–573
  dicarboxylic acids, 572
  halogenated, 572
  monocarboxylic acids, 571
Catalysis
  activation energy, 438
  energy diagram, 439
  heterogeneous, 441–443
  homogeneous, 441–442
  reaction course, 439–440
  reaction speed, 441
  theorems (Ostwald), 437, 438
Catalyst
  on fires, 479
  in fuel cells, 292
  motor vehicle, in, 437, 443, 445
  poisons for, 445
  promoter, 441
  selectivity, 440
  state of aggregation, 441
Catalyst poisons, 445
Cathode, 81, 82, 91, 93, 276, 284, 288–292, 294, 296, 300–304, 313, 319, 320
Cathode rays, 81, 82, 91, 93, 313, 319, 320
Cathode ray tube, 313, 319
C atom
  electron structure, 553
  hybridization, 554–555
Cavendish, H., 59
CBRNE, 613
Cell, galvanic, *see* Galvanic cell
Cell voltage
  alkaline manganese cell, 289
  Daniell element, 284, 285, 288
  H_2 fuel cell, 292
  lead-acid battery, 290
  Li-ion accumulator, 291–292
  Ni-Cd accumulator, 291
  Ni-MH accumulator, 291
  Zinc-carbon battery, 289
  Zn-Hg battery, 290
Celsius, A., 21
Cetane number (CN), 548
Chadwick, J., 94, 343
Chain explosion, 497
Chain propagation, 493
Chain reaction
  branched, 492, 494–496
  course, 496–497
  unbranched, 494
Chain termination, 444, 493, 527
Chalcogens, 118
Change of state of aggregation
  extinguishing agents usage, 53–57
  water, 53–56
Charge, formal, 130–131
Charging, 290, 295, 304
Charles, J., 63
Checkerboard method, 106, 107
Chemical agents
  fire service hazard groups, 616
  operational measures, 615–620, 656–657
Chemical reactions, *see* Reactions, chemical
Chemical warfare agents (CWA), 246, 393, 394, 621–645, 669, 674
Chemical Weapons Convention (CWC), 622, 623, 660
Cherenkov, P.A., 347
Cherenkov radiation, 347–348
Chimney fire, 54, 481, 521
Chlorine, 49, 72, 74, 75, 100, 101, 109, 110, 112, 118, 126, 132, 138, 139, 165, 186, 187, 222, 228, 229, 247, 249, 264, 271, 300, 317, 394, 465, 494, 506, 561, 601, 611, 616, 621, 622, 627, 632, 633, 685, 692
Chlorine-alkali electrolysis, 300–301
Chloroacetophenone, 630
Cholera, 648, 650, 662, 663, 666
Cis compounds, 549
Citric acid, 219, 259, 260, 571, 572
Clark I, 630, 631
Clark II, 630, 631
Close order, 152
CN (CWA), 630

CO_2, *see* Carbon dioxide (CO_2); Extinguishing agent carbon dioxide
Coefficient of volume expansion
  examples, 26
Cohesion, 20, 33, 94, 138, 139, 146, 332
Coin metals, 121
Combustibility
  influence of molecular structure on, 457–458
  influence of non-combustible elements on, 458–459
Combustible substances
  chemical-physical classification, 459
  classification into fire classes, 459–460
Combustion
  reaction steps, 494
Combustion equation, 44, 470–471, 488, 505
Combustion rate, 464, 467, 468
Combustion temperature, 55, 467–469, 479, 507, 508, 516, 517
Complex
  anionic, 198
  cationic, 197
  colourfulness, 197
  complexing agent, 194
  complex-ions, 195, 269
  denticity, 196
  nomenclature, 197–198
  stability, 196–197
  structure of, 195–196
Complex salts, 169, 194–199, 289
Composition
  ionic compounds, 141
Compounds
  naming, 168, 169, 558
Compressed air foam (CAF)
  advantages, 526
  disadvantages, 526
  notes, 526
Compressed air foam system (CAFS), 525–526
Comproportionation, 269
Compton, A.H., 350
Compton effect, 349, 350, 390
Compton electron, 350
Computed tomography, 380, 397
Concentration chain, 285–286
Concentration data, 157, 184, 187
Concentration-time product
  incapacitating, 628
  median lethal, 628
Condensation, 31, 34, 70, 427, 508
Condensation heat, 31
Condensation point, 31
Conduction electrons, 138, 275
Conductivity, *see* Conductivity, electrical
Conductivity, electrical
  and electron pair bonding, 142
  of metals, 138
  of saline solutions, 141
  of surfactant solutions, 610, 611
Conductor, electrical
  1st order, 138
  2nd order, 141
Constitutional formula, 133, 134
Container rupture, 73, 782
Contamination detection equipment, 373, 376
Contamination, detection of, 372, 376
Convection, 25–29
Conversion
  Bq in Ci, 367
  cal in J, 49
  Ci in Bq, 367
  °C in °F, 22
  °C in K, 22
  °dH in mmol CaCO3/L, 198
  eV in J, 312
  °F in °C, 22
  Gy in rad, 368
  H_i in H_s, 507–508
  H_s in H_i, 507–508
  H_x in H^*10, 377
  J in cal, 49
  J in eV, 312
  K in °C, 22
  kWh in J, 508
  L^* in *w*, *w*%, 182
  rad in Gy, 368
  rem in Sv, 368
  Sv in rem, 368
  Torr in mbar, 625
  $V_{s,m}$ in $V_{s,V}$, 471
  w, *w*% in L^*, 182
C-operations
  body protection, 619
  hazard groups, 619
  human life in danger, 619
Corpuscular radiation, 315
Corrosion
  electrochemical, 293–294
  element, 293
  phenomena, 293
  protection, 293, 302, 592
Corrosion protection, cathodic, 294
Counts per second (cps), unit, 376
Crimean-Congo fever, 653, 667
Criticality safety index (CSI), 406, 407

Critical point, 70
Crown ether, 569, 570
Crystal lattice
  diamond, 155, 556
  graphite, 155
  salts, 139
Crystallites, 11
Crystal water, 174, 176–177
CS (CWA), 628
Cubic law, 489
Curd soap, 608
Curie (unit), 367
Curie, M., 91, 320
Curie, P., 320
Cutaneous anthrax, *see* Anthrax
CWA detection paper, 246, 629, 630
C-warfare agents (CWA)
  blood warfare agents, 624, 633–634
  definition, 621
  division, 624
  evidence of, 625
  eye irritants, 630
  irritants, 630, 631
  latency, 627
  nasopharyngeal irritants, 630–632
  nerve agents, 624, 629, 637–641
  nerve agents, effect, 638
  properties of, 625–630
  psychotoxic, 624, 641–643
  pulmonary agents, 624, 632–633
  resistance, 627
  sedentariness, 626–627
  skin warfare agents, 624, 629, 634–637
  threshold value, 628
  tolerability limit, 628, 631
  vapor pressure, 625
C-weapons
  binary, 645
  prohibitions, 622
Cyanogen chloride, 633
Cyano group, 581
Cycloalkanes, 546
Cycloalkenes, 545
Cycloalkines, 545

D

Dalton, J.
  atomic model, 80–81, 91
  concept of atomic compounds, 125
  symbols for elements, 126
Dancing angels, 501
Danger area, 47, 675
Daniell element, 284, 285, 288
Daniell, J.F., 288
Davy, H., 205
de Broglie, L.V., 86
Decay, radioactive, 117, 319, 331, 339–352, 354, 359
Decomposition, 12, 159, 228, 295–297, 302, 355, 425, 435, 438, 441, 442, 484, 501, 502, 528, 535, 536, 550, 574, 575, 590, 591, 599, 603
Decomposition voltage, 296–297, 302
Deflagration, 228, 481, 483–484
Deformation
  of ion crystals, 142
  of metals, 138
Degree of dissociation, 239
Degree of protolysis, 239, 240
d-elements, 119–120
Democritus of Abdera, 79
Dengue fever, 649, 667
Density
  of extinguishing agents, 53, 528
  of solvents, 53
Dermal cancer, 383
Designations, systematic, 168
Desublimation, 32
Detection powder, 246, 629
Detonation, 481, 483, 484, 613, 671, 675
Diamond lattice, 155, 556
Diaphragm, 276, 300
Diene, 550
Diffusion
  definition, 67
Diffusion capacity, 68
Diffusion coefficient, 68
Diffusion rate, 67
Dioxins
  origin, 505
  and protective clothing, 506
Dipole-dipole interaction
  boiling point variation, 148
  effects, 148
Dipoles
  charge distribution, 147
  fluctuating, 151
  molecular geometry, 147
  permanent, 147, 148, 151, 153
  temporary, 151
Dirty bomb, 403
Dirty dozen, 561–562, 659, 662
Dirty dozen (bioweapons), 659, 662
Dirty dozen (halogen compounds), 561–562
Dischargeability series, 297, 298

Dispersed substance, 10, 199
Dispersing agent, 199
Dispersion
  categories of, 199
  forces, 151, 153
  paint, 200
Disproportionation, 268, 269
Dissociation, 206, 234, 238, 239, 492, 493, 521, 574
Dissociation, thermal
  water, temperatures, 55
Dissolving behaviour
  of substances, basic, 171
  in use, 178
Dissous gas, 72
Disulfides, 586, 698
Diver's disease, 74
Döbereiner, J.W., 101
d-orbital, 89, 105, 119, 130
Dose rate, 364–366, 372, 373, 375–377, 406, 409, 414–418, 682
Dose rate constant, 365–366, 682
Dose rate meters
  notes on working with, 375–376
Dose rate warning devices
  procedure, 375
Dose warning devices
  notes to, 375
Double bond
  conjugated, 550, 553, 558
  cumulative, 550
  display, 138
  influence on flammability, 457
  isolated, 550
  σ-/π-binding, 554
Double magic cores, 333
Double salts, 193–201
Drift gas, *see* Ion mobility spectrometer (IMS)
Droplet model, 331–332, 334
Dual use
  and B weapons, 657
  and C weapons, 624
  example substances, 623, 632, 633
Duration of stay, 412, 415–416, 418
Dust explosion, 482, 485, 486, 489, 521
Dynamite, 581

E

Ebola, 649, 650, 653, 662, 663, 667
EC sensors, service life, 445
Educts, 159, 165
EEE, *see* Equine encephalitis (EEE)
Einstein, A., 160, 308, 312, 335
Elastomers, 589, 591–593, 599, 601
Electrochemical voltage series, *see* Voltage series
Electrochemistry
  terms, 303–304
Electrodynamics, 308
Electrolysis
  applications, 300–303
  aqueous solutions, 297–300
  chlorine production, 229
  copper production, 302
  decomposition voltage, 296–297, 302
  electrode processes, 295–297
  electroplating, 302–303
  electro-refining, 302
  sodium hydroxide production, 230
  terms, 303–304
Electrolyte, 188, 191, 203, 238, 241, 276, 277, 289, 291, 293, 297
Electromagnetic spectrum, 310
Electromotive force (EMF), 283–294
Electron, 91
Electron affinity, 110, 112–113, 115
Electron capture, 347, 355
Electron conductor, 138, 303
Electron configuration
  display, 106–110
  orbital diagram, 107, 108
  term notation, 109–110
Electron deficiency compound, 129
Electron distribution, notation, 85
Electronegativity
  definition, 113
  difference in electron pair bonding, 142
  examples, 114
Electron flow, 138, 275, 287
Electron gas, 137, 138, 146
Electron pair bond (EPB)
  and atomic lattice, 153–156
  transition to ionic bonding, 142
  transition to metal binding, 138, 146
Electron pair bond, polarized
  symbolic representation, 144
Electron pair, free, 130, 135, 149, 196, 212, 578
Electron radiation, 312, 313, 315
Electrons, delocalized, 137, 558
Electron volt
  definition, 310
Electrophoresis, 296
Electro-refining, 302
Element, Galvanic, *see* Galvanic element
Elements

earth, fire, water, air, 79, 80
important, 126, 270
mononuclides, 328
Element symbols
Berzelius, 126, 127
Dalton, 125–127
Element, voltaic, *see* Voltaic element
Eloxal, 302
Emergency Response Planning Guidelines (ERPG), 689, 690
EMF, *see* Electromotive force (EMF)
Emulsion, 10, 199, 200, 372
Energy
formation energy, 425, 426
Gamma radiation, 310, 348
internal energy U, 425
kinetic, 20, 21, 32, 33, 61–63, 345, 362
and mass, 312–313
photon, 309, 315, 316, 350
reaction energy *E,* 425
thermal energy, 20, 21, 24, 29, 49, 292, 345, 395, 425, 521
turnover in chem. reaction, 425, 506–507
Energy transport
through waves, 307
Energy unit
electron volt, 310–312
Joule, 48, 70, 72, 310, 336, 342, 362, 474, 479, 609, 681
Newton meter, 310, 609, 681
Watt second, 310
Enthalpy
enthalpy of reaction, free, 432, 437
formation of enthalpy, 426, 432, 685
Enthalpy of reaction, 426–428, 430
Entropy
as a concept of order, 430
examples, 431
Environmental radioactivity, 379–380
EPB, *see* Electron pair bond (EPB)
Epicurus
Tenets, 79
Epoxies, 534, 592
Equine encephalitis (EEE), 667
Equivalent dose
effective, 362, 363
rate, 364–366, 377
Ester, 227, 441, 529, 538, 563, 567, 571, 573–574, 581, 592, 598, 606, 623, 629, 637, 639–642, 644, 698
Esterification, 573, 607
Ether
asymmetrical, 569–570
crown ether, 569, 570
cyclic, 568, 570
symmetrical, 569–570
ETW, 187, 689
Evaporation, 4, 12, 31, 32, 34–37, 47, 49–50, 52, 53, 149, 174, 176, 200, 220, 223, 479, 483, 513, 517, 518, 625
Evaporation number
examples, 36, 37
meaning, 36
Evaporative cooling, 32
Exclusion principle, *see* Pauli principle
Exemption limit, 408
Explosion, 37, 41–44, 47, 53, 56, 67, 73, 221, 222, 247, 451, 468, 471, 474–476, 479, 481–489, 496, 500, 521, 528, 607, 627, 671, 672, 686, 693, 694
chemical, 482, 483
differentiation, chem.-phys., 481–486
distinction medium, 485
and explosion range, 41–44, 47, 468, 474, 479
physical, 482, 483
Explosion hazard
exceeding the UEL, 43
Explosion indicators, 486–489
Explosion limits
and addition of inert gases, 42
and environmental conditions, 42
lower, 37, 41–45, 55, 468, 474–477, 479, 483, 485
lower, definition, 42
measurement of the, 43
and oxygen concentration, 42, 465–469
and pressure reduction, 43
and temperature change, 43
upper, 41–45, 468, 474–476, 479
upper, definition, 42
Explosion points, 43, 476–477, 496
Explosion pressure, maximum, 475, 487–488
Explosion range
definition, 41
Explosives
dangers due to, 675
Explosive substances
human life in danger, 674–675
operational measures, 674–675
operation sequence, 675
Extensive properties, 424
Extinguishing agent carbon dioxide, 56, 526
boundaries, 527
Extinguishing agent foam
compressed air foam, 525

Extinguishing agent foam (*cont.*)
diffusion of flammable vapours, 524
environmental protection, 524
foam destruction, 522
operation notes, 522
special foams, 525
water film forming, 525
Extinguishing agent Halons, 445, 530, 531
Extinguishing agents
definition, 520
density of, 53
Extinguishing agents for grease fires
mode of action, 529
Extinguishing agents other
halon substitutes, 529–531
hollow glass granules, 529
spare extinguishing agent, 529–531
Extinguishing agent water
advantages, 521
boundaries, 519
disadvantages, 519
Extinguishing effects
cooling, 516, 521
cover effect, 517, 522
diluting, 518
displacing effect, 518
insulating effect, 57, 517
powder, type ABC, 517, 526–528
powder, type BC, 444, 478, 526–528
powder, type D, 526–528
reaction inhibition, 518–520
reducing, 516
separating, 516–517
suffocating, 515–518, 522
Extinguishing method
cooling, 515, 516
reaction inhibition, 518–520
suffocation, 516–518
Extinguishing powder
particle size, 478
Extinguishing water demand, 510, 513
Extinguishing water demand, efficiency, 510

F

Fahrenheit, D.G., 21
Faraday constant, 299, 679
Faraday, M., 298
Faraday's Law, 297–299
Fat hydrogenation, 609
Fats, 10, 180, 461, 462, 481, 483, 516, 517, 521, 524, 525, 528, 529, 564, 571, 606, 607, 695
Fatty acids, 529, 571, 606–608
FCI steam explosion, 54, 56, 521, 528
f-elements, 120
Fermi, E., 321, 322, 335
Field, electromagnetic, 308
Film dosimeter
carrying, 373
function, 373
structure, 373
Fire, 449
and heat transport, 19–29
Fireball, estimation of size at BLEVE, 73
Fire classes, 57, 459–462, 478, 524–529, 674, 675
A, 57, 461, 525, 526, 528, 529
B, 461, 462, 478, 524–529
C, 461, 478, 526, 527
D, 461, 526, 529
F, 461, 462, 524, 528, 529
Fire course
phenomena, 499–501
Fire-fighter's rope, 597
Fire gases
definition, 503
Fire gas volume, 505, 506, 694
Fire house, 456
Fire load
and heating value, 507–508
Fire load, area-related, *see* Fire load density
Fire load density, 509, 511, 512
Fire pentacle, 456
Fire progression curve, 497–498
Fire protection equipment, 529
Fire quadrangle, 454
Fire-resistance class, 459–462
Fire service hazard groups
biochemicals, 655
chemical agents, 618
radioactive substances, 408
Fire smoke
composition, 506
definition, 506
Fire spread
danger of, 40
by heat transport, 29
by overflowing a container, 40
rate of, 498–499
Fire table, 456
Fire triangle, 454
Fissile material, 406–407
Fission, 321–322, 331, 337, 354, 399–402, 492
Flame-over, 500
Flame retardant

application of halogenated, 534
application of inorganic, 535
effect, 536
effect of halogenated, 534
effect of inorganic, 534
halogenated, 534
inorganic, 533, 534
inorganic P compounds, 536
organic N compounds, 535
organic P compounds, 536
Flame retardant clothing, 468, 533, 536
Flame retardants (FR), 117, 118, 468, 533–538, 599, 601, 603, 697
Flammability, 5, 15, 457–458, 461, 467, 470, 533, 599, 691, 692
Flash-over, 29, 498–500, 599
Flash point
comparison with ignition temperature, 37, 39
definition, 37
examples, 39
and fire hazard, 39
Flue gas cooling, 521–522
Flue gases
breathing protection, 503
colour, 502
definition, 502
quantities, emerging, 503
smoke density, 503
spread, 503
Flue gas ignition, 500, 522
Flue gas layer, 500, 501, 521, 522
Fluid, supercritical, 70, 71, 73, 152, 153
Fluorescence, 313–315, 319, 320, 350
Fluorine, 109, 112, 113, 115, 118, 264, 269, 458, 465, 611, 612, 641
Foaming agent
alcohol resistance, 611
fluorine-containing, 611–612
fluorine-free, 612
PFOS-containing, 611–612
Foaming number, 525
Foam, mixture, 10
f-orbital, 90, 106, 120
Force, electromotive, *see* Electromotive force (EMF)
Formal charge, 130–131
Formic acid, 149, 179, 205, 227, 259, 260, 502, 504, 571
Formula, chemical
constitutional formula, 134
Lewis formula, 131
mesomerism formulas, 131–132
molecular formula, 133, 134
partial sum formula, 134
semi-structural formula, 134
skeleton formula, 135
structural formula, 134
valence stroke formula, 133–135
Francium, 115, 121
Free Gibbs energy, 432
Freeze-drying, 32
Freezing point, 21, 30
Frequency, 83, 102, 295, 307–315, 333, 651, 681, 682
Frisch, O.R., 322, 331
Fuel cell, 287, 292
Functional group(s), 134, 543, 544, 563, 565, 567, 571, 577, 586, 594, 605, 698
Functional models, 6
Fused-salt electrolysis, 299, 301
Fusion, 4, 49, 50, 52, 322, 334, 337, 402

G

Gallium, 102, 116
Galvanic cell, 283, 287–294
Galvanic element, 275, 287, 293
Galvani, L., 287
Galvanization, 302
Gamma dose rate constant
examples, 365
Gamma radiation
emergence, 348–349
interactions, 349–351
ionization density, 390, 391
properties, 349
Gamma transition, 348–351
Gamow, G.A., 331
Gas
critical
data (examples), 71
formation of, 188, 190, 191, 501
ideal, 60, 66, 69
liquefied, 34, 71–73
permanent, 72
real, 69–75
solubility of
in emergency operations, 74–75
Gas chromatograph (GC)
with PID, 317
Gas cloud
condensation, 75
control, 75
Gas constant *R*, 65, 679
Gas detector array (GDA), 247

Gas equation
  general, 64–65, 69, 70
  universal, 65–66, 69
Gas explosion, 481, 484, 485, 489
Gas laws
  Amontons, 62, 64
  Boyle-Mariotte, 61
  Charles, 64
  Dalton, 66, 67
  Gay-Lussac, 62–64
  Humboldt's, 162
  partial pressures, 66–67
Gas, real, 69–75
Gassendi, P., 80
Gauss, C.F., 232
Gay-Lussac, J.L., 63, 162
GC, *see* Gas chromatograph (GC)
GDA, *see* Gas detector array (GDA)
Geiger, H., 371
Geiger-Müller counter, 365, 369, 370
Geissler, H., 93
Geissler tube, 93
Gel, 10, 442
Genetic engineering, 647, 655
Geneva Protocol, 622, 660
Germanium, 102, 117, 121
Ghost flames, 524
Gibbs energy, 432
Gibbs-Helmholtz equation, 432
Gibbs, J.W., 432
Glanders, 662, 663, 666
Gloss, metallic, 138
Glow, 119, 315, 347, 348, 472, 503
Gmelin, L., 102, 542
Goldstein, E., 93
Graphite
  in the nuclear reactor, 322
Graphite lattice, 155
Gray (unit), 362, 364, 366, 368, 384, 386
Green cross, 624
Groups of measures
  explosive substances (MG 1), 675
  toxic substances (MG 6), 620
Guerike, O. von, 59

H

Haber, F., 428
Haemorrhagic fevers, viral
  overview, 663
Hague Conventions, 622
Hahn, O., 321, 322
Half-cell, 275–277, 283–288
Half-layer thickness, 417
Half-life (HL)
  calculation with, 359
  examples, 359, 397, 408
  illustration, 360–361
  and spec. activity, 360
Hallucinogens, 641, 643
Halogenated flame retardants (HFR), 533, 534
Halogen compounds, organic, 561–562
Halogen-free flame retardants (HFFR), 533, 534, 536
Halogens, 111–113, 115, 118, 270, 444, 457–459, 465, 470, 472, 534, 543, 551, 561–562, 572, 585, 596, 611
Hard and soft acids and bases (HSAB), 213–215
Harkins, W.D., 94
Hasselbalch, K.A., 255
HD-PE, *see* Polyethylene (PE)
Heat
  accumulation, 473, 574
  capacity, 4, 48–57, 72, 526, 681
  conduction, 22–25, 29, 55, 473
  latent, 48–57
  length change, 24
Heat capacity, specific
  definition, 48
Heat conductor
  bad, 24
  good, 24
Heat explosion, 496, 497
Heat flow
  and fire spread, 29
Heating value, 507–509, 511, 599–601, 681, 696
Heating value, lower (Hl), *see* Heating value
Heating value, specific, 508, 511, 696
Heating value, upper (Hu), *see* Calorific value
Heat movement, 60, 146
Heat of evaporation, 4, 49–50, 53
Heat of evaporation, specific
  examples, 50
Heat of fusion, 50, 52
Heat of fusion, specific
  examples, 49
Heat of sublimation, 50
Heat quantity, calculation, 48, 50–52
Heat radiation
  and fire spread, 29, 40
Heat release rate (HRR), 509–513
  from the height of the flame, 513
Heat transport
  at fires, 29

Heavy gas effect, 72, 479
Heisenberg, W., 87
Helium, 49, 71, 105, 109, 110, 119, 126, 128, 129, 320, 322, 341, 342, 371, 379, 399, 402, 682
Helmholtz, H.L.F. von, 423
Helmont, J.B. van, 59
Henderson-Hasselbalch equation, 255
Henderson, L.J., 255
Hertz, H., 308
Hess, H.H., 428
Hess's theorem, 428
Heteroatoms, 135, 543, 558
Heterogeneous mixtures, 9–11, 14, 295
Heteropolar bond, 139
Hexogen, 580
HL, *see* Half-life (HL)
Homogeneous mixtures, 9–12, 14, 171, 179, 295, 431, 460, 479, 487
Homologous series, 545, 551
HSAB concept
  boundaries, 214
  overview, 213–214
Humboldt, A. von, 162
Humboldt's Gas Law, 162
Hund, F., 107
Hund's rule, 107, 108
Huygens, Ch., 308
Hybrid catalyst, 442
Hybridization, 554, 555
Hybridization, C atom, 554
Hydrocarbons, 68, 69, 152, 180, 222, 394, 444, 457, 458, 460, 501, 503, 506, 507, 518, 533, 544–561, 563, 586, 587, 605
  aliphatic, 545–553, 555–556
  aromatic, 501, 503, 506, 507, 545, 556–560, 566, 587
  bicyclic, 555
  cyclic, 555
  halogenated, 444, 518, 561
  polycyclic, 501, 507, 558, 560
  spiranes, 555
Hydrochloric acid, 121, 187, 189–192, 205, 206, 219, 221–223, 233, 240, 242, 245, 252–254, 257–259, 266
Hydrofluoric acid (HF), 191, 214, 224, 226
Hydrogenation
  on aromatics, 556
  fat hardening, 607
Hydrogen bomb, 305, 322, 323, 402
Hydrogen bond
  boiling point comparison, 150
  effects of this, 149–151
  illustration, 149
  intermolecular, 149
  intramolecular, 149
  and life processes, 150
Hydrogen cyanide, 187, 206, 303, 317, 394, 504, 542, 626, 633
Hydroiodic acid, 226–227
Hydrolysis
  definition according to Arrhenius, 208, 215
Hydrophilic, 172, 563, 570, 605, 609
Hydrophobic, 570, 605, 606, 609, 611
Hydroshields, use in gas clouds, 75
Hypothermia, 32
Hypoxia, 465

I

Ignition power, 438, 472–474, 498
Ignition temperature
  comparison with flash point, 37–41, 457–459, 691
  definition, 38
  examples (liquids), 39
Immediately Dangerous to Life and Health (IDLH), 689, 690
IMS, *see* Ion mobility spectrometer (IMS)
Incorporation, 218, 380, 387, 390, 408, 409, 411, 420, 459, 620, 657
Incrustation, 199
Incubation time
  bacteria, viruses, 663
Indicator, pH
  functionality, 255
Indium, 116
Indoor air quality, 688
Induction effect, 151
Infection dose
  bacteria, viruses, 666
  examples, 650
Infection rate, 650, 667
Infectivity, 650
Inflammation point
  compared to the flash point, 38
  definition, 38
  examples, 39
Infrared radiation, 28, 314
Ingestion, 380, 387, 616, 617, 628, 668, 669
Inhalation, 218, 380, 381, 387, 617, 618, 620, 627, 628, 633, 634, 636, 639, 640, 643, 663, 664, 668, 670, 689
Inhibition
  heterogeneous, 519–520

Inhibition (*cont.*)
homogeneous, 444, 518–519
Inhibitor, 444, 493–494, 638, 639
Initiator, 492–494
Injection anthrax, *see* Anthrax
Inorganic chemistry, distinction from organic chemistry, 542
Intensive properties, 424
International Atomic Energy Agency (IAEA), 409
Intestinal anthrax, *see* Anthrax
Intumescence
applications, 537
effect, 538
Inverse square law, 413, 414
Iodine, 32, 50, 101, 102, 118, 154, 226, 434
Ion binding fraction, 143
Ion character, 142–144, 210, 683
Ion charge, 140
Ion conductor, 141
Ion crystals
deformation of, 142
Ion dose, 364–365, 368, 682
Ionic bond
distorted, 143
transition to electron pair bonding, 143–145
Ionic compounds
composition, 141
Ionization chamber, 370
Ionization density, 390–391
Ionization detector, 395
Ionization energy
photoionization detector, 247, 316
Ionizing capacity, strength, 364
Ionizing radiation
avoidance contamination, 412
biological effect, 362, 383–392, 613
features, 352
half-layer thickness, 417
hazard groups, 420
interactions, 357
inverse square law, 413
measuring instruments, 372–378
prevention incorporation, 411
prevention of contamination transfer, 411, 412
protection from, 411–420, 613
protective measure distance, 413
protective measure duration of stay, 415
protective measures (combination), 412–420
protective measure shielding, 416–418
protective measure shutdown, 413
protective measures, STICKS worksheet, 420
Ion lattice, 140, 154, 210
Ion mobility, 141
Ion mobility spectrometer (IMS)
functionality, 393
possibilities of proof, 394
principle, 393, 394
reactant ions, 393, 394
substances CWA-mode, 394
substances ITOX-mode, 394
Ion radius, 110, 140, 213, 214
Ions
frequent, 140
Irradiation
external, 387–388
internal, 387–388, 390
Irritants
decontamination, 632
eye irritants, 628, 630
industrial chemicals, 632
nasopharyngeal irritants, 630–632
overview, 631
protective measures, 632
Isobar, 329
Isobar rule, 329
Isocyanates, 581–582, 602, 698
Isomerism (org. chemistry)
cis-trans, 549–550, 552
(Z)-(E), 550, 552
Isotonic, 329
Isotope battery, 395
Isotopes
hydrogen, 95, 402

J

Joliot-Curie, I., 321
Joliot-Curie, J.F., 321
Joule-Thomson effect, 72
Jungner battery, *see* Nickel-cadmium accumulator/battery

K

K-capture, 339, 347, 348, 355
Kelvin, 22, 64, 505, 626, 679
Ketones, 503, 531, 551, 563, 565, 567, 568, 611, 698
Key figures, safety-related, 35

L

Lanthanoids, 120, 127
Lassa fever, 649, 667
Latency, 627, 633, 634, 668
Lattice structure, 139, 154, 155
  diamond, 155, 156, 556
  graphite, 155
  salts, 139
Lavoisier, A. L. de, 160, 205
Law of conservation of energy, 423
Laws, chemical
  conservation of mass, 160, 161, 335
  constant proportions, 161, 162
  equivalent proportions, 161
  Humboldt's Gas Law, 162
  multiple proportions, 161–162
  Proust's, 161
LD-PE, *see* Polyethylene (PE)
Lead (Pb), 24, 25, 50, 110, 117, 120, 126, 127, 161, 270, 278, 279, 290, 349, 417
Lead accumulator, 117, 290
Lead-acid battery
  sulphuric acid concentration, 290
Leclanché element, 284–289
Leclanché, G., 288
Lenard, P., 81, 91
Lethal concentration (LC_{50}), 627–628, 630, 669, 689
Lethal concentration (LD_{50}), 689
Lethal dose (LD_{50}), 392, 628, 668, 670, 689
Lethality
  bacteria, viruses, 663
Leucip of Miletus, 79
Lewis acids, 212, 213
Lewis bases, 196, 212, 213
Lewisite, 394, 629, 634, 636
Liebig, J. von, 205
Ligands, 194–197, 212
Light equivalent, 310, 311
Light quanta, 308, 309
Light, visible, 308, 310, 311, 314, 315, 473
Light water, 525
Lime soap, 608
Limiting oxygen concentration, 487
Linear expansion
  calculation, 23
  coefficient of, 23, 24
  examples, 23, 24
Liquid gas
  vapour pressure, 33–37, 47, 67, 71
Liquid oxygen (LOX), 464, 469
Liquids, flammable
  water miscible, 40, 518
Liquids, miscibility gap, 180
Lithium-ion battery, 529
Litmus paper, 217, 218, 246, 248, 577
Lives in danger
  B-operations, 656
  C-operations, 619
  explosive substances, 671
Local elements, 293, 294
Lomonosov, M.V., 160
London, F.
  dispersion forces, 151
Long-range order, 154
Loschmidt number, 100
Lower explosion limit (LEL)
  calculated estimate, 44–45
  definition, 42
Lowry, T.M., 209–212, 237
Lucretius, 80
Luminescence, 314–315
Lye
  corrosion effect, 219
  corrosivity, 217–218
  important, 228–229
  measures after release, 260
  neutralization (*see* Neutralization)
  strength of, 238–244
Lysergic acid diethylamide (LSD), 641

M

Macromolecules, 589–591, 593, 594
Magic numbers, 332, 333
Magnesium, 49, 57, 59, 109, 116, 121, 126, 191, 219, 220, 280, 294, 457, 460, 461, 464, 517, 534, 610, 693, 694
Magnetic quantum number, 85–90
Main group elements, 103, 119, 141
Main groups
  1st main group (overview), 114–115
  3rd main group (overview), 116–117
  4th main group (overview), 117
  5th main group (overview), 117–118
  6th main group (overview), 118
  8th main group (overview), 119
Main valence bond, 137
MAK, 689, 690
Manhattan Project, 322
Marburg virus, 649
Mariotte, E., 61
Marsden, E., 92
Mask breaker, 630
Mass, 157
Mass burning rate, 508

Mass concentration
conversion to volume concentration, 186
Mass defect, 335, 336
Mass fraction
conversion to solubility, 182
Mass number, 325–329, 336, 337, 342–344, 347–349
Mass spectrometer (MS), 247, 327
Mass unit, atomic, 96–100
Mattauch, J., 329
Matter, forms of state, 155
Matter waves, 86, 87
Maxwell, J.C., 308
Measuring, 43, 156, 235, 245–247, 249, 256, 290, 316, 357, 365, 369–378, 395, 412, 414, 420, 445, 465, 620, 629
Measuring instruments for ionizing radiation, 372–378
Medium expansion foam pipe, for use in case of gas leakage, 75
Meitner, L., 321, 322, 331
Melamine resins, 582, 597, 693
Melioidosis, 662, 666
Melting point, 4, 15, 30, 31, 49, 57, 110, 154, 156, 301, 517, 529, 549, 550, 607, 625, 630, 634, 638
Mendeleev, D.I., 102–103
Mercaptans, 585
Mesomerism
Mesomerism arrow, 131
meta-, 559
Metal binding
transition to electron pair bonding, 142
Metal fires
and water, 53, 55, 57
Metal lattice, 138
Metalloid, 120–121
Metal refining, 302
Metals
base, 11, 219, 227, 278, 280
coin metals, 121
light metals, 55, 114, 119, 121, 278, 457
noble, 219, 279, 281, 293, 294, 297, 456
noble metals, 121
non-ferrous metals, 120, 121
in the periodic table, 121
platinum metals, 120, 121
refining of, 302
Meyer, J.L., 102–103, 110
Microwave radiation, 315
Milikan, R., 91
Minimum combustion temperature, 507, 508, 516
Minimum ignition energy, 473–474, 498
Minimum oxygen concentration, 469, 518
Mixed elements, 327, 328
Mixed phases, composition, 181–187
Mixture gap, for liquids, 179
Mixtures
aerosol, 10, 502
alloys, 10, 11
dispersion, 10
emulsion, 10
foam, 10, 522
gel, 10
heterogeneous, 9–11, 14, 295
homogeneous, 9–12, 14, 171, 179, 295, 431, 460, 479, 487
smoke, 10, 15, 502, 671
suspension, 10
Mixtures of substances, 9, 14, 43, 457, 468, 606, 671
Models
balloon model, 89
electron spin, 85
functional models, 6
hedgehog model, 137
orbital model, 84, 86–91
representation models, 6
validity range, 7
Modification
diamond, 155
graphite, 155
phosphorus, 117, 155
quinacridone, 156
tin, 155
Mol, 46, 65, 66, 99, 100, 146, 149, 150, 153, 165, 183, 184, 186, 194, 219, 231–233, 235, 236, 239–242, 249, 252, 254–257, 277, 278, 284–286, 298, 311, 336, 426, 428, 432, 441, 442, 457, 479, 530, 552, 553, 556, 574, 626, 638, 668, 679, 681, 692
Molality, 182, 184
Molecular crystal, 154
Molecular formula
alkanes, general, 546
alkenes, general, 548
alkines, general, 551
Molecular lattice, 153, 154, 156, 173
Molecular mass, relative, 97–98
Molecular motion, Brownian, 68, 99
Molecules
presentation, 125
term, 125
Monitor, use for gas clouds, 75

Monomers, 501, 589, 593
Montreal Protocol, 529
Morbidity, 651, 662
Mortality, 389, 651, 662, 665–668
Moseley, H.G.J., 102, 103
Moseley's Law, 102
MS, *see* Mass spectrometer (MS)
Multiple bindings, 142, 593
Mushrooms, 649
Mustard gas, 634, 636, 640
Mycotoxin, T2-, 668–670

N
Natural background reduction (NBR), 377
Natural sciences
exact, 3
NBC operations, and states of matter, 57
NBR probe, 377
Neon, 100, 109, 110, 119, 129, 139
Nernst's equation, 284–286
Nernst, W., 284
Nerve agents
data, 637
decontamination, 640
protective measures, 640
structural formula comparison, 641, 642
structural formulas, 637, 641, 642
Neutralization
acetic acid, 253, 260
ammonia, 253, 254, 259
CO_2 generation, quantities, 259, 260
definition according to Arrhenius, 207
with diluted acids/alkalis, 258–259
by dilution, 257, 258
in emergency operations, 257–260
fundamentals, 251
hazard element, estimation, 259
hydrochloric acid, 252–254, 258
quantity estimation, 258
sodium lye, 259
with solid acids/alkalis, 259–260
strong acid/strong base, 252–253
strong acid/weak base, 253–254
strong base/weak acid, 253–254
sulphuric acid, 252
weak acid/weak base, 253
Neutralization, heat of, 258
Neutralizing agent, 258, 260
Neutral point, 233–235
Neutrino, 334, 335, 343, 344, 347
Neutron
decay, 343
thermal, 322
Neutron activation analysis, 372, 396, 623
Neutron counter, 371–372
Neutron dosimeter, 371
Neutron number, 329, 353
Neutron radiation, 315, 377
Neutron stars, 95
Newlands, J.A.R., 101
Newton, I., 308
Nickel-63, 393
Nickel-cadmium accumulator/battery, 291
Nickel-metal hydride accumulator, 291
Nitric acid, 117, 149, 205, 219, 220, 224, 225, 230, 259, 281, 443, 464, 473, 580, 616
Nitric acid ester, 581
Nitrile, 504, 581, 698
Nitro compounds, organic, 577–583, 692
Nitrogen, 13, 19, 42, 49, 59, 66, 67, 70, 74, 83, 86, 93, 109, 110, 114, 117, 126, 127, 129, 130, 132, 144, 148, 157, 165, 166, 171, 270, 317, 321, 340, 396, 399, 443, 457, 463, 464, 466, 470, 503, 504, 517, 535, 543, 577–579, 583, 636, 685, 692
Nitrogen compounds, organic, 577–583, 692
Nitrogen perite, 636
Nitroglycerin, 581
N-Lost, 394, 627, 634, 636, 637, 640
Nobel, A., 13
Nobel Prize, 13, 81–83, 85–87, 91, 93, 94, 113, 205, 308, 313, 319–321, 328, 333, 347, 350, 353, 428, 437
Noble gas configuration
Octet rule, 128
Noble gases, 102, 105, 110, 112–114, 119, 125, 128–130, 137–139, 146, 153, 167, 341, 456, 463
Noble gas rule
exceptions, 129–130
Noble gas shell, 138
Noble metals, 219, 279, 281, 293, 294, 297, 456
Nomenclature
complex, 197–198
Non-ferrous metals, 120, 121
Normal conditions, 66, 278, 283–285, 672
Normal hydrogen electrode, 277, 278
Normal potential
of redox pairs, 277–279
Novichok, 640–642
Nuclear binding energy, 335–337
Nuclear chemistry, notations, 325–326
Nuclear crime, 403

Nuclear energy phase-out, 401
Nuclear fission
  controlled, 400–402
  discovery, 321–322
  energy, released, 322, 400
  uncontrolled, 400, 401
Nuclear force
  strong, 332, 333, 341
  weak, 343, 347
Nuclear fusion, 322, 334, 337, 402
Nuclear isomers, 348
Nuclear photoelectric effect, 349, 351
Nuclear power plants (NPPs), 305, 381, 400–401
Nuclear reaction
  conversion scheme, 340
  conversion scheme, rules, 340
  K capture, 347, 348
  reaction equation, 339–341
  representation, 339
Nuclear reactor, first, 322
Nuclear transformation
  artificial, 399
Nuclear weapons, 322–323, 402–403, 482, 613
Nucleon, 325, 326, 329, 331–334, 336, 340, 341, 347, 352
Nuclide map
  detail, 354, 355
  overview nuclides, 353, 354
  types of decay, 353–355
Nuclide notation, 326
Nuclides, 325–329, 332, 335, 347, 353–355, 359–361, 365, 377–379, 387, 395–397, 399, 400, 403

O

Occupation order
  chessboard method, 106, 107
  of the orbitals, 107
Octane number (ON), 548
Octave Law, 101
Octet rule, 128
Odorant, 585, 586
Oil test paper, 180, 181, 246
Oleophobic, 525
Operational measures/procedure
  B-operations, 656
  chemical agents, 625–630
  explosive substances, 485, 486, 671–675
Orbital diagram, 107–108
Orbital model, 84, 86–91
Orbitals
  atomic model, 86
  casting order, 105–106
  d-orbital, 89, 119, 130, 543
  f-orbital, 90, 106, 120
  p-orbital, 88–90, 105, 107, 109, 112, 130, 553–555
  s-orbital, 88, 90, 105, 109, 112, 113, 553, 554
Organic chemistry, differentiation from inorganic
Organisation of the Prohibition of Chemical Weapons (OPCW), 622, 623
ortho-, 148, 559, 630, 696
Ostwald, F.W., 437, 438
Outer electrons, 104, 105, 109, 112, 115, 128–130, 137, 139, 350, 543, 554
Overvoltage, 280, 297
Oxidation
  electrolysis process, 296
  fast, 451
  primary alcohols, 565
  secondary alcohols, 565
  slow, 450–451
  term, 263
  tertiary alcohols, 565
Oxidation number
  determination, rules, 269–270
  and stoichiometry, 167–168
Oxidizing agents, 222, 248, 265, 268, 278, 279, 284–286, 465, 473, 551, 558, 563, 565
Oxygen
  combustion pattern, 467–469
  and combustion temperature, 467–469
  element, 12, 97, 114, 264, 463, 469, 470, 543
  measurement, 465–467
Oxygen compounds, organic, 563–575
Oxygen index (*OI*), 470, 699
Oxygen intoxication, 465
Oxygen poisoning, 465
Oxygen toxicosis, 465
Ozone, 249, 530
Ozone layer, 529–531

P

PAA, *see* Peroxoacetic acid (PAA)
Pair annihilation, 351, 397
Pair formation, 351
para-, 148, 506, 559, 694, 696
Paracelcus, 80, 615
Partial charge, with polarized EPB, 143

Partial pressure, 66–67
Partial sum formula, 134
Particle radiation, 315
Pascal, B., 59
Passivation, 280
Pathogenicity, 651
Pauling, L.C., 113
Pauling notation, *see* Orbital diagram
Pauli principle, 85
Pauli, W., 85, 335
Pearson, R.G., 213
Pelargonic acid morpholide, 631
Pepper spray, 632
Peptide binding, 580
Perfluorooctane sulfonic acid (PFOS), 611, 612
Perfluorooctanoic acid (PFOA), 611
Periodic table
  display, 103–105
  main groups, 114–119
  structure, 103–106
  subgroups, 119–120
Periodic table of the elements (PTE), *see* Periodic table
Peroxides
  organic, 228, 574, 687
Peroxoacetic acid (PAA), 228, 575
Peroxycarboxylic acids, 575
Persistent organic pollutants (POPs), 561
Pest, 396, 647, 669
PFOA, *see* Perfluorooctanoic acid (PFOA)
PFOS, *see* Perfluorooctane sulfonic acid (PFOS)
Phenols, 179, 180, 494, 566–567, 597
Phenoplastics, 566, 597
Phenyl, 558
Philon of Byzantium, 60
pH meter, 232, 235, 248, 249
Phosgene, 247, 504, 598, 616, 621, 623, 632, 633, 640
Phosphate buffer, 256
Phosphate glass dosimeter, 372
Phosphorescence, 313–315, 319
Phosphoric acid, 130, 179, 219, 225–226, 258, 259, 535, 536, 538, 623, 629, 637, 640, 644
Phosphorus
  as a flame retardant, 117, 533, 535, 536
  modifications, 117, 155
Photoelectric effect, 308, 345, 347, 349–351, 390
Photoionization detector (PID), 247, 316, 317
Photon energy, 309, 315, 316, 350
Photon equivalent dose, 377
Photon equivalent dose rate, 377
pH paper
  deployment tips, 248–249
  sources of error, 248
pH value
  calculation, 232, 233, 236, 241–243, 255
  determination, 243–249
  pH indicator, 244, 245
  of saline solutions, 236–238
Pi binding, 554, 555, 557
PID, *see* Photoionization detector (PID)
Piloted ignition, 472–473
Pi-mesons, *see* Pions
Pions, 312, 333, 334, 442
Plague, 648, 653, 659, 661–665
Plague sepsis, *see* Plague
Planck constant, 83, 309
Planck, M.
  action quantum, 83, 308, 309
Plastic fires, 602, 603
Plastics
  classification, 589–593
  educational mechanisms, 593–595
  fire behaviour of, 599–603
  fully synthetic, 592
  heating values, 600
  MF, 597
  overview important, 592–593
  PA, 535, 590, 592, 596–597, 602
  PBI, 592
  PC, 592, 598, 602
  PET, 460, 574, 592, 599, 602
  PMMA, 460, 590, 592, 598, 602
  POM, 592
  PS, 460, 493, 503, 506, 590, 593, 598, 602
  PTFE, 596
  PU/PUR, 503, 593, 597, 602
  PVC, 219, 460, 493, 506, 590, 593, 596, 602
  semi-synthetic, 593, 595
  UF, 597
Platinum metals, 120, 121
Plutonium bomb, 322
Pneumonic plague, *see* Plague
pOH value
  calculation, 236, 241
Point, critical, 70
Poison
  endogenous, 615, 616
  exogenous, 616
  general, 616
  Poison effect, 616
  Poison groups (examples), 616

Poisoning, therapy, 617
Poison strength (p*LD*), 629
Polarity
  of solvents, 171
Polarization, spontaneous, 151
Policyclic aromatic hydrocarbons (PAHs), 506, 560
Polonium, 102, 118, 121, 320
Polyaddition, 593–595, 597
Polyamide (PA), 502, 534, 535, 592, 594, 596–597, 699
Polybenzimidazole, 536, 592, 599
Polycarbonate (PC), 534, 592, 598, 599
Polycondensation, 593, 594, 596–598
Polyene, 550, 553
Polyethylene (PE), 371, 442, 460, 503, 506, 535, 590, 592, 594, 595, 600, 602, 694, 696
Polyethylene terephthalate (PET), 460, 534, 574, 592, 599, 602
Polyfluorinated chemicals (PFCs), 611
Polyine, 552
Polymerization
  alkines, 552
Polymers, 292, 468, 537, 589, 591, 593, 595, 598–599, 601, 602
Polymorphism, 154
Polyolefins, 535, 595
Polyoxymethylene, 592, 597
Polystyrene (PS), 442, 460, 493, 503, 506, 534, 558, 580, 590, 593, 598, 600, 602, 694, 699
Polyurethanes, 503, 581, 593, 595, 597
Polyvinyl chloride (PVC), 219, 460, 493, 506, 573, 590, 593, 596, 601, 602, 693, 694, 696, 699
p-orbital, 88–90, 105, 107, 109, 112, 130, 553–555
Positron, 344, 345, 351, 352, 397
Positron emission tomography (PET), 351, 397
Potassium, 101, 102, 109, 115, 121, 172, 179, 189, 193, 219, 229, 230, 249, 259, 263, 319, 343, 450, 457, 460, 608, 616, 644, 700
Potassium lye solution, 229
Potential difference, 277, 283, 288
Precipitation, formation of, 188–190
Pressure
  conversion bar to Pa, 60
  conversion Pa to bar, 60
  conversion Pa to Torr, 60
  conversion Torr to Pa, 60
  supercritical, 70, 73
  unit Pa, 60
  unit Torr, 60
Pressure rise, maximum, temporal, 488
Pressure vessel burst, 482
Pressure wave
  damage caused by, 486
Primary element, 287, 288, 292
Principal quantum number, 85, 87, 110
Principles of operation, B operation, 656–657
Prions, 649
Processes, chemical
  distinction from physical, 4
Products, 49, 61, 117, 118, 125, 159, 164, 165, 167, 192, 194, 222, 225, 234, 235, 238, 284, 293, 304, 341, 380, 393, 396, 425, 427, 432, 434, 437–440, 442, 443, 450, 495, 497, 501–506, 533, 541, 550, 561, 568, 573, 581, 582, 589, 595, 616, 621, 623, 627–628, 649, 693
Properties
  chemical, 5, 11, 101, 327, 544
  physical, 4, 5, 11, 155, 156, 544
  physiological, 5
Properties, periodic
  atomic radius, 110, 111, 115
  electron affinity, 110, 112–113, 115
  electronegativity, 110, 113–115
  ionization energy, 110–112, 115
Proportional counter, 370
Protection of property, 374, 416
Protolysis, 209, 210, 215, 233, 237, 239, 240, 267
Proton
  stability, 335
Proton number, 329
Proton radiation, 315, 362
Proust, J.-L., 161
Psychotoxic agents
  decontamination, 643
  protective measures, 643
Pulmonary agents
  examples, 632
  protective measures, 633
Pulmonary anthrax, *see* Anthrax
Pulse rate, 357, 358, 376
Pure elements, 328
Pure substances, 5, 11, 12, 14, 30, 31, 99, 247, 295, 606
Pyrolysis, 497, 498, 501–504, 506, 516, 535, 538, 552
Pyrolysis gases, 500, 503, 535

Q
Q fever, 662, 663, 665
Quality factor (Q), *see* Radiation weighting factor
Quantum numbers
magnetic quantum number, 85–90
principal quantum number, 85, 87, 110
secondary quantum number, 85, 87, 88
spin quantum number, 85, 86, 88
Quaternary ammonium salts, 578, 698
Query fever, 662, 663, 665
Quinacridone
β-modification, 156

R
Rabbit plague, *see* Tularaemia
Rad (unit), 368
Radiation
cosmic, 379
infrared, 28, 314, 315
ionizing, 305, 314–317, 320, 321, 352, 357, 362, 369–378
non-ionising, labelling, 316–317
non-ionizing, 315
terrestrial, 380
Radiation damage
division, 392
early damage, 383
genetic damage, 384
late damage, 384
types of, 383–384
Radiation effect
factor dependence, 389–392
overview factors, 389–390
Radiation, electromagnetic
characteristics, 309–310
propagation speed, 309
Radiation exposure
atomic bomb test, 381
civilizational, 380–381
mean value, 381
medical, 380
reactor accidents, 378, 381
technical, 381
Radiation meters
detection of contamination, 372
dose rate meters, 373, 375
dose rate warning devices, 375
dose warning devices, 373, 374
Film dosimeter, 373–374
in firefighting operations, 372–378
Geiger-Müller counter, 365, 369, 370
ionization chamber, 370
ionization counter, 369–370
measured variable $H^{*}10$, 377–378
measurement of dose rate, 372, 373, 375
measurement of personal dose, 372–374
measurement principles, 369–372
NBR probe, 377
neutron counter, 371–372
overview (fire service), 373
scintillation counter, 371, 377
semiconductor detector, 371
Radiation protection
units of measurement (overview), 366, 367
units of measurement in, 357–368
units of measurement, obsolete, 367
Radiation protection areas, 409
Radiation sensitivity
of animals, 392
of organs, 392
relative, 390, 392
Radiation sickness, 383, 389
Radiation sickness, acute, 389
Radiation weighting factor, 352, 362–363
Radical chain reaction, 491–496, 519, 534, 575, 594
Radicals, 384–386, 442, 444, 445, 479, 491–497, 506, 519, 520, 527, 534–536, 544, 547, 549, 552, 558, 563, 574, 575, 577, 578, 583, 585, 586, 594, 605
Radioactive dispersion devices (RDD), 403
Radioactive substances
application IMS, 393–394
applications, 393–398
enclosed, 408
medical applications, 397–398
medical radionuclides, 394, 397
open, 408
scientific applications, 396–397
technical applications, 394–396
Radioactive substances, labelling
fire service hazard group, 408–409
IAEA supplementary markings, 409–410
packages, 406–407
transportation law, 405–410
vehicles, 405
workplace, 407–410
Radioactivity
artificial, 321
historical, 319–323
terms, 325–329
Radiolysis, 385, 492
Radionuclides

Radionuclides (*cont.*)
ingestion of, 380
for medical use, 397
Radio waves, 310
Radium, 92, 116, 340, 367
Radon, 380, 381
Radon inhalation, 380
Rare earths, 120, 315
Rate of flame spread, 498–499
Ratio formula, 133
Reaction
exergonic, 133
exothermic, 427–429, 435, 439, 457, 673
spontaneous, 284, 430, 442
transition state, 439
Reaction chain, radiobiological
cellular processes, 386–388
molecular processes, 385–386
time course, 384–385
Reaction entropy, 432, 439
Reaction equation
coefficients, 166
conventions for, 165
indices, 167
nuclear reactions, 339–340
particle stoichiometry number, 167
reaction stoichiometry number, 167
redox reaction, 269–273
rules for, 165–167
set up, 165–167
Reactions, chemical
basic, 163–164
basic laws, 160–162
stoichiometry, 167–168
Rechargeable battery, 273, 290, 291
Redox
explanation, 264–265
term, 264
Redox pairs, 265, 266, 275–281, 284
Redox potential, 283, 293, 301
Redox reaction
charge balance, 272
comparison with acid-base reaction, 266
comproportionation, 269
determine oxidation number, 269–270
disproportionation, 268
electron exchange, 267
electron transfer, 267, 271
examples, 268
protolysis-coupled, 267–268
reaction equation, constructing, 269
rules for determining the oxidation number, 269–270
strikeout trick, 272–273
types, 267–268
water balance, 272
Reducing, 40, 265–266, 268, 278, 279, 284–286, 449, 464, 493, 516–518
Reducing agents, 265–266, 268, 278, 279, 284–286, 449, 464
Reduction
electrolysis process, 295
term, 263, 264
Rem (unit), 368
Remote gripper, 414
Resonance formulae, 119
Rest energy, 312, 313
Rest mass, 312, 313
Richter, J.B., 161, 167
Ricin, 661, 662, 668–669, 700
Rickettsia, 649
Risk groups, biological agents, 652–654
Rod dosimeter, 365
Roentgen (unit), 368
Roentgen absorbed dose (Rad), 368
Roentgen equivalent man (Rem), 368
Roentgen, W.C., 319
Room cooling, 499, 521–522
Rubidium, 115, 121
Rule of thumb
duration of stay and dose rate, 416
dust explosion hazard, 486
extinguishing water requirement-fire compartment area, 513
foam concentrate requirement, 524
HL and activity, 359
linear expansion, 24
liquid gas, 72
personnel and fire area, 514
volume expansion, 27
Rutherford, D., 59
Rutherford, E.
atomic model, 82, 84

S

Sabotage poisons
criteria for, 643
examples of, 643–644
Sacrificial anode, 294
SAD temperature, 574
Safety-related key figures, 35
Salt bridge, 277
Salts, definition
to Arrhenius, 208
to Brønsted and Lowry, 237

Saponification, 529, 573, 607
Sarin, 394, 621, 626, 627, 629, 637–640, 645, 668, 700
Saturation concentration c_S, 625–626
Saxitoxin, 643, 668, 670
Scatter test, Rutherford's, 82, 91–93
Scheele, C.W., 59
Schrödinger, E., 87
Schrödinger equation, 87
Scintigraphy, 397
Scintillation counter, 371, 377, 395
Scintillator, 371
SEB, *see* Staphylococcal enterotoxin B (SEB)
Secondary element, 287, 290
Secondary quantum number, 85, 87, 88
Secondary valence bond, 137
Sedentariness
  dependence on boiling point, 626
  examples, 626
  influence of temperature, 626, 627
Segrè, E., 353–355
Selenium, 101, 118, 121
Self-accelerating decomposition temperature (SADT), 574
Semiconductor detector, 371
Semi-noble metals, 278
Sennert, D., 125
Sensing, 247
Sensory perceptions, 247
Shellfish, 670
Shell model, 332–333
Shielding
  alpha radiation, 342
  beta radiation, 345
  gamma radiation, 349
Sievert (unit), 362, 363, 366, 368, 682
Sigma binding, 682
σ-bond, *see* Pi binding; Sigma binding
Silicon (Si), 11, 101, 109, 114, 117, 121, 127, 154, 270, 543
Silicone, 117, 593, 599, 601
Single bindings, 143
Single-photon emission computed tomography (SPECT), 397
Skeleton formula, 135, 546
Skin warfare agents
  decontamination, 636–637
  examples, 634–636
  protective measures, 636–637
S-Lost, 394, 626, 634, 636
Smallpox, 650, 653, 659, 662, 663, 666–667
Smoke, 10, 15, 29, 201, 395, 450, 460, 463, 486, 498–506, 523, 526, 602, 603, 607, 671
Smoke density, for plastic fires, 602
Smoke detector
  ionization detector, 395
  optical, 201
Smoke generation, for plastics, 602–603
Smoke point (oils), 607
Smoldering, 247, 498, 500–502
Soap, 115, 229, 523, 529, 566, 632, 637, 640, 643
Sodium, 84, 101, 109, 115, 121, 130, 138–140, 172, 175, 177, 179, 188, 190, 191, 207, 219, 228–230, 236, 237, 245, 252–255, 257–259, 264, 280, 300, 371, 457, 460, 461, 484, 553, 585, 608, 624, 627, 637, 640, 644, 685, 700
Sodium lye, 259
Soft soap, 608
Solidification point, 30–31
Solids, amorphous, 155
Solid solution, 11
Solubility
  conversion to mass fraction, 182
  definition, 172
  dependence on solvent, 175, 176
  dependence on temperature, 175–176
  salts in water, rules, 181
Solutions, 10, 11, 56. 87, 99, 141, 153, 157, 159, 171–194, 199, 200, 205, 208, 209, 218, 220–223, 225–242, 244, 245, 248, 249, 251–260, 266, 267, 275, 277, 279–281, 284–286, 288, 289, 295–300, 302, 303, 314, 430, 441, 442, 450, 492, 511, 522, 524, 525, 528, 529, 556, 568, 575, 608, 609, 611, 616, 627, 630, 632, 637, 640
Solvent
  density of, 53
Soman, 394, 626, 629, 637–640, 700
Sommerfeld, A., 84
Soot formation, for plastic fires, 602
s-Orbital, 90, 105, 112, 113, 553, 554
Spectrum
  electromagnetic, 310
Speed of light
  in vacuum, 309, 347
  in water, 347
sp hybridization, 555
sp^2 hybridization, 554
sp^3 hybridization, 554
Spin quantum number
  explanation, 85, 86, 88
Spiro connections, 555
Spontaneous combustion, 451, 473

Standard conditions, 66, 100, 426, 505, 694, 695
Standard enthalpy of combustion, 507
Standard potential, 278
Standard state, 66
Standard temperature and pressure (STP), 66
Staphylococcal enterotoxin B (SEB), 661, 662, 668, 669
State functions, 424
State of aggregation
  change liquid-gaseous, 31–32
  changes, 4, 29–35
  change solid-gaseous, 32
  change solid-liquid, 30–31
  gaseous, 10, 19, 20, 30
  indices, 19
  liquid, 10, 19, 20, 30, 153
  in NBC operations, 57
  solid, 10, 19, 20, 30, 153
Stereoisomers, 549
Sterilization, 394
STICKS (mnemonic), 420
Stockholm Convention, 561
Stoichiometry
  and oxidation number, 167–168
Stoney, G.J., 91
Storage battery, 290
Storage cell, 290
Straßmann, F.W. 321
Strontium, 116, 121
Structural formula, 127, 133–135, 152, 457, 458, 530, 533, 546, 549, 550, 552, 637, 642
Subgroup elements, 119–120, 141, 177
Subgroups, 103, 104, 127, 671
Sublimation
  of water, 32
Sublimation printing, 32
Substance amount concentration, 183
Substance, dispersed, 10, 199, 200
Substance, flammable, 37, 453, 457, 459–460, 472, 475, 483, 485, 487, 516, 520
Substances
  water vapour volatile, 35
Substance separation
  heterogeneous systems, 11
  homogeneous systems, 11, 12
Substitution alloy, 11
Sulfonamides, 587
Sulfonic acids, 224, 562, 585–587
Sulfoxides, 698
Sulphur, 109, 110, 118, 130, 132, 149, 154, 172, 457, 460, 464, 470, 471, 504, 543, 558, 585, 586, 621, 634, 640, 685, 694
Sulphur compounds, organic, 585–587
Sulphuric acid, 26, 49, 118, 149, 169, 179, 205, 223, 224, 227, 252, 259, 290, 443, 551, 586
Sunburn, 383
Supercritical fluid, 70, 71, 73, 152, 153
Supporting beacon, 601
Surface tension, and surfactants, 608
Surfactants
  firefighting operations, 610–612
  fluorine-containing, 611–612
  fluorine-free, 611–612
  foaming, 609, 610
  micelle formation, 608
  mode of action, 608–610
  structure, 605–606
  water solubility, 606
  as a wetting agent, 610
Suspension, 10, 199, 200
Suspicion of poisoning, 615
Synthesis, 12, 13, 159, 160, 167, 226, 230, 396, 440, 441, 443, 460, 464, 541, 559, 564, 573, 594, 636, 647
System
  closed, 424
  colloidal disperse, 200
  completed, 424
  disperses, 199–201
  enemy-dispersed, 200
  molecular disperses, 199, 200
  open, 424

T

Table salt, ion lattice, 144
Tabun, 394, 629, 637–640, 700
Tammelin, L.-E., 640
Tammelin's ester, 640–641
Telescopic probe, 414
Tellurium, 101, 102, 118, 121
Temperature
  absolute, 22, 64
  critical, 70–73
  increase due to neutralization, 254
  Kelvin, 22, 64, 505, 626
  length change, 23
  volume change, 27
Temperature check, 521–522
Temperature class
  overview, 39
Temperature gradient, 25

Temperature scale
Celsius, 21, 402
Fahrenheit, 21
Kelvin, 22, 64, 505, 626
Temporary Emergency Exposure Limits (TEEL), 690
Tenacity, 651–652
Term notation, 109–110
Terrorist attack, suspicion of, 676
Test tube, 247, 629, 630
Thallium, 117, 371, 439, 644
Thermal conductivity
examples, 25
of metals, 138
Thermal dissociation, 55, 483, 521
Thermal energy, 20, 21, 24, 29, 49, 292, 345, 395, 425, 510, 521
Thermal explosion, 482
Thermal imaging camera (TIC), 28, 57
Thermal runaway, 292
Thermal state, 21
Thermochromism, 156
Thermoluminescence dosimeters (TLDs), 372
Thermoplastics, 461, 462, 535, 589–591, 593, 597, 601
Thermosets, 461, 589–591, 597, 601
Thioether, 585, 586, 634, 698
Thiols, 585–586, 698
Thomson, J.J.
atomic model, 81–82
Thomson, W., 64
Threshold value, 43, 201, 374, 586, 628, 630, 631
TIC, *see* Thermal imaging camera (TIC)
Tin
modifications, 155
Tin can, 289, 293
Tinplate can, 293
Tissue weighting factor, 363
T2 mycotoxin, 668–670
Tolerability limit, 628, 631
Torricelli, E., 59
Toxic effect
acute, 177, 616
chronic, 616
irreversible, 616
reversible, 616
systemic, 616
Toxicity, 5, 177, 389, 465, 618, 629, 631, 640, 641, 643, 645, 652, 669
Toxicological values, 684, 689
Toxicology, 616
Toxic substances
classification according to ADR, 617, 618
classification according to GHS, 617, 618
labelling, 617–619
labelling according to ADR, 617
labelling according to GHS, 617
Toxins
overview, 668
Trans-compounds, 549
Transesterification, 573
Transition complex, 433, 434
Transport index
and dose rate, 406
Transport law
biochemicals, 654–655
explosive substances and, 673
toxic substances and, 617
Treatment, thermal, 499–501
Triads, 101
Trichothecenes, 669
Triene, 550
2,4,6-Trinitrotoluene (TNT), 322, 323, 558, 559, 580
Triple bond
display, 128
σ-/π-binding, 555
Trivial names, 168, 551, 559
TRK, 689, 690
Tularaemia, 661–663, 665
Tyndall effect, 200–201
Types of decay, 353–355

U
Uncertainty principle, Heisenberg's, 87
Unicorn, 557
Unloading process
alkaline manganese cell, 289
Daniell element, 284, 288
Lead-acid battery, 290
Leclanché element, 288–289
Li-ion battery, 291–292
Ni-Cd battery, 291
Ni-MH battery, 291
Zinc-carbon battery, 289
Zn-Hg battery, 290
Upper explosion limit (UEL)
definition, 41
Uranium bomb, 402
Uranium radiation, 320

Urea resins, 597
UV radiation, 310, 315, 530, 648

V

Vacuum distillation, 34
Valence, 91, 127–135, 137, 138, 142, 143, 146, 160, 313, 566
Valence bond, 137
Valence electrons, 128, 130, 131, 138, 142, 146
Valence stroke formula
  rules on education, 128
Valency, 132
Van der Waals forces
  close order, 152
  effects, 148, 151–153
  gecko, 151
  influencing the boiling point, 152
Vapour explosion
  aluminium melt, 483
  at the chimney fire, 54, 481
Vapour pressure
  and ambient pressure, 33, 34, 39, 44
  formation, 33
  gases, 34, 36, 37
  liquid gas, 34, 72
  as a safety-related key figure, 35
Vapour to density ratio
  definition, 46
VEE, *see* Equine encephalitis (EEE)
Verifying, 245
VHF, *see* Haemorrhagic fever, viral
Virulence, 652
Viruses
  human pathogens (examples), 649
  incubation time, 651, 667
  infectious dose, 650, 663
  lethality, 663, 666, 667
vis vitae, 542
Volta, A., 287, 288
Voltage series, 280, 296, 297
Voltaic element, 275, 288
Volta's column, 288
Volume binding energy, 332
Volume concentration, 183, 186, 475, 682
Volume expansion
  apparent, 27, 28
  calculation, 26
  completely filled container, 28
  and emergency operations, 28
Volume fraction, 182–183, 463, 682
Volume, molar, 66, 99–100, 186
Vomiting agent, 631
VR, 640, 641
VX, 394, 626, 629, 637, 638, 640, 641, 668, 700

W

Wall effect, 444, 493, 519, 520, 527
Warfare agents
  biological, 613, 645, 659, 660, 662, 665, 668
  BZ as, 641
  chemical
    blood warfare agents, 624, 633–634
    irritants, 621, 624, 628, 630–632
    nerve agents, 624, 628–630, 637–641
    nerve agents effect, 638
    skin warfare agents, 624, 629, 634–637
    See also C-warfare agents (CWA)
  CN, 625, 630, 631
  CS, 630–632
  Cyanogen chloride as, 633
  detection
    B warfare agents, 659
    C warfare agents, 629–630, 659
  hydrogen cyanide as, 626, 633
  Lewisite, 629, 634, 636
  LSD as, 641
  N-Lost, 634, 636, 637, 640
  Novichok, 640, 641
  psychotoxic, 624, 641–643
  S-Lost, 626, 634–636
  Soman, 626, 629, 637–640
  strategic, 624
  Tabun, 629, 637–640
  tactical, 624
  Tammelin's ester, 640–641
Water
  anomaly, 150, 153
  hardness ranges, 198
Water detection paste, 180, 181, 246
Water Half Time, 523
Water hardness
  and extinguishing water supply, 198–199
  hardness grades, 198
  permanent hardness, 198–199
  temporary hardness, 198–199
  total hardness, 198–199
Water miscibility
  of liquids, 178–180
Water steam distillation, 34
Water vapour volatility, 35

Wave, electromagnetic, 307–310, 349, 352
Wave equation, 87
Wave function, 87
Wavelength, 86, 87, 307, 309–311, 315, 316, 349, 350, 383, 682
Wave-particle duality, 308
Wave phenomena, 307, 308
Wave radiation, 307–315, 349, 352
Weapons, radiological, 403
WEE, *see* Equine encephalitis (EEE)
Weight force, 157, 182, 681
Weighting factor q, *see* Radiation weighting factor
Weizsäcker, C.F. von, 331
White cross, 624, 630
Whole body irradiation, 363
Whole Number rule, 328
Wien, W., 93
Wöhler, F., 542
Wood glue, 200
Workplace labeling
  biochemicals, 655
  explosive substances, 673

X
Xenon, 110, 119
X-ray camera, 371
X-ray examination, 380, 397
X-rays, 86, 102, 116, 117, 309, 310, 313–314, 319, 320, 339, 345, 349, 350, 362, 371, 373, 378, 380, 381, 388, 390, 391, 393, 397, 413, 682
  origin, 313
  properties, 313–314

Y
Yellow cross, 624
Yellow rain, 670
Yukawa, H., 333

Z
Zero point, absolute, 64

GPSR Compliance

The European Union's (EU) General Product Safety Regulation (GPSR) is a set of rules that requires consumer products to be safe and our obligations to ensure this.

If you have any concerns about our products, you can contact us on ProductSafety@springernature.com

In case Publisher is established outside the EU, the EU authorized representative is:

Springer Nature Customer Service Center GmbH
Europaplatz 3
69115 Heidelberg, Germany

Batch number: 10370501

Printed by Printforce, the Netherlands